AUSTRALIAN MINE VENTILATION CONFERENCE 2022

10–12 OCTOBER
GOLD COAST, AUSTRALIA

The Australasian Institute of Mining and Metallurgy
Publication Series No 6/2022

AusIMM

Published by:
The Australasian Institute of Mining and Metallurgy
Ground Floor, 204 Lygon Street, Carlton Victoria 3053, Australia

ISBN 978-1-922395-12-2

ORGANISING COMMITTEE

Duncan Chalmers
MAusIMM
Conference Chairperson UNSW

Bharath Belle
MAusIMM(CP)
Technical Editor

Basil Beamish
MAusIMM(CP)

Rick Brake
FAusIMM(CP)

Katie Manns
MAusIMM

John Rowland

Michael Shearer

Guangyao Si

Craig Stewart

Jerry Tien

Claudia Vejrazka
MAusIMM

AUSIMM

Amelia Lundstrom
Manager, Events

Samara Brown
Conference Program Manager

REVIEWERS

We would like to thank the following people for their contribution towards enhancing the quality of the papers included in this volume:

Ms Cheryl Allen, *Vale North Atlantic, Canada*

Dr Rao Balusu, *CSIRO Energy Flagship, Australia*

Dr Basil Beamish, *CB3 Mine Services Pty Ltd, Australia*

Dr Bharath Belle, *Australia*

Dr Sekhar Bhattacharyya, *USA*

Dr Rick Brake, *Mine Ventilation Australia*

Dr Jurgen Brune, *CSM, USA*

Dr Aleksandar Bugarski, *USA*

Mr David Carey, *Mines Rescue Services, Australia*

Mr Duncan Chalmers, *UNSW, Australia*

Prof David Cliff, *University of Queensland, Australia*

Mr Andrew Derrington, *Ozvent Consulting Pty Ltd, Australia*

Dr Gerrit Goodman, *Office of Mine Safety and Health Research, USA*

Dr Adrian Halim, *Luleå University of Technology, Sweden*

Dr Gerald Joy, *National Institute for Occupational Safety and Health, USA*

Dr Gareth Kennedy, *SIMTARS, Australia*

Ass Prof Mehmet Kizil, *The University of Queensland, USA*

Mr Bob Leeming, *Health and Safety Executive, UK*

Mr Kevin Lownie, *Australia*

Dr Pierre Mousset-Jones, *Professor Emeritus, USA*

Mr Paul O'Grady, *Australia*

Late Professor Emeritus Huw Phillips, *South Africa*

Dr Qingdong Qu, *Australia*

Mr John Rowland, *Dallas Mining Services Pty Ltd, Australia*

Dr Emily Server

Dr Guangyao Si, *UNSW, Australia*

Prof Jerry Tien, *Australia*

Dr Purushotham Tukkaraja, *USA*

Mr Michael Webber, *Australia*

Dr Hsin-Wei Wu, *Gillies Wu Mining Technology Pty Ltd, Australia*

FOREWORD

Welcome to the Australian Mine Ventilation Conference 2022. As chairman I feel honoured to lead such a dedicated team that has worked hard to bring this conference to fruition. I know that there are many more people who would have stepped up if they were asked and I thank them too for being available. In these uncertain times, we were hopeful that we would be able to continue this series of conferences.

Thank you to those that submitted abstracts and those who formed them into full papers. Without these contributions the conference would be a lot of awkward silences. Eleanor Roosevelt is attributed as saying that we should learn from the mistakes of others as we won't live long enough to make them all ourselves. However, this is only one side of the coin. We should also learn from the success of others as that will also enable us to avoid making mistakes.

Our industry uses risk assessment tools to help provide the world class safe working environments that we expect of our mines. Risk assessments have to be underpinned by the broadest base of knowledge or otherwise they will become flawed. The speedy dissemination of and access to knowledge is imperative to provide this base.

This conference provides a platform for the collection and presentation of some of this information. we can only present the contributions that have been made and encourage all to consider providing papers and case studies for future conferences so that the learning can continue.

I encourage you all to share freely as much as you can so that we can have a safe productive industry that meets the needs of society.

Again welcome and I hope that you gain as much as you can from this conference.

Yours faithfully,

Duncan Chalmers
Conference Chairperson UNSW

EDITOR'S FOREWORD

Greetings to you all and welcome to the Gold Coast, Queensland, Australia.

It's been three years as a result of the COVID pandemic, since last we met in Perth for the 2019 Mine Vent biennial conference. As the Technical Editor, I would like to highlight our journey in bringing this conference together and various challenges faced along the way. Let me sincerely thank you all, and our sponsors in particular, for participating in this biennial Australian Mine Ventilation Conference series. Mine ventilation engineering is a critical safety and health function in various spheres. Our continued aim of this gathering has not changed, which is by co-creating and improving the health and safety of Australian and international mine workers who are contributing to the well-being of our fellow citizens. Thus, your individual contribution through participation, discussions, presenting and publishing in this proceeding is appreciated.

The proceedings of the 2022 Australian Mine Ventilation Conference 'Mine Vent 2022' focuses on the theme of Improving the productivity and safety of mines. The Organising Committee, chaired by Mr Duncan Chalmers, aims to share the latest developments and challenges faced by the industry over the last three years, by providing engineering solutions with embedded operational risk management (ORM) processes so that the Australian and global mining industry can improve the productivity and safety of our people. Over 50 papers were submitted this year, with cross-sections of commodities from Australia and around the world. Amongst these, 40 full papers have been accepted for presentations in person on mine ventilation engineering, a health and safety enabler.

The three-day conference sessions include: explosions and air blasts, coal and metal mine ventilation, main and auxiliary fans, mine gases, monitoring and control, electric vehicles, dust and DPM, heat and refrigeration engineering, safety and health hazard management, emergency egress, detection and control of spontaneous combustion. This conference is supported by featured keynote speakers from regulator, academia and the industry, who continually support this unique ventilation engineering profession over the years, namely, Mr Peter Newman (RSHQ, Qld), Prof Jim Galvin (UNSW) and Hope Mulvihill (NSW).

In addition, as an Australian representative and Chair of the IMVC Committee, it is with pleasure to inform you that Australia (AusIMM/UNSW) has won the bid to host the 12th International Mine Ventilation Congress in Sydney in 2024. This recognition to host the next IMVC is a significant acknowledgement of the Australian mine ventilation engineering community of mine ventilation engineering professionals, academics, researchers, suppliers and regulatory bodies. And it was made possible with every one of your contributions. I will be approaching people in the very near future to make it a great success.

I trust that you will enjoy the next three days and make use of this opportunity to enhance your networks in improving worker safety and operational productivity. Finally, thanks and appreciation go to all the technical peer reviewers (local and overseas) for their precious time and assistance in making these proceedings a reality. Also, special mention goes Amelia Lundstrom of AusIMM, who has been with us since the journey started as a constant link between the organising and technical committee along with Dr Guangyao Si of UNSW in ensuring the smooth flow of reviewed papers and primary authors.

It is hoped that the Mine Vent 2022 technical conference proceedings presented herein will be a valuable reference material. Best wishes to you all for a successful Mine Vent 2022 conference, and I look forward to meeting you all at the 12th International Mine Ventilation Congress in Sydney in 2024.

Yours faithfully,

Dr B Belle, MAusIMM(CP)
Proceedings Editor, Mine Vent 2022 Conference
Australian Representative and Chair IMVC

SPONSORS

Platinum Sponsor

Howden

Gold Sponsors

BBE GROUP

gordon

MTV Mine & Tunnel Ventilation

TLT-Turbo

zitrón

Silver Sponsor

OreTeck Mining Solutions

Conference Dinner Sponsor

Howden

Exhibition Lounge Sponsor

HATCH

Name Badge and Lanyard Sponsor

MINETEK

Session Sponsor

COAL FOAM

CONTENTS

Design

Health and safety

Operations

Technology

Design

Highlighting duct constraints in duct-fan systems operating under negative pressure

M Francoeur[1]

1. Ventilation Consultant, BBE Consulting Canada, Sudbury Canada P3E 5S1.
   Email: myriam.francoeur@bbegroup.ca

ABSTRACT

In the last decade, North American mines have witnessed the advent of extensive push duct-fan systems requiring in-line booster fans for ventilating greenfield underground project development. Several of these mines also resort to or explore the use of lengthy pull systems to swiftly exhaust contaminants from blind development tunnels. The first type of duct-fan system may be subjected to a partial vacuum, ie negative static pressure, upstream of the booster fan; the latter operates entirely under a partial vacuum. Both designs are sensitive to system effects that may further decrease their operational static pressure.

Generally, ducting is characterised by its geometrical properties (length, cross-sectional dimensions and aspect), its Atkinson friction factor and ad hoc leakage description. This is a rather incomplete property list for duct-fan systems that operate under a partial vacuum: thickness, stiffness and cross-section aspect determine ducting structural integrity as well as its coupling airtightness and section length. Then, the quality of the installation of a duct-fan system, a function of a duct stiffness, junction design and mine or contractor installation practices influence system effects.

However, ducting mechanical properties are often overlooked when designing a forwarding duct-fan system with negative static pressure potential or any exhausting duct-fan system. Failure to account for such properties may create unforeseen system effects that can lead to operational issues for both systems. In these proceedings, the author explores mechanical ducting properties as limiting factors in the design of duct-fan systems operating under negative static pressure. The author then highlights the role of a duct's thickness, wall stiffness, cross-sectional aspect and junction type for maintaining system integrity when subjected to atmospheric compression. An empirical method for determining negative static pressure ratings of ducting is also presented.

INTRODUCTION

Underground mine development relies massively on extensive and often complex forwarding duct-fan systems, though pull systems of equal complexity are being set-up as well. Francoeur, Bowling and McFadden (2022) speculate that the increasing prevalence of extensive duct-fan systems arises from technical and economic considerations, including:

- Postponing vertical development.

- Promoting long single-heading development.

- Favouring truck or conveyor haulage.

Such duct-fan systems, of considerable lengths – from 500 m to several thousand metres –, must deliver airflows between 30 m^3/s and 60 m^3/s in 1060 mm to 1830 mm ducting and present with significant Atkinson resistance R (Equation 1), where k is the system Atkinson friction factor, L, its length and d its diameter.

$$R = \frac{64kL}{\pi^2 d^5} \tag{1}$$

Today's increasing airflow requirements due to larger airway dimensions and more powerful diesel truck motors, combined with substantial Atkinson resistance, increase the high static pressure that must be countered by the fan(s) in the duct-fan system as per Square Law (Equation 2).

$$p_S = RQ^2 \tag{2}$$

In Equation 2, p_S is the static pressure and Q is the airflow.

Despite the latest improvements made to auxiliary vane axial fans to boost airflow and pressure outputs, generally, their pressure envelope is maxed at about 4 kPa for a 1520 mm diameter and single-stage arrangement. For this reason, auxiliary mine ventilation designers opt to either group two or more fans in series to increase their pressure output or divide the static pressure load between several fans spaced along the duct-fan system length. While simple, the former strategy imposes high pressures on the early ducting in push systems, thus increasing the risk of bursting – or collapse in the case of strictly exhausting systems. Moreover, fan grouping in series has been seldom studied, especially with three or more vane axial fans (Francoeur, Bowling and McFadden, 2022).

Distributing fans along the duct-fan system is a more common approach and can be done in three ways:

- The duct-fan system, made of canvas-based ducting is sectioned into several, equal segments, each with its own fan at its intake. Each section is separated by a 150 mm gap between the duct and the following fan, ie the in-line booster fan, to prevent the occurrence of negative static pressure at the latter fan's inlet. However, it may lead to uncontrolled recirculation if fan duty points are not set judiciously, ie if any downstream fan pulls more airflow than available upstream (Brake, 2008).

- A first, long canvas-based duct-fan system discharges fresh air from the outside or an underground fresh air source into a plenum space where it is picked up by a secondary canvas-based system that forwards it to a heading or another plenum space. Once again, it may lead to uncontrolled recirculation if the plenum-space fan's duty point is set at a similar or greater airflow than that of the primary duct-fan system (De Souza, 2015).

- The duct-fan system is made of a single, gapless section, ie in-line booster fans are connected to the ducting along with the entire system, from its intake to its outlet. Such a system is made of rigid or semi-rigid ducting, whether steel, glass-reinforced plastic (GRP) or thermoplastic, to account for a partial vacuum (Francoeur, Bowling and McFadden, 2022).

The third solution is being increasingly implemented in North American mines and is operated in South African deep mines (Holtzhausen, June 2021, personal communication). Typically, the design of such duct-fan systems relies on the rigid or semi-rigid ducting capacity to withstand negative static pressure (NSP). Yet, steel, GRP and thermoplastic ducting all have different mechanical and geometrical properties that determine their NSP ratings, including wall stiffness and thickness, cross-sectional profile, coupling type and length – if applicable; all steel and GRP ducting are considered NSP-rated while not every thermoplastic ducting can withstand a partial vacuum.

Failure to account for the abovementioned properties may create unforeseen system effects that can lead to a series of operational issues. Past its negative static pressure rating, ducting collapses in a swift and catastrophic manner with the following results, in order of seriousness:

1. Permanent deformation of ducting.

2. Rapid stalling of system fans, leading to blades and drive failure.

3. Disassembly or breakage of ducting segments, or both.

4. Propelling of ducting parts both near and far of the failure location.

The subsequent downtime to proceed with repairs ultimately leads to the loss of time and inevitable delay of development.

As ducting is the last and often the only line of defence against a duct-fan system collapse, it is critical to select one with a suitable negative static pressure rating for a given application. While Francoeur, Bowling and McFadden (2022) have broached the topic of ducting NSP ratings, they have limited themselves to wall stiffness and duct geometry.

In these proceedings, the author revisits the underlying principles behind the effect of negative static pressure on ducting, this time through the theory of the buckling of thin-walled cylinders subjected to external pressure. Then, the author examines the impact of the mechanical and geometrical properties on the ducting under atmospheric compression, including wall stiffness and thickness, duct diameter and profile as well as the effect of stiffeners and coupling design. Finally, the author presents a methodology for testing negative static pressure ratings of rigid or semi-rigid ducting and practical considerations for designing reliable systems operating partially or fully under a partial vacuum.

DUCTING – A THIN-WALLED CYLINDER

The topic of thin-walled cylinders subjected to external pressure, especially their *failure*, has been studied extensively in mechanical engineering (Moss and Basic, 2013). Examples of thin-walled cylinders include distillation towers, freight pressurised gas storage tanks and even space shuttles. While these vessels are made to contain gases – or liquids – at a pressure greater than that of the atmosphere, they are also designed to withstand *some* compressive forces without failure; a failure may arise from an accidentally-induced internal vacuum from unloading a pressurised gas vessel or a sufficient temperature drop within the vessel that forces condensation. The failing is both sudden and generally destructive as the cylinder walls crumple beyond recovery.

Mine ventilation ducting is also a type of thin-walled cylinder, which may be subjected to the compressive force of the atmosphere when its internal static pressure drops below zero. When a duct-fan system is powered off, its inner pressure is identical to the atmospheric pressure (Figure 1a). Then, upon the creation of a partial vacuum from the pulling force of an in-line fan, the atmosphere applies a uniform, circumferential inward force on the ducting. The ducting wall stresses react to such compression; the sum of the wall compressive strength and the outward force applied by the air within the ducting cancel out the atmosphere compressive force (Figure 1b) and the ducting maintains its profile. However, past a given partial vacuum – depending on the ducting properties and reinforcement – the compressive force of the atmosphere is greater than the combined forces of the wall compressive strength and the inner air. Therefore, the ducting can no longer maintain its profile and starts to cave in (Figure 1c).

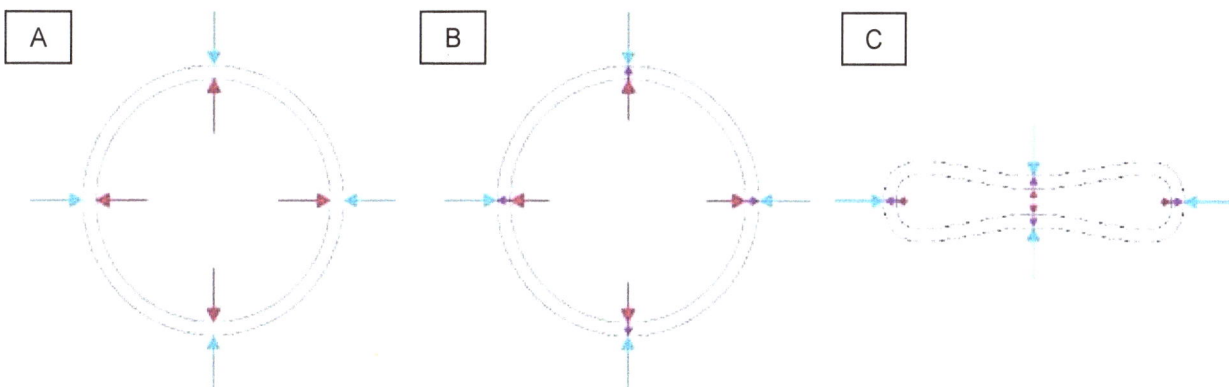

FIG 1 – Force diagrams of the different stages of ducting deformation due to the atmospheric compressive force: (a) No differential pressure is applied. (b) A partial vacuum is created and the ducting maintains its profile. (c) The ducting caves in as the compressive force of the atmosphere increases beyond the ducting rating. Blue: Outer atmosphere-induced inward force. Magenta: Inner air outward force. Purple: Ducting wall compressive strength.

The initial ducting deformation is generally elastic, ie the ducting recovers its profile as the partial vacuum disappears. However, at a given partial vacuum, the wall reaches its yield strength and any deformation becomes irreversible and permanent, resulting in buckling, with severe buckling leading to a collapse. In some cases, the ducting may even fracture locally.

The stiffer the ducting, the more likely it will undergo permanent rather than elastic deformation and collapse without early buckling. On the other hand, semi-rigid thermoplastic ducting will progressively buckle before collapsing. Figure 2 shows a pull duct-fan system made of 4.2 mm thermoplastic ducting in a Canadian mine. The ducting buckles without collapsing; the junction steel rings have been deformed as well.

FIG 2 – 762 mm thermoplastic ducting installation showing buckling.

As for any thin-walled cylinders subjected to external pressure, a ducting critical pressure p_{CE}, ie the maximum compressive pressure it can withstand, depends on the Young's modulus E of the wall material, which describes its stiffness and its thickness t; p_{CE} is inversely proportional to the cube of its diameter d (Equation 3). Equation 3 reveals that for a given material, larger ducting will exhibit smaller critical pressures unless made thicker.

$$p_{CE} \propto E \times \frac{t^3}{d^3} \tag{3}$$

Table 1 presents typical mine ventilation rigid and semi-rigid circular, 1067 mm ducting NSP rating, reproduced from Francoeur, Bowling and McFadden (2022). One may notice that while galvanised steel ducting is thinner, it presents with the greatest NSP rating because of its higher stiffness. Table 1 NSP ratings for galvanised steel and polyester-based GRP ducting include a 2.5-to-1 factor of safety; the high-density polyethylene (HDPE) values are critical pressures.

TABLE 1

Negative static pressure ratings for 1067 mm rigid and semi-rigid circular ducting based on their respective wall material types, average Young's moduli and thicknesses. *Maximum NSP ratings include the addition of stiffeners to improve ducting structural integrity.

Wall material	$E_{AVERAGE}$ (GPa)	t (mm)	NSP rating (kPa)
Galvanised steel	200	1.2	5.1
GRP (polyester matrix)	10	6.4	2.2–5.0*
Thermoplastic (HDPE)	1.5	5.3	1.8–4.3*

Critical pressures also depend on material nonlinearity and are largely sensitive to geometrical, material and manufacturing variations (Moss and Basic, 2013). The following variations can lower ducting NSP ratings due to localised effects such as:

- Longitudinal welds, which change ducting wall internal stresses.

- Thickness discrepancies that make ducting either stiffer or more flexible.

- Wall material imperfections that lead to different material properties.

- Indentations or other small deformations, which change ducting profile.

The first three sources of NSP rating variations are strongly dependent on the uniformity of ducting material; the last is inherently linked to handling, transportation and operational conditions.

Also, some ducting material properties may change according to temperature variations. This is especially true for thermoplastic ducting, which becomes more flexible with rising temperatures, effectively reducing their NSP ratings (SMACNA, 1995). Additionally, thermoplastic ducting diameters fluctuate with temperature because of thermal expansion. Besides, thermoplastic-based ducting cannot be employed at temperatures beyond their softening point.

A circular profile is the most desirable geometry for partial vacuum rating. Because of its symmetrical cross-section, all atmospheric compressive forces are applied circumferentially and therefore nullify. However, any deviation from a circular symmetry, whether an indentation, a flattened section, or an elongated cross-section, breaks the compressive force uniformity and results in lower ducting NSP ratings. For example, Figure 3 compares the atmospheric compressive force applied to a circular ducting of diameter d to that of an oval ducting of semi-major and semi-minor axes a and b respectively. F_C, the compressive force applied on the ducting outer surface, is given by Equation 4:

$$F_c = p_{ATM} \times A_{PRO} \tag{4}$$

where p_{ATM} is the local atmospheric pressure at the ducting location – a constant – and A_{PRO} is the projected surface on which the force is applied. In the case of circular ducting, $A_{PRO} = L \times d$, where L is the ducting length, is the same independently of the compressive force orientation and so does the force ratio. As for an oval ducting, $A_{PRO,a} = L \times 2a$ whereas $A_{PRO,b} = L \times 2b$; evidently, as $A_{PRO,a} > A_{PRO,b}$, the elongated surfaces of the ducting are therefore subjected to greater compressive forces, effectively decreasing its NSP rating. The previous analysis also applies to flat oval (oblong) ducting to describe their inferior partial vacuum ratings compared to those of circular ducting.

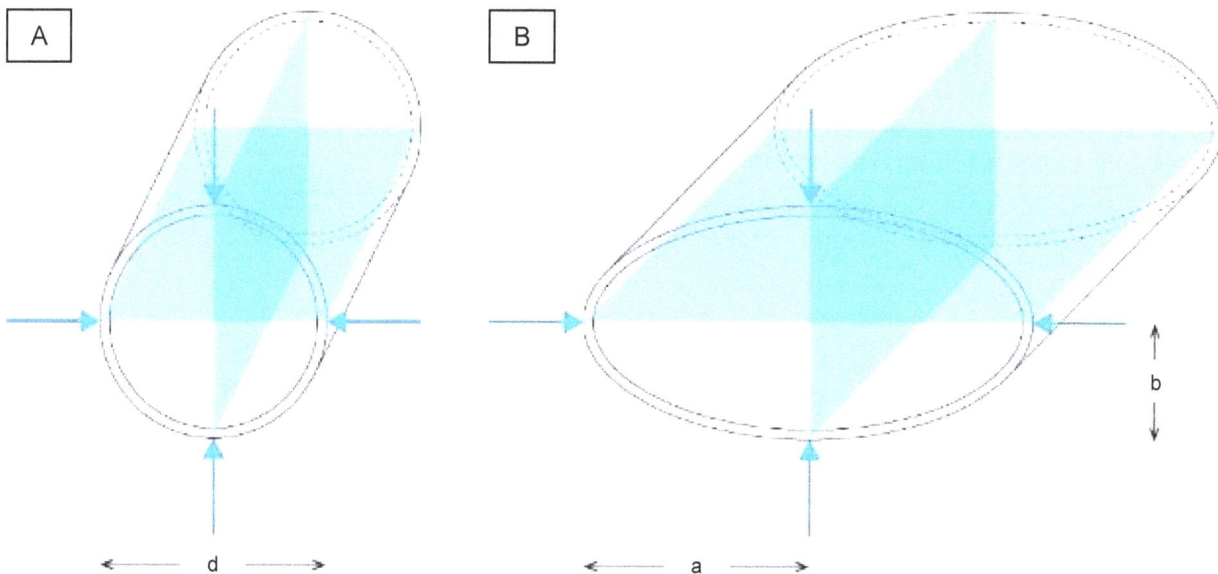

FIG 3 – Force diagram of the compressive force of the atmosphere on circular (a) and oval ducting (b). The blue arrows represent the compressive force of the atmosphere; the planes indicate the projected surface on which the compressive force is applied.

One must recall that a duct-fan system is made of numerous duct sections, each being a thin-walled cylinder. Static pressure profile is not homogeneous along the duct-fan system's length because of airflow losses and shock losses. Consequently, no duct is subjected to the same partial vacuum; furthermore, the closest ducts to a fan inlet are submitted to the greatest compressive forces in the duct-fan system.

DUCTING REINFORCEMENT

Ducting can be reinforced to raise its NSP rating regardless of its wall material, thickness and diameter. Reinforcement is achieved with circumferential stiffeners, either welded or detachable, that can be added to ducting of different geometries although circular ducting still offers better NSP ratings for a given thickness and material. Ducting stiffening is also a more practical means to achieve a predetermined rating for a given application without increasing thickness significantly.

Thicker ducts are also heavier and require stronger back support and specialised lifting vehicle (fan handler, scissor lift (while fan handlers and scissor lifts are common equipment in Canadian underground mines, there are seldom used outside North America)). Besides, rigid and semi-rigid ducting is already manufactured in short sections – relative to canvas-based ducting – with typical lengths between 1.8 m and 3.1 m to ease handling and transportation. Still, NSP-rated ducting is generally 10 to 20 more expensive than its canvas-based counterpart (G+ Plastics, 2020, personal communications).

A ducting section length is a critical component of its NSP rating. As seen in Figure 3, the compressive force of the atmosphere applied to the ducting depends on the projected area from the said ducting surface, which is proportional to its length; as a duct length increases, so does the compressive force of the atmosphere. As a result, shorter duct lengths are also preferred in partial vacuum operations.

Rigid and semi-rigid duct couplings are either of the flanges, bell-and-spigot, or insertion type; they strengthen the ducting as they induce additional stresses in the material and therefore act as stiffeners. By design, flanged junctions produce the greatest stiffening of all couplings, they are thus preferred for negative pressure applications. Additionally, the grooved spiralised lockseams of galvanised steel ducting contribute to its negative pressure rating, albeit modestly.

Couplings alone are insufficient to reinforce ducting and so NSP ducts are often braced with circumferential stiffeners (stiffening rings). These can be welded to the exterior of the ducting or later added; they may be either screwed onto the ducting wall or tightened enough to compress the wall to locally increase its stiffness. Stiffening rings essentially reduce the length of AP_RO. Their effect is optimum when they are evenly spaced along the ducting length (Figure 4). The more stiffening rings are added, the better the ducting NSP rating as spacing is reduced and critical pressures are inversely proportional to the spacing (Annaratone, 2009).

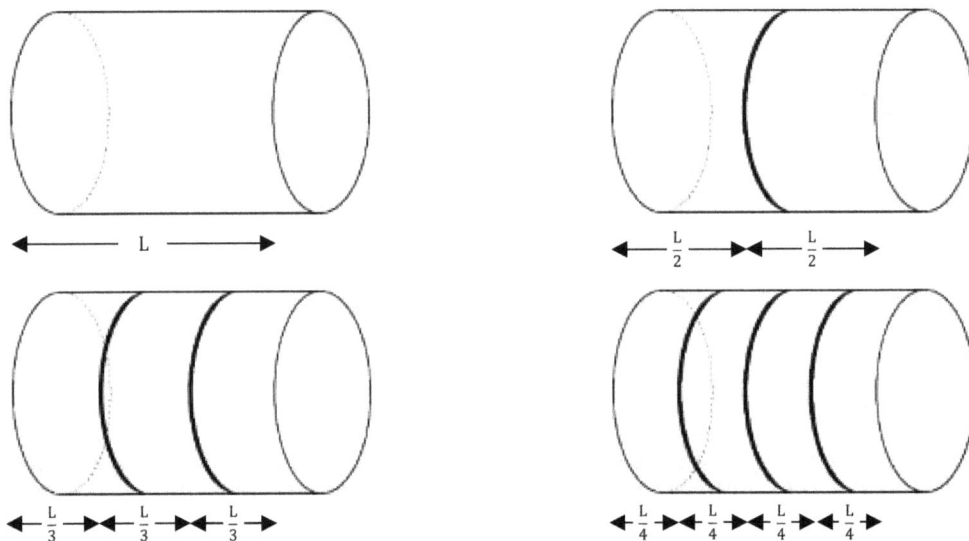

FIG 4 – Various stiffening ring configurations on a single duct of length *L*. Stiffeners are represented by bold contours.

If internal stiffeners have been tested (G+ Plastics, 2019, personal communications), they however create additional pressure losses in a duct-fan system as they obstruct the airflow and generate unwanted turbulence; external stiffeners are therefore preferred.

STANDARDISED DUCTING CONSTRUCTION

Designers can compute ducting NSP ratings to satisfy a given application. The Sheet Metal and Air Conditioning Contractors' National Association (SMACNA) presents guidelines to design ducting made of either steel, glass-reinforced plastic, or thermoplastic – specifically polyvinyl chloride (PVC) – and provides their NSP ratings based on specific material mechanical properties, thickness, geometry, coupling type and stiffening ring design and spacing (SMACNA, 1995, 1997, 1999). SMACNA assumes reasonable material and production quality controls from ducting manufacturers. Table 2 indicates the maximum design NSP ratings for galvanised steel, glass-reinforced plastic and thermoplastic ducting as per SMACNA for a given temperature of 37.8°C.

TABLE 2

Round duct maximum NSP ratings as per SMACNA, including stiffening. *Excludes furan-, phenolic- and acrylic-based matrices. ** With flanged couplings. † At 37.8°C.

Wall material	Maximum NSP rating (kPa)	Maximum diameter (mm)
Galvanised steel	7.5	2440
GRP* **	7.5	1800
Thermoplastic (PVC)†	2.5	2134

SMACNA guidelines are laid out for industrial ducted ventilation, not ducted auxiliary mine ventilation: they present methods to compute the different structural loads to which the ducting will be subjected, including reinforcement, ice, wind, heat, hanging, maintenance, corrosion and earthquakes. SMACNA does not consider underground blast waves loads and mobile equipment impacts. The *Thermoplastic Duct (PVC) Construction Manual* does consider the long-term effect of temperature cycling and creep – specifically the ducting sagging as a result of gravity – on such materials because of their lower stiffness compared to that of galvanised steel and GRP.

DETERMINING DUCTING NSP RATINGS EMPIRICALLY

It may be best to either establish or validate a ducting negative static pressure rating empirically for mine ventilation applications. Refined numerical models yield NSP ratings with a ±5 per cent error for industrial thin-walled cylinder applications (NASA, 2020); however, they do not account for the various loads a ducting is subjected to in the underground environment besides atmospheric compression. It is still possible to build a crude stress analysis model to predict ducting behaviour under external pressure. Yet one must bear in mind that novel ducting materials properties may not be adequately characterised and the ducting stress analysis model can only be based on approximate values.

To determine ducting NSP ratings empirically, one must build a closed pressure vessel in which a vacuum can be created and then measure p_{CE} at the maximum allowed wall deformation – or proceed to collapse the ducting and assess the consequences of a rapid failure.

Figure 5a illustrates an ideal critical pressure testing set-up; Figure 5b shows the 2018 set-up used by G+ Plastics, a thermoplastic duct manufacturer out of Rouyn-Noranda, Canada, to determine its 762 mm duct NSP rating for different thicknesses and stiffening ring configurations. Such a set-up should include an assembly of at least three ducts with the vessel's outer ends sealed and all couplings made airtight during the installation to achieve a stable partial vacuum – the goal is to assess the ducting behaviour under compression and leaky junction only stall the testing. If stiffening rings are added, they must be tightly secured and spaced evenly along the length of every ducting section in the set-up (see Figure 4) to optimise stiffening.

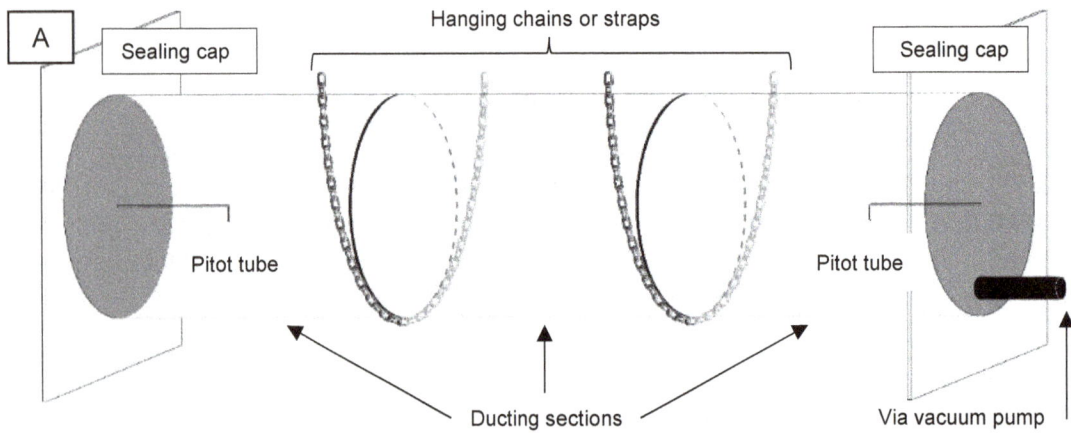

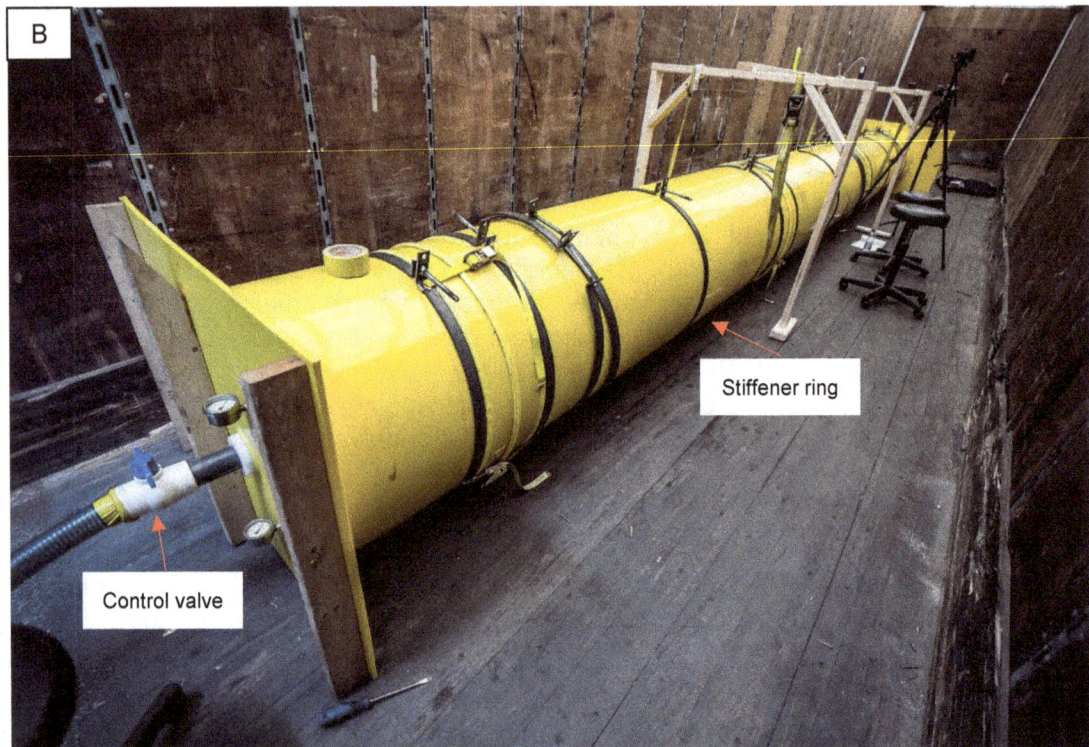

FIG 5 – (a) Conceptual critical pressure measurement set-up. (b) G+ Plastics 2018 critical pressure measurement set-up.

A valve-controlled vacuum pump is connected to one of the vessel's ends to create a partial vacuum. The valve regulates the system's emptying rate: a high rate expedites the emptying, which can lead to the collapse of the ducting, while a slow rate ensures better characterisation of the ducting deformation although extending the testing duration. The negative static pressure is probed with Pitot-static tubes on both sides of the vessel to ensure the uniformity of the partial vacuum and subsequently determine a duct's critical pressure – NSP ratings are afterwards computed by dividing the critical pressure by the FOS.

As shown in Figure 5, the ducts are hung above the ground to both reproduce the underground mine installation and prevent them from flattening, which would change the ducting's geometry and reduce its NSP rating. Chains or straps are to be positioned at the couplings, which have greater structural strength.

Determining a ducting critical pressure can withstand before collapsing can prove a delicate and challenging task. One must first set an upper limit to the ducting deflection resulting from the partial vacuum and then link it to the negative static pressure reading, keeping in mind that once flattened, the ducting is prone to a rapid collapse. However, if thermoplastic-based ducting deflections can be visible to the naked eye, GRP or steel variations are minute – in the order of the millimetre – and therefore hard to measure unless using a deflectometer.

Under compression, the ducting deflects following the buckling modes n of thin-walled cylinders subjected to external pressure illustrated in Figure 6, which shows modes 2, 3 and 4.

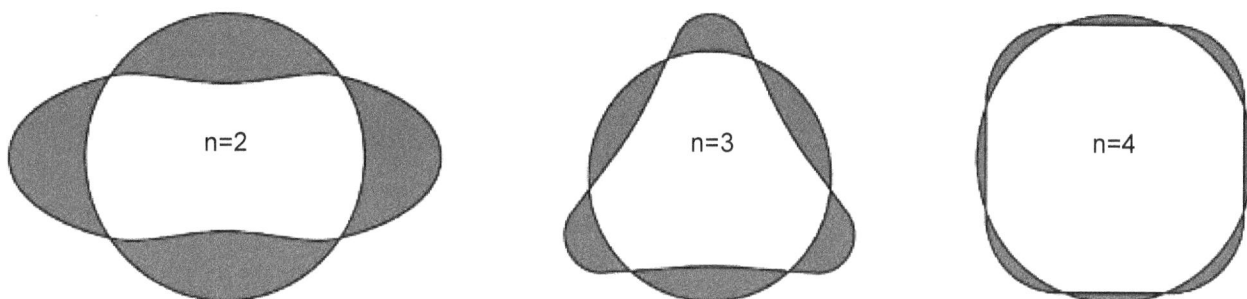

FIG 6 – Thin-walled cylinders buckling modes 2, 3 and 4.

The second mode, also illustrated in Figure 1c, is typical for non-stiffened buckled ducting whereas the third and higher modes are encountered with an increasing number of stiffening rings (De Paor *et al*, 2012). Readers must note that the angle at which the buckling modes appear is unpredictable and it is therefore recommended to assign observers to identify the location of the initial buckling.

For thermoplastic-based ducting, critical pressure tests must be conducted in a temperature-controlled room and the ducts must thermalised prior to testing. One shall also ensure that the ducting temperature is uniform along its length and diameter or semi-axes, thus requiring good air mixing in the room. Besides, critical pressures are to be measured for several temperatures to account for thermoplastic temperature-dependent stiffness.

DESIGN CONSIDERATIONS

In these proceedings, the author proposes a new set of parameters to evaluate when designing a duct-fan system in order to increase its reliability in the eventuality of a partial vacuum. All parameters are to be weighted against the risk of operating a system under negative pressure, even partially. They can be grouped into five categories, illustrated in Figure 7 and described below:

- Performance: airflow delivery, system resistance minimisation.

- Operations: power, maintenance, system lifetime.

- Procurement: duct, fan, transport, production and/or delivery delays.

- Installation: shock loss minimisation, junction airtightness, duration, lifting equipment availability, clearance compliance.

- Reliability: NSP ducting, static pressure sensors, live telemetering, fan interlocking, vacuum-relief system.

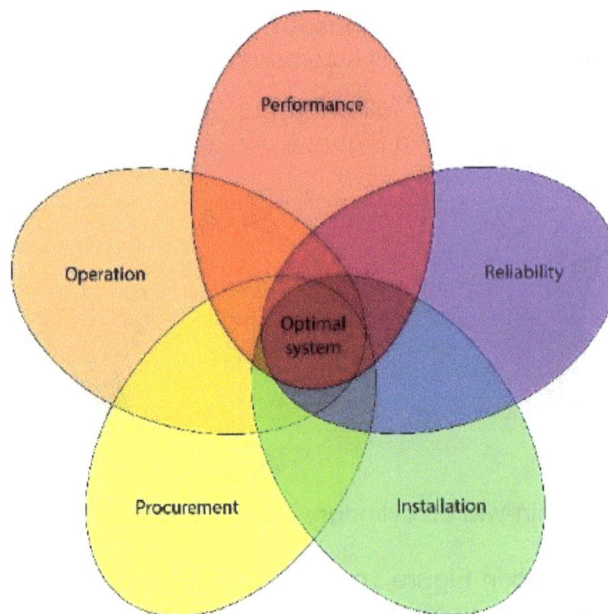

FIG 7 – Parameters to consider for the design of an optimal duct-fan system.

The design should also include the type of auxiliary ventilation system (strict push or pull or a combination of these, the use of plenums), the maximum length of development, the maximum allowed duct diameter, the fan(s) performance and their location along the system – as demonstrated by Francoeur, Bowling and McFadden (2022), the position of in-line booster fans in gapless system is correlated with the extent and amplitude of a partial vacuum. Finally, if planning for a long development, designers should weight in the aging of the duct-fan system, which generally manifest as an increase in the ducting friction factor and leakage.

CONCLUSIONS

In these proceedings, the author has presented the foundations of ducting critical pressure and negative static pressure ratings. Special attention was paid to wall material properties, particularly stiffness and uniformity, cross-sectional profile and section length, which limit the ability of a given ducting to withstand negative static pressure. The author has also examined the effects of stiffening rings to achieve higher ducting NSP ratings while optimising its weight and cost. Then, these proceedings conclude with the presentation of an empirical method to measure ducting critical pressures with which ducting NSP ratings can be computed.

With mine ventilation increasingly relying on duct-fan systems subjected to a partial vacuum, it is important to remember that there is no single, all-purpose ducting that fully satisfies techno-economical constraints and limitations, especially operational pressures. Any ducting, even the toughest, can collapse 'unexpectedly' if the system modelling and/or operation disregards its NSP rating. Its reliability is of utmost significance for the completion of development in a timely manner as well as the safety of underground personnel.

ACKNOWLEDGEMENTS

The author has performed ducting critical pressure tests during her time as Ventilation Specialist at G+ Plastics. The testing methodology and set-up was developed by Samuel Chartré-Bérubé, former Mechanical Engineering Technician at G+ Plastics and Pierre Trudel, G+ Plastics Vice-President and Production Manager; the methodology was further improved by the author. The author would like to thank G+ Plastics for its contribution to these proceedings and for providing confidential reports and pictures.

REFERENCES

Annaratone, D, 2009. *Pressure Vessel Design*, p 140 (Springer-Verlag: Berlin).

Brake, D J, 2008. *Mine Ventilation: A Practitioner's Manual, Revision 33–02*, pp 460–461 (Mine Ventilation Australia: Brisbane).

De Paor, C, Kelliher, D, Cronin, K, Wright, W M D and McSweeney, S G, 2012. Prediction of vacuum-induced buckling pressures on thin-walled cylinders. *Thin-Walled Structures*, 55:1–10.

De Souza, E, 2015. *Auxiliary mine ventilation manual, Version 1.6*, pp 38–40 (Workplace Safety North: North Bay).

Francoeur, M, Bowling, J R R and McFadden, A, 2022. Extensive Duct-Fan Systems with In-line Booster Fans: Planning for Partial Vacuum and an Algorithm for In-line Booster Fan Positioning [online]. *Mining, Metallurgy & Exploration*. Available from <https://doi.org/10.1007/s42461–022–00575–3> [Accessed: 08 April 2022].

Moss, D R and Basic, M, 2013. *Pressure Vessel Design Manual*, 4th edition, pp 3–7 (Elsevier: Oxford).

National Aeronautics and Space Administration (NASA), 2020. Buckling of Thin-Walled Circular Cylinders, NASA/SP-8007–2020/REV 2 [online], Available from: <https://ntrs.nasa.gov/citations/20205011530 > [Accessed: 20 April 2022].

Sheet Metal and Air Conditioning Contractors' National Association (SMACNA), 1995. *Thermoplastic Duct (PVC) Construction Manual*, p 2.3, p A.1 (Chantilly: Virginia).

Sheet Metal and Air Conditioning Contractors' National Association (SMACNA), 1997. *Thermoset FRP Duct Construction Manual*, p C.3 (Chantilly: Virginia).

Sheet Metal and Air Conditioning Contractors' National Association (SMACNA), 1999. *ANSI/SMACNA 005–2003. Round Industrial Duct Construction Standards*, p 11.29 (Chantilly: Virginia).

Coalmine ventilation – examining the past, looking to the future

J M Galvin[1]

1. Emeritus Professor, University of New South Wales, Sydney NSW 2000.
   Email: j.galvin@bigpond.net.au

ABSTRACT

History attests to the catastrophic consequences that fire, explosion and contaminated, noxious and irrespirable atmospheres can have for workplace health and safety in underground coal mining. It also confirms that fires and explosions present the greatest risk of partial or total loss of underground mines and equipment and that this risk continues to materialise. The design, implementation, operation and maintenance of the mine ventilation system are core controls for managing the hazards associated with these risks, both for preventing them from materialising and for mitigating their consequences if they do materialise. These are most effective when embedded in a risk management approach.

This paper provides a brief summary of the origins and basics of a risk management approach to mine ventilation and reviews select case studies of coal mining mishaps over the last 200 years that have some association with mine ventilation. It identifies insights, learnings and initiatives from these relevant to the effective management of mine ventilation risk. Learnings relate to:

- The need for ventilation management to be integrated into mine planning over the full life cycle of the operation.

- The difficulties in identifying the ignition source for an explosion in an underground coalmine.

- The importance of contingency planning to manage consequences.

- The basis for accepting risk.

An important initiative in the late 1990s in Australia was the establishment of the statutory position of *ventilation officer*. Ventilation officers have vital contributions to make in the design of mine ventilation systems and the development and execution of risk-based management plans for these systems. This paper notes the need for ventilation officers to be closely engaged and consulted in day-to-day operational decisions that may impact on mine ventilation and the importance of undertaking continuing professional development to maintain the currency of their competencies. The paper looks back as a guide to going forward.

INTRODUCTION

History attests to fire, explosion and contaminated, noxious and irrespirable atmospheres being very high risks to workplace health and safety in underground coal mining. All three core contributors to determining the magnitude of these risks, namely likelihood, exposure and consequence, are elevated in the confined space of an underground coalmine, giving rise to the potential for the total loss of a mine and all persons within it when control is lost.

Up until about 1980, a prescriptive legislative approach was generally adopted in an endeavour to prevent these risks from materialising and to control them should they still materialise. The legislation required the appointment of a *mine manager*, being a statutory official who was accountable for all aspects of the mine. It evolved over time in a reactive manner in response to adverse incidents, with each revision stipulating additional actions that a mine manager was required to take in an endeavour to eliminate the identified causes of these incidents. Subsequently, many jurisdictions including Britain, Australia and New Zealand, moved to a proactive risk-based legislative approach to managing risk in the workplace. The review presented in this paper of select case studies of mining mishaps associated with mine ventilation over the last 200 years reinforces the aptness a risk management approach to mine ventilation. The statutory ventilation officer has a vital role to play in bringing knowledge and competency to risk assessment and management processes, in implementing the outcomes of these processes, and in monitoring and improving their effectiveness.

EVOLUTION OF A RISK BASED APPROACH TO MINE VENTILATION

The invention of the steam engine and improvements to its design during the 1700s were a boom for underground coal mining from a production perspective but were to bring great misery and suffering to those who worked in the mines. Not only did the steam engine create a huge demand for coal but it also provided a practical means for pumping water, thereby enabling this demand to be met by mining at much greater depths and over much larger areas. However, the greater depth of mining also resulted in the gas (methane) content of the extracted coal seams being higher and in the introduction of pillar extraction and longwall mining techniques to compensate for reduced rates of production and productivity. Both of these changes were to have devasting impacts on the safety of mineworkers well into the 1900s, as reflected in Table 1. While explosions involving high loss of life during that period have received a lot of press, the statistics reveal that falls of ground actually accounted for more lives lost than any other cause. Fatal fall of ground events were much more likely to occur than fatal fire and explosion events but generally had much lower death tolls.

TABLE 1
Historic fatality rates for underground coal mining in Britain (Galvin and Hocking, 1994).

| | | | | | | FATAL ACCIDENT RATES | | | |
| | | | | | | Per 1000 employed | | Per million tonnes | |
Year	Total Fatal Accidents Below Ground	Total Deaths Due to Falls of Ground	% Due to Falls of Ground	Total Persons Employed Below Ground	Annual Output (million tonnes)	Other than Falls of Ground	Falls of Ground	Other than Falls of Ground	Falls of Ground
1851	790	346	43.8%	171,893	53.0	2.58	2.01	8.39	6.5
1911	927	639	68.9%	863,512	271.9	0.33	0.74	1.06	2.3
*NSW	23	10	43.5%	43,830	201.0	0.30	0.23	0.07	0.05

*NSW Underground 1990-1993 (incl.)

While the earliest recorded date of death by an explosion of methane in an underground coalmine in Britain is probably October 1621, very little is known about fatal accidents that occurred in British coalmines during the 1700s and early 1800s (Duckham and Duckham, 1973). The requirement to record fatalities on an annual basis only came into effect in Britain in 1850, not long after the passing of the first Coal Mines Act in 1843.

Up until 1843, girls and boys as young as six were employed in underground coalmines in Britain. Of the 453 persons employed in Howgill Colliery in 1820, for example, 39 were girls and 63 were boys between the ages of 7 and 14 (Devlin, 1988). The youngest children were employed as *trappers* and played a critical role in controlling coalmine ventilation. They had the task of sitting beside one of the ventilation doors associated with *'coursing the air'* around the working area as shown in Figure 1, opening and closing it every time someone or something needed to pass through it. Shift length was up to 14 hours and they could spend much of that time in the dark. Explosions of methane are known to have occurred as a result of trappers leaving a ventilation door open (Devlin, 1988).

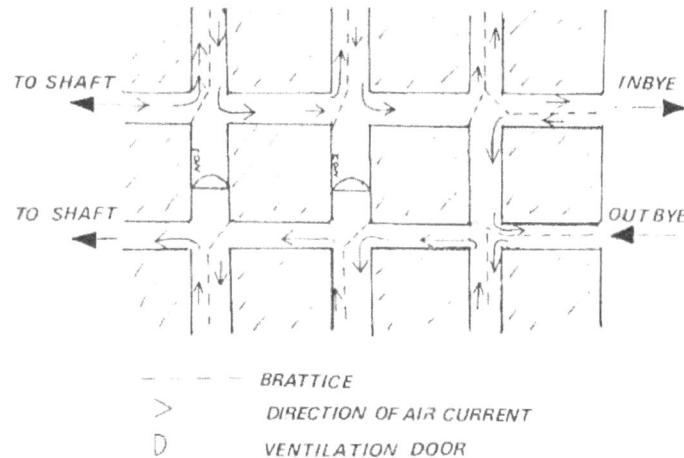

FIG 1 – Ventilation circuit for bord and pillar mining in the early 1800s (Devlin, 1988).

The first legislation to regulate coalmines in Australia was passed in 1851 and mirrored that passed in Britain in 1843. The primary objective of the legislation was to improve health and safety in mines. The legislation subsequently evolved in a highly prescriptive manner in both jurisdictions as it attempted to respond to adverse incidents.

Over time, advances in mining technology, production methods and mining legislation resulted in some of the traditional functions of a mine manager, such as mine surveying, being devolved to specialists. However, the duties associated with mine ventilation remained embedded with and jealously guarded by coalmine managers. This progressively changed from around 1998 in Australia, with the primary functions, accountabilities and responsibilities associated with mine ventilation being transferred from the mine manager (or *manager of mining engineering* as is now the case in NSW) to the statutory function of mine ventilation officer and with the ventilation competency requirements to fulfill this role being significantly higher than those required of a mine manager. These changes were driven by a number of serious mine ventilation related incidents in Australian coalmines and a government move away from a prescriptive legislative approach to workplace health and safety to a risk management approach.

By the late 1960s, the effectiveness of prescriptive legislative for achieving a safe place of work was being called into question in Britain, leading to the establishment of the Robens Inquiry. The Inquiry concluded (Robens *et al*, 1972) that prescriptive style legislation was only partially effective because it:

- is reactive

- does not cover all circumstances

- does not keep up to date with evolving knowledge, technology and practice

- does not encourage owners and management to seek out risks and develop their own set of controls

- can create a mindset amongst management that, because it is complying with the prescribed legislative requirements, health and safety are being properly managed.

The Robens Inquiry recommended a shift from prescriptive occupational health and safety legislation to so-called enabling legislation in which, in theory, the regulator sets the performance standards and leaves it up to the owner of the risk to decide how to best achieve these standards. This recommendation has been adopted in many countries whose legal system is based on the British system, including Australia. It is founded on formalised risk management processes that have regard to ISO 31000:2009 Risk Management – Principles and Guidelines (ISO 31000, 2009), an international standard that applies to controlling all forms of risk, not just risk to health and safety. These processes have been adopted by the International Labour Organisation in its Safety and Health in Mines Convention, 1995 (No 176).

In practice, however, some controls required to manage health and safety in the workplace are considered so critical and essential that any latitude for the risk owner to change performance criteria or dispense with the controls altogether has been removed by continuing to prescribe them in the new style of enabling legislation. This applies in particular to the ventilation of underground coalmines.

Figure 2 summarises the fundamental steps involved in the risk management process. These steps are premised on the philosophy that uncertainty gives rise to *risk*, which is a combined measure of the *likelihood* of an event occurring (taking into account the level of *exposure* to the event) and the *consequences* should the event occur. A source of potential harm or loss which could result in an event is referred as a *hazard*, with any means by which the hazard could materialise being referred to as a *threat*. The consequences of an event influence the level of uncertainty that can be tolerated. In Australian mining legislation, a hazard with the potential to result in the loss of two or more lives is referred to as a *principal hazard*.

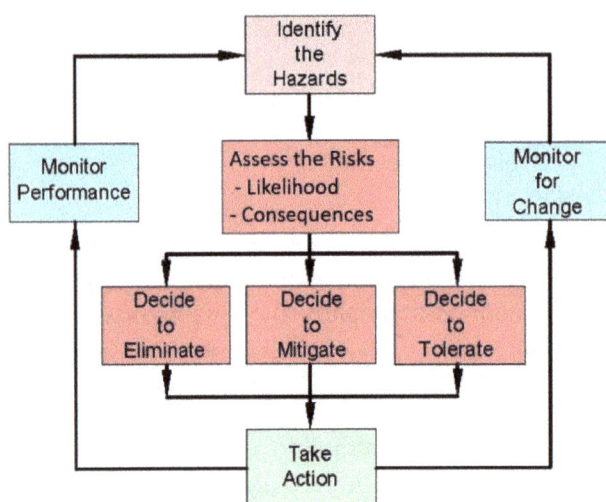

FIG 2 – The basic framework for managing risk (adapted from Joy, 1998).

The risk management process commences with identifying hazards and then assessing associated likelihood and consequences in order to determine the risk presented by each hazard. Next, controls are devised to eliminate each hazard where possible, or otherwise to reduce the risk associated with it to an acceptable level. These controls need to be risk assessed in their own right to confirm their likely effectiveness, to verify that they will not give rise to higher risks than those they are intended to address, and to determine residual risk levels. Then, having implemented the controls, it is essential that performance is monitored to verify the effectiveness of the risk assessment process. It is also essential that monitoring for change is undertaken to identify any deviations from the conditions and circumstances on which the risk management process was based and to intervene in a timely manner before a hazard materialises. Hence, the risk management framework equates to a continuous improvement process of Plan-Do-Check-Act.

Risk management has added significance for the ventilation of underground coalmines because mitigation and recovery measures also need to be pre-planned to minimise the consequences of any hazard that materialises. A bow tie diagram, illustrated in Figure 3, is a particularly useful tool for analysing risk. It provides a powerful graphical representation of upstream threats and downstream consequences and facilitates the identification of preventative controls and, should the unwanted event still occur, contingencies for managing and mitigating the consequences.

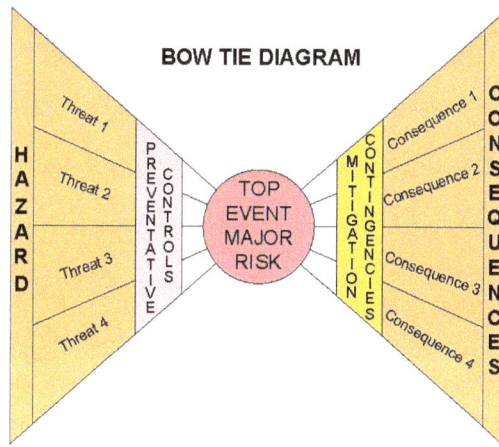

FIG 3 – The concept of a 'bow tie' diagram for analysing risk.

The outcomes of risk analysis and risk assessments inform the development of Trigger Action Response Plans (TARPs). A TARP is a plan designed to prevent a hazard from developing into a *top event* by proactively identifying precursory signs; assigning a series of staged threshold limits, or trigger levels, to each precursor; and specifying responses and response accountabilities when a trigger level is reached (Galvin, 2016). They can also be used to manage the consequences of a top event should it still occur, in which case they may place a high focus on contingency planning. Hence, TARPs are a very important elements of principal hazard management plans, such as Mine Ventilation Management Plans and Spontaneous Combustion Management Plans.

The effectiveness of a risk management approach as described is highly dependent on the knowledge and competence of the persons who input into the risk analysis and assessment processes, who develop the management plans and who are responsible for implementing the management plans. Ventilation officers have vital roles to play in all those aspects.

The Australian mining industry began to adopt a risk management approach to health and safety in the mid-1980s and to support it with a range of guidelines and research and development initiatives. The benefits of this approach, supported by technological innovation, are reflected, for example, in trends and improvements in the safety performance of the NSW coal sector, shown in Figure 4. Since the early 1980s, this sector has experienced a 15-fold decrease in fatalities, with a number of fatality free years, and a 10-fold decrease in lost time injuries per one million employee hours worked (LTIFR).

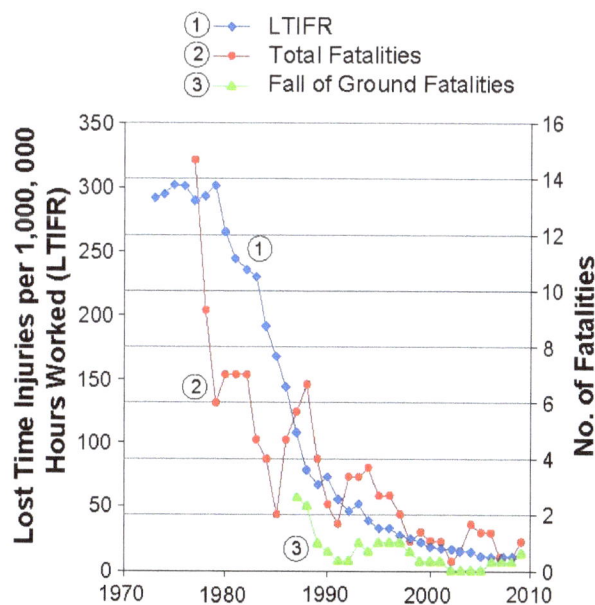

FIG 4 – Trends in safety performance measures for the NSW coal mining sector, expressed as three year rolling averages (Galvin, 2016).

CASE STUDIES RELATED TO VENTILATION INCIDENTS

Case studies of incidents related to mine ventilation provide insight into the broad range of factors and circumstances that can impact on the design, implementation, operation and maintenance of mine ventilation systems and provide the opportunity to refresh and reinforce learnings from past experience. The following case studies represent a snapshot over 200 years of major incidents in the western world, with a focus on those of particular relevance to underground coal mining operations and practices in Britain and Australia, which tended to evolve in parallel.

Felling Colliery, England; 25 May 1812

An explosion resulted in the deaths of 92 mineworkers. The cause of the explosion was not definitively established but most probably an ignition of methane.

Outcome: George Stephenson, an engine wright employed in the local collieries, designed a safety lamp known as the Geordie Lamp which was the forerunner to the development of the Davy safety lamp.

Hartley Colliery, England; 16 January 1862

Half of the 42 t cast iron beam of a pumping engine snapped off and fell down the colliery's only shaft, which was partitioned for furnace ventilation. The debris chocked off the shaft, blocking egress as shown in Figure 5. The furnace was extinguished by those trapped underground but it continued to smoulder. It took six days for rescuers to clear and secure the shaft. They found 199 men and boys assembled and awaiting rescue but dead from carbon monoxide poisoning, A further five had been killed in the shaft.

Outcome: Legislation was passed requiring every mine to have not less than two means of egress separated by not less than 45 feet (~14 m).

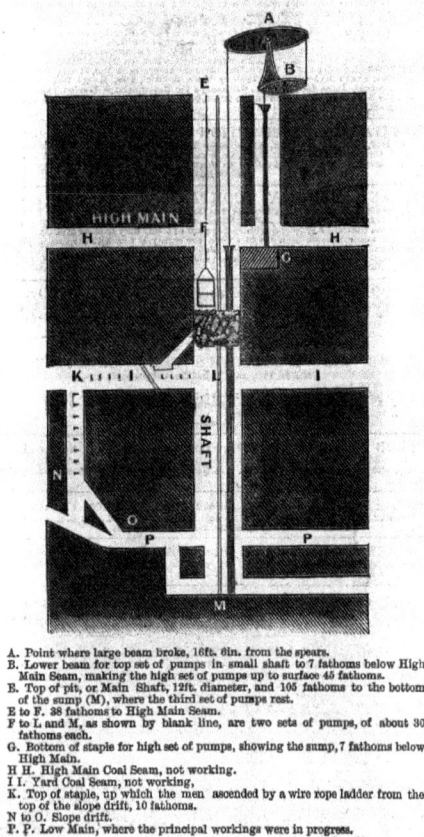

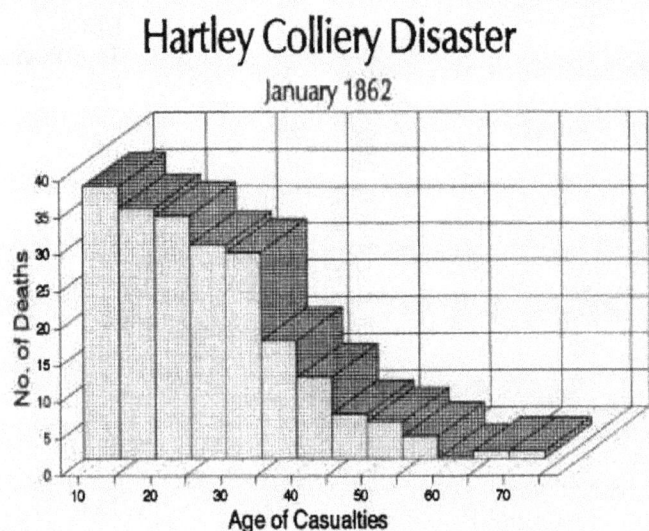

A. Point where large beam broke, 16ft. 6in. from the spears.
B. Lower beam for top set of pumps in small shaft to 7 fathoms below High Main Seam, making the high set of pumps up to surface 45 fathoms.
E. Top of pit, or Main Shaft, 12ft. diameter, and 105 fathoms to the bottom of the sump (M), where the third set of pumps rest.
E to F. 38 fathoms to High Main Seam.
F to L and M, as shown by blank line, are two sets of pumps, of about 30 fathoms each.
G. Bottom of staple for high set of pumps, showing the sump, 7 fathoms below High Main.
H H. High Main Coal Seam, not working.
I I. Yard Coal Seam, not working.
K. Top of staple, up which the men ascended by a wire rope ladder from the top of the slope drift, 10 fathoms.
N to O. Slope drift.
P. P. Low Main, where the principal workings were in progress.

FIG 5 – Cross-section through the single egress shaft and worked seams at Hartley Colliery (left; The Illustrated London News, 1862); and a profile of the age of the victims (right; Galvin and Hocking, 1993).

Mount Kembla Colliery, NSW, Australia; 31 July 1902

Aside: This case study is presented in a little more detail than others due to the additional relevance and interest it still has to ventilation management in Australia and New Zealand, particularly from practical, political and sociological perspectives. The reader is referred to Piggin and Lee (1992) for an in-depth account of aspects of these.

An explosion occurred at about 2:00 pm when 261 boys and men were underground. It was heard up to 20 km away and claimed the lives of 96, including two rescuers. Up until the Black Saturday bushfires which claimed 174 lives in Victoria in 2009, the disaster ranked as the highest loss of life on land in Australia and still ranks as Australia's most lethal mining accident.

Coal was extracted as shown in Figure 6 by pairs of miners forming up bord and pillar workings, with pillars subsequently being extracted. Some areas had also been extracted by longwall mining. Large areas of goaf remained open after pillar extraction. Miners referred to as *wheelers* conveyed coal from the face in track mounted tubs using horses and endless rope haulage systems. The mine manager at the time was William Rogers, an experience Welsh undermanager whose appointment as mine manager was based on him being in possession of a *Certificate of Service* obtained on the basis of time spent as an undermanager.

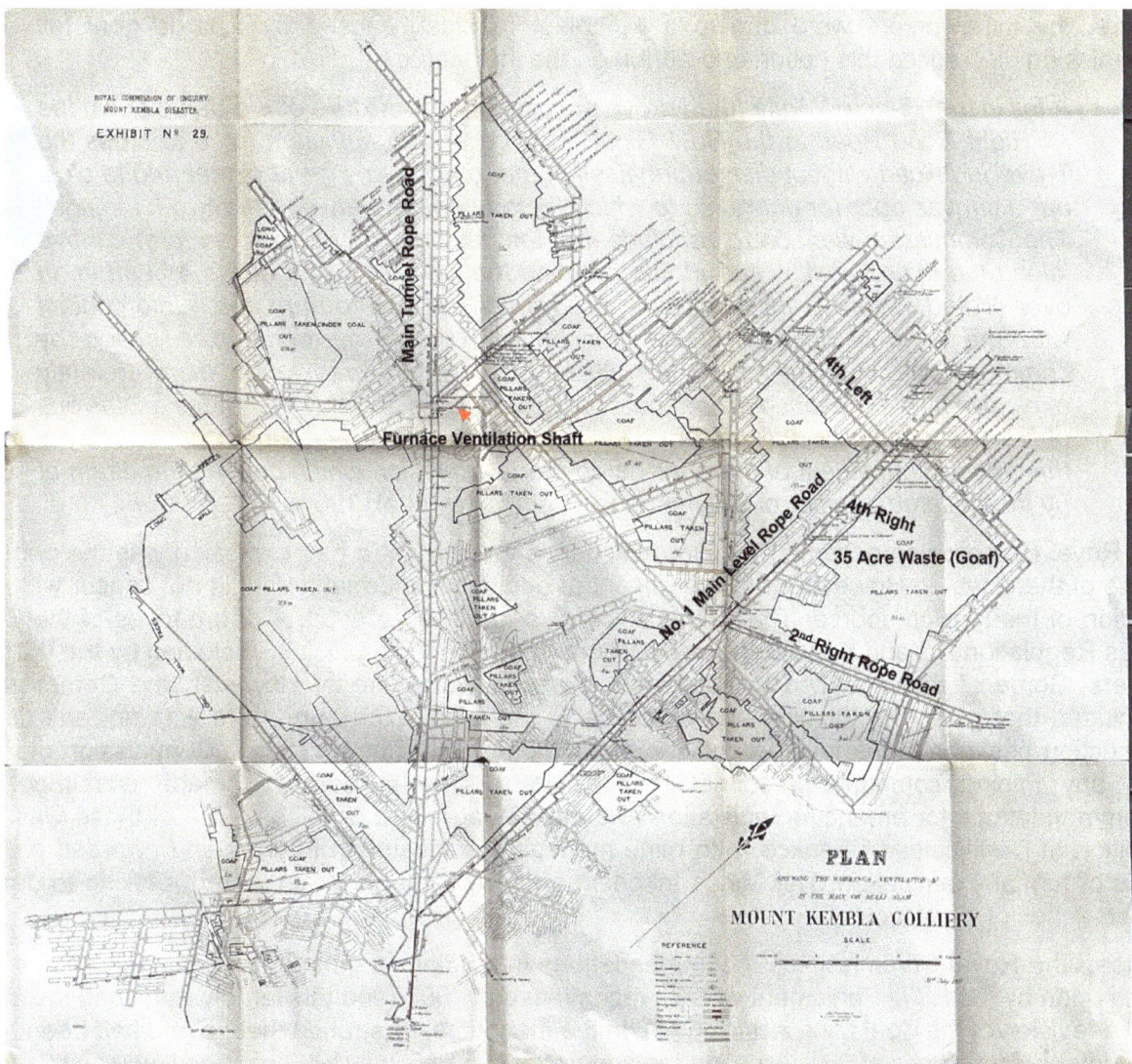

FIG 6 – Mine Plan of Mount Kembla Colliery at the time of the Disaster (Murray *et al*, 1903).

A ventilation furnace was situated adjacent to the Main Tunnel Rope Road, about 800 m from the mine mouth and generated between 40 and 60 m³/s (80 000 to 100 000 cubic feet per minute) of air flow. This by all accounts provided good ventilation. At the time of the incident, senior management held the conviction that the mine was not gassy. This was despite a mineworker having been severely burnt by a gas ignition some 10 years earlier and a previous mine manager advising a Royal Commission seven years earlier that methane could be given off from fissures in all sections of the mine. No reports of methane in the mine were made in the years immediately leading up to the explosion. The conviction that the mine was a non-gassy mine permitted the use of naked flames for illumination underground instead of safety lamps, which were not favoured by the miners. Subsequent inquiries into the disaster established that the presence of gas had been reported to mining officials by the workforce on a number of occasions in the years leading up to the disaster but the mindset apparently prevailed that the mine was not gassy.

Effectively, four inquiries were held into the incident (Coronial, Royal Commission, Judicial Commission, Parliamentary debate). The Coronial Inquiry focused on the cause of death of three of the victims, with the Coroner's Jury concluding (Piggin and Lee, 1992) that they died:

> From carbon-monoxide poisoning produced by an explosion of fire-damp ignited by the naked lights in use in the mine and accelerated by a series of coal-dust explosions, starting at a point in or about the number-one main level back headings, and extending in a westerly direction to the small goaf marked 11 perches [~0.07 acre] on the mine plan.

A company official argued strongly before the Royal Commission that the incident and injuries suffered by mineworkers were due to a windblast (airblast) caused by a large goaf fall. The Commission discredited this notion and attributed the incident to:

> A fall in the 35-acre Waste [that] drove an inflammable mixture of fire-damp down the 4th Right Rope Road to the No.1 Right Main Level with sufficient force to cross the Travelling Road without distributing itself in that road to any great extent and to blow out a canvas door (or doors, as to which there is doubt) between the two headings'. The Commission went on to conclude that the mixture was then driven further inbye 'along the Main Level,[where it] first reached the wheeler's light at the 4th Left in an over-diluted state; but as soon as a mixture rich enough to burn came into contact with the light, a flash of flame ran back, starting the destructive action by communicating the ignition to the whole body (which was made more violently explosive by the presence of coal-dust raised by the first blast). ....these explosions of fire-damp and coal-dust generated a large quantity of carbon-monoxide; and it was this deadly constituent of the after-damp of the explosion which caused the death of by far the large number of victims (p. xxxv of Murray et al, 1903).

The Royal Commission accepted the Coronial Jury's verdict that a gas explosion was the primary cause of the disaster. However, with the benefit of additional information, it did not concur with the location of the ignition source. The Commission also identified a number of breaches of the Coal Mines Regulation Act and Special Rules by Mount Kembla mining officials, including by the William Rogers. Some of these related to failures to examine waste areas. However, the Commission concluded that given the location of the source of the initial explosion, there was not any direct connection between these failures and the catastrophe itself. Therefore, the Commission did not hold any mining company personnel responsible for the incident but made a number of recommendations for amending legislation and the management of mines. One of these was the abolition of Certificates of Service, with mine manager certificates in future being awarded on the basis of formal examination. The Mines Inspectorate also came in for criticism for failing to detect the breaches of legislation.

None of the Royal Commission's recommendations in relation to amendments to legislation were acted upon by the NSW Parliament when it reconvened in 1903 (and this largely remained the case until a revision of the Coal Mine Act in 1943). In the meantime, frustration that no one had been held accountable for the loss of 96 lives led to the appointment on 15 July 1903 of a third inquiry, this time a Judicial Inquiry conducted by a District Court judge (Charles Heydon) who had the power to impose penalties under the Coal Mines Regulation Act. Judge Heydon heard evidence and visited the mine before issuing his decision on 14 September 1903 that Rodger's certificate as a mine manager should be suspended for 12 months. Heydon stated that he could not make a finding against Rogers that he was unfit to carry out his duties as a mine manager and that Rogers failures were a 'a partial *or temporary unfairness that might be fairly expected to be cured by the censure necessarily accompanying temporary suspension of his certificate'*, an action he took with *'great regret'*.

Judge Heydon's finding resulted in a petition to the Minister of Mines, John Fegan, to exercise his prerogative under the CMRA to restore Roger's certificate. However, Fegan was an ex-miner and along with some of his political colleagues who were also ex-miners, was vehemently opposed. This resulted in a motion for Parliament to debate *'the excessive and disproportionate punishment inflicted upon Mr Rogers....and the desirability, on the basis of equity and justice, of a remission of the balance of punishment'*. Parliament accepted that the penalty was adequate. Rodger's 12 months suspension stood, whereafter he resumed his post as the Mine Manager of Mt Kembla Colliery, retiring in 1911.

Fairmont Coal No. 6 and No. 8 Mines, West Virginia, USA; 6 December 1907

An explosion resulted in the deaths of 363 miners, making it the worst mining disaster in US history. The cause of the explosion could not be established but investigations suggested that it was an electrical spark or the flame of a miner's lamp that ignited methane or coal dust.

Outcome: The incident led to the establishment of the United States Bureau of Mines (USBM).

Hutton Colliery Bank Pit No. 3, England; 21 December 1910

An explosion resulting in 344 deaths. The cause of the explosion was not definitively established by the coronial inquiry, with the jury concluding that an accidental ignition of gas and coal dust had occurred in some manner unknown but probably from a defective or over-heated safety lamp and that this subsequently caused an explosion that affected the whole mine. It is reported to be the worst mining disaster on one day in England.

Senghenydd Colliery, Wales; 14 October 1913

An explosion resulted in the deaths of 439 miners, 60 of whom were younger than 20, and one rescuer. The cause of the explosion could not be established but an inquiry concluded that the most likely cause was an ignition of methane by a spark from underground signalling equipment that was then fuelled by coal dust. It was the worst mining disaster in the history of the United Kingdom.

Bulli Colliery Fire, NSW, Australia; 9 November 1965

On 9 November 1965, a fire developed in 8 Right Panel at approximately 9:00 am, engulfing the intersection outbye of the continuous miner and trapping five men at the face. One man was able to escape by running through the flames, the other four did not survive. The fire commenced near the shuttle car shunt at the intersection, which was separated from the goaf by a brattice stopping. An accumulation of flammable gas developed in the shunt. The Inquiry into the incident concluded that a piece of timber jammed between the disc brake and the drive coupling of the shuttle car was the logical ignition source.

Two auxiliary fans in series supplying air to the face were shutdown. Smoke immediately started to backup in the working panel. The Southern Mines Rescue Station was alerted and rescue teams were deployed using BGI74 model breathing apparatus for the first time in the NSW Mines Rescue Service. The transport intake road to the panel was bratticed off some four pillars outbye of the fire to contain the smoke. Many rescue teams sourced from local mines and supported by the Southern Mines Rescue Station were required to deal with the fire, which was brought under control in about 24 hours by fighting it with water.

Outcomes: Legislation was introduced requiring:

- improved underground firefighting facilities
- appropriate ventilation
- stringent and continuous gas monitoring and detection
- the carrying of carbon monoxide filter self rescuers.

Wanki No. 2 Colliery, Rhodesia (Zimbabwe); 6 June 1972

A series of explosions over two days resulted in the deaths of 427 mineworkers, with 423 being entombed in the mine. It is the worst mining disaster to have occurred in Zimbabwe.

Box Flat No. 5 Mine, Qld, Australia; 31 July 1972

A spontaneous combustion heating was detected in a pile of fallen coal. The heating coincided with an 11-hour fan stoppage, during which time the reduced airflow allowed a build-up of heat sufficient to cause the coal to self-ignite. The heating developed into a large fire, which was assisted by the increased airflow when the mine fan was restarted. Efforts to extinguish the fire were unsuccessful.

Recirculation of the ventilation occurred, causing an explosive mixture of gas generated by the fire to ignite. The explosion claimed 17 lives, including seven members of a rescue team who had entered the mine in an attempt to control the fire by sealing a main intake airway.

Kianga No. 1 Colliery, Qld, Australia; 20 September 1975

At 7:30 am on 20 September 1975, a heating was detected in the goaf area of 4 North Section of the mine. A decision was made to seal off the area with permanent brick stoppings after the setting of brattice sheets. Construction of the brick stoppings began at 11:30 am. At 5:10 pm an explosion occurred which claimed 13 lives. The bodies were never recovered.

The cause of the explosion was attributed to a barometric pressure drop during the sealing operation causing an overflow from a large body of methane in the goaf area. The resulting gas mixture was ignited by contact with the fire and was followed by a coal dust explosion.

Outcomes: Inquiry recommendations included:

- The establishment of an autonomous 'Safety in Mines Research Organisation' (which was to become the Safety in Mines Testing and Research Station – SIMTARS).

- All mines have available at short notice the means of analysing the air samples obtained whilst dealing with an outbreak of fire below ground.

- No person should enter an area on the return side of a suspected heating or fire or on the intake side where smoke is present unless they have the instruments and knowledge to ensure their own safety.

Appin Colliery, NSW, Australia; 24 July 1979

Appin Colliery extracts the Bulli Seam, which is noted to be gassy. At approximately 11:00 pm on 24 July 1979, an explosion occurred in K Panel, which was a three heading longwall gate road development. On the day of the explosion a ventilation changeover was carried out to provide a return airway on each side of the panel. The changeover was intended to allow for the use of an auxiliary fan in each of the two returns. Evidence indicated that a breakdown of one of the auxiliary fans occurred on the shift of the explosion, resulting in a large accumulation of gas in the right-hand heading, which had been advanced some 50 m inbye the last intersection. Rescue teams recovered 14 bodies.

Two possible sources of ignition were seriously considered without resolution. They were the deputy's (foreman's) flame safety lamp, which was found in a damaged condition, and the flameproof enclosure on the fan starter, which was found to be in a non-flameproof condition. The violence indicated that the explosion had travelled outward from the face area and it was considered that the initial ignition probably travelled up the fan ducting to the face before developing its major force as it travelled out from the face area. The explosive force extended 600 m outbye of its source, damaging roof supports, machinery and belt structures and destroying overcasts and brick stoppings.

Mitsui Mike Coal Mine, Japan; January 18, 1984

458 mineworkers were killed by an explosion and associated carbon monoxide, with 839 of the 939 survivors suffering carbon monoxide poisoning.

Moura No 4 Colliery, Qld, Australia; 16 July 1986

Moura No. 4 Mine was nearby to Kianga No. 1 Mine. At about 11:05 am on 16 July 1986, a thick cloud of dark grey dust was observed rising above the spoil pile near the vicinity of the main ventilation fan serving the No. 4 underground mine. The underground power was interrupted and an inspection of the fan revealed substantial damage. These effects were initially attributed to a windblast associated with a large goaf fall in the Main Dip Section.

There were 20 men underground at the time comprising 12 extracting coal from pillars in the Main Dips Section, five in 3 South Section and three who were out of contact in the access roadways in the mine. The five men in 3 South Section were contacted by telephone and told to start making their way to the surface and that they would be met on the way by a deputy and an experienced miner. This occurred and all reached the surface safely by about 11:55 am. The three men in the access roadways had also made their own way to the surface by this time. There was no contact with the remaining 12 men.

The next day conditions permitted an inspection underground and a rescue team operating within severe time constraints sighted the bodies of most of the missing miners. The positions of the machines and the bodies of the men suggested that immediately prior to the explosion, a goaf fall was anticipated and that an orderly withdrawal was initiated. During the withdrawal a young miner may have been pinned beneath a shuttle car, with the explosion occurring while he was being assisted. The body of the deputy, who carried a flame safety lamp, was close by the young miner. All bodies were recovered by mines rescue teams on Wednesday 23 July 1986. The explosion was believed to have been initiated by frictional ignition or the deputy's flame safety lamp. It was concluded that all victims had died at the time that the incident occurred.

Outcomes: The Inquiry expressed serious concerns re:

- Those responsible failed for some considerable time to consider alternative causes for the disaster other than a windblast.

- The failure of any of the eight survivors to don their SCSR when evacuating the mine.

Recommendations of the inquiry included the banning of flame safety lamps (which was implemented in Qld).

Moura No. 2 Colliery, Qld; 7 August 1994

Moura No. 2 mine extracted the No. 2 Seam immediately beneath the No. 4 Seam involved in the 1986 explosion. The No. 2 seam was susceptible to spontaneous combustion. At about 11:35 pm on Sunday, 7 August 1994, an explosion occurred in 512 Panel, almost directly beneath where the Moura No. 4 explosion has occurred. There were 21 persons working underground at the time. Ten men from the northern area of the mine escaped within 30 minutes of the explosion by using filter self rescuers. The remaining 11 men failed to return to the surface. A second and more violent explosion occurred at 12:20 pm on Tuesday, 9 August 1994. Rescue and recovery attempts were thereafter abandoned and the mine was permanently sealed at the surface.

Mining height in 512 Panel was typically 4.5 m. Figure 7 shows the partial pillar extraction devised for the panel in order not to disturb the overlying site of the Moura No. 4 explosion. The layout is unique and had three features that were conducive to promoting spontaneous combustion, in both their own and collective right, these being:

- A lot of loose coal was left on the floor in the panel.

- A staggered layout of the pillars, which obstructed ventilation flow-through the panel.

- A large mining height, which resulted in low ventilation velocities through the goaf.

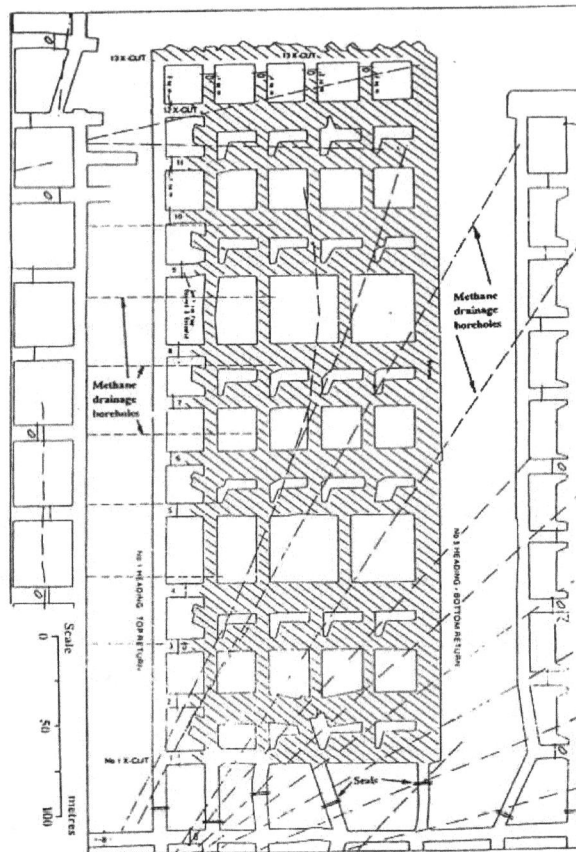

FIG 7 – Plan of 512 Panel, Moura No. 2 Mine.

The resulting Warden's Inquiry (Windridge *et al*, 1996) found that the first explosion originated in the 512 Panel of the mine and resulted from a failure to recognise and effectively treat a heating of coal in the panel. This, in turn, ignited methane gas, which had accumulated in the panel after it had been sealed. The 21 recommendations from the inquiry included:

- The position of Ventilation Officer should be established at all underground mines.

- Spontaneous Combustion Management Plans to be developed and implemented at mine sites to provide effective long-term control of the risk.

- All mines to put in place Mine Safety Management Plans to cater for key risk areas. Such plans to be based on detailed risk/hazard analysis and to be subject to regular audit.

- Emergency procedures should be exercised at each mine on a systematic basis, the minimum requirement being on an annual basis for each mine.

- Persons holding statutory certificates of competency should be required to demonstrate their fitness to retain the certificate on a regular basis (three to five years).

- Re-entry for the purposes of recovering bodies and collecting evidence must be weighed against the risks to those involved in these tasks.

Endeavour Colliery, NSW, Australia; 28 June 1995

On 28 June 1995, a gas explosion associated with a goaf fall and windblast occurred in a cut-through immediately outbye of the face in the 300 Panel pillar extraction unit, which was at the extremity of the mine workings as shown in Figure 8. Panel width was restricted in order to prevent inundation from the overlying tidal lake (Lake Macquarie). The coal was not prone to spontaneous combustion and like most other pillar extraction panels in the mine, 300 Panel was planned to have bleeder roadways to clear gas from the goaf and help dissipate windblasts associated with goaf falls. However, this did not eventuate due to an accumulation of water at the site of the proposed holing site to the main return airway. There were eight mineworkers in the panel at the time of the explosion.

They felt their way out of the panel in zero visibility in three independent groups while a rescue team was still being assembled to enter the mine.

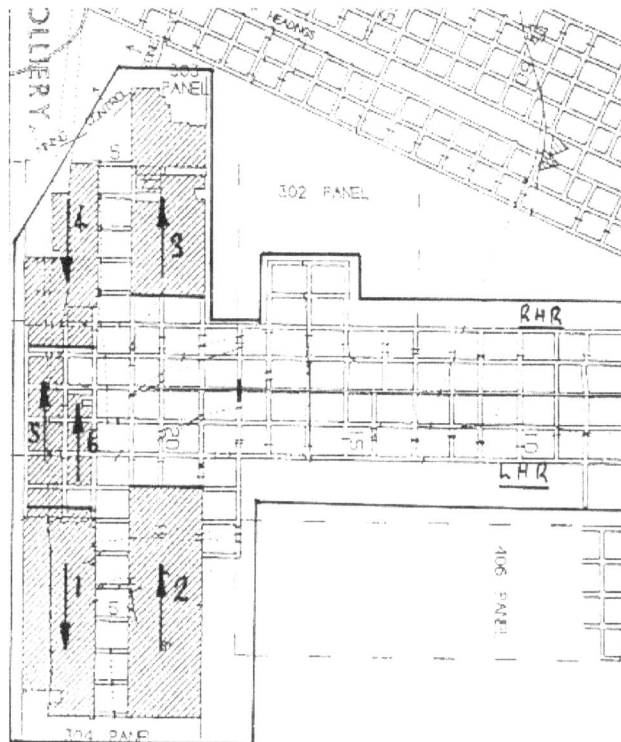

FIG 8 – Extract of mine plan showing the layout and extraction sequence for 300 Panel at Endeavour Colliery (MDG-1007, 1996).

A fall of ground in the goaf is believed to have pushed an explosive mixture of methane into the working face area but the source of ignition could not be definitively identified. It was initially suspected to be a non-flameproof back-to-back shuttle car cable connector but subsequent laboratory testing of the connector failed to confirm this.

Outcomes:

- The Chief Inspector of Mines (Bruce McKensey) approached the author, through the School of Mining Engineering at the University of New South Wales to develop and deliver a Ventilation Officer Training Course.

- The Coal Mines (Underground) Regulation enacted in August 1999 included a provision that a mine manager must appoint a person as a ventilation officer for a mine and that any person appointed to that role *'on or after 1 September 2000 must have successfully completed the Ventilation Officers Training Course conducted by the School of Mining Engineering at the University of New South Wales (or a course specified as an equivalent course)'*.

North Goonyella, Qld, Australia; September 1999

A spontaneous combustion event resulted in the sealing of Longwall Panel 5 South and the loss of the majority of the face equipment.

Pike River Coal Mine, New Zealand; 19 November 2010

An explosion followed by three more over the next nine days resulted in the deaths of 29 mineworkers whose bodies have not been recovered. Very soon after the initial explosion, which is believed to account for all fatalities, two mineworkers escaped from the drift. The second egress through the main ventilation shaft, shown in Figure 9, was not functional due to a fall of ground that had occurred some months earlier.

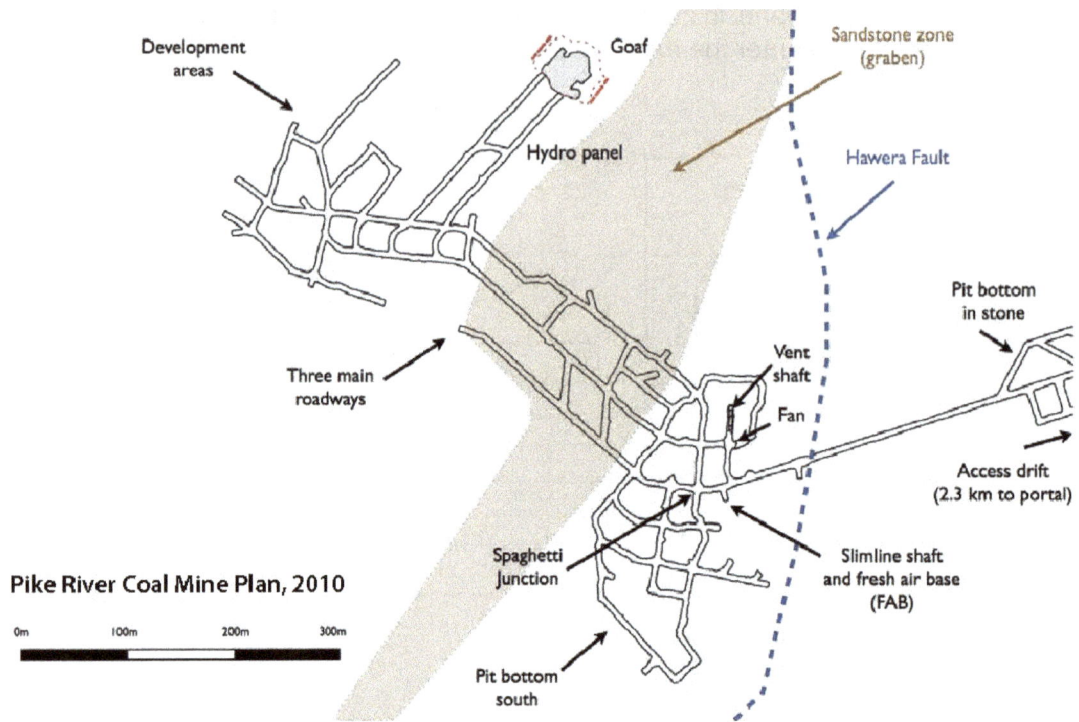

FIG 9 – Pike River Coal Mine Plan as at 2010 (Panckhurst *et al*, 2012).

The mine was regarded as gassy and the ventilation monitoring system was not fully commissioned at the time of the incident. Surface constraints resulted in the unusual practice of the main ventilation fan, which was not flameproof, being installed underground. Nominal roadway dimensions were 5.2 m wide by 3.6 m high. Hydraulic mining was adopted for secondary extraction of the full seam, which was up to 13 m thick. Panel width was restricted in order to limit surface subsidence. The first hydraulic mining panel was in the process of being extracted at the time of the incident and roof falls from the goaf void were expected, with a significant fall having already occurred.

There were multiple possible sources for the initial ignition. The Royal Commission into the disaster (Panckhurst *et al*, 2012) and subsequent re-entry up to the site of a roof fall near the end of the drift, which had been driven in stone, could not establish a definitive cause. Families of the deceased have campaigned aggressively for the recovery of their loved ones and for someone to be held accountable for the incident, with their actions leading to the sealing of the mine being delayed. Controversy and legal debate continue regarding the approach taken by government to prosecution. In December 2013, charges were dropped against the mine manager, Peter Whittall, with prosecutors claiming that there was a lack of evidence. Instead, agreement was reached with WorkSafe for Whittall and Pike River Coal to make a payment of $3.41 million to the families of the deceased and the two survivors. On 23 November 2017 the NZ Supreme Court ruled that this agreement was unlawful, stating that *'If accepted this proposal would undoubtedly have constituted a bargain to stifle prosecution'*. The matter is ongoing.

Outcome: Worksafe New Zealand was established in 2013 and workplace health and safety legislation was updated with the passing of the Health and Safety at Work Act 2015.

North Goonyella, Qld, Australia; September 2018

A heating developed behind the longwall face in early September during the relocation of Longwall 9 North to Longwall 10 North Panel. Nitrogen injection failed to stop gas levels rising and the mine was evacuated towards the end of September. Jet engine inertisation technology (GAG) was then employed to suppress the fire. The mine was sealed and mining is yet to resume.

Grosvenor Mine, Qld, Australia; 6 May 2020

Construction of Grosvenor Mine commenced in 2012, with extraction of its first longwall panel (LW 101) commencing in May 2016. Extraction of LW 104 commenced on 9 March 2020. On 6 May

2020, five miners working towards the tailgate end of the 300 m wide longwall face were injured by an event described in the report of the Board of Inquiry (Martin and Clough, 2021) as comprising *'two consecutive pressure waves (wind blasts) separated by about 10 to 15 seconds. Each occurred without warning. These pressure waves proceeded from the tailgate end of the longwall. Both were of considerable force. A flame front, which burnt the five workers closest to the tailgate, accompanied the second pressure wave'.* Subsequently there was a second event on the 8 June 2020, with no one injured as the mine had been evacuated. As a result, the mine was not re-entered until April 2021 after the longwall equipment and face were abandoned and permanently sealed. A new set of longwall equipment was installed in LW 104A and longwall operations re-commenced in February 2022.

The Board of Inquiry was tasked with performing its function at the same time that the regulator, Resources Safety and Health Queensland (RSHQ), was investigating the 6 May 2020 incident and determining whether a prosecution process may follow. Consequently, the Board did not have access to the RSHQ investigation report or to employees or contractors of the mine owner, Anglo American Metallurgical Coal (AAMC). Notwithstanding that RSHQ and AAMC did not fully participate in the Inquiry, they were cooperative. Select findings of the Board of Inquiry are noted in this paper, bearing in mind the scope of the information available to the Board.

At the time, Queensland regulations defined a methane exceedance high potential incident (HPI) as an occasion when methane is present in air in an underground coalmine in a concentration of at least 2.5 per cent. Points noted in the Executive Summary of the Board of Inquiry's report (Martin and Clough, 2021) included:

- During the period between 1 July 2019 and 5 May 2020 there were 27 methane related HPI incidents at Grosvenor Mine.

  'Another substantial part of Grosvenor's gas management strategy for LW 104 was to increase gas drainage capacity by doubling the number of goaf wells and reducing their spacing from 50 metres to 25 metres.' In noting this, the Inquiry reported that *'Greater extraction of methane from the goaf carries with it an unacceptable risk of spontaneous combustion if the resultant oxygen ingress is not well managed.'*

  'The efficiency of emission capture by Grosvenor's post-drainage system could accommodate production on LW 104 of around 70,000 tonnes per week. This was regularly far exceeded.'

  In the case of a number of HPIs, *'The systemic cause was that the gas emissions being generated by the mine's rate of production were in excess of the capacity of the mine's gas drainage system'.*

  'In early May 2020, LW 104 was progressing through unstable strata, and gas emissions were causing elevated methane levels in the tailgate, with both issues resulting in production delays and stoppages.'

  'Gas Incident Management Teams were formed on 28 April and 2 May to try to address the gas delays. Subsequently, a number of goaf holes were put on venturi to maximise gas flow rate.'

  'On 3 May 2020, approximately 5,600 litres of PUR was injected into 35 holes in the longwall face and roof between shields #97 and #132' (each shield being approximately 2 m wide).

The Board of Inquiry considered that the first pressure wave could only have been a result of (a) a strata collapse, or (b) a methane explosion. It concluded that *'Having found that a strata fall in the goaf is an unlikely explanation of the first pressure wave, the Board reviewed the evidence indicating a methane explosion in the goaf, ultimately concluding that it is the likely explanation.'* The Inquiry's report (Martin and Clough, 2021) goes on to include statements that:

  'Careful analysis of the gas data reveals the presence of a number of subtle indicators of a small but intense heating in the goaf';

'In the circumstances, the Board concludes that a spontaneous combustion-initiated methane explosion was the probable cause of the first pressure wave';

'In the circumstances, the Board considers that the probable ignition source for the methane deflagration on the longwall face was the PUR-initiated heating of coal to thermal runaway, which ignited an explosible atmosphere behind the longwall in the vicinity of shield #111, resulting in a flame propagating onto the longwall face';

'The level of stone dust maintained in the first 100 metres of the longwall return outbye the face was sufficient to prevent the methane ignition from initiating a coal dust explosion that could have propagated to other parts of the mine.'

APPLYING LEARNINGS TO VENTILATION RISK MANAGEMENT

The review of historical incidents attests to the catastrophic consequences that fire, explosion and contaminated, noxious and irrespirable atmospheres can have for workplace health and safety in underground coal mining. It also confirms that fires and explosions present the greatest risk of partial or total loss of underground mines and equipment and that this risk continues to materialise. A number of contributing factors are common to many of the incidents (for example, ground control and management decisions). The incidents give insight into a range of matters relevant to managing mine ventilation, with four of particular note being:

- The need for ventilation management to be integrated into mine planning over the full life cycle of the operation.

- The difficulties in identifying the ignition source for an explosion in an underground coalmine.

- The importance of contingency planning to manage consequences.

- The basis for accepting risk.

Life cycle risk management

Many of the case studies illustrate how ventilation system design and performance can be impacted and, as is often the case, compromised by competing factors which also have to be taken into account in mine design. Ground control is one of these factors, as reflected for example in the case studies involving open goaves that are a consequence of having to restrict panel dimensions for safety, environmental or practical mining reasons. Careful consideration should be to be given to the stability of goaves as part of selecting the mining system and mining dimensions at the conceptual mine design stage and the risk assessment of the concept plan. A goaf that does not fully cave and choke off can give rise to a range of hazards including:

- the accumulation and concentration of flammable and noxious gases.

- rock-on-rock frictional ignition sources.

- access of oxygen that promotes oxidation and spontaneous combustion of residual coal in the goaf.

- significant reduction in the velocity and quantity of air as it passes through the goaf, thus possibly resulting in inadequate removal of heat produced by the oxidation of residual coal.

- dead ventilation spots within the goaf.

- airblast impacts such as over-pressurisation, flying projectiles, expulsion of flammable and noxious gases into the workplace, creation of flammable clouds of coal dust, suction of personnel back into the goaf, and destruction of mine ventilation devices and circuits.

Similar considerations apply to longwall mining, where the following types of factors and how they may impact on the mine ventilation system need to be carefully assessed from the conceptual mine design stage forward:

- dip and mining direction.

- the extent of fracturing of the floor and roof strata and the gas content of all coal seams that fall within this zone.

- longwall panel dimensions that are compatible with retreat rates that result in residual coal being buried within its spontaneous combustion incubation period.

- production targets that are matched to the capacity of gas drainage system, ground destressing operations and water management systems.

Often competing demands do not make it possible to achieve mine design outcomes that are optimum from all perspectives (ventilation, ground control, water management, economic performance etc). In particular, mining environments that are both gassy and prone to spontaneous combustion are particularly challenging because they can have diametrically opposed controls. For example, bleeder ventilation can be a very effective control for gas management but very conducive to the promotion of spontaneous combustion.

In a mature risk management environment, these types of factors need to be identified at the conceptual stage of mine design and factored into prefeasibility studies. It is often too late to resolve and optimise them after production has commenced, Permits to Mine need to be signed off and operators are required to achieve targets set by project feasibility studies. They need to be addressed by a multidisciplinary team throughout the full life cycle of a mine, from developing the concept mine plans that support feasibility studies, through to planning for mine closure.

The determination of cause

The case studies reveal that while contributory factors to explosions in underground coalmines may be relatively easy and quick to identify, it is usually very difficult and unlikely to be able to determine their initiating ignition source. This has a number of significant implications, especially for risk assessment; managing and satisfying the expectations of the families of victims and of community; and for legal proceedings.

In the case of mine explosions, the question arises as to how likelihood can be quantified for risk assessment purposes if one cannot be reasonably sure of the status of all the potential sources of ignition. While many barriers can be put in on the left-side of the bow tie to reduce the likelihood of explosion, experience shows that these may not always consider and/or effectively address all potential threats and that the assumptions on which they are based can be deficient; for example, as the case studies show, it cannot be assumed that all electrical enclosures relied upon to be flameproof will be maintained in a flameproof condition. Therefore, the determination of risk needs to give careful consideration to the confidence that can be placed in the reliability of likelihood estimates against a backdrop of a poor track record in identifying the sources of gas ignitions.

Identifying the cause of a critical incident can have great significance to families and communities affected by the incident and be very distressing for them if this is unsuccessful. For more than ten years, the families of the victims of the Pike River Mine disaster have been lobbying for the mine to be searched for evidence of what caused the disaster. This human response to tragedy is natural and consistent with human responses to other tragic events. Unfortunately, learnings from other catastrophic mine explosions suggest that the likelihood of categorically identifying the initiating source of an explosive event is very low in most cases, let alone when the scene of the incident has been impacted and altered by subsequent explosions. These circumstances, in turn, can have significant implications for legal proceedings.

The control of consequences

The case studies show that there has been a dramatic reduction in the number of deaths caused by ventilation related incidents. This is only to be expected due to mechanisation and technology resulting in fewer persons being employed underground at a mine. However, there is a more important reason for this reduction and that is the implementation of effective measures to mitigate the consequences when a hazard materialises; that is, measures focused on the right-hand side of the risk analysis bow tie. Stonedusting to prevent coal dust explosions, water barriers to suppress flame fronts, self-contained self-rescuers, and self-escape systems are some of the important and effective mitigation measures in this regard.

The determination and acceptance of risk

Basically, risk assessment is concerned with providing a rational for deciding whether to accept or reject a situation based on its risk profile. That profile, in turn, is determined by identifying and weighting the consequences if the situation arose and by evaluating the likelihood that the situation could arise.

When dealing with situations that have the potential for catastrophic consequences, such as mine explosions, a fundamental question that needs to be asked is: Does the owner of the risk, or government, or community accept that, irrespective of likelihood, the potential consequences are intolerable? If the answer is 'yes', then obviously that should be the end of the matter – the activity should not be undertaken.

However, a common pitfall in risk management is that the potential severity of a catastrophic event can be masked when likelihood is factored into the risk equation. This is not necessarily evident to end-users that were not involved in the risk assessment process. End-users can include executive management and directors who also have accountability for managing all forms of risk. This is a recognised pitfall in risk assessment and why, as per risk assessment guidelines such as MDG-1010, 2011 – *Minerals Industry Safety and Health Risk Management Guideline*, risk assessments and the reporting of risk assessment outcomes should always include a ranking of risk based only on the consequence ratings for hazards.

VENTILATION OFFICERS – LOOKING TO THE FUTURE

The Endeavour Colliery explosion and the findings of the Wardens Inquiry into the Moura No. 2 Mine explosion were not the only motivators for the NSW Chief Inspector Mines requesting UNSW to develop a Ventilation Officer Training Course in 1997. These incidents were just two of 16 mine ventilation related incidents to have occurred in the Australian coal mining industry since 1990. The other 14 events included gas outbursts in two collieries (four fatalities), three underground fires (resulting in two mines being sealed), several frictional ignitions of gas, a number of face gas outs and three incidents involving excessive concentrations of carbon dioxide in mine workings (Galvin *et al*, 1998).

Subsequently, the UNSW Ventilation Officer Training Course was accredited for delivery into Queensland coal sector. Some 450 persons have successfully completed it to date, including a number who did so as part of a UNSW Graduate Diploma in Mine Ventilation. A number of Registered Training Organisations have also delivered accredited ventilation officers training programs during the same period.

In the 25 years since the statutory position of ventilation officer was established in the Australian coal mining industry, the status of ventilation officers has increased, along with a corresponding increase in their accountabilities and responsibilities. Under managers and mine managers (since replaced in NSW by the statutory role of *'manager of mining engineering')* who previously fulfilled the role of ventilation officer are still required to have a sound practical understanding of mine ventilation, but more specialist and advanced knowledge and skills in mine ventilation have been brought into the industry through the establishment of the role of statutory ventilation officer.

Historically, the statutory mine manager ruled, with legislation holding the mine manager accountable for control of all persons and operations. This included determining the location and settings for ventilation devices and the overseeing of ventilation changes, both being activities which can be closely related to and influenced by coal production related decisions. Australian mining legislation prior to 1999 did provide for the appointment of ventilation officers to assist in these regards, but those appointments were not statutory appointments. Given the functions and accountabilities of a ventilation officer today and the catastrophic consequences that can arise if mining production practices and goals and ventilation practices and goals are not aligned, it is very important that the ventilation officer and production manager (mine manager, manager of mining engineering) at a mine site work together closely on a day-to-day basis and not just during the management of a critical incident.

Since the creation of this role, there have been significant technological advances in the mining industry that impact on ventilation officers. Computing capabilities, more sophisticated software packages and more advanced real time and remote monitoring technologies present important opportunities for improving the design, implementation, operation and management of ventilation systems. These include collecting a larger amount of data, collecting data from areas otherwise inaccessible, and utilising algorithms and artificial intelligence to process data faster and to make more appropriate decisions. These advances are particularly beneficial when dealing with critical incidents where control is threatened or may have been lost, such as the outbreak of fire underground.

Given the vital contributions that ventilation officers are expected to make to the design of mine ventilation systems and the development and execution of risk-based management plans for these systems, it is essential that ventilation officers undertake continuing professional development (CPD). This not just because it is a requirement for retaining their statutorily endorsed qualification but also because there is a need for the ongoing enhancement of skills and innovation in mine ventilation. Attendance at conferences such as this Mine Ventilation Conference provide an important opportunity to undertake CPD through both attending presentations and networking with colleagues.

Looking to the future, the most effective control for eliminating risk to the health and safety of persons underground is not to have anyone employed underground. The underground coal mining industry is moving in that direction, with a number of longwall faces now having the capacity to be operated for extended periods of time from the surface. Achieving the aspiration of an unmanned operating underground mine will require ventilation officers to be engaged in considerable change management and innovation.

An emerging challenging issue is the capture and processing of fugitive gas emissions from underground (and surface) operations. Regulators and planning approval authorities have an increased focus on this source of greenhouse gases. Research into a range of options to limit or to capture and utilise these emissions has shown that there are no simple or quick solutions. It remains to be seen if the underground coal sector, especially the thermal coal sector, will continue to exist long enough for ventilation officers to play a significant role in solving that problem.

SUMMARY KEY POINTS

- Fire, explosion and contaminated, noxious and irrespirable atmospheres continue to have a very high potential to result in catastrophic consequences in underground coal mining, including multiple fatalities and partial or total loss of the mine workings and infrastructure.

- Risk management approaches to coalmine ventilation have proven to be effective in many cases in preventing situations from escalating to a point where control is lost, in removing mineworkers from exposure sites before control is lost and in mitigating the consequences of ventilation related failures.

- The ventilation management process needs to commence at the pre-feasibility stage and be integrated into mine planning over the full life cycle of a mine.

- Often, competing demands do not make it possible to implement the optimum mine ventilation system.

- History shows that the identification of the initiating source for an explosion is seldom able to be determined.

- When dealing with situations that have the potential for catastrophic consequences, such as mine explosions, a fundamental question that needs to be asked is: Does the owner of the risk, or government, or community accept that, irrespective of likelihood, the potential consequences are intolerable? Obviously, if the answer is 'yes', then the activity should not be undertaken. In those circumstances, care is required to avoid the level of risk being masked or diluted by including likelihood in the risk assessment equation.

- It is very important in regard to mine ventilation that the practices and goals of the ventilation officer and production manager are aligned and that they work together closely on a day-to-day basis and not just during the management of a critical incident.

- Ventilation officers need to undertake continuing professional development (CPD) not only because it is a requirement for retaining their statutorily endorsed qualification but also because there is a need for the ongoing enhancement of skills and innovation in mine ventilation.

REFERENCES

Devlin, R, 1988. *Children of the Pits. Child Labour and Child Fatality in the Coalmines of Whitehaven and District*, Whitehaven: The Friends of the Whitehaven Museum.

Duckham, H and Duckham, B, 1973. *Great Pit Disasters*, David and Charles Abbot.

Galvin, J M, 2016. *Ground Engineering: Principles and Practices for Underground Coal Mining,* Switzerland: Springer.

Galvin, J M and Hocking, G, 1993. Strata Control for Coal Mine Design Research Project Technology Transfer Newsletter, Issue 1, Principal Theme: The Coal Pillar System, 8 p (UNSW School of Mines: Sydney).

Galvin, J M and Hocking, G, 1994. Strata Control for Coal Mine Design Research Project Technology Transfer Newsletter, Issue 5, Principal Theme: Coal Pillar Performance, 8 p (UNSW School of Mines: Sydney).

Galvin, J M, Moreby, R and Chalmers, D R, 1998. *Mine Ventilation, Ventilation Officers Training Course*, UMRC9808/1 (School of Mining Engineering, University of New South Wales: Sydney).

ISO 31000, 2009. Risk Management – Principles and Guidelines, 38 p (International Standards Organisation: Geneva).

Joy, J, 1998. *Introduction to Risk Management (Unpublished Lecture Notes),* School of Mining Engineering, University of New South Wales.

Martin, T and Clough, A, 2021. Queensland Coal Mining Board of Inquiry Report, Part II, 562 p.

MDG-1007, 1996. *Explosion at Endeavour Colliery, 28 June 1995, Summary of Investigation* (NSW State Government: Sydney).

MDG-1010, 2011. *Minerals Industry Safety and Health Risk Management Guideline* (NSW State Government: Sydney).

Murray, C E R, Robertson, D A W and Ritchie, R, 1903. Mount Kembla Colliery Disaster, 31 July 1902, Report of the Royal Commission together with Minutes of Evidence and Exhibits, 1050 p (Legislative Assembly, NSW Government: Sydney).

Panckhurst, C, Bell, S and Henry, D, 2012. Royal Commission on the Pike River Coal Mne Tragedy (New Zealand Government: Wellington).

Piggin, S and Lee, H, 1992. *The Mt Kembla Disaster,* Oxford University Press in Association with Sydney University Press.

Robens, A, Beeby, G H, Pike, M, Robinson, S A, Shaw, A, Windeyer, B and Wood, J C, 1972. Safety and Health at Work, Report of the Committee 1970–72 (Her Majesty's Stationery Office: London).

The Illustrated London News, 1862, 25 January. The Hartlet Pit Disaster, *The Illustrated London News*.

Windridge, F W, Parkin, R J, Neilson, P J, Roxborough, F F and Ellicott, C W, 1996. Report on Accident at Moura No. 2 Underground Mine on Sunday, 7 August 1994 (Queensland State Government: Brisbane).

Automatic ventilation model calibration using measured survey data

M D Griffith[1] and C M Stewart[2]

1. Software Team Leader, Howden Ventsim, Brisbane Qld 4101.
   Email: martin.griffith@howden.com
2. Principal Consultant, Minware, Brisbane Qld 4163. Email: craig@minware.com.au

ABSTRACT

A crucial part of maintaining a useful mine ventilation model is calibrating it to real mine ventilation data. A common way to do this is using pressure and quantity surveys (PQ) to set the correct resistances to the model airways. However, it is often impractical to do this for an entire mine due to limited time and resources, or restrictions in access to some locations. Furthermore, the generalisation of PQ surveys from measured airways to other similar but unmeasured airways can introduce errors. An additional worry is that imposing a measured resistance corresponding to a PQ survey will not necessarily produce the measured flow in the model, due to the interdependence of other airway resistances and flows across the model. This study builds upon a numerical method that uses a model-wide sensitivity matrix to calculate the response of quantities and pressures to changes in resistances elsewhere in the mine. The method then automatically modifies the resistances throughout the model so that the airflow solution returns the pressures and airflows measured at the PQ survey sites. This method can potentially help calibrate an entire model based on an incomplete set of PQ survey data, or recommend further PQ surveys required to achieve a successful calibration. In this study, the method will be applied to a test model, but with variations on the algorithm, such that some airways can be preferred for resistance variation over others. This allows the calibration method to account for variations in user confidence in some airways, or to focus the method on a particular subset of airways, such as regulators.

INTRODUCTION

A model of a mine ventilation system can be useful for the design, evaluation, visualisation and future planning of the system. All models are an approximation of real ventilation systems, with inevitable inaccuracies (Griffith and Stewart, 2019); but a suitable level of accuracy is required for the model to be useful to the ventilation engineer. This study examines a computational method able to assist in calibrating models from limited PQ surveys, highlight airways where significant resistances errors may exist, or to calibrate resistances such as regulators to achieve known flows in the model.

In this study, the Atkinson resistance equation is used as the basis for ventilation simulation (McPherson, 1993):

$$\Delta P = rQ^n \text{ [Pa]} \tag{1}$$

where:

ΔP	is the difference in pressure from one end of the airway to the other
Q	is the air volume flow rate [m³/s]
n	is the flow coefficient, normally equal to 2 for turbulent flow
r	is the Atkinson resistance, corrected for density

The Atkinson resistance is:

$$r = \frac{\rho_{actual}}{\rho_{ref}} R = \frac{\rho_{actual}}{\rho_{ref}} \frac{kLS}{A^3} \text{ [Pa·s}^2\text{/m}^6] \tag{2}$$

where:

L	is the length of the airway [m]
S	is the perimeter of the airway cross-section [m]
A	is the cross-sectional area of the airway [m²]
ρ_{actual}	is the air density in the airway [kg/m³]

ρ_{ref} is a reference air density [kg/m³]

k is the Atkinson friction factor at reference air density [kg/m³]

To calibrate a model, the engineer can survey the mine airways using the PQ survey technique, measuring the quantity flow rate with an anemometer, and pressure loss with either a trailing tube with manometer or differences in barometric pressure adjusted for elevation and temperature (Prosser and Loomis, 2004). The calculated resistance from Equation (1) can then be used to assign a more accurate resistance to the airway. If the surveyor also has confidence that the model airway dimensions adequately represent the true airway size, the engineer can instead calibrate the friction factor (Michelin *et al*, 2019).

A correctly performed PQ survey can assign the correct resistance to individual or common airways, but there are some limitations. One limitation is that the flow-through an airway depends not only on the resistance in the airway, but on the flow and resistances in other parts of the mine. This is demonstrated in Figure 1, which shows an airflow quantity of 40 m³/s evenly split between two branches. A PQ survey is carried out in the right hand branch revealing a pressure drop of 10 Pa and a flow rate of 15 m³/s, giving a resistance of R = 0.0444 Pa·s²/m⁶. If this resistance is applied to the model, the simulation result in the airway concerned changes to 18.4 m³/s instead of 20.0 m³/s as measured in the PQ survey.

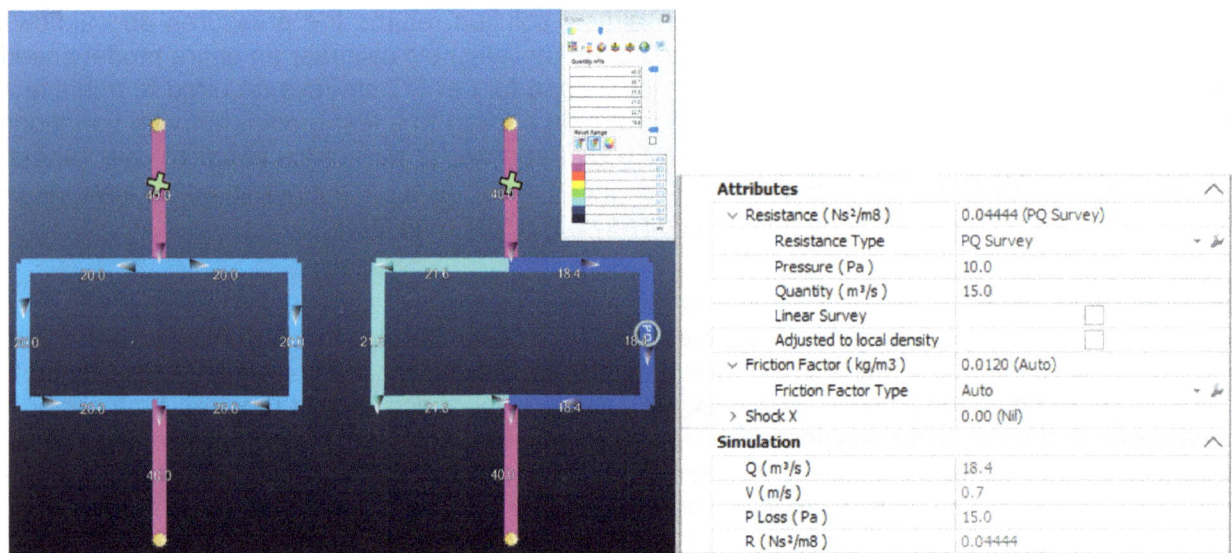

FIG 1 – A comparison of airflow with resistance change.

To obtain the correct flow rate, all other airways in the network require measured resistances. However, in reality this might be difficult. Firstly, there may be too many airways to measure in a real mine with the available time and resources. Additionally, it is common for some airways to be inaccessible to the surveyor due to limited access or mine activities. A way to help calibrate the model based on a limited number of surveys would be useful.

Some previous work has been done on this subject. Danko, Bahrami and Lu (2015) used a least-square-fit numerical algorithm to match the simulated data to a limited subset of PQ surveys. The method uses a Jacobian sensitivity matrix of the ventilation system relative to airway resistance to adjust the Atkinson friction factors across the model to minimise the root-mean-square (RMS) error between the measured and simulated results. Xu *et al* (2018) present a method that identifies circuits in the model and then adjusts resistances within constraints to match the simulated airflows with surveyed airflows.

This study uses a similar method to Danko, Bahrami and Lu (2015). The objective is to calibrate the model from a limited set of PQ surveys, or known flow rates. In the simple case in Figure 1, the method will deduce how much to change the resistance in the airway in the left branch to make the flow agree with the PQ survey in the right branch. The paper will present the method; then further options will be presented where the user can limit or induce the calibration on certain airways to achieve a more practical result. For example, the user may have more confidence in existing model resistances in some airways than others. Further to this, the application of the method to calibrating regulator resistances will be shown.

METHODS

Base calibration method

We define a dimensionless resistance calibration factor, C, which is applied to the airway resistance, such that Equation (1) is replaced with:

$$\Delta P = CrQ^n. [\text{Pa}] \tag{3}$$

The mine ventilation model with no calibration applied will have $C = 1.0$ applied on every airway; the method will adjust these coefficients so that the airflow quantities at the PQ survey sites agree with the surveyed quantity. On the airways that form the subset of airways with known flows, r is calculated directly from the PQ survey result, so no adjustment is applied and C remains unchanged at 1 throughout the process. Hence our numerical algorithm is adjusting the values of C throughout the rest of the model where the resistance is not specified with a PQ survey, searching towards the condition:

$$\sum_{i=0}^{N}|Q_{S_i} - Q_{PQ_i}| = 0 \tag{4}$$

where:

N	is the number of airway segments with known flows throughout the model
S	subscript denotes the simulated value
PQ	subscript denotes a known flow rate from a PQ survey

The simulated values are obtained using the air simulation function from Ventsim DESIGN 5.4, which solves airflows and pressures in the ventilation model using the gradient method of Todini and Pilati (1987).

To adjust the coefficients C_j, a Jacobian sensitivity matrix is built; this matrix is effectively a record of how sensitive the airflow quantity in each known flow site is to changes of the resistance on every other airway segment in the model. It is defined as:

$$[J_Q] = \begin{bmatrix} \frac{\partial Q_1}{\partial C_1} & \cdots & \frac{\partial Q_N}{\partial C_1} \\ \vdots & \ddots & \vdots \\ \frac{\partial Q_1}{\partial C_M} & \cdots & \frac{\partial Q_N}{\partial C_M} \end{bmatrix} \tag{5}$$

Where M is the number of airway segments without a known flow, on which we will change the values of C_j. Note, that an airway segment refers to a series of one or more connected airways between two junctions or nodes of the network. In a model, a segment will often comprise several airways, such as a spiral decline between two intersections which uses several connected airways to account for the curvature between junctions. Segments are more efficient to calculate as they may be substantially fewer in the model than the number of individual airways.

Each element of this matrix gives the sensitivity of the quantity on a known flow airway to a change in resistance on a single airway segment elsewhere in the model. Constructing this matrix on a moderately sized mine model can take a long time; each element of the matrix requires an air simulation to calculate the individual sensitivity on each known flow airway, meaning constructing this matrix will require M air simulations. For example, on a model of 3600 airway segments, if each simulation takes 1 second to complete, constructing the matrix will take 3600 seconds, or 1 hour, to complete.

Once the matrix is calculated it provides a guide as to where to adjust, or nudge, the coefficients, C_j, across the model to drive the difference between measured and simulated quantity values on the known flow or PQ survey airways to zero.

A column vector called the step magnitude vector is defined:

$$[S] = \begin{bmatrix} \partial s_1 \\ \vdots \\ \partial s_N \end{bmatrix} \qquad (6)$$

which tracks the adjustment or nudge that each airway segment's resistance coefficient, C, is receiving. Each element is initialised at an arbitrary value of 0.4, although this initialisation value could be optimised in the future.

The next step is to obtain the adjustment to the resistance coefficients by multiplying the step magnitude vector by the values in the Jacobian sensitivity matrix, such that:

$$[A]_a = \begin{bmatrix} \frac{\partial Q_1}{\partial C_1} & \cdots & \frac{\partial Q_N}{\partial C_1} \\ \vdots & \ddots & \vdots \\ \frac{\partial Q_1}{\partial C_M} & \cdots & \frac{\partial Q_N}{\partial C_M} \end{bmatrix} \begin{bmatrix} \partial s_1 \\ \vdots \\ \partial s_N \end{bmatrix} \qquad (7)$$

This means the resistances will be adjusted according to the sensitivities set by the matrix. The airway resistance coefficients are then incremented by A_a:

$$[C] = [C] + [A]_a \qquad (8)$$

At each iteration, this nudges the airway segment resistances towards the values suggested by the Jacobian. At the end of the iteration, a record is made of whether the simulated quantity at each of the known flow airways has moved away or towards the measured value, or if it has crossed over from less than to greater than, or vice versa. If the sign of the error is switching back and forth every iteration, this is an indication that the algorithm has moved as close to the target as it can with the current step magnitude, so the increment is halved, $\partial s_i = \partial s_i \times 0.5$. If the error between the simulated and measured quantity has increased or decreased, then the sign of the step is adjusted accordingly. In this way, the algorithm steps progressively towards the corrected resistance solution. At a set period of iterations, the nonlinearity of the process is accounted for by recalculating the Jacobian in Equation 5. Danko, Bahrami and Lu (2015) suggest every ten iterations, however this can be adjusted on a case-by-case basis, considering the rate of convergence of the algorithm as the iterations progress and the length of time it may take to rebuild the Jacobian.

The result is a set of coefficients, C, which when applied in Equation 3 across the model gives simulated airflows acceptably close to those at the known flow sites.

Smoothed method

Adjusting the resistance coefficients using the method outlined in the base calibration method leads to a tendency to prioritise adjustments on the airway segments with the strongest effect on the quantities at the known flow sites. This behaviour is not desirable if a more even and holistic adjustment is required. Another option is available which spreads the adjustment across a wider selection of airways, instead of focusing changes on airways which have the greatest effect on the calibrator airways. To smooth the coefficients, each element of the Jacobian matrix in Equation 7 is inspected. If the magnitude of the element is negligible (<0.0001) then it is set to zero, and -1 or 1 otherwise depending on the original sign, giving a simplified matrix, $\overline{J_Q}$, with elements of -1, 0 or 1. The two matrices are then multiplied to give an adjustment vector A_b:

$$[A]_b = \begin{bmatrix} \overline{\frac{\partial Q_1}{\partial C_1}} & \cdots & \overline{\frac{\partial Q_N}{\partial C_1}} \\ \vdots & \ddots & \vdots \\ \overline{\frac{\partial Q_1}{\partial C_M}} & \cdots & \overline{\frac{\partial Q_N}{\partial C_M}} \end{bmatrix} \begin{bmatrix} \partial s_1 \\ \vdots \\ \partial s_N \end{bmatrix} \qquad (9)$$

$$[C] = [C] + [A]_b \qquad (10)$$

This method effectively damps the sensitivity matrix, meaning that the resistance adjustments are spread more evenly through the model, and the airway segments with the most effect on the known flow site quantities are not so strongly prioritised. In the case of a model with a single calibration point, this leads to a model with airways with either a calibration factor of 1 or of plus or minus some standard amount across the whole model; this may appear to be too blunt a tool, but on only slightly more complex models with several known flows, it quickly leads to greater variation of calibration factors across the model.

Designated airways method

Many airway segments may already have known or more likely accurate resistances. Another option can be applied whereby a greater confidence in the existing model resistances can be assigned to certain airways. This could be done to reflect a greater confidence in a certain airway given a recent calibration in that airway; or conversely to favour adjustment in other parts of the model where the engineer has low confidence in the current resistance.

When airways segments are denoted as accurate, we restrict the change in the coefficient to (for example) 5 per cent. This means during the calculation of the coefficients in Equation 10, the resulting coefficient is restricted to the range $0.95 < C < 1.05$. Furthermore, during subsequent constructions of the sensitivity matrix of Equation 5, if the coefficient of the accurate airway has reached 0.95 or 1.05, the airway is given a zero sensitivity. This restriction typically means that to reach a calibrated state, the algorithm will apply greater adjustments to other 'non-accurate' airways.

Control resistance only method

Control resistances are additional resistances in mine ventilation systems used to direct or block airflow, such as bulkheads, doors, seals and regulators. Another option tested is where the element of the sensitivity matrix of Equation 5 is set to zero if the associated airway of the element is not a pre-defined control resistance. This means that adjustment of coefficients will only occur on airways with a control resistance and not on open airways. This has the effect of calibrating the model only with the regulators and seals.

This option repurposes the method somewhat to act as a calibration tool for regulator openings. Typically during a manual model design process, a user identifies the desired flow amounts at locations in the model, then iteratively adjusts regulator resistances (often at different locations) to achieve those flows. The proposed calibration method will automatically achieve a solution for this problem.

USE CASES

Figure 2 shows the base case for the various use cases of this section in Ventsim™ software. The flow is evenly split between left and right in the model.

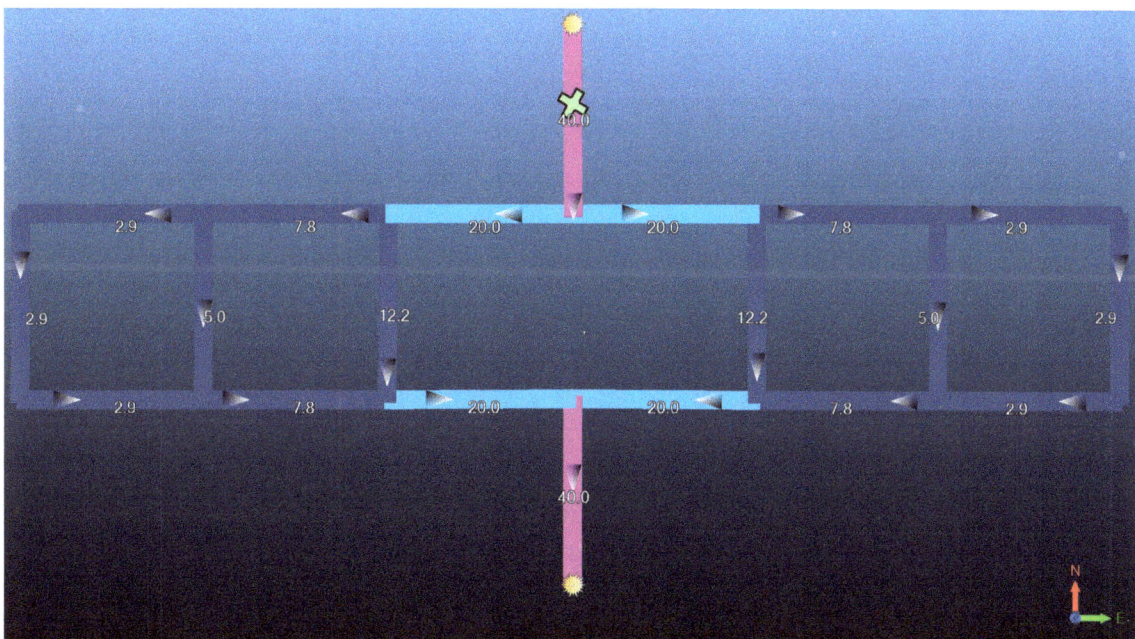

FIG 2 – Base model, showing airflow rates in m³/s.

A resistance corresponding to a theoretical PQ survey measured pressure loss of 6 Pa and a flow rate of 13 m³/s is applied to one of the central airways, as indicated by the icon and red circle shown in Figure 3.

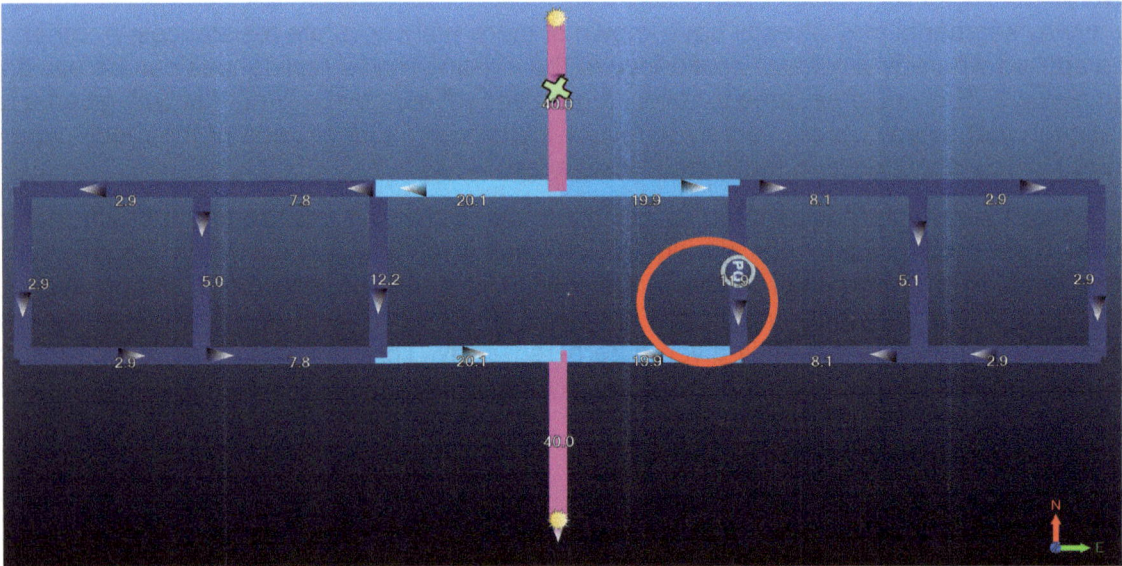

FIG 3 – Base model, showing PQ survey and airflow rates in m³/s, (pressure loss 6 Pa, flow rate of 13 m³/s) applied.

This produces a higher resistance than the previous model and simulation reduces the flow rate from 12.2 to 11.9 m³/s. The objective of the following test cases is to adjust the other resistance in the model to make the flow in this calibrated airway equal to the measured flow rate of 13 m³/s.

Base calibration method

Running the calibration tool with the base method yields the following results in Figure 4, after 27 iterations. The figure shows on each airway the resistance coefficient C and the airflow rate; the airways are coloured by contours of the resistance coefficient multiplier. The method follows the sensitivity amounts contained in the sensitivity matrix of Equation 5, so the change in resistance is focused on the airways closest to, and with the biggest effect on, the calibrated airway.

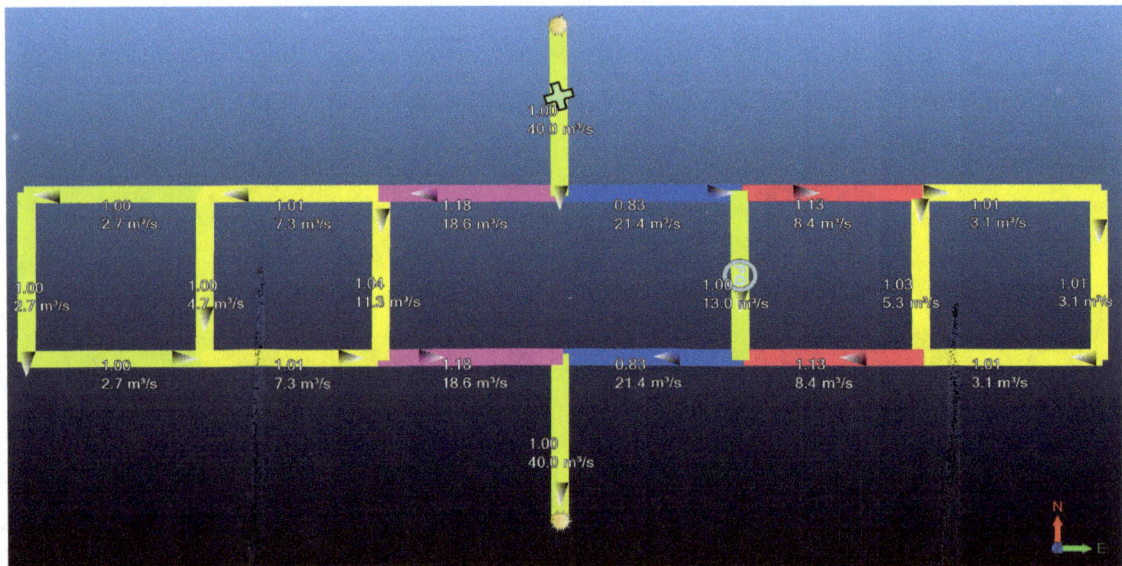

FIG 4 – Base Calibration Method, coloured by contours of the calibration factor, and showing in text the calibration factor and the airflow rate. Red and pink colours indicate higher calibration factors.

Smoothed method

For the smoothed method, the change to resistance factor does not follow the sensitivity level, but rather the sensitivity sign, so the effect is spread around the model as shown in Figure 5.

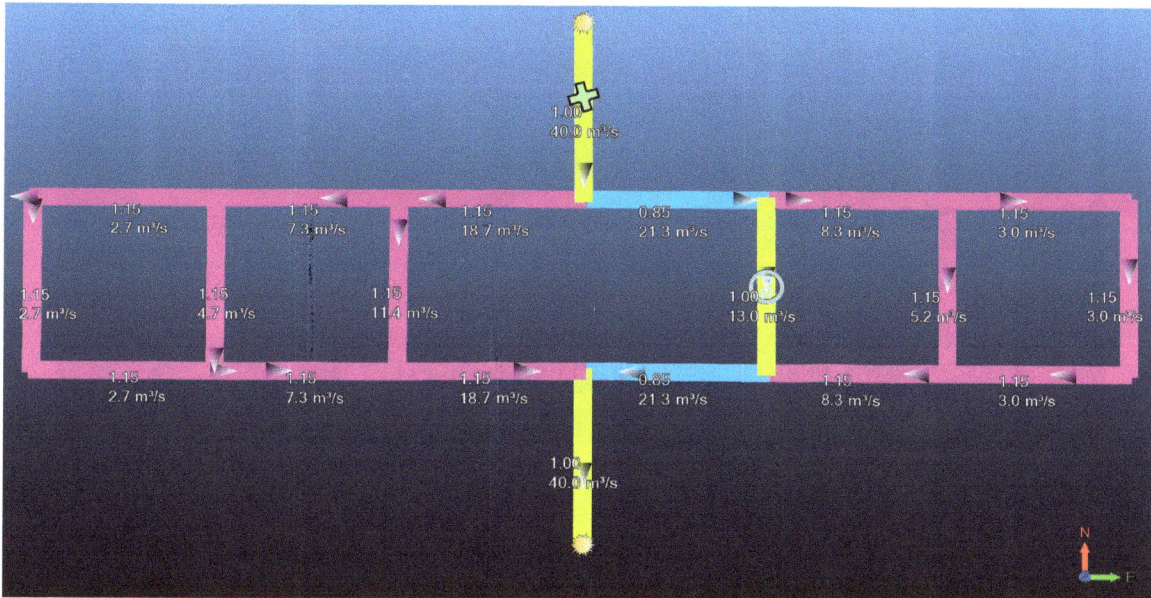

FIG 5 – Calibration factors and airflows using the smoothed method. Pink colours indicate higher calibration factors.

In this case, airways have only three values of calibration factor, $C = 1.0$, or $C = 1.0 \pm 0.15$, reflecting the blunt nature of the tool, where the calibration factor is adjusted by the same amount across the entire model to achieve the known flows. This is the opposite extreme to the base method, which only strongly adjusts sensitive airways. This method is more applicable to a model where most resistances are considered reasonably accurate, and any adjustments are likely lower in magnitude and made more evenly across the entire model.

Designated airways method

In the example shown in Figure 6, the horizontally aligned airways are assumed to have been measured with a high degree of confidence. The calibration focuses on the vertically aligned 'inaccurate' airways. Accordingly, the horizontally aligned airways are restricted to a 5 per cent change in resistance, and we see very large variation in the vertically orientated airways.

This method could be employed to represent a range of scenarios. For example, the horizontally aligned airways in the example may have been surveyed and calibrated, so a high level of accuracy to the resistances there is assumed. By contrast, the vertically aligned airways (other than the one with PQ survey) could be inaccessible and the surveyor, through their knowledge of the mine and experience, might strongly suspect that the model resistances are wrong.

The model shown, and the set of calibration factors found, is simplified and has been purposefully designed to show the potential use of the tool. In a real mine with all its complexity, the solution may not be as obvious, and the calibration method could highlight areas of the model requiring further work or investigation.

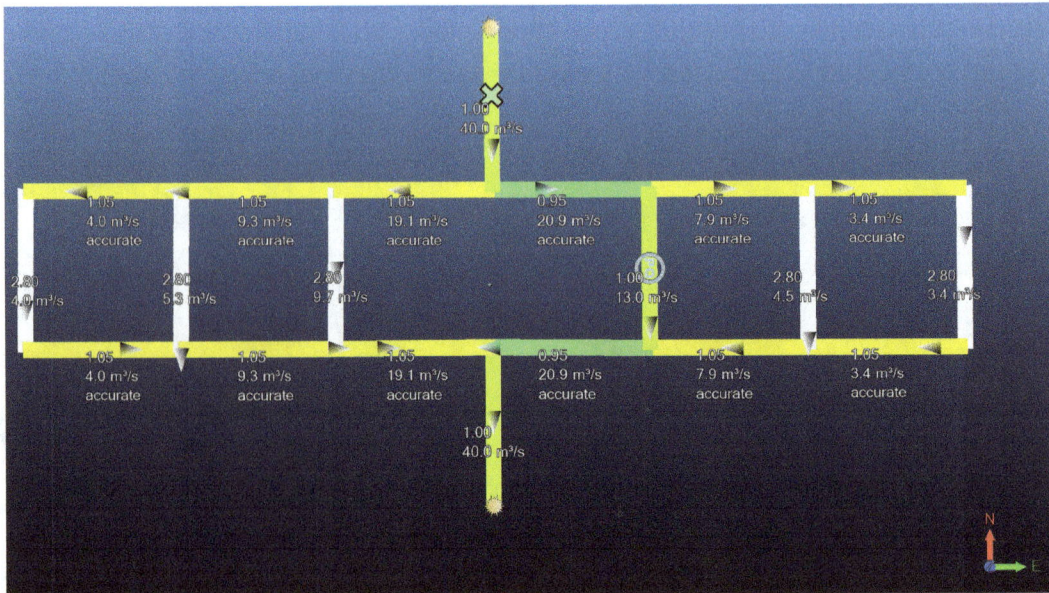

FIG 6 – Calibration with the Accurate Airways method (white colour shows areas of high calibration).

Regulator only method

This method effectively re-proposes the tool to calibrate a set of regulators to achieve flows in a different part of the mine. Calibrating a set of regulators to achieve set flows elsewhere in the model is a fairly common task, but one often done by user iteration, or by a different system of calibration. One method is to take a known, or targeted, flow and detect the regulator with the greatest effect and adjust it accordingly. If fully opening that regulator cannot achieve the flow, the next most influential regulator is adjusted and so on. If we employ the method described here to this task, then it does something similar, but will also handle multiple set points. If nothing else, employing the current method would be likely to produce a different, possibly more efficient, solution.

In this example, the vertically aligned airways have been replaced with custom regulating resistances. The calibration method can now be restricted to these resistances/regulators, giving the result shown in Figures 7 and 8, whereby the calibration has entirely occurred on the resistances, leaving the open airways untouched. This is performed for both the base and smoothed method. Again, while this is a simple example, the method could be useful in more complex problems where there are multiple set points affected by multiple regulators.

The situation could also arise where there is not sufficient fan pressure to achieve the set flows, even though regulators are fully open. This would manifest in the algorithm as a calibration factor reducing to zero. A different but similar calibration could be done which would increase fan pressures across one or a range of fans to increase the flow to the set point.

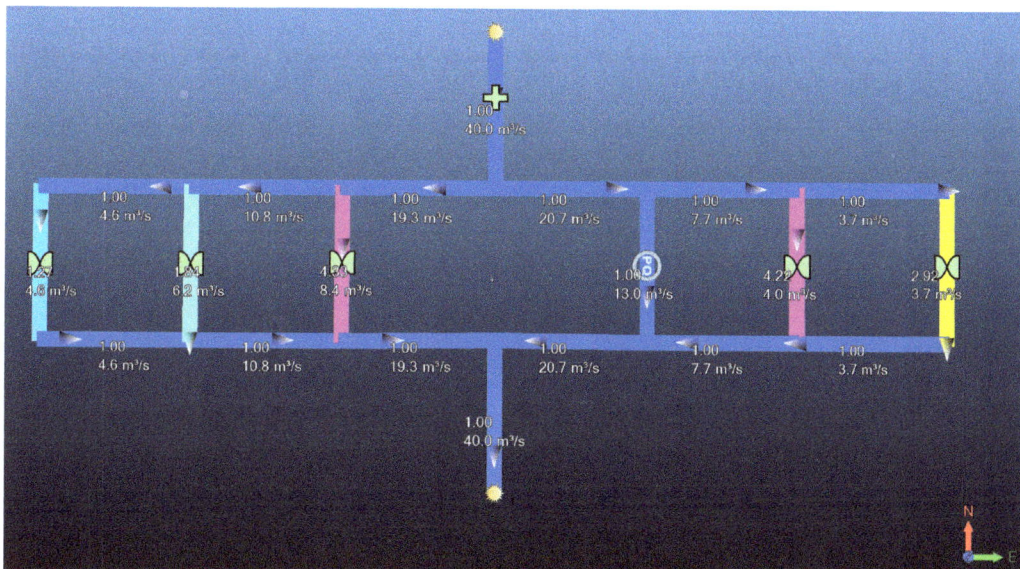

FIG 7 – Regulator only calibration, where resistances are targeted to achieve the flow – base method.

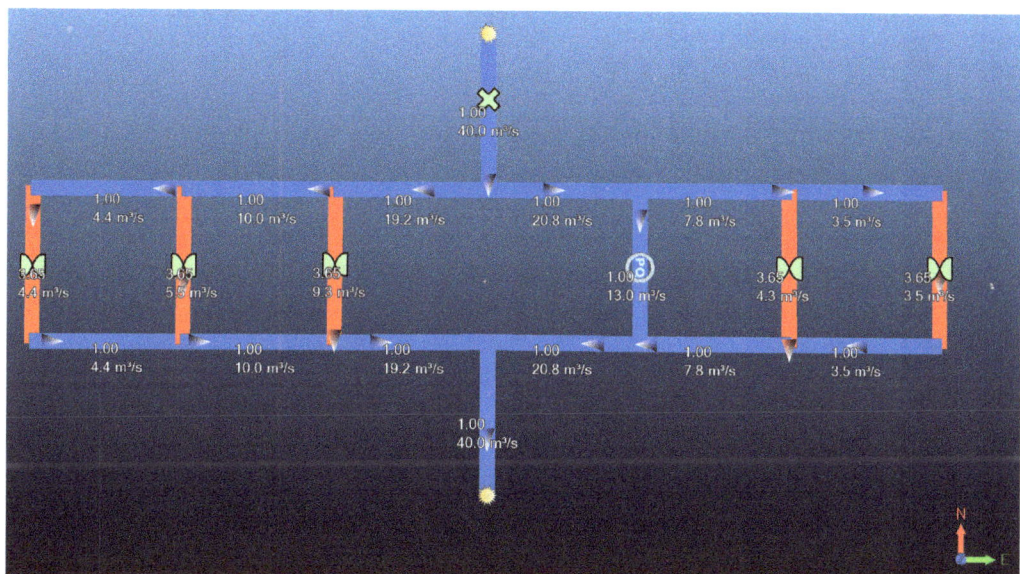

FIG 8 – Regulator only calibration, where resistances are targeted to achieve the flow – smoother method.

CONCLUSIONS

A range of use cases has been presented for a mine ventilation network calibration tool, based on a limited set of known flow rates. Several adjustments to the algorithm can be made to reflect a user's preference and knowledge of particular aspects of the mine and model. For example greater confidence in the model resistance of certain airways can be included. The tool has the potential to greatly improve the speed and accuracy of calibrating a model, and help users locate problem areas of the model, motivating either further model improvement or reinspection and investigation of parts of the mine.

REFERENCES

Danko, L D, Bahrami, D and Lu, C, 2015. Mine Ventilation Model Calibration with Numerical Optimization, *Proceedings of the 15th North American Mine Ventilation Symposium 2015*, Blacksburg, USA.

Griffith, M D and Stewart, C M, 2019. Sensitivity Analysis of Ventilation Models – Where not to trust your simulation, *Proceedings of the 17th North American Mine Ventilation Symposium 2019*, Montreal, Canada, pp 654–660.

McPherson, M J, 1993. *Subsurface Ventilation and Environmental Engineering*, Chapman & Hill.

Michelin, F, Stewart, C M, Griffith, M D and Andreatidis, T, 2019. Calibrating Model Airway Size and Resistance with Survey Asbuilt Data, *Proceedings of the AusIMM Mine Ventilation Conference 2019* (The Australasian Institute of Mining and Metallurgy: Melbourne).

Prosser, B S and Loomis, I M, 2004. Measurement of Frictional Pressure Differentials during a Ventilation Survey, *Proceedings of the 10th North American Mine Ventilation Symposium*, Anchorage, USA, pp 59–66.

Todini, E and Pilati, S, 1987. A Gradient Method for the Analysis of Pipe Networks, *International Conference on Computer Applications for Water Supply and Distribution*, Leicester, UK.

Xu, G, Huang, J, Nie, B, Chalmers, D and Yang, Z, 2018. Calibration of Mine Ventilation Network Models using the Non-Linear Optimization Algorithm, *Energies*.

Ventilation contaminant characterisation for a greenfield project

A Hatt[1]

1. MAusIMM, Sr Underground Mining Engineer, Newcrest Mining Limited, Melbourne Vic 3006. Email: alex.hatt@newcrest.com.au

ABSTRACT

Ventilation demand estimating is a multi-faceted problem requiring understanding of the ambient environment that the mine operates/will operate in. Characterisation work tends to focus on the requirements to dilute the by-products of machine operation (gases, heat, diesel particulate matter (DPM)) and can neglect other risks that use ventilation as a control. Failing to completely characterise (or mischaracterising) the mine environment can result in a ventilation system that is inefficient or ineffective at controlling contaminants of concern.

This paper presents a case study of characterisation work undertaken at a greenfield exploration project to provide engineering and OH&S inputs for the design of a ventilation system. The paper discusses how comprehensive early characterisation of the ambient surface and underground environment including aspects related to heat, dusts, strata gas, and geologic/mineralogical factors enables incorporation of the right ventilation and non-ventilation controls at a study level to mitigate the risks identified.

INTRODUCTION

The Havieron Project is located in the Paterson Province, 45 km east of Telfer on the traditional lands of the Martu people (Figure 1). The Project is operated by Newcrest under a Joint Venture Agreement with Greatland Gold.

FIG 1 – Havieron location map.

The project is overlain by approximately 420 m of sediments of the Paterson Formation and Quaternary aeolian sediments. Gold and copper mineralisation consists of breccia, vein, and massive sulfide replacement typical of intrusion-related and skarn styles of mineralisation. Mineralisation is hosted by metasedimentary rocks (meta-sandstones, meta-siltstones, and metacarbonates) and intrusive rocks of an undetermined age. Drilling has partially defined the extents of mineralisation within an arcuate shaped zone to depths of up to 1400 m below surface.

As a greenfield project, Havieron is in an ongoing state of exploration and development with concurrent technical studies to determine preferred mining options. In this context, the design of the ventilation system to manage contaminants from underground mining is a key safety consideration. During early study phases a typical (and frequently valid) approach is to adopt and modify inputs from adjacent or similar operating mines with consideration of the similarity of the existing operation to the one under study.

In the case of Havieron there are several potential contaminants that require site specific characterisation driven by the style of mineralisation and minerology. Put simply, a 'copy/paste' of parameters related to the thermodynamics of the ventilation system and contaminants from adjacent operations would not provide the correct inputs to engineering controls.

IDENTIFICATION OF CONTAMINANTS AND RISKS

Concept ventilation studies for the project identified multiple potential contaminants and associated risks based on information from exploration geology and early metallurgical test work.

Massive sulfides

The SE Crescent Zone is a geologic domain characterised by massive sulfide accumulations and contains well developed pyrrhotite-chalcopyrite and pyrite sulfide mineral assemblages as breccia and vein infill, and massive sulfide lenses. Massive sulfides present risks for self-heating, with potential impacts to both the underground environment as well as surface stockpiling and processing. More specifically, mixed sulfides present a risk of self-heating with many recorded instances from Canadian base metal operations (Farnsworth, 1977; McCutcheon and Walker, 2020). Nesset and Rosenblum (2020) suggest a series of electrochemical reactions leading to the evolution of gases (H_2S, SO_2) and heat. They also provide key factors controlling the occurrence and rate of the reaction are:

- A permeable pile of rocks to allow introduction of moisture and oxygen (blasted muck, stockpiles).

- Mixtures of electrochemically dissimilar sulfide minerals in sufficient quantity to sustain the reaction, balanced by presence of buffering (acid-neutralising) minerals.

- Fragmentation/particle size, with self-heating being proportional to the surface area of the sulfide particles (Payant *et al* 2007).

- Time for the self-heating reaction to initiate and develop into a self-sustaining reaction.

Another identified risk from high sulfide content is sulfide dust explosions. Experimental data presented by Enright (1988) and others gives a minimum 18 to 20 per cent total sulfur as a lower bound for explosible sulfide dusts. Enright also includes some case histories of sulfide dust explosions from the early 1900s through 1980s. Factors that affect the explosibility of sulfide dusts include:

- Particle size distribution, with finer dusts requiring lower ignition temperatures/energy.

- Dust concentrations, with increasing concentrations leading to higher explosive pressures and rates of pressure rise.

- Minerology of the dust, with inert components (non-sulfides) inhibiting the explosion reaction by a combination of methods.

Respirable crystalline silica

The SE Crescent Zone includes units of 'Cloudy Quartz Veins' and 'Pyrrhotite Quartz Rock' characterised by white to grey euhedral quartz (Wilson, 2019). Respirable Crystalline Silica (RCS) dust is a known source of various lung diseases (silicosis, cancers, emphysema) in miners and the construction industry. Australian states have adopted a recently reduced Workplace Exposure

Standard (WES) of 0.05 mg/m³ (Safe Work Australia, 2019) leading to increased focus on management.

Fibrous minerals

Core logging and early minerology identified the presence of Actinolite and Tremolite in the metasedimentary packages and breccias surrounding the SE Crescent and within the cloudy quartz veins. These minerals can exhibit a fibrous (asbestiform) habit however their presence alone is not sufficient to understand the risk. A key consideration in the regulation of exposure to fibrous minerals are the dimensions of fibres collected from airborne sampling and what constitutes a 'countable fibre', with slightly differing requirements based on the applicable regulation or guidance. The current WES for Asbestos from Safe Work Australia is 0.1 fibres/mL (Safe Work Australia, 2019).

Strata gases

Exploration drilling has encountered Carbon Monoxide (CO), Sulfur Dioxide (SO_2), Hydrogen Sulfide (H_2S), and flammable gases at the collar of drill holes. Approximately 5 per cent of holes and wedge holes drilled at the time of review had gas intercepts with typically only one intercept in each hole.

Heat

The Havieron Project is located in the Great Sandy Desert, a physiographic region with hot summers and minimal rainfall. The adjacent Telfer mine employs refrigeration for the mine ventilation system during the summer months to manage the effects of the ambient climate, and similar weather conditions are expected for the Havieron project. With the difference in geologic setting and style of mineralisation from Telfer the need to establish site specific rock parameters for modelling of heat loads from strata was identified.

CHARACTERISATION

Based on the contaminants and risk identified a characterisation program was incorporated into the Pre-Feasibility Study (PFS). The intent of the program was to establish the site-specific engineering inputs/factors and to validate the identified risks to inform engineering level controls. The following sections describe the laboratory and data collection efforts undertaken as well as some of the results to date.

Self-heating of sulfides

Testing was completed for a suite of 7 existing metallurgical composites during the PFS for reactivity of sulfides using the FR-2 protocol, with composites also submitted for quantitative minerology and analytical head assays. The FR-2 protocol is a two-stage test for self-heating that incorporates a simulated weathering phase followed by a higher temperature stage that test the overall reactivity of the sample. Samples of pulverised material are moistened and then heated to 70°C for Stage A of the test. Air is sparged into the sample regularly and the resulting increase in sample temperature is recorded. After Stage A the temperature is increased to 140°C and the sparging process is repeated (Rosenblum, Nesset and Spira, 2001).

Inclusion of a simulated weathering phase differentiates the protocol from the United Nations (UN) Self-Heating Substances Test by allowing time and conditions for the formation of acid in the initial stages of the reaction and reduces the rate of false negatives that reactive sulfides can produce using only the UN test (Moon *et al*, 2019). This more closely resembles the conditions expected to be found in an underground mine by the inclusion of heat and moisture to the sample, whereas the UN test is designed to test conditions for the determination of dangerous good packaging which can have tightly controlled conditions for moisture and temperature. An example of the resulting thermogram (BBA Labs, 2021) is shown in Figure 2.

Test 2021-027 : Newcrest HAD 018 : Unit 1 : 2021/04/12

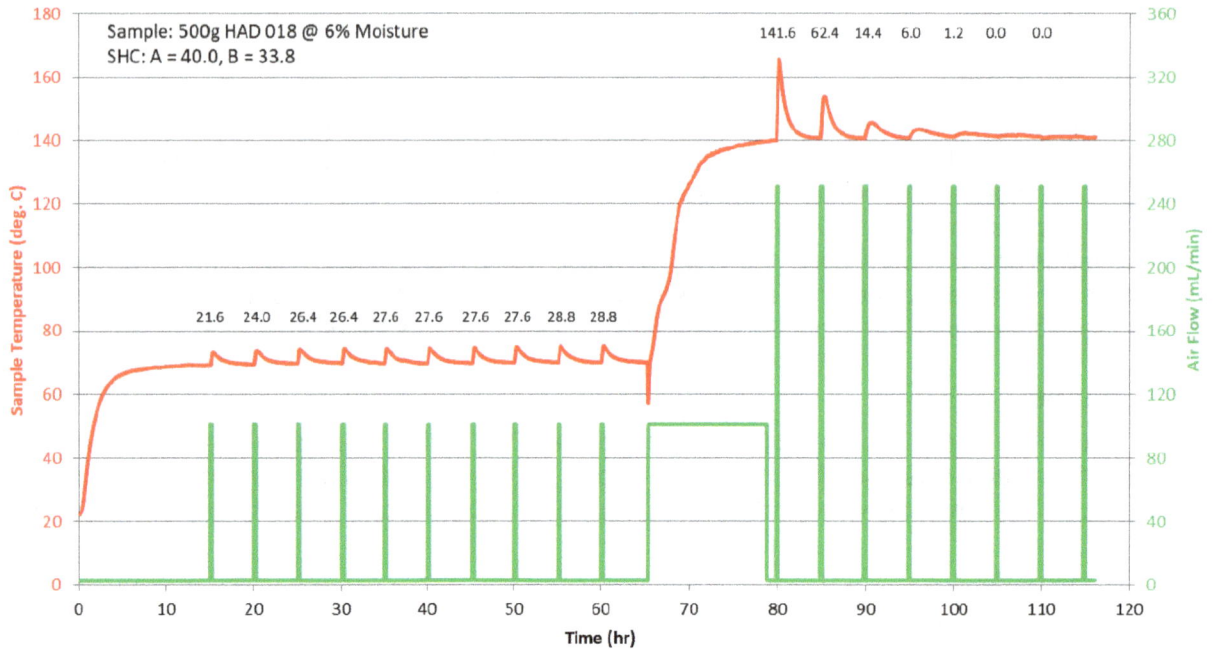

FIG 2 – FR-2 protocol thermogram (BBA Labs, 2021)

The data recorded from each stage of the test is used to calculate a 'self-heating capacity' and plotted on an empirically derived risk chart. The results of the work completed for the PFS (BBA Labs, 2021) are shown in Figure 3.

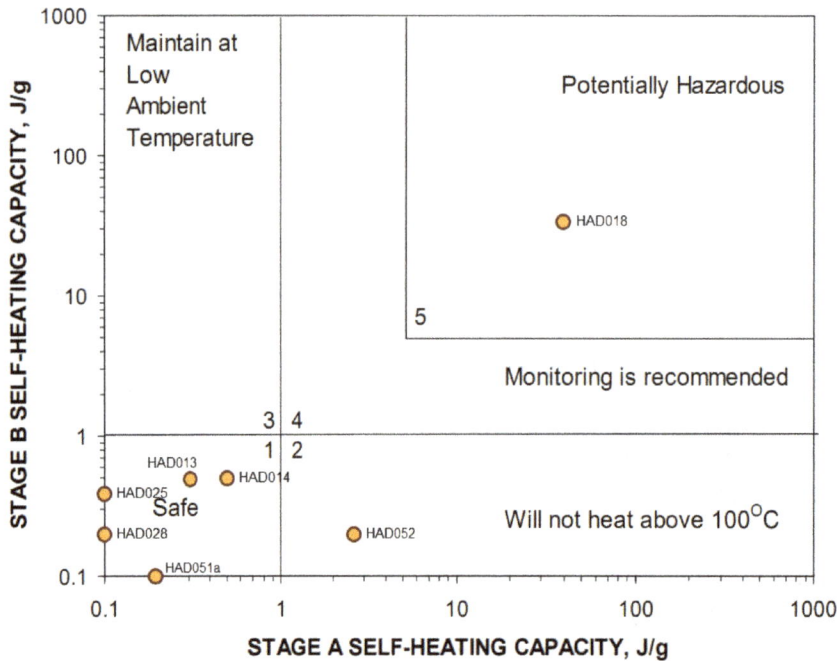

FIG 3 – Self-heating risk chart.

Analysis of results indicated that samples with elevated levels of pyrite and pyrrhotite had increased self-heating capacity as expected, and that where significant amounts of carbonate minerals were present the self-heating capacity was less than expected. Conceptually this is due to the acid neutralising nature of the carbonates interfering with the initial steps of the self-heating reaction and provides a key input to any risk classification scheme.

Sulfide dust explosibility

Sulfide dust explosions were identified as a risk in the Concept Study due to the elevated sulfur assays in areas of massive sulfide mineralisation. To identify geologic units of concern boxplots were generated from raw assays of total Sulfur as shown in Figure 4.

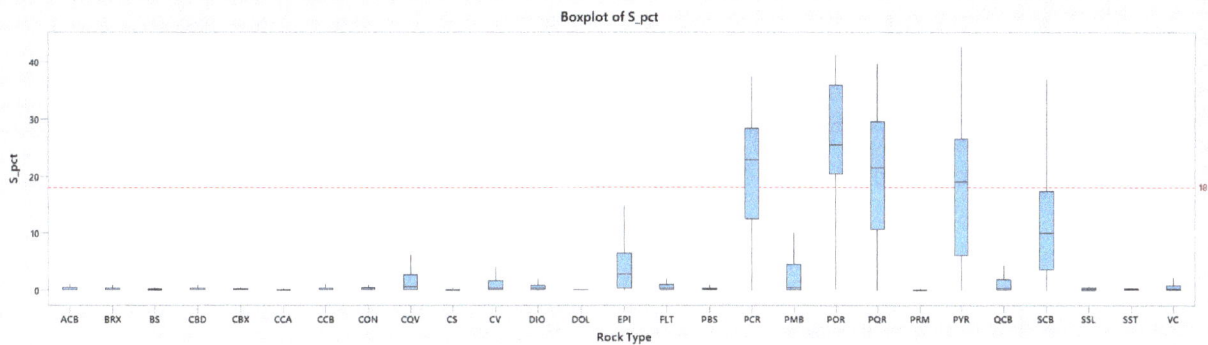

FIG 4 – Boxplots of % sulfur by geologic unit.

The boxplots indicate the four massive sulfide units (Pyrrhotite Calcite Rock (PCR), Pyrrhotite Quartz Rock (PQR), Massive Pyrrhotite (POR), and Massive Pyrite (PYR)) and the Sulfide Cemented Breccia (SCB) that comprise the Crescent Zone have mean total sulfur >18 per cent which exceeds the accepted minimum sulfur content for an explosible sulfide dust. Explosibility Go/No Go (Standards Australia, 2016) was completed as part of the PFS to determine if an explosive/ combustible dust could be formed.

Composites were selected from each of the units above as well as the Cemented Calcite Breccia (CCB) and Crackle Breccia (CBX) units as they had extreme values exceeding 18 per cent and are of economic interest. Samples for compositing were selected by the exploration geology team to represent the 'worst case' example with high levels of sulfides, with samples submitted for quantitative minerology and analytical head assays. Sample preparation was conducted dry with coarse crushing of samples followed by staged pulverisation and screening to P_{100} of 212 µm. Dry processing was used to minimise self-heating risk during preparation and transport, with samples oven dried and vacuumed sealed in nitrogen purged bags before shipping to labs. Abridged minerology, analytical assays, and testing results are shown in Table 1.

TABLE 1

Dust explosibility minerology, assays and results.

		PCR	POR	PQR	PYR	SCB	CBX	CCB
Powdered XRD Minerology (% Mass)	Pyrrhotite	48	92	71	0	36	1	17
	Chalcopyrite	3	3	3	1	2	0	1
	Pyrite	0	0	0	53	0	0	5
	Annite – biotite – phlogopite	0	0	0	0	8	9	0
	Calcic amphibole	0	<1	2	0	5	24	0
	Plagioclase	1	1	2	<1	32	29	1
	Quartz	3	4	18	10	9	<1	4
	Dolomite – ankerite	45	0	0	31	1	0	69
	Calcite	0	0	2	4	4	34	1
	Siderite type carbonate	0	0	3	0	1	0	0
Head Assay	Fe – (%)	45.7	56.8	46.3	32.1	33.2	3.53	29.9
	S – (%)	30.5	38.3	30.0	36.0	20.9	0.93	19.7
	S-2 – (%)	28.3	26.9	27.3	35.7	18.8	0.54	17.3
Explosibility	Explosibility	Yes	Yes	Yes	Yes	No	No	No
	P_{50} (µm)	66.2	68.1	77.5	86.4	65.5	58.4	78.6
	Moisture (%)	0.5	0.3	0.6	0.5	0.5	0.3	0.5
	Min. Explosible Concentration (g/m^3)	500	500	750	500	-	-	-
	Pmax (bar)	2.3	2.6	2.2	2.1	-	-	-
	dP/dt (bar/s)	39	63	36	32	-	-	-

Testing confirmed that an explosible dust could be formed in the massive sulfide units where total sulfur exceeded 18 to 20 per cent. In addition to validating the risk, some 'academic' points of interest were also observed:

- The SCB and CCB samples were classified as non-combustible, however they are approximately at the limit of the accepted sulfur content.

- The rate of pressure rise (dP/dt) and maximum pressure (Pmax) of the Massive Pyrrhotite was appreciably higher than that of the massive pyrite unit. While both samples had similar concentrations of total sulfur, the PYR was only 57 per cent pyrite by mass, while the POR unit was 92 per cent Pyrrhotite by mass. This suggests some influence of inert minerals on severity of the explosibility.

Respirable crystalline silica

Quartz content from quantitative minerology of 23 metallurgical composites was used to generate boxplots for the SE Crescent and Breccia domains shown in Figure 5. The composites cut the full width of the domains are a 'mining scale' composite of the material that is likely to generate dust during the mining process. Mineral Liberation Analysis of the composites identified the quartz as crystalline α-quartz. The distinction is important at early-stage evaluation as not all methods of estimating quartz content draw the distinction which is critical to the 'crystalline' aspect of the risk.

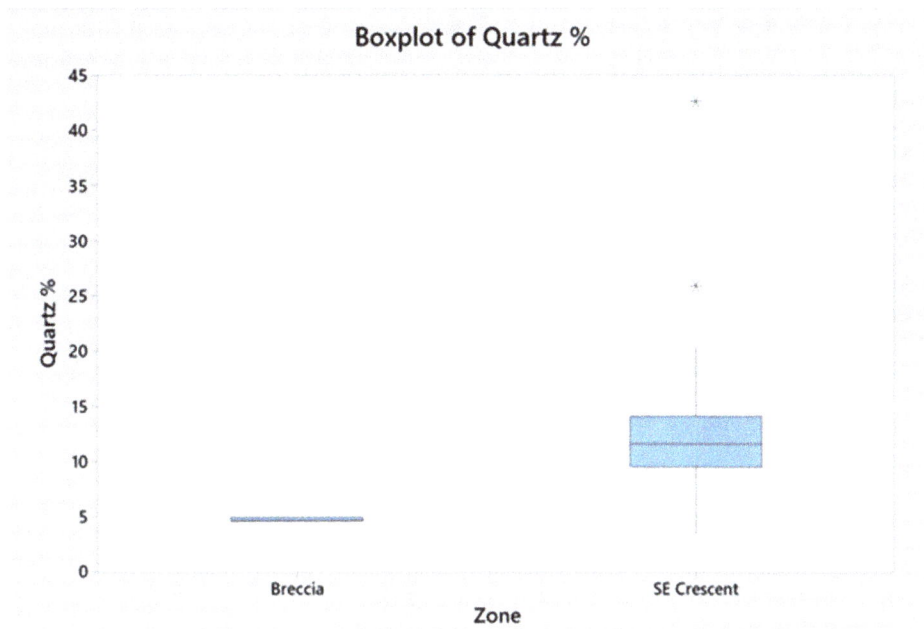

FIG 5 – Boxplot of % quartz by geologic unit.

Using the recently introduced 0.05 mg/m^3 WES from Safe Work Australia (2019) (adjusted to 0.034 mg/m^3 for a 12-hour shift), an equivalent total respirable dust concentration was calculated for each domain. For the SE crescent percentiles were generated to assist with understanding the distribution. Since silica content varies between mine sites the use of total respirable dust allows for a better comparison to trends in dust exposure and management within the industry as opposed to RCS exposures which are, all other things being the same, a function of site minerology.

The trend from (Department of Mines, Industry Regulation and Safety, 2019) indicates average total respirable dust exposures in the WA mining industry for the period of 2006–2018 of ~0.4 mg/m^3. Assuming dust generated in the mining process has the same composition as the metallurgical samples, interpretation suggests that:

- Management of RCS from the breccia domain should be achievable using standard dust management techniques. Total respirable dust exposures would need to be ~2× the industry average for an exceedance.

- Management of RCS in the SE Crescent will require 'above average' dust management practices, with quartz concentrations requiring total respirable dust exposures to be less than ¼ the average exposure previously mentioned.

This comparison represents a 'worst case' scenario and does not consider differential liberation/sizing of dust generated and the subsequent impacts to respirability or retention in the lungs. While the mechanisms of breakage and dust generation are understood well at a conceptual level for mining activities, the ability to predict particle size distributions at the micron scale is limited.

Fibrous minerals

Previous Scanning Electron Microscope (SEM) work from the development of the Fibrous Minerals Management Plan for mineral exploration identified the presence of Actinolite and Tremolite fibres in Havieron rock samples. The samples, as observed, were within the specified length, diameter, and aspect ratio as defined in the NOHSC:3030 (NOHSC, 2005) to be counted as a fibre for the purposes of quantifying respirable fibres. Note that the work completed to identify the material is not a measure of exposure, but uses the definitions established from exposure monitoring standards to determine if the observed fibre would qualify in the instance that it was captured during exposure monitoring. This is an indication that the risk exists and requires management.

Characterisation work completed by Wilson (2019) indicates the presence of Actinolite in surrounding metasedimentary rocks, breccias, and cloudy quartz veins. Decline development and level accesses are expected in the metasediments and breccias. Stoping activities are expected to focus on the massive sulfide and breccia units. This is a key distinction as it dictates the nature of the nature of the fragmentation (stope versus development blasting) and relative planned exposure.

The NOHSC:3030 method is based on manually counting fibres in a gridded microscope field from airborne dust sampling and is centred on exposure rather than predictive measures. Quantitative minerology would only indicate the presence of the minerals and not their form/nature of fibres generated by breakage. Accordingly, predicting the concentrations of fibrous minerals in a spatially representative manner is effectively impossible due to the lack of a repeatable quantitative sampling method for *in situ* material. For the purposes of managing the hazard, knowledge of which units the fibrous minerals exist in and that a countable fibre can be generated is sufficient.

Strata gases

A small number of gas intercepts from drilling were logged in exploration drill database with the downhole position and maximum gas concentrations measured from logging gas meters on the drill. No robust association between gas and geologic structures or units has been identified to date, mainly due to the sparsity of gas intercepts. Attempts have been made to further characterise the gas with the use of gas speciation sampling tubes to resolve questions of cross-sensitivity of some of the gas sensors however there has been no success at the time of writing due to the practicalities of getting close enough to the emission point to safely obtain a sample. Consideration was given to gas bag sampling for chromatograph analysis however transportation times were not short enough to guarantee sample integrity.

Heat

Ambient climate

Climate data was sourced from the Australian Bureau of Metrology Telfer Aerodrome weather station (Australian Bureau of Metrology, 2020) located at the Telfer airport. Inspection of station details provided by the Bureau and Google Earth indicates no conditions that would lead to biased/non-representative data of the regional climate (eg nearby bodies of water, shading, wind shadowing, mechanical heat sources). The weather station is ~45 km NW of Havieron with the next closest weather station located in Nullagine ~260 km NW.

The data set consists of hourly measurements of dry-bulb temperature (°C), relative humidity (%), average 10 m elevation wind speed for 10 minutes prior (km/hr, wind direction), and local barometric pressure (hPa) from 1 January 1996 to 31 December 2019 (24 years). Wet-bulb temperatures were calculated by iterative solution to standard psychrometric equations for the given dry-bulb, relative humidity, and barometric pressure.

Data exhibits typical seasonal variation with occasional extreme events. Design criteria for surface conditions shown in Table 2 is based on 95th percentile wet-bulb temperatures for each time period, and the mean co-incident dry-bulb temperatures and barometric pressures (±0.25°C).

TABLE 2

Monthly and annual 95th percentile climate data.

	Jan	Feb	Mar	Apr	May	Jun	Jul	Aug	Sep	Oct	Nov	Dec	Year
Mean Coincident Dry Bulb (°C)	31.6	30.6	29.9	29.9	27.1	23.4	24.7	27.3	31.0	32.5	33.6	31.7	30.8
Wet Bulb (°C)	24.5	24.6	24.2	21.5	18.6	17.2	15.9	15.9	17.2	19.8	21.5	23.5	23.4
Pressure (kPa)	97.1	97.1	97.4	97.8	97.9	98.1	98.1	97.9	97.8	97.7	97.5	97.2	97.3

Subsurface thermal conditions

Surface virgin rock temperature

Surface virgin rock temperature (VRT) was estimated at 32°C using the average annual temperature method with an offset of +5 degrees as given by Gerner and Budd (2015), with temperature data supplied from the Telfer Aerodrome weather station. The calculated value agrees with the surface temperature model for the continent given in Haynes *et al* (2018).

Thermal gradient

Data for the determination of the thermal gradient was sourced from downhole survey measurements of exploration drill holes and compared to other temperature data including aquifer pump tests and downhole geophysical tools. Temperatures are taken along the length of the hole using the survey tool and are stored in the drill database. Downhole temperature data from survey instruments has many limitations that require quality control of the raw data before incorporation into any estimates of the gradient. Some of these limitations are:

- Survey instruments are typically left in direct sunlight before being lowered down the hole, or in the instance of underground drilling in the prevailing airflow. This can significantly change the starting temperature of the instrument and lead to inaccurate or erratic readings until it stabilises in the hole.

- Holes on survey control for orientation/deviation frequently have single point survey measurements taken where the survey tool will not have sufficient time to temperature stabilise. The result is significant scatter in the data sets or conflicting information when compared to a final continuous survey.

- As the instrument is moved down the hole water/mud circulates past. Depending on the speed the instrument is lowered at and the response of temperature sensor this can cause 'smearing' of temperatures.

- Depending on the survey instrument technology the tool may be deployed downhole in a continuous or a 'station observation' approach. This can have a significant effect on the amount of time the sensor has to stabilise and the validity of the temperature measurements.

- Water bearings structures and aquifers can locally skew the data, particularly if they are recharged with surface or rainwater that is lower than surrounding rock temperature.

- The temperature sensor is internal to the survey tool and in some instances is located adjacent to 'hot spots' in the tool casing. This can cause a bias in the reported temperature compared to the true temperature.

- In the instance of holes where the instrument is pumped in (rather than lowered) the temperature of the water used to pump the instrument can be significantly lower than the rock temperature or the temperature of the stagnant water in the hole, rendering the temperature data useless.

Data from individual holes was quality checked though the use of scatterplots and time series plots. Holes with no temperature trend or clear effects from the limitations listed above were removed from the data set. Of 102 unique hole identifiers only six were retained for estimation of the thermal gradient. Many of the reject holes had conflicting temperature data, with differences on the order of 10–20°C between sequential measurements indicating some of the potential quality issues above.

Heat from strata and broken rock is estimated to account for ~10 per cent of the heat source underground based on initial desktop estimates, with auto-compression and diesel equipment accounting for the majority. Given the minor nature of the estimated heat loads from strata and broken rock, the method and amount of data collected to estimate the thermal gradient is considered appropriate. In instances where heat loads driven by thermal gradients (such as hot water inflows) are a more substantial portion of the heat sources (say, 20 per cent or more) a dedicated measurement program may be justified to generate a larger body of representative measurements.

For the rock below the permian individual fit lines for each hole were drawn with highest and lowest calculated gradients from the six holes excluded. An average gradient of 2.2°C/100 m was calculated. This agrees with the visually fit gradient when all holes are plotted on a single scatterplot shown in Figure 7. In the cover sequence, the range and variability of fit lines along with the noise in the data set made comparison with measured data difficult. The gradient was set by using the calculated surface VRT and the fit line VRT at the unconformity between the Permian and Metasediments/SE Crescent, ensuring 'thermal continuity' of the VRT profile and resulting in a gradient of 3.9°C/100 m. Point measurements from groundwater pump tests in the cover sequence were used as a second point of confirmation for the approach with reasonable agreement, and the difference in the rock thermal parameters in the next section supports a difference in gradient.

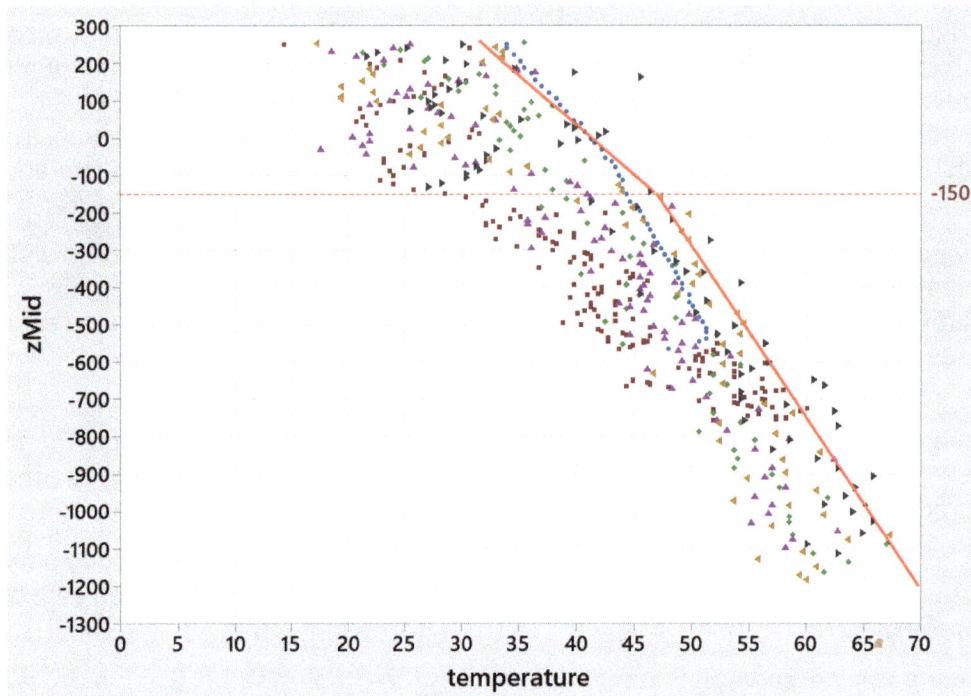

FIG 7 – Downhole temperature data with thermal gradient.

Rock thermal parameters

Lab testing was completed across dominant seven lithologies (Permian, Bedded Sediments, Calc-Silicate, PQR, PYR, CBX, CCB) with three samples each for rock thermal properties (conductivity, specific heat, density) using a divided bar apparatus (Antriasian and Beardsmore, 2014). The number of samples tested was largely driven by the availability of core that had not already been destructively tested. Samples were tested 'as received' as there was no information to indicate any degree of *in situ* saturation and the samples had been previously stored in core trays. The results indicated significant deviations from textbook thermal properties assumed in the concept study (a uniform marble covered by sandstone) due to the massive sulfide minerology of the SE Crescent. The test results were combined into their applicable mine planning geology units and averaged for development of design criteria shown in Table 3.

TABLE 3

Rock thermal parameters.

	Conductivity (W/mK)	Specific Heat (J/kgK)	Density (kg/m³)
Permian	1.7	860	2030
Metasediments	3.1	790	2690
Crescent	6.4	675	3940
Breccias	2.9	810	2770

CONTROLS AND IMPACTS TO VENTILATION SYSTEM DESIGN

Self-heating of sulfides

The risks posed by self-heating of sulfides arise from the release of gases and the complications introduced in the handling of hot material in the mine. Given the demonstrated ability for the material to self-heat through lab scale testing, the following controls have been brought into the study process:

- Additional ventilation capacity for ducted exhausting of stopes allowing the stope to be put under negative pressure with a small diameter auxiliary fan. This reduces the risk of outgassing at the drawpoint and minimises the risk of the exhausted air further contaminating the other working levels of the mine.

- Additional ventilation capacity for top exhausting of orepasses. Like stopes, orepasses present a confined volume where gas generation and ultimately build-up can be released as material is pulled, potentially at very high concentrations.

- Inclusion of required lateral development to support handling of self-heating material 'on the level' if use of orepasses is problematic.

- Based on the results, a second round of testing has been initiated for the FS stage with the intent to develop a comparative risk classification scheme driven by standard analytical assays as proxy for quantitative minerology. A conceptual classification scheme for risk based on threshold assay values and ratios (specific to the SE Crescent mineralisation) has been hypothesised, aligned to the series of reactions suggested by Nesset and Rosenblum (2020) for acid generation, electrochemically dissimilar sulfides, and acid neutralising capacity from carbonate minerals. The intent of the testing is to further refine/validate the classification scheme to support identifying high risk stopes to apply the controls above.

Sulfide dust explosions

Having demonstrated that an explosible dust can be generated, the following controls have been brought into the study process:

- Intrinsically safe electrical switching equipment where located in the exhaust path. This is in line with current sulfide dust explosion controls at the Telfer underground mine.

- Additional ventilation capacity for top exhausting of orepasses, the same as for self-heating. The orepass is, in effect, a large Hartmann tube apparatus and preventing development of an explosive concentration of dust is seen as a critical element.

- Use of non-sparking fans for top exhausting of stopes and orepasses.

- Fogging water sprays for stope blasts in elevated risk areas.

- Stemming of stope blastholes to reduce ejection of fines from blasting.

Additional operational controls related to charging, blasting mine clearances, and wash-down practices will be incorporated into management plans if the project proceeds to execution.

RCS and fibrous minerals

Due to the difficulty of reliably predicting the location and concentration of fibrous mineral dusts a precautionary approach assuming that all areas affected by dusts may have fibrous mineral dusts has been adopted for the purposes of technical studies. This paired with potential high levels of RCS indicate that aggressive dust management will be required. Controls integrated into the study phase focus on capture of dust where practicable, minimising exposures through filtration, and prevention of the spread of dust to less controlled areas of the operation such as camps and offices. The controls include:

- Dedicated capture ventilation for major dust generation and release points (orepasses, load points).

- High Efficiency Particulate Air filters (HEPA) for underground crib rooms, control rooms, and equipment cabins.

- Parallel (single pass) ventilation distribution.

- Supplied uniforms with on-site laundry and change/shower blocks.

- Prohibition on 'dirty' clothing being worn back to camp/off-site.

Additional operational controls related to wash-downs, water application to broken material, and PPE will be incorporated into management plans if the project proceeds to execution. A challenging conflict exists for the management of dusts, as the typical approach of wet suppression goes against the approach of minimising water addition for management of self-heating. The minimal effective water application will need to be determined in practice, and selection of spray nozzle type will be key in achieving the correct balance of water application while still adequately wetting the material to manage dust levels.

Strata gases

The sparse data on the location of gas at Havieron limits the ability to design or plan controls at the study stage. Current thinking is that there are 'pockets' of gas that may be encountered during development and production, however no clear geologic or geometric relationship that has been identified to date. Single pass ventilation in development can minimise the impacts to operations however the lack of data makes preventative and engineering controls challenging to develop at a study level. If/as more information on gas intercepts becomes available the question of geologic correlation will be revisited in later stages of study.

Heat

The results of the characterisation of the surface climate and the subsurface thermal conditions have confirmed expectations that the site will require refrigeration during the summer months to manage the heat in the underground environment. The inputs established have been incorporated into studies and have allowed for higher confidence in marginal sources of heat in the underground (heat from strata and broken rock) and seasonal and day/night variations in cooling demand. In particular, the high conductivity of the SE Crescent indicates high rates of heat release and has emphasised the importance of time dependant effects in heat loads. This has supported further refinement of the cooling loads over initial conceptual studies and informed timing of implementation.

CONCLUSIONS

The Havieron Project is a greenfield development with a significantly different style of mineralisation and geologic context to that of the adjacent Telfer Mine. Identification of these differences at an early stage of study allowed the necessary data collection and laboratory testing to be undertaken. The comprehensive nature of the characterisation early in the study process has assisted in setting expectations with respect to safety and risks, ultimately resulting in the inclusion of the right controls at a study level.

ACKNOWLEDGEMENTS

The author acknowledges the assistance and effort by the various contributors to the Havieron project who have assisted with the characterisation process:

- The Havieron Mining Studies team for their willingness and appreciation of the need to undertake this scope of work.

- Ben Ackerman and the Havieron Exploration Geology team for their assistance with supplying the necessary drill core samples, gas intercept data, and SEM data.

- Quentin Broadbent for his ongoing assistance with the challenges of transporting samples and willingness to share his data, knowledge, and metallurgical composites.

- Michael MacDonald for his assistance in understanding the intricacies and limitations of the different analytical methods in quantitative minerology.

Lastly, the author thanks Newcrest Mining Limited for permission to publish this paper.

REFERENCES

Antriasian, A and Beardsmore, G, 2014. Longitudinal Heat Flow Calorimetry: A Method for Measuring the Heat Capacity of Rock Specimens Using a Divided Bar, *Geotechnical Testing Journal*, 37(5):859–868.

Australian Bureau of Metrology, 2020. Climate Data Online – Long Term Temperature Data, TELFER AERO, Station number 013030, 1 January 1996 to 27 March 2020.

BBA Labs, June 2021. Technical Report: Self-Heating Tests – Newcrest Havieron Ore Samples.

Department of Mines, Industry Regulation and Safety, 2019. *Mines Safety Bulletin No. 163 – Reducing exposure to respirable crystalline silica (quartz)*, March 2019.

Enright, R J, 1988. *Practical and Theoretical Aspects of Sulphide Dust Explosions*. Canada Centre for Mineral and Energy Technology, Mining Research Laboratories, 1988.

Farnsworth, D J M, 1977. Introduction To and Background Of Sulphide Fires In Pillar Mining at the Sullivan Mine, *CIM Bulletin*, 70(782):65–71.

Gerner, E and Budd, A, 2015. Australian Surface Temperature Corrections for Thermal Modelling, in *Proceedings Word Geothermal Congress 2015*, International Geothermal Association, Netherlands, 8 p.

Haynes, M W, Horowitz, F G, Sambridge, M, Gerner, E J and Beardsmore, G R, 2018. Australian mean land-surface temperature. *Geothermics*, 72:156–162.

McCutcheon, S R and Walker, J A, 2020. Great Mining Camps of Canada 8. The Bathurst Mining Camp, New Brunswick, Part 2: Mining History and Contributions to Society, *Geoscience Canada*, 47(3)143–166.

Moon, S, Rosenblum, F, Tan, Y H, Nesset, J E, Waters, K E and Finch, J A, 2019. Examination of the United Nations self-heating test for sulphides, *Canadian Metallurgical Quarterly*, 58(4):438–444.

National Occupational Health and Safety Commission (NOHSC), 2005. NOHSC: 3003 (2005) – Guidance note on the membrane filter method for estimating airborne asbestos fibres, April 2005.

Nesset, J E and Rosenblum, F, 2020. Basics of Self-heating of Sulphide Mineral Mixtures, in *27th Annual BC MEND Metal Leaching/Acid Rock Drainage Workshop*, Vancouver, BC, December 1–3.

Payant, R, Rosenblum, F, Nesset, J E and Finch, J A, 2007. *Galvanic Interaction and Particle Size Effects in Self-Heating of Sulphide Mixtures in Separation Technologies for Minerals, Coal, and Earth Resources*, C A Young and G H Luttrell (eds.), pp. 419–431 (The Society for Mining, Metallurgy, and Exploration: Littleton).

Rosenblum, F, Nesset, J and Spira, P, 2001. Evaluation and control of self-heating in sulphide concentrates, *CIM Bulletin*, 94(1056):92–99.

Safe Work Australia, 2019. Workplace Exposure Standards For Airborne Contaminants, Canberra, Act: Safe Work Australia.

Standards Australia, 2016. *AS ISO/IEC 80079.20.2:2016 – Explosive atmospheres – Part 20.2: Material characteristics – Combustible dusts test methods*.

Wilson, A J, 2019. Appendix: Rock, Alteration and Mineralisation Assemblages, Havieron Project, WA, GeoAqua Consultants Limited, October 2019.

Analysis of dust behaviour in the face zone utilising CFD modelling tools

O Joneydi[1], J Viljoen[2] and R Funnell[3]

1. Mechanical Engineer, BBE Consulting, Perth WA 6000. Email: ojoneydi@bbegroup.com.au
2. Senior Engineer, BBE Consulting, Perth WA 6000. Email: joeline.viljoen@bbegroup.com.au
3. Principal Engineer, BBE Consulting, Johannesburg, South Africa. Email: rfunnell@bbe.co.za

ABSTRACT

Computational Fluid Dynamics (CFD) modelling is a useful tool that can assist engineers in designing effective ventilation systems for the control of airborne dust generated during mining operations. This technology can save time, effort and the costs associated with re-work or the traditional on-site experimental field trials by running desktop 3D simulations of the various proposed scenarios using state-of-the-art CFD analysis tools. Before implementing a dust management system, an understanding of the behaviour of airborne dust particles in the face zone can be achieved, which can improve the confidence in the secondary ventilation and dust management design. Using advanced numerical models, suitable dust extraction strategies can be foreseen, and multiple scenarios can be simulated and investigated quickly. As a case study, dry drilling operations in Kimberlite orebodies produce large quantities of fine dust which poses health and safety risks. An effective dust management system is required to achieve acceptable face conditions. A possible practical solution is a ducted exhaust ventilation system with a force fan in the face zone. This paper demonstrates how CFD analysis can be used as a tool to assist the design engineer with validating and optimising the proposed ventilation design by tracking the trajectories of different size dust particles, to determine the best equipment layout and appropriate flow rates which minimises airborne dust.

INTRODUCTION

This paper presents a case study on dust control for a Kimberlite diamond mine using a Sub Level Caving mining method. Access to the orebody will be through a series of parallel extraction drives. Production from the extraction drives will involve longhole drilling, blasting, and mucking activities in a retreat direction.

It is anticipated that the dry method of drilling for the longhole drilling operation will produce large quantities of fine dust. Dust in the underground workings can lead to worker health issues, specifically lung diseases, which relate to the inhalation of airborne dust particles in the critical respirable range. Dust can also pose safety risks associated with operating mining equipment in conditions with poor visibility.

Depending on the aerodynamic diameter, the particles can reach various regions of the respiratory tract and are assigned to the inhalable, thoracic, or respirable dust fraction. Inhalable dust consists of particles with an aerodynamic diameter of up to 100 µm, as classified by various international standards like ISO 7708:1995 (2017) and SIST EN 13205–2:2014 (2014). The smaller particles can reach the gas-exchange region of the lungs and form the respirable dust fraction. The limit for entering the alveolar region is between 10 µm and 15 µm, as published in WHO (1999) and SIST EN 13205–2:2014 (2014).

The three major airborne dust control and management methods, as described in NIOSH (2003) are ventilation, water, and dust collectors. The ventilation methods provide the best use of air in the vicinity of workers and in the vicinity of dust sources, and thus a suitable ventilation system is required to mitigate the risks. This paper's focus is on the ventilation component of dust management.

To contain the dust in the face zone, each extraction drive will be ventilated with an exhaust duct. The exhaust flow quantity will be regulated with a high flow for the loading operation (diesel dilution issue) and a reduced flow for all the other activities such as the longhole drilling operation. The exhaust duct will not extend fully to the face due to the risk of damage from blasting activities. Therefore, the face zone is planned to be force ventilated with a jet fan positioned at a safe distance from the blasting activities at the face. The purpose of the jet fan is to ensure that there will be adequate fresh ventilation delivered to the face for effective clearing of blast fumes, diesel fumes and dust. To minimise the risk of recirculation, the jet fan is generally selected at a lower flow than the exhaust duct. The bypass flow in the force-exhaust overlap zone is calculated as the difference between exhaust duct flow and the jet fan flow.

Van Niekerk (1990) gives the dust particle size distribution by mass for typical Kimberlite dust, in which the particles with a diameter of less than 1 μm and up to 45 μm make up 80 per cent of the mass as follows:

- 26.9 per cent of particles with a diameter of less than 1 μm

- 9.3 per cent of particles with a diameter between 1 μm and 4 μm

- 13 per cent of particles with a diameter between 5 μm and 12 μm

- 16.8 per cent of particles with a diameter between 13 μm and 25 μm

- 14.3 per cent of particles with a diameter between 25 μm and 45 μm.

Prior to mine-wide implementation, it is important to validate the effectiveness of the proposed ventilation design. Traditionally this was achieved through on-site experimental field trials, which are generally difficult, expensive, and time-consuming. As an alternative, this paper demonstrates how state-of-the-art CFD analysis tools can be used to model dust behaviour in the face zone for different configurations and flows. The results of these desktop 3D simulations provide valuable information to assist the design engineers to make a more informed judgement regarding the final ventilation design including the layout, equipment selection and whether further dust mitigating measures may be required. This technology can provide a rapid and cost-effective solution that can greatly reduce the reliance on field trials once findings have been proven to be comparable with field trial results.

CFD MODELLING PROCESS

The modelling process, inputs and results are described in this section.

Workflow

As an overview, performing a CFD analysis generally involves the following main steps:

1. Information gathering.

2. Generate a 3D model of geometry.

3. CFD analysis:

   o Definition of inputs and assumptions.

   o Select models and solvers.

   o Definition of geometry and boundary conditions.

   o Designing the computational mesh.

   o Solving and monitoring.

   o Post-processing of the results and review.

4. Discussion with design engineers.

Case study inputs and assumptions

The following inputs and assumptions apply to the case study:

1. Force flow of 7.5 m³/s from the Ø570 mm jet fan duct.

2. As a simplification, the jet fan is modelled as a force duct, ignoring the inlet side.

3. Model two scenarios for the location of the jet fan, namely:

   o Scenario 1 – jet fan discharge at 20 m from the face.

○ Scenario 2 – jet fan discharge at 30 m from the face.

4. Exhaust duct with size Ø1200 mm and located 20 m from the face.

5. Model two flow cases for the exhaust duct, namely:

    ○ Loading case – higher exhaust flow of 20.7 m³/s.

    ○ Drilling case – reduced exhaust flow of 12.5 m³/s.

6. The intake air has a dry-bulb temperature of 30°C at a density of 1.06 kg/m³.

7. Assume one point location used for particles injectors and located 1 m from the face.

8. Dust particle sizes to be modelled in discrete sizes of 1 μm, 5 μm, 10 μm and 50 μm.

9. The dust particle density is 2950 kg/m³.

10. As a simplification, the face zone is modelled without any equipment.

11. As a simplification, the face wall is modelled as a solid flat surface.

CFD software

The software used for the CFD analysis is Simcenter's STAR-CCM+ (2021), which is a desktop Computational Aided Engineering (CAE) tool for solving problems involving flow (of fluids or solids), heat transfer, and stress within a single integrated user interface.

The Simcenter STAR-CCM+ (2021) simulation software offers a full suite of analysis tools, including import and creation of geometries, mesh generation, solution of the governing equations, analysis of the results, and demonstrating a wide range of graphical data for interpretation of results.

Models and solvers

The face zone will involve turbulent flow discharging from a jet fan with dust particles entrained in the air stream. Simulations are to be conducted in steady state with particle tracking selected.

The physics, as described in the software user manual, is modelled using several built-in solvers in the STAR-CCM+ software, namely:

- Non-reacting multi-component gas solver.

- Reynolds-Averaged Navier–Stokes equations for mass, momentum and energy were solved in steady-state for each discrete cell of the computational mesh.

- Turbulence was modelled using the k-epsilon turbulence model equations, and the boundary layer at the solid interfaces was modelled with logarithmic wall functions.

- Particles were modelled as Lagrangian particles. The Lagrangian particle model is employed for multiphase flow modelling involving fluid interaction with solid particles with actual weight, density, and diameter. The particle track function was engaged in steady-state simulations to evaluate the dust particle behaviour.

Geometry and boundary conditions

Two geometries were modelled, namely Scenario 1 and Scenario 2, with the jet fan discharge either at 20 m or 30 m from the face, respectively.

For Scenario 1, the face zone model was set-up with the jet fan discharge and the exhaust duct inlet boundaries at 20 m from the face. The model has a total length of 25 m and a cross-sectional area of about 25 m² (5.0 m × 5.0 m). The jet fan discharge diameter is Ø570 mm, and the exhaust duct is Ø1200 mm, with their position inside the extraction drive as shown in Figure 1.

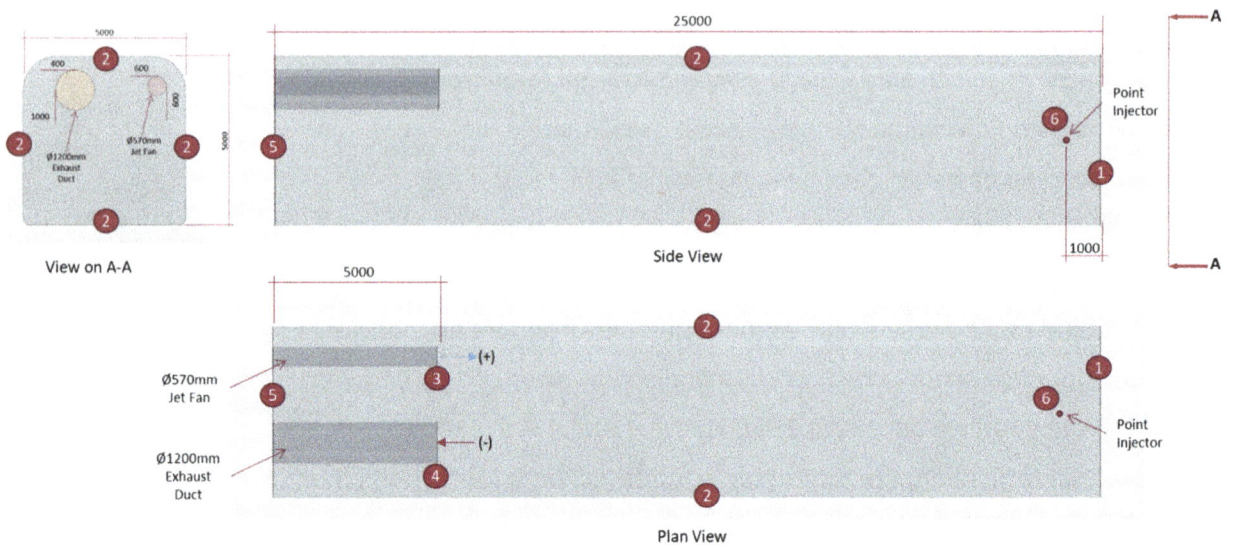

FIG 1 – Side, plan, and front view of face zone model for Scenario 1.

For Scenario 2, the face zone was set-up with the force jet fan discharge at 30 m from the face and the exhaust duct remaining at 20 m from the face to evaluate the effect of a reduced jet stream discharge velocity in the face. The model has a total length of 35 m and a cross-sectional area of about 25 m² (5.0 m × 5.0 m). The jet fan discharge diameter is Ø570 mm, and the exhaust duct is Ø1200 mm, with their position inside the extraction drive as shown in Figure 2.

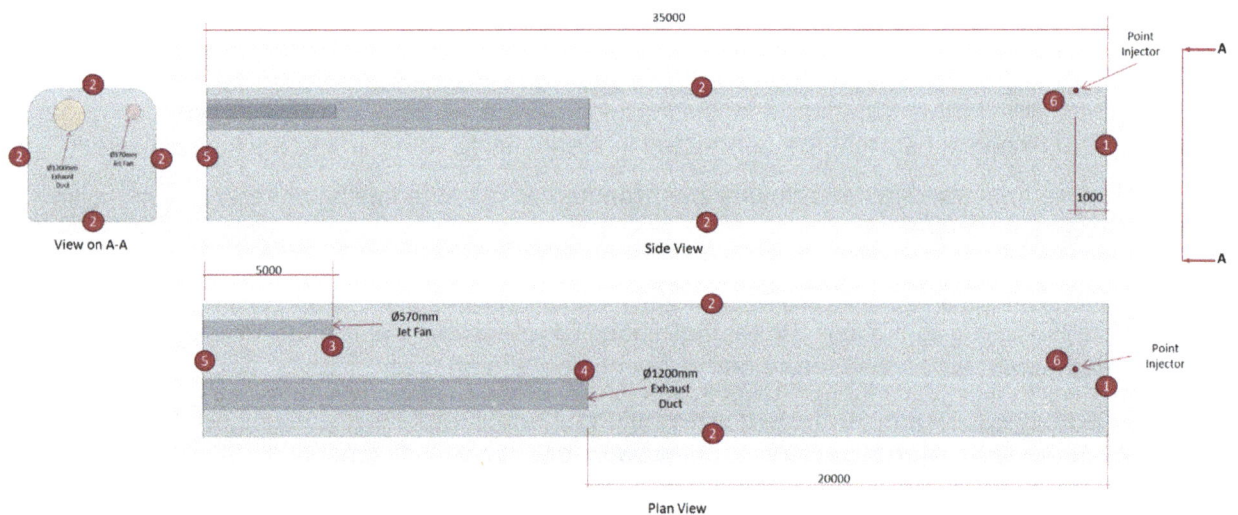

FIG 2 – Side, plan, and front view of face zone model for Scenario 2.

The numbers in red in Figures 1 and 2 indicate the model boundary conditions, namely:

1. Face with a rough wall (assume 50 mm surface roughness).

2. Sidewalls, roof, and floor with a rough wall (assume 50 mm surface roughness).

3. Velocity inlet boundary for the Ø570 mm jet fan inlet at 29.4 m/s (7.5 m³/s) at a temperature of 30°C and air density of 1.06 kg/m³.

4. Velocity outlet boundary for Ø1200 mm exhaust duct:

    o Loading case at 18.3 m/s (20.7 m³/s).

    o Drilling case at 11.0 m/s (12.5 m³/s).

5. Back wall with zero pressure boundary.

6. Point particle injector with a nominal mass flow of 0.5 kg/s and initial vertical downward velocity of 0.5 m/s. The particle density is 2950 kg/m^3. Discrete particle sizes that will be modelled are 1 μm, 5 μm, 10 μm and 50 μm.

The jet fan discharge is set as a velocity inlet boundary and the exhaust duct boundary is set as a velocity outlet boundary (a negative sign indicates an outlet flow). The back wall is set as a zero-pressure boundary to allow airflow into or out of the region as required due to pressure changes in the face zone. All other boundaries are set as walls, either smooth or rough. A portion of the intake air from boundary 5 will bypass the jet flow and flow directly to the exhaust duct.

Mesh and mesh refinement

A polyhedral cell mesh is used as the core mesh model as it provides a balanced solution for complex 3D mesh generation problems. A boundary layer mesh next to the wall surfaces and boundaries was applied to improve the accuracy of the near-wall flow solution. A mesh refinement process was done near the exhaust and force duct openings as well as around the area where particles are injected into the face zone to ensure that there are enough cells in those areas to accurately capture the flow characteristics of the air as well as the behaviour of the dust particles.

Figure 3 demonstrates the meshed model of the face zone for Scenario 1. The same mesh strategy was applied for Scenario 2.

Refined Mesh at Duct Inlet and Exhaust

Refined Mesh at Proposed Location of Particle Injection

FIG 3 – Meshed model of face zone for Scenario 1.

Solving and monitoring

To initialise and get the simulations to start solving, the model geometry and the mesh must have no errors. Secondly, the physics models and solvers must be correctly defined and be appropriate for the multi-phase flow, turbulence, energy etc as defined in the Star CCM+ user guide. It is the solvers that compute the solution during simulation runs and perform tasks of various levels of complexity.

The software has the functionality to post-process the solution after each iteration or time-step, (while the simulation is running), as well as when the simulation completes. For this reason, postprocessing objects (plots, scenes, and reports) are created before running the simulation, which can then be monitored while the simulation is running. It is important to monitor these postprocessing objects for any divergence of the solution. While closely monitoring the data, it can be established which solvers are not converging and changes to the physics models, solvers, and boundary conditions can be applied to assist the simulations to run more stable.

CFD RESULTS

Scenario 1 – results for model with jet fan discharge at 20 m from face

Airflow only solution

Before evaluating the dust distribution patterns, an airflow only solution is run to confirm that the flow modelling yields realistic flow patterns and behaviours. Figure 4 shows the velocity scalars on a vertical plane through the centre of the jet fan discharge for the Loading Case, which has an average airstream velocity of 29 m/s. The jet fan airstream profile compares well with the results from the work done by Beyer *et al* (2016) in which the performance of jet fans in tunnels is evaluated. This high-velocity airstream causes a face velocity between 5 m/s and 6 m/s.

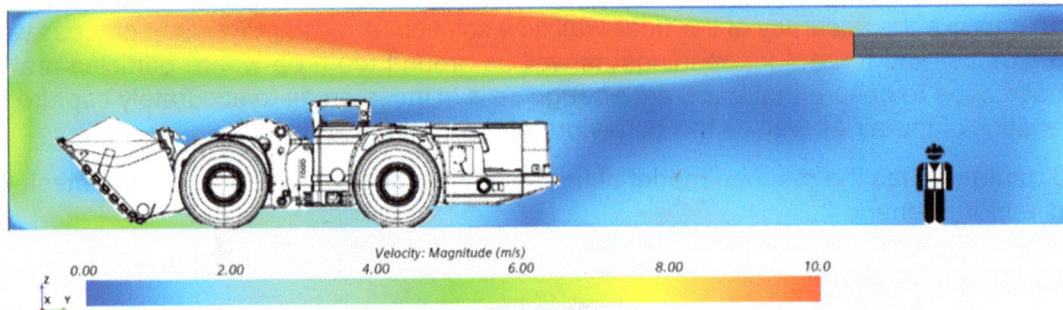

FIG 4 – Velocity scalars on a vertical plane through jet fan with discharge at 20 m from face for loading case.

Figure 5 shows the velocity streamlines of the air discharging from the jet fan, sweeping the face and returning to the exhaust duct for the (a) Loading Case (20.7 m³/s) and (b) Drilling Case (12.5 m³/s). It is noted that there is a significant amount of turbulence and swirling flow in the face zone, especially close to the entrance of the exhaust duct. The high-velocity discharge from the jet fan entrains a large quantity of surrounding air. In the current geometry, the jet fan discharge is near the exhaust duct entrance, which aggravates recirculation of dust particles, which might result in the smaller dust particles not settling, but rather having an increased residence time in the face before being exhausted. To reduce the airborne dust, it may be beneficial to locate the jet fan further away from the exhaust duct, within practical constraints, thus reducing the velocity at which the jet fan airstream sweeps the face and returns to the exhaust duct. This modelling is carried out in Scenario 2.

The observed flow patterns are understood, and the trends can be explained through flow dynamics. The jet fan flow pattern is consistent with those by Beyer *et al* (2016).

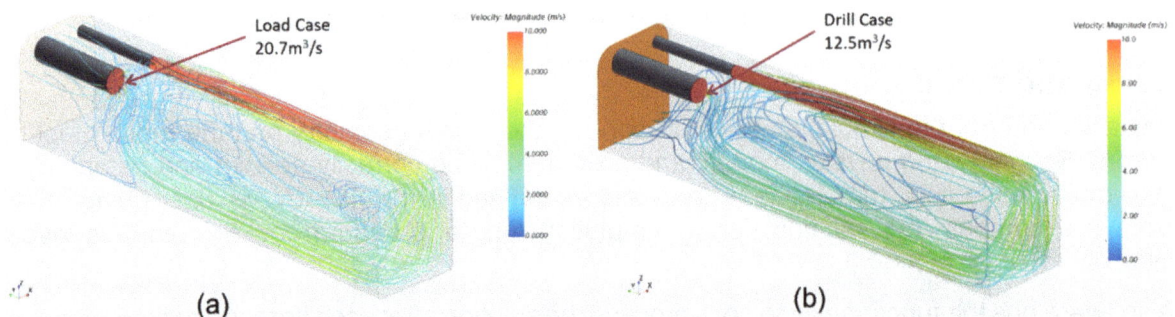

FIG 5 – Velocity streamlines with jet fan discharge at 20 m from face for (a) loading case and (b) drilling case.

Particle injection solution for loading case

Figure 6 shows the tracking and monitoring of various size particles in the face zone for the Loading Case. A nominal mass flow of 0.5 kg/s of dust particles is introduced at the injector point but is not necessarily an indication of actual dust loading. The dots provide a visualisation of the location of dust particles and the colour of the dots indicates the particle residence time. The percentage of particles that enters the exhaust duct in the elapsed time is reported for each particle size, with a maximum residence time of 120 seconds. The results are summarised below.

Figure 6a shows that 100 per cent of the 50 µm particles are exhausted within about seven seconds. The weight of the particles causes the particles to be less affected by the turbulent airflow and thus fall into the area where more laminar flow is observed. The velocity in this more laminar region is still sufficient to carry the larger particles toward the exhaust duct.

Figure 6b shows that 100 per cent of the 10 µm particles are exhausted within about 25 seconds. The smaller particles' reduced weight makes them more sensitive to the turbulent airstream and they will get carried around in the face zone for a longer period than the 50 µm particles.

Figure 6c shows that about 40 per cent to 60 per cent of the 1 µm particles are exhausted within 120 seconds. These small particles are highly affected by the turbulent airstream and will remain in the face zone for an extended period. If the dust production rate exceeds the exhaust rate, the face will get filled with small dust particles, decreasing visibility, and also exposing the working to higher concentrations of respirable dust.

Thus, in conclusion, the general trend is that the smaller particles (less than 10 µm) are greatly influenced by the air currents and tend to remain within the face zone for longer periods compared to the heavier particles.

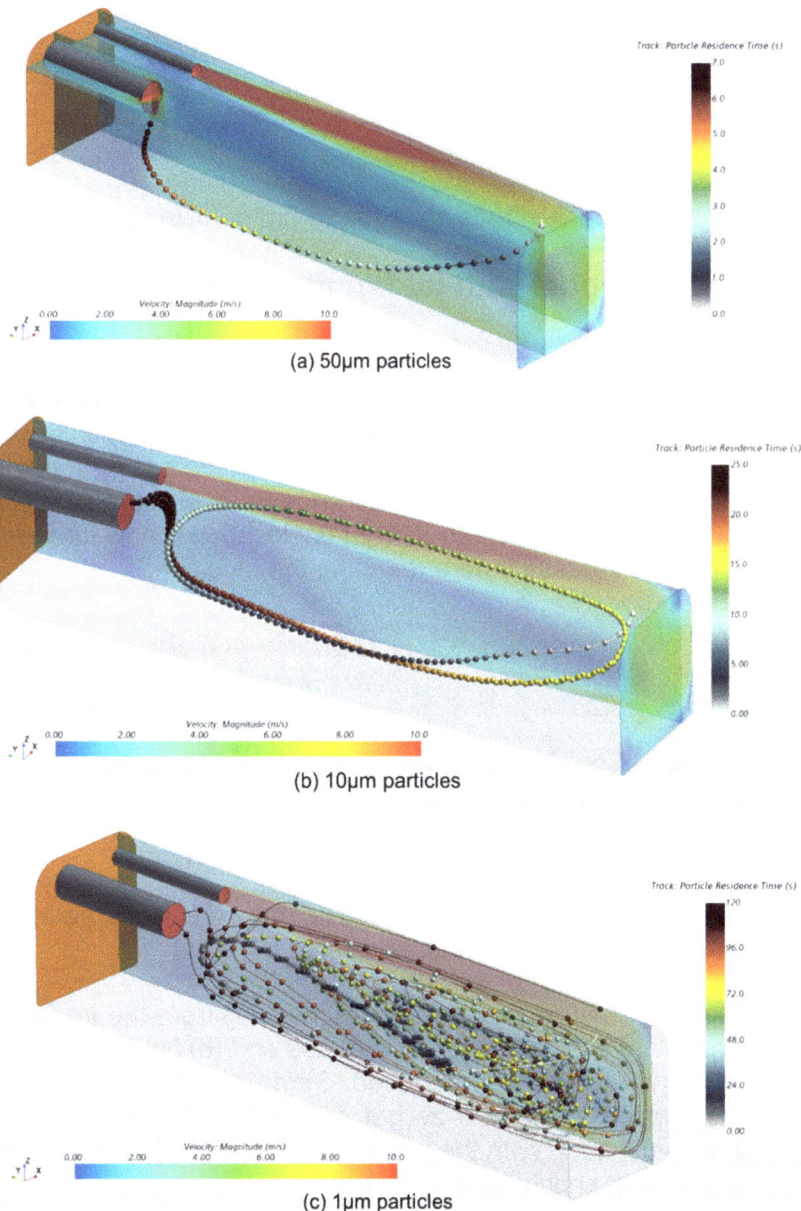

(a) 50µm particles

(b) 10µm particles

(c) 1µm particles

FIG 6 – Dust particle track in face zone for various particle sizes (a) 50 µm, (b) 10 µm, (c) 1 µm.

Particle injection solution for drilling case

In Figure 7, the behaviour of the 10 µm particles for the Drilling Case be observed. It can be seen that about 60 per cent to 80 per cent of the 10 µm particles are exhausted within 135 seconds for the Drilling Case. This is a significantly longer residence time compared to the Loading Case with 100 per cent of particles exhausted in 25 seconds. At the reduced exhaust flow, the percentage of recirculation flow increases which leads to longer residence time. Therefore, the general trend is that higher exhaust flows will result in reduced airborne dust in the face zone.

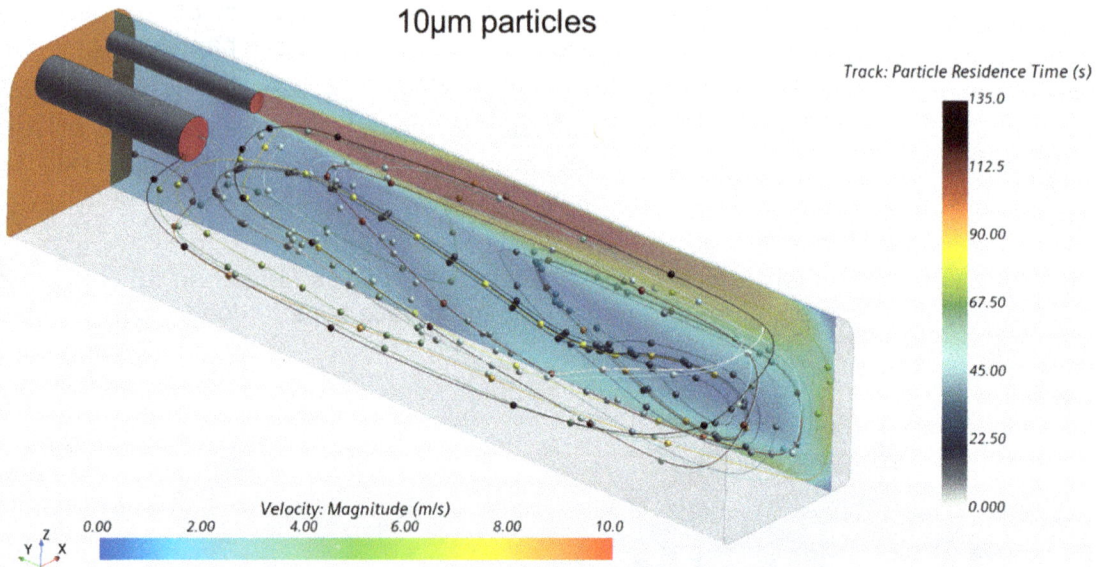

FIG 7 – Dust particle track in face zone for 10 µm particle size for drilling case.

Scenario 2 – results for model with jet fan discharge at 30 m from face

Airflow only solution

As previously discussed, for the Scenario 2 set-up, the exhaust duct boundary remains in the same position relative to the face and the jet fan boundary is moved back to discharge 30 m from the face. Figure 8 shows the velocity scalars on a vertical plane through the centre of the jet fan discharge for the Loading Case (exhaust flow of 20.7 m³/s), which has an average airstream velocity of 29 m/s and face air velocity between 0.3 m/s to 2 m/s. The jet fan location is now further away from the face; therefore, the airstream gets dispersed before reaching the face.

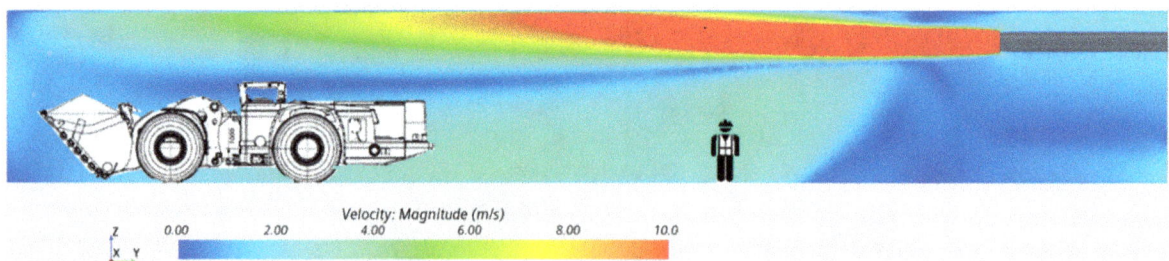

FIG 8 – Velocity scalars on a vertical plane through jet fan with discharge at 30 m from face for loading case.

Figure 9 shows the velocity streamlines of the air discharging from the jet fan, sweeping the face and returning to the exhaust duct for the (a) Loading Case and (b) Drilling Case. Due to the reduced face velocity, the swirl in the face zone is reduced. Some of the air will pass the return duct to combine again with the bypass air as well as get entrained in the force airstream before being exhausted through the exhaust duct. With the jet fan position being recessed from the exhaust duct entrance, the recirculation effect should be minimised.

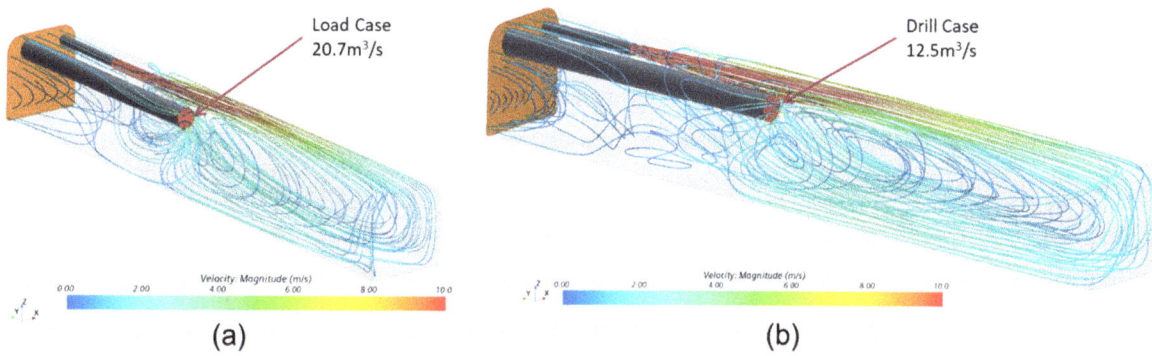

FIG 9 – Velocity streamlines from jet fan to exhaust duct during (a) loading case and (b) drilling case.

Particle injection solution for loading case

For the jet fan discharge at 30 m from the face, the behaviour of the 10 µm dust particles was modelled. The simulation indicated a transient solution with two modes of operation; at times, the dust particles might remain in the face zone for some time and not get exhausted (mode 1) and then later get picked up by the exhaust airstream (mode 2).

As a reference, modelling of the jet fan discharge at 20 m from the face, 100 per cent of the 10 µm particles were exhausted within 25 seconds for the Loading Case. Therefore, moving the jet fan further back (to 30 m) would indicate an improved face condition for the Loading Case due to settling of dust particles.

Particle injection solution for drilling case

Figure 10 shows that 100 per cent of the 10 µm particles are exhausted within about 30 seconds for the jet fan discharge at 30 m from the face. The particle tracks indicate that the 10 µm particles will tend to settle towards the floor because of the reduced velocities in the face zone and tend not to be recirculated. The particles are light enough to be picked up by the suction of the exhaust duct.

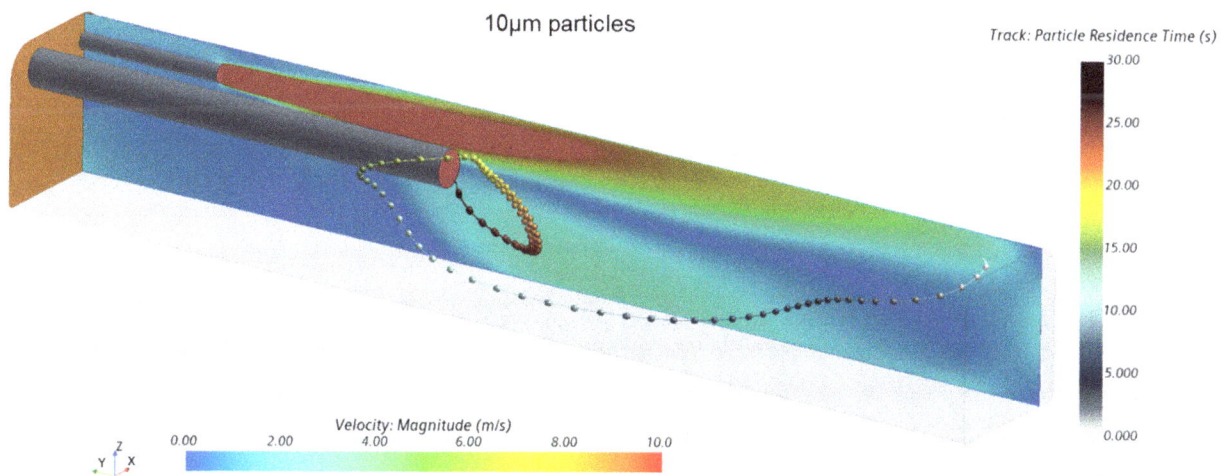

FIG 10 – Dust particle track in face zone with jet fan discharge at 30 m for drilling case.

In comparison, in modelling the jet fan discharge at 20 m from the face, only 60 per cent to 80 per cent of the 10 µm were exhausted within 135 seconds for the Drilling Case. Therefore, moving the jet fan further back (to 30 m) indicates an improved face condition for the Drilling Case.

Validation of CFD results

The mine under consideration is still in the development phase, and thus no field trials have been conducted to validate and calibrate CFD results. Once operational, data from field trials will be collected, to compare to CFD results and to further improve on modelling inputs and assumptions.

CONCLUSION

The CFD study results indicates that locating the jet fan discharge at 20 m from the face results in a relatively high face velocity of 5 m/s to 6 m/s with a significant amount of swirl in the face zone. As a result, the small respirable dust particles are readily re-entrained into the jet stream and recirculated which increases the residence time of the dust particles in the face zone. Moving the jet fan discharge to 30 m from the face would still provide adequate face velocity for clearing of blast fumes, with face velocities in the range of 0.3 m/s to 2 m/s. Dust tracking simulations indicate that this arrangement will result in significantly less recirculation in the face zone and thus more effective removal of the dust by the exhaust duct. Modelling also shows that the residence time of dust particles can be reduced by increasing the exhaust duct flow, as demonstrated by the higher flow in the Loading Case compared to the Drilling Case. The proposed ventilation arrangement looks promising to minimise the amount of airborne respirable dust particles in the face zone, although it still falls short of eliminating the dust problem and further mitigation measures may be required.

The results, and findings of CFD simulations conducted for the various secondary ventilation arrangements in the face zone, were presented in this paper. The secondary ventilation layouts proposed, for the effective management of dust in the face zone, were simulated and evaluated, prior to any implementation or field trial testing. The findings from these simulations can now serve as a valuable input into the secondary ventilation design. The findings further aid in a more in-depth understanding of the dust behaviour by means of visual representations, by plotting the airflow as streamlines and by tracking the dust particle paths as trajectories. It is concluded that CFD can be used as a tool, to quickly demonstrate and quantify changes in airflow and dust particle behaviour, when changes are applied to the ventilation system layout.

Simulation results can be validated by minimal field trials, and once models are calibrated, more simulations can be conducted, requiring less expensive fieldwork.

REFERENCES

Beyer, M, Sturm, P J, Saurwein, M and Bacher, M, 2016. Evaluation of Jet Fan Performance in Tunnels, *Proceedings of the 8th International Conference 'Tunnel Safety and Ventilation'*, Graz, Austria.

International Organization for Standardization (ISO), 2017. ISO 7708:1995 Air quality – Particle size fraction definitions for health-related sampling.

NIOSH, 2003. Handbook for Dust Control in Mining, Publication No. 2003–147.

Simcenter STAR-CCM+ User Guide, Version 2021.3. www.siemens.com/simcenterccm.

Slovenian Institute For Standardization (SIST) and European Standards (EN), 2014. SIST EN 13205–2:2014 – Workplace exposure – Assessment of sampler performance for measurement of airborne particle concentrations – Part 2: Laboratory performance test based on determination of sampling efficiency.

Van Niekerk, G H C, 1990. Dust Handling Methods in the Kimberley Diamond Mines, *Journal of the Mine Ventilation Society of South Africa*, Oct 1990, pp 182–190, South Africa.

World Health Organization (WHO), 1999. Hazard prevention and control in the work environment: Airborne Dust, Geneva, Switzerland, Publication Reference Number: WHO/SDE/OEH/99.14.

Geothermal gradient determination for ventilation and air conditioning modelling at Malmberget Mine

F K R Klose[1], T H Jones[2] and A L Martikainen[3]

1. Graduate Student, University of Leoben, Leoben Austria 8700.
   Email: frederic-karl-rene.klose@stud.unileoben.ac.at
2. Senior Research Engineer, LKAB, Malmberget Sweden 98381. Email: tristan.jones@lkab.com
3. Ventilation Specialist, LKAB, Malmberget Sweden 98381. Email: anu.martikainen@lkab.com

ABSTRACT

Malmberget is a sublevel caving mine owned and operated by Luossavaara-Kiirunavaara Aktiebolag (LKAB), a Swedish state-owned iron ore mining company, in northern Sweden. The mine has an annual iron ore production of approximately 15 Mtpa. The current strategy foresees an expansion which will increase the mine's depth considerably. Ventilation and air conditioning modelling for the planned expansion is ongoing. One of the most influential parameters for ventilation planning and design for a deep mine is geothermal gradient, which is crucial for the design of the refrigeration system. It is determined through virgin rock temperature (VRT) measurements carried out at various depth below the surface. This paper details these measurements and the determination of geothermal gradient for Malmberget Mine for the expansion study, which were necessitated by the lack of information for the focus depth and area.

There are two methods used for VRT measurements. The first method is placing static temperature sensors in a short, purpose-drilled borehole located on several levels inside an underground mine. The second method is lowering a temperature logging device into a long borehole such as surface and underground exploration boreholes. Both methods were used in this work.

Short borehole instruments using Type-K thermocouples connected to dataloggers were constructed based on modified methods found in literature. These were installed in six boreholes on three different levels. Locations were chosen based on literature recommendations, but under operational restrictions imposed by the mine.

For deep borehole measurements, a commercial logger was used and attached to a winch. The holes were selected based on dip, intersecting geology, and depth. The selected holes yielded data from up to 2820 m below surface and included the target level.

Data analysis showed that VRT varied only marginally across the mine, and best-fit was used to identify the geothermal gradient. It was found that length of the short hole measurements wasn't sufficient to reach steady state, thus results cannot be regarded as VRT. The installation of longer instruments using the same construction method was unsuccessful. In conclusion, deep borehole measurements are preferable to short borehole measurements to define the geothermal gradient. Also, measurements in deep boreholes are far easier to obtain and have fewer variables to consider.

INTRODUCTION

The Malmberget iron ore mine is a sublevel caving mine with an output of 15 Mt per annum (Mtpa) located above the arctic circle (Figure 1). The mining lease comprises several magnetite orebodies of varying thickness covering an area of approximately 12.5 km^2. About half of the known orebodies are currently being mined.

The ongoing expansion study of the mine has raised a question on whether the current ventilation practices need to be adjusted to accommodate the expansion of the mine from the main level 1250 potentially down to level 2000. Current mine ventilation practices include heating of intake air for approximately half the year, followed by local cooling of critical areas during the summer months. To estimate the extent of changes required due to the increase in rock temperature and auto-compression at depth, climate simulations are being conducted with mine-specific input parameters. The geothermal gradient is an essential parameter in estimating the potential cooling needs at the new target level.

FIG 1 – Location of LKAB operations in northern Sweden and Norway (modified from LKAB, 2019).

The geothermal gradient, which is the change in temperature per depth increase, is typically derived from measurements of the virgin rock temperature (VRT) at various depths below surface. VRT can be defined as the temperature of the undisturbed rock at a particular location, in equilibrium with its surroundings. The inverse of the geothermal gradient, or the change in-depth necessary per temperature increase, is the geothermal step.

Variations in gradient used in simulations can quickly lead to over- or underestimation of the cooling requirements at depth. This is illustrated by Russell and von Glehn (2019). Their simulations for typical South African reef mining scenarios show that for a variation in VRT of 10°C, a potential error of 10 per cent occurs in heat load estimation for the case of a production site and up to 70 per cent for a 1000 m haulage where broken rock is transported. While this is a highly specific example, it does illustrate the importance of accurate VRT inputs. As VRT at depths are estimated using the geothermal gradient, an accurate gradient is particularly important for large, deep and high production mines such as the Malmberget Mine.

Unfortunately, there is only limited information about geothermal gradient of Malmberget region. Using estimated value derived from literature in climate simulations may not reflect the true situation of the mine and hence could produce significant error in heat load estimation. The only specific geothermal gradients for part of the current study area were documented by Parasnis (1982), who observed the geothermal gradient for three surface boreholes located near the Fabian (FA) orebody to be 1.378°C/100 m (average from three sections), 1.535°C/100 m and 1.381°C/100 m. These gradients are the reported measured values and are uncorrected for uplift. The boreholes covered depths shallower than the current main level and can no longer be considered accessible. Measurements in the mine itself had never been conducted.

Due to the importance of having correct geothermal gradient in climate simulations, it is therefore crucial that the gradient of Malmberget Mine is determined properly. The work that was done for this purpose is outlined in this paper.

MEASURING VIRGIN ROCK TEMPERATURE

VRT can be obtained using measurements in short holes drilled within existing openings on different levels of a mine, or from measurements within a long borehole that is drilled from the surface or from an underground level, eg an exploration borehole (Fenton, 1973; Calizaya and Marks, 2011).

The practice of using short hole measurements is well established (Duckworth, 1999; Oender and Güyagüler, 1995; Ashworth, Ashworth and Mahajan, 1983; Cleland, 1933). To measure temperature, an instrument with temperature sensors attached is inserted into the borehole. The data is collected either manually or automatically, with the instrument either left in place or advanced. Different instrument designs have been described in literature and are based primarily on the use of thermocouple readings or resistive measurements.

Location of the boreholes is very important to obtain good measurement results. General recommendations on-site selection procedures and VRT measurements were formulated decades ago, for example by Morris (1953) as cited by Fenton (1973):

- *The boreholes should be drilled towards the end of the week, since the face may receive additional cooling during the weekend.*

- *The borehole should be drilled in development faces, which have been regularly advancing for at least 150 feet (45.72 m).*

- *The holes should be drilled at approximately right angles to the face to obtain maximum thickness of rock.*

- *Holes should have a slight upward inclination.*

- *Known faults, dykes, water fissures, or any other disturbances should be avoided.*

- *Holes should not be nearer than 300 feet (91.44 m) from through ventilation.*

These recommendations are to ensure that the temperatures can be interpreted as VRT. This requires the location to have had little opportunity to deviate from its former state of thermal equilibrium. Development headings are more likely to fulfill this criterion than other locations and reduce the need for deeper boreholes while improving comparability of measurements.

Particular borehole depths are not included in the recommendations and the choice varies significantly between sites where the method has been documented. Fenton (1973) cites Morris (1953) as saying that approximately 3 m boreholes would be sufficient to reach VRT when the above guidelines are adhered to, but also gives the example of Robertson and Bossard (1970, as cited in Fenton (1973)), who reported that a steady state was not reached after 15 m encountered in Butte, MT, United States. Another example is Cleland (1933), who used boreholes between approximately 1.5 m and 6 m in length in the Sudbury area in Canada. Misener (1949) on the other hand, sought to extend Cleland's results, deemed the necessary length in the same mines at greater depth to be between around 12 m and 27.5 m. The gradients are in good agreement (Misener, 1949).

In comparison to short borehole measurements, few specific examples exist for the method of determining geothermal gradients from VRT measurements conducted in long exploration boreholes (Parasnis, 1975, 1982; Jones, 2016, 2018). The instruments in use for borehole measurements are typically based on thermistors, thermocouples, or resistance temperature detectors, with longhole measurements taken either continuously or in intervals. Data is transferred from the sensor in the borehole to the borehole surface via a data cable or onto an onboard memory for later extraction. It is possible that boreholes experience collapse or blockage over time, limiting their measurable length.

For these long boreholes, letting temperatures revert to equilibrium after drilling and before measurements is crucial. Few specific recommendations exist with regards to the required duration for exploration boreholes. Fenton (1973) suggests that one hour is sufficient, and continues to give the example of Mullins and Hinsley (1958, as cited in Fenton (1973)), who compared the bottom-hole temperature method with that of interval measurements taken several days apart. The results of both approaches were in agreement.

SITE SELECTION

The long borehole method was selected as the main method in this study because it allows measurements to intersect the target depth. Short boreholes, on the other hand, enabled measurements in areas outside the coverage of long borehole method and served as supplementary method. In this work they primarily assist to gather additional data in order to create a complete picture of the thermal characteristics of the mine, and to improve interpretation of long borehole temperature measurement results.

The work was focused in the eastern section of the mine because this area contains the largest current and future reserves and its orebodies are mined at a greater depth than those in the western section of the mine. Nevertheless, few measurements were done in the western section because their result could provide important information to assist the interpolation and extrapolation of geothermal gradient from measurement results in the eastern section. The intent for the selected sites was to, as much as possible, gather thermal gradient information that represented the current and future extent of the mine, both laterally and vertically.

It was found that each method has different site availability. Performing short borehole measurements at or near fresh development headings in line with the recommended approach was not always possible due to safety and operational concerns. Recently developed areas with little or no activity or ventilation were considered as alternatives. Long borehole measurements, on the other hand, were limited to those areas and levels that included exploration core drilling sites. Boreholes completed at the time of the study were generally of historical nature and not deep enough to intersect the target depth or extend significantly beyond the current mined depth. Poor accessibility due to obstruction and caving had to be expected. Therefore, and as virtually all expansion-relevant drilling had been completed underground, only underground sites were considered.

For the short borehole measurements, three different sites were identified in or near future production areas on levels 755 in the Baron (BA) orebody, 1023 in the Gunilla orebody (GN) and 1112 in the Alliansen (AL) orebody, with level numbering referring to vertical distance from the reference surface elevation according to the local mine grid. Long borehole measurements were done in four different locations, exploration drilling sites on levels 600, 970, and at various depths centred on level 1250. These boreholes target orebodies Johannes (JH), Parta (PA), Printzsköld (PR) and Fabian (FA). These locations were spread across the mine as shown in Figure 2.

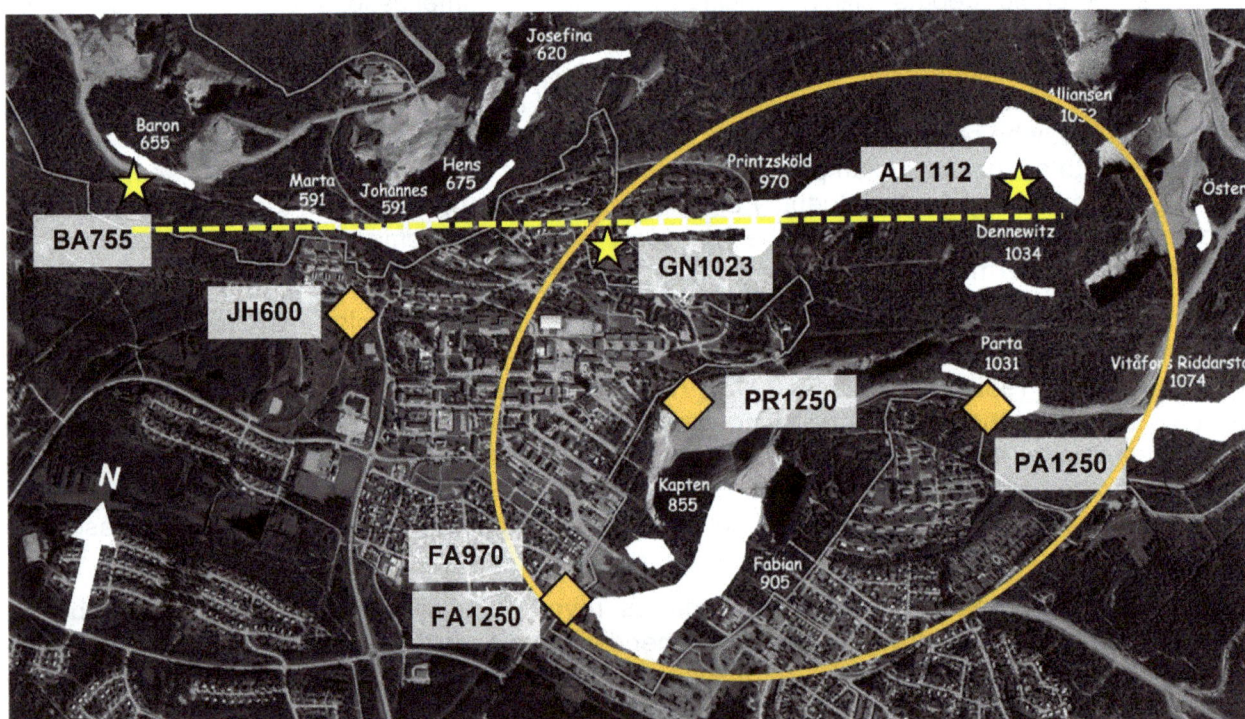

FIG 2 – Approximate location of short borehole (star) and long borehole (diamond) measurement sites, by level, modified from LKAB (2018). The focus area is circled.

All boreholes inspected at the chosen locations prior to the measurements featured water, which can be explained by the location of the exploration drilling sites away from and/or below current active mining areas. This is preferable to dry boreholes for the purpose of temperature measurements, provided that no active groundwater inflow disturbs the measurements (Jones, 2016). The water acts as a superior thermal conductor between the borehole wall and the sensor.

A borehole database was utilised to select specific boreholes for the deep measurements. Candidate boreholes were required to have a steep inclination and their deviation was checked using geological software. Boreholes with potentially fragile sections (defined by the presence of biotite schist) were avoided to minimise the risk of instrument loss and replacement. The distance from the collar of the borehole until a section of biotite or significant biotite alteration was judged to be the measurable distance. Absence of significant biotite sections was a main selection criterion. Clayey and potentially fractured zones as indicated by RQD logging information were also avoided. Newer boreholes can generally be expected to be better quality than older ones.

INSTRUMENTATION

Short borehole instrumentation

Instrumentation for the short holes consisted of different types of pipes, either PVC or stainless steel, in which multiple Type K thermocouples were attached. These were non-factory calibrated but came with a Class 1 tolerance as in accordance with the relevant standard set out by the International Electrotechnical Commission (IEC), IEC 605841:2013. Instruments were 9 m or 15 m long, and pairs of thermocouples were installed along the length of the pipes in 2 m intervals so that redundant data was collected along the entire length of the borehole. The first and last half-metre were left without instruments in order to provide an insertion handle on the shallow end and a buffer to protect instruments during insertion of the pipe on the deeper end. Temperature data was necessary along the length of the hole because to be able to say with confidence that VRT has been reached, the temperature values of the two deepest measurement points must converge to the same value. The thermocouples were connected to a Campbell Scientific CR1000X datalogger with a multiplexer for simultaneous hourly recording of temperature into an Excel readable format.

Sufficient coupling of the sensors with the rock mass must be ensured to achieve accurate measurements, and convection must be minimised. This can be achieved with physical barriers such as plugs or baffles, or by replacing the air in the borehole with a different medium such as water or cement grout. Rybach and Busslinger (2013) performed measurements in boreholes purposely filled with water for thermal coupling. To examine the impact of these methods and gather information for future work, the current study drilled two holes side by side, filling one with pumped grout and separating the instruments in the other with foam plugs. The design of the instruments varied slightly for each method to suit their different purposes and were primarily based on the instrument used by Duckworth (1999).

Ultimately, three different types of collapsible instruments were constructed, allowing for easy transportation to the sites (Figure 3). Instruments A and C were constructed from 3 m sections of 32 mm diameter PVC pipe with pipe couplings, while instrument B was constructed using a set of stainless-steel setting rods for stress cells. Instrument A was designed for grouted holes and attached the thermocouples directly to the outside of the pipes, while instruments B and C were removable and were used in ungrouted holes. On these instruments foam pipe insulation was used to construct plugs that both pushed the thermocouples against the rock and blocked airflow in the holes, preventing convection.

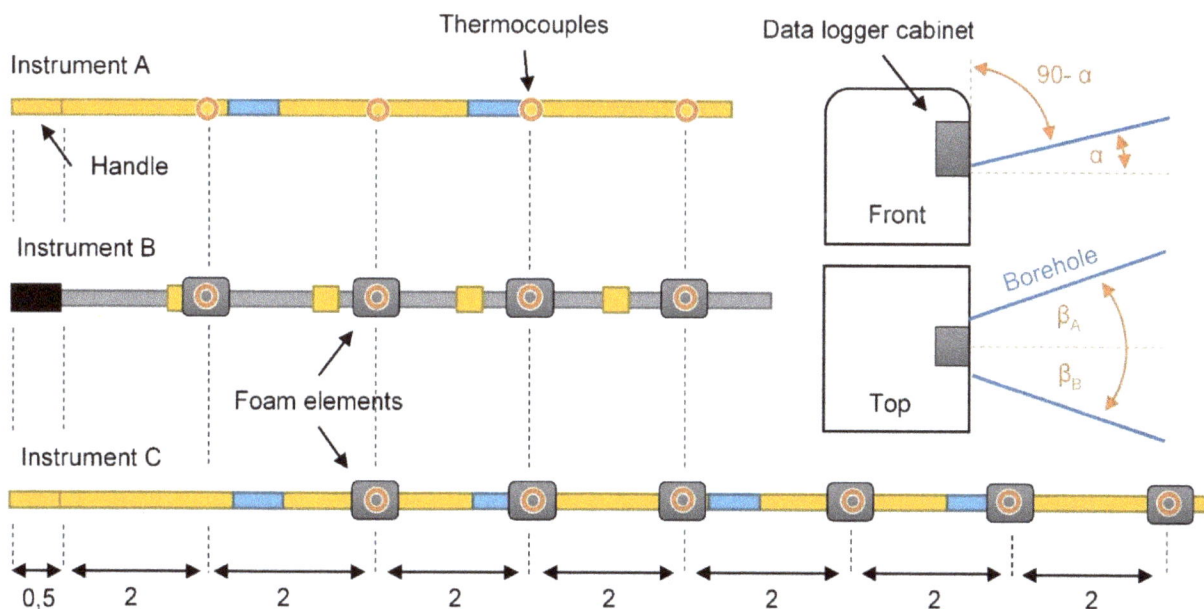

FIG 3 – Schematics of instruments A, B and C (top to bottom) and installation concept (right).

Long borehole instrumentation

The instrument needed to be capable of measuring about 1000 m from the collar in a water-filled borehole to reach the target depth while accounting for borehole inclination. Conventional borehole logging tools with temperature sensors were discarded as an option due to lack of immediate availability. As a flexible alternative, commercial ocean temperature loggers were chosen.

This type of instrument is battery-powered and features on-board memory storage, which simplifies its application as no data cable was necessary. An aluminium housing contains the electronic components, while the temperature and pressure sensors are located at the bottom and are protected by a cage-like construction (Figure 4). Depth data is obtained via the pressure sensor simultaneously with temperature measurements, allowing easy creation of depth-temperature profiles.

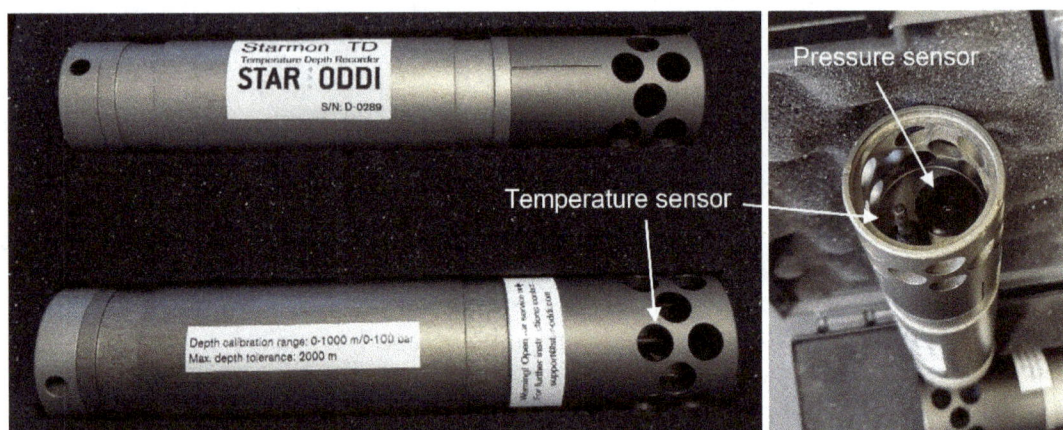

FIG 4 – Depth-temperature logging unit for deep borehole measurements.

Corresponding to its commercial application, the calibrated temperature range was from -2°C to +40°C with an accuracy of ±0.025°C. The maximum depth rating guaranteed by the manufacturer was 2000 m, with an accuracy of ±0.3 per cent of the full range. The data was transferred and processed first through the software of the manufacturer and then exported into a spreadsheet format.

A borehole logging winch that was modified to feature 1000 m of steel wire and a battery-powered electric motor with an onboard winching program enabled lowering and retrieving of the instrument at a constant speed.

Instrument calibration

The instruments selected for use underground were calibrated to better understand their limitations, to help build confidence in the readings returned by the sensors, and to help ensure that the data returned was of high quality. The completed calibration tests provided information about the behaviour of the various instruments under different circumstances, which helped with interpretation of their results after the main test campaigns in the mine.

First, randomly selected thermocouples were connected to a datalogger and allowed to collect temperature data from a single point over a period of days. This showed that the individual thermocouples were in good agreement with one another. Next, the factory-calibrated downhole logger was evaluated along with the thermocouples, both in air and in water. This showed that the instruments agreed very well with one another in water but took a long time to stabilise in air. This matched the findings of Jones (2016).

Finally, the longhole instrument was evaluated in a water-filled borehole underground to assess whether water in the borehole was being disrupted or mixed in a way that required temperature stabilisation to achieve high-quality measurements. The instrument was lowered into the borehole and was held stable for several minutes at intervals of 25 m, 50 m and 100 m. The temperatures collected were compared against those collected during a second run, where the logger was lowered at a continuous rate down into the borehole with a sampling rate of 1 second. The results showed that there were no identifiable changes in the recorded temperatures between the two methods.

INSTRUMENT INSTALLATION

Short borehole measurements

The first instruments to be installed were at AL1112 (see Figure 2). These included one grouted type A and the ungrouted type B instrument. They were both connected to the datalogger immediately after installation. This allowed the instruments to capture the full temperature profile during and after curing of the cementitious grout. An example of the instrument installation is shown in Figure 5.

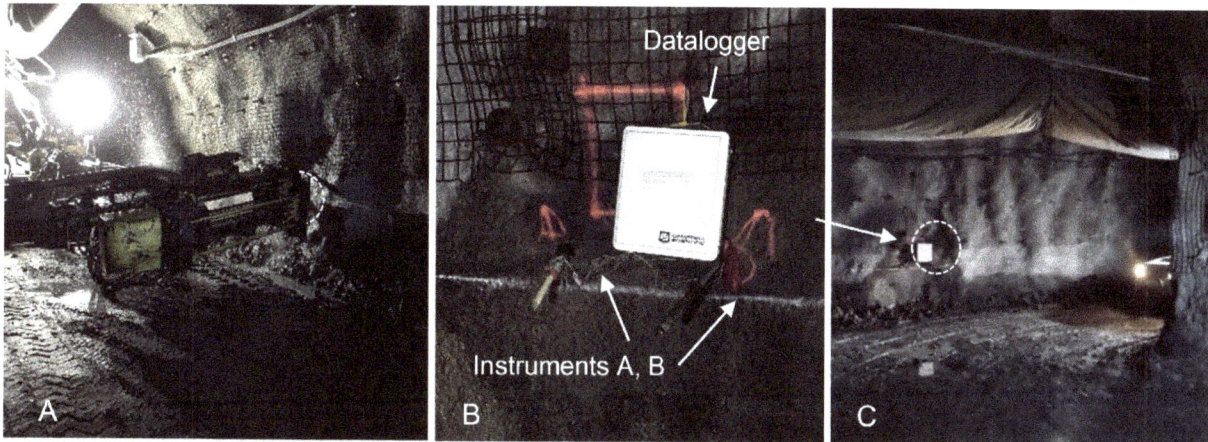

FIG 5 – Summary of on-site procedure: drilling and grouting (A), installation and connection of instruments and datalogger (B), site overview (C).

Installation proceeded to GN1023 (see Figure 2). At this location the holes for both instruments were drilled, the grout hole was filled, and a type A instrument was installed and connected to the datalogger. The ungrouted borehole was plugged for a later installation of instrument type B after temperatures had stabilised at Site I.

Installation at BA755 (see Figure 2) followed the same pattern as at GN1023. The instrument holes were drilled, and the grouted type A instrument was installed. The second drill hole was made longer (15 m), and the Type C instrument was installed. Unfortunately, the installation went poorly, and part of the instrument was lost in the borehole, preventing measurements at the target depths. To make use of the remaining measurement point at the 10 m mark, the installation plan was changed, and the remaining type C instrument was moved to GN1023 and later to AL1112.

Long borehole measurements

The selected exploration boreholes were not always available due to damage, poor marking, or other reasons. In these cases, the steepest borehole in the same area was chosen as the next best alternative. In virtually all cases, the water level in the borehole was at or near the collar.

The standard procedure for measurement in the exploration drill holes commenced by placing a tripod with feeding wheel over the borehole, securing the downhole winch to a parked truck, and installing the winch cable through the feeding wheel. The instrument was attached to the carabiner on the end using a cable binder. As a safety feature for the winch, the cable binder was notched to provide a breakage point. This way, the chances of wire recovery in case of the instrument becoming lodged in the borehole would be improved. The recovery would be at the loss of the instrument and the acquired data, but time-consuming repairs and delays to the winch would be avoided. The procedure is summarised in Figure 6.

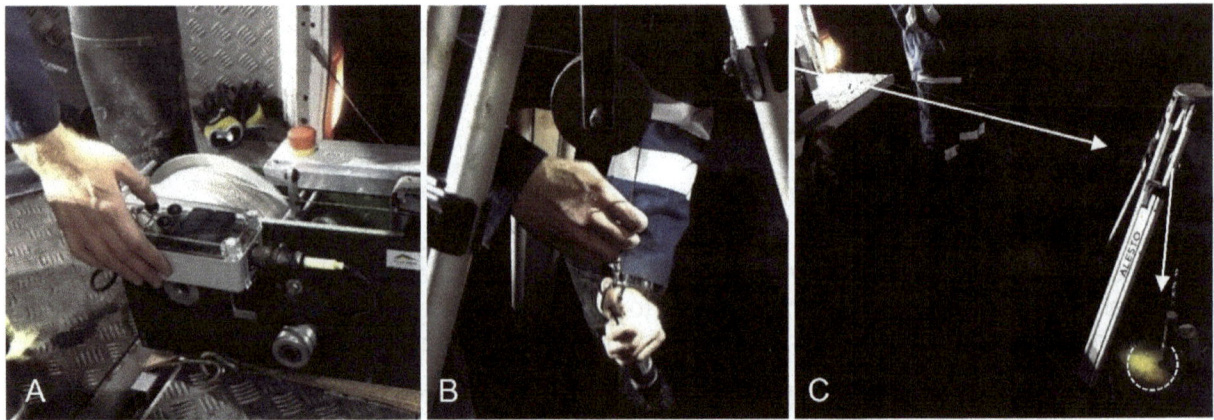

FIG 6 – Summary of on-site procedure: set-up of winch (A), set-up of tripod and attaching of instrument (B), measurement (C, borehole circled).

Due to the light weight of the instrument, the winch gears were set to free-wheel and the instrument was fed manually down the borehole until weight of the wire plus instrument became heavy enough to slide down the hole on its own. Upon reaching the bottom of the hole, a blockage, or the end of the winch cable, the automated winch was set to retrieve the instrument at a rate of around 10 m per minute. Upon reaching the surface the data was downloaded immediately to a computer and checked for inconsistencies.

On one occasion it was possible to achieve even deeper temperature measurements than were possible with the standard procedure and winch equipment. An opportunity presented itself to work with one of the drilling contractors and use their drilling rig and its winch for lowering the temperature sensor down a recently completed borehole. The borehole had sat quiet and untouched for some days and the contractor built a special attachment to allow lowering the temperature logger to the bottom. This was an exceptionally deep hole and would not have been possible without the full-size winch available on the drill rig. In total it allowed measurement of temperature from a depth of 2820 m, relative to the hole collar location on level 1250. This was well beyond the target depth but provided an excellent opportunity to record an extended temperature profile of the mine.

RESULTS – SHORT BOREHOLE INSTRUMENTATION

For the short boreholes, the raw data was graphed and reviewed for temperature steadiness and convergence. Temperature profiles along the length were created from the position of each measurement point, corrected for the dip and angle of the borehole so that the results reflected the true depth in the rock mass. For each sensor pair along the hole, daily temperatures were obtained by averaging the data from the previous 24 hours of measurements. The temperature development and the final profile for each of the boreholes is shown in Figure 7.

The data from the short holes showed that it took only a matter of hours for the air-filled holes to stabilise in temperature, while it took the full 28 days of curing time for the grouted holes to stabilise (Figure 7). Both sets of measurements returned stable values and didn't show signs of fluctuation as one might expect due to convection within the borehole. After the curing of the grout the temperatures of both holes were at worst within 0.3°C of each other, and the temperatures could be seen to get colder as the length into the boreholes increased in the cases of BA755 and GN1023, and warmer in the case of AL1112.

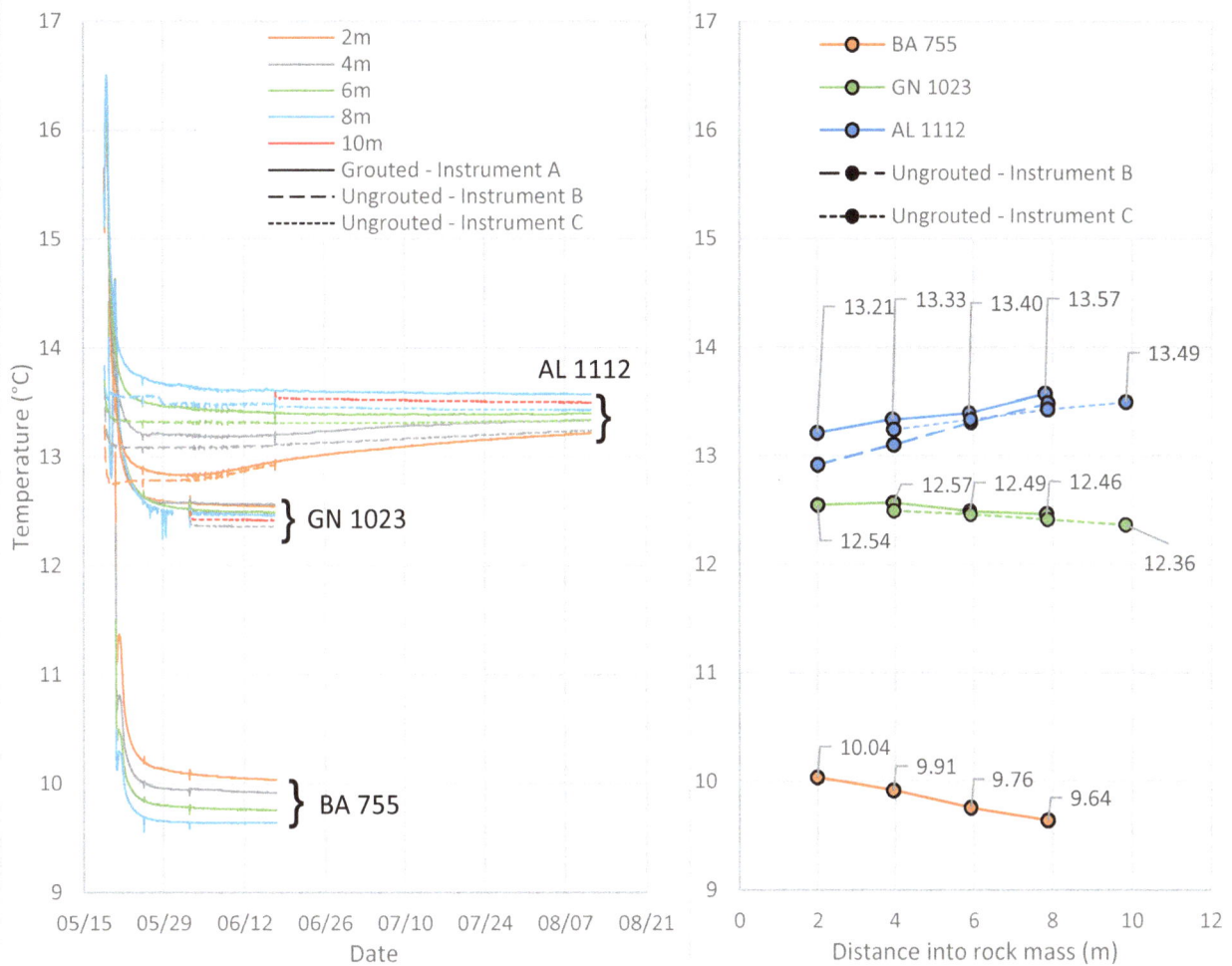

FIG 7 – Temperature development (left) and final profiles for grouted and ungrouted boreholes (right).

The performance of the non-grouted holes and the grouted holes were similar enough that grouting of holes is likely unnecessary to achieve temperature measurements that are close enough for calculating gradients and VRT, and that grouting may be seen as unnecessary. The foam plugs into which the sensors were installed did an adequate job of pressing the thermocouple against the rock. In this configuration, only some thin layers of tape were between the sensor and the rock, likely contributing to their precision with respect to the grouted sensors.

However, when it comes to the practicalities of instrument construction, installation, and success rate, the grouted instruments were far superior to the foam-plug instruments. They were easier and faster to construct, more robust, easier to install, and more likely to result in a successful installation.

Because the temperatures along the borehole length never converged upon a single point, the short holes were not deep enough to reach VRT. As such, the results couldn't be used for thermal gradient calculation.

Analysis and discussion

Temperatures measured on level 1112 are observed to rise with increasing distance into the rock mass, with the upward trend being more pronounced close to the wall. As the deepest two sets of thermocouples did not record the same temperature, it can be said with certainty that they did not reach a deep enough depth to record the VRT. Because these holes were roughly horizontal, there should be relatively little influence of depth increase due to depth change. The temperature flux along the hole length is clearly due to absorption and dissipation of heat from the air into the rock mass.

The same trend was found in all the short hole readings, regardless of their installation method, grouted or ungrouted. The trend only varied in direction, positive or negative. All of the short holes from levels 1023 and 755 show that the rock temperature decreases, rather than increases, as the depth of the measurement sensors increase. This is shown in Figure 7.

These results indicate that the average air temperature is actually higher than the VRT at the 755 and 1023 depths. Additionally, the rate of temperature decrease (°C/m) into the rock is faster on level 755 than it is on level 1023. The rate of temperature change is an indicator of the temperature difference magnitude between VRT and the average air temperature, indicating that there is a bigger difference on 755 than on 1023.

The overall conditions are reasonable for a mine in a subarctic region that is built in very old, stable rocks. The average surface air temperature is quite low on an annual basis and there are no significant heat sources other than depth in the rock mass. As such, the mine air, which is heated to a minimum of 1°C in the winter, has a higher average temperature than the unheated air, which is the primary cooling source of the rock in the region. There exists a depth, though, at which the average mine air temperature is equal to the VRT at that depth. At that depth there should be nearly zero temperature change in a hole drilled horizontally into a tunnel wall. In Malmberget, that depth appears to occur between the depths of 1023 and 1112 m on the mine coordinate system. That depth would tend to vary throughout the year as the average surface air temperature fluctuates seasonally, and potentially varies from place to place around the mine depending on the ventilation system design.

RESULTS – LONG BOREHOLE INSTRUMENTATION

In total there were 13 exploration boreholes from which temperature data was collected. Of these, 12 provided high-quality temperature and depth data that was of sufficient length for further processing. Unlike other boreholes, Borehole 1 had noticeable water inflow which prevented the water column from reaching temperature stability. In Boreholes 8, 10 and 11 the temperature logger encountered difficulty proceeding down the hole and was retrieved after only a short distance. The pertinent data for each hole and their location is shown in Table 1 and Figure 2, respectively.

TABLE 1

Summary of data for measured deep boreholes and gradients at 40 m below collar.

Borehole	Level	Collar level	Hole length (m)	Gradient (°C/100 m)	Comment
1	JH600	586	446	1.027	Excluded due to groundwater inflow
2	JH600	586	108	1.213	-
3	JH600	586	340	1.356	-
4	FA970	976	240	1.508	-
5	FA970	976	259	1.465	-
6	FA1250	1255	1565	1.627	-
7	FA1250	1255	279	1.479	-
8	PA1250	1247	57	1.051	Excluded due to insufficient length
9	PA1250	1247	303	1.603	-
10	PR1250	1237	51	1.442	Excluded due to insufficient length
11	PR1250	1237	35	-	Excluded due to insufficient length
12	PR1250	1233	109	1.419	-
13	PR1250	1233	152	1.438	-

The VRT in Malmberget is often lower than the mine air temperature in most areas. In these situations, the rock around the tunnel acts as a heat sink and absorbs heat from the air. The depth to which the rock is heated depends upon the temperature difference between VRT and the air temperature, and the length of time the rock has been exposed. In every longhole case the strata around the tunnel cools as the temperature probe moves deeper until the distance from the collar is reached where the heating influence of the air minimises and the VRT takes over as the dominant rock temperature driver. At this point the rock begins to heat as the probe travels deeper into the borehole. This is shown by the inflection point in Figure 8.

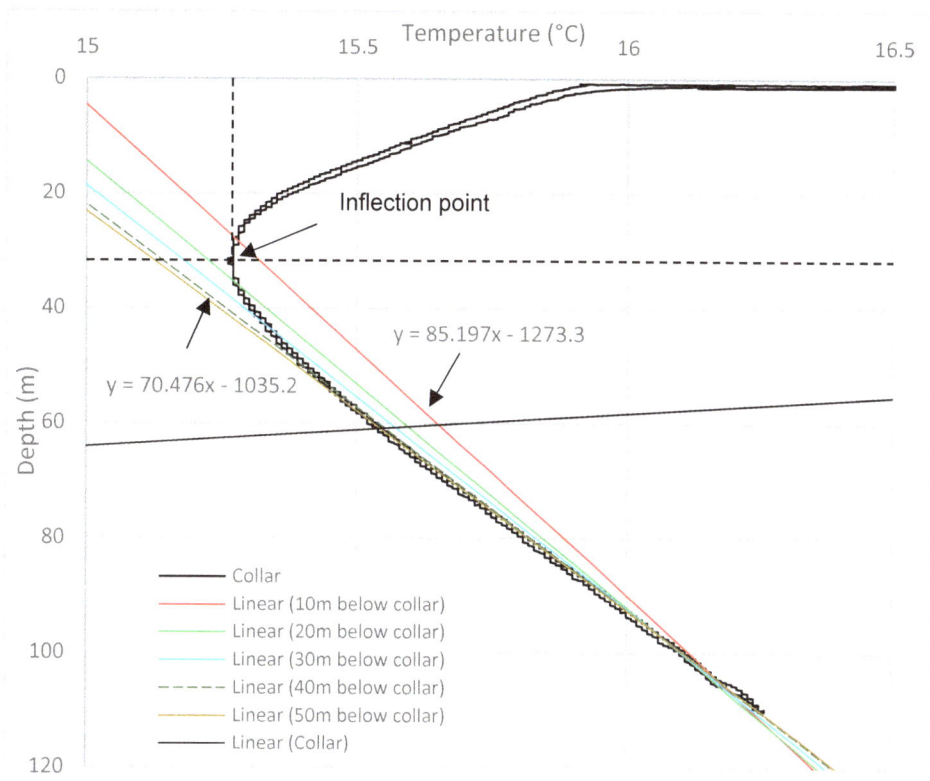

FIG 8 – Example of geothermal steps for different starting points below collar for Borehole 12.

To obtain an accurate thermal gradient from the data acquired during the long borehole measurements, the depth data was first corrected to reflect the elevation according to the mine grid using the survey coordinates of the borehole collar. A cut-off length had to be determined to eliminate the influence of the drift conditions and the heating of the rock by the air. This effectively truncated the data record, disregarding the temperature data from the first part of each hole and leaving the remaining data that exhibited a steady gradient.

A spreadsheet program was used to calculate an individual gradient for each borehole while successively excluding data in 10 m intervals, starting at the borehole collar. After each 10 m interval of data was removed, the absolute change in gradient relative to the previous scenario was calculated. If the gradient change exceeded a subjectively selected ±0.2°C/100 m, it was deemed that the influence from the drift was still too strong and an additional 10 m was excluded. With each successive removal of 10 m intervals, the slope of the trendline of the remaining data approached the measured, nearly linear trend of the VRT. In Figure 8, which illustrates the method, the linear trend is observed at depths greater than 55 m below the hole collar.

This method automated the process of identifying the VRT trend allowed a uniform calculation to be used on all data sets. The calculated trends can be extrapolated upwards to estimate the presumed VRT at the collar location. In general, it was determined that a cut-off length of 40 m was appropriate for eliminating the influence of the heating of the rock by the air.

While this approach functioned well and produced good results, it led to Boreholes 8, 10 and 11 being excluded from further processing as the measured length proved too close to the cut-off length and resulted in too little data remaining for the determination of a reliable geothermal gradient. In addition, Borehole 1 was also excluded, as the non-linear temperature profile was a clear outlier to all other data collected even within the same general area. This odd profile was caused by clearly identified fresh water inflow into the hole. A summary of all measured boreholes, including those not considered during gradient calculation, is given in Table 1.

Boreholes 2 and 3 were considered separately as they are consistent with one another but are clearly offset from the remaining data. A line of best fit was plotted through each data set to determine gradients for above level 1000 and as shown in Figure 9, which were calculated from the geothermal step to be 1.349°C/100 m and 1.591°C/100 m, respectively.

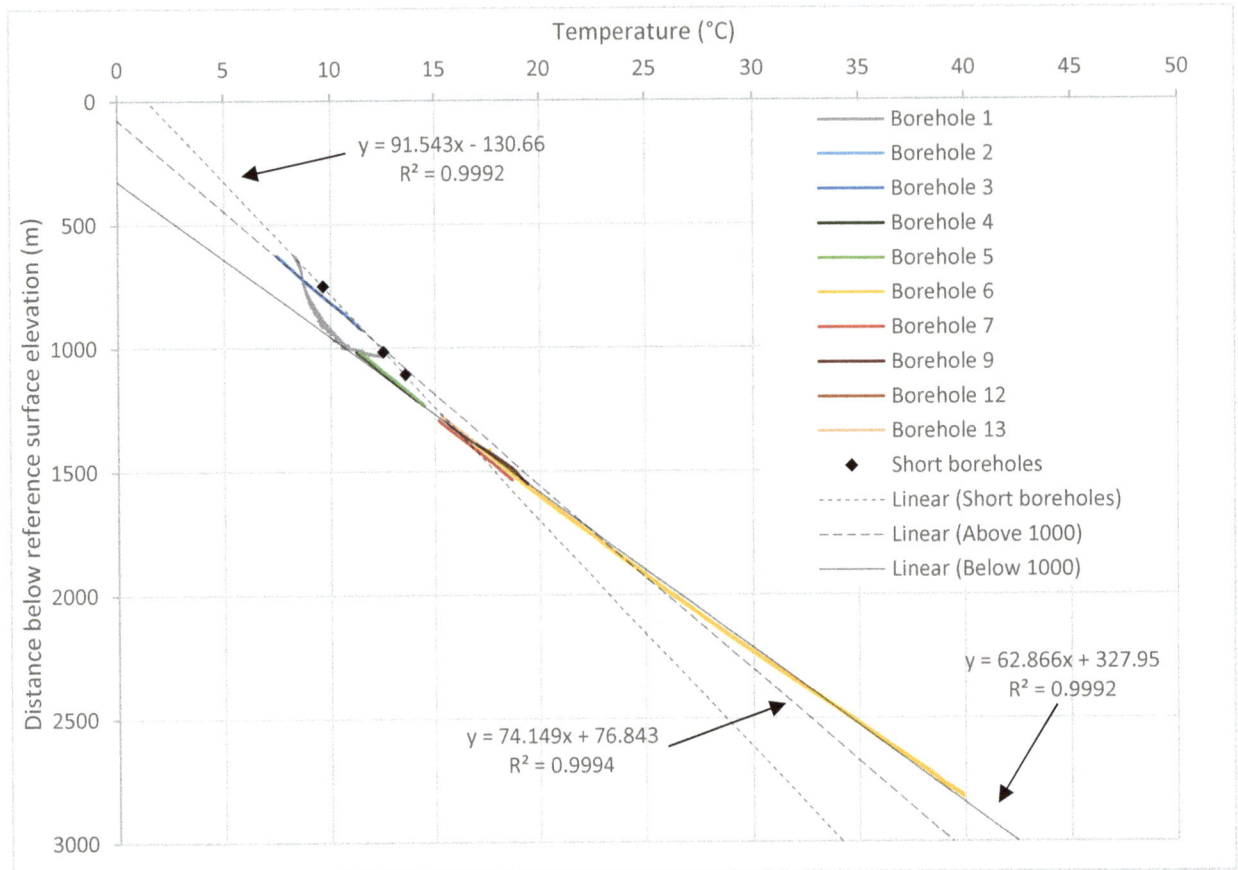

FIG 9 – Geothermal steps derived from selected boreholes for above and below level 1000 (short borehole data and Borehole 1 included for comparison).

Analysis and discussion

The temperature profiles graphed from raw collected data show that the upper borehole sections have been influenced by the prevailing local conditions since the time of excavation. A representative example of this is given in Figure 8. Following an inflection point at around 20–40 m depth, temperatures quickly establish a nearly linear trend for boreholes without water inflow. While the conditions at the short borehole sites are not directly comparable to the situation at the long borehole locations, the position of the inflection point nevertheless indicates that short borehole measurements would have to take place much deeper in the rock mass than they did in this study. The increased lengths and additional measurements necessary to record VRT is considered impractical due to the operational constraints of the mine.

Linear temperature profiles are not always encountered. As seen in Figure 9, temperature fluctuations are clearly visible in Borehole 1 and groundwater outflow was observed on-site. This highlights the importance of examining how boreholes can potentially be affected prior to measurements. In locations with consistent water inflow, it is likely that grouted instrumentation would be necessary to achieve accurate temperature measurements.

Depth-temperature profiles and gradients derived from data collected in the other boreholes follow each other closely, including the very deep Borehole 6. The depth of this borehole was within the logger's depth range but greater than the manufacturer's calibrated range. However, the instrument showed no signs of damage and the data obtained did not show any deviations from previous measurements at other sites on the same level and followed the other data closely. As drilling had been completed only days prior, this implies that thermal recovery of long exploration boreholes is relatively quick. Given the thermal transmissivity of water, this is not surprising.

The gradients obtained by Parasnis (1982) appear to mostly support the decision to divide thermal gradients above and below level 1000. The Parasnis gradients were determined for borehole sections known to have different geologies and are shown in Table 2. They are in a similar depth and gradient range to Boreholes 2 and 3 of this study as shown in Table 1.

TABLE 2

Summary of measured and historical gradient data for Malmberget Mine.

Location	Sections (m)	Gradient (°C/100 m)	Comment	Source
Borehole 4092	654–1008	1.390	Gneiss + magnetite	Parasnis (1982)
	710–896	1.362	Magnetite only	
	915–1027	1.382	Gneiss only	
Borehole 4000	711–1096	1.535	Gneiss only	
Borehole 3946	696–1092	1.381	Blended rock types	
Above level 1000	640–940	1.349	2 measurements	This study
Below level 1000	1010–2820	1.591	7 measurements	

When the depth range is viewed from an overall standpoint, the arithmetic average gradient of the Parasnis (1982) data is 1.410°C/100 m. The overall arithmetic average gradient for the measurements taken in this study, not including excluded holes, is 1.537°C/100 m. This is a difference of 0.127°C/100 m, or 9.0 per cent. If one uses the data in Table 1 to normalise the gradient data by hole length, so that longer holes have a proportionally greater influence than shorter holes, then the normalised average gradient for this study is 1.479°C/100 m. The same can be done with the Parasnis (1982) data in Table 2, leading to a normalised average of 1.422°C/100 m. These normalised values show differences of only 0.057°C/100 m, or 4.0 per cent overall. The clear agreement of the two data sets is seen as validation of the utilisation of available exploration boreholes, the approach of determining single gradients from available data, and the method of eliminating the cooling or heating of the rock nearest to an opening due to differences in mine air temperature.

The comparison between the results of this study and that of Parasnis (1982) is a complicated one. The variables at play when determining geothermal gradients include questions of depth, geology, orebody, mining history, age of the excavation, water inflow, and more. The holes measured in the Parasnis (1982) study were from similar areas of the mine as those measured in this study, and from similar depths, but it is difficult to identify precisely why different holes have varying gradients. The best course of action for any mine is to attempt to get temperature measurements from representative areas of the mine, and then develop thermal gradients according to the needs of the mine and the characteristics of the site.

As the geothermal gradient information collected during this study is utilised for modelling of the mine extension below the current main level 1250, the decision to separate the gradient based on depth was considered a more conservative approach than relying on the normalised averaged gradient over the entire depth of the mine. A too-low gradient value could easily lead to underestimation of future cooling requirements. Implementing multiple gradients in climate simulation programs is possible, for example in Ventsim by applying rock type pre-sets customised using the different gradient. Refinement of the near surface gradients could be performed by considering boreholes drilled from the surface following the conclusion of the study. This would allow the examination of the validity of the chosen cut-off depth of 1000 m for the two gradients and the evaluation of the possibly rising trend of the geothermal gradient in Malmberget Mine.

CONCLUSION

Shallow and deep borehole measurements were used to determine geothermal gradients in the Malmberget Mine. The focus was on the eastern section of the mine, although measurements were also collected in certain areas of the western section. Collected data had a good spread about the mine despite some locations no longer being accessible due to ongoing production or caving. Of the three types of measurements attempted, air-filled, grouted and water-filled, the last was shown to be the best as long as there was limited water inflow.

When comparing short holes, grouted temperature sensors are easier to install, though air-filled and grouted holes showed similar results. However, short boreholes were determined to be generally inadequate for VRT measurement because the temperature of the surrounding rock mass is affected by the air temperature inside the tunnel where the boreholes are located. The extent of the rock mass after which the measured VRT trend is approached, and thus the influence of the air sufficiently eliminated, was determined to be around 40 m from the borehole collar. This makes even the longest 'short' holes insufficient.

A better method was shown to be temperature measurement in water-filled exploration holes, again, controlling for water inflow affecting the measurements as seen in Borehole 1. These yielded more reliable data down to the target depths and resulted in gradients of 1.381°C/100 m above level 1000 and 1.591°C/100 m below level 1000, as determined by line of best fit. The results are judged to be reliable based on the consistency between the results obtained for different boreholes at different locations and the normalised average gradient of 1.479°C/100 m, which is consistent with historical data, even though that data did not reach the target depths.

ACKNOWLEDGEMENTS

LKAB is gratefully acknowledged for providing means to conduct these activities and for permitting this paper to be published. We extend our thanks to our colleague Joakim Douhan and the cable bolter operators, without whom no short hole measurements would exist. Thank you also to our colleague Simon Engström for providing geological insights helping with borehole selection and for enabling data to be collected in Borehole 6, in agreement with the drilling contractor. Furthermore, we wish to extend our thanks to our colleagues at the exploration department, Niklas Juhojuntti and Anton Lejon, for lending us the winch that allowed measurements in deep boreholes to take place. Our appreciation goes out to all our colleagues within LKAB, who provided information for this study as well as for their genuine interest and good suggestions.

REFERENCES

Ashworth, E, Ashworth, T and Mahajan, E, 1983. *Study of temperature profiles in rock using a new thermistor probe*, Society of Mining Engineers of AIME, Preprint 83–24.

Duckworth, I, 1999. Rapid evaluation of rock thermal parameters at the Lucky Friday Mine, in *Proceedings of the 8th US Mine Ventilation Symposium* (ed: J C Tien), pp 337–346 (University of Missouri-Rolla Press: Rolla).

Cleland, R H, 1933. Rock temperatures and some ventilation conditions in the mines of northern Ontario, *The Canadian Mining and Metallurgical Bulletin,* 256:379–407.

Calizaya, F and Marks, J, 2011. Heat, humidity and air conditioning, in *SME Mining Engineering Handbook* (ed: P Darling), third edition, chapter 15.5, pp 1611–1624 (Society of Mining, Metallurgy and Exploration: Englewood).

Fenton, J L, 1973. A study of underground mine heat sources, United States Bureau of Mines report OFR 19(1)-74.

International Electrotechnical Commission, 2013. IEC 60584-1:2013 – Thermocouples – Part 1: EMF specifications and tolerances, August 2013.

Jones, M Q W, 2016. Virgin rock temperature study of Venetia diamond mine, in *J S Afr Inst Min Metall,* 116(1):85–92.

Jones, M Q W, 2018. Virgin rock temperatures and geothermal gradients in the Bushveld Complex, in *J S Afr Inst Min Metall,* 118(7):671–680.

LKAB, 2018. Blasting in Malmberget [online]. Available from: <https://www.lkab.com/en/sustainability/environment/land-and-remediation/blasting-in-malmberget/> [Accessed: 6 March, 2022].

LKAB, 2019. This is LKAB [online]. Available from: <https://www.lkab.com/en/SysSiteAssets/documents/publikationer/broschyrer/this-is-lkab.pdf> [Accessed: 6 March, 2022].

Misener, A D, 1949. Temperature gradients in the Canadian shield, in *The Canadian Mining and Metallurgical Bulletin* 446:280–287.

Oender, U Y and Güyagüler, T, 1995. Determination of thermal properties and heat emission of coal measure rocks in Zonguldak coal region, Turkey, in *Proceedings 7th US Mine Ventilation Symposium* (ed: A M Wala), pp 449–454 (Society for Mining, Metallurgy, and Exploration: Littleton).

Parasnis, D S, 1975. Temperature phenomena and heat flow estimates in two Precambrian ore-bearing areas in north Sweden, in *Geophys J R Astr Soc,* 43(2):531–554.

Parasnis, D S, 1982. Geothermal flow and phenomena in two Swedish localities north of the Arctic circle, in *Geophys J R Astr Soc,* 71(3):545–554.

Russell, A J and von Glehn, F H, 2019. The importance of accurate rock property and virgin rock temperature data in ventilation and cooling planning requirements, in *NAMVS: Proceedings of the 17th North American Mine Ventilation Symposium* (eds: A Sasmito, A Madiseh, F Hassani and J Stachulak), pp 536–544 (Canadian Institute of Mining, Metallurgy and Petroleum: Westmount).

Rybach, L and Busslinger, A, 2013. Verification of rock temperature prediction along the Gotthard base tunnel – a prospect for coming tunnel projects, in *Underground. The Way to the Future: Proceedings of the World Tunnel Congress 2013* (eds: G Anagnostou and H Ehrbar), pp 1714–1721 (CRC Press: London).

Calibration of post blast gas simulation using mine site readings

F C D Michelin[1], J Norris[2] and T Andreatidis[3]

1. Managing Director, Howden, Brisbane Qld 4169. Email: florian.michelin@howden.com
2. Senior Mine Ventilation Engineer, Kirkland Lake Gold, Fosterville Vic 3557.
   Email: jnorris@kl.gold
3. Ventilation Officer, Glencore, Cloncurry Qld 4824. Email: tim.andreatidis@glencore.com.au

ABSTRACT

Blast fumes are a significant health risk to underground workers, containing several dangerous gases such as carbon monoxide (CO), nitrogen oxides (NOx) and ammonia. To ensure safe working conditions, workers are not normally allowed to re-enter a mine until the gases are cleared. To predict if gases can be cleared efficiently, simulations can be performed to estimate the gas levels following a blast.

For accurate modelling, a calibrated airflow simulation is required in addition to blast-specific parameters such as the yield rate (representing the amount of gas released per kg of explosive) and the dispersion factor (representing how quickly the gases can enter the air stream from the explosive site). Those parameters are not easily determined, as they will vary from blast to blast.

Many post-blast gas recordings on mine sites were studied and the correct parameters were determined to ensure an accurate gas simulation. The variability of the results was then analysed and correlation to known parameters is suggested. Finally, recommendations on how to determine the correct parameter are presented.

BACKGROUND

Why simulate blast fumes?

To establish safe re-entry times, a common method is to calculate the number of air changes required to guarantee fresh air in a heading (Kocsis *et al*, 2019). While this approach is valid, it does not take into consideration the gas fumes that could be entering from other locations, or if recirculation is present. Simulation of blast fumes allows for a greater understanding of the impact of all blasts across the entire network.

Using visual simulation tools also allows 'what if' scenarios to be analysed, to try and find the best configuration for quick blast clearance. For example, increasing the main fan speed, opening regulators near blasts or closing regulators in other areas could improve clearance times. The simulations can suggest airflow control changes to maximise blast clearance effectiveness.

While implementation of those scenarios is possible by manually setting up timers, a scheduling tool allowing the creation of scripts that are run automatically daily decreases the time required to operate the equipment while also decreasing the risk of human errors in the implementation (Pinedo and Torres Espinoza, 2019). In addition to the use as a daily prediction tool, blast clearance simulation should also be used in the planning stage to ensure that the mine ventilation design will allow efficient blast clearing.

Initial model requirement

An accurate ventilation model is required to perform a blast clearance simulation. Models should typically have simulated values within 10 per cent of the actual readings of airflow quantities and velocities. For blast modelling, the airflow velocity is particularly important as it affects the assumed speed and distribution of fumes.

A common error that prevents this is the use of original design dimensions to build ventilation models. Actual airway sizes often differ by 20 per cent or more due to overbreak of the tunnels (Michelin *et al*, 2019), and even if the airflow is correctly calculated in the model (through the modification of friction or resistance values), the air velocity (the component used to calculate the speed of contaminant and gas removal), can be significantly in error. Accurate velocities are required to simulate the timed spread of fumes from fires, gas sources, heat, diesel particulate matter (DPM) and blasting fumes.

Where possible, resistance surveys (pressure and flow measurements) combined with correctly measured airway dimensions should be performed to ensure friction factors and shock losses are accurate, and that future development and model predictions use realistic values.

For blast fume analysis, auxiliary ventilation and duct should also be included in the model to better represent airflow distribution, as the movement of air through these pathways will affect distribution and clearance times. The ventilation duct should also consider leakage, as high leakage will reduce flow delivered to the blasting site, increasing the fume clearance time.

Blasting simulation in Ventsim design

Simulation of blasting fumes is based on calculating the gas yield (l/kg), which is the total volume of gas released compared to the explosive mass blasted. The gas released is calculated by multiplying the yield and the explosive quantity, and this amount is then gradually released into the passing airstream during the simulation based on the dispersion factor.

The yield in Ventsim is set as part of the explosive presets list while the explosive amount and dispersion are set per airway through the EDIT form.

Recording monitors to store simulated gas results over time can be placed throughout the model. For this study, the monitors are placed in the same place as the sensors.

Gas yield

The gas yields provided by explosive manufacturers are rarely accurate enough for simulation and should be calibrated to site conditions if possible. The yields are measured in controlled environments (Table 1), which do not represent the variation in the product and conditions experienced underground. Some factors influencing the gas yield include oxygen balance, water content and confinement (Rowland III and Mainiero, 2000), with variations in CO yield from ammonium nitrate/fuel oil (ANFO) and emulsion ranging from 5 L/kg to 50 L/kg.

TABLE 1

Ventsim Explosive presets.

Name	# in use (total all stage)	Yield generic L/kg	Density kg/m³	YieldCO L/kg	YieldNO L/kg	YieldNO2 L/kg	YieldNH3 L/kg
Contaminant only		12.5	1.2	0	0	0	0
ANFO 96/4		0	1.3511765	4	2	4	0.2
ANFO 94/6	4	0	1.21483...	16	0.5	1.5	0.2
ANFO 92/8		0	1.1907618	38	1.5	1.5	1
70/30 ANFO/Emulsion		0	1.26674...	22	3.1	5.6	0
50/50 ANFO/Emulsion		0	1.27447...	18	2.7	5.2	0
Emulsion		0	1.2073586	14	1.25	0.65	0

Dispersion factor

The dispersion factor was introduced by (Stewart, 2014) and represents how quickly the blast gases are removed and enter the clearing air stream. Some default values are available in Ventsim (Table 2), however, the offered factors are somewhat arbitrary, and selection often comes down to guesswork or calibration with observed results.

TABLE 2

Default dispersion factor in Ventsim.

Description	Dispersion factors
Very Slow	1
Slow	2
Moderately Slow	3
Moderate	4
Moderately Fast	5
Fast	7
Very Fast	10

METHOD

Measurement conditions

The data from 34 blasts across two sites were studied, with both sites using emulsion explosives. All measurements were taken underground and measured using the site's gas detectors and airflow measurement equipment. Multiple factors can affect the results including airflow measurements, the post-blast mixing of fumes, and possible discrepancies in calibration between the different sensors, however, despite some likely inaccuracies, the values are the best available to help decide on re-entry and are therefore relevant to the study.

To make the analysis easier, it is preferable to have each sensor measure the fumes from one location only. Similarly, all the fumes from that blast should pass in front of the sensor. The distance between the blast and the sensor should be sufficient to ensure the fumes are fully mixed with the air. As most sensors are located close to the exhaust raises, this condition is met by most development blasts.

The sensors used were electrochemical sensors which measure the concentration of the gas by oxidising the gas at an electrode and measuring the resulting current. These types of sensors usually require regular calibration every three months. Poorly performing or inaccurate sensors can often be easily determined through nonsensical values or by comparison to manual ventilation surveys.

Calculation of yield

To calculate the yield gas release, the concentration for each time step is multiplied by the time step, summed and divided by the mass of explosives. The concentration was also adjusted based on pre-existing gases.

$$Y = \frac{\sum \left(\frac{(C_{n+1} + C_n)}{2} - Ce \right) Q_{sensor} \Delta t}{1000\,E} \tag{1}$$

where:

Y	is the Yield inL/kg
E	is the amount of explosive in kg
C_n	is the nth concentration of gas in ppm
Ce	is the concentration existing before the blast in ppm
Q_{sensor}	is the amount of airflow at the sensor in m³/s
Δt	is the time between two readings in seconds

The pre-existing concentration is taken by looking at concentration values before the blast time. To calculate yield accurately, it is important to sum the gas volume from the time the gas level exceeds the pre-existing concentration to when the gas level pass below the pre-existing concentration.

Calculation of the dispersion factor

The dispersion is calculated using the peak concentration of gas and the total amount of gas generated.

$$D = -\frac{50E \ln \left(1 - \frac{C_{max}Q_{sensor}}{1200YE}\right)}{1.2Q_{face}\sqrt{\frac{1.2YE}{1000}}} \tag{2}$$

where:

C_{max} is the maximum gas concentration in ppm

Q_{face} is the airflow at the face

EXAMPLE

Key data required

The main information required is the gas measurements over time from the blast. To ensure effective calibration and keep the calculation simple, the sensor should only be located to detect the gases from one blast and all gases from this blast should pass by the sensor.

The first step is to identify if there are any background gases present before the blast (from normal mine activities) so that all gas readings can be reduced by the number of gases already present. Time step and maximum concentration can be found in the gas readings as well.

The quantity of explosives is usually easily obtained, and the airflow at the sensor and the face is usually measured or assumed from the Ventsim model (which should have values typically within 10 per cent of the measurements).

Yield

An example of calculation of the calculation volume of gas passing in front of the sensor can be seen below, considering the two following concentrations of 21 ppm and 19.5 ppm measured 60 seconds apart, an existing concentration of 2 ppm before blasting, airflow at the sensor (104 m³/s) and 500 kg of explosives:

$$V = \frac{\left(\frac{(21+19.5)}{2}-2\right)104 \times 60}{1000 \times 500} = 0.22776\,l \tag{3}$$

The gas volumes would then need to be summed from when the gases exceed 2 ppm until return back to 2 ppm.

Dispersion factor

The dispersions factor calculation requires the yield previously found (29 L/kg), airflow at the face (20 m³/s), airflow at the sensor (104 m³/s) and the mass of explosives (500 kg). The result of this example is 0.41.

$$D = -\frac{50 \times 500 \times \ln\left(1 - \frac{272 \times 104}{1200 \times 29 \times 500}\right)}{1.2 \times 20\sqrt{\frac{1.2 \times 29 \times 500}{1000}}} = 0.41 \tag{4}$$

Method of calculation

An application in the programming language C# was developed to calculate the yield and dispersion of many blasts efficiently. This also serves as a first step to integrating this calculation as part of an automated system analysing blast fumes automatically. It is also possible to do this in excel, the workings of the previous example can be seen in excel in Table 3.

TABLE 3

Example of blast fumes readings and calculation of litres of CO created per minutes.

Data	Unit	Value	Source
Airflow at sensor	m³/s	104	From Ventsim
Airflow at face	m³/s	20	From Ventsim
Explosive quantity	kg	500	From charging report
Total CO	litre	14 514	Calculated from gas readings
Max CO	ppm	272	Calculated from gas readings
Yield	L/kg	29	Calculated
Dispersion Factor		0.41	Calculated

RESULTS

Simulation results

Once the factors are calculated from the available measurements, the simulation results using the adjusted calibration factors in Ventsim align closely with the readings, with similar peak and blast clearance times (Figure 1). Despite two different test sites being used, the results were similar, likely due to the use of emulsion at both sites. For this reason, the results are presented combined as single graphs.

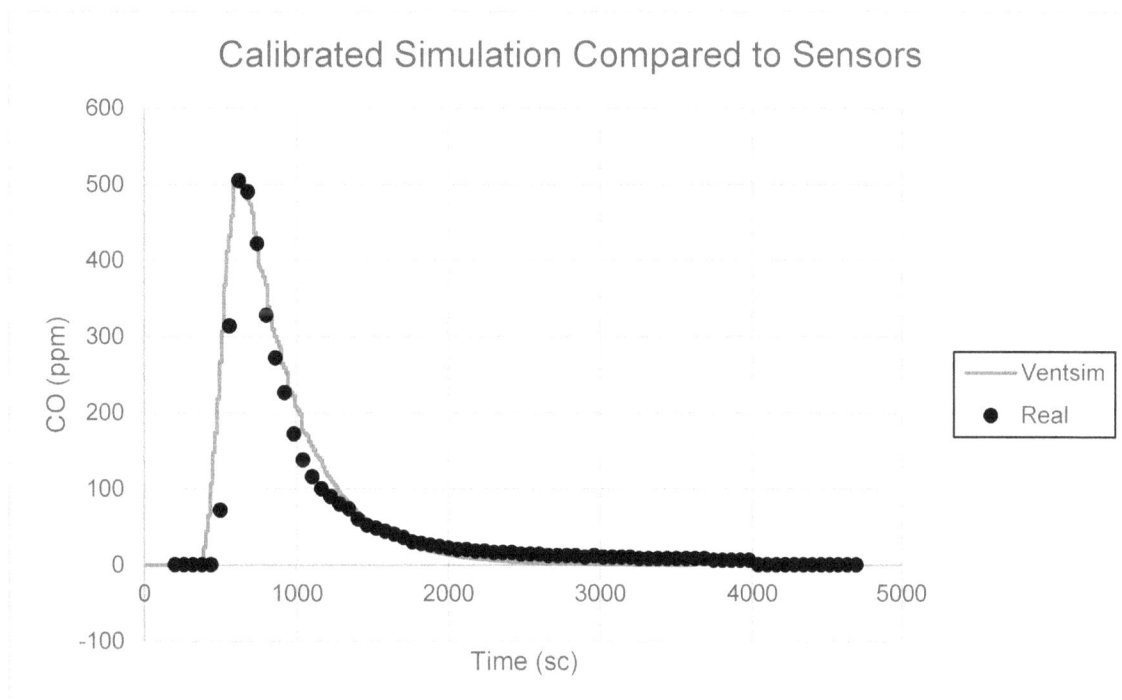

FIG 1 – Calibrated simulation compared to sensors.

Yield

The yield of carbon monoxide to explosive mass varies significantly (from 5 L/kg to 60 L/kg) depending on the blast, and the results are presented in Figure 2. The mean was approximately 26.5 L/kg. The Ventsim preset value of 16 L/kg does represent a reasonable value however higher yields on some mine sites may be likely. Simulation of a worst-case scenario would require the use of a higher yield value, 50 L/kg representing the 80th percentile.

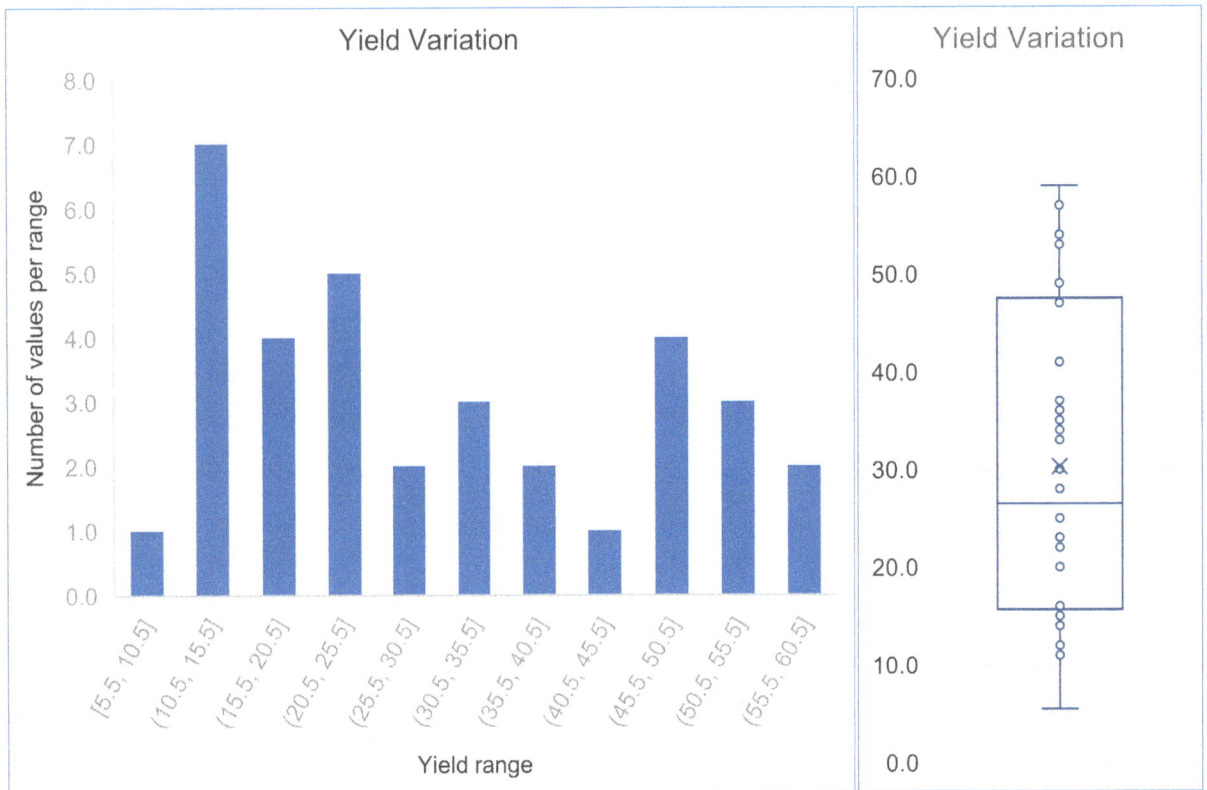

FIG 2 – Yield variation.

Dispersion factor

The dispersion factor had a mean of 0.22 with 70 per cent of the readings below 0.345 (Figure 3). This result suggests the use of a lower factor than the Ventsim default factor and it is recommended to use a custom dispersion factor of 0.20 in most cases for blind development headings.

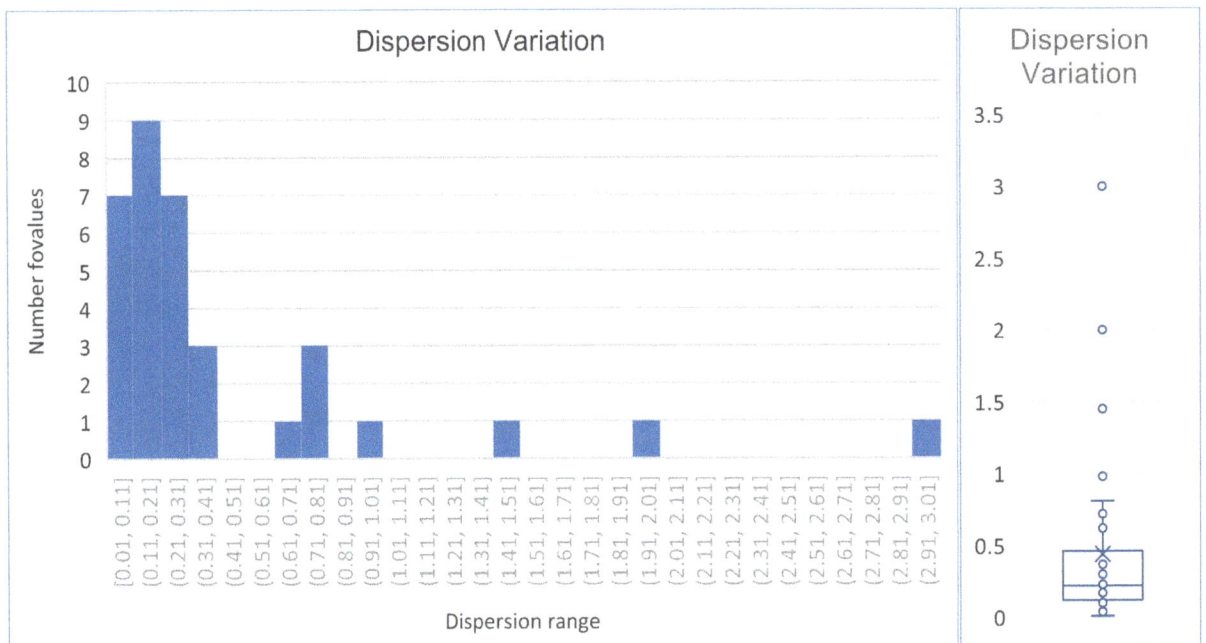

FIG 3 – Dispersion variation.

FURTHER RESEARCH

Factor affecting dispersion factors

While the dispersion factor is often somewhat arbitrary, it is influenced by measurable values like the airflow at the face or the distance between the ventilation duct and the face. Unfortunately, no records are kept of the distance to ventilation ducts making it hard to study. Regular measurement of those parameters would need to be taken to study the impact of those factors. The use of those factors will remove significant guesswork from the blasting simulation.

Production blasts

Production blasts present unique challenges to the prediction of blast fume clearance due to close void and lack of flow-through ventilation. Calculation of dispersion and resistance of broken rock would be the key parameters to study and compare to real-time measurements.

Live data analysis

As mines are becoming more connected, more data is available for research and analysis. Large-scale data gathering could be implemented on-site providing regular statistics on past blasts and estimates for future ones.

Purely statistical analysis can be performed where only the gases are analysed. Blast clearance time could be recorded every day, building a database which can be used to predict blast clearance time and highlight problem areas.

A better approach would be to couple gas readings with blast information and simulations. By combining the gas, airflow and explosive data, yield and dispersion can be calculated. Blast simulation could then be run automatically pre-blast using mean and 90th percentile values to estimate the time of re-entry and automatically combine it in a daily report.

CONCLUSION

This study highlights the high variation of explosive yield factors and to a lesser extent the dispersion factors. To be used efficiently on-site, the use of manufacturer information is a good place to start but calibration of explosive yields through comparison with measured data is recommended.

The results for the dimensionless dispersion factor used in the Ventsim presets are suggested in this study at 0.20, well below the default slow setting (1.0). More work is ongoing to determine the factors influencing the dispersion. This initial study suggests a correlation between airflow at the face and distance to the ventilation duct., however, the number of examples studied is not yet sufficient to make further conclusions.

Considering the high variation in parameters, larger sets of blast data analysed directly on-site through an automated data collection system such as SCADA or Ventsim CONTROL would be beneficial. A higher number of data points could be processed automatically leading to greater confidence in the likelihood of the different yields and dispersion factors. It would then be possible to run a daily blast clearance time simulation automatically and report on the re-entry time probability. This data could form part of a KPI measurement and improvement system, indicating the efficiency of the ventilation and clearance, and the productivity improvements through faster re-entry times.

Due to the variability of data, computer-modelled reentry times must not be considered a sole source of information to determine safe re-entry, which should still be backed up with remote sensor recordings or manual inspection with gas sensors.

REFERENCES

Kocsis, K C, Parrish, J, Teixeira, M and Scalise, K, 2019. Reducing Heat Induced Health and Safety Problems In Unddeground Metal Mines By Means of New Technologies and Renewable Energy Based Cooling System, *Canadian Institute of Mining, Metallurgy and Petroleum*, pp 336–345.

Michelin, F, Stewart, C, Griffith, M D and Andreatidis, T, 2019. Calibrating model airway size and resistance with survey asbuilt data, in *Proceedings Australian Mine Ventilation Conference 2019*, pp 363–368 (The Australasian Institute of Mining and Metallurgy: Melbourne).

Pinedo, J and Torres Espinoza, D, 2019. Implementation of Advanced Control Strategies Using Ventsim Control at San Julian Mine in Chihuahua, Mexico, *Canadian Institute of Mining, Metallurgy and Petroleum*, pp 435–441.

Rowland III, J H and Mainiero, R J, 2000. *Factors Affecting Anfo Fumes Production*. International Society of Explosives Engineers, pp 163–174.

Stewart, C, 2014. Practical prediction of blast fume clearance and workplace re-entry times in development headings, *International Mine Ventilation Congress, IMVC2014*, The Mine Ventilation Society of South Africa.

Cooling on demand – the next step in ventilation automation

F C D Michelin[1], C M Stewart[2] and A R Loudon[3]

1. Managing Director – Howden Ventsim, Howden, Brisbane Qld 4169.
   Email: Florian.michelin@howden.com
2. Principal Engineer, Howden, Brisbane Qld 4169. Email: Craig.stewart@howden.com
3. Mine Cooling Specialist, Howden, Rydalmere NSW 2116. Email: alan.loudon@howden.com

ABSTRACT

As mines get deeper, the critical underground temperature limits will be exceeded, and an increasing number of mines will need to implement, expand or modify cooling systems. Surface cooling plants have many advantages over underground cooling such as efficient heat rejection, and lower relative capital cost; however the delivery of cooling through long underground distances can result in lost refrigeration capacity before the chilled air reaches the areas it is needed most. Refrigeration plants are therefore often overdesigned in size and capacity to compensate for underground losses.

More mines are now utilising real-time monitoring and automated control of ventilation airflow and fan systems to ensure the conditions underground are safe and productive, and power consumption is efficiently managed. There may be significant opportunities to extend ventilation monitoring and control into the refrigeration delivery and control system.

Energy reduction is an effective way to reduce costs and CO_2 emissions, which is an increasingly important part of corporate governance. More efficient refrigeration control is a further step toward reducing energy consumption. This paper looks at an added step in how real-time monitoring can assist in controlling the output and chilled air delivery of a cooling plant. The requirements, implementation, benefits and challenges will be discussed.

INTRODUCTION

Most cooling plants are designed based on worst-case mining scenarios and operated at constant air and temperature output. While this process takes into consideration the change in temperature outside of the mine it does not cater for underground changes, such as the current level of diesel activity or the daily location of different mining cycles.

Ensuring cooling is efficiently distributed has many safety and financial benefits. Energy efficiency in coal or gas-powered mines plays a key part in the level of CO_2 emissions and the impact of the operation on the environment.

Several operational strategies can be implemented to adjust the cooling output and resultant load on the powerplant based on underground conditions. For example, the air used in an area can be selected to come from either a chilled air shaft when personnel would be openly exposed to hot conditions or from non-refrigerated sources when work is done remotely or operators are protected in airconditioned vehicles.

BENEFITS

Safety

Working in a hot and humid environment creates a risk of impaired performance, heat illnesses and heat strokes. Working in hot conditions has been demonstrated (Kampe *et al*, 2016) to contribute to unsafe acts and poor judgement in the workplace, leading to injury.

Risk management procedures can help reduce the risk of heat related incidents such as regular workplace measurements, ensuring hydration, acclimatisation, rest periods (Figure 1) and wearing appropriate clothing. However, those measures ultimately have an impact on productivity and temperatures increase.

Work-Rest Regimen	Work Load		
	Light	Moderate	Heavy
Continuous work	30.0	26.7	25.0
75% Work / 25% Rest, each hour	30.6	28.0	25.9
50% Work / 50% Rest, each hour	31.4	29.4	27.9
25% Work / 75% Rest, each hour	32.2	31.1	30.0

FIG 1 –Typical heat management procedure resulting in lower productivity.

While a lower temperature environment underground is generally desirable, activities conducted in protected areas or air-conditioned vehicles do not require as low temperatures. This is even more relevant when considering remotely operated machines, where the operator is separated from the hot environment.

Most cooling plant design and underground operating activity does not currently consider the option of operating different areas at different limiting temperatures at different times. Implementing this design philosophy would encourage a reduction in average cooling plant output as well perhaps as size.

Cost

Most mines still operate their ventilation fans and controls on a fixed and unchanging basis. The equipment is set to specific set points and operates that way regardless of short-term changes in activities. It has been demonstrated that automation of fans and regulators in mines can result in large power savings (Costa and Margarida da Silva, 2020; Chatterjee *et al*, 2015).

Cooling plants are often operated on a fixed temperature output, regardless of the requirement underground. This process takes into consideration the change in temperature outside of the mine but does not cater for underground changes.

A growing number of mines have invested in costly communication and tracking of underground equipment. The utilisation of this data to help automate ventilation can be a very cost-effective expansion of capability with a rapid return on investment.

Environment

Reducing power usage is a cost-effective way to combat global warming and reduce CO_2 emissions where the mine supply is by coal or gas-fired power stations. CO_2 emissions are now a standard reporting performance indicator for many mining organisations, with both a corporate and public expectation of reducing carbon footprints where possible (Fakoya and Chitepo, 2019). Shareholder valuation, community expectations and even the level of government support are increasingly dependent on expectations of a company's efforts to reduce carbon output.

REQUIREMENTS FOR AN AUTOMATED SYSTEM

Communication

The first element of automation is a reliable digital communication system to allow centralised monitoring and control of the various parts of a mining system. For ventilation, the communication network is the backbone of sensor reading, control of underground ventilation fans and controls, and vehicle tracking to flag the need and location of ventilation. The presence of an existing communication makes the implementation of automation a much simpler process, and if not present, must be included in any ventilation automation proposal.

Sensors

Control of the underground ventilation environment can only be performed if the current conditions are measured and connected to the communication backbone that transfers real-time data to a centralised control location. Measured data should include airflow, temperature, relevant gas concentration, and if possible, location of equipment and activities. As well as instantaneous ventilation data, sensors also offer trends in historical data which can assist in ventilation planning and improvement. Access to live underground readings does not replace regular ventilation surveys which are still required to confirm the accuracy of the readings and highlight any faulty sensors.

Controllers

Ventilation automation requires the use of a local control device, connected to a centralised control communication system. Programmable logical controllers (PLCs) and microcontrollers are the preferred methods of interfacing a control to sensors and a communication system, assisting in standardising control communication protocols, and providing a level of local intelligence. PLCs are generally considered a robust industry standard for logic controlled applications, whereas microcontrollers can achieve more complex tasks and can be programmed to offer standalone computer-like capabilities (such as displaying a web interface page for settings and values for example).

Controllers enable the equipment to be controlled from the surface and allow the status of the equipment to be read. Without an intelligent local PLC or microcontroller, if there is a loss of communication, the equipment will normally either switch off or remain at its last status unless changed manually.

The use of a PLC offers additional redundancy in case of loss of communication. The PLC can take over control and set the equipment to a default position or an optimum position based on local sensor inputs. This adds a layer of redundancy to the system.

PLC programming for the equipment typically includes:

- Specific tags for control.

- A default setting in case of loss of communication.

- Possibility to lock the device from external control for maintenance.

- Potential to maintain flow-based control.

Regulators

Many mines are equipped with drop-board or manually louvred regulators. While it is possible to automate a ventilation system without automated regulators (for example by adjusting the secondary ventilation or circuit fans), manual controls will still require constant adjustment to achieve optimum ventilation conditions. While more expensive, automated louvres have the following advantages:

- The current status is visible from the surface, ensuring that activities are effectively ventilated.

- Adjustment is performed without the need to travel to the location, saving man-hours.

Automated actuators can be directly controlled from a communication switch, or preferably with a PLC or microcontroller that can be used to provide some local intelligence, particularly if the communication system fails.

Primary and secondary fans

Primary fans are responsible for delivering air and cooling to mining activity regions, while secondary fans are responsible for delivering the cool air to the working face or location. Together they usually form the largest power consumer in a ventilation system, often in equal amounts to each other. Remote control of fans can be performed by connecting controllers to the fan starters. Control of secondary fans in particular offers a high degree of local control and the potential for significant power savings. Secondary fans are often left turned on even when there is no mining activity, and therefore result in considerable power consumption, heat generation and unnecessary use of chilled air.

A ventilation control automation system has the opportunity to link secondary fan demands to primary fan demands. Cooling systems are intrinsically linked to primary ventilation with the overall flow often dictating the amount of cooling that can be transported underground. Changing the airflow available to a cooling plant influences cooling output (if a set discharge temperature is adopted). The control of primary and secondary fans, therefore, allows the cooling plant to effectively become a 'cooling on demand' system without any additional control methodology.

Cooling plant

Cooling plants are usually already highly automated and controlled with minimal manual intervention again using microcontrollers and PLCs to deliver optimum plant performance. However, the targets for plant output are often simply a target output temperature or scheduled control of daily and nightly outputs. Demand output is normally restricted to respond to variances in atmospheric temperatures and changes to airflow through the cooler due to underground demand. A controlled system integrated into other underground devices through a control system permits much more sophisticated control of output, including predictive analysis of future cooling demand, responsiveness to underground activities generating additional heat, and adjustment of cooling output depending on where the demand is required (for example higher or lower in the mine).

Control system

A control system is typically a server-based software system interfacing all control components into an intelligent decision-making system that both reads and control all devices to achieve optimal outcomes (Figure 2). Systems typically employ different algorithms such as PID (proportional integral derivative) loops to achieve targets, energy optimising and predictive analysis of real-time data.

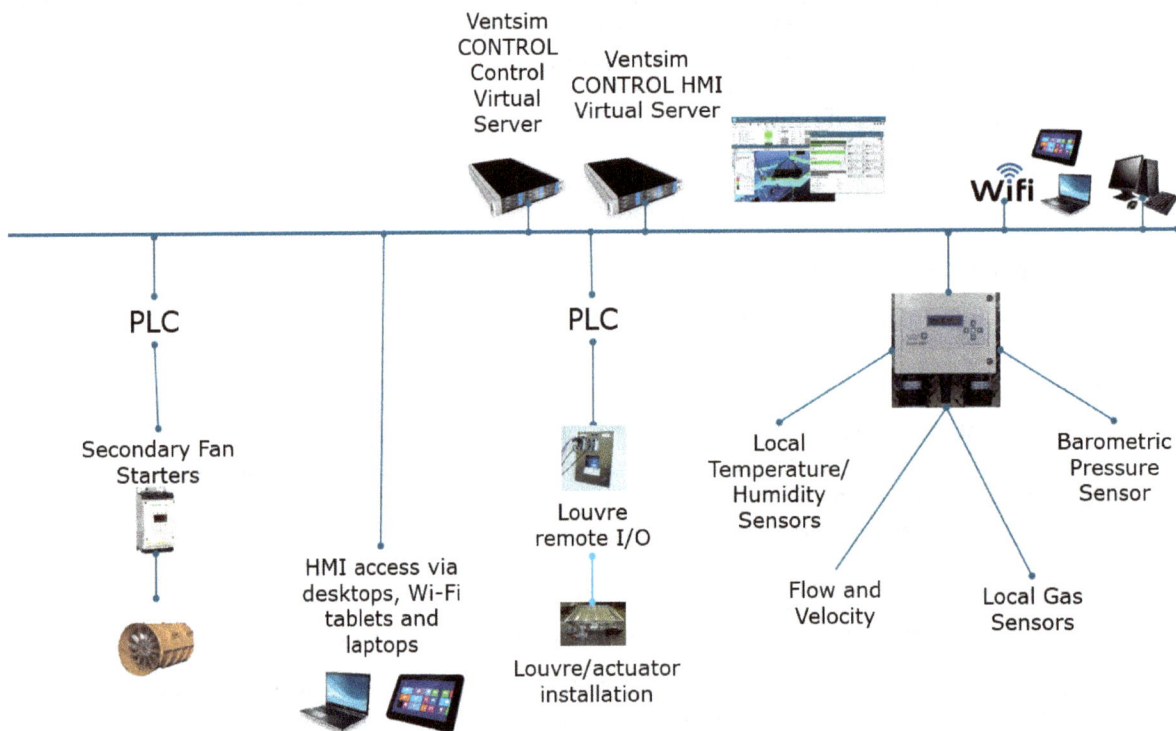

FIG 2 – An example of a control system architecture.

EXAMPLES OF ADVANCED AUTOMATED COOLING STRATEGIES

Direct chilled air supply to auxiliary fans

A simple way to implement cooling on demand to ventilate each level is by using an auxiliary fan sourcing fresh chilled air from a refrigerated shaft through a sealed bulkhead. This method results in only the air required by each fan being cooled. Turning off or reducing the speed of an auxiliary fan in the circuit will impact the total airflow drawn through the shaft and at the cooling plant. If the cooling plant is set to a fixed temperature output, this will automatically reduce the power required to cool the air to the required temperature.

This system would be especially efficient where auxiliary fan automation is controlled by activity schedules or equipment tracking, where the airflow at the face is set to the minimum required for each type of activity.

Pause supply during inactivity

Savings can be achieved by turning off the cooling plant during blasting or shift changes, providing the thermal stored mass and delay of cooling delivery is considered. While the ventilation system is still required to clear blast fumes and remove contaminants, the requirement for chilled air when personnel are not present is greatly reduced.

Table 1 above considers the plant being fully off or idled at minimum power for two hours when active hot faces no longer have personnel present and personnel are either travelling or have retreated from the mine. While the time taken for resumed cold air to reach the face needs to be considered as well as the loss of some chilled thermal rock mass, the delay can be factored to allow the chiller to be reduced in advance of the mine being evacuated for shift change and blasting. As a result, the strategy may involve reducing or turning off refrigeration (for example) one hour before evacuation and resuming refrigeration one hour before workplace faces are re-entered. The strategy may be limited in some mines with extreme depth or heat problems where cooling limits are being reached.

TABLE 1

An example of turning off the cooling plant during inactive hours.

Cooling plant power MW	Blasting/shift change hours	Shifts/blasts per day	Savings in kWh (COP = 4, 50% util)	Power cost $/kWh	Saving in $ per annum
10	2	2	1 825 000	0.15	$273 750

Dynamic simulation can be used to assess the transient effects of such a strategy. Figure 3 shows the simulated result in an example mine of turning off the cooling plant for two hours during blasting twice a day at the base of the chilled air delivery shaft. It must be noted that these results are likely to differ from other mine models which may have different configurations. The following observation can be made when looking at the temperature in the fresh air shaft at 1100 m depth:

- The immediate temperature change from turning off the cooling plant is relatively low in the afternoon (from 17°C to 19°C WB) due to the surrounding cooled airway rock continuing to provide some temporary thermal mass cooling even without the cooling plant running.

- When the cooling plant restarts it must rechill the warmed thermal rock mass, delaying the return to the slightly cooler temperatures of a continuously running plant. In this example, it takes 2 hrs for the air to cool back down.

- The simulated reject temperature showed little variation over the period studied.

- Shallower mines will see a more rapid change in underground conditions due to shorter chilled air delivery times.

The small temperature change in this example may still be too much for mines experiencing limiting temperature conditions, however, in many cases, the variance may be acceptable and barely noticeable.

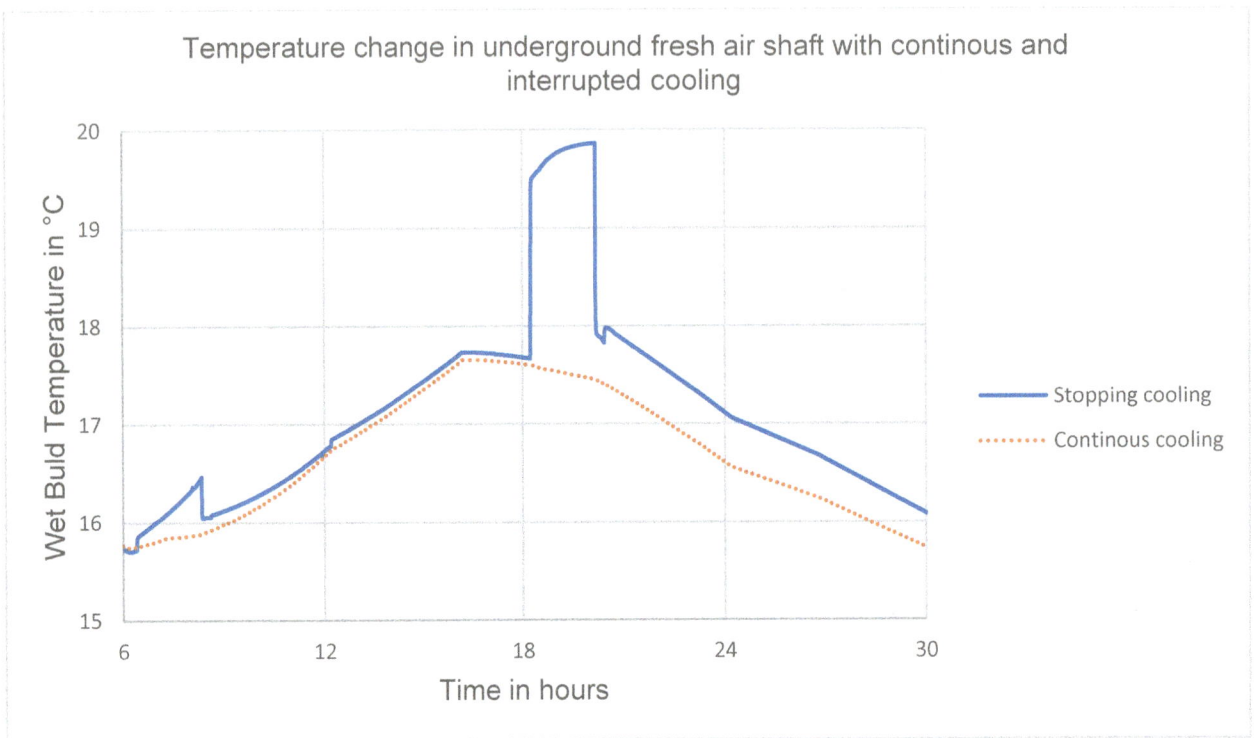

FIG 3 – Change in wet bulb temperature following stopping of the cooling plant during blasting.

Chilled air selectivity

While all activity requires ventilation to clear fumes, some activities may accommodate different temperatures and levels of cooling. For example, when people are working in air-conditioned cabins, or only travelling through warm areas and not actively working, the outside air temperature may not affect or impact workers' health and productivity.

This type of arrangement may be possible by designing a system using multiple sources of air. For example, a working level with an auxiliary fan could draw air from a chilled air shaft during activities requiring cooling, or from warmer air on the main ramp during airconditioned activities such as stope production. The system could incorporate a dual duct and fan system (from each source) into a common duct, or a more complex damper system utilising only a single fan to draw from either source. A blend of sources may also be possible to achieve the desired temperature.

Temperature sensors can also be used on the exhaust side to modulate the damper and achieved the required temperature of air drawn through a region. This becomes especially relevant if chilled air is cooler than required and airflow is more than what is required for current activities. When temperatures rise, dampers can be opened and airflow (and cooling) increased.

For the above scenarios, it is important to keep in mind the risk of workers being exposed to hot conditions in unexpected circumstances – for example, if they leave their airconditioned vehicle or are required to travel by foot in cases of emergency.

Controlled recirculation

Controlled recirculation is the process of reusing a portion of exhaust air underground (Hall *et al*, 1990). It offers the advantages of lower resistance in the system compared to exhausting the full flow of air to and from the surface and may also assist with cooling requirements if the air does not need to be fully cooled again. Dust, diesel particulate matter (DPM) and gases such as carbon monoxide (CO) and nitrogen oxides (NOx) still need to be controlled to ensure a safe environment.

A control system is essential to ensure safe conditions are maintained by decreasing recirculation and increasing fresh air intake if gases are high (during blasting for example). Similarly, the control system could recirculate a greater portion of air while it is cool but increase the chilled air intake if the exhaust air is too warm or not of suitable quality.

Thermal stored rock mass cooling

Once a cooling plant is stopped, the rock mass surrounding the immediate airway remains cool for some time and can continue to provide some cooling (Stewart *et al*, 2017). This phenomenon can be used to 'over' cool the mine at night when the outside temperatures are colder, and the plant can run more efficiently. In addition, lower power tariffs are often available during nighttime periods. During the day, the cooled thermal rock mass can help maintain cooling even if plant output is reduced. Some cooling systems are also designed to build a supply of chilled water at night in tanks, for additional cooling during the day, another form of thermal stored mass cooling.

Reject temperature control

While cooling plants are often designed on the concept of an assumed workplace reject temperature, the plants are rarely controlled on this basis. As a result, while the reject temperatures are maintained during peak hot periods, for significant other periods the reject temperatures will be substantially less than design conditions, effectively wasting cooling. In the case of an overdesigned cooling system (or one that does not require full output all year round), the cooling plant output could therefore be based on the underground reject temperatures, allowing underground reject air temperature sensors to control output instead of using set chiller output delivery temperatures.

While thermal rock mass and airflow delivery times created significant delays in sensor feedback, this can be overcome with schedules, and trends in temperature measurements to provide triggers for varying cooling output. Maintaining a more constant reject temperature can lead to power savings due to the reduction in wasted chilled air output.

While it is recognised that reject temperatures will be different in different mining zones, the limiting temperature in the warmest zone can be prioritised for the controlled chiller output. When coupled with other cooling on-demand strategies mentioned above, overall mine cooling output can be further optimised to achieve a goal of similar reject temperatures in all mining zones (and hence the lowest potential cooling plant output).

CONCLUSION

Cooling on demand is a concept that already exists in many refrigerated mining environments, however it is often simply the resultant outcome of other unrelated ventilation strategies and is rarely pursued as a design concept in its own right. While varying flow and controlling chilled air distribution is commonplace, taking the next step in evaluating the instantaneous need for cooling and dynamically adjusting the cooling system to suit the immediate local requirements has not been generally considered.

In the past when mines were poorly monitored and ventilation automation did not exist (beyond manual control), a 'set and forget' design philosophy for cooling was understandable. With modern mines embracing improved communication, monitoring and control automation technology, cooling on demand as a design goal becomes more attractive and can potentially result in significant power savings and cost reductions, while promising improved conditions underground, particularly in mines constrained by hot conditions.

REFERENCES

Chatterjee, A, Zhang, L and Xia, X, 2015. Optimization of mine ventilation fan speeds according to ventilation on demand and time of use tariff, *Applied Energy*, 146:65–73.

Costa, L D V and Margarida Da Silva, J, 2020. Cost-saving electrical energy consumption in underground ventilation by the use of ventilation on demand, *Mining Technology*, 129:1–8.

Fakoya, M B and Chitepo, K T, 2019. Effect of Corporate Environmental Investments on Financial Performance in Mining and Manufacturing Companies Listed on the Johannesburg Stock Exchange Social Responsibility Index, Acta Universitatis Danubius: Oeconomica, 15 p.

Hall, A, Mchaina, D and Hardcastle, S, 1990. Controlled recirculation in Canadian underground potash mines, *Mining Science and Technology*, 10:305–314.

Kampe, E O, Kovats, S and Hajat, S, 2016. Impact of high ambient temperature on unintentional injuries in high-income countries: a narrative systematic literature review, *BMJ Open*, 6:e010399.

Stewart, C, Aminossadati, S M, Kizil, M S and Andreatidis, T, 2017. Diurnal thermal flywheel influence on ventilation temperatures in large underground mines, *The 16th North American Mine Ventilation Symposium*, 8 p.

Transition of mechanical cooling refrigerants

T Roberts[1] and K McLean[2]

1. Mining Operations Manager, Gordon Brothers Industries P/L, Brunswick Vic 3056.
   Email: travis.roberts@gordonbrothers.com.au
2. MAusIMM, Mining Director, Gordon Brothers Industries P/L, Brunswick Vic 3056.
   Email: keith.mclean@gordonbrothers.com.au

ABSTRACT

The transition of mechanical cooling refrigerants from their inception in the 1800s through to the current day has been driven by several factors.

Original refrigerants such as sulfur and carbon dioxide, chloromethane and ammonia, whilst effective came with toxicity and/or flammability issues. In 1930 DuPont trademarked a new synthetic refrigerant known as Freon® 12, this development was the introduction of 'chlorofluorocarbons' (CFCs).

Throughout the 1930s and 1940s additional synthetic refrigerants were developed, namely Freon® 11 and Freon® 22. Due to their non-toxic and non-flammable characteristics, synthetic refrigerants were extensively used for domestic, commercial and industrial applications for the next 60 years.

In 1985, one of the first scientific papers suggested a potential mechanism linking CFCs and a reduction in the density of the ozone layer. In 1987 the Montreal protocol is ratified; it is the first UN treaty to be ratified by every country on earth. Due to CFC's ozone depletion potential (ODP), the treaty required the consumption of CFC's to be reduced by 50 per cent over the next ten years. By 1996 the phase out of CFC's (Freon® 12) is completed, the phase out of HCFC (Freon® 22) refrigerants comes into effect.

On 1 January 2018 the phase down of HFC's begins. Whilst having zero impact on the ozone layer HFC refrigerants have high global warming potential (GWP). As we transition through the HFC phase down (2018–2035) we are beginning to see the emergence of what some consider to be the fourth synthetic wave, hydrofluoro-olefins (HFO's).

HFO's and their blends have zero ODP and low GWP's, ranging from 150 to less than 1. The reduction in GWP with HFO's typically comes at the expense of flammability, which is not ideal. Each refrigerant has its optimal application, therefor when considering refrigerant choices for mining systems, a complete assessment of the environmental, health and safety, efficiency and commercial impacts should be considered now more than ever.

INTRODUCTION

All modern refrigeration systems use some form of working fluid, a refrigerant which facilitates the refrigeration process. The benefits of refrigeration are all around us, it is applied in the production and storage of many of the foods and beverages we consume. Refrigeration is used to air-condition our homes, our vehicles, offices and the sleeping quarters at mining camps. In the situation where underground mines experience heat issues, the appropriate application of refrigeration is critical to facilitating a safe and productive work environment.

There is no such thing as the perfect refrigerant. The selection of the most appropriate refrigerant for any application requires careful consideration of a range of factors. NASA scientists reviewed over 860 refrigerants and despite its toxicity, selected ammonia as the best and most appropriate for use in the space station (www.iiar.com).

Synthetic refrigerants have been associated with some of the world's greatest environmental issues, causing damage to the ozone layer (CFC's and HCFCs) and directly contributing to global warming due to their high global warming factors (CFC's, HCFC's and HFC's their replacement). The latest synthetic refrigerants HFO's, are due to replace HFC's and while having zero ozone depletion and low global warming factors, HFO's are not a perfect solution and come with their own challenges. Some in the refrigeration industry fear that as with CFC's and HFC's, it's only a matter of time before HFO's are also found to present with environmental issues of concern.

Natural refrigerants (water, ammonia, carbon dioxide, hydrocarbons) have been used across the 100+ years that refrigeration systems have been applied. Ammonia has been used continuously and is still the refrigerant of choice for the vast majority of industrial applications across the world. It has also been used extensively in many surface mine cooling systems. All forms of natural refrigerants are being promoted and are being used in an ever increasing number of applications as a replacement to synthetic refrigerants. However natural refrigerants are not perfect and have a range of issues that must be suitably managed, such as high pressures, toxicity and flammability.

The best refrigerant for any application requires consideration and should take into account those directly impacted as well as the effects on the wider environment. Despite the broader environmental benefits that come with natural refrigerants, some may be unsuitable for use in specific applications where local factors of toxicity or flammability are an issue. In many applications these local aspects can be suitably managed. It is the consideration of this broad range of factors, that facilitates the best overall refrigerant selection.

A BRIEF HISTORY OF REFRIGERATION AND REFRIGERANTS

Air could be viewed as the first refrigerant, particularly in cold climates. The harvesting of ice and snow were some of the earliest methods of 'refrigerating' food and goods, however these primitive methods were somewhat limited by geographical locations, coupled with seasonal availability of ice/snow (Pearson, 2003).

Throughout the early 1800s particularly in the USA, natural ice was becoming an ever increasing commodity. As demand increased, prices were driven down as more and more ice became accessible. By the early 1900s around 40 000 tons of ice were consumed in the USA daily.

Due to the increase in demand, water quality and subsequent ice quality began to decline. Food and product contamination followed. Pollution and contamination of natural water/ice sourcing started a decrease in natural ice production.

The introduction of refrigeration with a 'working fluid refrigerant' provided significant benefit. Advancements in food production, storage and availability helped create healthier balanced diets as meat, produce, dairy and seafood became more widespread.

'The primary driving force for most new inventions is a need'. Although the new form of refrigeration provided significant benefits, the original vapour compression refrigerants came with a considerable negative effect. Numerous refrigerants were trialled with various success. Ethyl chloride, methyl chloride, carbon dioxide, sulfur dioxide, propane, iso-butane and ammonia. These can be described as the first generation refrigerants. Propane was used as an alternative refrigerant to ammonia from as early as the 1900s (Pearson, 2003).

Water and air were also used as refrigerants, however at the time the low thermodynamic efficiency of air and water's limitations due to low pressure and high volumetric flow rates, made these unviable.

The creation and introduction of the second and third generation synthetic refrigerants heralded a quantum change in safety, due to their stable, non-toxic and non-flammable characteristics (McLinden and Huber, 2020).

In the late 1970s, scientists discovered that the ozone layer had experienced an overall reduction in concentration of approximately 4 per cent per decade and by 1985 ozone levels had dropped significantly. The main cause of the depletion was traced back to the use of man-made substances containing CFC's that were released into the atmosphere from the use of aerosols and refrigerants (Baum, 2017).

TABLE 1

The challenge the refrigerant industry has with balancing safety and environmental impact at a local and global level.

Properties of early refrigerants (first generation)									
Substance	R Number	Chemical Formula	M Kg/kmol	NBP °C	CRT °C	CRP Bar	Safety Group	ODP	GWP$_{100}$
Carbon dioxide	R-744	CO_2	44.01	-55.6[1]	31.6	73.77	A1	0	1
Ammonia	R-717	NH_3	17.03	-33.3	132.25	113.33	B2(B2L[2])	0	0
Sulfur dioxide	R-764	SO_2	64.06	-10.0	157.49	78.84	B1	0	0
Ethyl ether	R-610	C_4H_1O	74.12	35	194.0	36	-	0	0
Dimethyl ether	E170	C_2H_6O	46.07	-25	126.9	53.7	A3	0	0
Methyl chloride	R-40	CH_3Cl	50.49	-24.2	143.1	66.77	B2	0.02	16

1 – triple point

2 – new class introduced since 2010

Properties of CFC and HCFC refrigerants dominant in 20th century (second and third generation)									
Substance	R Number	Chemical Formula	M Kg/kmol	NBP °C	CRT °C	CRP Bar	Safety Group	ODP	GWP$_{100}$
Trichlorofluoro-methane	R-11	CCl_3F	137.4	23.71	197.96	44.1	A1	1	4000
Dichlorodifluoro-methane	R-12	CCl_2F_2	120.91	-29.75	111.97	41.4	A1	1	8500
Chlorotrifluoro-methane	R-13	$CClF_3$	104.5	-81.3	29.2	39.2	A1	1	11 700
Chlorodifluoro-methane	R-22	$CHClF_2$	86.47	-40.81	96.15	49.9	A1	0.055	1700
R22/R15	R-502	$CHClF_2 + CF_3CClF_2$	111.6	-45.3	80.73	10.2	A1	0.33	5600

The synthetic refrigerants (CFC's second generation, HCFC's third generation) that were created to improve safety, subsequently were found to damage the environment. In late 1987 Montreal protocol came into effect with the aim of phasing out ozone depleting substances. This saw the phase out of R11 and R12, which typically were the most prolific synthetic refrigerants of the time due to their versatility across domestic, commercial, and Industrial applications.

As part of the phasing out of ODP refrigerants, another group of substances, hydrofluorocarbons (HFCs) were introduced as non-ozone depleting alternatives to support the timely phase out of CFCs and HCFCs. HFC blends such as R404A, R407C, R410A quickly expanded into the industry replacing CFC and HCFC refrigerants in predominately domestic air-conditioning and commercial systems.

The Kyoto Protocol and Kigali amendment came into effect with the focus of phasing down HFC refrigerants with high global warming potential, or GWP. The GWP of a gas refers to the total contribution to global warming resulting from the emission of one unit of that gas, relative to one unit of the reference gas, CO_2, which is assigned a value of 1. Figure 1 defines the impact greenhouse gases will have on global warming over different time periods or time horizons. These are usually 20 years, 100 years, and 500 years. A time horizon of 100 years is typically used as a reference point (Environmental Protection Agency, Victoria, www.epa.vic.gov.au).

Whilst not in the same league as R13 or R23 (11 700 and 14 800 GWP respectively), the third generation refrigerants present a significant risk to the environment due to their considered high GWP's.

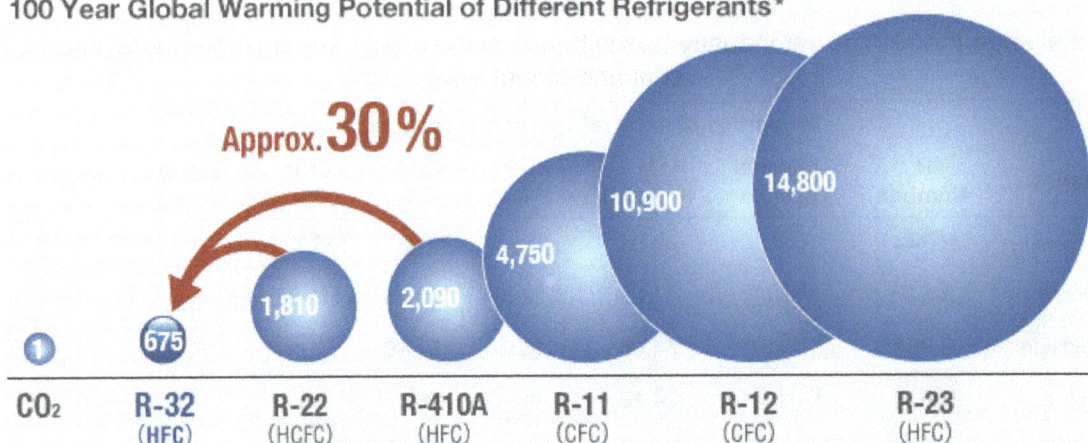

FIG 1 – Global warming potentials of various refrigerants (Daikin Industries Ltd, 2022).

Throughout the journey, the industry has been presented with what some may say is 'Newton's third law of motion'. That is, any action to improve the safety at a local level has an equal and opposite reaction at a global level.

Now the industry finds itself transitioning to what some consider the fourth generation of refrigerants, hydrofluoro-olefins (HFO's) and their blends. HFO refrigerants are categorised as having zero ODP and very low GWP, offering in principle a more environmentally friendly alterative to the previous CFC, HCFC and HFC refrigerants.

The atmospheric lifetime of HFO-1234yf is estimated to be about 11 days compared to HFC-134a (R134a), which has an atmospheric lifetime of 14 years. There is already debate over the rapid 100 per cent breakdown of some HFO's in the atmosphere and potential for localised higher concentrations of trifluoroacetate (TFA) in rainwater due to the shorter degradation time (Dudita and Kauffeld, 2021).

History has shown that safety and protection of the environment are for all attempts, somewhat mutually exclusive. A compromise on safety results in a reduction to the environmental impact and vice versa.

Unlike the past where single refrigerants were versatile and could be used in domestic fridges, room and office air-conditioners through to industrial applications, we are seeing a narrowing of applications which refrigerants are suitable for. A1 classification HFO blends such as R1233zd are more suitable for large centrifugal chillers. On the other hand, there are currently limited candidates from the HFO group with similar volumetric refrigerating capacity, such as R22/R407C, R404A/R507A and R410A available for Industrial use.

THE FUTURE AND WHERE WE ARE HEADED

Figure 2 and Table 2 provide an indication of the timeline associated with the development and use of various refrigerants and the associated driving forces for change. Note the return to greater use of natural refrigerants almost 200 years from when they were first widely used.

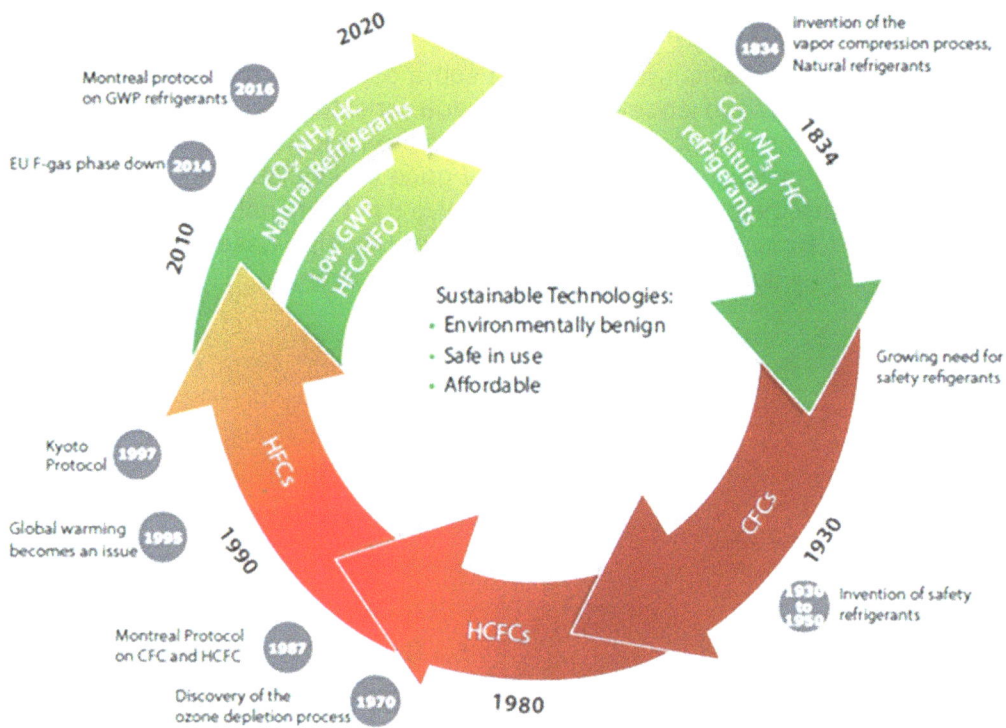

FIG 2 – Timeline of refrigerant development and use (Environmental Sustainability Rotary Action Group, 2017–2022, www.esrag.org).

TABLE 2

Changes in use of synthetic and natural refrigerants from the 1980s.

	Synthetic refrigerants	**Natural refrigerants (most still require manufacturing)**
1980s	CFC and HCFC	NH_3 and HD
1990s	HCFC and HFC	NH_3 and CO_2 and HC
2000s	HFC and HCFC	NH_3 and CO_2 and HC
2010s	HFC and HFO	NH_3 and CO_2 and HC and H_2O
2020s	HFO and `	NH_3 and CO_2 and HC and H_2O
2030s	?HFO?	?NH_3 and CO_2 and HC and H_2O?

Grey indicates disappearing from general use and red shows that general use is increasing

(HC= HYDROCARBON, HFO = hydrofluoroolefin, NH_3 = ammonia)

In the mining industry R134A would be considered the most prevalent refrigerant in recent times. The non-flammable, non-toxic characteristics enable R134A to be used close to an intake shaft or for underground localised cooling applications.

R134A is currently being phased down in Australia, as part of the Kigali agreement. As we approach the year 2035, the availability of R134A will reduce, resulting in a price increase therefore providing financial incentive to retrofit existing systems with an alternative refrigerant or replace with new equipment.

Carbon pricing policy's adopted by previous Australian governments, led to large increases in HFC refrigerant pricing. This resulted in some mine cooling chillers having refrigerant charges with values up to $500 000. While these policies have been reversed, there is always the possibility they may be reintroduced in some form.

WHAT IS THE BEST ALTERNATIVE TO R134A AS WE APPROACH 2035?

The decision is complex and inevitably comes with some compromise and a great deal of thought given to efficiency, safety, cost and longevity. Without due consideration, transitioning back to natural refrigerants, in particular ammonia for mining may seem the logical choice. As a natural refrigerant it ticks the ODP and GWP boxes, its efficiency is very good and its longevity is proven. However its safety and system cost are the primary consideration. We can't eliminate the toxicity or flammability yet it can be managed through comprehensive site-specific risk assessments, well-engineered systems and maintained by suitably trained and competent personnel.

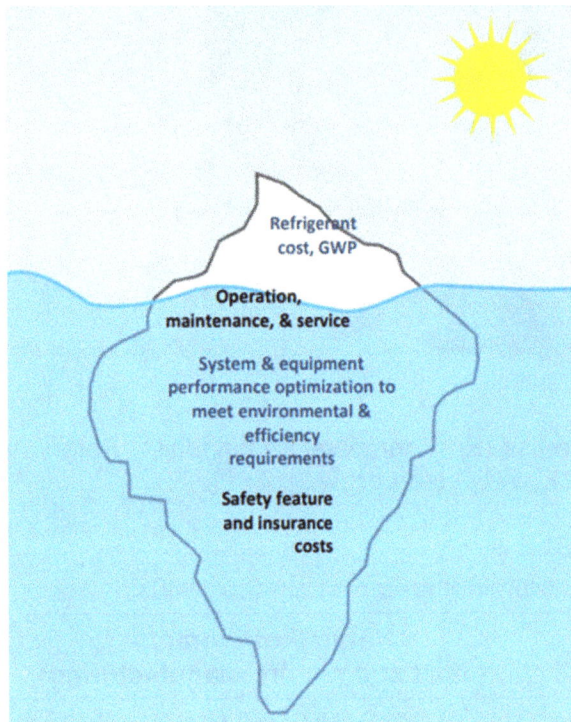

FIG 3 – Hidden and total costs of refrigerant and associated system.

For synthetic refrigerants, the direct global effect is its GWP rating. This only occurs when refrigerant leaks into the atmosphere. Improvements in design, manufacturing, technology and regulations, along with changes to maintenance practices and training, have significantly reduced the risk of leakage.

A typical large, water-cooled chiller has a R134A refrigerant charge of 1100 kgs (3000–4000 kW). Based on Table 1, the direct impact of the refrigerant could be around 100 000 kg of CO_2 if 7 per cent typical, (note the authors believe this can and should be much less) of the chiller refrigerant charge leaked to atmosphere each year of its life cycle. In isolation, 100 000 kgs of equivalent of CO_2 appears very concerning however the indirect effect of refrigerants, ie emissions that result from generating the electricity used by the chiller, is significantly more. The energy efficiency of the chiller and system overall can be an even more important consideration.

TABLE 3

Industry annual leak rates by equipment class/application (IIAR, no date).

Annual leak rates by equipment class/application

Equipment class/application	Annual leak rates (% pa)		
	Lower	Typical	Upper
Refrigeration applications			
Centralised system (ie supermarket rack)	5%	Maintained = 12.5%, otherwise 15%	23%
Chillers (ie cold storage facilities, process refrigeration)	5%		16%
Condensing units (ie Walk-in coolrooms)	5%	Maintained = 12.5%, otherwise 15%	23%
Self-contained refrigeration system	-	2%	-
Road transport	15%	20%	23%
Marine	20%	30%	40%
Air conditioning applications			
Chillers	5%	7%	9%
Chillers (HCFC-123)	-	2%	-
Roof top packaged system	4%	5%	9%
Split systems (single and multi)	3%	4%	9%
Window/wall units and portable	-	2%	-

NOTE – The authors believe that an optimally designed, configured and maintained system should achieve leakage rates much less than the figures in Table 3. Many mining cooling systems that are optimally configured and maintained, have leakage rates approaching zero.

Coefficient of performance (COP) is a measure of chiller efficiency and is the ratio of refrigeration kW provided divided by the power input kW.

A typical water-cooled chiller designed to operate at 'standard' conditions, generally has a COP of around 6.4, (6.7°C leaving chilled water temperature LCWT, 29.5°C entering cooling water temperature ECWT).

With a detailed analysis of the geographical ambient conditions, cooling capacity required, water flows/temperatures at full and part load efficiency, it's not unreasonable to improve the chiller COP to 9 or higher (11°C LCWT, 25.5°C ECWT).

Excluding any ancillary electrical loads (pumps, fans etc), the higher efficiency chiller over a typical 5000 hr annual run time could consume around 1.1 M kWh creating a CO_2 footprint of around 753 tons/a. The typical chiller based on the same operating time and diversification (0.7) is around 1.6 M kWh and 1 100 000 kgs of CO_2.

Over a typical life cycle of 20 years with an assumed efficiency loss of zero, the additional CO_2 contribution of the typical chiller could be as high as 16 000 000 kgs of CO_2 at the end of its expected useful life (based on the carbon intensity look out below). The indirect impact of a 7 per cent refrigerant leak rate over the same period is 1540 kgs of R134A or around 2 200 000 kgs of equivalent CO_2.

The higher efficiency chiller with the same zero efficiency loss assumption is anticipated to contribute around 11 000 000 kgs of CO_2, 27 per cent less over the same 20 year period.

The calculation of CO_2 emissions associated with power consumption is based on the outlook for the power sector in Australia from 2020 to 2050 with a reduction in grams of CO_2 per kWh reducing from around 664 to 294 g/kWh over the period 2022 to 2041 (Tiseo, 2021). The reduction in CO_2

being based on lower emission forms of power generation. The reduction in CO_2 emissions from power generation, particularly from renewable sources with no emissions, provides opportunities to use refrigeration system configurations that can also considerably reduce water consumption, which is a very important consideration for some remote mine sites.

TABLE 4

Properties of current and next-generation refrigerants (Osachi, 2019).

Refrigerant for HVAC		Low Pressure			Medium Pressure				High Pressure				NH3
This table compares various properties of both current and next-generation refrigerants. The efficiencies and capacity changes shown are based on the theoretical properties of the refrigerant alone, with all design variables held constant for objective comparison													
		R-123	R-1233zd	R-154A	R-134a	R513-A	R-1234 ze³	R-1234yf	R-22	R-410a	R-452B	R-32	R717
Flammability	ASHRAE Class BV (cm/s)	Non(1) n/a	Non(1) n/a	Non(1) n/a	Non(1) n/a	Non(1) n/a	Slight (2L) 0.0	Slight (2L) 1.5	Non(1) n/a	Non(1) n/a	Slight (2L) 3.0	Slight (2L) 6.7	Slight (B2L) 7.0
Toxicity	ASHRAE Class OEL	Higher (B) 50	Lower (A) 800	Higher (B) 320	Lower (A) 1000	Lower (A) 650	Lower (A) 800	Lower (A) 500	Lower (A) 1000	Lower (A 1000)	Lower (A) 870	Lower (A) 1000	B2 25
Efficiency (COP)		8.95	8.85	8.91	8.47	8.28	8.45	8.17	8.48	7.99	8.14	8.22	8.73
GWP²		79	1	2	1300	573	1	1	1760	1924	675	677	0
Atmospheric Life		1.3 years	26 days	22 days	13.4 years	5.9 years	16 days	11 days	11.9 years	17 years	5.5 years	5.2 years	1 week

ASHRAE – The American Society of Heating, Refrigerating and Air-Conditioning Engineers

[1] None of the synthetic refrigerants shown in the table are considered 'toxic' or 'highly toxic' as defined by the IF, UFC, NFPA 1 or OSHA regulations. Ammonia is toxic can cause serious or permanent injury.

[2] GWP values reported are per the fifth assessment report (R5) of the IPCC (intergovernmental panel on climate change).

[3] R-1234ze is not flammable at room temperature, so it's BV is zero by definition. It does, however become flammable at temperatures above 30°C (86° F).

Existing HFC refrigerants, shaded blue, and their replacement lower GWP HFC's or HFO's, shaded green, ammonia shaded yellow

TABLE 5

The American Society of Heating, Refrigerating and Air-Conditioning Engineers (ASHRAE) safety classifications of refrigerants. R-134a is A1 and ammonia R-717 is B2L. Refrigerants close to intakes or underground must be A1. Ammonia due to its B2 rating must not be used underground or within 200 m minimum of intakes.

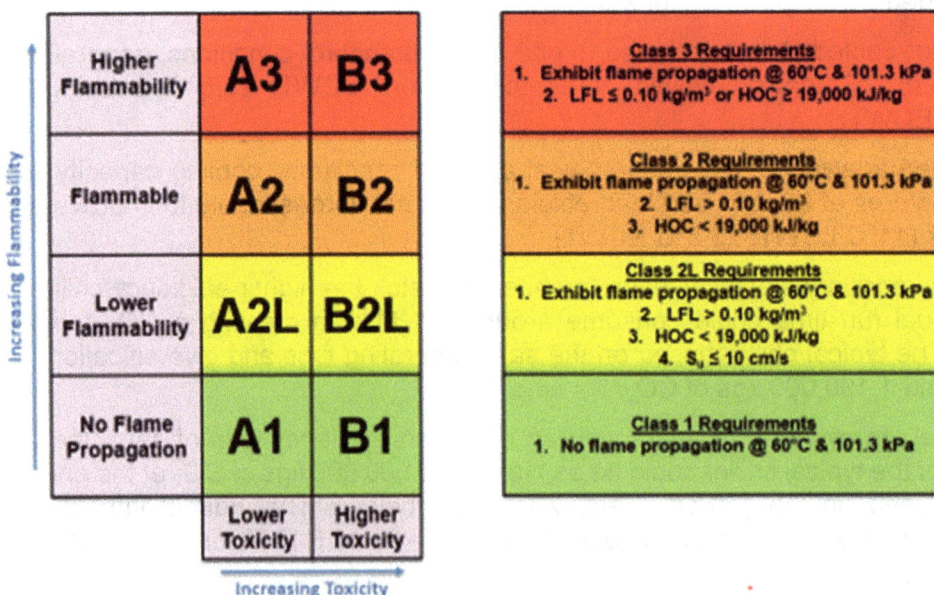

REFRIGERANT SELECTION CONSIDERATIONS FOR MINE COOLING SYSTEM

In the case of synthetic medium pressure refrigerants which cover most mine cooling systems we can see that two out of the three alternatives (R1234ze/R1234yf) have an ASHRAE classification of

A2L (mildly flammable). R513A (blend of R134a and R1234yf) which shares the same classification as R134A (A1) is currently considered a 'transitional' refrigerant due to its intermediate GWP (573).

Without playing down the safety considerations of A2L refrigerants, they typically share the same flammability characteristics as ammonia (B2L) however minus the toxicity of ammonia. In simplified terms they are not easy to ignite, require specific concentrations and have a relatively low flame speed. In well ventilated installations the flammability concerns can be largely mitigated.

A2L refrigerants do provide a good alternative to R134A for the reduction of 'indirect' equivalent CO_2 emissions for leakage. Their medium pressure also provides flexibility as these refrigerants can be utilised in either air-cooled or water-cooled chillers.

With the use of A2L refrigerants, it is essential to understand that A2L refrigerants must only be used in systems designed specifically to consider their flammability characteristics. They should never be used to replace non-flammable refrigerants (A1 classification) in retrofit situations without a full risk assessment and system modifications by a 'competent person' in consultation with the OEM.

R1233zd is currently an A1 classification alternative. R1233zd is limited to large, water-cooled chillers (+1MWr) and new installations. R1233zd is a low-pressure refrigerant, at typical HVAC operating conditions the chiller would be expected to operate in a significant vacuum (-60 kPa).

R717 – Ammonia is a high pressure natural refrigerant and is also used extensively in surface mine cooling plants. Through proper design, installation, servicing and with the suitable controls it has an enviable safety record and has no impact on global warming or the ozone layer. A principal use of ammonia is as a fertiliser. Ammonia is used extensively in industrial refrigeration systems with charges into the hundreds of tonnes. There are numerous storage facilities throughout Australia that contain thousands of tonnes of ammonia. The ammonia is kept at safe pressures via use of appropriately designed refrigeration systems.

Modern ammonia chillers can be designed with low ammonia charges around 0.1 kg/kW. Therefore a 5000 kW chiller should only have a charge around 500 kg, a synthetic refrigerant chiller will typically have a much higher refrigerant charge.

CONCLUSION

To reduce mine cooling system CO_2 emissions, the chiller and overall system efficiency is critical. A system that uses less power has a lower operating cost, along with a reduced CO_2 emissions associated with the reduced power consumption. Tailored solutions which cater to the variability in operating conditions and which suit each specific system, will provide the greatest increase or improvement in efficiency.

Some considerations that form a critical part of the design process are:

- geographical ambient conditions (peak dry and wet bulb temperatures)
- power generation (power grid, site power station, gas or diesel)
- availability of water (good quality)
- intake airflow
- intake dry/wet bulb temperatures required
- preventative/predictive maintenance.

Refrigerant selection can form part of the decision-making process on the optimum configuration of the refrigerant plant, however indirect emissions via power consumption will in most cases be greatest cause of CO_2 emissions. Irrespective of the refrigerant used, if a low environmental footprint is a principal design criteria, then the focus needs to be on ensuring the overall system design is as energy efficient as possible. A more energy efficient plant will also use less water if it is water cooled.

The implementation of engineering controls as per the automotive industry, can provide considerable efficiency improvements for mine cooling systems. Most of us know or have experienced firsthand, the ability of late model cars that switch off the engine when stationary or those with variable displacement systems (engine technology that allows the engine displacement to change, usually *by deactivating cylinders, for improved fuel economy).* Similar based initiatives of reducing power consumption can be implemented that allow cooling systems to operate at the upper end of their efficiency most of the time. These can include floating chilled water temperatures and smart use of VSD's on pumps and fans.

An often misunderstood aspect of refrigerant selection is what is the best refrigerant to ensure optimum efficiency. That is, does the cooling duty require a very low chilled water temperature, in which case ammonia may be an optimum solution providing it can be safely applied. If chilled water temperatures can be raised or are highly variable, then the type of the chiller may be much more important to ensure optimum efficiencies. There are chillers that will provide optimum overall efficiency if water temperatures can be floated and greater with variable loads. Consequently to achieve a solution that provides optimum capital and the lowest power and water consumption, may involve a mixture of refrigerants and chiller technology. The selection of the best refrigerant is just one consideration, albeit an important one when deciding on the optimal design solutions to mine heat problems, which also need to be cost-effective in terms of operational and capital costs and leave the smallest carbon footprint possible.

Reducing CO_2 emissions requires the adoption of the best possible engineered solutions. The best solution requires an open mind and may necessitate discarding previous solutions and a rethink on how things have been done in the past. Many mine cooling plants are amongst the largest capacity refrigeration plants, so even small improvements in efficiency can result in relatively large savings in power consumption and associated CO_2 emissions.

The longer the design life of the refrigeration plant, the more important the refrigerant selection and striving for maximum possible efficiency is.

ACKNOWLEDGEMENTS

The authors gratefully acknowledge Gordon Brothers Industries Pty Ltd for their support and permission to publish this paper.

REFERENCES

Baum, R, 2017. *Chlorofluorocarbons and Ozone Depletion*, American Chemical Society, University of California, http://www.acs.org/content/acs/en/education/whatischemistry/landmarks/cfcs-ozone.html

Daikin Industries Ltd, 2022. R32 Next Generation Refrigerants, www.daikin.com

Dudita, M and Kauffeld, M, 2021. Environmental impact of HFO refrigerants and alternatives for the future, www.openaccessgovernment.org/hfo-refrigerants

International Institute of Ammonia Refrigeration (IIAR), no date. *About Ammonia Refrigeration,* https://www.iiar.org/IIAR/About_Ammonia_Refrigeration/About_Ammonia_Refrigeration_Home_Page

McLinden, M O and Huber, M L, 2020. (R)Evolution of Refrigerants, *Journal of Chemical & Engineering Data*, 65(9):4176–4193.

Osachi, H F, 2019. Refrigerant Choices for Chillers Remain Complex, *ACHR News*, https://www.achrnews.com/articles/141178-refrigerant-choices-for-chillers-remain-complex

Pearson, S F, 2003. Refrigerants Past, Present and Future, *Proceedings of the 21st IIR International Congress of Refrigeration: Serving the Needs of Mankind.*

Tiseo, I, 2021. Carbon Intensity Outlook of the Power Sector in Australia from 2020–2050, Statista Inc.

The influence of roadway configurations on VCD and bulkhead design

M Salu[1] and D Pateman[2]

1. Director – Salu Structural Engineering, Brisbane Qld 4178. Email: msalu@gmx.com
2. MAusIMM, Business Development and Marketing Lead, Minova Australia, Nowra NSW 2541. Email: douglas.pateman@minovaglobal.com

ABSTRACT

Ventilation control devices (VCD's) and bulkheads are critical controls used for managing risk in underground coalmines. Following the Moura No. 2 mine explosion in 1994, the Queensland Mines Department required that the design of VCDs be based on live testing (such as in a mine testing explosion gallery) and more recently on numerical modelling using measured seal material properties. While the stipulated methods represent an idealistic solution, they are not practical to implement as solutions to daily challenges presented in underground coalmines.

Not all VCD's and bulkheads are placed in areas which have been mined to standard roadway dimensions. Operationally, VCD's may be required to be placed close to intersections or require large dimensionally different areas (eg overcasts and pre-driven recovery road shutters) compared to the live testing or numerical modelling originally undertaken.

This paper describes the various structural engineering methods utilised to account for the design of VCD's where the operational requirements are non-standard. Siting and construction of VCD's near the entry of cut throughs and geological structures requires additional design considerations. While VCDs constructed from a single material (eg sprayed mining plaster) are relatively straightforward to design, the addition of multiple openings (doors) or large pipes will complicate the structural analysis significantly.

Temporary bulkheads used for the containment of grout such as for the construction of pre-driven recovery roads in coal mining have a significantly different risk profile to VCDs and should be designed accordingly. Recent failures and near-misses in such bulkheads have highlighted this issue for the authors.

The existence of an Australian Standard for Underground Mine VCDs and Bulkheads would go a long way to addressing many of the issues raised in this paper. Without a single, industry-accepted reference for VCD design, individual mines will have to continue to exercise their best judgements using whatever local expertise is available to them.

INTRODUCTION

Ventilation Control Devices (VCDs) are used in underground mines to control and adjust ventilation air flow, as well as to provided protection against gas explosions and air blasts caused by roof falls (Mutton and Salu, 2013). They can be constructed from a wide variety of structural materials, be intended for short – or long-term use and rely on a variety of structural engineering design theories for their performance and safety.

Bulkhead is the term used for underground devices intended to hold water or other liquids (eg slurries or grouts). They are often constructed using the same materials and similar techniques to those used for constructing VCDs. Similar to VCDs, they can be intended for short – or long-term use but unlike VCDs they will almost always be subjected to their full design pressure when used to contain backfill slurries or grouts.

STRUCTURAL BEHAVIOUR

It is important to understand the different types of structural behaviour that VCDs and bulkheads can use to resist forces from gas or liquid pressures, which are largely dependent on the material of construction used for any particular device. Four distinct behaviours were identified by Salu and Mutton (2017), building on earlier work presented in Pearson et al (2000). These behaviours were labelled: sail, plate, flat arch, and plug. The sequence described form a continuum of behaviour from most flexible to least flexible (ie most rigid). Each is described briefly below.

- **Sail** – Flexible, so-called 'sail' stoppings, are VCDs that act purely in tension just like a sailboat sail (Figure 1). These are typically constructed from geotextiles or other similar heavy-duty fabrics and are typically used for low-pressure applications (up to 35 kPa overpressure). Their strength is often limited by the challenge of reliably fixing the fabric to the surrounding strata. Research is currently underway on developing flexible VCDs rated up to 140 kPa.

- **Plate** – VCDs and bulkheads constructed from materials such as sprayed mining plaster or reinforced concrete carry loads using a bending mechanism (Figure 2). Such devices require significant flexural tensile strength to carry these loads. In the case of reinforced concrete VCDs steel reinforcing is provided for that purpose.

- **Flat Arch** – As devices increase in thickness, the bending mechanism starts to transition to an arching mechanism (Figure 3), where the loads are carried by direct compression to the surrounding strata. While Australian structural design codes limit the span to thickness (s:t) ratio for so-called 'arching action' to 3 or 4:1, full-scale explosion testing (Salu and Kay, 2007) have shown that this behaviour can start to occur with s:t ratios as high as 15:1.

- **Plug** – Relatively low compression (and tensile) strength materials can be used for VCD and bulkhead construction if they are thick enough. These extremely thick devices form a solid block of material, which resists loads through direct compression and internal shear strength as well as shear resistance along the perimeter interface (Figure 4).

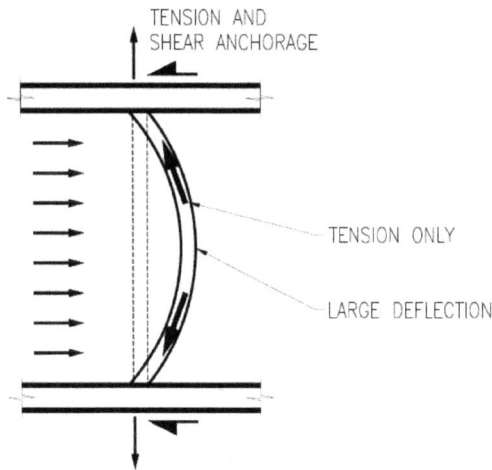

FIG 1 – Sail mechanism (Salu).

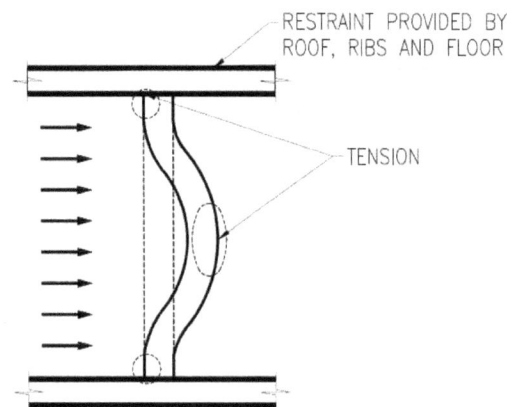

FIG 2 – Plate mechanism (Pearson).

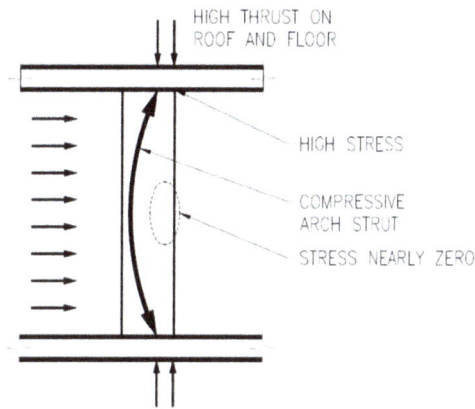

FIG 3 – Arch mechanism (Pearson).

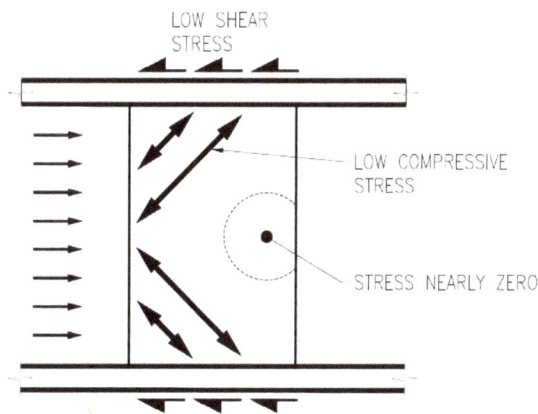

FIG 4 – Plug mechanism (Salu).

Structural analysis of the first three mechanism types is often simplified by assuming that the cross-sections shown in Figures 1 to 3 represent the actual load-carrying mechanisms in full. In other words, that the full pressure load is transmitted *only* to the floor and roof strata (50 per cent to each). The reality is significantly different.

Figure 5 shows that approximately 75 per cent of the load on the device will be transmitted to the roof and floor, compared with 50 per cent for Figure 6. As the W:H ratio increases, an increasing proportion of the pressure load will be transmitted to the roof and floor and the simplified structural assumption known as 'one-way action' will become more accurate.

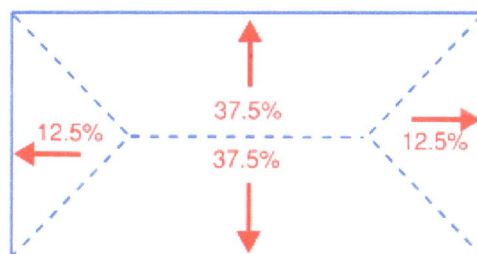

FIG 5 – Typical coal mining roadway (W:H = 2:1).

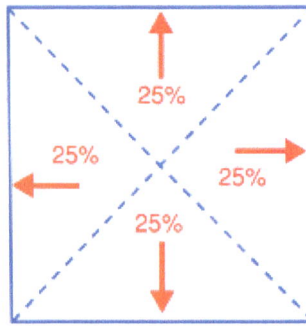

FIG 6 – Typical metal mine drive (W:H = 1:1).

Australian design code requirements

There are no Australian Standards covering the design of VCDs or bulkheads for underground mining. To the Authors' knowledge, there are no Overseas or International Codes or Standards for underground mining that could be applied or used for guidance.

This deficiency was examined in a paper by Salu and Mutton (2017) and will not be further discussed except to the extent of describing the Authors' design approach and methodologies.

Australian Standard (2002) AS/NZS1170.0 Structural design actions – General principles, is intended 'to provide designers with general procedures and criteria for the structural design of structures. It outlines a design methodology that is applied in accordance with established engineering principles.' Of relevance to this paper is that structures must be able to safely resist the loads imposed on them. Therefore, if a designed structure fails, AS/NZS1170.0 provides a starting point for investigation and potentially prosecution of a structural failure.

There are Australian materials design codes covering the design of structures made from concrete, steel and other materials. These Codes can be used for guidance, although their scope is often specifically limited to a narrow range of buildings and some infrastructure elements. Confusingly, there are a number of steel and concrete codes, for example, that overlap in scope. An unscrupulous designer could arguably 'pick and choose' a preferred Code to suit a particular VCD or bulkhead design to provide a commercial advantage over a competitor's product.

PITFALLS IN THE STANDARDISATION OF UNDERGROUND VCD DESIGNS

Most mines will be developed using a Mine Plan that is based on standard size roadways or drives. It follows that standard designs will be developed for VCDs, based on the standard roadway geometries.

A structural design engineer will adopt certain assumptions for the VCD design in order to simplify and speed up the design process, but these are not always fully communicated to the mines Ventilation Officer (VO) or to the workers responsible for its construction. The underground environment is non-uniform and continually changes as mining proceeds, which is not an ideal situation to be using standard designs.

A challenge always arises when a non-standard situation is encountered and a solution is required quickly to minimise additional costs resulting from potential lost production while a solution is found and implemented. The following mini case-studies provide examples where the Authors have observed such situations arising and have participated in developing solutions.

VCD placement and influence on design

As noted above, not all locations underground are mined to standard dimensions and operationally, not all VCD's are able to be placed in ideal locations. This results in VCD's being constructed which are much larger than the original live testing or numerical modelling used to determine the standard designs. Practically, VCD's may require placement close to geological structures or the entrance to cut-throughs, which require refinements in the design methods to ensure robust VCD's are built. The following section provides an overview of three examples where placement of a VCD has influenced the design.

Construction of seals near intersections

When placing seals in cut-throughs, best practice is to locate the seals away from the intersection with individual mines specifying their standard seal location, however this may not always be possible to achieve. Near intersections, the strata experiences increased stresses from the larger spans mined, which causes a reduction in the ability of the ribs to carry explosion loads experienced by the seal. Additionally, as the coal ribs fret near intersections, the ability of the seal to remain airtight decreases without further treatment of the ribs through either grout or polyurethane strata injection. Positioning of seals should also allow for sufficient room to construct a new seal if required (Qld Mines Inspectorate, 2011).

In some instances, it is not possible to construct the seal in any location except at the entrance to a cut-through, as is shown in this case study. This seal was placed a distance inbye the cut-through entrance with sufficient room to replace the seal if it was required, however this positioned the seal at the direct entrance to the inbye end of the cut-through where it would experience an explosion load, shown in Figure 7.

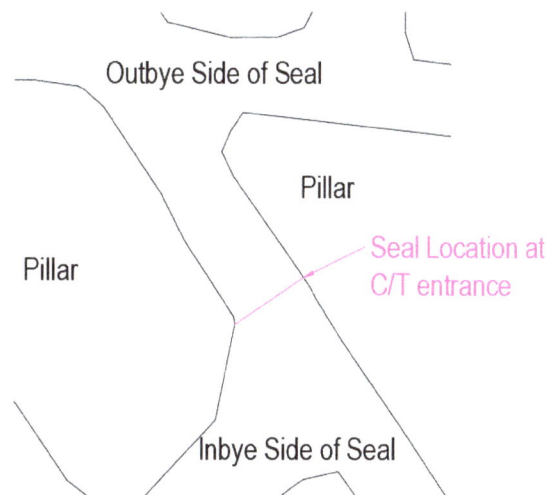

FIG 7 – Location of seal at far inbye end of cut-through.

For this seal design, any influence from the ribs to provide sufficient confinement to the seal when experiencing an explosion load were ignored, which increased the minimum sprayed thickness of the seal. The increase in seal thickness of approximately 23 per cent allowed for the floor and roof strata to be fully mobilised in the event of an explosion. When constructing seals at the entrance to intersections, additional care also needs to be taken with spraying the ribs to minimise leakage paths which are more prevalent in this highly fractured area.

Construction of seals in front of existing seals

Ongoing management of ventilation in a mine sometimes requires upgrading of existing VCD's to a higher rating. Depending on the seal design and operational requirements of the mine, this may require the existing structure to be used as formwork for the construction of a new VCD.

When sealing a section of a mine, the proposed construction method of an 827 kPa (120 psi) rated seal used the current seal as formwork for the front wall of the new seal. The current seal was constructed six metres inbye the intersection of the cut-through, which allowed sufficient room in front of the seal for future repair or replacement of the seal. Placing the 827 kPa seal on the inbye side of this existing seal would have retained this ability. The proposed seal was designed as a plug seal (Figure 4) constructed using Minova's FB200 high yield grout and utilising Sprayplast gypsum plaster walls as grout retaining formwork.

During inspection of the seal locations, the following operational concerns were identified when using the current seal as an existing front wall:

- Spraying a dry-mix plaster stopping on the inbye side of the current seal for use as a back formwork wall would result in unacceptable exposure of operators to dust due to the limited ventilation available.

- Placement of pumping ports at the top of the existing seal to allow filling and pressurisation of the seal would require operators to jack pick holes at the top of the seal. From a safety and time management perspective this was not suitable.

A discussion was undertaken about the suitability of utilising the current seal as a back wall for the formwork instead of a front wall, with concerns raised about the high-pressure rating of the seal and its location closer to the intersection. It is preferable for seals with a high rating such as 827 kPa to be a minimum of 8 to 10 m from an intersection to reduce the risk of the strata failing under a 827 kPa blast pressure. Significant forces are generated within the seal core material which are transferred into the strata and become a concern where destressed and damaged ribs occur, potentially requiring alternate design methods.

The site geotechnical engineer undertook a review into the suitability of the strata in front of the existing seal location as an alternative location for the new seal, with the review incorporated into the structural engineering certificate for this seal. The review determined that a new seal would be able to be placed in front of the existing seal as long as it was at least four metres from the intersection as:

- Roof conditions had been mapped as good with additional support already installed at the seal location to provide long-term stability.

- There were no long-term plans for mining within the seals location which would potentially cause re-loading of the seals or the surrounding strata.

- The planned seal was designed at 1.1 m thickness which would still provide room for repair or construction of an additional seal, if required at a later date.

With the geotechnical review showing the location was suitable for construction of the seal on the outbye side of the existing seal (Figure 8), the certifying structural engineer was able to be sign off on the seal being constructed in the revised location.

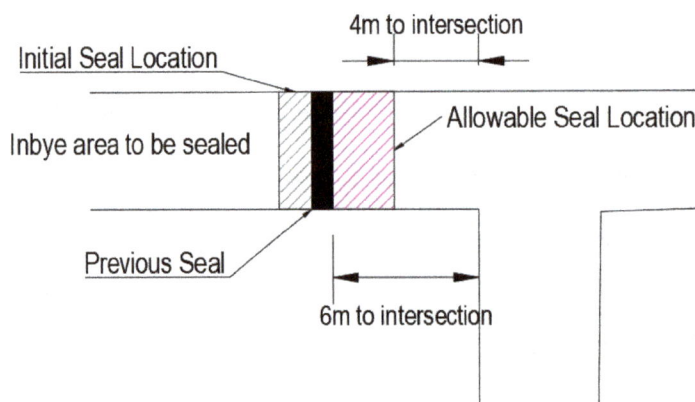

FIG 8 – Seal location showing current seal and proposed construction locations.

Non-standard stoppings near intersections

Over the life of a mine, the requirements of ventilation throughout an area can change requiring the construction of VCD's for unique applications. One such application was the construction of a rated stopping to withstand an explosion pressure of 35 kPa (5 psi) with an additional fan pressure of 3.5 kPa. The construction of this device was designed to span across two headings and a pillar, for a total length of 40 m with heights up to 4.2 m as shown in Figure 9.

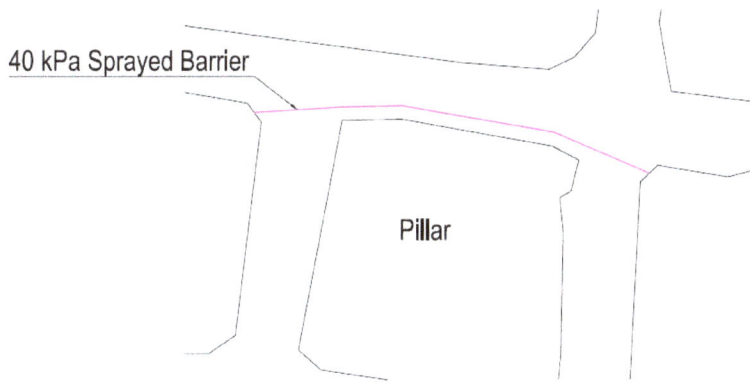

FIG 9 – Overview of 40 m long rated stopping.

Additional challenges in developing this VCD design were the poor conditions of the ribs at the intersections where the stopping ended, as well as the stopping becoming inaccessible for repairs post construction requiring it to be a robust structure.

The engineering design was undertaken for a stopping rated to 40 kPa with a factor of safety of 1.5. Sprayed stoppings are designed using plate theory (Figure 2) whereby as the seal shape becomes more rectangular (W/H > 3), the seal begins to experience one-way loading. For this design, the W/H ratio was 9.5, meaning that the loads were designed to be completely carried through the roof and floor, with no load distributed to the ribs.

This design method allows for stoppings to be designed where rib conditions are questionable and are not expected to transmit any loads from the stopping into the strata due to their poor condition. Further design refinements included spraying buttresses (Figure 10) at the locations where the stopping intersected the ribs to locally strengthen the stopping, increasing its robustness.

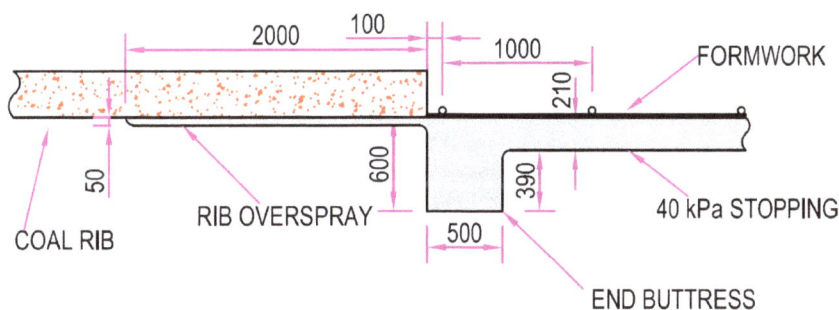

FIG 10 – Buttress details at the stopping ends.

This case study shows the versatility available when using sprayed plaster for the construction of VCD's. Design thicknesses can be optimised for different applications to the nearest 10 mm and large structures are able to be built with confidence, with localised modifications to designs as required. Further, the design methodologies used and the consideration of robustness during the design process demonstrate compliance with the intentions of Australian Standard (2002) AS/NZS 1170.0.

Potential gaps in design of underground VCDs

Access doors and hatches

Many stoppings and some seals are fitted with doors or hatches to provide access between ventilated compartments. It is logical that a door fitted to an explosion rated VCD should be rated at least equal to the VCD rating. However, it is critically important that the rated VCD must be designed to be structurally adequate despite having a discontinuity at the door or hatch location.

Some VCDs have been full-scale explosion tested with hatches installed. More commonly, testing has been performed on 'basic' VCDs with no hatches or any other embedded fittings. In either case, the VCD designer must ensure that the opening provided for a door or hatch does not compromise the intended VCD rating. One way to do this is to increase the VCD thickness, when a door or hatch is fitted, compared with the basic thickness.

Most explosion rated door designers will assume that the door frame provides unform, rigid support to the perimeter of the door. This is not always the case. The authors are aware of many door designs that are much stronger in one direction than in the opposite direction. Often, this is due to the use of a single locking lug in the centre of one door edge and two or three hinges on the opposite side. Instead of the door being a 'plate' supported uniformly around its edge, the forces in the door due to an overpressure event will be concentrated in just a few locations. For a 1 m × 1 m hatch in a 35 kPa rated stopping, this will increase the local design load from approximately 2 kN per 100 mm length of door perimeter frame to 17 kN at the section of door frame adjacent to the catch, a local force increase of 850 per cent. Refer to Figure 11 for a typical example of an explosion rated door.

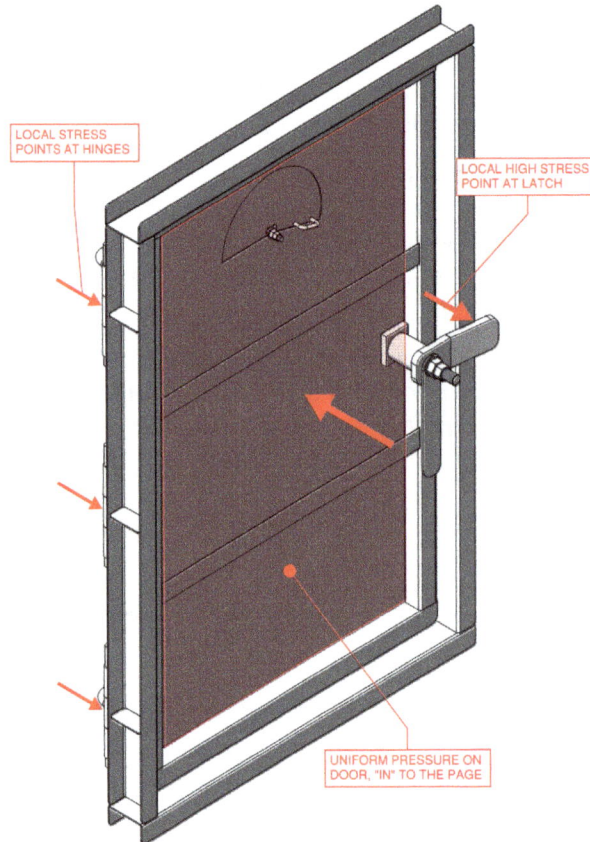

FIG 11 – Example explosion rated door.

To close the potential gap between designers of rated VCDs and designers of doors and hatches, the authors recommend that VCD designers should be made explicitly responsible for the certification of the 'VCD system', including any inclusions. This is not intended to let VCD door designers 'off the hook'. They would remain responsible for the structural performance of their devices and in addition, they would be required to stipulate the structural assumptions that they have used during their design process to fully inform the VCD designer.

Doors in flexible VCDs

Doors or hatches in flexible VCDs present similar issues to those described in the preceding section. As previously described, a flexible stopping relies on its characteristic 'sail' shape to support a uniform pressure load purely by tension. If a large, rigid flat section (eg a steel door) is inserted into the stopping then its structural behaviour may change significantly. Ideally, doors or hatches in flexible stoppings should be suspended from steel cables to carry their weight while allowing the door to move with the flexible stopping. Where a door is supported on a structural steel frame fixed to the floor, the fabric/door frame interface must be checked to ensure that excessive stress concentrations do not occur at the corners of the door frames.

Usually, a single door or hatch positioned centrally and up to 1 m wide × 1.5 m high can be accommodated in most flexible stoppings rated at up to 35 kPa. The authors have found that when two doors are required in a flexible stopping, the location and relative spacing of those doors will be limited to a few configurations due to stress concentration effects. Figure 12 shows a Finite Element computer model of a steel door frame attached to a flexible sail stopping. The relatively low to moderate stresses in the fabric show up as blue and green colouration, while immediately adjacent to the door frame the red colouring indicates extremely high local stresses in the stopping.

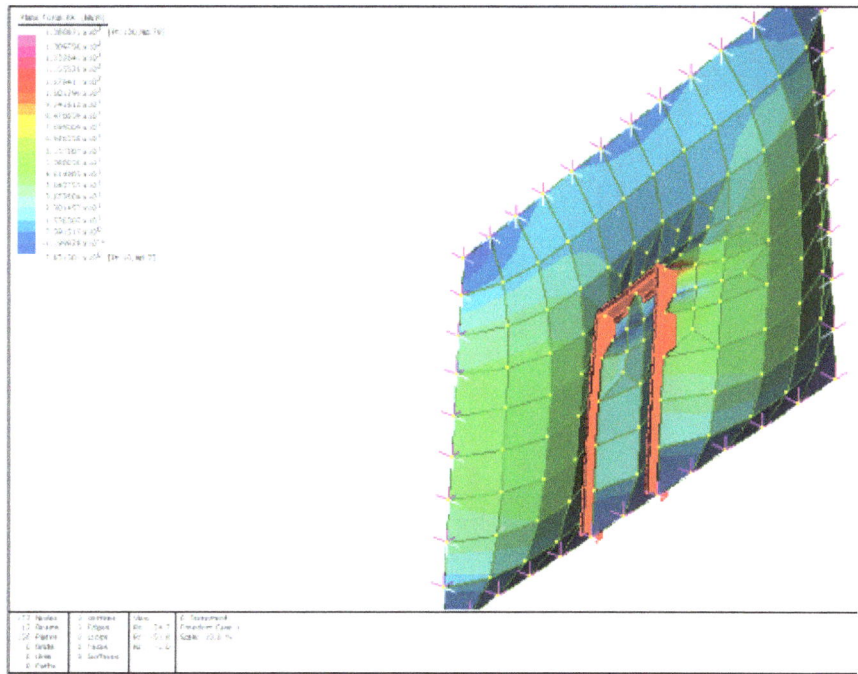

FIG 12 – FEA model of flexible VCD showing high stresses at door frame.

If multiple doors are required in a flexible stopping, a simple rule of thumb is to allow for a double thickness of rated cloth over the entire stopping. Doubling up the thickness in this case has the advantage of doubling the stiffness of the stopping overall. This is likely to bring it closer to the stiffness of the steel door frames, which in turn will tend to reduce the stress concentration due to differences in stiffnesses between stopping structural elements.

Overcasts

Overcasts are used to separate fresh intake air flow from stale air being extracted from a coalmine. In the past 20 years, overcasts have been increasingly pre-fabricated in standardised sizes, based on standardised components. These standardised overcasts are intended to be installed in mine roadways that generally have very consistent dimensions. However, at an overcast location, the roadway roof must be raised in one direction to allow extracted air to flow over the overcast. By definition, this work will result in a non-standard roadway size and due to normal construction tolerances and the natural variability of a coal seam the overall dimensions of a roadway intersection can vary significantly from 'standard'.

As an example, a typical pre-fabricated steel overcast size for a Qld coalmine is 2.8 m high × 5 m wide × 4.8 m long. To provide a fully airtight seal, sprayed mining plaster wing walls and headwalls are typically installed between the overcast structure and the ribs and roof. Rated overcasts are typically designed for an overpressure of 35 kPa, but the overcast designer rarely provides any information regarding the infill wing walls and headwalls with their design certification.

This is a significant omission, as the headwalls and wingwalls of a rated overcast also need to be rated and neither the overcast nor the infills can be designed in isolation.

The larger the infill dimensions, the higher the load that will be transferred back to the overcast structure. Designing either component in isolation can lead to problems, as the following example shows. Refer to the overcast illustrated in Figure 13.

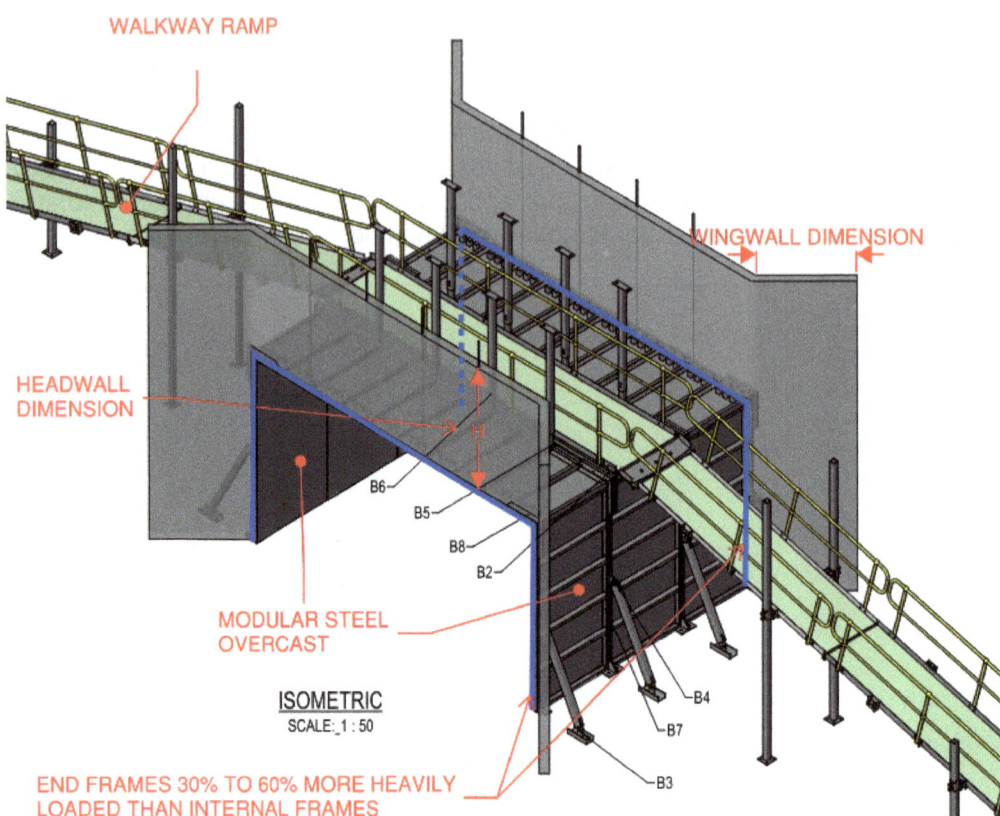

Labels in figure:
- WALKWAY RAMP
- WINGWALL DIMENSION
- HEADWALL DIMENSION
- H
- B6
- B5
- B8
- B2
- MODULAR STEEL OVERCAST
- ISOMETRIC SCALE: 1 : 50
- END FRAMES 30% TO 60% MORE HEAVILY LOADED THAN INTERNAL FRAMES
- B4
- B7
- B3

FIG 13 – Example overcast showing sprayed infill walls.

For the 2.8 m × 5 m × 4.8 m overcast described earlier, the load on a typical roof beam assuming 4 × 1.2 m wide panels, due to 35 kPa overpressure, is 35 × 0.6 = 21 kN/m. If a sprayed headwall 150 mm thick is to be supported by the end frame, the additional load will vary from 6.6 kN/m at 2 m high up to 13.2 kN/m at 4 m high. That is an additional 30 per cent – 60 per cent loading and could be higher subject to the finished excavation size. A similar calculation applies for the wing walls: the wider the wing walls, the more load is exerted on the end frames.

Bulkheads – failure and near-miss case study

As mentioned earlier, bulkheads that are designed to contain backfilling grouts or slurries will always experience their full design loads, until the backfill sets. This means that any structural deficiency *will* be exposed, unlike many water-retaining bulkheads which may never be fully loaded. A structurally sound VCD or bulkhead must be both designed correctly and then constructed in accordance with engineer-approved drawings and methodology. If supervision of construction is poor or non-existent, then the risk of failure of a bulkhead will be high, as illustrated in the following example.

A Qld mine was using sprayed plaster bulkheads (shutters) to contain flyash grout being used for construction of pre-driven recovery roads (PDRR). The shutters were designed to be 100 mm thick mining plaster sprayed onto a plasterboard backing, in turn supported by timber posts hitched into the roof and floor. The design depth of grout was 1 m, which was never achieved due to the size of the PDRR compartments and the slow rate of grout filling. At some point during construction of the backing walls and subsequent spraying of the shutters, the construction sequence was inverted. The backing boards were placed first and then the vertical posts were attached onto the front face and then the shutter was sprayed. Once the spray covered the posts (by approximately 10 mm), construction was deemed to be completed. Figure 14 shows the difference between the 'as designed' and 'as constructed' shutters.

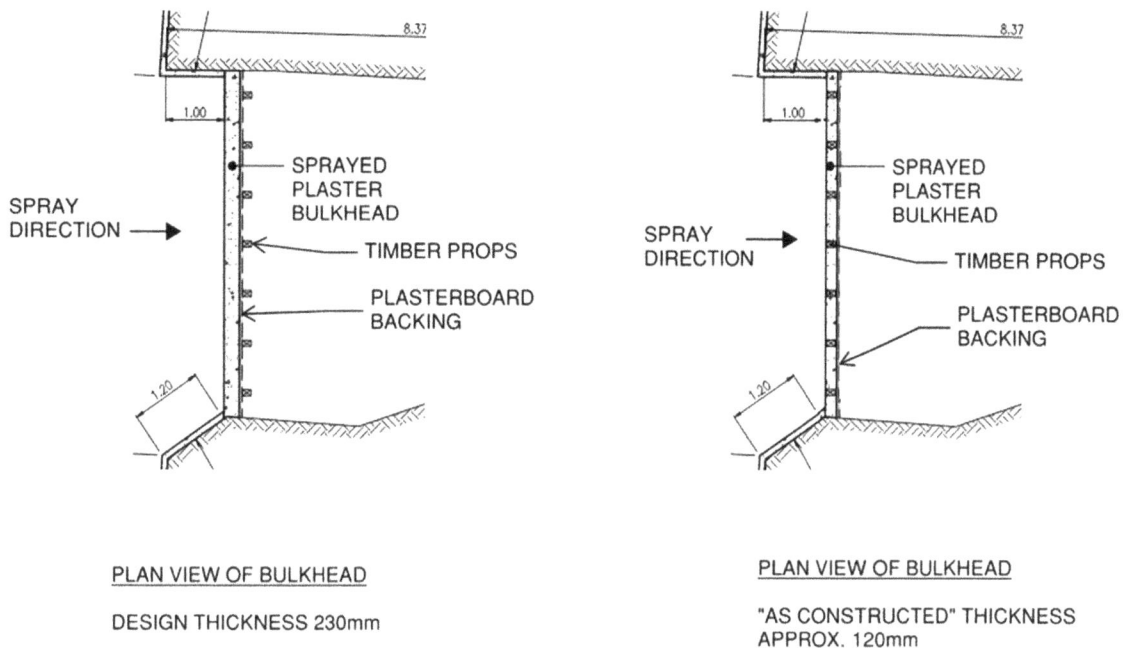

FIG 14 – Comparison between As Designed and As Constructed bulkheads. Note the reduced strength of the as constructed bulkhead due to timber props sprayed into bulkhead plus overall reduced bulkhead thickness.

None of these incorrectly constructed shutters failed, almost certainly because the depth of liquid grout never exceeded approximately 300 mm during the PDRR compartment filling process. Unfortunately, this inverted method of shutter construction became the norm and would shortly lead to a serious shutter failure.

At the same mine, a redundant niche was identified as requiring backfilling and an engineer-approved design was procured. This shutter was 230 mm thick and designed to contain liquid grout up to 3.2 m deep. Unlike the PDRR shutters, the small size of the niche meant that it could be grout filled in a single shift. Subsequent investigations of the shutter failure revealed a litany of other construction and procedural deficiencies, but all could be linked back to a root cause of lack of effective construction supervision.

CONCLUSIONS

Ventilation is a critical aspect of operating safe and productive coalmines, especially as mines age and become deeper, hotter and gassier. Ensuring that VCD's are designed and constructed to a high standard and in accordance with industry best practice will minimise the risk of ventilation failure due to an unexpected overpressure event.

VCD's are typically designed for standard roadway dimensions and placement towards the middle of a cut-through, however due to the variabilities of the underground mining environment this is not always possible. Operationally, VCD's may need to be sited at the entrance of a cut-through or near geological structures which reduce the ability of the seal to transmit loads into the strata. It is important for the designer to understand these influences and ensure that the correct design method has been utilised to design a fit for purpose VCD.

Three short case studies have been presented which outline different situations that have arisen along with the processes, both from a design and operational perspective, to develop a timely and cost-effective solution.

Three further VCD design examples have been presented, where there is potential for VCD failures to arise due to lack of coordination between designers of different elements of VCDs.

Finally, a brief case study highlighting the importance of on-site supervision of construction of bulkheads and by extension of VCDs was presented to demonstrate how failures are often caused by a combination of factors rather than a single, simple event.

In conclusion, the authors believe that Mine Management in general and Ventilation Officers in particular must continue to pay close attention to the design and construction of VCDs and bulkheads, despite the majority usually being constructed in accordance with standard designs. Complacency with routine installations can lead to carelessness or errors when a non-standard solution is required for a particular location. In such cases, it is important to obtain sound technical advice from relevant subject matter experts. It is also critically important to ensure that any geotechnical, structural or operational requirements developed to address a non-standard situation are fully implemented and/or complied with on-site and that such compliance is physically verified prior to acceptance of the device.

ACKNOWLEDGEMENTS

The authors would like to thank Minova Australia for giving permission to publish this conference paper. Special thanks go to those people who gave valuable advice and encouragement to write this paper.

REFERENCES

Australian Standard, 2002. AS/NZS1170.0–2002 Structural design actions – General principles.

Mutton, V and Salu, M, 2013. Full scale explosion testing and design of gypsum plaster ventilation seals, in *Proceedings of the 2013 Coal Operators' Conference* (eds: N Aziz, B Kininmonth, J Nemcik, T Ren and J Hoelle), pp 199-208, (University of Wollongong – Mining Engineering: Wollongong; The Australasian Institute of Mining and Metallurgy: Melbourne; The Mine Managers Association of Australia: Toronto).

Pearson, R D, Gillies, A D S, Green, A R, Day, R and Dux, P, 2000. Evaluation of a full scale pressure test for ventilation control devices, *ACARP Project No: C8006*.

Qld Mines Inspectorate, 2011. Construction of Seals, *Mines Safety Bulletin 107*, https://www.rshq.qld.gov.au/safety-notices/mines/construction-of-seals.

Salu, M and Kay, G, 2007. Assessment of full scale explosion testing data to develop a design methodology for underground mine VCDs, Presentation to Bowen Basin Coal Conference, (Coppabella, Qld).

Salu, M and Mutton, V, 2017. The case for developing an Australian technical specification for structural design of ventilation control devices, in *Proceedings of the 2017 Coal Operators' Conference* (eds: N Aziz and B Kininmonth), pp 411-425 (University of Wollongong – Mining Engineering: Wollongong).

The impact of ventilation on demand on impeller fatigue life

L Salvestro[1] and T Rafferty[2]

1. Sales and Applications Director, Howden Australia, Norwest NSW 2153.
   Email: livio.salvestro@howden.com
2. Chief Engineer, Howden Australia, Norwest NSW 2153. Email: tom.rafferty@howden.com

ABSTRACT

Ventilation on demand (VOD) has been demonstrated to significantly reduce operating costs for ventilation fans, by reducing fan output to meet demand. The most effective method for reducing operating costs is to use variable speed drives to adjust the speed of ventilation fans as required – this achieves significantly larger power savings that damper or vane control alone.

However, this mode of control will significantly increase the number of speed cycles the ventilation fans will experience. Fabricated impellers have a limited number of start stop cycles based on the materials and method of construction, and the operating stresses due to centrifugal forces. This means that fans experiencing constant or regular speed changes will experience a reduced fatigue life compared with fans that are run at constant speed.

The fatigue life of fans operated under various speeds using VOD is difficult to predict, as it is dependent on the magnitude of speed changes, and the frequency, which will be varied to meet demand. It is important to be able to estimate remaining fatigue life to assist in maintenance planning and avoid unplanned maintenance activities.

This paper will review extensive VOD data on fan operating conditions from two primary mine ventilation fans. This paper will demonstrate how this data can be used to calculate accumulated fatigue and predict the remaining fatigue life, and how this compares with fixed speed or non-VOD controlled fans.

The increased maintenance costs for the impellers will also be estimated, and compared to the operating cost savings using VOD, to determine if the savings outweigh the additional maintenance costs.

INTRODUCTION

Electric motors are the heart of most mining and industrial process plants. The three largest consumers are electric motors driving fan, pump and compressed air systems. Motors account for 60 per cent of the industry's electricity consumption.

Eskom Holdings Limited (2010) showed that fans are subject to more misuse and faulty applications than virtually any other type of equipment. They are one of the largest energy consumers in that they operate for the entire 24-hour cycle.

To reduce operating costs and associated carbon emissions, ventilation on demand is being used more frequently by mine operators.

Ventilation-on-Demand (VOD) is a system capable of real-time adjustments to mine ventilation equipment to meet underground atmospheric quality requirements for personnel and equipment based on status and location. Through active monitoring and equipment modulation, VOD systems are able to reduce the amount of 'wasted air' supplied to mine workings, thereby reducing overall operational costs, as shown by Sanftenberg et al (2019) and Pinedo and Torres Espinoza (2019).

For primary ventilation fans the most effective method to achieve this is through the use of a variable speed drive, adjusting the speed of the fan up and down as require to meet the ventilation demand. Other methods of ventilation control include adjusting the inlet louvre damper, or the inlet radial vane control, but these methods are not as efficient as motor speed control.

While VOD has been shown to result in significant operating costs savings in papers by Sanftenberg et al (2019) and Pinedo and Torres Espinoza (2019), the high number of speed changes on a fabricated impeller is a concern, as this could lead to early fatigue failure of an impeller. Fatigue could result in a reduced design life of the impeller, and would require additional maintenance costs to inspect, repair or replace the impeller more often than required for a fixed speed design. One concern that has been raised is that the increased maintenance cost could offset the operating cost savings, reducing the appeal of installing a VOD system.

To evaluate this issue, this paper reviews six years of operating data from two primary exhaust fans installed at a Nickel mine in North America, that have been exclusively controlled by ventilation on demand during this period. The exhaust fans are controlled using a control strategy called speed optimiser. The control strategy is as follows: the exhaust fans adjust their speed in order to maintain the open most regulator at an 80 per cent opening position (80 per cent was set to be a balance between energy savings and smooth control response). More specifically, if the regulator reaches an opening position greater than 80 per cent (lower resistance), the exhaust fans will increase their speed. Conversely, if the regulator position is lower than 80 per cent (higher resistance), their speed will decrease. Adopting such a strategy optimises the operating speed of the exhaust fans while maintaining current airflow demands. Consequently, this control strategy leads to a decrease in total mine airflow, which results in reduced speed on the intake fans and lower gas consumption through the mine air heaters during winter conditions. The cost savings associated with the VOD speed optimiser control strategy is calculated later in the paper.

By calculating the stress at each operating point, it is possible to process the data to calculate the Miners sum of fatigue life, as defined by Miner (1945), which can be used to calculate the design life of the impeller (in years).

FAN DETAILS

The fans are centrifugal double inlet plate bladed, with a diameter of 3263 mm and a design speed of 710 rev/min with 4000 hp (2983 kW) electric motors. The fan is a welded steel construction, made from quenched and tempered steel. The blades are not a casting or a hollow aerofoil, but are plate blades, made from ½" thick steel plate which is then welded to the centreplate and side plates. Wear liners are installed on the blades and centreplate.

The fans have been controlled by ventilation on demand using the speed optimiser control strategy for the entire six years of operating data. The data is from 1 January 2016 until 15 December 2021, and has a sampling interval of 10 minutes, resulting in over 313 000 data points for each fan. Figure 1 show the general arrangement of the fan, and Figure 2 shows a typical week of speed change cycles.

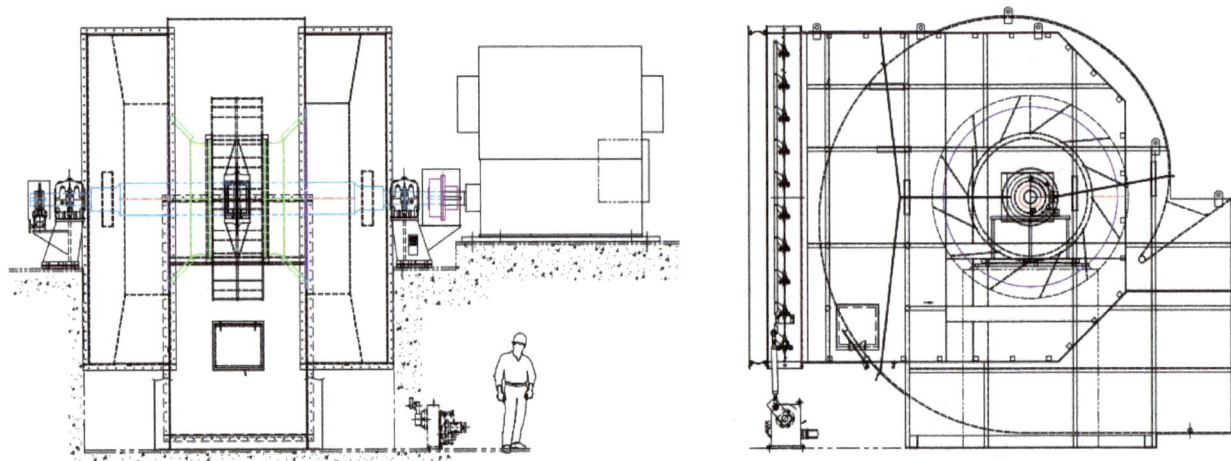

FIG 1 – General arrangement of fan.

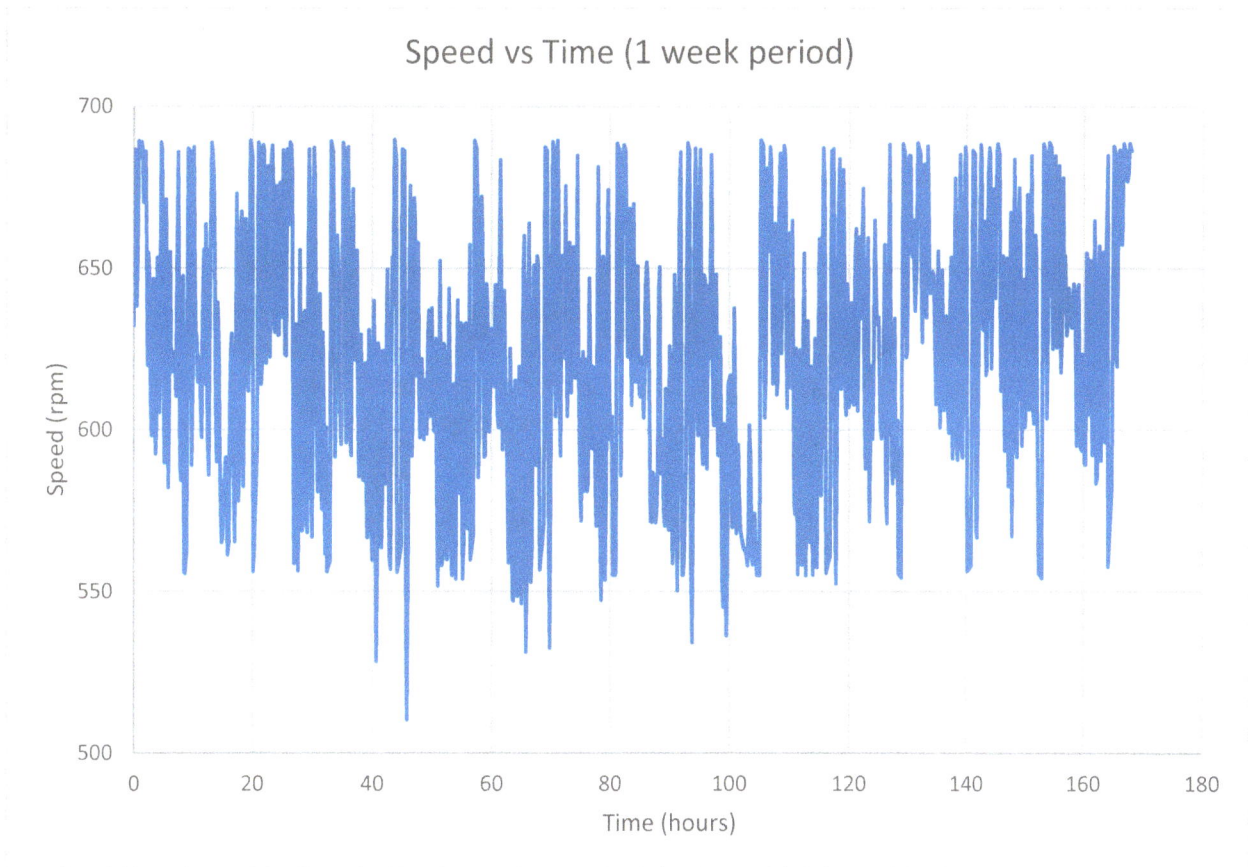

FIG 2 – Typical speed cycles in one week.

It can be seen from Figure 2 that the speed range is normally in the 550–700 rev/min range, and there are no fan stops in a typical week. There are some daily variations in the speed required, and within each day there are multiple speed cycles fluctuating the fan speed.

IMPELLER FATIGUE ANALSIS

To determine the fatigue life of the impeller, the standard BS 7608 (2014) is used, which provides detailed methods for calculating the fatigue life of welded steel structures such as impellers. The calculations are based on SN curves which graph stress range against fatigue life in stress cycles. Constants are provided for various welded connections that allow the fatigue life (in number of cycles) to be determined for a given stress range. The authors use the standard C_2 constant, which is applicable for two standard deviations below the mean line (d=2), with a probability of failure of 2.3 per cent. This is the nominal probability of failure used in the standard and throughout industry.

Note that for this analysis, the stresses remain within the elastic range and typically have more than 10^4 cycles to failure, which is considered high cycle fatigue. Low cycle fatigue is associated with repeated plastic deformation, and a much lower number of cycles to failure.

The stress at design speed is calculated using a finite element analysis (FEA) of the impeller, to determine the maximum hot spot stress at the weld toe. FEA was carried out using Nastran In-CAD, and the maximum hot spot stress (determined in accordance with BS 7608 appendix C) was found to be 400 MPa at 710 rev/min design speed. This peak stress was located at the blade to centreplate weld, which is a perpendicular plate connection with a single partial penetration weld and two fillet welds. Figure 3 shows an image of the impeller FEA, showing the area of maximum stress.

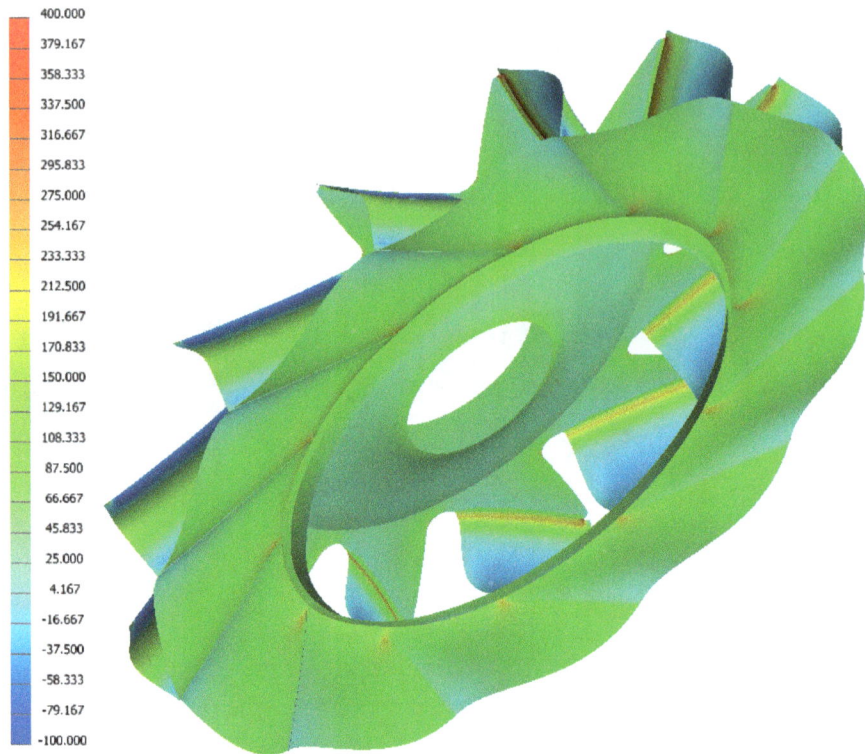

FIG 3 – FEA of impeller.

This connection is a class D weld according to BS 7608 *Table 7*, so the applicable fatigue life constants from BS 7608 *Table 18* are m = 3 and C_2 = 1.52 × 10^{12}. Using these constants, the fatigue life in cycles is calculated as follows using equation 4 from BS 7608:

$$S_r{}^m N = C_d \tag{1}$$

$$N = \frac{C_d}{S_r{}^m} \tag{2}$$

$$N = \frac{1.52 x 10^{12}}{400^3}$$

$$N = 23\ 750 \text{ cycles}$$

For the example above the stress range is equal to the stress at full speed minus stress at zero speed. As the stress at zero speed is 0 MPa, the stress range is equal to the stress at full speed. Note that BS 7608 contains additional factors for the stress range, but these have not been included in this analysis for simplicity. Refer to the assumptions discussion for further details.

Using the formula above, the fatigue life of the impeller is calculated to be 23 750 full speed cycles (0 rev/min ↗ 710 rev/min ↘ 0 rev/min). This same formula can also be used to determine the allowable number of fatigue cycles for any stress range. The impeller stress varies directly with the speed squared, so the stress can be calculated for any speed using the formula below:

$$\sigma_x = \sigma_d \times \left(\frac{z_x}{z_d}\right)^2 \tag{3}$$

$$\sigma_x = 400 \times \left(\frac{z_x}{710}\right)^2 \tag{4}$$

where:

σ_x	= Stress at speed x (MPa)
σ_d	= Stress at design speed (MPa) = 400
Z_x	= Speed x (rev/min)
Z_d	= Design speed (rev/min) = 710

This formula can be used to convert the speed data into stress data, which can now be used for stress range and fatigue life calculations.

It is noted that calculating the speed range first then converting this to a stress range is not possible, due to the square law relationship. A 100 rev/min speed change from 400 rev/min to 500 rev/min gives a much higher stress range than a 0 rev/min to 100 rev/min speed change (71 MPa versus 8 MPa). For this reason it is essential to calculate the stress at every speed in the data set (>313 000 data points) before further analysis is carried out.

ASSUMPTIONS

Several assumptions made in the analysis are discussed below.

BS 7608 has modifiers based on thickness, bending, temperature and sea water. These modifiers were not included, so the fatigue life is based on the 'standard basic SN curves'. The modifiers were not used for the following reasons:

- Bending and thickness modifier was not used to simplify the calculations, and to make them applicable for all fans regardless of thickness or fan dimensions. This modifier is always ≥1, so assuming 1 for this modifier is a conservative approach.

- Temperature modifier is only applicable above 150°C, and the authors assume all mining fans will operate below this temperature.

- The sea water modifier applies to welds without protection submerged in sea water. Although many mines may have wet/saline/corrosive conditions, this factor was not applied as the authors assume the welds are protected from corrosion (painted), and not in contact with sea water.

The sampling rate was one sample every 10 minutes. This means that any speed change cycles shorter than 10 minutes will not be captured – ie if the fan is stopped and re-started within 10 minutes, this start/stop cycle may not appear in the data if it occurs between two measurement points. This is not expected to significantly affect the results for the following reasons:

- Speed changes are gradual, and if the fan changes speed it then holds the new speed for a period of time before changing again.

- Reviewing the data and control algorithm it does not appear there are many rapid speed changes shorter than 10 minutes.

- Fan stops are assumed to exceed 10 minutes duration (so will be recorded).

The authors have therefore assumed a 10 minute sampling rate to be sufficient but agree a reducing sampling period would improve accuracy and may increase the number of cycles recorded.

The authors ignore the effects of erosion, and assume the material thicknesses are constant as per the original manufacturing drawings. Erosion can reduce thicknesses and therefore increase hot-spot stresses at the blade welds. The analysed impeller is fitted with liners, and the authors understand the blade is not experiencing significant erosion in service. Erosion of the liner could actually decrease the blade weld stresses (due to the reduced mass), but was not considered in the analysis for simplicity.

Stresses on the impeller due to aerodynamic loads (pressure on the blade) are assumed to be negligible compared to stresses due to centrifugal loads and have been ignored. This assumption is consistent with calculations, FEA and strain gauge testing carried out by Howden.

RAINFLOW ANALYSIS

To calculate the accumulated fatigue damage for over 313 000 data points, it is necessary to process the data. ASTM E1049 (1985) 'Standard Practices for Cycle Counting in Fatigue Analysis' describes a rain flow method, which was used by the authors to calculate the accumulated fatigue damage. This method consists of determining peaks and valleys, assigning 'buckets' for stress ranges, and then assigning each speed cycle to a relevant bucket, as described below.

As stress ranges are required for fatigue analysis, intermediate speeds are not required – ie if the data goes from a low of 0 rev/min through 100 rev/min, 200 rev/min, 400 rev/min, up to a max of 500 rev/min, the only data points that are required are 0 rev/min (the valley) and 500 rev/min (the peak). The intermediate points can be removed, reducing the number of data points significantly.

A series of 'buckets' is then created, each with a given stress range. In this analysis, each bucket was for a stress range of 10 MPa, so bucket 1 was for 0–10 MPa stress range, bucket 2 was for 10–20 MPa etc. This was continued up to a maximum of 420 MPa, so there were 42 buckets created.

Each valley to peak and peak to valley is then analysed to determine the stress range. Stress range = stress at peak – stress at valley. This stress range was then assigned to the appropriate bucket and a cycle count of 0.5 is added for each half cycle (valley to peak or peak to valley). For example, if the calculated stress range was 196 MPa, then this would be added to bucket 20 (for 190–200 MPa stress range).

This process was repeated for all the cycles in the data set, adding up the number of cycles (n) in each bucket. For each bucket, the allowable number of fatigue cycles (N) was calculated using Equation 4, based on the maximum stress range for each bucket. For example, for bucket 20 (190–200 MPa stress range), the maximum stress range was 200 MPa, and the allowable number of fatigue cycles (N) was calculated to be 190 000 cycles.

The ratio of recorded number of stress cycles to allowable fatigue cycles (n/N) was then calculated for each bucket. The summation of these ratios $\left(\Sigma \frac{n}{N} \right)$ gives the Miner summation, which is a linear cumulative damage summation based on 'Miners Rule', devised by Miner (1945). The Miners sum is a damage fraction. This is a ratio of the fatigue life that has been consumed so far based on all speed cycles. A value of 0.5 means 50 per cent of the fatigue life has been consumed. This is the most widely used cumulative damage model for fatigue failures, and is the damage model used in BS 7608.

The resulting fatigue data from two fans can be seen in Table 1 which shows the stress ranges, actual number of cycles (n), allowable number of fatigue cycles (N), and the ratio of these two numbers (n/N). At the foot of the table, the Miner sum is calculated. The number of cycles are also shown in graphical form in Figure 4.

TABLE 1

Rain flow results.

Bucket stress range		Fatigue life in cycles (N)	West fan		East fan	
Start	End		No. cycles (n)	n/N	No. cycles (n)	n/N
0	20	190 000 000	16 318.5	0.01%	51 200.5	0.03%
20	40	23 750 000	8399	0.04%	8720	0.04%
40	60	7 037 037	6413	0.09%	6774	0.10%
60	80	2 968 750	4133	0.14%	4526	0.15%
80	100	1 520 000	2506	0.16%	2680	0.18%
100	120	879 630	1541	0.18%	1682	0.19%
120	140	553 936	3450	0.62%	3478	0.63%
140	160	371 094	607	0.16%	716	0.19%
160	180	260 631	370	0.14%	415	0.16%
180	200	190 000	173	0.09%	190	0.10%
200	220	142 750	71	0.05%	88	0.06%
220	240	109 954	44	0.04%	49	0.04%
240	260	86 482	50.5	0.06%	48.5	0.06%
260	280	69 242	36	0.05%	30	0.04%
280	300	56 296	50	0.09%	67	0.12%
300	320	46 387	230	0.50%	223	0.48%
320	340	38 673	26	0.07%	18	0.05%
340	360	32 579	23	0.07%	25	0.08%
360	380	27 701	39	0.14%	33	0.12%
380	400	23 750	62	0.26%	82	0.35%
400	420	20 516	2	0.01%	2	0.01%
Miner's Sum:			2.97%		3.16%	

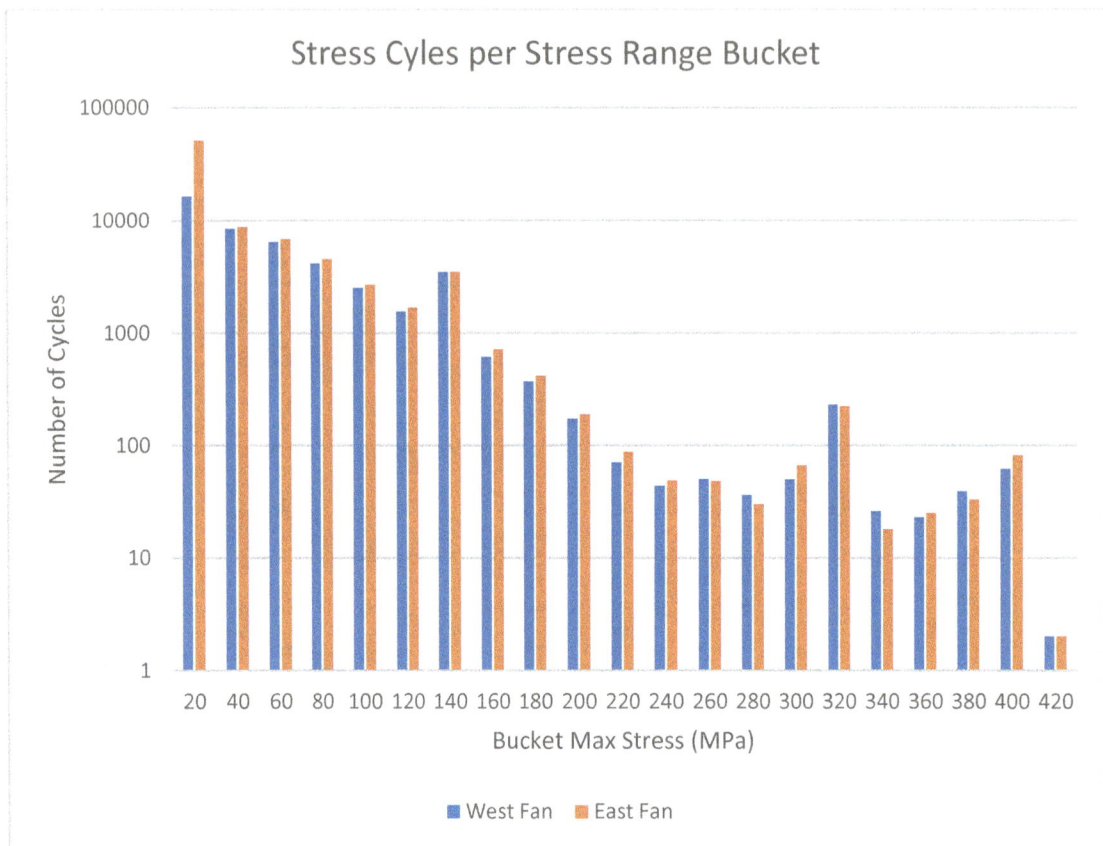

FIG 4 – Number of cycles versus bucket maximum stress.

DISCUSSION ON RESULTS

The west fan has around 45 000 speed cycles in 5.96 years of operation, with an average of 20.5 per day. The east fan has substantially more, with around 81 000 cycles in 5.96 years of operation and an average of 37.3 per day.

Figure 4 shows that most cycles are at very low stress ranges, with the number of cycles substantially dropping off as the stress range increases. The Y axis is a logarithmic axis, so the difference in cycle count between the low stress and high stress is significant.

The Miner sum results show that the west fan has consumed 2.97 per cent of the total fatigue life in ~6 years of operation, and the east fan has consumed 3.16 per cent. This can be used to estimate the total and remaining fatigue life (assuming similar operational characteristics). For the west fan, this is 219 years fatigue life, with 213 years remaining. For the east fan, this is 206 years fatigue life, with 200 years remaining. The results are very similar for the two fans, despite a significantly lower number of speed cycles. The reason for this is that the majority of the additional cycles are in the 0–10 MPa stress range, so do not have any significant impact on the Miner sum.

For both fans it can be seen the Miner sum is very low, and the predicted fatigue life is very high. This is due to several reasons

- Maximum stress in the impeller is reasonably low.

- Most cycles have a very low stress range, where the allowable number of fatigue cycles is very high.

- The fan is run continuously, with a minimal number of start/stops experienced in the six year period (~100 start stops, or ~17/annum).

- Fan speed is consistently in the 550–700 rev/min range and not regularly used at low speeds, so stress cycles are low.

- The operating regime for the fans is not aggressive. Damping in the control loop prohibits rapid speed changes and hunting (where the speed is constantly increasing and decreasing speed to try and match a target ventilation output).

CONSIDERATION OF HIGHLY STRESSED FAN DESIGNS

The authors note that the two fans studied in this paper have relatively low operating stresses, which contributes to a reasonably high fatigue life. Fan blade quality and type are factors that are part of fatigue analysis. Fans using aerofoil bladed impellers and quenched and tempered steel materials have significantly higher operating stresses, up to 700 MPa peak stress. The authors believe the use of VOD on a highly stressed fan could have a more significant impact on fatigue life and therefore increased maintenance costs.

To compare blade types, the same base speed data was used, but with a higher design stress of 700 MPa at 710 rev/min design speed. This increased all the stress values, and therefore reduced the fatigue life for each stress range bucket. The results of this analysis are shown in Table 2.

TABLE 2

High design stress results.

Bucket stress range		Fatigue life in cycles (N)	West fan		East fan	
			No. Cycles		No. Cycles	
Start	End		(n)	n/N	(n)	n/N
0	35	35 527 972	16 318.5	0.05%	51 200.5	0.14%
35	70	4 440 997	8399	0.19%	8720	0.20%
70	105	1 315 851	6413	0.49%	6774	0.51%
105	140	555 125	4133	0.74%	4526	0.82%
140	175	284 224	2506	0.88%	2680	0.94%
175	210	164 481	1541	0.94%	1682	1.02%
210	245	103 580	3450	3.33%	3478	3.36%
245	280	69 391	607	0.87%	716	1.03%
280	315	48 735	370	0.76%	415	0.85%
315	350	35 528	173	0.49%	190	0.53%
350	385	26 693	71	0.27%	88	0.33%
385	420	20 560	44	0.21%	49	0.24%
420	455	16 171	50.5	0.31%	48.5	0.30%
455	490	12 948	36	0.28%	30	0.23%
490	525	10 527	50	0.47%	67	0.64%
525	560	8674	230	2.65%	223	2.57%
560	595	7231	26	0.36%	18	0.25%
595	630	6092	23	0.38%	25	0.41%
630	665	5180	39	0.75%	33	0.64%
665	700	4441	62	1.40%	82	1.85%
700	734	3836	2	0.05%	2	0.05%
Miner's Sum:			15.87%		16.91%	

This change in design stress has resulted in a larger Miner sum of 15.9 per cent and 16.9 per cent for the west and east fans respectively. The resultant predicted life is 37 years (31 years remaining), and 35 years (29 years remaining) for the two fans.

While these reduced fatigue lives are substantially lower than the original calculated values, they are still in excess of the fan design life of 25 years, so would not be expected to contribute to increased maintenance costs with the current operating regime. If a more aggressive control regime is combined with a high design stress impeller, this combination could result in increased maintenance costs.

OPERATING COST SAVINGS DUE TO VOD

The two exhaust fans discussed are presently operating under speed optimiser control as described in the introduction of this paper with energy savings being achieved. In addition to speed optimiser, an event is scheduled (scheduling control) at the end of each shift to decrease the fan speed for about 40 minutes, and thus, further reducing its energy consumption. These two components are included within the savings calculation.

To establish a business case, a base case must be established. The base case considers the VOD system not operational, the exhaust fans will operate at a constant speed of 690 rev/min during the colder months and 695 rev/min during the warmer months of the year. The exhaust fans' operating power and the electric motors' speed is recorded and used to calculate the energy savings. Data from 5 January 2012 to 9 December 2012 has been used for the base case, as it represents almost a full year where the exhaust fans were mostly operating at 690 rev/min or above, prior to the use of speed optimiser control. The data verifies if the exhaust fans' speed are above 690 rev/min, and used to calculate the average operating power at that speed. It is important to note that there are some instances where the data power readings are erroneous; for example, the power reading states 0 kW, even though the fan is operating. The data was cleaned to avoid these instances in the calculations. The total average operating power (for both fans) at speeds of 690 rev/min (or higher) was calculated at 5821 kW.

For the VOD energy savings evaluation, 20 September 2017 to 20 September 2018 was chosen as a data range as it provides a full year and deemed sufficiently representative of the typical operation. Within the selected period of time, the speed optimiser was operational with the exception of an underground regulator communication failure, which can change the exhaust fans to manual mode for several hours or days. The power consumption during communication failure is still included within the calculations (failures will always occur).

For the energy savings period of time, the average power consumption of the exhaust fans, using the recorded data, is 4580 kW; 1241 kW less power consumed than the base case. Assuming that the power savings are constant throughout the year, that is 365 days a year and 24 hours a day, the estimated annual energy savings would be 10 874 MWh. At an electrical energy cost of $0.075/kWh, this is equivalent to an annual cost savings of $816 000 CAD (~$870 000 AUD) due to speed changes of the exhaust fans and a reduction of 21 per cent compared to the base case. Applying the savings over six years of operation, that the fatigue calculations are based, results in an estimated savings of $4.9M CAD (~$5.2M AUD).

Additional maintenance costs due to VOD

Given the low Miner sum and the very high predicted fatigue life, it is very unlikely that fatigue cracking will occur in the lifetime of the fan installation, so the impeller life will be limited by other normal operational reasons (erosion, corrosion, or mine operating life). For this specific example, the use of VOD is not expected to result in a requirement for any additional impeller replacements or maintenance costs. There will be costs associated with the implementation, operation and maintenance of the VOD system itself, but these costs are not reviewed in this paper.

The authors note the operating cost saving is substantial, and if the fatigue accumulation was much higher with the impeller being replaced every two to five years, due to fatigue damage, the net power savings would still be significant. Based on this analysis of the data and the cost savings achieved, the additional maintenance costs due to fatigue would not exceed the VOD savings.

CONCLUSION

For the given mine ventilation fan's speed controlled by VOD, the consumed fatigue life during a six year period is approximately 3 per cent, and the estimated fatigue life of the impellers based on the current operating regime is over 200 years which significantly exceeds the design life of the fans. The frequency and magnitude of the speed changes are not excessive when compared to industrial or power station fans which typically have significantly greater frequency and magnitudes changes of speed cycles.

It was shown that no additional impeller replacements or associated maintenance costs should be required due to the use of VOD, and the full operating cost saving due to the use of VOD can be realised by the end user, less the costs to implement the VOD system, such as controls, sensors and licensing fees.

The results in this paper are only applicable to the specific fans analysed, and results may vary for other fans that have multiple start/stops in one day, or more significant speed variations. However, this information is useful in that it is not based on theoretical numbers, but on actual data from VOD controlled fans installed in a working mine primary ventilation system.

The authors note this approach can be used to calculate the consumed and remaining fatigue life on any fan where sufficient speed data exists. The opportunity exists to incorporate a real time monitoring system, to estimate the remaining fatigue life of the impeller (based on the previous 1 month, 3 months, 6 months or 1 year of operating data). This information would be very useful to understand impeller life to predict and schedule maintenance or if possible to make changes to reduce the frequency and magnitude of stress cycles increasing the remaining fatigue life. This analysis approach is incorporated into the Howden Uptime system.

ACKNOWLEDGEMENTS

The authors would like to acknowledge Muhammad Shariq Khan, for carrying out the finite element analysis, Pavel Bartak, Florian Michelin and Leo Botha for their assistance in sourcing the raw data and the operating cost analysis, and Stephan Bergh, for his assistance on ventilation cost savings. The authors also acknowledge Howden approval to publish this paper.

REFERENCES

American Society for Testing and Materials, 1985. ASTM E1049–85 – Standard Practices for Cycle Counting in Fatigue Analysis, 1985.

British Standards Institution, 2014. BS 7608:2014 – Guide to fatigue design and assessment of steel products, March 2014.

Eskom Holdings Limited, 2010. The Energy Efficient Series: Towards an energy efficient mining sector, February 2010.

Miner, M A, 1945. Cumulative Damage in Fatigue, Journal of Applied Mechanics, 12(3):A159-A164.

Pinedo, J and Torres Espinoza, D, 2019. Implementation of Advanced Control Strategies Using Ventsim Control at San Julian Mine in Chihuahua, Mexico, 17th North American Mine Ventilation Symposium (Montreal), 2019:435–441.

Sanftenberg, J, Cameron, C, McLaren, E and Dello Sbarba, H, 2019. Mine Ventilation on Demand System at Nickel Rim South Mine; Testing Procedures, System Maintenance and Results, 17th North American Mine Ventilation Symposium (Montreal), 2019:104–113.

Comparative performance of single and tandem stator arrangements in a one stage ventilation fan

G C Snedden[1,2], P Rossouw[3] and J C F Martins[4]

1. Professor of Aerospace Systems, University of KwaZulu Natal, Durban, KZN, 4001 South Africa. Email: sneddeng@ukzn.ac.za
2. Director of Research, Air Blow Fans, Pretoria, Gauteng, 0133, South Africa.
3. Senior Design Engineer, Air Blow Fans, Pretoria, Gauteng, 0133, South Africa. Email: pieter@aiblowfans.co.za
4. Engineering Manager, Air Blow Fans, Pretoria, Gauteng, 0133, South Africa. Email: jose@aiblowfans.co.za

ABSTRACT

Tandem row stator arrangements are common in the last stage of multirow, gas turbine compressors. Tandem row stators, or stator and de-swirler vane combinations are used to relieve the loading on the stator but ensure axial flow leaving the compressor. In this paper a standard 40" (1016 mm) ventilation fan in the Air Blow Fan's range is modified to utilise the plate support struts on the barrel to relieve the aerodynamic loading on the aerofoil stator. Performance maps for both versions of the fan were generated using computational fluid dynamics and are compared across a wide range of rotor setting angles showing a clear advantage of approximately 2 per cent in efficiency for the tandem blade design. In addition, the area of high efficiency that represents the practically usable area of the map extends over a greater area in terms of flow rate and total pressure rise.

INTRODUCTION

Two new aerodynamic features where experimented with at Air Blow Fans during the design process of a new fan range. The first was bowing of the stator vanes which allowed for far greater turning in the stator vane without corner separation developing at the endwall junction. This however raises concerns over the manufacture of these bowed vanes. An alternate option was sought to relieve the turning in the aerofoil stator and it was decided to investigate the use of the plate support strut downstream on the barrel to de-swirl the flow. The aerofoil stator in this case was slightly twisted but not bowed. A fan representing this design with the slightly twisted aerofoil stator and cambered plate support strut as a second stator row has been fabricated and tested, Snedden and Rossouw (2021).

These aerodynamic features were applied in the design of a standard 1016 mm diameter secondary fan. This paper describes the results of both bowed stators and tandem blade rows in a standard 40" (1016 mm) diameter secondary fan (Figure 1).

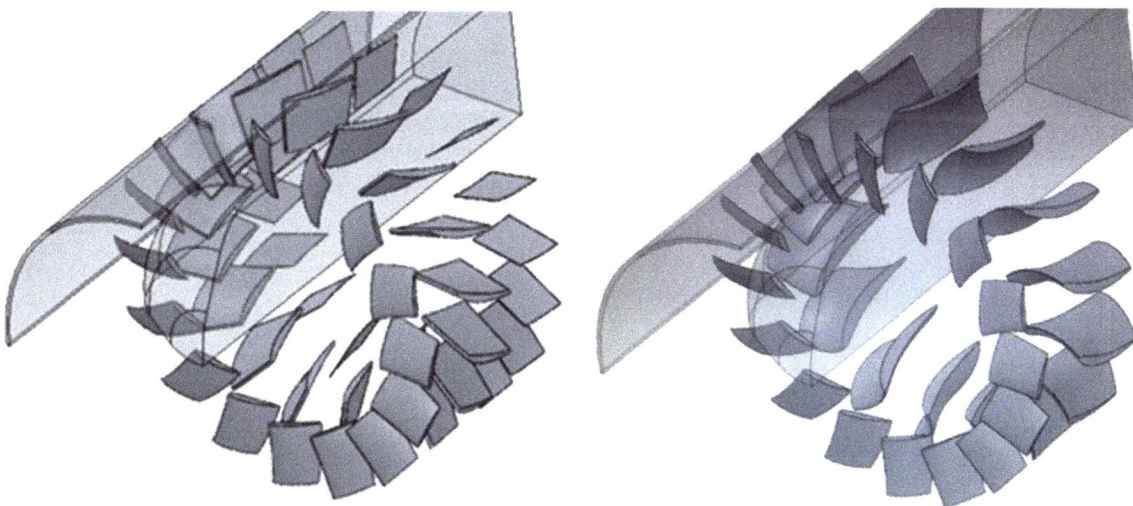

FIG 1 – Tandem stator design (left), bowed stator design (right).

Benefits of bowing and twisting stator vanes

It is common to adjust the localised twist of the rotor blade in order to alter the work distribution produced in the rotor or increase the product of radius and outlet absolute tangential velocity at midspan and to alleviate secondary or loss generating flows in both turbine stators and rotors (Lampart *et al*, 1999; Watanabe and Harada, 1999).

In compressors or fans, where the pressure gradients adversely affect the onset of secondary flows the turning angles are milder to compensate, localised twist is used to reduce corner separation (see Figure 2), otherwise known as corner stall at the hub which had only been identified in the 1980s. In the stator row localised twist can be used in the same manner, but in addition the twist can be used to compensate for any off-design radial variation in the flow angle emerging from the rotor row.

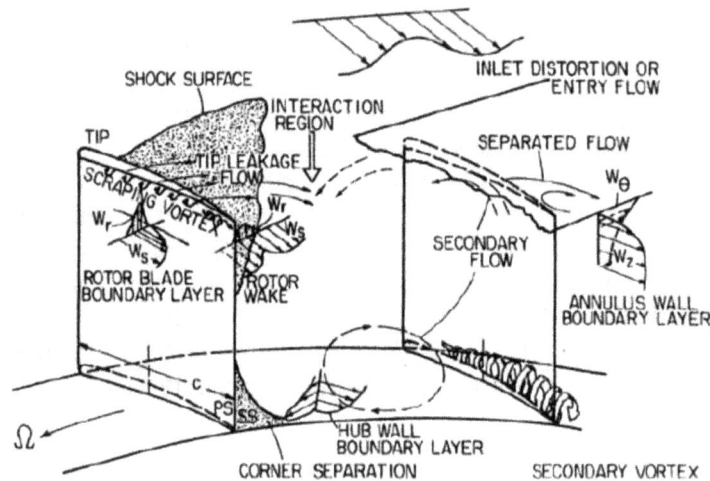

FIG 2 – The nature of flow in an axial compressor rotor passage (Lakshminarayana, 1996).

Bow, is also known as dihedral or compound lean in the literature and has been studied in compressors since the 1990s to address secondary flows. Weingold *et al* (1997) describes the design of a compound bowed stator designed to reduce corner stall and the resulting flow blockage in the compressor achieving more than 1 per cent efficiency improvement. They describe bowing as reducing diffusion rates in the suction surface corner and delaying or eliminating the formation of corner separation.

Sasaki and Bruegelmans (1998) showed that bowing increased loading at midspan and that dihedral made the blade row relatively insensitive to incidence compared to sweep. Gallimore *et al* (2002a, 2002b) report a 5 per cent pressure increase over the entire speed line for a 3D optimised multistage compressor utilising free form sweep and bow with the rotors largely utilising sweep and the stators, compound bow limited to close to the hub and tip. Fischer *et al* (2004) applied strong bowing, 30° at tip and 35° at the hub, to the stators of a four-stage compressor and achieved a 1.4 per cent increase in pressure rise which was ascribed to the shift of flow towards the midspan and the reduction of corner separation as a result of unloading the blade in the proximity of the endwall (see Figure 3).

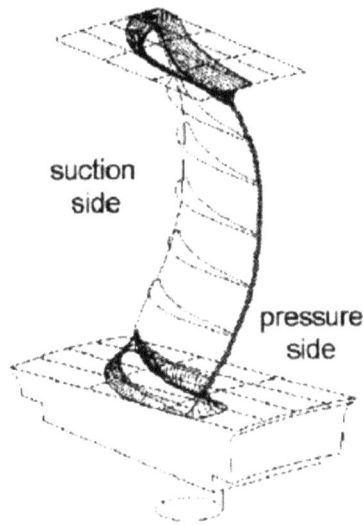

FIG 3 – A heavily bowed Stator after Fischer *et al* (2004).

The Air Blow Fans standard 40" (1016 mm) fan

This fan, with a 23" barrel was selected as the test model and as such was used for the analysis of these aerodynamic features applied to the stator vanes. The current design for this axial flow fan features C4 rotor blading with twisted aerofoil profiles and radially optimised camber. The blades are forward swept, and bowed below 50 per cent span and have irregularly spaced tubercles. The stator blading is of a tandem nature with the plate struts taking half the turning at the design point and are straight with a constant rolled camber. The aerofoil stator vanes are modified NACA65 profiles with a slight twist from hub to tip. The design is very similar to that pictured in Figure 1 (left).

The performance of this design is shown in Figure 4. Maximum total to total efficiency is in the region of 85.5 per cent at a rotor blade tip angle of 58° after accounting for the expansion behind the barrel without a cone. The rotor blade setting angles are defined as the angle of the chord line at the tip of the blade, adjacent to the fan casing, relative to the axis of rotation of the rotor. This fan configuration was tested on a test duct and the maximum efficiency achieved for the test curve was 3.5 per cent lower than predicted and the efficiency at the duty point was 5 per cent lower than predicted. This can be attributed to inaccuracies in the manufacturing methods and also due to differences in the geometry of the impeller dome. The CFD results were predicted with an impeller dome with an ideal bullet nose geometry whereas the actual impeller dome used in the test results was a 10 per cent torospherical spun dome which has a much flatter profile than the impeller dome geometry used in the CFD analysis.

1016 Straight Tandem Stator

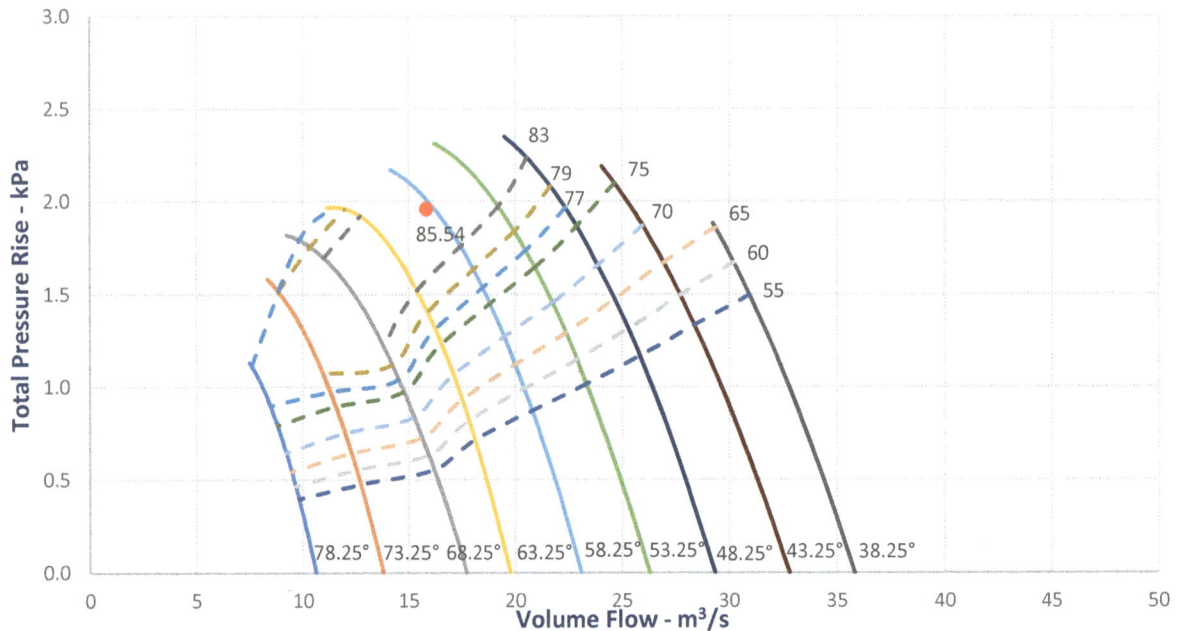

FIG 4 – The standard 1016 fan performance map, straight vaned tandem row stators.

The modified geometries under investigation

The aim of this paper is to investigate whether a new bowed stator arrangement for this fan would still benefit from a tandem row design. For this purpose, two new geometries have been created. The first being a single bowed and leant stator capable of removing swirl to less than 3.5°. Also included in this model is the simple straight support strut included for completeness. The second model has the aerofoil stator reduced in camber by half with the same bow, lean and chord as the first case. Outlet swirl is roughly the same as in the previous case. The strut is increased in chord and rolled to accept the incoming swirling flow but is straight otherwise. The same blade counts of eight rotors and nine stator vanes is retained as is the original rotor blade design. The only other difference between these configurations and the standard design mapped in Figure 4 is a modification to the hub, eliminating the convex contour.

Axial extension plates to the stators are not considered here as the stators are bowed and this practise while reducing the barrel length does not allow for indexing of the second row to improve the de-swirl efficiency of the second stator row.

COMPUTATIONAL FLUID DYNAMIC (CFD) ANALYSIS

Table 1 indicates the details of the modelling approach taken in the NUMECA FineTURBO® package for the purposes of this study. Between five and ten points are run per setting angle and these take approximately 3.5 hours on a standard i7 laptop with 8 GB per core. The runs are undertaken in parallel on three cores. The rotor is re-meshed at each setting angle.

TABLE 1

CFD Parameters.

CFD model	Parameters	Setting
Package	Numeca	FineTurbo meshed with Autogrid 5 release 17.1
Mesh	No. of cells	3 332 700
	Type	Structured hexahedral
	Tip gap	2 mm
	Fillets	5 mm (Stators only)
	Near wall cell	0.002 mm yields y^+ in the region of 1
	Geometry	Single passage per blade row, inlet bulb and tandem row enabled
Solver settings	Gas	Air (Perfect Gas Law)
	Turbulence	Spalart-Allmaras
	Multigrid levels	3
	Course grid iterations	100
	Fine mesh iterations	1000 (CPU Booster is not enabled)
	Inlet boundary	Total pressure set to standard temperature and pressure. Flow normal to boundary.
	Outlet boundary	Pressure adaption massflow with radial equilibrium and backflow control enabled.
	Rotor/stator mixing plane	Steady

Figures 5 and 6 are included to show the overall geometries in question and at the same time to indicate the density and quality of the mesh. The mesh shown is however the medium coarseness multigrid level and not the final mesh. The Inlet bulb is included in the simulation but the mesh ends in the annulus behind the strut and therefore the sudden expansion behind the barrel is not modelled in the CFD.

All runs were performed at 1775 rev/min equivalent to 60 Hz four pole and the performance maps generated reflect this speed. Inlet conditions are near 1.2 kg/m³ but are normalised to this condition as the maps can be easily scaled to a given speed and density.

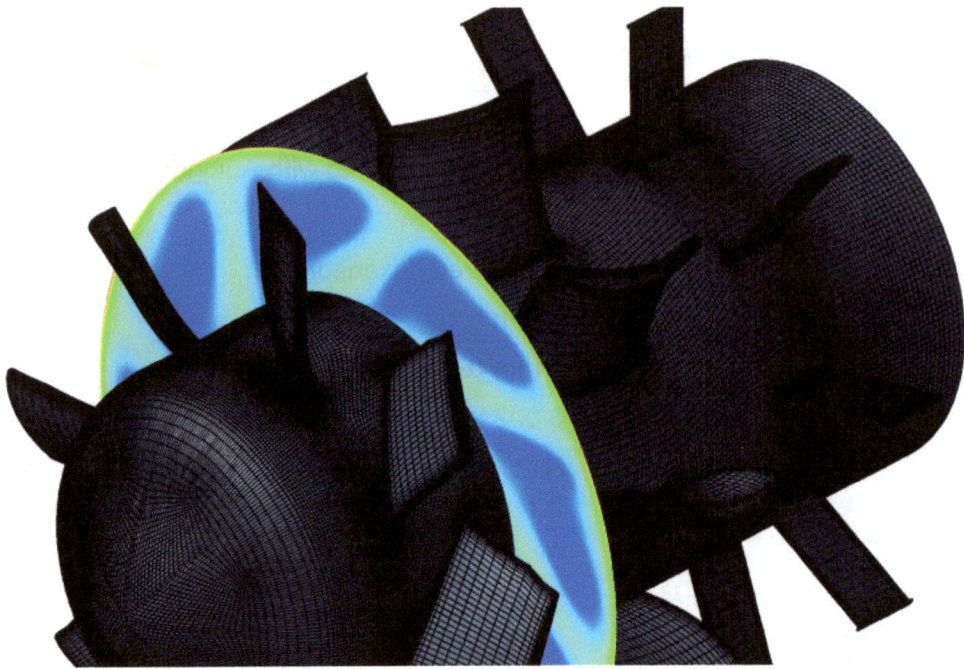

FIG 5 – The 1016 fan with a bowed and leant aerofoil stator designed to de-swirl the flow in a single row and a simple straight support strut for structural purposes.

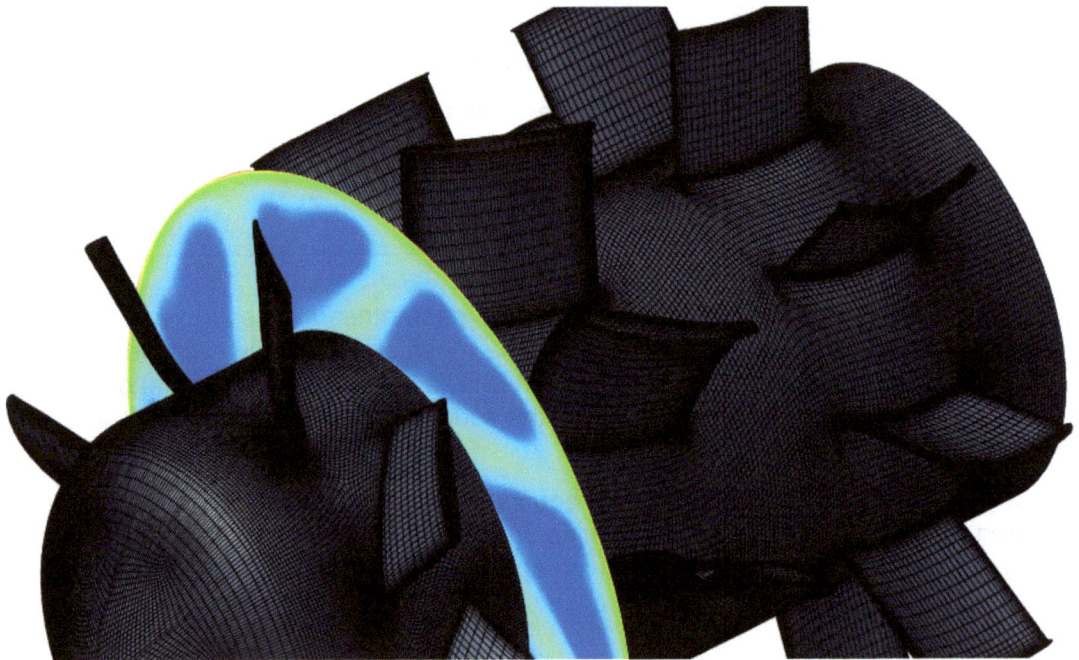

FIG 6 – The 1016 fan with a tandem stator arrangement featuring the same bow and lean as the design above in the aerofoil stator but with half the turning and a straight, rolled plate tandem de-swirl vane/strut.

The sudden expansion is accounted for in the mapping by including a total pressure loss and, if required, a static pressure increase, based on the Carnot-Borda equations for frictionless sudden expansions.

Each set of points generated by CFD at a given setting angle has a quadratic curve fitted to it and this is used to extend the curve to zero pressure. A similar approach is used to find the peak efficiency for each curve and again through the peaks to locate an approximate maximum efficiency point. Efficiency bands are created by interpolating points of the same efficiency on each of the setting angle curves. This process is performed in Excel and is automated as far as possible.

Figure 7 shows quite clearly in the reduction slow moving flow, how the reduction in loading allows the flow to stay attached over a far greater volume of the stator passage in the tandem design at the design point.

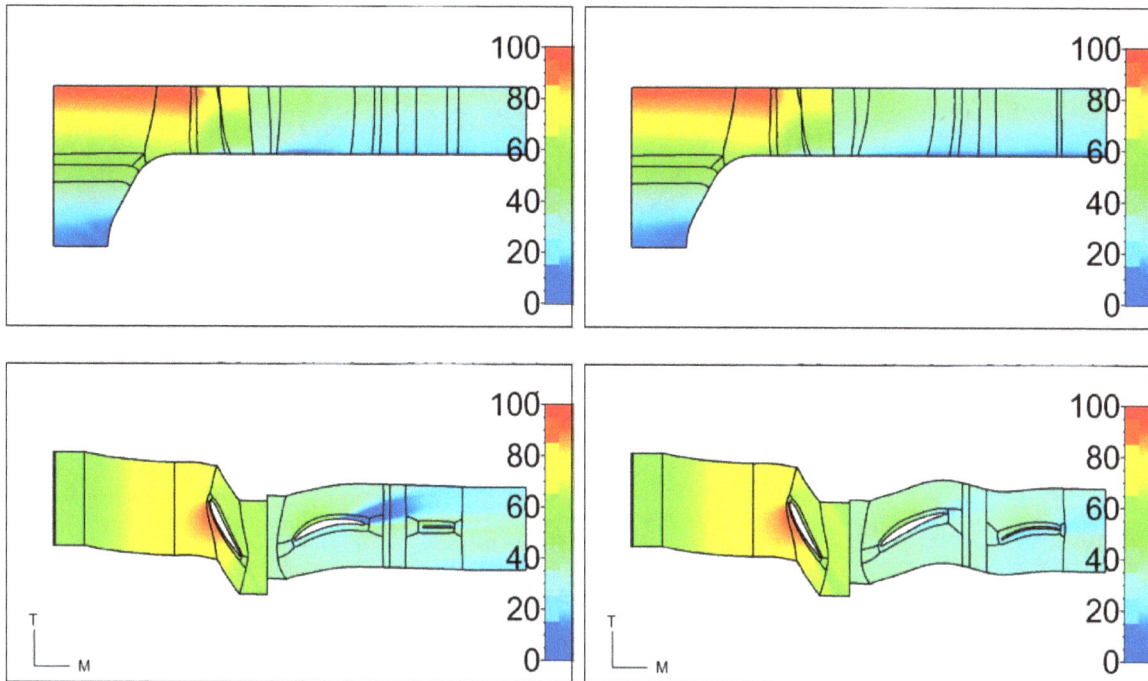

FIG 7 – Meridional and Azimuthal averaged relative velocity contours (m/s) at design for Single (left) and Tandem (right) fans.

RESULTS FROM THE UPDATED DESIGN

The results generated using the CFD and mapping approach described in the previous sections are given in Figures 8 and 9.

The first thing to note is that the maximum efficiency of the tandem arrangement is nearly 2 per cent higher than the single stator model, reaching 86.48 per cent. This is 1 per cent greater than that for the original straight stator. Therefore, despite the increased wetted area and consequent skin friction drag of the tandem stator arrangement it is clearly a better choice from an efficiency point of view than a single row of stators, even when comparing a tandem straight stator to a single bowed stator from a maximum attainable efficiency point of view.

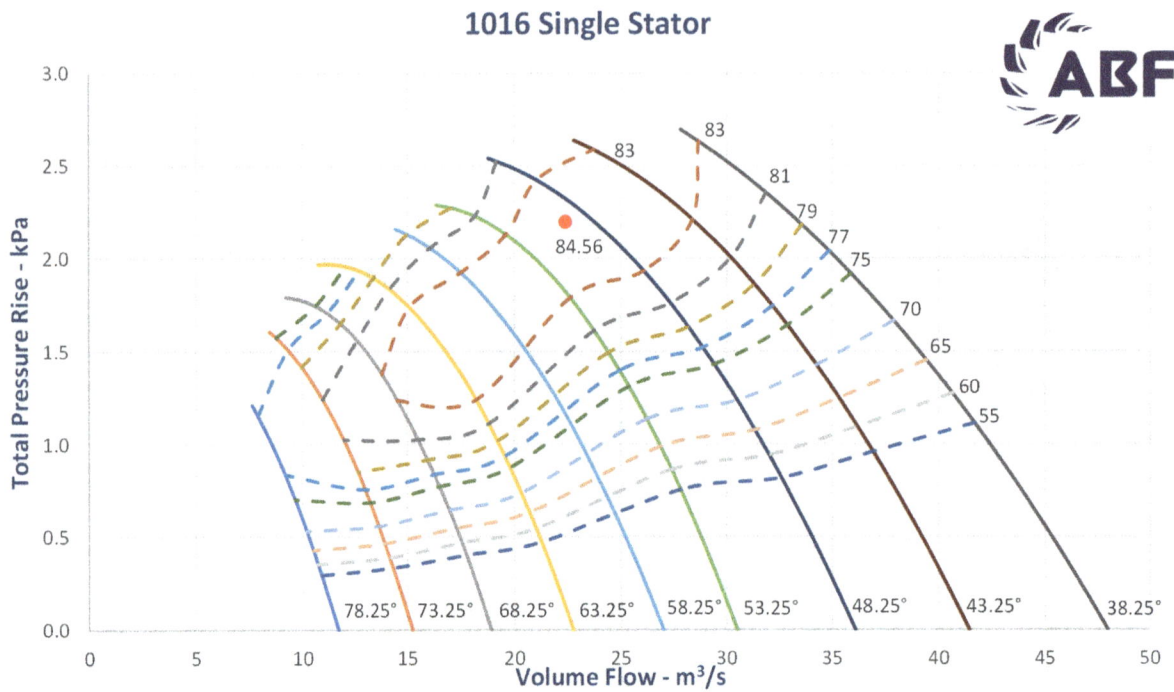

FIG 8 – The standard 1016 fan performance map, straight vaned tandem row stators.

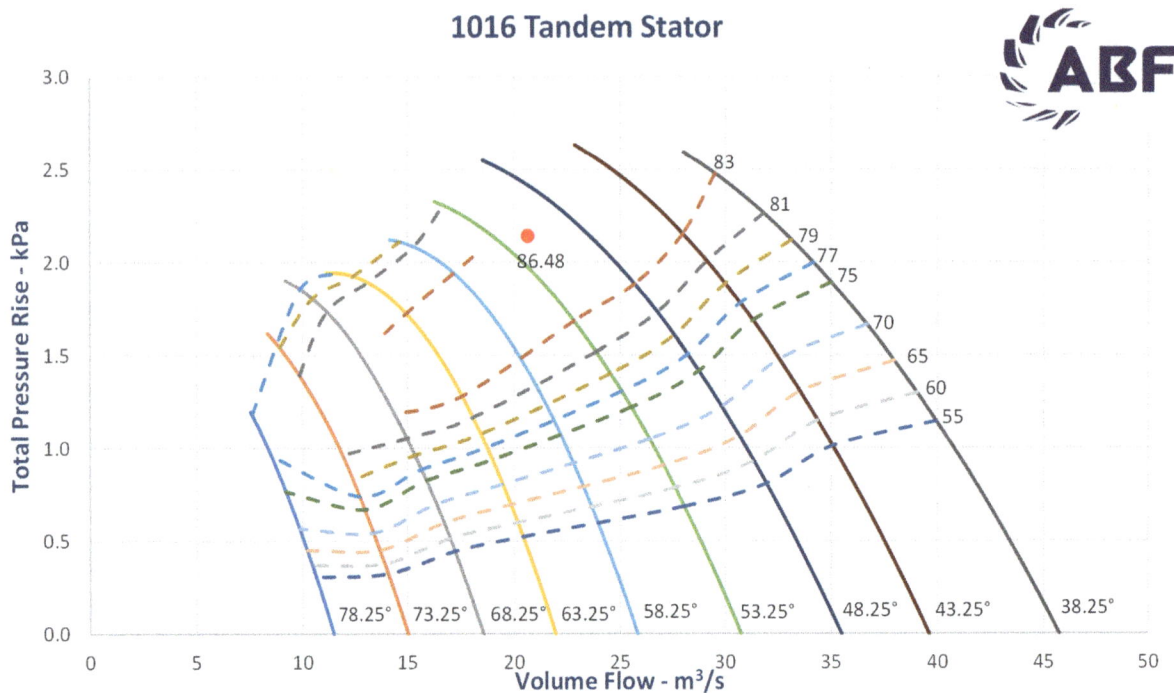

FIG 9 – The standard 1016 fan performance map, straight vaned tandem row stators.

The point of maximum efficiency has moved from approximately 58° in the case of the straight stators to between 48° and 53° which is an unexpected result, but may be due the incidence setting selected for the aerofoil stator. The efficiency at 58° remains good however, in fact the ridge of high pressure on the map is broad and relatively flat.

For both cases with bowed and leant aerofoil stators the band of high efficiency extends from 38° to 68° rotor setting angle starting well below surge or stall at the low angles and moving closer to surge at the high setting angles. The tandem design however maintains a consistently higher value of efficiency over this entire area. As the curves drop to lower pressure rise values however the tandem stators loose efficiency at a higher rate that the single stator case. But this effect is marginal in terms of efficiency value given the higher starting point.

The most notable effect of bowing the stators when compared to the straight stators is the overall extent of the performance map, and here it is important to note that the three maps provided all have the same axes with the bowed blades filling the graph and the straight stator map limited to values below 2.5 kPa pressure rise and 37 m³/s. Maps for the bowed blades stretch from the same starting point a 68° rotor setting angle to well over 45 m³/s and 2.5 kPa at 38° rotor setting angle indicating an improved delivery and not just an improved efficiency.

Bowing the blades has arguably shown to have a greater improvement on the performance map of the fan by extending the usable range of the fan across a broad range of rotor setting angles and expanding the range of the fan performance. Without utilising the tandem stator concept however, the fan efficiency would lose 1 per cent efficiency against that of the existing tandem stator design with straight stators.

STRUCTURAL CONSIDERATIONS

Although the aerodynamic benefits of the bowed stators were demonstrated above, the concern exists that bowed stators may be detrimental to the structural integrity of the fan and especially to the stiffness. It is not the purpose of this paper to provide a full structural analysis of these designs but it can be stated that although the structural stiffness of the fan with bowed stators is lower than that of the straight stators not of such a magnitude to cause a concern for this fan.

A modal analysis was carried out of this 40" fan with bowed stators and straight stators. The lowest natural frequency for the straight stators was 88.8 Hz while the lowest natural frequency for the bowed stators was 80 Hz. Both of these frequencies are well above the operating speeds of these fans, even with two pole motors so both designs are structurally sound.

CONCLUSIONS

The introduction of bowed stators into a low-pressure ventilation fan has the effect of expanding the operational range of the fan at high efficiency levels to a band ranging from 38° to 68° rotor setting angles. This is a significant improvement when compared to the straight stator design which shows a more typical bubble of high efficiency between 68° and 53° setting angle. This is due to the unloading of the stator hub due to the bowing which delays the stall of the stator in the hub region.

By further unloading the aerofoil stator utilising the aft strut on the barrel as a de-swirl vane the operational range of the fan is maintained and despite the increased wetted area of the struts, the efficiency is again improved, by as much as 2 per cent over an equivalent single stator design. The drawback might be the proximity of this high efficiency area to surge or stall at the higher setting angles.

ACKNOWLEDGEMENTS

The management of Air Blow Fans, in particular Mr Gavin Ratner for funding this work and allowing its publication.

REFERENCES

Fischer, A, Riess, W and Seume, J R, 2004. Performance of Strongly Bowed Stators in a Four-Stage High-Speed Compressor, *Transactions of the ASME Journal of Turbomachinery*, 126:333–338, July 2004.

Gallimore, S J, Bolger, J J, Cumpsty, N A, Taylor, M J, Wright, P I and Place, J M M, 2002a. The Use of Sweep and Dihedral in Multistage Axial Flow Compressor Blading – Part I: University Research and Methods Development, *Transactions of the ASME Journal of Turbomachinery*, 124:521–532, October 2002.

Gallimore, S J, Bolger, J J, Cumpsty, N A, Taylor, M J, Wright, P I and Place, J M M, 2002b. The Use of Sweep and Dihedral in Multistage Axial Flow Compressor Blading – Part II: Low and High-Speed Designs and Test Verification, *Transactions of the ASME Journal of Turbomachinery*, 124:533–541, October 2002.

Lakshminarayana, B, 1996. *Fluid Dynamics and Heat Transfer of Turbomachinery*, John Wiley and Sons Inc.

Lampart, P, Gardzilewicz, A, Rusanov, A and Yershov, S, 1999. Effect of stator blade compound lean and compound twist on flow characteristics of a turbine stage – numerical study based on NS simulations, *Trans-actions of ASME, Pressure Vessels and Piping*, 397:195–204.

Sasaki, T and Bruegelmans, F, 1998. Comparison of Sweep and Dihedral Effects on Compressor Cascade Performance, *Transactions of the ASME Journal of Turbomachinery*, 120:454–463, July 1998.

Snedden, G C and Rossouw, P, 2021. The Case for Advanced 3D design of Ventilation Fans – Overcoming the Cost of Capital, *MVSSA Conference 2021*.

Watanabe, H and Harada, H, 1999. Suppression of secondary flows in turbine nozzle with controlled stacking shape and exit circulation by 3D inverse design method, *ASME 99* GT-72.

Weingold, H D, Neubert, R J, Behlke, R F and Potter, G E, 1997. Bowed Stators: An Example of CFD Applied to Improve Multistage Compressor Efficiency, *Transactions of the ASME Journal of Turbomachinery*, 119(2):161–168, April 1997.

Quantitative airblast risk assessments for block and sublevel caves

C Vejrazka[1]

1. Principal Caving and Ventilation Consultant, Resolve Mining Solutions, Mountain Gate Vic 3156. Email: claudia@resolvemining.com.au

ABSTRACT

The occurrence of airblasts is one of the principal mining hazards in caving operations between the time of cave establishment and breakthrough of the cave to surface. As such, it needs to be managed continuously and risk assessments need to be reviewed on a regular basis for applicability to changing conditions. The likelihood of an airblast occurring is based on a range of geotechnical factors over some of which engineers on-site only have limited control. However, the impact of an airblast to the underground mine, if it does occur, can be managed by following established safety guidelines for the production from the cave.

This paper outlines how the Caving Airblast Simulation Tool in Ventsim was used to determine a risk management strategy for a sublevel and a block cave at a mining operation in Australia. The block cave has a footprint of 14 500 m^2 with the extraction level located 1250 m below surface while the sublevel cave currently has a footprint of around 18 000 m^2 operating between 660 m and 800 m below surface. Both caves are currently in their early operational periods. The risk analysis provided valuable information for the production planning of the caves including confirmation of extraction ratios for the block cave undercut, locations of airblast walls and maximum allowable air gaps for different muck pile heights to ensure safe operation is achieved.

INTRODUCTION

The block cave assessed as part of this case study is in an existing mining operation below a sublevel cave that was extracted in the past. The sublevel cave is situated in another orebody at the same mine site. In July 2022, both the block cave and sublevel cave were still in development, with sublevel cave mining its seventh successive level and the block cave having approximately 75 per cent of the undercut completed. Current underground workings extend to 1350 m below surface and laterally extend approximately 3 km between the two orebodies (Figure 1).

This paper details the process whereby the potential airblast impact for both of these caves was assessed.

FIG 1 – Long section of the mine.

VENTSIM CAVING AIRBLAST SIMULATION TOOL (CAST)

The Caving Airblast tool available in Ventsim estimates the impact of an airblast within the mine network. The main benefit is that downstream mine resistances can be modelled dynamically as the airblast progresses. Additionally, the impact of cave connections at differing levels can be modelled as they connect to the cave.

Prior to the tool being available, airblasts were simulated using a combination of excel spreadsheets and the VentFire tool within Ventsim (Vejrazka, 2016).

The CAST modelling is based on the following features:

- Isentropic compression of the air within the air gap of the cave assuming a remaining air gap of 5 m at the end of modelling. This is a limitation of the isentropic compression model as pressures approach infinity as volume approaches zero.

- Air velocities in the mine network are calculated at each timestep based on the pressure differentials within the mine network.

- Network pressures are based on the interaction with the cave at each time step.

- At each timestep, the full mine network is modelled to a steady state. This might lead to an overestimation of the impact extent of very small events such as stope failures but, in the authors' experience, has not been found to impact sublevel and block caving models.

- All cave connections can be included in the models. Drives higher up in the cave are blocked at the time step when caving material passes the drive elevation. It is thus important that the vertical extent of the model reflects mine conditions as closely as possible.

- Airblast plug loads are determined at each timestep and represent the simulated pressure only. Design safety factors need to be applied to these loads based on the mine's risk management framework to determine design criteria for the civil engineers.

When modelling airblasts in Ventsim, the following limitations are currently applicable:

- Ventsim creates a circular cave model based on an input of cave diameter, muck pile height and caveback height.

- When connecting drives to the cave, only z-coordinates are taken into consideration. If there are significant lateral differences, these must be accounted for manually as it does not use northing or easting as spatial inputs.

- The lowest connected drive in the model is assumed to be the base of the muck pile. This means that muck pile height might need to be adjusted based on what the mine uses as the base of the muck pile.

- The connections to the airblast model were completed in intervals based on user input. The intervals start from the lowest RL. Care should be taken when selecting interval size as this has the potential to affect local airflows significantly. Alternatively, the elevation of the cave connections affected can be adjusted to suit.

- The model assumes that an air gap of 5 m remains at the end of the modelling process. By using adiabatic compression, the pressure in the cave will approach infinity as the air gap approaches zero. In reality, the top part of the muck pile would pressurise as the void pressurises, leading to residual volume in the muck pile. Additionally, it is highly likely that the rock will breakup as the material falls in the cave allowing backflow upwards in the muck pile.

The principal layout of the cave created as part of the simulation is shown in Figure 2.

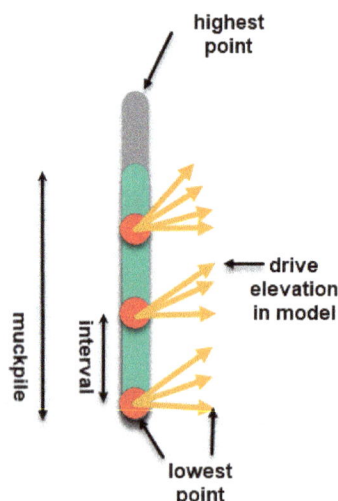

FIG 2 – Graphical input parameters for caving airblast model (cross-sectional view).

Acceptance criteria for maximum air velocities should be determined by the mines based on risk assessments and suitable controls. In absence of risk criteria defined by the mine the following two criteria can be applied and were used in this case:

- 15 m/s for areas with a large amount of foot traffic, and
- 20 m/s for areas with mainly vehicle traffic and return airways.

The basis of these criteria is on research completed on wind blasts in the coal mining industry (Fowler *et al*, 2003) that found that the likelihood of serious injury to a person on foot significantly increases with air blast peaking above 15 m/s, and damage to vehicles is more likely when air velocities exceed 20 m/s.

MODELLING OF BLOCK CAVE AIRBLAST RISK

At the time of the review the mine had completed half of the undercut for the block cave and it was agreed that three stages were to be modelled for the block cave:

1. Half of the undercut excavated. (geometry at time of analysis).
2. The full undercut excavated, and the first half of the drawbells have been established.
3. All drawbells have been established.

Table 1 shows the Ventsim input parameters used for each of the scenario. The range of muck pile resistance is based on experience from previous modelling to create a range from a young developing cave (0.04 Ns^2/m^8 per 10 000 m^2) with low porosity to a more mature cave (0.02 Ns^2/m^8 per 10 000 m^2) with higher porosity. Vejrazka (2016) contains a detailed discussion on muck pile resistance factors for airblast modelling.

TABLE 1

Model simulation parameters for Ventsim.

Input parameters	Half undercut	Full undercut	All drawbells established
Model undercut area (m^2)	7740	14 500	14 500
Min void (m)	5.5	7	10
Max void (m)	8	20	70
Min caveback height (m)	18 (above UC – blasted)	37 (above EL – blasted)	65 (above EL)
Max caveback height (m)	28 (above UC)	65 (above EL)	255 (above EL)
Min muck pile height (m)	12.5	30	55
Max muck pile height (m)	22.5	55	185
Min muck pile resistance (Ns^2/m^8 per 10 000 m^2)	0.02	0.02	0.02
Max muck pile resistance (Ns^2/m^8 per 10 000 m^2)	0.04	-	-

Half undercut simulations

For the first scenario the parameters were set based on the undercut ring geometries and the swell material extracted at the time. Without any drawpoints in place caving would stall once the swell has filled the void. As the CAST simulation assumes a void of 5 m remains to avoid unrealistic pressures, the minimum void was limited to 5.5 m. Figures 3 and 4 show the apex and undercut level at the time of the simulations, respectively. Rather than connecting each individual undercut drive, the drives are connected to the cave inside the ore pillar with the resistance of half the pillar added to the connection. This is due to the assumption in Ventsim that all cave connections on one level are being connected to the same cave node. With direct connections this can lead to unintended flow between adjacent undercut drives.

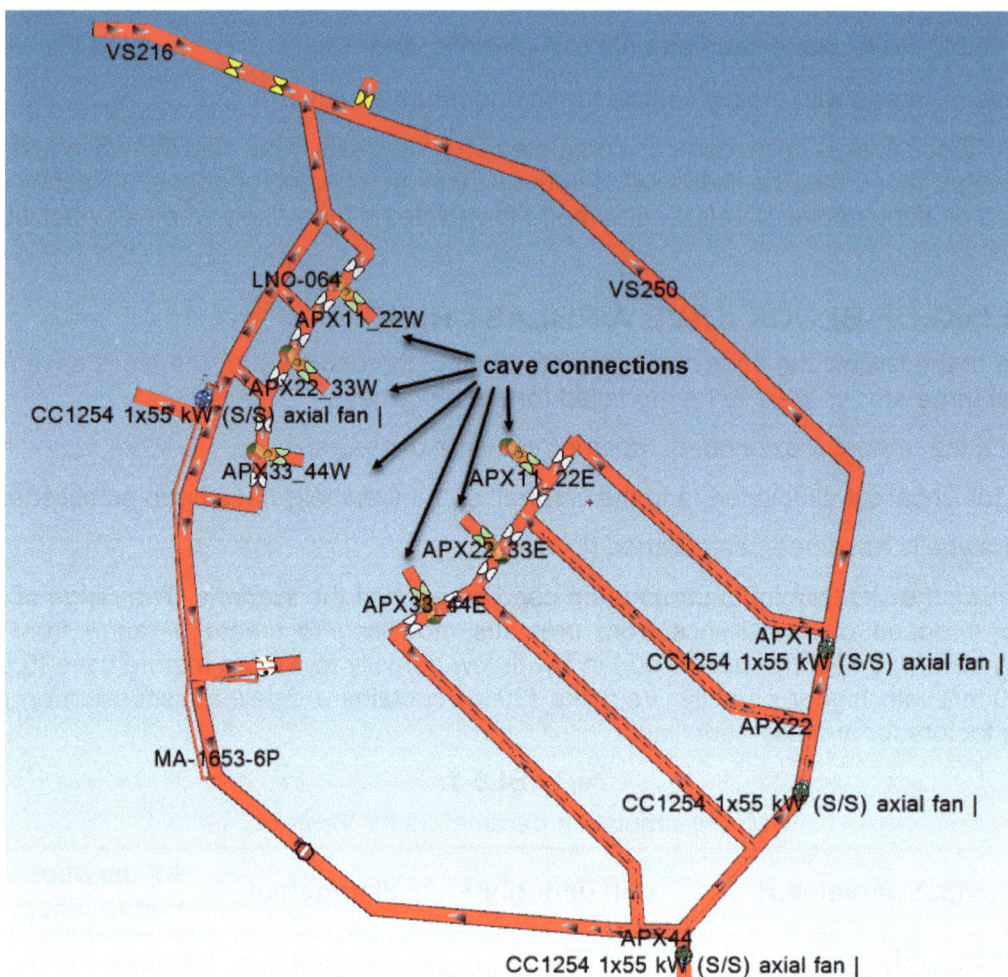

FIG 3 – Plan view of the apex level showing the cave connections with names for first stage.

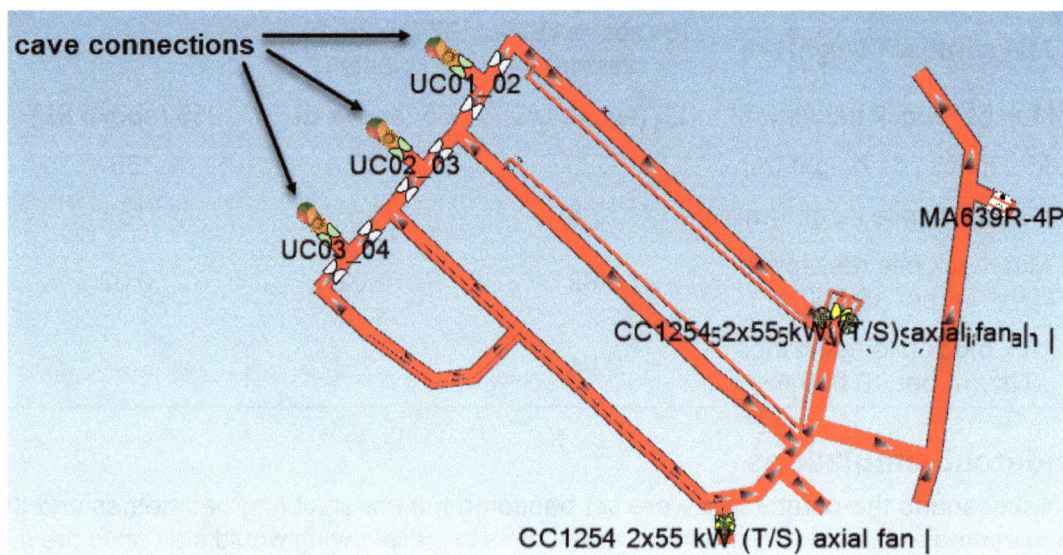

FIG 4 – Plan view of the undercut level showing the cave connections with names for first stage.

During undercutting, the maximum simulated air velocities could always be found in the Extraction Level Access to the block cave. Air from the airblast would vent out mainly from the north-western apex drives and travel up the 9300–9280 VR as well as the R2 vent rises. However, air will also travel down the 9280–9250 OPA and the 9250 to 9280 RAD3 rises and then flow along the extraction level perimeter up the block cave access (Figure 5).

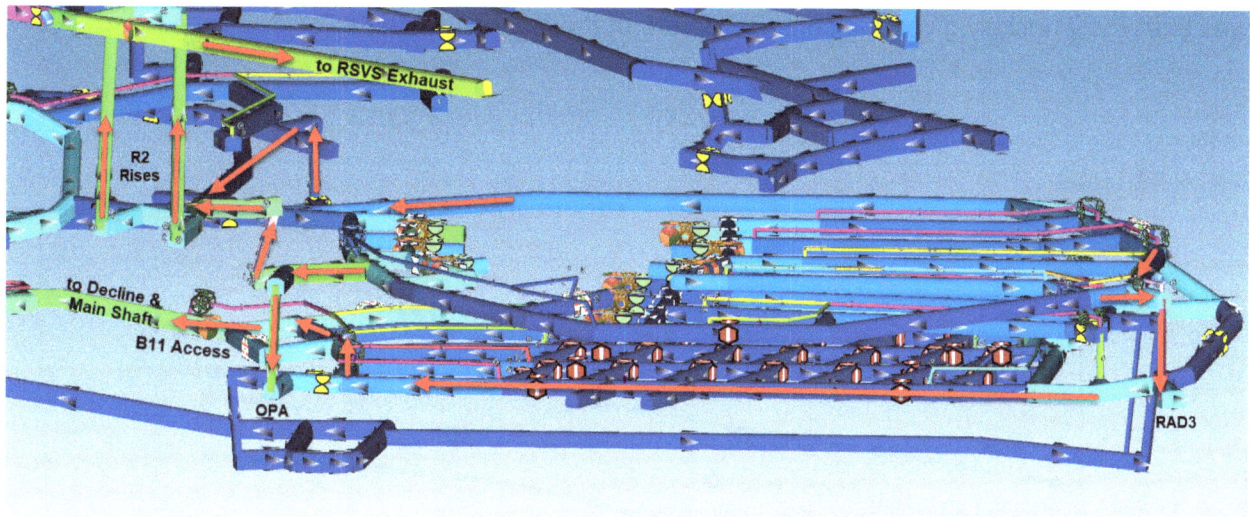

FIG 5 – Exhaust paths around the extraction level in the event of an airblast.

The severity of the airblast, ie peak velocities, was primarily influenced by the muck pile height and, to a lesser extent, by muck pile resistance (Figure 6). Muck pile resistance of 0.02 Ns^2/m^8 per 10 000 m^2 can be expected in mature caves and typical blasted material corresponding to about 20 to 30 per cent porosity. The lower the porosity, the higher the resistance with 0.04 Ns^2/m^8 per 10 000 m^2, more representative of developing caves and confined blasted material corresponding to about 5 to 10 per cent porosity. As porosity changes as the cave matures, generally lower resistances are used to determine the overall safety limits. This is due to the uncertainty of actual porosity in the cave. As the graph shows, the increasing extraction ratio has a much more severe impact on the air velocities.

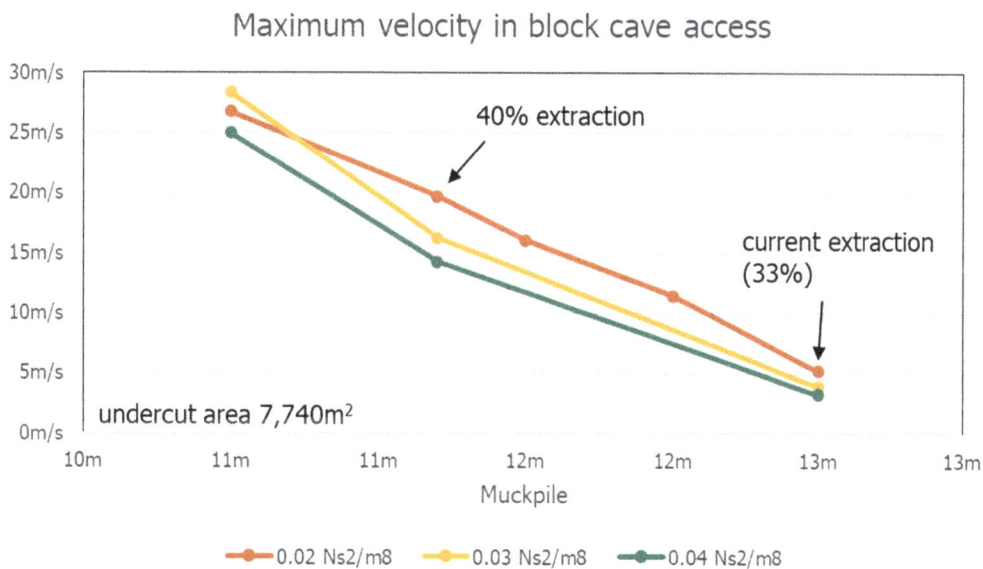

FIG 6 – Maximum air velocities in block cave access for current undercut at different muck pile heights and muck pile resistances.

At the time of the analysis the mine had an extraction ratio of 33 per cent and were investigating changes to the extraction ratio. As muck pile resistance is also dependent on undercut area, an analysis was carried out using 0.02 Ns^2/m^8 per 10 000 m^2 for an increasing undercut area with different extraction ratios. The results can be found in Figure 7. The graph shows that peak velocities increase as undercut size increases due to the resistance decreasing according to the square law of ventilation. If the density of material decreases from 2.8 t/m^3 to 2.7 t/m^3 (4 per cent swell) during blasting, the extraction ratio of 33 per cent was adequate. To increase the extraction ratio to 40 per cent a swell factor of 15 per cent would need to be achieved consistently when blasting, which was extremely unlikely given the confined nature of the blasts.

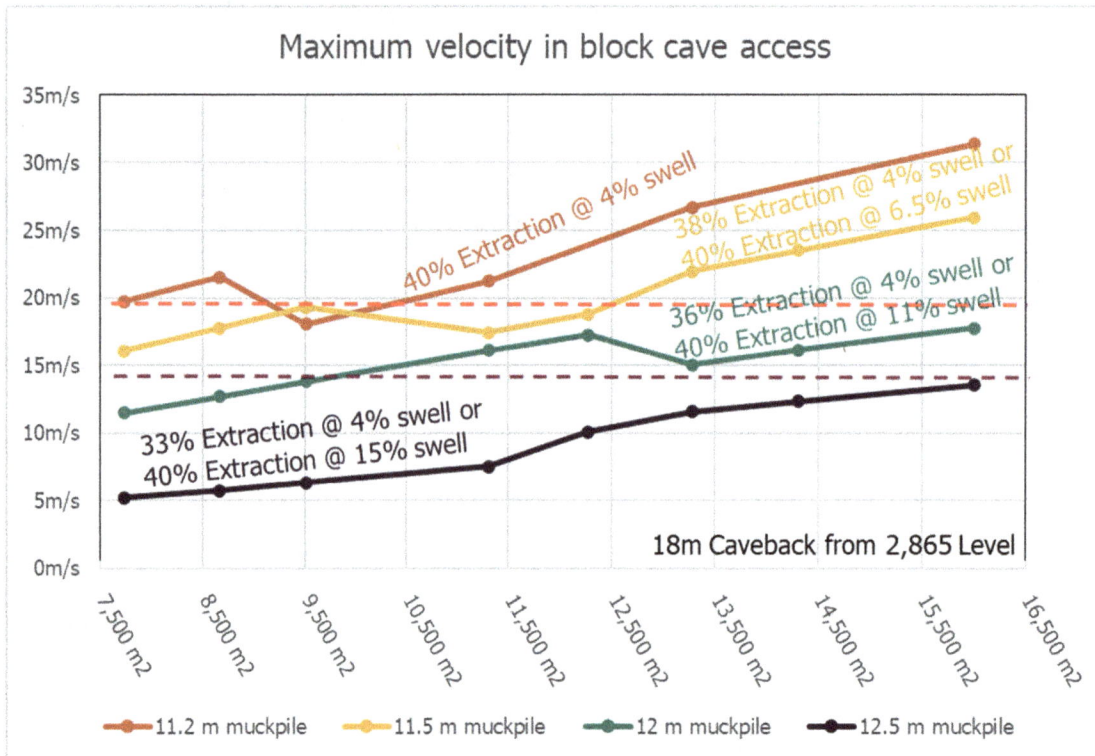

FIG 7 – Maximum air velocities in block cave access for increasing undercut areas with extraction ratios and blast swell factors.

Complete undercut simulation

The simulations for the remaining two scenarios are discussed in combination as it was found that the number of drawbells connected on the extraction level did not impact the air velocities around the area in the event of an airblast. In either case most of the air exits the cave from the undercut and apex levels.

Figures 8 and 9 show the apex and undercut levels after the completion of the undercut, respectively. Figure 10 shows the drawbell connections to the cave after all bells have been completely established. Again, all cave connections are done inside to ore pillar.

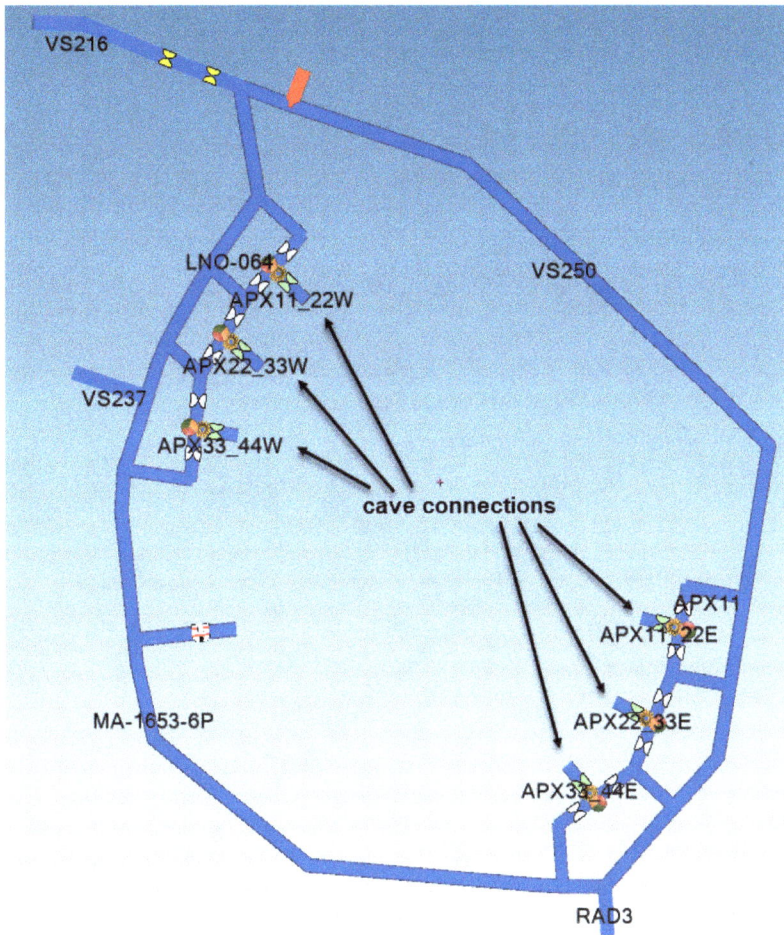

FIG 8 – Plan view of the apex level showing the cave connections with names for second and third stage.

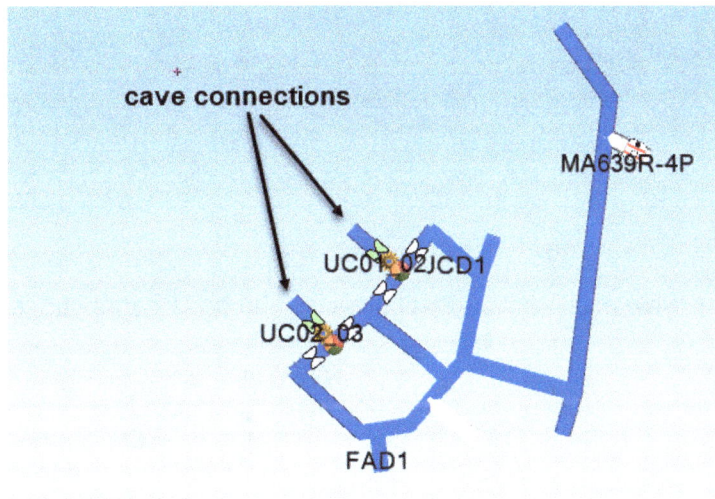

FIG 9 – Plan view of the undercut level showing the cave connections with names for second and third stage.

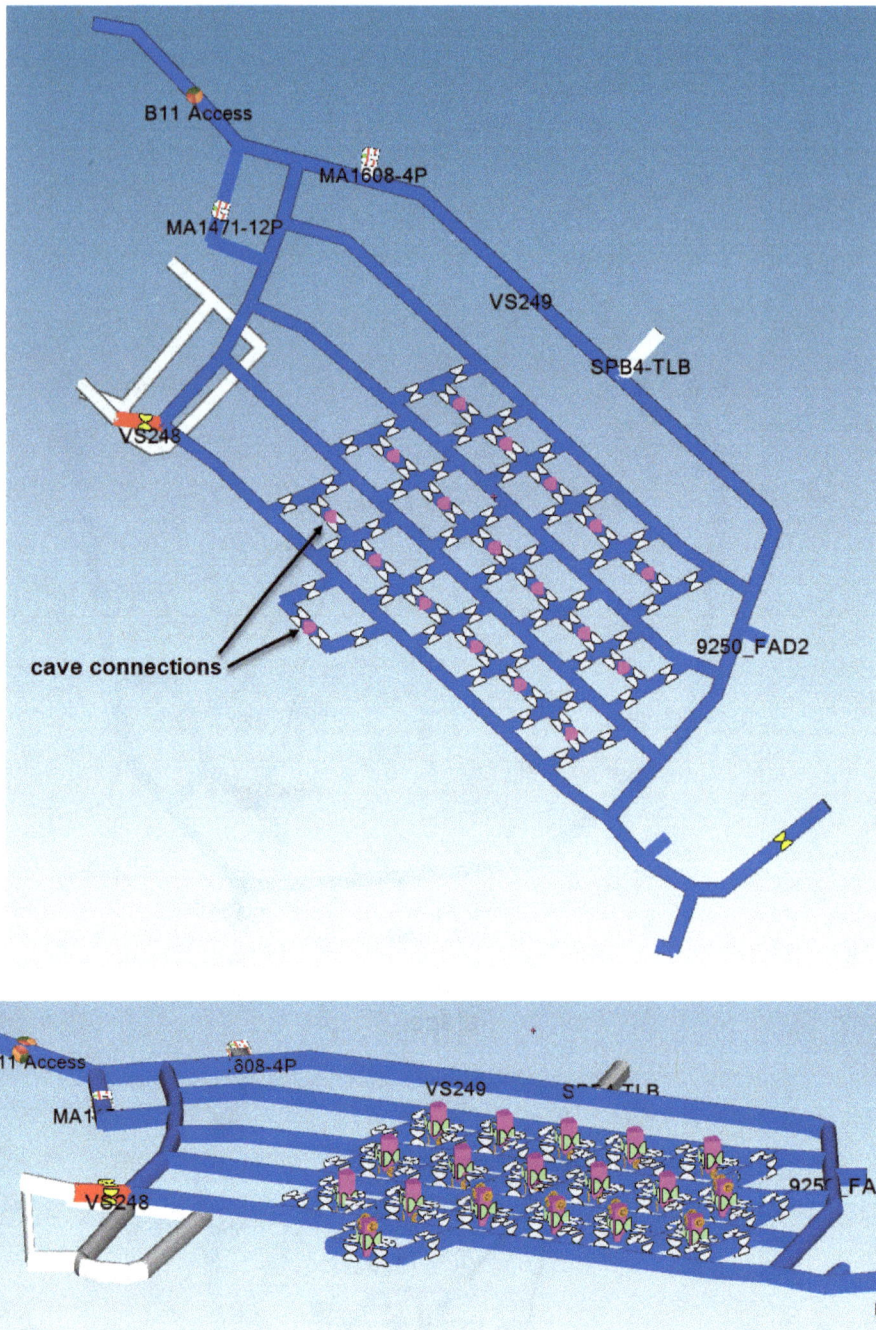

FIG 10 – Plan and iso view of extraction level showing the cave connections after completion of cave establishment.

The simulations were carried out with the extraction level, undercut level and apex level connected initially. Levels that had the potential to be exposed to the cave were connected as the caveback went past to understand the pressures an airblast plug would be exposed to if the respective levels were connected. However, as they do not allow the venting of air, they do not have an impact on air velocities in other parts of the mine. The levels were added at the following caveback heights which corresponded to their elevation in the model rather than nominal level elevation:

- 9325 Level – 71 m
- 9350 Level – 93 m
- 9390 Level – 118 m
- 9415 Level – 153 m
- 9440 Level – 181 m

- 9465 Level – 207 m
- 9490 Level – 234 m.

Figure 11 shows an example of historically developed levels above the block cave footprint including the predicted cave shape. It also highlights additional areas that will need to be monitored for cave intercepts and which will potentially require the installation of an airblast wall.

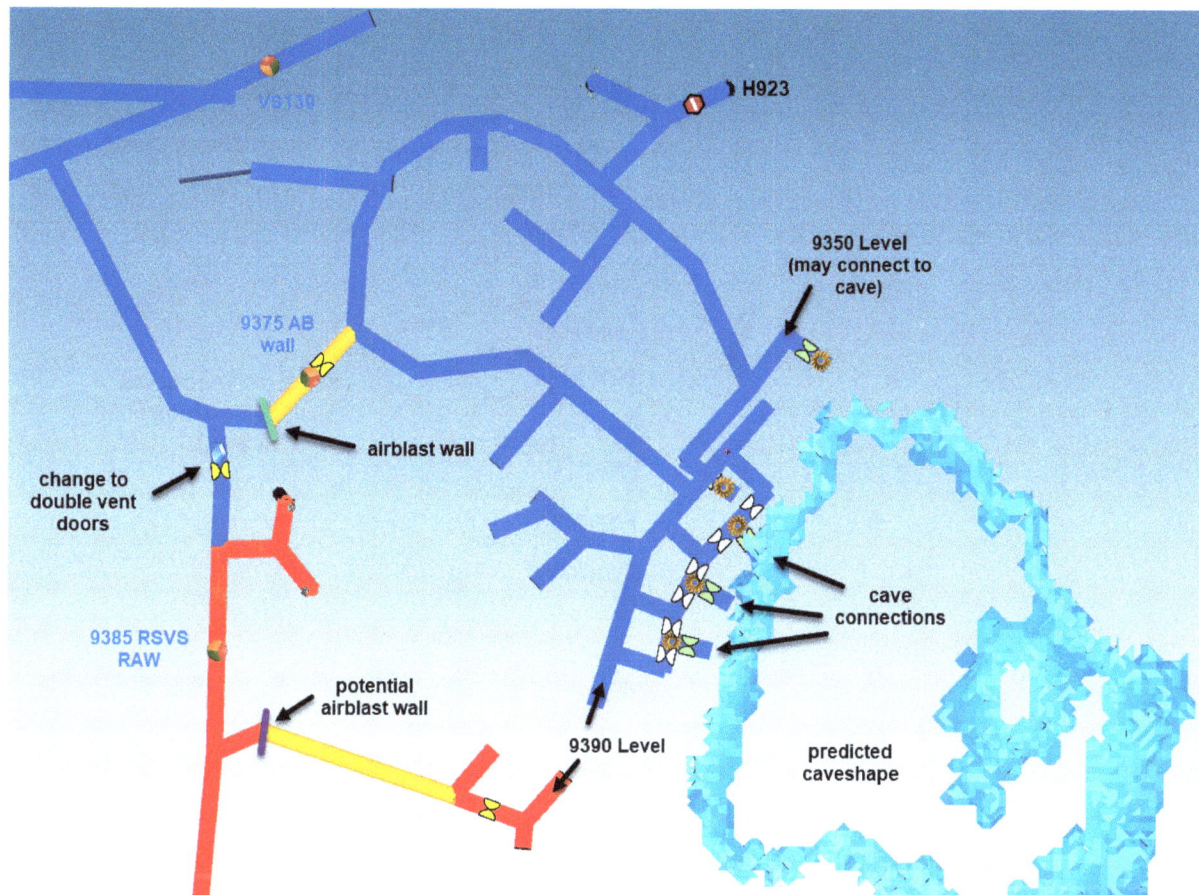

FIG 11 – 9350 and 9390 Level showing planned airblast wall location (green) and additional suitable location (purple).

It was predicted that the block cave will intercept the old sublevel cave 153 m above the block cave extraction level. It is likely that the cave material in the sublevel cave has consolidated since the cessation of production resulting in a much higher muck pile resistance in that cave. As the resistance is extremely difficult to quantify and the bogging in the adjacent block cave could potentially loosen up some of the sublevel caving material, it was decided to include the sublevels into the airblast assessment. Figure 12 shows the 9490 Level as an example of the connected sublevel cave.

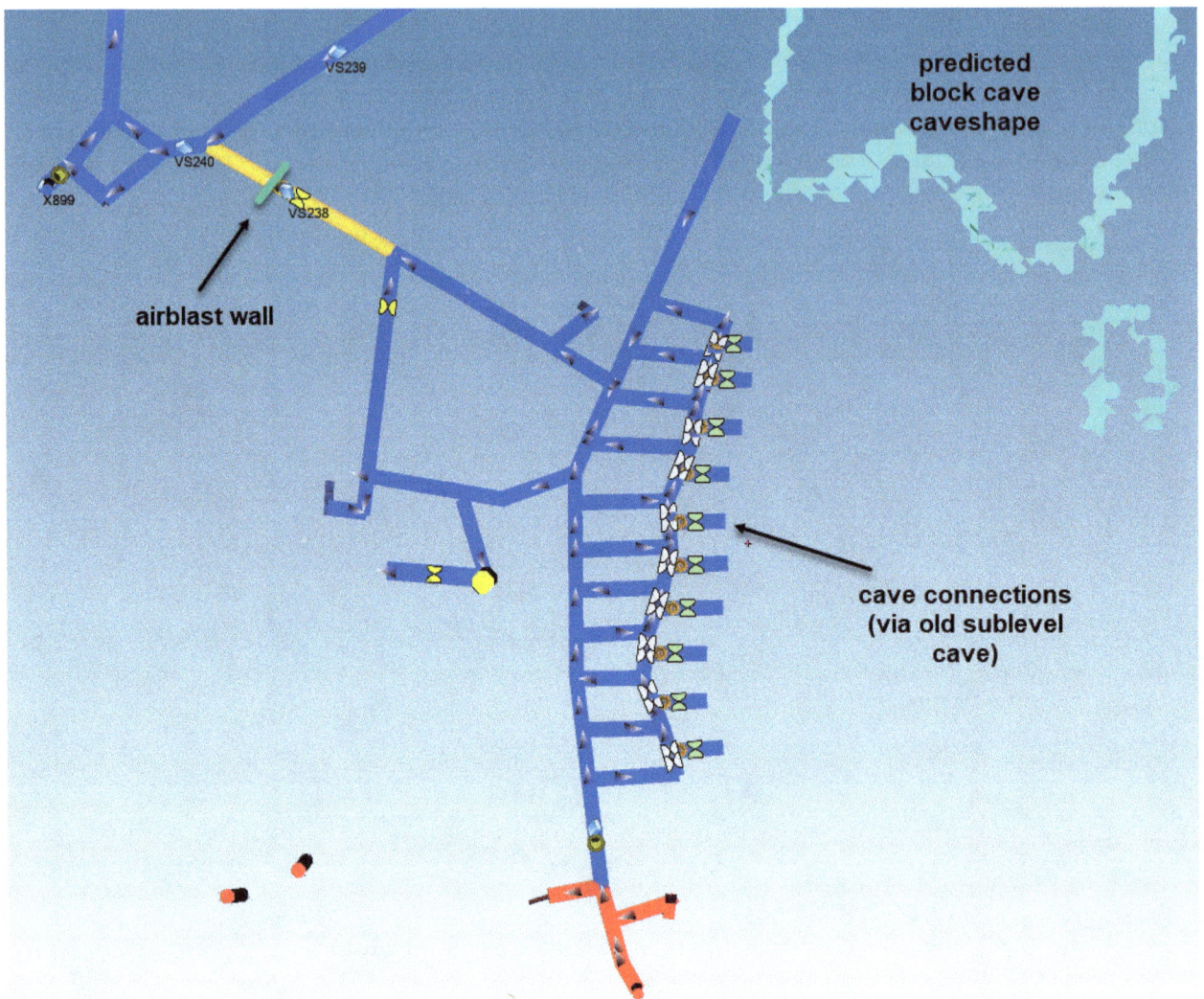

FIG 12 – 9490 Level showing planned airblast wall location (green).

Based on all the simulations a nomogram was interpolated using the two mentioned ventilation criteria. Figure 13 shows the nomogram for the block cave growth in terms of height over extraction level. As can be seen from the graph the 15 m/s air velocity graph follows a linear trendline very well with the 20 m/s air velocity graph following a linear trendline overall except for when the muck pile is just above the 9415 Sublevel.

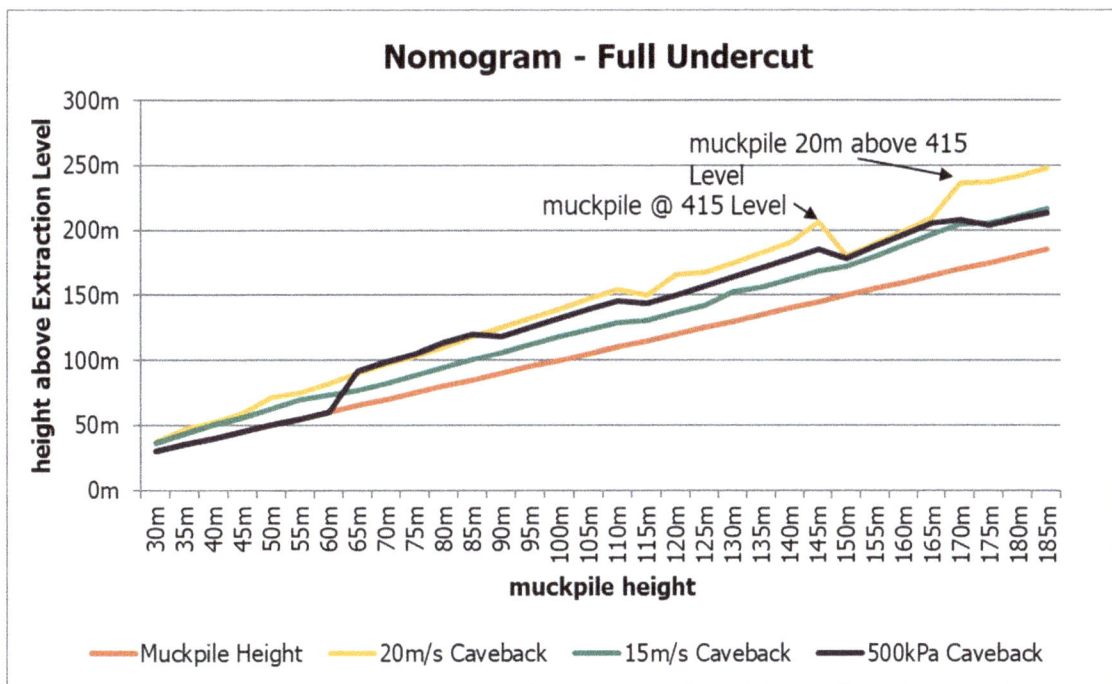

FIG 13 – Block cave Nomogram based on predicted growth using height above extraction level.

The simulations found that the proposed blast walls had to withstand significant pressure if the velocity criteria were used as the only ones to limit the size of the air gap (Figure 14). After discussions with the civil engineers designing the blast walls, it was decided to add an additional pressure criterion limiting the overpressure to 500 kPa on the inside of the blast walls. This resulted an allowable air gap of around 30 m over most of the simulated growth period once the first level with an airblast wall was connected to the cave 71 m above the extraction level (Figure 13).

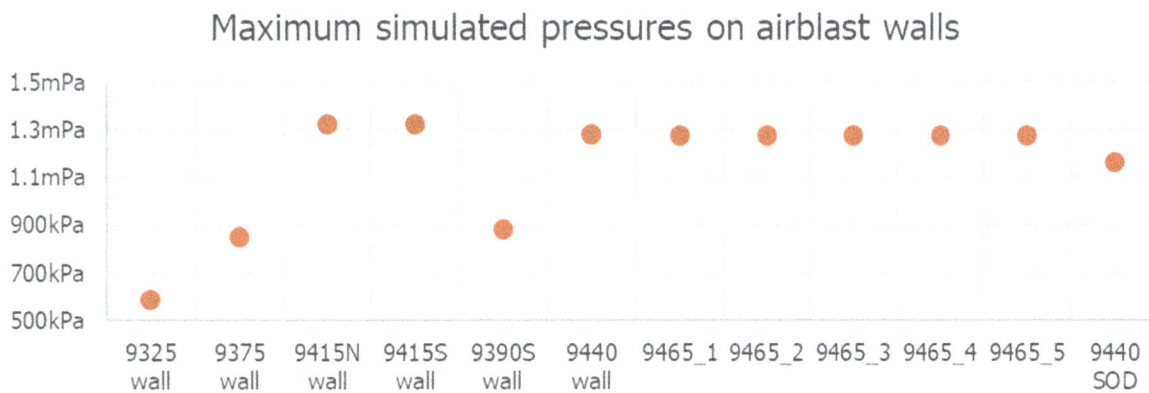

FIG 14 – Graph showing the maximum simulated pressures for the proposed blast wall locations (1 MPa = 102 t/m^2).

SUBLEVEL CAVE SIMULATIONS

The sublevel cave was divided into the southern and northern areas with a total of three scenarios investigated for both:

1. Southern cave with 9835, 9810 and 9785 Sublevels connected.

2. Northern cave with 9785, 9760 and 9735 Sublevels connected.

3. Northern cave with 9785, 9760, 9735 and 9705 Sublevels as well as southern cave connected.

To create the models, all ore drives connected to a particular model were cut back to 20 m stumps assuming finishing of production. As the ore drives in the upper levels of the sublevel caving area are relatively short no horizontal offsets were included in the model. As the orebody dips to the north-east offsets will be necessary in the lower sublevels (9575 RL and below) in the future depending on the location of the cave air gap in relation to the sublevels.

Southern sublevel cave model

At the time of the simulation the southern cave was estimated to be 436 502 m³, with three levels connected to the cave. Figure 15 shows an overview of the South cave area, including some monitoring points used to monitor simulations.

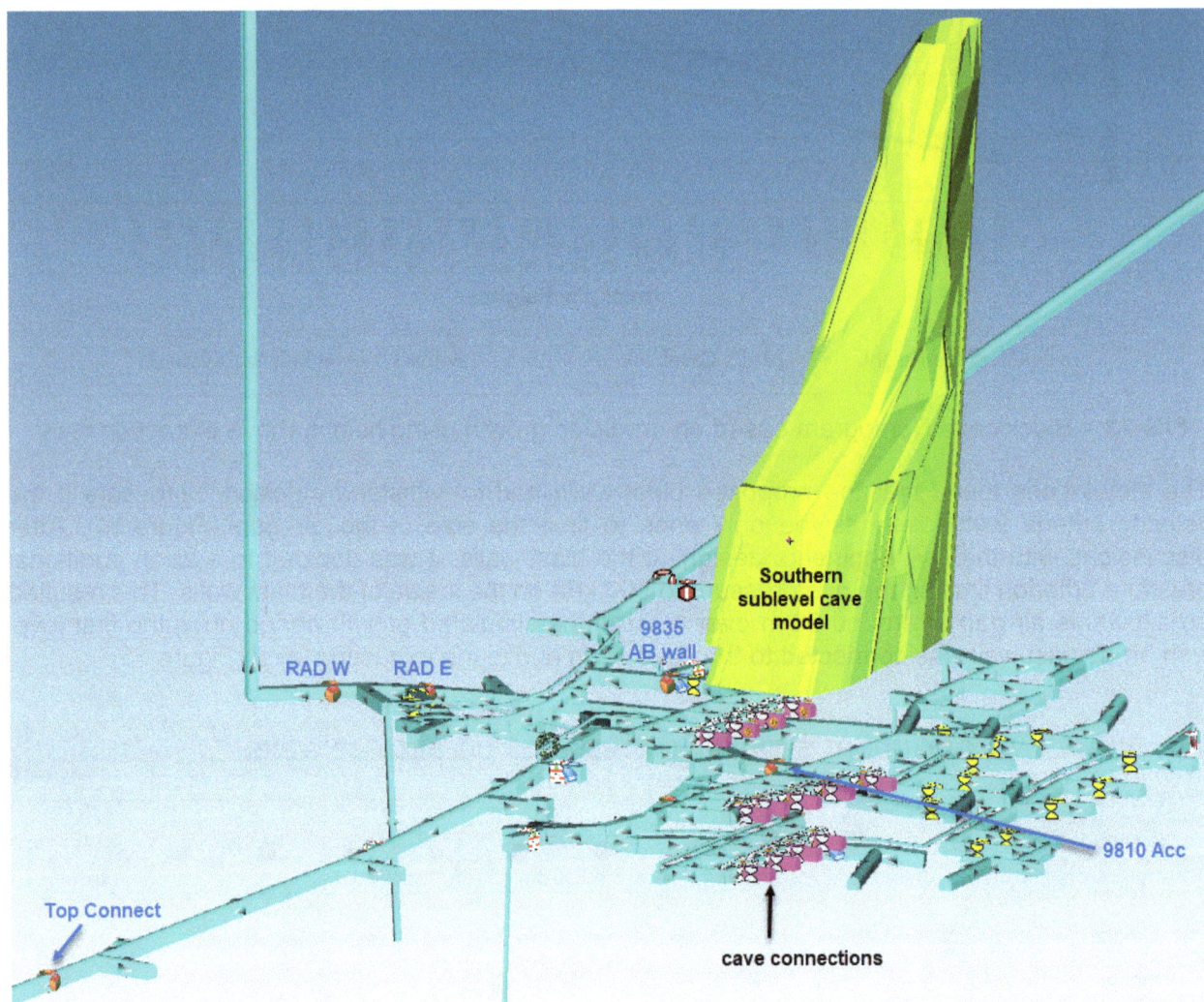

FIG 15 – Isometric view looking north-west – southern sublevel caving area model in Ventsim showing cave connections (pink) and monitoring points (blue labels).

The mine had already installed an airblast wall in the 9835 Level, which was included in the simulations.

As the cave was smaller in cross-sectional area further up, the cave was divided into 20 m slices and respective volumes determined for those slices. These volumes were then converted to an equivalent height based on the cross-sectional area at the base of the cave.

The mine estimated the material density after firing to be between 2.0 t/m³ and 2.7 t/m³. Swell factors of caved material are likely going to be between 5 to 10 per cent for such a young cave, or 2.67 to 2.55 t/m³. For the calculation of the muck pile height, it was assumed that the swell from firing was bogged in the cross-cuts after each firing and that only swell from cave growth contributed to muck pile growth. Based on production data provided by the mine the assumptions resulted in an estimated muck pile height range between 109 m and 115 m.

Simulations were run for a muck pile height of between 75 m and 135 m with the results shown in Figure 16. The highest air velocities are in the 9810 Access and the Decline just below the

9875 Sublevel. There is some reversal of air in the connections between this orebody and the orebody containing the block caving operation. However, even in the worst case modelled, with only a 75 m muck pile the overpressure from the airblast does not reach the other orebody (Figure 17). The reversal decreases with an increasing muck pile height.

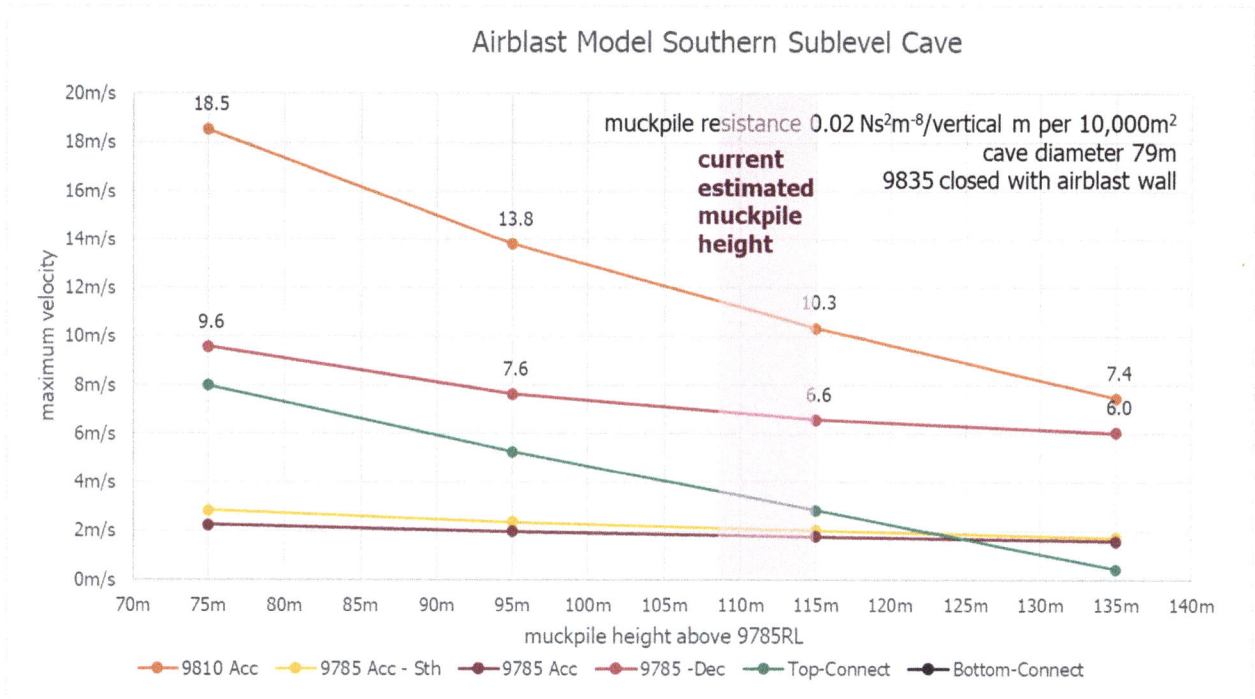

FIG 16 – Airblast assessment of current southern sublevel cave configuration.

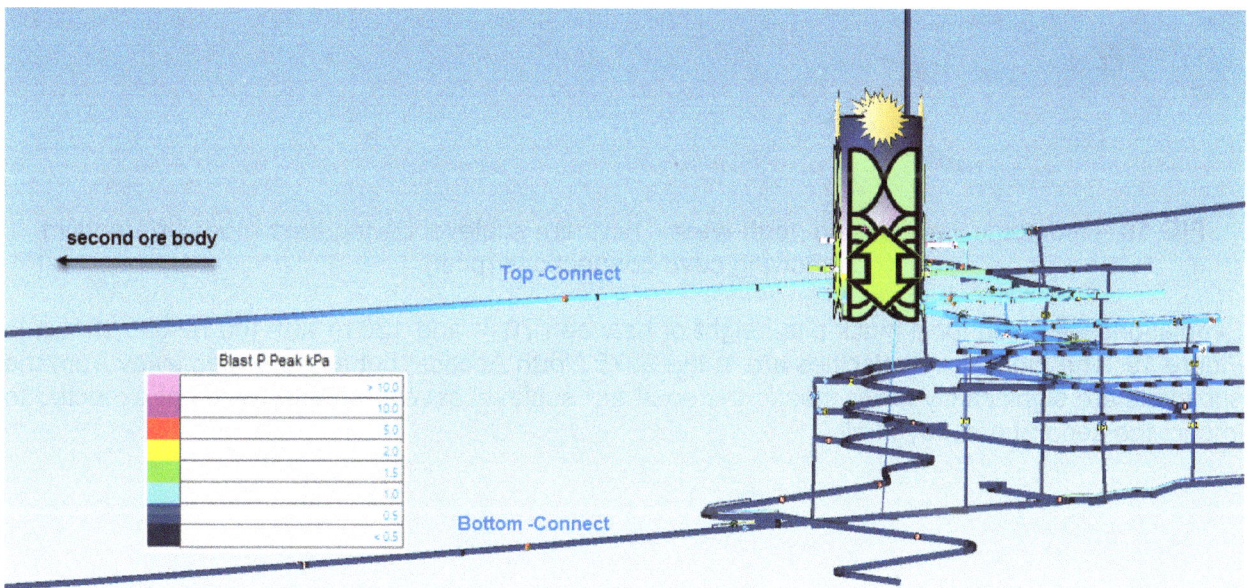

FIG 17 – Peak blast pressure distribution for 75 m muck pile in the southern sublevel cave.

Northern sublevel cave model

The northern sublevel cave was estimated to be 201 568 m³ in size at the time of the simulation with the 9785, 9760 and start of 9735 Sublevels connected to the cave. Figure 18 shows an overview of the northern sublevel cave area. Based on the production data provided by the mine and using the same parameters as the southern sublevel cave the muck pile is estimated to be between 105 m and 111 m high.

Again, the cave gets smaller in cross-sectional area further up and was divided into 20 m slices and respective volumes determined for those slices. These volumes were then converted to an equivalent height based on the cross-sectional area at the base of the cave.

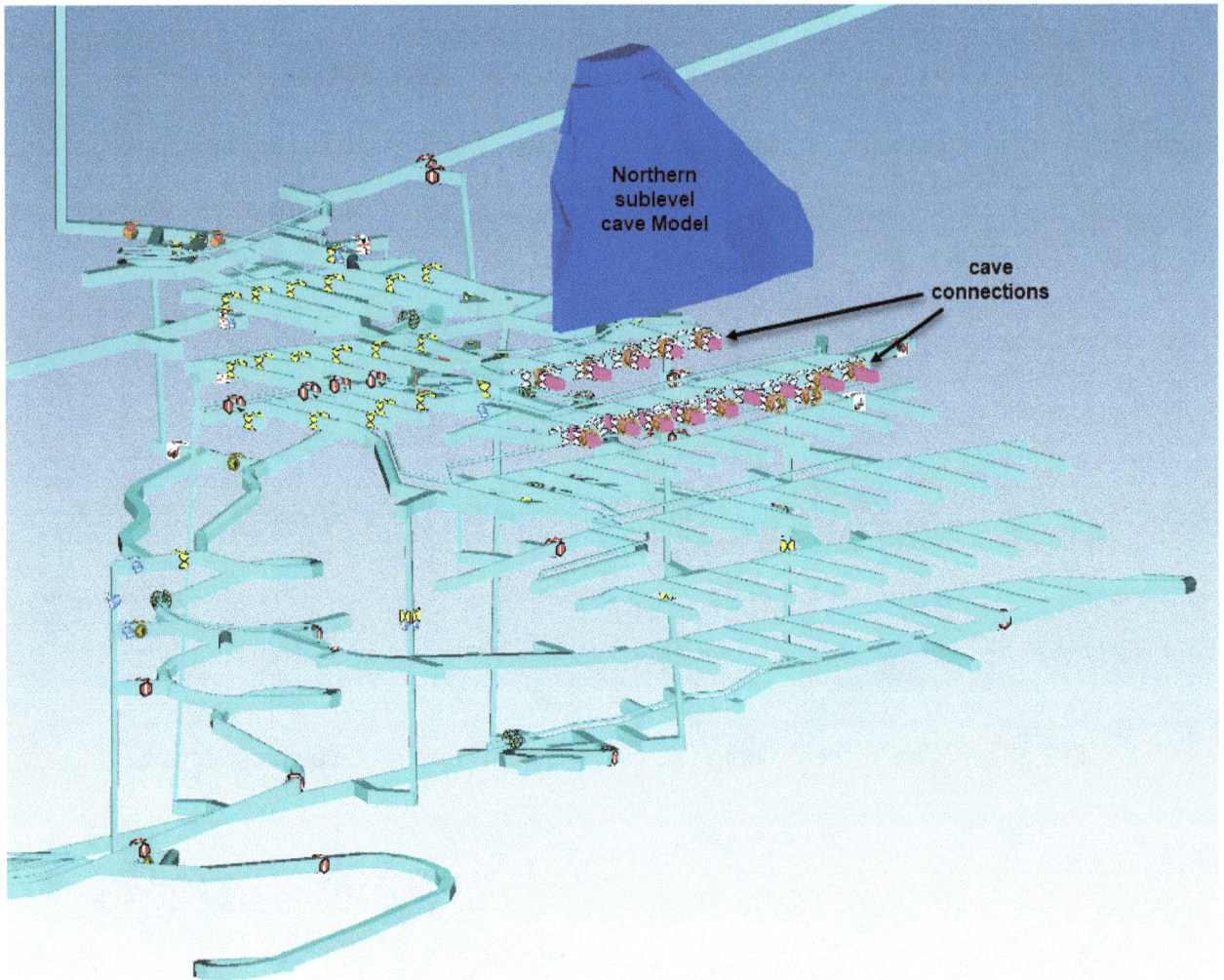

FIG 18 – Isometric view looking north-west – northern sublevel caving area model in Ventsim showing cave connections (pink).

Simulations were run for a muck pile height of between 76 m and 136 m with the results shown in Figure 19. The highest air velocities are in the 9875 North Access, but air velocities away from the sublevels are significantly lower than in the southern sublevel cave case and were not expected to impact the second orebody at all.

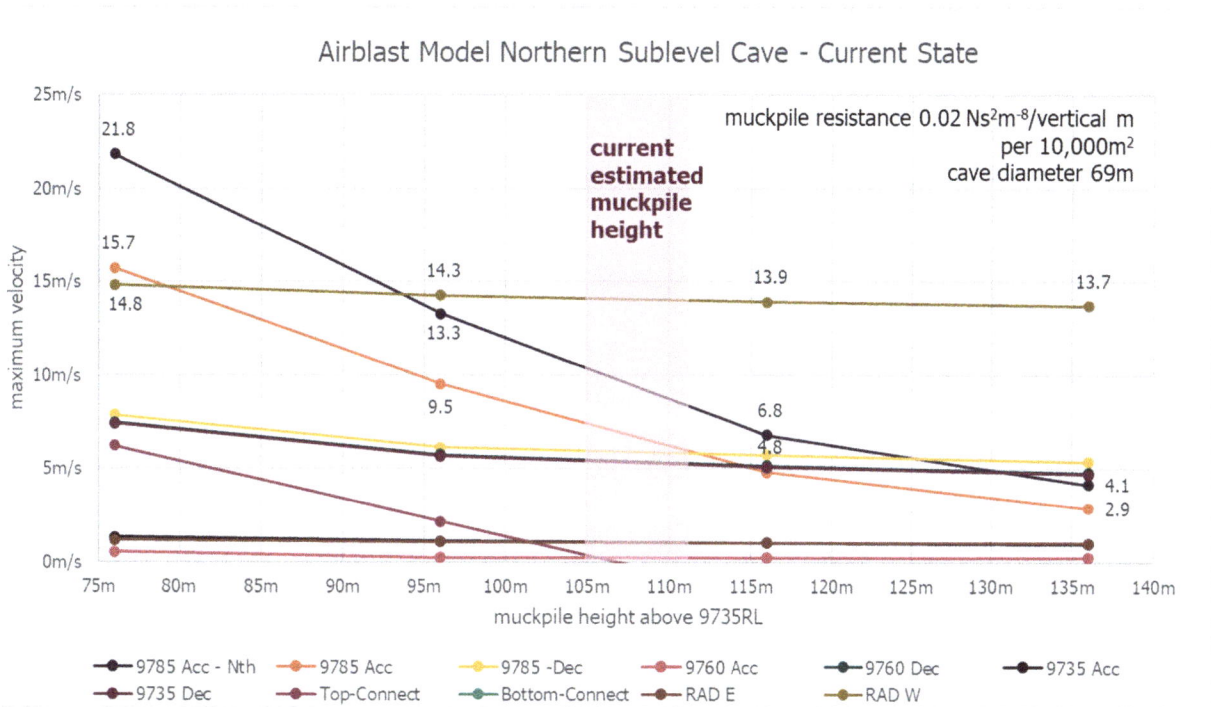

FIG 19 – Airblast assessment of current northern sublevel cave configuration.

Northern sublevel cave model – adding future 9705 Sublevel

With the orebody dipping to the north-east cave growth north of the current North Cave Model is expected. For the simulations, the footprint of the first four levels of the north cave was projected vertical up to estimate the maximum strike length of the cave (Figure 20). The same was done with the cave's width, resulting in a footprint of 300 m by 60 m or 18 000 m². As there is a distinct drop in cross-sectional area of the predicted cave above the 9950 RL modelling was only extended to that level. The airblast risk assessment will require updating in the future based on actual cave growth as differences are likely based on the actual cave growth of the South cave. While the South cave is not assumed to contribute to the overall airblast risk area, the ore cross-cuts in the South Cave area are used as potential exit points for the airblast from the cave.

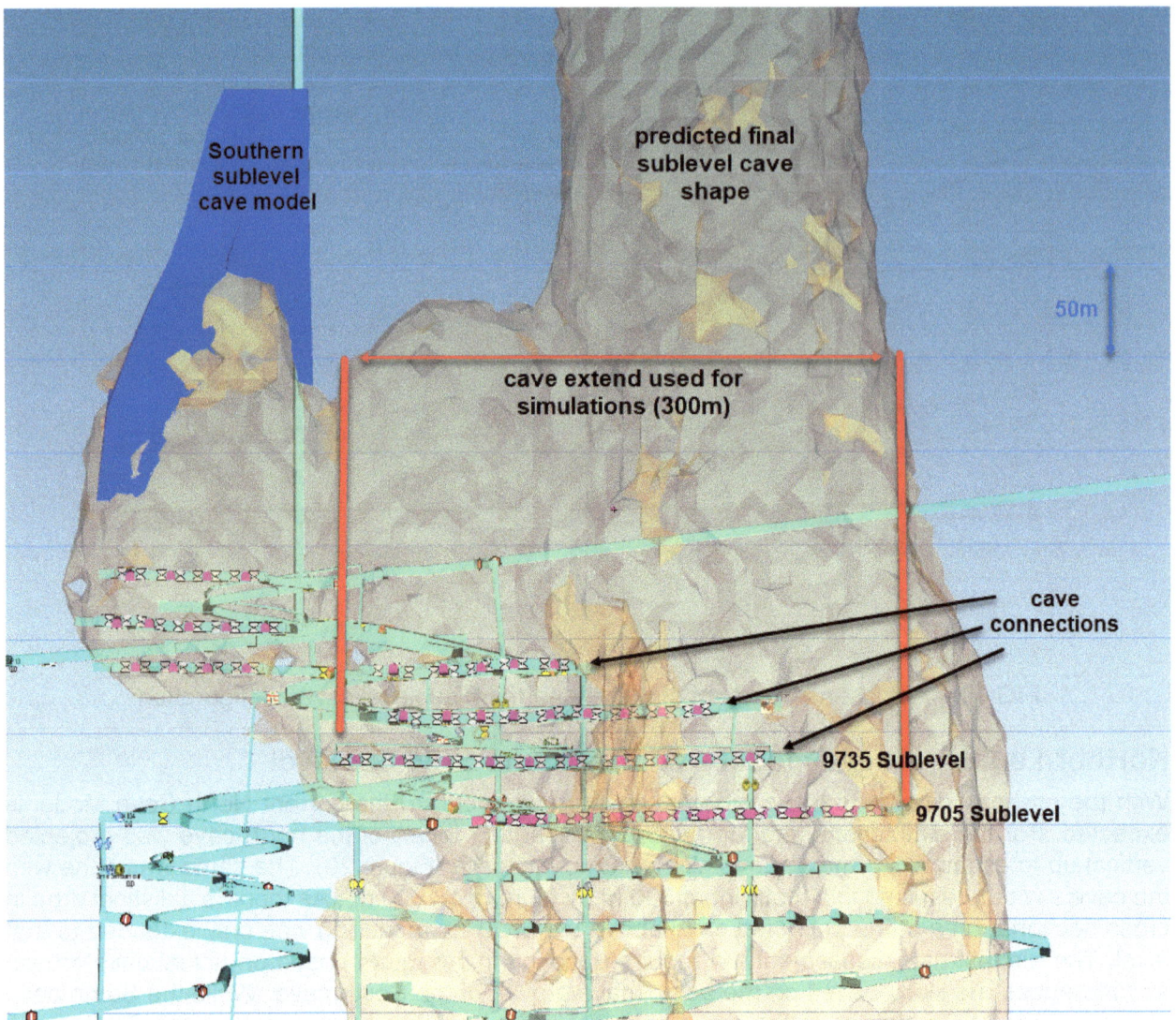

FIG 20 – Isometric view looking west – Final predicted sublevel cave shape in Ventsim showing cave connections (pink) and extends for forth level simulations.

All three sublevels of the southern sublevel cave and the 9785 Sublevel North Access were expected to have airblast walls installed at this point (Figure 21).

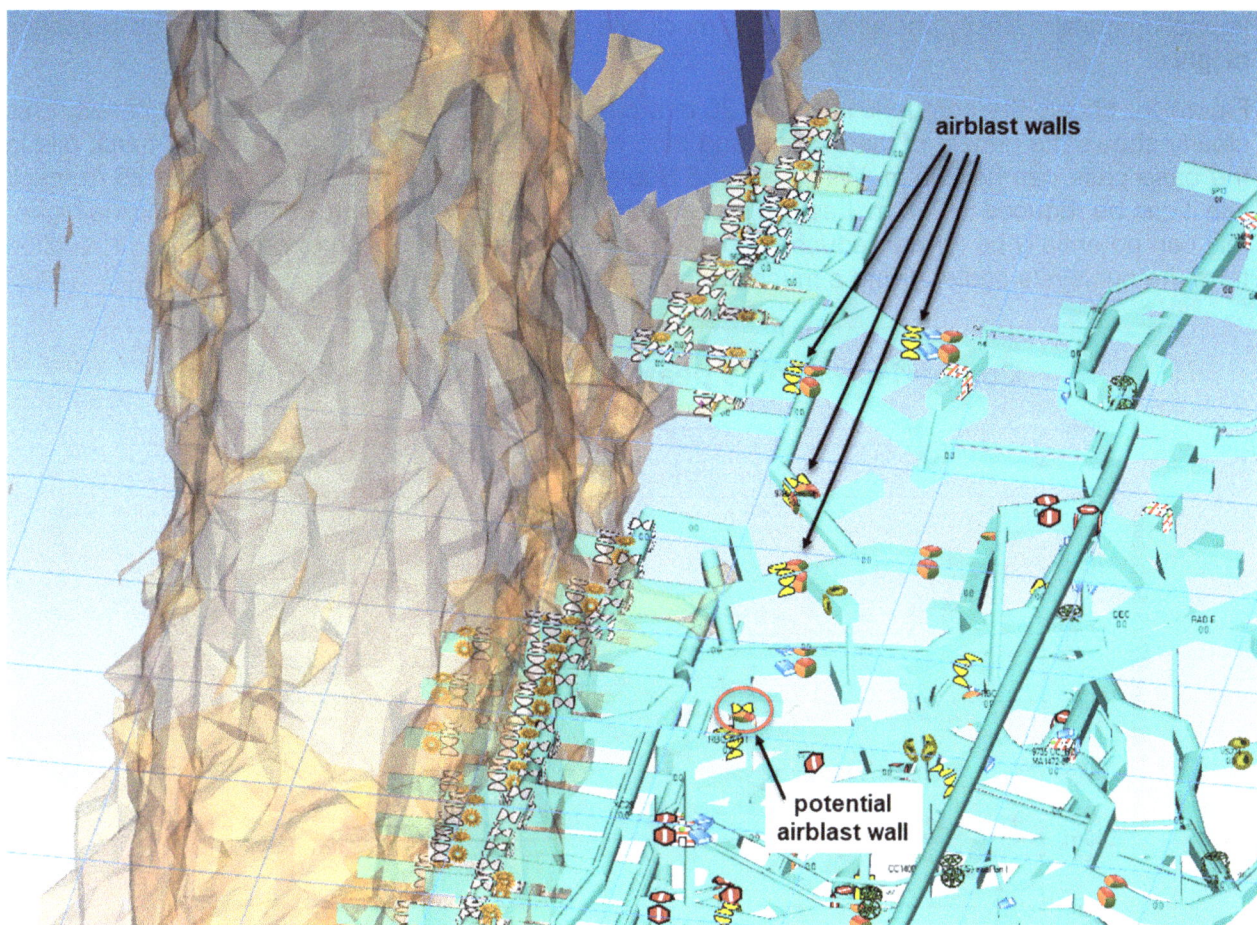

FIG 21 – Sublevel cave looking south-east showing simulated airblast walls and optional airblast wall in 9785 RAD.

With the opening of the cave footprint, the resistance of the muck pile will decrease due to the larger cross-sectional area. This means that air from an airblast will flow more freely through the muck pile when compared to a smaller cross-sectional area. As a result, the allowable air gap to maintain the same velocity in the mine will decrease.

Initially, no airblast wall was installed on the 9785 Return Airway side of the level, effectively using this sublevel and the return airway as a designated exhaust in the event of an airblast. While this will limit exposure for personnel, air velocities were still capped at 20 m/s to minimise likelihood of damaging the surface fans on top of the R115 raise.

In a second set of simulations, an airblast wall was installed on the return airway side of the 9785 Sublevel. This effectively means that the 9760 Sublevel is now the closest cave exit from the air gap. With the added 25 m muck pile a larger air gap would be possible.

While overall the second option allows for larger air gaps it was found that the air velocities in the top connection drive to the Perseverance deposit are higher in this set of simulations. The pressure wave from the airblast does not extend to the second orebody for any of the cases modelled. This means that while the simulation shows air flow to that deposit it is likely overstated due to Ventsim doing a steady state simulation after each timestep, which will increase the extend of the high air velocity area.

Given the significant reduction in air gap between the current cave size and the larger footprint, a third set of simulations was completed using a muck pile resistance of 0.04 Ns^2 m^{-8}/vertical m per 10 000 m^2. While this resistance value is suitable for a young cave with active growth and a muck pile porosity of less than 10 per cent, unconfined blasted material will have a higher porosity and thus a lower resistance. To use this nomogram for the mine muck pile resistance of the actual sublevel cave material needed to be confirmed. The increase in resistance has a significant positive impact on the air velocity and allows for the muck pile to be drawn further down at the same caveback height.

Figure 22 shows the comparison of the 20 m/s caveback line for the three simulated cases. This clearly shows the benefit of the second and third scenario. For a 250 m caveback the muck pile in the base case can be 230 m. With the 9785 Sublevel completely sealed and an extra 26 m muck pile it can be reduced to 216 m. Lastly if measurements can show that muck pile resistance is higher this can potentially be further reduced to 196 m. The benefit is the option of ramping up production from the sublevels earlier giving a much better return on investment.

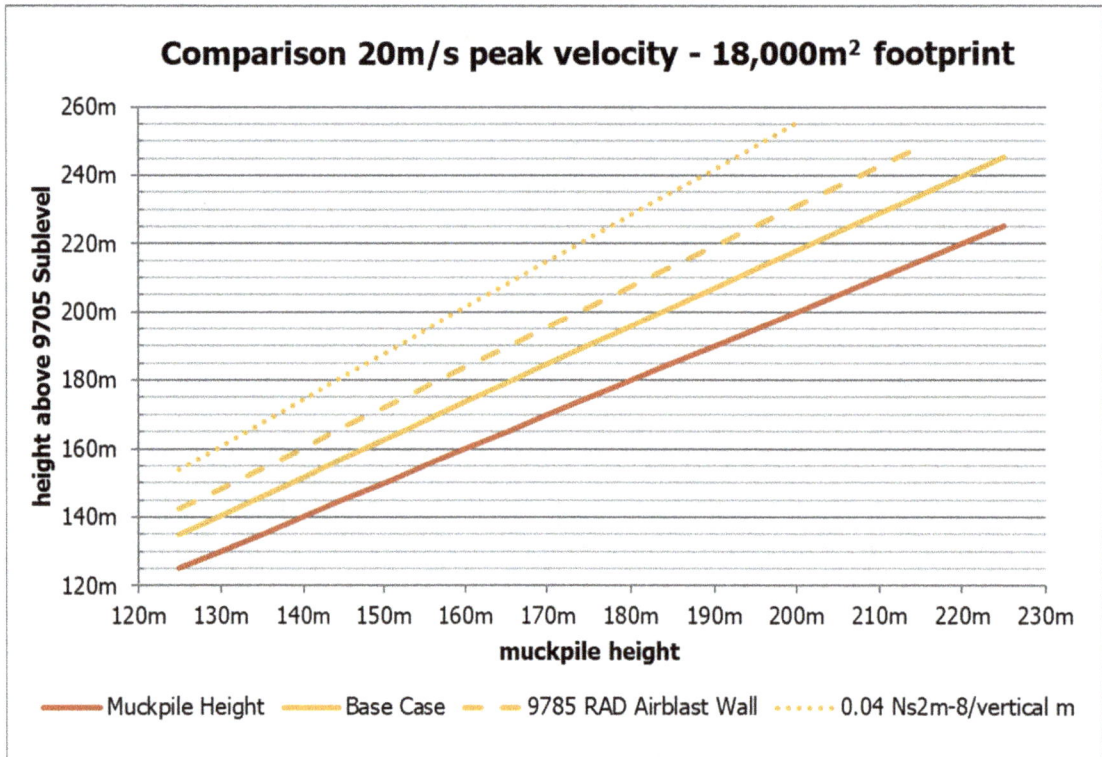

FIG 22 – Comparison of 20 m/s nomograms for sublevels down to 9705 RL connected to cave for different simulated cases.

The pressure on the airblast walls for all the cases simulated did not exceed 200 kPa. The pressure would increase, however, as the muck pile is drawn down and the height of the muck pile between the respective levels and the air gap diminishes. An airblast simulation can then determine the maximum pressure for that muck pile level to inform the design criteria for the airblast walls. Subsequent production decisions need to be made to ensure this pressure is never exceeded. Fundamentally the airblast walls on the top sublevels must be substantially heavier than on the sublevels further down as they will experience much higher pressures. A modular design to the airblast walls could be considered to allow for increasing capacity of the walls later on if required while minimising costs upfront.

Once production on the 9735 Sublevel has completed the 9760 Sublevel can be sealed and the 9735 Sublevel opened to the exhaust (Figure 23). To prevent air short circuiting a set of double steel vent doors needed to be installed in the access. They need to be able to withstand reasonable pressure determined by predicted cave growth until the completion of production on the 9675 Sublevel until which the 9735 Sublevel will be designated exhaust.

FIG 23 – Sublevel cave looking south-west showing 9735 Sublevel as designated exhaust with all levels above sealed with airblast walls.

CONCLUSIONS

The CAST module in Ventsim can be used to help mines understanding potential impacts of airblasts on the mine ventilation network and help with decision-making for adequate mitigating controls. It has a place both in the planning of future caving operations as well as adjusting risk controls as information becomes available during production from caving operations.

It offers an additional tool to help manage the safe establishment of caves until such time as they break through to surface or other caves above them. It is imperative, however, that the limitations of the module are well understood and form part of the decision-making process.

ACKNOWLEDGEMENTS

The author would like to acknowledge the ongoing support by the team at Howden Ventsim to further develop and improve the Ventsim CAST module.

REFERENCES

Fowler, J C W, Hebblewhite, B K and Sharma, P, 2003. Managing the hazard of wind blast/air blasts in caving operations in underground mines, in *ISRM 2003 – Technology roadmap for rock mechanics*, SAIMM 2003.

Vejrazka, C, 2016. Northparkes Mines' current air blast risk assessment practices for block caving operations, in *MassMin 2016*.

Health and safety

Examining the toxicity of respirable coal dust from interaction with moisture

S Azam[1], S Liu[2], S Bhattacharyya[3] and B J Arnold[4]

1. PhD Student, The Pennsylvania State University, University Park, PA 16802. Email: sma6414@psu.edu
2. Associate Professor of Energy and Mineral Engineering, The Pennsylvania State University, University Park, PA 16802. Email: szl3@psu.edu
3. Program Chair and Associate Teaching Professor of Mining Engineering, The Pennsylvania State University, University Park, PA 16802. Email: sxb1029@psu.edu
4. Professor of Practice in Mining Engineering, The Pennsylvania State University, University Park, PA 16802. Email: bja4@nb.net

ABSTRACT

Coal is a vital energy source and source of carbon for metallurgical applications, contributing significantly to the world economy. Coal mining and processing involve multiple processes that generate Respirable Coal Mine Dust (RCMD), one of the primary sources of health hazards for coal workers. Since 2000, the rate of Coal Workers' Pneumoconiosis (CWP) has continually increased in the U.S, with severity of the cases also worsening. Increasing evidence shows that exposure to RCMD in coal workers is a high-risk factor for lung cancer, which causes more deaths than the following three major cancers combined in the US Recent observations show an increased incidence of CWP or other lung diseases, particularly among young miners in the Appalachian basin. The CWP prevalence in Appalachia is four times higher than the national average. Although various regulations are in place, the resurgence of such disease among coalminers demands additional efforts to better understand all the possible facets of coal dust toxicity, taking insights from current interdisciplinary research efforts. One such important field is the interaction of coal dust with moisture in underground mines. Although moisture is prevalent and dynamic in an underground mining environment, it has never been evaluated for its impact on modifying the toxicity of coal dust and transport or flow behaviour. However, research conducted in atmospheric sciences in the past decade makes it clear that moisture in the atmosphere can significantly alter the behaviour of mineral dust. This paper is a part of research aimed at understanding how moisture interacts and alters the characteristics of coal dust. Such insight is crucial in understanding the potential toxicity of coal dust and developing mitigation technologies.

INTRODUCTION

Coal is the most abundant energy source for electricity production worldwide and currently accounts for over one-third of global electricity generation. In addition, coal is utilised as a source of carbon for coke making and anodes for aluminium production, among other applications. Several coal extraction methods are practiced worldwide depending upon the situation (National Research Council (NRC), 2007). Almost all coal mining processes produce airborne respirable dust, identified as one of the major occupational health hazards impacting workers' health. Most of these coalmine workers work in conditions with prolonged interaction with coal dust. Because of such prolonged exposure, miners are at risk of developing a lung disease called *pneumoconiosis*. Chronic inhalation of coal dust could cause several lung diseases, such as coal workers' pneumoconiosis (CWP), progressive massive fibrosis (PMF), chronic bronchitis, lung function loss, and emphysema (Schins and Borm, 1999). Various regulations are in place in almost all major coal mining economies to tackle the coal dust issue. Still, it remains one of the major global problems the mining industry faces. Moreover, recent observations suggest an increased incidence of CWP or other lung diseases in the U.S, particularly among young miners in the Appalachian basin. It has been recorded that the CWP prevalence in Appalachia is four times higher than the national age-adjusted prevalence rates in the US (5.9 per cent) (Croft *et al*, 2018). It has been identified that CWP was either the primary or contributing cause of death of more than 4000 miners between 2007 and 2018.

The coal dust problem impacts the miners as well as nearby populations as dust can travel great distances, as evident from recent studies (Ghosh and Cebula, 2021). Recently, a group of researchers found that, although the large coal dust particles almost all settle inside the open pit mining region, small particles escape from the mining pit (Wu *et al*, 2021). Despite various regulations in practice, the resurgence of such disease among coalminers demands additional efforts to enhance our understanding of the possible facets of coal dust toxicity, taking insights from ongoing interdisciplinary research efforts. One such important field is the interaction of coal dust with moisture in underground mines. Many studies have been conducted in previous decades on different kinds of mineral dust, whether natural, anthropogenic, or engineered, released into the atmosphere. The fundamental idea is that as the dust particle is liberated from the primary source, it can stay in the atmosphere for months depending on particulate size and other factors and get transported long distances before settling out of the atmosphere or even staying indefinitely. Depending upon their size and surface characteristics, surface reactions can be extremely fast, as shown by several studies (Jimenez *et al*, 2009; Rubasinghege *et al*, 2010; Zhang *et al*, 2010). The characteristics of modified dust particles resemble the source and the atmosphere in which they travel. So far, we are making excellent progress on research on primary coal dust particles, whether understanding their toxicity (Liu and Liu, 2020; Schins and Borm, 1999) or flow behaviour in a constrained underground mine environment (Wei *et al*, 2020; Yueze *et al*, 2017). However, there is a critical missing link in our research, which should account for the environmental changes associated with the characteristics of the coal dust particles. Also, the atmosphere in an underground mine is much more dynamic and constrained. Therefore, there is an intensified interaction of coal dust produced during different operations within the underground mine environment.

The relative humidity is a crucial parameter that could impact the characteristics and flow behaviour of the particulates in the mine atmosphere. Although moisture or humidity is prevalent and dynamic in a mining environment its impact on modifying the toxicity of coal dust and transport or flow behaviour has not been well understood. However, research conducted in atmospheric sciences in the past decade makes it clear that moisture in the atmosphere can significantly alter the behaviour of mineral dust (Chen *et al*, 2020; Joshi *et al*, 2017; Seisel *et al*, 2005; Tang *et al*, 2016). A recent paper mentioned that the water in the atmosphere could: (i) alter reaction pathways and surface speciation relative to the dry surface; (ii) hydrolyze reactants, intermediates, and products; (iii) enhance surface reactivity by providing a medium for ionic dissociation; (iv) inhibit surface reactivity by blocking sites; (v) solvate ions; (vi) enhance ion mobility on surfaces, and (vii) alter the stability of surface adsorbed species (Rubasinghege and Grassian, 2013). The relative humidity of underground coalmines depends upon many factors, such as strata heat release, ventilation, outside atmosphere, atmospheric pressure, etc. It might be possible that the coal dust released into the underground mine atmosphere could interact with moisture, and the characteristics could change without being noticed. In such a case, if we implement controls that target the primary characteristics of the coal dust, they might not be very useful. Depending on the chemical and physical attributes of the coal dust, it can react with the humidity; and accordingly, its properties might get modified.

The Free Radical (FR) content of coal dust plays a prominent role in determining its toxicity (Dalal *et al*, 1989; Huang *et al*, 2009). Free radicals are species that have unpaired electrons that make them more reactive. Earlier studies have shown that coal mostly have carbon centred free radicals (Huang *et al*, 2009; Petrakis and Grandy, 1978). Free radicals are introduced into the coal during the coalification process and are modified over time. During different coal mining operations, such as drilling, blasting, crushing, or cutting etc, these free radicals are present on the surface of the coal dust particles and are further modified because of their fast-reacting nature. These free radicals impart toxic properties to the coal dust. Once the coal dust interacts with the cells of the human body, they react very quickly, leading to the generation of unwanted species and the destruction of cells. These F.R.s could interact with anything they contact, not only with the cells. Based on the available literature and reasoning, it was thought that coal dust could interact with the humidity in the environment; and, because of that, it's free-radical properties could change.

Spray jet treatment is a common technique applied in the mining industry to tackle the problem of coal dust and prevent workers' exposure. Although spray jet or misting technology effectively knocks down coal dust particles, it has certain limitations. They are found to be less effective for very small or submicron coal dust particles (Wang *et al*, 2019). The humidity might adsorb on the surface of these submicron-sized coal dust and enhance surface reactivity by providing a medium for ionic dissociation.

To analyse this behaviour, we conducted Electron Paramagnetic Resonance (EPR) analysis to understand the FR changes in the coal dust as it interacts with moisture or water vapour. This paper presents a preliminary understanding of the F.R. changes occurring in different types of submicron coal dust when they interact with moisture. This work will detail the experimental procedure and discuss the implications. The process and the knowledge generated in this work are unique and could provide a better understanding of the toxicity modification of the coal dust.

MATERIALS AND METHODS

Coal dust preparation

Moisture interaction behaviour of coal dust was performed using nano-size coal dust prepared in the lab from lignite and bituminous coal samples. The demonstration of the step-by-step procedure followed in this study is represented in Figure 1. Each step will be detailed further in this section.

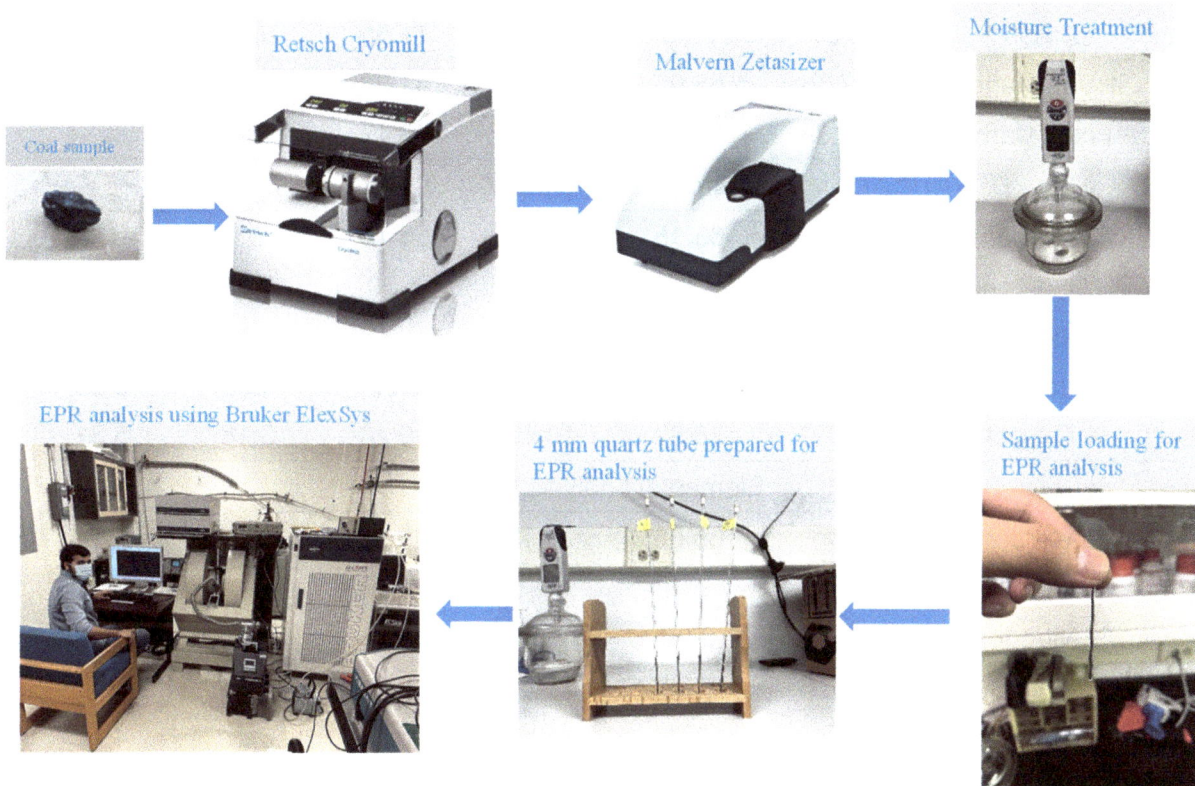

FIG 1 – Demonstration of sample preparation and analysis steps.

A bituminous and two lignite coal samples were selected for the study. It is known that lignite is hydrophilic and bituminous is less hydrophilic/more hydrophobic. Hence, we selected these two for our present analysis. All the coal samples were obtained from The Pennsylvania State University coal sample bank. Penn State is maintaining a suite of 38 well-preserved Department of energy coal samples collected in an ongoing effort starting in 1989. As soon as feasible after collection, processing was performed to obtain representative subsamples. These were sealed under argon in foil multilaminate bags to preserve samples very well and are kept in refrigerated storage (3°C). The lignite coal samples Lig1 and Lig2 were collected from Beulah Seam (Mercer County, North Dakota) and Pust seam (Richland County, Montana), respectively. The bituminous coal sample, Bit1, was collected from Upper Banner Seam (Dickenson County, Virginia). Table 1 shows the proximate, ultimate, and elemental analysis data of the coal samples as obtained from the coal sample bank.

TABLE 1

Ultimate and proximate analysis data for coal samples.

Sample Name	Lig1		Lig2		Bit1	
Location	Beulah Seam		Pust Seam		Upper Banner Seam	
	Mercer County, ND		Richland County, MT		Dickenson County, VA	
Coal Rank	ligA		ligA		hvAb	
Proximate analysis (wt%)						
	as received	dry	as received	dry	as received	dry
% Moisture	33.38	-	34.91	-	2.44	-
% Ash	6.37	9.56	7.71	11.85	6.20	6.36
% Vol. Matter	37.36	56.08	27.32	41.98	33.46	34.30
% Fixed Carbon	22.89	34.36	30.05	46.17	57.89	59.34
Ultimate analysis (wt%)						
% Carbon	44.07	66.15	42.80	65.76	79.10	81.08
% Hydrogen	2.68	4.03	2.99	4.60	5.05	5.18
% Nitrogen	0.60	0.90	0.61	0.94	1.44	1.48
% Total Sulfur	0.49	0.74	0.47	0.72	0.58	0.59
% Oxygen	12.40	18.62	10.50	16.13	5.18	5.31
Elemental analysis (wt%)						
% Carbon	-	66.05	-	65.38	-	80.91
% Hydrogen	-	3.91	-	4.45	-	5.10
% Nitrogen	-	0.90	-	0.94	-	1.48
% Organic Sulfur	-	0.39	-	0.35	-	0.50
% Oxygen	-	17.76	-	15.32	-	4.78
% Mineral Matter	-	10.99	-	13.56	-	7.23

As indicated, the two lignite samples, Lig1 and Lig2, have low fixed carbon contents of 22.89 and 30.5 per cent, respectively. In comparison, the bituminous coal sample, Bit1, has a much higher fixed carbon content of 57.89 per cent. Another important feature is that the proximate analysis shows high moisture content in these two lignites at 33.38 and 34.91 per cent, compared to the bituminous coal with 2.44 per cent moisture. Similarly, the ultimate analysis of the samples shows a much higher oxygen content of 12.40 and 10.50 per cent for the two lignites compared to the bituminous, which only has 5.18 per cent oxygen. In terms of elemental carbon and oxygen, we again observe the same trend that the lignites have lower elemental carbon and higher elemental oxygen content. Several other elements are present in the coal samples, such as nitrogen and sulfur. The amount of nitrogen is slightly higher in the bituminous coal samples as compared to the lignite coal sample. The difference in the carbon and oxygen content of the bituminous and lignite coal samples is expected because of the differences in the degree of coalification. Bituminous coal is more mature, and because of the longer coalification process compared to lignite, it has become more dense, dry, and rich in carbon (Liu *et al*, 2021).

Initially, each coal sample was hand crushed (HC) by a pestle and mortar to a top size of -80 mesh (-177 μm). This HC sample was suffixed as -HC for all the coal dust samples. This HC sample was further crushed to prepare nano-sized coal dust through cryogenic ball milling using a CryoMill

(Retsch Inc.) located at the Materials Research Institute at Penn State. During the whole cryomill processing, liquid nitrogen was continuously circulated through the autofill system to keep the temperature at -196°C during the ball milling. The use of a liquid nitrogen environment prevents the heating of the powder sample, which makes the process faster and maintains the original nanopore structure of the coal dust. A maximum of 20 ml of the HC coal dust powders were put into the sample cell. A vibrational frequency of 5 Hz was used for pre-cooling, and two cycles with 2 min in each cycle and 30 Hz of vibrational frequency were used. One minute of intermediate cooling time was used between each cryo cycle. The coal samples crushed for two and four cycles are suffixed as Cr2 and Cr4 samples, respectively. So, this makes three samples for each of the coal dust based on the type of crushing for a total of nine samples. We kept the sample in the high-grade air-tight plastic bag continuously and flushed it with helium to minimise oxidation. It was taken out only during crushing and immediately stored back in the bag. During the cooling time in the cryo-milling process, the sample was in a liquid nitrogen atmosphere.

Table 2 shows the important atomic ratios and vitrinite reflectance of the coal samples. As mentioned in the previous studies, the O/C ratios could represent the surface hydrophilicity of the coal matrix of the three coal samples (Buchner *et al*, 2016; Kovtun *et al*, 2019). Lignites generally have higher atomic H/C and O/C ratios than the bituminous coal samples. The greater maturity of the bituminous coal is depicted by higher vitrinite reflectance values of 1 compared to the two lignites at 0.35 and 0.23, as mentioned in Table 2. As per the Van Krevelen diagram, the rank of the studied coal samples is Lig1 is lignite-A, Lig2 is lignite-A; whereas, Bit1 is of the high volatile-A bituminous rank.

TABLE 2
Atomic ratios and reflectance of coal samples.

Sample Name	Lig1	Lig2	Bit2
Atomic Ratios			
Atomic H/C	0.732	0.840	0.733
Atomic O/C	0.206	0.177	0.038
Vitrinite Reflectance (%)			
Mean-max Ro	0.35	0.23	1

Chemical characterisation of coal samples suggests that the molecular structures of the two lignites are much more heteroatomic and aliphatic or have more side chains. In contrast, the bituminous coal sample is much more aromatic and less heteroatomic. We will discuss the implications of these observations in further sections.

Particle size analysis of the crushed coal dust samples

As described earlier, each of the three coal samples was initially hand crushed, and then cryo milled for four cycles. The particle size distribution for these samples was measured by the Malvern Zetasizer® Nano ZS located at the Materials Characterization Lab at Penn State. The Zetasizer Nano ZS uses the laser diffraction technique.

Before each measurement, each coal dust sample (dispersant) and isopropanol (solvent) were mixed inside a small beaker and put in an ultra-sonic oscillator for about 5 min for preliminary dispersion purposes. The solvent with dispersant was then put into a flow cell containing pure solvent only for the measurement. The samples were placed in a 12 mm OD (outer diameter) square polystyrene cuvette for aqueous solvents for measurement. Zetasizer software (v8.01) was chosen for data collection and particle size analysis. The data collected from the Zetasizer Nano is plotted as a graph and shown in Figures 2 and 3. We can see that most of the volume and the number size distribution for all the three coal dust samples, whether 2-cycle or 4-cycle cryomilled, are less than 1000 nm. This means that we successfully generated submicron or nano-sized coal dust samples. These samples were further used for the EPR analyses.

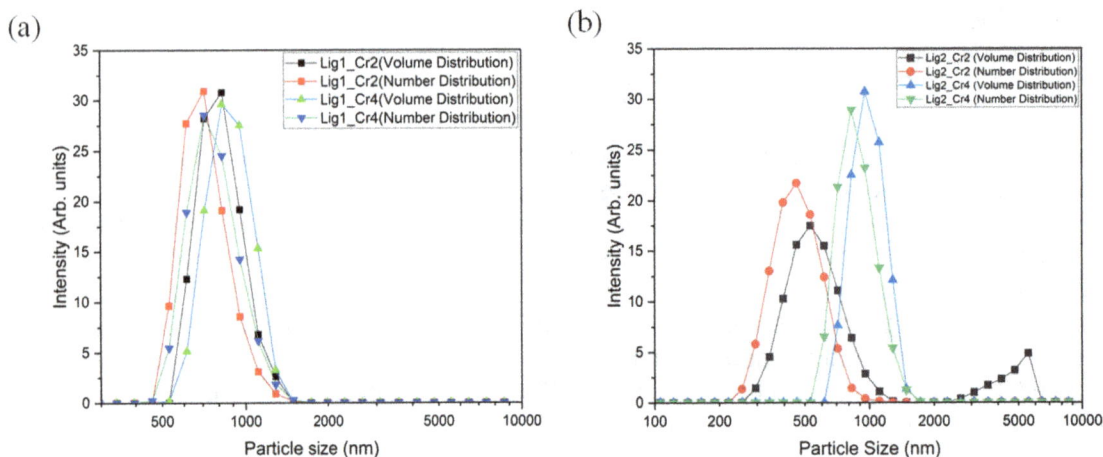

FIG 2 – Volume and number size distribution of the lignite coal dust samples prepared determined using Malvern Zetasizer-nano.

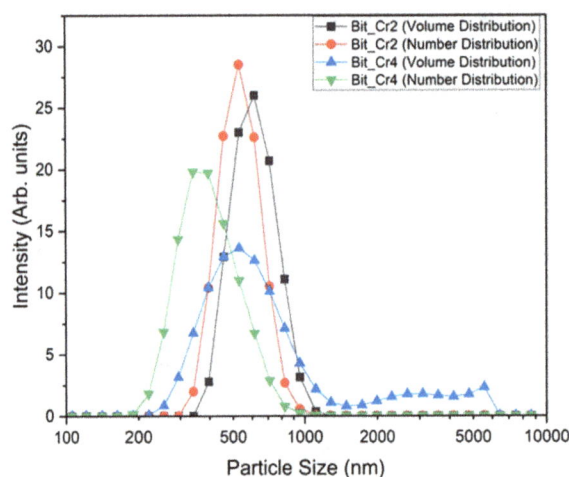

FIG 3 – Volume and number size distribution of the bituminous coal dust sample measured using Malvern ZetaSizer-nano.

Moisture treatment of the prepared coal dust samples

Constant relative humidity (RH) environments in relatively small containers were generated utilising the standard ASTM E104–02 (2012). This work used saturated sodium chloride solution as a humidity control in the desiccator, as shown in Figure 1. The equilibrium relative humidity value for NaCl solution is 75.1±0.2 at 30°C. An RH sensor was fitted on the top of the desiccator to continuously monitor the internal temperature and the RH A small beaker containing saturated NaCl solution was kept at the bottom of the desiccator, and all the coal dust samples were placed over a stage in small containers. After adding the coal dust samples, the desiccator was completely sealed using glycerine gel. This set-up was left for seven days. The RH and the temperature were continuously monitored for the experimental days. The mass of the coal dust sample was taken before and after moisture treatment, and the difference was the moisture adsorbed on the coal dust samples. This moisture was then converted into μmol of moisture per gram of each coal dust sample. Figure 4 shows the moisture adsorbed by different types of coal dust. It clearly indicates that the lignite coal dust has a much higher tendency to adsorb or interact with humidity.

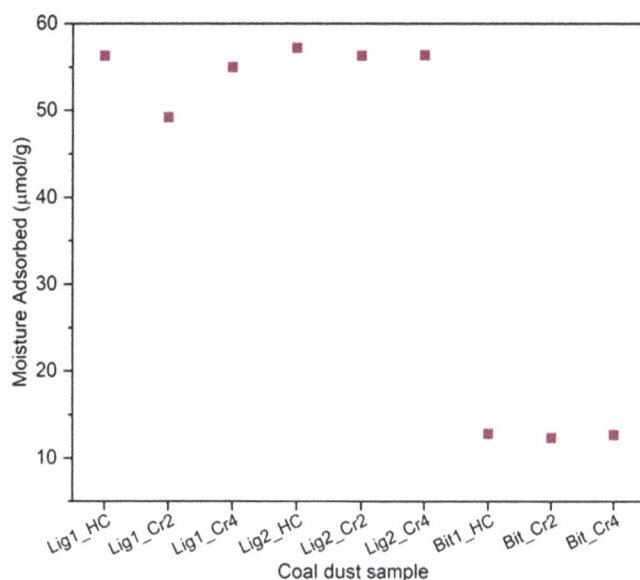

FIG 4 – Measurement of moisture adsorbed by different coal dust samples using saturated NaCl solution in a glass desiccator.

Continuous Wave EPR analysis

CW-EPR (continuous wave EPR) spectra were recorded using an X-band E500 Elexsys Bruker spectrometer operating at 9.0–9.5 GHz. EPR scan parameters were kept the same during all the experiments: modulation frequency, 100 kHz, X-band; microwave frequency, about 9.87 GHz; attenuation 40 dB; time constant-50 ms and number of scanning points-2048. All the observations were taken in air at ambient temperature and atmospheric pressure. EPR spectroscopy scans were performed with the Bruker computer software WinEPR Acquisition. The parameters recorded and further calculated from EPR spectra are g-value and spin concentration (radical intensity, ie concentrations of paramagnetic centres). The CW-EPR experiments were conducted for each coal dust sample before and after moisture treatment. The experiments were repeated three times, and the obtained values are the average of these two sets of experiments. A detailed analysis of these features provides information about the free radical characteristics of coal, as will be discussed in the later section.

RESULTS AND DISCUSSION

Change in FR quantity of coal dust due to moisture

EPR Intensities were recorded for all the coal dust samples. The peak area was obtained by double integrating the EPR absorption curve spectra. It was then converted into the spin count by calibrating and comparing the signal with a standard substance. The standard substance has known free radical type and quantity. In our analysis, we used Tempol ($C_9H_{18}NO_2$, 2,2,6,6-tetramethyl piperidine-1-oxyl) to compare the spectra and count the spins in the coal dust samples. The spin quantity is often further converted into a spin or stable radical concentration (spins/g). Tempol is a substance of known spin counts, and a calibration curve could be set between the required sample and the Tempol to define a best-fit equation. This best-fit equation then provided the multiplier to find the spin count in the required samples. Figure 5 shows the stable FR concentration of coal dust samples with a different type of crushing. Sample names are in the format '*sample(mois)*', with *(mois)* indicating that the sample was moisture treated. It can be seen that the higher rank bituminous coal dust (Bit1) has considerably higher FR radical content than the lower rank lignite coal dust (Lig1 and Lig2) for all crushing levels. Lignites have a lower rank than bituminous coal dust, as evident from the reflectance in Table 2. Some previous studies have also observed that the higher rank coal has higher free radical content (Zhou *et al*, 2019). We know that the coal's carbon content represents the rank of the coal. Also, it is widely accepted that the EPR signals mainly arise from coal macerals. Retcofsky *et al* (1968) observed that the FR concentration is directly proportional to the carbon content of the coal dust sample found using EPR analysis. Bit1 coal dust used in this study contains around 80.91 per cent elemental carbon, which is much higher than both the lignite dust samples Lig1 and

Lig2 (which have 66.05 per cent and 65.38 per cent elemental carbon) for cryomill 2 and 4 cycles, respectively.

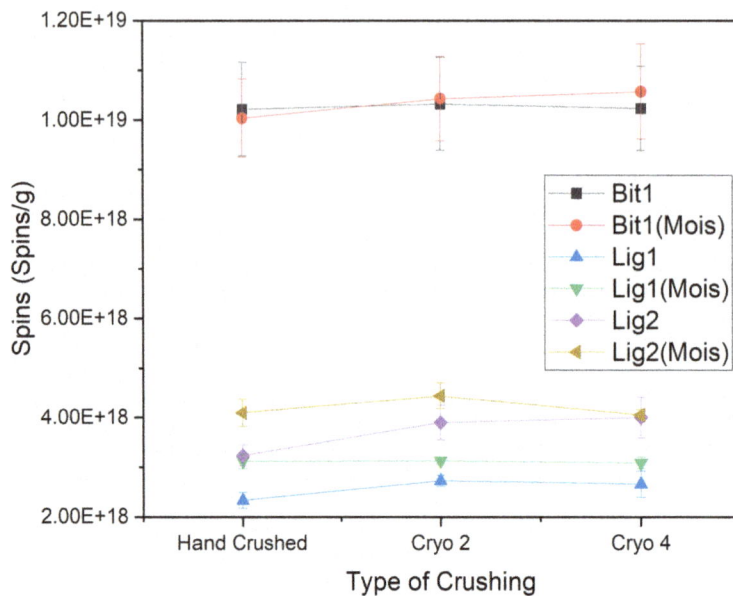

FIG 5 – Free radical spin count in different coal dust samples before and after moisture treatment (Error bars shows the standard deviation for the spin count measurements).

In Figure 5, we see that the bituminous coal dust has a much higher FR radical content than the lignite coal dust samples for any crushing level. Moreover, the FR concentration of Lig1 and Lig2 coal is of the order of 10^{18} spins/g, around ten times lower than the Bit1 samples for all the crushing. This might be because of the longer coalification undergone by the bituminous coal. This trend suggests that the coalification process involved cracking of covalent bonds that resulted in the release and coupling of small free radicals and confinement of unpaired electrons in large structures, especially the solid, ie the stable radicals. In Table 1, we see that the carbon content increases with the rank of the coal, which suggests that Bit1 coal dust has a much more aromatic structure when compared to the lignite. This trend was attributed to the bond breakage and the subsequent structural aromatisation and condensation during the longer coalification process undergone by the bituminous coal (Zhou *et al*, 2019). That's why the observed trend of EPR spin counts/g is consistent with the understanding that the aromatic rings show exclusive radical stabilising mechanisms, such as resonance and hyper-conjugation, which is lacking in less aromatic or aliphatic structures. So, it is reasonable to conclude that the stable free radical concentration increases with the increasing aromatic carbon content of coal dust samples.

Another important observation was that the FR content has increased for all coal dust samples, for any level of crushing (HC, Cr2 and Cr4). That increase is related to the amount of water the coal dust samples have absorbed. Figure 4 shows that the moisture adsorbed/absorbed by the lignite coal dust samples is greater than that adsorbed by the bituminous ones. Several other studies have suggested that the lignite coal dust can interact more with moisture than bituminous coal dust (Gosiewska *et al*, 2002; Xu *et al*, 2017). The moisture adsorption by the Lig1 and Lig2 coal dust samples is well beyond 40 μmol/g; whereas, for Bit1, it is just above 10 μmol/g. Consequently, the FR content increased in the Bit1 sample when compared to the non-moisture treated sample. There is an exception for the hand-crushed sample of the Bit1 as it shows that the FR count has decreased, which is almost negligible. However, the increase in the FR count is not well-pronounced for the Bit1 samples. For the Lig1 and Lig2 samples, the increase in the FR content is well pronounced and can be clearly seen in Figure 5. The increase in the FR content is more evident for the lignite HC samples. This is because of the dependence of the wettability of the coal dust on size. As mentioned earlier, the hand-crushed coal dust samples have larger sizes and contain most of the micron-sized particles reaching up to 177 μm. At the same time, the cryomilled samples are mostly submicron-sized. It has been found that the coal dust particle size is inversely proportional to its wettability. As the particle size decreases, more and more carbon gets exposed to the surface as the constitution of all the other elements is much lower than carbon in the coal, and the C/O ratio increases (Wang *et al*,

2019). The exposed area of coal dust increases as the coal dust particle size decreases. Meanwhile, the atomic exposure inside the coal dust becomes the surface atoms, so the proportion of carbon in the surface layer of coal dust increases while the proportion of oxygen elements decreases. The dust that has more hydrophobic groups increases its hydrophobicity (Wang *et al*, 2019). Carbon is hydrophobic, and more carbon in the smaller particles could mean less wettability (Xu *et al*, 2017). That is why the Lig1 and Lig2 cryomilled samples have slightly lower water interaction, and consequently, the change in the spin count is slightly less as compared to the HC sample. Overall, we observed that the FR count in the coal dust changes because of its interaction with moisture. This preliminary observation indicates that the higher FR in the moisture-treated samples suggests increased toxicity of these coal dust particles.

Change in FR type of coal dust due to moisture

This section will discuss the changes in the FR type of the coal dust samples because of moisture adsorption. The type of the FR is characterised by the Lande '*g-value*'. Lande g-value is constant for a specific compound. In the case of coal, we know that it is a hybrid structure; therefore, it has many types of FR of different g-values, and its g-value reflects the hybrid nature of all the stable FR present. The g-value, which reflects the spin-orbit coupling directly, can provide information on the FR's locations and environment in the paramagnetic molecules. It is particularly sensitive to the chemical environment of the unpaired electron, which is useful for obtaining structural information on the radicals. Previous studies have found that the g-values positively relate to coal maturity. As for lower-ranking coal, higher g-values indicate that an unpaired electron has some spin partially localised on heteroatoms, such as oxygen, sulfur, and nitrogen, which have a strong spin-orbit coupling. In general, the g-values of coal are higher than that of free electrons (gE = 2.0023) caused by the spin-orbit coupling. With increasing coal rank, oxygen and other heteroatoms decrease, which causes the decrease of g-values. If the unpaired electrons are located on aromatic hydrocarbon molecules, the g-values decrease further (Zhou *et al*, 2019). Figure 6 shows a distinct difference between the g-values of the coal dust. Coal dust from the Lig1 and Lig2, which has a higher oxygen content (atomic O/C = 0.206 and 0.177, respectively) and lower carbon content, exhibits g-values of around 2.0035, which is a characteristic of carbon-centred FR with a nearby oxygen atom. However, the dust from Bit1 (higher rank coal) exhibits g-values around 2.0030, characteristic of carbon-centred radicals because of the much higher carbon content in this coal dust. Figure 6 also shows that the g-value increases with the increase in the reflectance of the coal dust (Table 2). This trend is also attributed to the loss of heteroatom-containing radicals, especially oxygen, and the formation of aromatic radicals due to the increase in the rank of coal during the coalification process.

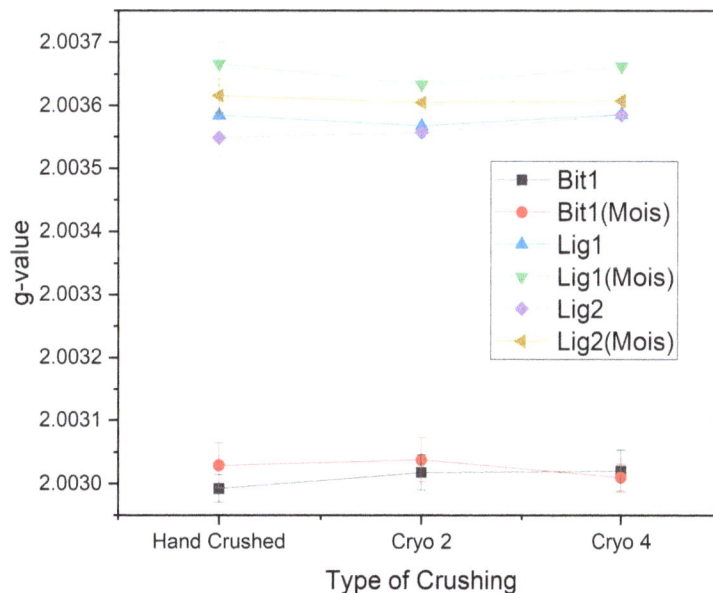

FIG 6 – Free radical g-values in different coal dust samples before and after moisture treatment.

Figure 6 also shows that the g-values have increased slightly for almost all coal dust samples for each crushing step. Again, the increase in the g-value is more pronounced for the Lig1 and Lig2 coal dust samples than the Bit1 samples. The reason for such a small increase in the g-value for the Bit1 coal dust samples is that these samples didn't adsorb much moisture (Figure 4) compared to the Lig1 and Lig2. There is an almost unnoticeable anomaly in the g-value of the Bit1-Cr2 sample. This might be because of the experimental error in recording the EPR spectra. For Lig1 and Lig2, the g-values have increased due to the higher interaction with the moisture. The primary adsorption of the moisture on the coal dust surfaces is dominated by the degree of oxidation; and, combined with the cumulative pore volume, this determines the secondary adsorption (Liu et al, 2021). Lig1 and lig2 coal dust samples have higher oxygen and low carbon content, resulting in more hydrophilic sites. Because of that, they adsorbed more moisture than the Bit1 coal dust (Figure 4). The H and O content increases in the coal dust samples as the water molecules attach to the hydrophilic sites. This means that the hydrogen (H) and oxygen (O) content will increase more in the coal dust sample with more hydrophilic sites. Lig1 and Lig2 have more hydrophilic sites than the Bit1 sample because of their composition (Table 1). Consequently, the heteroatom content of the lignite coal dust samples increases due to moisture treatment, and we already discussed that the higher g-values are related to the more heteroatomic nature of the material. Bituminous coal dust has more aromatic structures and less hydrophilic sites and will not adsorb/attach as many water molecules as lignite coal dust, which is reflected in a low increase in the g-value after moisture treatment. The noticeable increase in the g-values on moisture treatment of the lignite coal dust samples shows that the carbon-centred free radicals are getting more surrounded by oxygen and hydrogen. Whereas bituminous coal dust samples adsorbed less moisture and although it has higher spins/gm they are getting comparatively less surrounded by the O and H atoms around them. Overall, this EPR analysis shows that the type and surrounding of FR in the coal dust also change due to interaction with the moisture.

CONCLUSION

This study was performed to characterise the FR of the respirable coal dust produced from two different coal ranks and examine the toxicity from interaction with moisture. As discussed above, coal has a complex structure, and coal dust also exhibits complex free radical characteristics. Persistent FRs present in coal dust are directly related to coal dust toxicity, which is why it is essential to characterise FR, for which EPR is an excellent technique. Through this preliminary study, we found that the coal dust generated from the bituminous coal has a higher FR concentration than the lignite coal dust samples. The g-values using EPR spectroscopy were lower for the bituminous coal dust than the lignite coal dust. This is due to the higher carbon content and lower heteroatom nature of coal dust produced from the bituminous coal dust. Similarly, we also observed that the FRs are extremely sensitive to the environment from our observation of EPR analysis of the moisture-treated sample. It was observed that the spin counts/g, which is representative of the FR quantity in the coal dust, could increase due to its interaction with the moisture. This observation suggests that the coal dust released into the high moisture environment could have higher carbon centred free radical content. This also suggests that if the submicron coal dust particles are not efficiently diminished by the spray jets used in the underground mines, their carbon centred free radical could increase because of the high-moisture environment generated by the spray jets. A number of studies performed in the past few years have shown that the free radicals which are similar to what has been observed in coal dust can affect the toxic characteristics of particulates (Phaniendra et al, 2015; Vejerano et al, 2018; Xu et al, 2019).To avoid this, fine water droplet misters or foggers and surfactants or other chemicals that could effectively settle the submicron-sized coal dust should be utilised. This study also suggested that the type and distribution of FR on the coal dust could also change due to moisture interaction. The attachment of an increasing number of water molecules on the coal dust surfaces could increase the heteroatomic nature of the FR found on coal dust. We already know that different types of FR can have different stability and toxicity depending on their type and surrounding composition.

This study demonstrated that EPR spectrometry is an excellent technique for characterising the FR characteristics of coal dust. The changes in the FR characteristics due to environmental conditions can be understood with the help of EPR spectroscopy. This study further showed that the FR characteristics of the coal dust change as it interacts with the environmental moisture that could govern their toxicity. The experimental procedure and the results presented in this study are unique and crucial for understanding the coal dust toxicity.

ACKNOWLEDGEMENT

This study was sponsored by National Institute for Occupational Safety and Health (NIOSH), USA. The authors would also like to thank Dr. Shimin Liu (Associate Professor, Penn State) and Dr. Golbeck (Professor, Penn State) for their suggestions on the EPR analysis. We also want to thank Dr. Golbeck for the use his Chemical and Biomedical Engineering lab to conduct EPR analyses. We also thank Dr. Vasily Kurashov (Associate Research Professor, Penn State) for supervising the EPR experiments.

REFERENCES

Buchner, F, Forster-Tonigold, K, Bozorgchenani, M, Gross, A and Behm, R J, 2016. Interaction of a Self-Assembled Ionic Liquid Layer with Graphite(0001): A Combined Experimental and Theoretical Study, *Journal of Physical Chemistry Letters*, 7(2):226–233. https://doi.org/10.1021/ACS.JPCLETT.5B02449/SUPPL_FILE/JZ5B02449_SI_001.PDF

Chen, L, Peng, C, Gu, W, Fu, H, Jian, X, Zhang, H, Zhang, G, Zhu, J, Wang, X and Tang, M, 2020. On mineral dust aerosol hygroscopicity, *Atmospheric Chemistry and Physics*, 20(21):13611–13626, https://doi.org/10.5194/ACP-20-13611-2020

Croft, J B, Wheaton, A G, Liu, Y, Xu, F, Lu, H, Matthews, K A, Cunningham, T J, Wang, Y and Holt, J B, 2018. Urban-Rural County and State Differences in Chronic Obstructive Pulmonary Disease - United States, 2015, *MMWR – Morbidity and Mortality Weekly Report*, 67(7):205–211, https://doi.org/10.15585/MMWR.MM6707A1

Dalal, N S, Suryan, M M, Vallyathan, V, Green, F H Y, Jafari, B and Wheeler, R, 1989. Detection of reactive free radicals in fresh coal mine dust and their implication for pulmonary injury, *Annals of Occupational Hygiene*, 33(1):79–84, https://doi.org/10.1093/annhyg/33.1.79

Ghosh, S and Cebula, R J, 2021. Proximity to coal mines and mortality rates in the Appalachian Region of the United States: a spatial econometric analysis, *Regional Studies, Regional Science*, 8(1):130–142. https://doi.org/10.1080/21681376.2021.1906311

Gosiewska, A, Drelich, J, Laskowski, J S and Pawlik, M, 2002. Mineral Matter Distribution on Coal Surface and Its Effect on Coal Wettability, *Journal of Colloid and Interface Science*, 247(1):107–116, https://doi.org/10.1006/JCIS.2001.8130

Huang, X, Zalma, R and Pezerat, H, 2009. Chemical reactivity of the carbon-centered free radicals and ferrous iron in coals: Role of bioavailable Fe2+ in coal workers' pneumoconiosis, http://dx.doi.org/10.1080/10715769900300481, 30(6):439–451, https://doi.org/10.1080/10715769900300481

Jimenez, J L, Canagaratna, M R, Donahue, N M, Prevot, A S H, Zhang, Q, Kroll, J H, DeCarlo, P F, Allan, J D, Coe, H, Ng, N L, Aiken, A C, Docherty, K S, Ulbrich, I M, Grieshop, A P, Robinson, A L, Duplissy, J, Smith, J D, Wilson, K R, Lanz, V A, ... Worsnop, D R, 2009. Evolution of organic aerosols in the atmosphere, *Science*, 326(5959):1525–1529, https://doi.org/10.1126/science.1180353

Joshi, N, Romanias, M N, Riffault, V and Thevenet, F, 2017. Investigating water adsorption onto natural mineral dust particles: Linking DRIFTS experiments and BET theory, *Aeolian Research*, 27:35–45, https://doi.org/10.1016/J.AEOLIA.2017.06.001

Kovtun, A, Jones, D, Dell'Elce, S, Treossi, E, Liscio, A and Palermo, V, 2019. Accurate chemical analysis of oxygenated graphene-based materials using X-ray photoelectron spectroscopy, *Carbon*, 143:268–275. https://doi.org/10.1016/J.CARBON.2018.11.012

Liu, A, Liu, S, Liu, P and Wang, K, 2021. Water sorption on coal: effects of oxygen-containing function groups and pore structure, *International Journal of Coal Science and Technology*, pp 1–20, https://doi.org/10.1007/s40789-021-00424-6

Liu, T and Liu, S, 2020. The impacts of coal dust on miners' health: A review, *Environmental Research*, 190:109849, https://doi.org/10.1016/j.envres.2020.109849

National Research Council (NRC), 2007. *Coal: Research and Development to Support National Energy Policy*, The National Academies Press, https://doi.org/10.17226/11977.

Petrakis, L and Grandy, D W, 1978. Electron spin resonance spectrometric study of free radicals in coals, *Analytical Chemistry*, 50(2):303–308, https://doi.org/10.1021/AC50024A034

Phaniendra, A, Jestadi, D B and Periyasamy, L, 2015. Free radicals: properties, sources, targets, and their implication in various diseases, *Indian Journal of Clinical Biochemistry*, 30(1):11, https://doi.org/10.1007/S12291-014-0446-0

Retcofsky, H L, Stark, J M and Friedel, R A, 1968. Electron Spin Resonance in American Coals, *Analytical Chemistry*, 40(11):1699–1704, https://doi.org/10.1021/ac60267a011

Rubasinghege, G, Elzey, S, Baltrusaitis, J, Jayaweera, P M and Grassian, V H, 2010. Reactions on atmospheric dust particles: Surface photochemistry and size-dependent nanoscale redox chemistry, *Journal of Physical Chemistry Letters*, 1(11):1729–1737, https://doi.org/10.1021/jz100371d

Rubasinghege, G and Grassian, V H, 2013. Role(s) of adsorbed water in the surface chemistry of environmental interfaces, *Chemical Communications*, 49(30):3071–3094, https://doi.org/10.1039/c3cc38872g

Schins, R P and Borm, P J, 1999. Mechanisms and mediators in coal dust induced toxicity: a review, *The Annals of Occupational Hygiene*, 43(1):7–33. doi: 10.1016/s0003-4878(98)00069-6

Seisel, S, Pashkova, A, Lian, Y and Zellner, R, 2005. Water uptake on mineral dust and soot: A fundamental view of the hydrophilicity of atmospheric particles?, *Faraday Discussions*, 130(0):437–451, https://doi.org/10.1039/B417449F

Tang, M, Cziczo, D J and Grassian, V H, 2016. Interactions of Water with Mineral Dust Aerosol: Water Adsorption, Hygroscopicity, Cloud Condensation, and Ice Nucleation, *Chemical Reviews,* 116(7):4205–4259, https://doi.org/10.1021/acs.chemrev.5b00529

Vejerano, E P, Rao, G, Khachatryan, L, Cormier, S A and Lomnicki, S, 2018. Environmentally Persistent Free Radicals: Insights on a New Class of Pollutants, *Environmental Science & Technology*, 52(5):2468, https://doi.org/10.1021/ACS.EST.7B04439

Wang, P, Tan, X, Zhang, L, Li, Y and Liu, R, 2019. Influence of particle diameter on the wettability of coal dust and the dust suppression efficiency via spraying, *Process Safety and Environmental Protection*, 132:189–199, https://doi.org/10.1016/J.PSEP.2019.09.031

Wei, J, Xu, X and Jiang, W, 2020. Influences of ventilation parameters on flow field and dust migration in an underground coal mine heading, *Scientific Reports 2020*, 10(1):1–17, https://doi.org/10.1038/s41598-020-65373-7

Wu, T, Yang, Z, Wang, A, Zhang, K and Wang, B, 2021. A study on movement characteristics and distribution law of dust particles in open-pit coal mine, *Scientific Reports 2021*, 11(1):1–10, https://doi.org/10.1038/s41598-021-94131-6

Xu, C, Wang, D, Wang, H, Xin, H, Ma, L, Zhu, X, Zhang, Y and Wang, Q, 2017. Effects of chemical properties of coal dust on its wettability, *Powder Technology*, 318:33–39, https://doi.org/10.1016/J.POWTEC.2017.05.028

Xu, M, Wu, T, Tang, Y T, Chen, T, Khachatryan, L, Iyer, P R, Guo, D, Chen, A, Lyu, M, Li, J, Liu, J, Li, D, Zuo, Y, Zhang, S, Wang, Y, Meng, Y and Qi, F, 2019. Environmentally persistent free radicals in PM2.5: a review, *Waste Disposal and Sustainable Energy*, 1(3):177–197, https://doi.org/10.1007/S42768-019-00021-Z/FIGURES/7

Yueze, L, Akhtar, S, Sasmito, A P and Kurnia, J C, 2017. Prediction of air flow, methane, and coal dust dispersion in a room and pillar mining face, *International Journal of Mining Science and Technology*, 27(4):657–662, https://doi.org/10.1016/J.IJMST.2017.05.019

Zhang, R, Khalizov, A, Wang, L, Hu, M and Xu, W, 2010. Nucleation and Growth of Nanoparticles in the Atmosphere, https://doi.org/10.1021/cr2001756

Zhou, B, Liu, Q, Shi, L and Liu, Z, 2019. Electron spin resonance studies of coals and coal conversion processes: A review, *Fuel Processing Technology*, 188:212–227, https://doi.org/10.1016/j.fuproc.2019.01.011

Laboratory evaluation of ventilation effects on the self-heating incubation behaviour of high volatile bituminous coals

B Beamish[1] and J Theiler[2]

1. Managing Director, B3 Mining Services Pty Ltd, Darra Qld 4076.
   Email: basil@b3miningservices.com
2. Senior Mining Engineer, B3 Mining Services Pty Ltd, Darra Qld 4076.
   Email: jan@b3miningservices.com

ABSTRACT

The concept of a 'critical velocity zone' in the longwall goaf environment for the development of a spontaneous combustion event has been supported by numerical modelling. However, there is no experimental data available for Australian coals to show ventilation effects on the self-heating incubation behaviour of broken coal in the longwall goaf. A preliminary study using incubation testing has been completed on three high volatile bituminous coals (A, B and C) with R_{70} self-heating rate values of 3.86°C/h, 3.15°C/h and 2.09°C/h and moisture contents of 11.3 per cent, 3.9 per cent and 4.7 per cent, respectively. Mass to flow rates used in the adiabatic oven reaction vessel have been selected to be indicative of a sluggish, natural air leakage and medium ventilation environment. Under a sluggish ventilation flow rate all three coals incubate to thermal runaway in minimum time frames of 2.8–3.1 years, 34–42 days and 79–95 days, respectively. When the ventilation flow rate is increased to natural air leakage, coal A with the highest R_{70} value and highest moisture content self-heats to 28°C above mine ambient temperature before heat loss takes over due to moisture evaporation. However, both Coals B and C at this flow rate incubate to thermal runaway in a shorter time frame than under sluggish flow. At the medium ventilation flow rate, only Coal B incubates to thermal runaway, albeit after a significant extended period of time. These results show how important it is to determine the heat balance between the coal intrinsic reactivity (as indicated by the R_{70} value) and moisture content. The results also have implications for the possible location of the 'critical velocity zone' and subsequent hotspot development and migration.

INTRODUCTION

The influence of mine ventilation on the possible location for the development of a spontaneous combustion event in a longwall mining environment is illustrated in Figure 1 in terms of the presence of a 'critical velocity zone' (Smith *et al*, 1994). This concept highlights that in the immediate face area adjacent to the goaf fringe the air velocity is too high for heat to accumulate, but as the distance increases into the goaf the air velocity decreases to a critical zone where there is insufficient heat dissipation and a sufficient supply of oxygen to support self-heating. Consequently, in the event of a prolonged face stoppage ideal conditions may be present to allow a caved coal pile to incubate to thermal runaway. Deeper into the goaf the atmosphere becomes too oxygen deficient to support self-heating. While numerical modelling (principally CFD modelling – Yuan and Smith, 2008; Song *et al*, 2017) has often been used in support of this concept, there is no experimental data available for Australian coals to show ventilation effects on the self-heating incubation behaviour of broken coal in the longwall goaf environment.

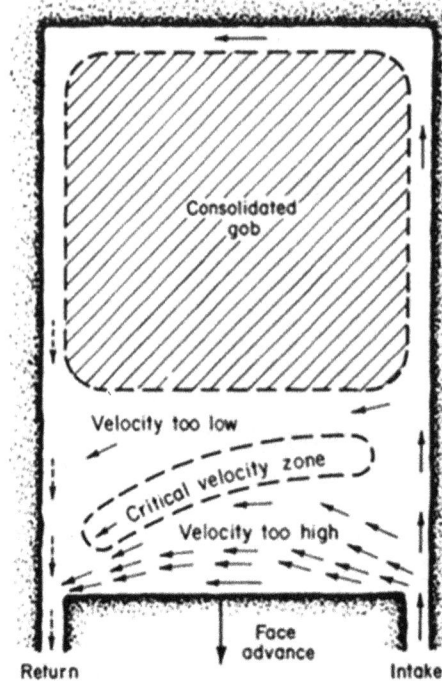

FIG 1 – Schematic diagram of ventilation flow in the vicinity of the goaf fringe on a longwall face (from Smith *et al*, 1994).

Laboratory experiments have been conducted on US coal samples using flow rates ranging from 100 to 500 mL/min and a sample mass of 150 g (Yuan and Smith, 2012). The flow to mass ratio used in these experiments is much too high to replicate site conditions, and causes evaporation to dominate the heat balance in favour of heat loss. Hence, they partially dried the coal samples prior to testing and heated the coal to a temperature in excess of 100°C to create a heat balance where heat gain could be achieved to produce thermal runaway. For the three US coals tested a different combination of flow rate and applied temperature increase was necessary for each coal to induce heat gain to thermal runaway. Consequently, practical demonstration and characterisation of ventilation flow rate effects are lacking, particularly with respect to the range of coals being mined in Australia.

Until recently, existing spontaneous combustion index tests have produced relative ratings of spontaneous combustion propensity. These tests do not provide any context of the self-heating behaviour in an actual mine environment and they do not indicate any time frame for an event to occur under mine site conditions. However, a new adiabatic Incubation Test method is now available that overcomes these deficiencies (Beamish and Theiler, 2019) and makes it possible to replicate site-specific conditions including different mine ventilation flow rate scenarios. This paper presents the Incubation Test results for three high volatile bituminous coals using three different flow rates indicative of sluggish ventilation, natural air leakage ventilation and medium ventilation, where the natural air leakage flow rate is double the sluggish flow rate and the medium ventilation flow rate is double that again. These coal samples are part of a more extended research program on Australian coals from different basins covering a wide range of reactivities and moisture contents.

DESCRIPTION OF SAMPLES AND ANALYTICAL DATA

The samples used in this laboratory evaluation are from Australian coalmines and are high volatile bituminous in rank. Two of these mines have recorded minor heating events in the past, with minimal disruption to production. Coal analytical and ranking data for the three samples are contained in Table 1. There is a significant variation in coal type as indicated by the location of each sample on a Suggate rank diagram (Figure 2). Coals A and B plot within the low-medium vitrinite coal band, whereas Coal C plots within the high vitrinite coal band. All three of these coals have low total sulfur contents and therefore the presence of pyrite is negligible. The as-mined moisture content of coal A is very high at 11.3 per cent, whereas Coals B and C have much lower as-mined moisture contents at 3.9 per cent and 4.7 per cent respectively.

TABLE 1

Analytical data for high volatile bituminous coal samples.

Proximate analysis (air-dried basis)	Coal A	Coal B	Coal C
Moisture (%)	6.1	1.3	2.6
Ash (%)	5.2	15.4	30.1
Volatile Matter (%)	24.0	25.7	27.5
Fixed Carbon (%)	64.7	57.6	39.8
Total Sulfur (%)	0.32	0.32	0.32
Calorific Value (MJ/kg)	30.42	28.23	22.38
Ultimate analysis (dry ash-free basis)			
Carbon (%)	85.7	84.2	80.5
Hydrogen (%)	4.43	4.86	5.69
Nitrogen (%)	1.92	1.55	1.59
Sulfur (%)	0.36	0.38	0.46
Oxygen (%)	7.6	9.0	11.8
ASTM rank	hvBb	hvAb	hvBb

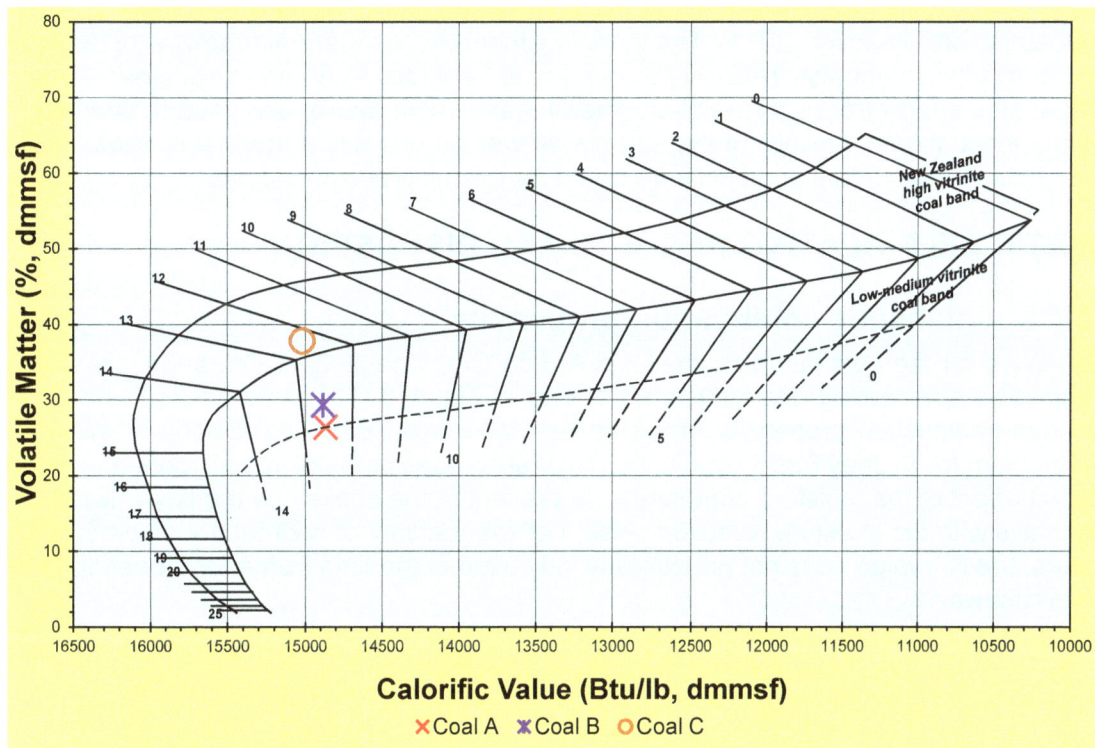

FIG 2 – Suggate rank plot for high volatile bituminous coal samples showing volatile matter and calorific value on a dry mineral matter sulfur free basis.

SELF-HEATING TEST PROCEDURES

Adiabatic oven R$_{70}$ self-heating rate

Full details of the adiabatic oven are given in Beamish, Barakat and St George (2000). The sample to be tested is crushed and sieved to <212 μm in as short a time as possible to minimise the effects of oxidation on fresh surfaces created by the grinding of the coal. A 150 g sample is placed in a 750 mL volumetric flask and a unidirectional flow of nitrogen at 250 mL/min applied to the flask inside a drying oven. Precautions are taken to ensure the exclusion of oxygen from the vessel prior to heating the coal for drying. Hence, the air is flushed from the system at room temperature for a period of one hour. After one hour, the oven is ramped up to 110°C and the coal is dried under nitrogen for at least 16 h to ensure complete drying of the sample. All R_{70} tests are performed on a dry basis to standardise the test results.

At the completion of drying, the coal is transferred into the reaction vessel and left to stabilise at 40°C in the adiabatic oven with nitrogen passing through it. The reaction vessel is a 450 mL thermos flask inner. When the sample temperature has stabilised, the oven is switched to remote monitoring mode. This enables the oven to track and match the coal temperature rise due to oxidation. The gas selection switch is turned to oxygen with a constant flow rate of 50 mL/min. The temperature change of the coal with time is recorded by a datalogging system for later analysis. The oven limit switch is set at 160°C to cut-off the power to the oven, and stop the oxygen flowing when the sample reaches this temperature. When the oven cools down, the sample is removed from the reaction vessel, which is then cleaned in preparation for the next test. The results are used to classify the intrinsic spontaneous combustion propensity of the sample according to the rating scheme published by Beamish and Beamish (2011).

Adiabatic oven self-heating incubation

This test is designed to replicate true self-heating behaviour from low ambient temperature. As such, the normal in-mine temperature is used as the starting point for the test. The nature of the test also assumes that in the real operational situation there is a critical pile thickness present that minimises any heat dissipation (represented by the adiabatic oven testing environment) and there is a sufficient supply of oxygen present to maintain the oxidation reaction. A larger sample mass and lower oxygen flow rate is used, compared to the R_{70} test method, to produce conditions that more closely match reality (Beamish and Beamish, 2011). The sample either reaches thermal runaway, or begins to lose heat due to insufficient intrinsic reactivity to overcome heat loss from moisture release/evaporation and/or heat sink effects from non-reactive mineral matter. The results are used to characterise the self-heating incubation behaviour of the sample as well as quantify if thermal runaway is possible and if so, does this occur in a practical time frame for the mine site conditions.

ADIABATIC SELF-HEATING RESULTS AND DISCUSSION

Intrinsic spontaneous combustion propensity

The R_{70} values for the coal samples are Coal A 3.86°C/h, Coal B 3.15°C/h and Coal C 2.09°C/h. Their respective self-heating rate curves are shown in Figure 3. These results indicate an intrinsic spontaneous combustion propensity rating of high for Bowen Basin conditions or alternatively a rating of medium for Sydney Basin conditions. This rating does not take into account any moderating self-heating effect of the moisture content that is present in the coal since the R_{70} value is obtained on a dry basis with the moisture removed. Also, like the majority of spontaneous combustion index parameters, the R_{70} value does not provide any indication of the time frame for a heating to develop to thermal runaway.

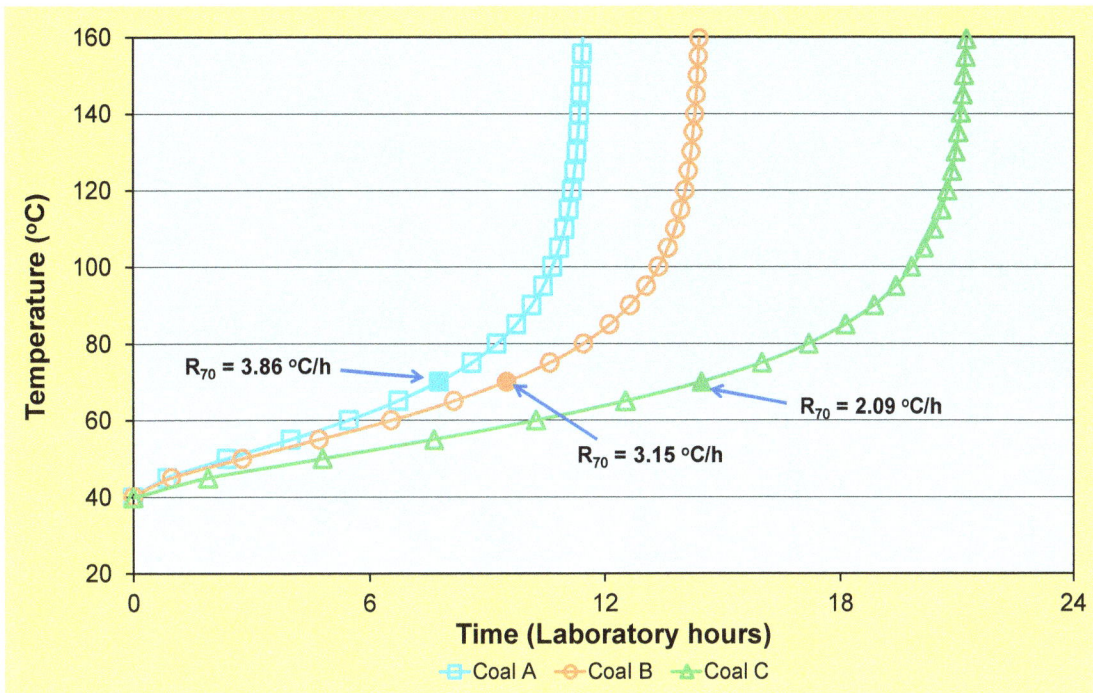

FIG 3 – Adiabatic R_{70} self-heating rate results for high volatile bituminous coal samples.

Self-heating incubation behaviour under different flow rate conditions

Using a sluggish ventilation flow rate of 5 mL/min (3.85×10^{-5} m/s), all three coals incubate to thermal runaway in time frames that reflect the heat balance between their respective intrinsic reactivities and moisture contents as-mined for the mine site ambient start temperature condition. The self-heating curves are shown in Figures 4 and 5 with the laboratory hours obtained from the test converted to a site equivalent time frame based on case study benchmark results.

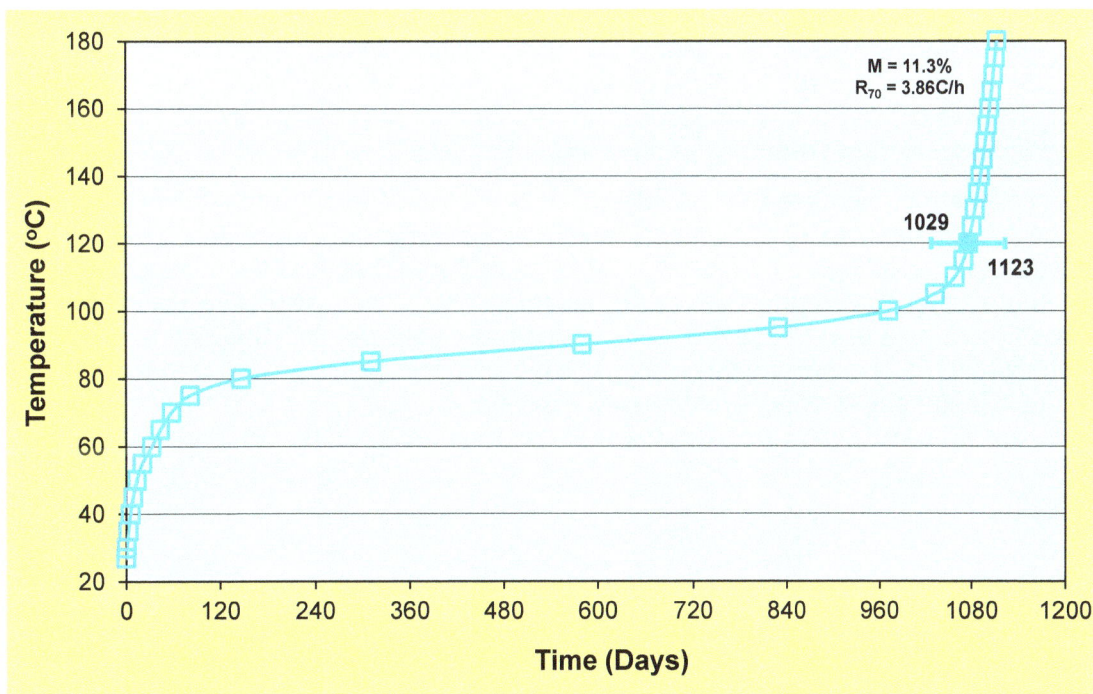

FIG 4 – Adiabatic Incubation Test results for coal A under a sluggish ventilation flow rate.

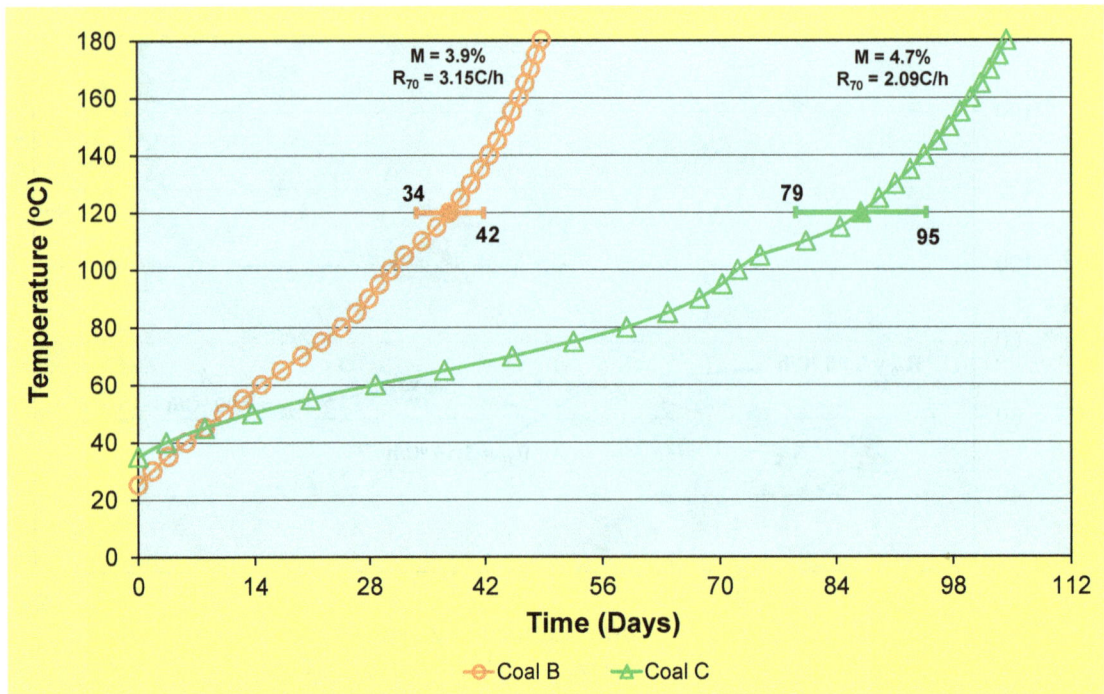

FIG 5 – Adiabatic Incubation Test results for coals B and C under a sluggish ventilation flow rate.

It can be seen that the high moisture content of Coal A has a significant moderating effect on the coal self-heating rate despite this coal having the highest R_{70} value. Initially, the coal self-heats to reach a temperature of 80.0°C in a site equivalent time frame of approximately 147 days as shown in Figure 4. Evaporative heat loss then reduces the self-heating rate for an extended period of time in the order of 890 days till the coal reaches 105.0°C. Over the next 40 days the coal locally dries out and a well-defined hotspot develops that is then capable of self-heating and migrating to thermal runaway. The time frame error bars shown in Figure 4 are the minimum incubation period lower and upper limits (1029–1123 days) for the coal under the sluggish ventilation flow rate conditions. The lower moisture contents present in Coals B and C produce minor evaporative heat loss, which can be seen in the shape of their self-heating curves where there is virtually no moisture shoulder present (Figure 5). Their respective minimum incubation periods are 34–42 days and 79–95 days, which is consistent with their intrinsic reactivity and moisture contents.

When the flow rate is increased by double to 10 mL/min (7.7 × 10^{-5} m/s), replicating natural air leakage, coal A self-heats to 28°C above mine ambient temperature before heat loss takes over due to moisture evaporation (Figure 6). However, both Coals B and C at this flow rate incubate to thermal runaway in a shorter time frame than under sluggish flow. Their respective minimum incubation periods are 26–33 days and 52–63 days, which again is consistent with their intrinsic reactivity and moisture contents. At this flow rate the supply of oxygen to the coal is greater and therefore the self-heating rate is faster due to the greater oxygen availability.

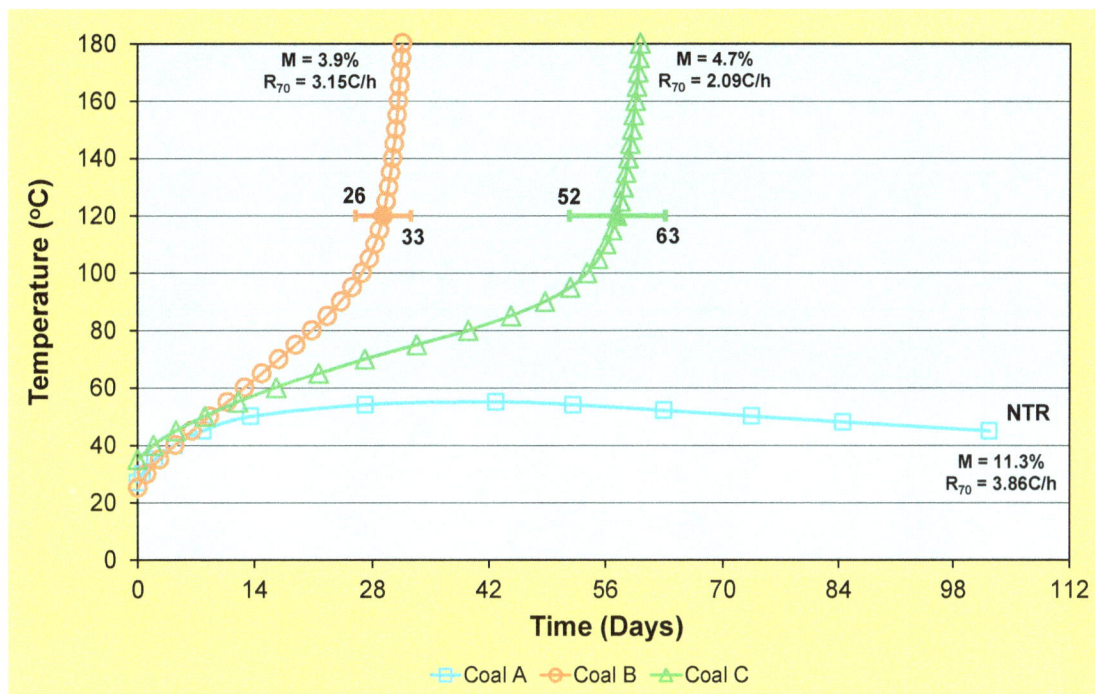

FIG 6 – Adiabatic Incubation Test results for coals A, B and C under a natural air leakage flow rate.

When the flow rate is doubled again to 20 mL/min (1.54 × 10^{-4} m/s), designated as a medium ventilation flow rate, Coal A only reaches a maximum temperature of 35.2°C before evaporative heat loss overcomes the heat released from oxidation (Figure 7). Coal C also only reaches a maximum temperature of 49°C before evaporative heat loss takes over and the coal temperature decreases. However, Coal B gradually self-heats over an extended period of time as shown in Figure 8, which is due to the coal having a low moisture content that even at this higher flow rate creates minimal evaporative heat loss.

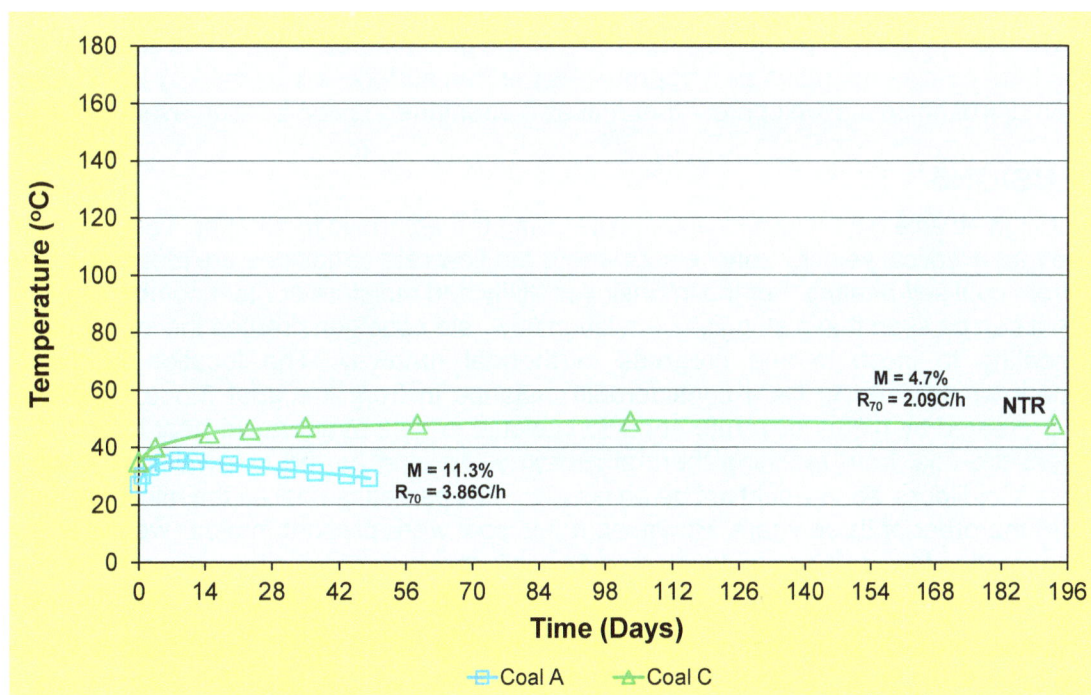

FIG 7 – Adiabatic Incubation Test results for coals A and C under a medium ventilation flow rate.

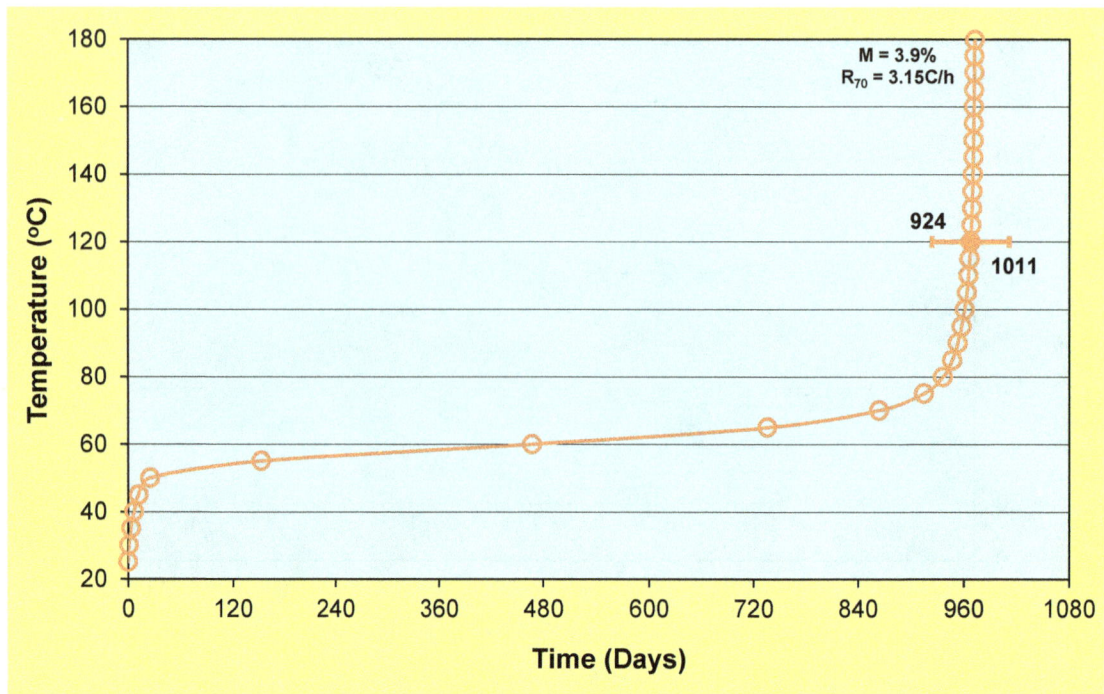

FIG 8 – Adiabatic Incubation Test results for coal B under a medium ventilation flow rate.

From the incubation test results it can be inferred that the critical velocity zone for a hotspot to develop into a heating event for Coal A is where sluggish ventilation conditions occur. This would be at some distance in from the goaf fringe. In addition, once the hotspot has developed and reached a temperature in the order of 120°C in the broken coal pile it would begin to migrate upwind and progressively move into a higher flow rate environment until such time as it reaches a free surface where ignition becomes possible. The temperature of the hotspot also rapidly escalates as it migrates due to the greater oxygen availability.

The faster self-heating shown by Coals B and C under natural air leakage indicates that the critical zone for a hotspot to develop would be shallower into the goaf for these coals. The same temperature escalation from hotspot migration as it approaches the free surface is indicated by both these coals. It is this feature of hotspot development that makes a spontaneous combustion event so dangerous.

CONCLUSIONS

Testing of high volatile bituminous coal samples using the adiabatic Incubation Test procedure has confirmed that a 'critical velocity zone' exists where the flow rate conditions are ideal for a hotspot to develop from coal self-heating. For the intrinsic reactivity and moisture content combination obtained for Coal A it can be seen that a sluggish ventilation flow rate condition creates the ideal environment for self-heating to incubate and progress to thermal runaway. The location of the hotspot development would need to be a considerable distance in from the goaf fringe, otherwise heat dissipation created by higher flow rate conditions would enable evaporative heat loss to dominate and prevent the coal from reaching thermal runaway. As long as the longwall face continues to retreat, this 'Goldilocks' zone would not be an issue for this particular coal as the minimum incubation period is in the order of three years. However, if this coal were present in older workings of a bord and pillar operation the likelihood of developing a spontaneous combustion event would be increased if the workings were kept open for an extended period of time under a sluggish ventilation regime.

For the intrinsic reactivity and moisture content combinations obtained for Coals B and C, the location of the 'critical velocity zone' would be closer to the goaf fringe. In fact, Coal B is able to generate the shallowest hotspot of the three coals. In addition, as the hotspot migrates towards the free surface the temperature of the coal rapidly escalates due to the greater oxygen availability.

ACKNOWLEDGEMENTS

The authors would like to thank ACARP for providing funding for the research presented in this paper under ACARP Project C33025. In addition, the mines who supplied the coal samples for testing in this project are gratefully acknowledged for their contribution.

REFERENCES

Beamish, B and Beamish, R, 2011. Testing and sampling requirements for input to spontaneous combustion risk assessment, in *Proceedings of the Australian Mine Ventilation Conference*, pp 15–21 (The Australasian Institute of Mining and Metallurgy: Melbourne).

Beamish, B B and Theiler, J, 2019. Coal spontaneous combustion: Examples of the self-heating incubation process, *International Journal of Coal Geology*, 215:103297.

Beamish, B B, Barakat, M A and St George, J D, 2000. Adiabatic testing procedures for determining the self-heating propensity of coal and sample ageing effects, *Thermochimica Acta*, 362(1–2):79–87.

Smith, A C, Diamond, W P, Mucho, T P and Organiscak, J A, 1994. Bleederless ventilation systems as a spontaneous combustion control measure in US coal mines, US Bureau of Mines Information Circular IC9377.

Song, S, Wang, S, Liang, Y, Li, X and Lin, Q, 2017. Influence of air supply velocity on temperature field in the self-heating process of coal, *Sains Malaysiana*, 45(11):2143–2148.

Yuan, L and Smith, A C, 2008. Effects of ventilation and gob characteristics on spontaneous heating in longwall gob areas, in *Proceedings of the 12th US/North American Mine Ventilation Symposium*, The Society of Mining, Metallurgy and Exploration Inc., Littleton, pp 141–147.

Yuan, L and Smith, A C, 2012. The effect of ventilation on spontaneous heating of coal, *Journal of Loss Prevention in the Process Industries*, 25:131–137.

Developing and maintaining mine site ventilation capability and new trends in mine ventilation career paths

D J Brake[1]

1. Principal Consultant and Director, Mine Ventilation Australia, Brisbane Qld 4017.
   Email: rick.brake@mvaust.com.au

ABSTRACT

The importance of mine ventilation to health and safety in both coal and metalliferous mines has been long recognised. The primary and secondary ventilation systems in most metalliferous mines incur 50 per cent or more of the total mine electrical power consumption and frequently about 10 per cent of total underground mine operating costs. The requirements for ventilation-related horizontal and vertical development, fans and cooling have a significant impact on mine capital costs and also on mine schedules and start-up times and hence cash flows. There are increasing pressures on regulators and miners to reduce airborne contaminants such as diesel particulate matter, nitrogen dioxide and heat (wet bulb temperatures), and significant pressures to reduce greenhouse gases such as fugitive methane emissions from coalmines and the carbon dioxide footprint of both coal and metalliferous mines. Many previously Australian miners are now global operators and have extensive overseas mining operations where local ventilation standards and expertise may not be consistent with Australian practices. The role of the site ventilation officer is also changing with strict new requirements for the statutory position being introduced into Queensland metalliferous mines and the potential for this to become more widespread. Recent regulator and parliamentary review committees have identified systematic shortcomings in mine ventilation compliance, and statutory certificates in both Queensland and NSW now require continuing professional development to maintain currency. Public and investor support for mining are also increasingly contingent on safe, healthy and environmentally responsible practices and the standards for these are frequently tightening. Good ventilation design and operating practices that are consistently well executed therefore contribute significantly not only to a safe and healthy mine but also to a productive and cost-effective operation and the ongoing social licence to operate. There are a variety of ways of providing technical design and oversight into a mine's ventilation systems; this paper explores some current trends in ventilation standards and statutory requirements, as well as ways to support and develop mine ventilation staff for both domestic and international operations and also the emerging senior technical roles of Group/Principal/Corporate mine ventilation engineer.

INTRODUCTION

Very little has been written directly about the structure or management of ventilation departments (Rose, 1979; Shaw, 1979; Self, 1999; Marks, 2012). However, after working more than 45 years in the underground mining industry, most of which has been in the area of mine ventilation, the author has seen a large number of different approaches to managing mine ventilation at individual mine sites and at the corporate level. Some of these have been quite effective but many have left the operation struggling with persistent ventilation problems, unwanted surprises and unexpected and sometimes major operating and capital cost shocks. Some operations seem to be consistently on the front foot but all-too-many seem to be consistently the opposite. This paper summarises the author's observations, reflections and opinions about the direction in which the mining industry should be proceeding with respect to managing mine ventilation, and options to get there.

DOES MINE VENTILATION ADD VALUE TO THE BOTTOM LINE?

There is no doubt that the future of mining is moving inexorably towards low impact underground mines. And as underground mines get deeper and/or more productive, they are getting hotter and more difficult and expensive to ventilate.

Mine owners, managers and regulators are all concerned about ventilation-related OH&S issues such as the resurgence of coal workers' pneumoconiosis, silica dust, DPM, NO_2 and heat stress as well as the negative publicity and reputational damage from major ventilation-related disasters or incidents such as fires or explosions eg Pike River (Panckhurst, Bell and Henry, 2012) and

Grosvenor (Australasian Mine Safety Journal, 2020). There is also a groundswell of public and investor concern about the environment and especially 'climate change'.

Partially offsetting but also complicating this trend towards deeper, more productive and higher cost underground mines is the technology trend including the digital revolution, automation, tele-remote and mobile equipment engine developments as well as environmental, societal and employment trends. All these trends will have significant impacts on future mine ventilation system design and operation. It would therefore be expected that the mine ventilation profession should be seen by mining companies as an important and even key element in successful, future mine operations. This emerging recognition is certainly the case for the larger mining corporations, if sometimes only begrudgingly, but is yet to be seen in most of the smaller players.

In practice, competent (ie well designed and operated) mine ventilation systems add value in at least the following ways:

- Better safety and health outcomes on a day-to-day basis (reduced potential for acute medical conditions and/or improved safety) as well as a long-term basis (reduced incidence of chronic disease) and reduced incidence of emergency issues (fewer or less severe mine fires and explosions and other ventilation-related negative events).

- Lower health costs and insurances (eg 'workers comp').

- Less stress and better mental health (less burnout) for workers, supervisors and managers.

- Reduced short-term operational delays and downtime.

- Improved reliability and achievability of development and production schedules, and hence improved reliability of cost and profitability forecasts.

- Minimised wastage of capital and operating costs.

- Improved productivity due to better working conditions (improved temperatures, improved visibility etc).

- Improved workforce morale and attitudes to management, and even reduced absenteeism and turnover in some cases.

- Reduced environmental footprint and surface discharges.

- Increased social and investor licence to operate for the mine, the mine owner and the industry.

Unfortunately, in our industry, there continues to be a minority of workers, supervisors, middle managers and sometimes even senior mining executives who believe that ventilation is more of an annoying impediment and unnecessary constraint to the flexibility required by them to run their operation. These individuals see ventilation standards as adding another burden of red tape and compliance costs. This poor attitude to mine ventilation is no different to the poor attitude of similar groups in the past towards mine safety or towards more topical issues such as discrimination or inclusiveness. There are many studies at the individual mine site level as well as the industry level which have consistently shown that mines can and have simultaneously improved their safety performance and improved productivity and reduced operating costs (Ramjack, 2019). It is essential for senior management on the mine site, and at the corporate level, to recognise both in word and in deed the role that good mine ventilation has in achieving its overall business objectives.

So how can competent ventilation design and management be consistently achieved in our industry?

THE VENTILATION PERFORMANCE PLATFORM

In the author's view, there are three critical supports that need to be in place for a mine to consistently deliver operationally reliable, cost-effective and safe ventilation outcomes at an individual site (Figure 1). These are (with further discussions later):

- The **ventilation control plan** (VCP), which is the single most important structure needed to support an effective ventilation system. The VCP must cover all elements that need to be in place at the mine to safely and effectively manage the ventilation, both on a day-to-day basis and in terms of short, medium and long-term planning. The VCP needs to cover the what, how, when, where and why. In this respect, it not only contains what standards are to be achieved, but also sets out how management will ensure these standards are achieved as well as ensuring regular audits of the ventilation system. The corporate version of a mine site VCP is usually a Group ventilation standard or similar.

- **Competent ventilation support** which is mostly the responsibility of the ventilation officer (VO) on-site, often assisted by in-house or external advisers in specialist areas. In larger or more complex operations, this ventilation support is provided by a team rather than an individual. The ventilation officer needs to not only be technically competent, but also needs to have the personal effectiveness and interpersonal skills required to engage with all the stakeholders needed to discharge the VCP. In this paper, the author refers to the 'ventilation officer', but in larger mines, this reference to the VO should be read as the ventilation team.

- **Consistent management support** is the third pillar for an effective ventilation system in a mine. Without consistent management support for both the VCP and the ventilation team, the ventilation system on the site, or group-wide, will not support the business objectives to the extent that it should.

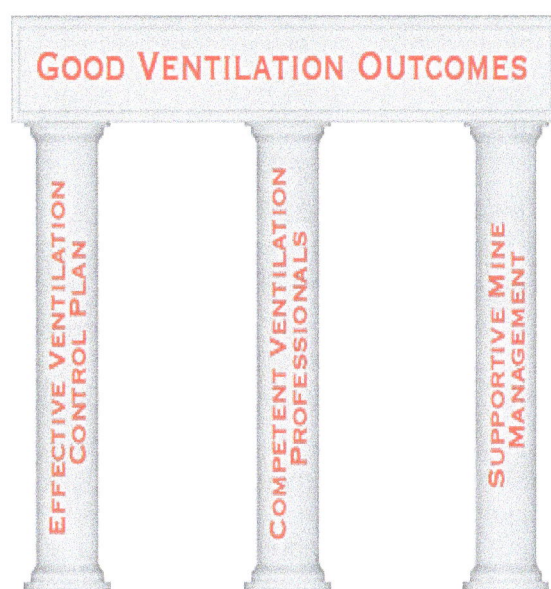

FIG 1 – The three pillars for effective mine ventilation outcomes.

Ventilation control plan

All mines should have a VCP even if there is no legal requirement to do so. The VCP is the blueprint for the successful design and operation of the ventilation system at the mine. It contains the standards, the overall strategy and the detailed game plan for managing the ventilation to meet the business objectives. It should incorporate all statutory requirements, corporate requirements, good practice industry standards and also any site-specific requirements that have been identified in earlier studies (eg consultants' reports) or on-site or off-site risk assessments etc There should be no item of importance with respect to planning or operating the ventilation system that is not in the VCP. Mine personnel should not have to forage through numerous other documents, some of which may not be able to be found or which no one even knows about any more, to establish some important ventilation matter – it should all be in the VCP.

The VCP should be developed after an extensive risk identification and assessment process to understand any ventilation-related safety, health or operational aspects of the ventilation system. It needs to include a review of regulatory requirements on the mine as well as good practices and standards used in other jurisdictions. It should be auditable, and must be audited on some defined, regular basis. The VCP should be a live document in the sense that it is updated as frequently as needed to continue to support the mining operation as it evolves and in the light of changes in any of the underlying assumptions, knowledge or practices elsewhere. In Australia, the minimum requirements for a VCP are laid out in regulation (eg NSW Government, 2014) and also in guidance notes (eg NSW Resources Regulator, 2021). Change management, document control and initial and refresher training for all the stakeholders as to their responsibilities under the VCP are all essential and should also be set out in the VCP. Key performance indicators (KPIs), risk assessments (RAs) and the need for Trigger Action Response Plans (TARPs) and supporting or related standards and procedures should likewise be included or identified in the VCP. Not all Australian states have a legal requirement to produce and operate under a VCP, but all states require risks to be managed in accordance with the ALARA principle via documented safety management systems, and it is therefore at least prudent for all mines to have a VCP and to develop and operate these VCPs in accordance with good practices or legislated requirements applicable in other jurisdictions in Australia.

Competent ventilation support

In addressing the question of how to provide competent ventilation support on a mine site, the key questions to be answered include:

1. What, if any, are the legal requirements for the ventilation officer position on-site?

2. What is the size and complexity of the operation? Physically larger mines or those with more complex ventilation circuits will need more than one VO and therefore some structure to the ventilation organisation to provide the necessary coordination.

3. Is support needed 24/7, or day shift only seven days per week, or day shift only five days per week, or on a roster of (say) seven on and seven off, in which case the site has no on-site ventilation support for seven days out of every 14?

4. What are the ventilation-related hazards at the operation? Operations with significant hazards (toxic or flammable gases, toxic or explosive dusts, radon, temperature etc) will need more experienced and better trained ventilation staff.

5. What is the role of the VO in the event of specific mine emergencies such as an underground fire, gas or dust explosion etc? If the incident management team protocols require competent on-site ventilation support during these incidents, then a VO must always be available.

6. What is the stage of the mine life? A mine that is starting up or embarking on a major development or construction program will require more ventilation support due to the important circuit changes being implemented than one where the mine is in a steady state mode. A mine that is very old with extensive underground ventilation circuits and ventilated old workings or with very dispersed remnant mining may also need significant ventilation support.

7. Does the operation need the ability to independently complete full ventilation designs or only need the ability to complete minor redesigns, incremental designs or perhaps only needs the ability to implement someone else's designs?

8. How much of this ventilation <u>design</u> support needs to or should happen on-site versus off-site?

9. How much ventilation <u>operational</u> support is required by site and how frequently? For example, does the operation need daily fault finding of the ventilation system, daily adjustments to the circuits or daily attendance at production or other meetings.

10. Does the workforce live at a nearby town (residential operation) or is it a FIFO operation? If FIFO, then some person on-site usually needs to provide ventilation support over the full roster coverage. If it is a residential operation, then perhaps having the ventilation staff working five days per week with an on-call roster over the weekend may be sufficient.

11. What are the mine's recruitment and retention objectives for the VO role? Recruiting and retaining capable and experienced ventilation staff is expensive not only in terms of the salary package and FIFO costs, but also because such staff will want (and need) to attend professionally relevant conferences, peer networking opportunities and training to stay up to date with their skills. They are also likely to want new technical challenges from time to time and also a career path, otherwise they will inevitably be poached by other organisations and some of the mine's investment will be wasted. This need for continuous professional development is also rapidly becoming a legal requirement.

12. What role does management see for the Ventilation Control Plan, if any? If the mine has a comprehensive and effective VCP, then it can generally operate with less ventilation support on-site.

13. Is the position of VO seen as a short-term rotational role or as a longer-term career role? How long will the incumbent be expected to remain in the job? What further career opportunities will be made available? How will the incumbent be mentored, if at all? What formal and informal training is needed or will be offered? Where the VO is part of a site team which includes more experienced ventilation staff or has the support of a group ventilation engineer (GVE), then mentoring and training should be more straightforward.

14. Will the VO role be:

   o A dedicated ventilation role.

   o A dual role (eg a single engineer tasked with ventilation and backfill or services or drilling).

   o A cross-training role, ie where a small group of engineers continuously and repeatedly rotate through several technical roles including ventilation. For example, a mining engineer might be the fill engineer on one 'swing', the drill and blast engineer on the next and the ventilation officer on the third, and then cycle back through these. The concept here is to have more than one engineer who can 'do' ventilation to cover FIFO issues, annual holidays, people attending training courses or people leaving the operation. In practice, this system tends to result in frustrated and uninspired engineers as none of them can really take ownership for any of these technical fields or they take ownership of one of these areas and neglect the others.

15. In third-world countries, a key issue or even a legal requirement is for the company to train up local staff to assume all possible roles in the operation. This will impact on the structure and focus of the ventilation department.

In addressing the question of whether a mining company should have a group (or principal) ventilation engineer role, the key questions to be answered include:

- What is the size of the mining company and how many underground mines does it have?

- How complex are the ventilation-related issues across the various mine sites?

- How much site and corporate ventilation-related risk is there across the various mine sites?

- Does the mining company have group policies relating to ventilation? If so, who is the owner of this policy or policies? Who will ensure it is being complied with? Who will lead any updating or auditing process for it?

- How much autonomy is provided to individual mine sites? If considerable, then that may impact on the relevance of a group ventilation engineer?

- How much mentoring needs to be done for the site-based VOs and will this be part of the role of the group ventilation engineer?

Management support

It is easy to be critical of management but ultimately management does set the priorities, the standards and the 'ethos' or tone for a mine operation. Therefore, the ventilation standard achieved in practice on a mine site does depend on the importance that management gives to this area. Specifically, management needs to address the following:

- Properly resourcing development of an effective and comprehensive VCP.

- Properly resourcing the ventilation function on the mine site as discussed above.

- Supporting and 'backing up' the VO in implementing and complying with the VCP.

- Providing proper equipment for the VO. Sometimes the VO has the wrong or outdated (or no) monitoring equipment for a particular measurement or no vehicle to actually get underground and has to rely on begging others for a lift.

- Providing a position description with responsibilities and authorities that empower the VO to do the job. All too often, the VO has a position description that has vaguely defined responsibilities or too much overlap with other roles and therefore lacks clarity as to who is responsible for what.

- Providing time for the VO to do the job. In some cases, the VO simply hasn't got the time to do any real engineering in the role. This will always result in a reactive (rather than proactive) 'whack a mole' approach which produces a substandard, ineffective, costly, risky and frequently unsafe or unhealthy ventilation operation.

- Providing <u>opportunities</u> for technical training and skills development. For younger engineers, this needs to include two years as an underground worker 'on the tools' and as shift supervisor or foreman, as well as at least a further two years in a graduate rotation program.

- Providing <u>opportunities</u> for soft and professional skills training and development.

- Providing <u>opportunities</u> for mentoring and networking, ie making sure the VO gets the technical and soft skills support and has the network available to draw upon to 'be a winner'.

These last three (the 'opportunities') are essential and, in the author's view, professional staff on a mine site should be provided with between two weeks per annum (for younger technical staff) and four weeks per annum (for senior technical staff) for professional development, soft skills development, technical training, writing papers, attending conferences, and supporting professional organisations by contributions as members on committees and the like.

In particular, it is highly regrettable that many mining companies no longer support publication of technical papers or no longer support local chapters of the AusIMM or similar professional bodies. The bigger companies, in particular, need to take a longer term and more holistic view of their contributions to the industry than most of them are currently doing.

It should also be noted that one of the critical global constraints on mining operations at present (and has been many times in the past) is the shortage of people with critical technical skills (Zakharia, 2021) and as discussed below by Scanimetrics (2021):

> The National Research Council recently released a report explaining that that the U.S. mining and energy sectors will experience a shortage in skilled workers over the next five years, according to AP. This will impact coal mining, gas drilling, solar and wind power and other manufacturing industries. A skills shortage could mean that several companies will struggle to fill vacancies, which could inhibit overall productivity levels.

> The Federal Mine Safety and Health Administration estimates that 46 percent of the mining workforce will be eligible to retire within the next five years. In the oil and gas industry, most workers are older and nearing retirement age or are fairly new to the industry. This has created a "gap in experience and maturity" that will need to be addressed as soon as possible by prioritizing the development of mentoring relationships between senior employees and new workers, explained AP.

> Although the report encouraged schools to do a better job of training workers, it warned that programs aren't currently designed to train students for these industries and additional training programs or educational modifications may be needed. The report encouraged companies to form partnerships with colleges and universities. It also emphasized that there is a "bright present and future" for those in the energy and mining industries, since demand and pay will remain high for the foreseeable future.

Taylor, Coulter and Espitia (2020) states with respect to the current shortage of mining engineers generally:

> No house can be built without foundation and no industry can be built without forward thinking and the succession planning needed to keep a constant flow of necessary skills required to attain the productivity to run an efficient asset or operation….

> One of the key factors is the best way for mining companies to equip graduates for careers in mining, and how to attract, retain and engage the engineers that have been in the industry from 0–5 years that will become tomorrow's managers and leaders.

Legislative support

As stated earlier, not all mine workers or management believe that good ventilation is important for health and safety, or that it truly adds value to the operation's bottom line. Larger mining companies can identify and manage these sceptics out of their organisation, but smaller companies may struggle to do this. This is where legislative support, in particular, can make a significant contribution to good ventilation outcomes. In fact, it could be argued that legislative support should be a fourth pillar in the effective mine ventilation outcomes system (Figure 1).

In several Australian states, the position of ventilation officer at an underground mine is a statutory appointment. This means that the position itself is established by legislation and puts specific legal responsibilities onto the appointment holder over and above any responsibilities they may have to their employer. There are of course other statutory appointment holders in underground mines including positions as diverse as surveyor, electrician, winder driver, mechanical engineer and mine manager. In all such cases, the intent behind making these positions a legal appointment is to ensure these 'mission critical' roles are filled by competent persons and that those persons discharge their responsibilities as required under the legislation rather than to any lower standard as allowed by their employer or even their own opinion.

The concept of the ventilation officer being a statutory appointment is not new. A recommendation from the Inquiry into the Appin Colliery disaster of 1979 which claimed 14 lines was that ventilation officers be appointed (NSW Resources Regulator, 2022a). The Inquiry into the Moura #2 disaster of 1994 stated (Windridge, 1996) (author's underlining):

> *Although a person with the title 'Ventilation and Fire Officer' was appointed at Moura No. 2, he did not have overriding responsibility, under the manager, for the mine ventilation system. Rather, the role of ventilation officer appears to have been one of taking statutory measurements, keeping records, and little else. The provision and maintenance of good ventilation is vital to the safety of underground coal mines and there must be a system in place to secure it. We believe that an essential requirement to that objective is to have a person who is in charge of ventilation at a mine and is <u>directly responsible to the manager</u> for the provision, maintenance, monitoring and control of ventilation.*

> *It is recommended that a position of ventilation officer be established <u>as a statutory position</u> at all underground coal mines. The ventilation officer appointed must have <u>demonstrated competencies</u> appropriate to the duties and responsibilities of the position and would be directly responsible to the mine manager for the planning, design and implementation of the mine ventilation system and for the establishment of effective standards of ventilation for the mine, methods for its control and protection, monitoring of performance, reporting procedures, maintenance of ventilation records and plans, and emergency action plans.*

The report from the Pike River disaster likewise recommended (Pike River Royal Commission, 2012):

> *Additional statutory roles and qualifications are required in new regulations, including a statutory ventilation officer and an agreed level of industry training and supervision for all new or inexperienced workers.*

The new (2021) Western Australian underground mining regulation continues its previous requirement for the position of ventilation officer to be a statutory appointment in metalliferous mines and the statutory position of ventilation officer in metalliferous mines was reintroduced into Queensland in 2019 (after being removed in 1999).

In Queensland, a recent review of the ventilation of both coal and metal mines found that (Brake, 2019):

> *Compliance activities have highlighted ongoing deficiencies in the competency and training of ventilation officers.*

> *To maintain high safety standards, persons appointed to ventilation officer positions must be sufficiently skilled. The proposed amendment will ensure there are people*

with sufficient experience, expertise, status and understanding of statutory obligations working at the operational level as ventilation officers, in the complex and hazardous mining process...

Other reforms will enable the development of a scheme of continuing professional education for holders of certificates of competency, helping to ensure the competency standards of those in safety-critical roles at mines over time.

The legislative intent is to improve the '...experience, expertise, status and understanding of statutory obligations.'

These observations about the importance of the VO role are not new and are not restricted to Australia. A survey completed for the USBM in the late 1970s (Hoskins, undated) described the following under a section titled 'deficiencies noted':

1. *Technical Supervision*

   A. *Full-time ventilation engineer rarely appointed. Part-time ventilation engineer, if appointed, spends less than ¼ of time on ventilation needs*

   B. *Dearth of basic mine ventilation measuring instruments*

   C. *Designated ventilation person receives little support from mine management. Often ignored by production supervisor. Sometimes involved in short/long term planning.*

4. *Ventilation planning: little consideration given*

5. *Education problems: undersupply of mining engineers with ventilation experience*

8. *Ventilation costs: little or no knowledge of true cost of ventilation system.*

13. *Fans: poorly monitored, poorly installed, nonexistent fan curves or operating point*

A Safety Bulletin from Western Australia (WA Resources Safety, 2011) titled 'Ventilation standards in underground mines' states:

Inspections of underground operations in the Goldfields have highlighted poor ventilation practices, particularly in the lower level workings of decline mines and in relation to the risks associated with excessive temperatures, the use of diesel equipment and dispersal of fumes after blasting

This same Safety Bulletin goes on to state that the 'contributory factors' to the poor standards were:

- *Inadequate planning and scheduling resulting in the main return shafts and airways being too far behind the decline and level development;*

- *Electrical infrastructure not properly planned to provide an adequate power supply for multiple fans;*

- *Fan characteristics not properly assessed for the diameter and length of ventilation ducting required;*

- *Inadequate consideration of the regulatory requirements for ventilation standards at truck loading stockpiles in declines and on operating levels;*

- *Shift supervisors not aware of ventilation standards with regard to velocities and quantities of air;*

- *Lack of appropriate equipment to monitor blasting fumes and noxious gases; and*

- *Use of single fans to ventilate multiple headings*

All of these issues are highly avoidable if a mine site has its trifecta of VCP, competent ventilation support and supportive management in place.

It is particularly interesting to note that the Qld government's comments in introducing the new legislation specifically refer (in two places) to the need to lift the status of ventilation officers. One of the clear intents of the legislation is that the input and concerns of site ventilation officers should be heard more clearly by senior management on-site and in their corporate offices. All three references above (Qld, USBM, WA) highlight government authorities' concerns that the role of the VO is not given enough gravitas in underground mines. This concern about ensuring that the voice of the ventilation officer is heard by senior management is also consistent with one of the key recommendations from the Inquiry into the Wankie mine disaster of 1960 (425 fatalities) which was that the position of ventilation officer should report directly to the General Manager of the mine and the same was a recommendation from the Moura #2 disaster as noted above.

One final point should be made in this section. It is impractical and perhaps even unsafe for regulators to not have their own technical specialists in 'mission critical' aspects of their own legislation. Therefore, it is this author's view that jurisdictions that have significant underground mining activity should have at least one inspector in their organisation who is competent to design and operate the ventilation system in a mine, at least to a basic level.

THE VENTILATION SUPPORT STRUCTURE

Fundamentally there are three satisfactory models to provide ventilation support for the planning and operations functions on a mine site.

Model A – minimum on-site competence with off-site support from group ventilation engineer

This model retains only the minimum possible ventilation support on-site backed up with a competent and experienced Group Ventilation Engineer, probably based in a major city and servicing all of the mines in that Group. The Group Ventilation Engineer needs to not only be called in as required by the site VO (ie on a reactive basis), but needs to take proactive technical oversight of the ventilation standards at the mine, as well as reviewing mine designs and schedules etc. It would be normal for him to visit the mine site every two to six months (depending on local factors) for an inspection, discussions, reviews of mine plans, mentoring of the site ventilation and planning staff, etc. The Group VE not only provides technical support but also helps with recruitment, internal transfer or development of new ventilation staff (as required) and provides the longer-term focus and coordination as ventilation staff come and go through the operation and the group. This model works well where three factors are present:

- where the mine is small and does not have or need large planning or operating teams on-site, *and*
- where the Group VE has sufficient time to fill this mentoring role, *and*
- where the Group VE has sufficient opportunities to get to site and does so regularly.

For metal mines, the site VO role needs at least three years' underground mining experience with at least one year of that in underground metal mines as well as completing a recognised competency-based and assessed ventilation training program.

For coalmines, the site VO role needs at least five years' underground mining experience with at least two years of that in underground coalmines as well as completing a recognised competency-based and assessed ventilation training program.

Model B – minimum on-site competence with off-site support from long-term consultant

This model is the same as model A but is backed up by a longer-term relationship with a suitable ventilation consultant instead of the Group Ventilation Engineer. This model works well where the company is too small to justify its own group ventilation engineer.

Model C – self-contained ventilation team on-site

This model sets up a mostly self-contained ventilation department on-site. As an example, Figure 2 considers the ventilation department in a large metalliferous mine which has three large production areas (or it could be three separate but nearby underground mines). The leader of the team (position A) is the ventilation superintendent (or chief ventilation engineer or senior ventilation engineer). This person could be tertiary trained (or not) and could report either to mine operations or mine technical services (MTS).

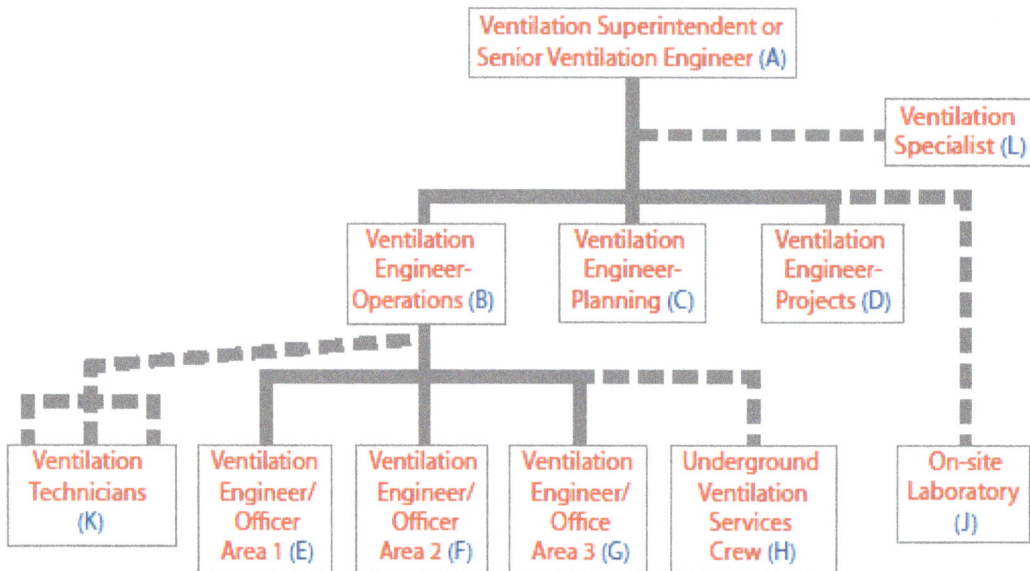

FIG 2 – Options for larger site-based ventilation teams.

In general, the 'owner' of the VCP would be the ventilation superintendent. However, in practice, this person may have limited authority in the operation's chain of command. This is a key reason why there needs to be a VCP. If there is no VCP, then any <u>instruction</u> from the ventilation superintendent or his team merely become a <u>recommendation</u>. Adoption of a VCP binds all persons in the organisation which significantly improves ventilation outcomes. The presence of a VCP on the site effectively empowers the ventilation department who otherwise may be something of a toothless tiger.

The Ventilation Engineer-Operations (position B) could be tertiary trained but often is not, as this is a long-term career position and this role needs extensive experience and local knowledge of the operation and all its key operating and planning personnel, as well as the ability to set priorities and manage subordinates E to G (and potentially K and H).

Positions E to G are the ventilation officers for particular mines or parts of a large mine. In uranium mines, the operation may need a VO on each shift, in which case four VOs are needed to cover a 24/7 operation. In some mines, positions A to D report to MTS and positions E to G/H report to mine operations. This so-called 'matrix reporting' rarely works effectively in the author's opinion and it is far better for the entire team to report to MTS for reasons discussed below. Positions E to G could be filled with graduate mining engineers or career ventilation technicians or a combination of the two.

In some mines where training of local people is important, it can be useful to have ventilation technicians (positions K) which are typically entry-level pathways into the ventilation team. These individuals go underground daily and take measurements in accordance with instructions from others and typically enter this collected data into a suitable database on their return to surface. From this position, they could potentially move into one of the VO roles (E to G) or into other roles in the organisation.

Mines with significant ventilation-related hazards (eg uranium operations) where damage to ventilation controls or secondary duct or other ventilation circuit changes need to be rectified with high priority may also have a dedicated ventilation services crew, or crews, to do this sort of work, in which case this crew (H) should report through to position B.

Large mines in remote locations may also have their own on-site ventilation laboratory (J) for instrument calibrations or gas, dust or DPM sampling measurements and analysis (Loomis *et al*, 2006).

All underground mining engineers should spend at least 6 to 12 months in the mine ventilation section. The best position is as one of the VOs (positions E to G) as this allows the VE-Operations (position B) to supervise and mentor the graduate engineer or 'trainee'. After their stint (tour of duty) in this role, they can then move into positions C or D if they have both the capability and interest in doing so or can move into other roles in the organisation in planning, services or operations. In most cases the progression would be initially be from E-G into D and then into C.

The VE Planning (C) is in constant communication with mine design, scheduling and other technical personnel and is basically the ventilation modeller for the mine. He or she should also be involved in assisting with liaising with internal and external specialists (eg consultants) and projecting future secondary fan requirements and the like. The VE Projects (D) is the team's 'gopher' and can be tasked with any of a number of different types of projects, including major projects such as new surface fans, evaluating new suppliers for ventilation duct or instruments etc.

Positions C and D can ultimately move into positions A or B, or into other senior technical or operating roles in the company or at other mines.

Technicians in K need at least a two-day technicians' course to be able to properly use the various ventilation instruments and obtain consistently reliable measurements.

Engineers moving into the E to G roles need at least two weeks of competency-based and assessed practical training. The rest of the team (positions A to D) will need additional supplementary technical training usually in specialised areas such as gas management, radiation, dusts, heat stress, cooling and refrigeration or ventilation modelling.

For global companies operating in third world countries, the ventilation department sometimes has an expatriate ventilation specialist (position L) who has no line authority in the ventilation department but who helps provide technical support and acts as an on-site mentor particularly for the ventilation superintendent, who in this case would be a national.

One advantage of all three of the above models is that most or all of the company's graduate mining engineers obtain experience in mine ventilation under the supervision and mentoring of experienced ventilation professionals, and therefore gain valuable skills and a positive attitude towards ventilation that will be valuable to them and their organisation in their subsequent career in mine planning, mine operations or other specialist roles.

The ventilation staff (or at least the ventilation superintendent) must also have sufficient confidence and maturity to recognise when the operation needs some external input (eg from specialists) or additional temporary resources and be prepared to negotiate with management to ensure these are provided in a timely manner.

Finally, it is very important that each person on the ventilation team has a clearly written job description so that accountabilities are clear, and that all mining regulations and company policies are covered within the range of job descriptions of the ventilation team.

Unsatisfactory models

Less satisfactory, or unsatisfactory, models for ventilation support on-sites include an inexperienced or ventilation-untrained person in the role of ventilation officer in structures as follows:

- Model D: with no competent technical support or technical oversight on or off-site.

- Model E: with no competent on-site support and who is expected, based solely on his own inexperience, to determine when to bring in outside specialist advice (and then justify this to management, who may not be sympathetic).

- Model F: backed up by a site senior mining engineer or operations manager, unless the senior mining engineer or manager is a competent ventilation engineer in his/her own right.

- Model G: operating in a dual role (eg ventilation and services) where it is not clearly established in word and in practice that the primary responsibility of that person is ventilation.

Models D to G are unsatisfactory because the inexperienced engineer is left to flounder in the role, with the only option of learning by his mistakes, which can be unsafe in some cases, expensive for the operation, result in morale problems for the workforce or unnecessary conflict with management. The result of these models is that the operation is left with poor ventilation standards and outcomes and the engineer is left with a poor understanding of ventilation, poor ventilation habits and a poor attitude towards ventilation which he takes with him as he then progresses into other roles.

Should the VO or team report to technical services or to operations?

Irrespective of how the ventilation support team is set-up or who the VO reports to, it is essential that the mine ventilation team has good input into both the mine planning and mine operations functions and has the respect of both. In turn, this means the ventilation department must have the required range and depth of skills and experience to handle the routine ventilation problems of the mine.

In the author's opinion, in almost all cases it is best if the VO or the ventilation superintendent reports through to mine technical services (MTS). The root cause of most of the failures in mine ventilation lies in planning and the operation of the ventilation system should proceed relatively smoothly if the mine has a quality VCP, sufficient competent people in the ventilation team and supportive operations management. It is important, however, that the MTS teams (including ventilation) are not physically located away from the operational teams. Both the MTS and operations teams should be located as close to the mine entry as realistically possible. There have been a few operations where the planning roles in the ventilation team (positions A, C and D above) report to MTS and position B has a dual reporting relationship (in technical respects to A but operationally to the mine manager). This splits the team, creates inevitable friction and this author has not seen this arrangement work any better than the entire team reporting to MTS, assuming the operation has a VCP.

Factors affecting the ability to provide ventilation support from off-site

It may appear attractive to outsource all but the essential daily operational ventilation support to off-site, but there are many factors to consider in this decision, particular in supporting operations in third-world countries:

- Language issues: especially local proficiency in oral and written English.
- Existing site practices: how big is the gap between current site practices and good practice?
- Difficulties travelling to site: visas, flight schedules, vaccinations, security, luggage restrictions.
- Time required to get to site.
- Time zone differences between site and off-site.
- Communication options for face-to-face discussions via video conferencing (internet bandwidth/speed).
- Company and government policies on using off-site labour.

THE COMPETENT SITE-BASED VENTILATION OFFICER

This author's view is that there are six key requirements needed to produce a competent VO. These are shown in Figure 3.

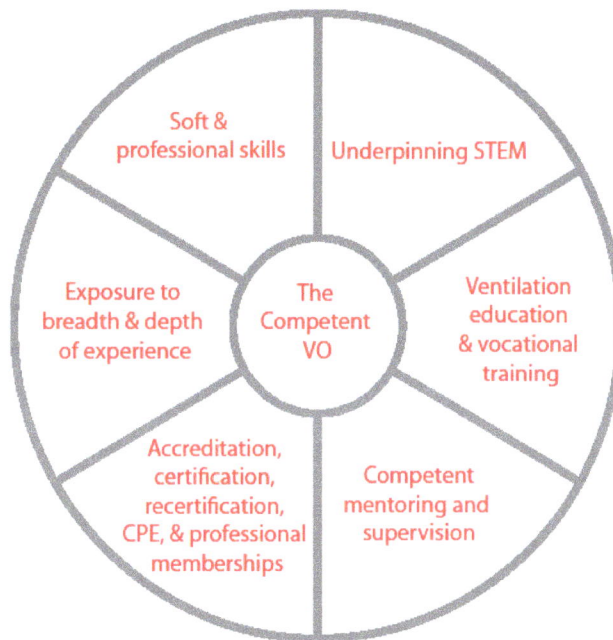

FIG 3 – The competent ventilation officer.

Underpinning knowledge in science, technology, engineering and mathematics (STEM)

There is no doubt that mine ventilation is a multi-disciplinary, science-based, technical field. A competent VO is going to need to understand the relevant aspects of a range of sciences as diverse as mathematics (including algebra), geometry, chemistry, physics, economics, electrical theory, physiology, occupational hygiene and statistics. There is no need to be an expert in any of these areas, but a strong grounding in and understanding of the underpinning knowledge is essential.

Ventilation education and vocational training

The role of tertiary (academic) education

In this author's experience, mine ventilation education at universities is mostly and in some cases entirely useless in terms of <u>practical application of mine ventilation</u>. This is a function of three factors:

- Lack of access to real mines and real mining people for site visits and real-world projects. Some universities do take considerable effort to provide site-visit opportunities for their students and obtain visiting speakers from industry, but many do not.

- Academics frequently have little or no currently relevant mine ventilation operating and planning experience. This is often a result of university emphasis on research.

- Insufficient time in an undergraduate program to cover mine ventilation in anything other than a very superficial or theoretical basis. There is great pressure on universities to shorten their engineering programs and also to broaden the range of topics covered. This inevitably leads to less time being available for mine ventilation.

However, none of these negates the essential value of a good university education. A good quality university education will provide graduate engineers with the underpinning STEM knowledge as well as many of the other skills they need for problem solving, researching information, critical thinking, writing reports and other professional skills, at least at a graduate level.

The corollary to this is that the concept of producing 'job-ready graduates' is a myth if 'job ready' means the individual can immediately enter a mine ventilation role and be competent. 'Job ready' individuals are produced by an apprenticeship system not an academic system. This is true even for the most specialised fields of applied human endeavour, such as or neurosurgeons or commercial pilots. A university education will never produce a 'professional' of any sort; it will never produce a competent neurosurgeon or commercial airline pilot or lawyer or mechanic; neither will it produce a competent mine ventilation officer; and it is entirely unreasonable and even dangerous for the industry, employers or the university sector to expect otherwise.

This position is reflected in most professional society's professional qualifications schemes, such as the AusIMM's Chartered Professional scheme which has requirements (in part) as follows (AusIMM, 2022a):

- At least five years of relevant work experience within the resources sector in at least one area of practice in the applied discipline.

- Demonstrate key competencies and a detailed written response with proof the applicant has worked proficiently in practice and the relevant discipline for five years.

- Nominate three sponsors who can verify the applicant's qualifications and experience.

A chartered professional is someone recognised as being competent in their field by a professional body representing professionals in that field. It is effectively a peer-to-peer recognition (Engineers Australia, 2022; Wikipedia, 2022; AusIMM, 2022a). Most professional organisations require a minimum of five years of high-quality experience after completing their academic qualification before being granted chartered status. This is entirely prudent and reasonable.

However, if by 'job ready', the industry means the graduates have the technical and personal skills to tackle a range of graduate-level engineering roles, and then develop competency over time in those roles if provided with the right support, supervision, experience and mentoring, then that certainly is an entirely reasonable expectation of a quality university undergraduate program.

The VO also needs a strong or at least sufficient grounding in safety science and safe design principles, risk management, mining regulations as they apply to ventilation, occupational hygiene and, importantly, some sort of personal moral compass. Often the VO may be the only person on-site who properly understands some risks and so there is an ethical responsibility on the VO to make sure these risks are properly understood by the decision-makers. It is unreasonable to expect this level of responsibility and possible confrontation to be taken up by a VO in a very junior position, which reinforces the view that the VO role needs to have increased gravitas and status in the mine organisation. Alternately, the group ventilation engineer may need to tackle senior management about such matters.

The role of vocational (on-the-job) training

As noted above, there is only so much that an academic education can contribute to a future professional, irrespective of the profession or the industry that person will enter.

Training for the professional VO requires, in fact, a combination of knowledge and skills. But professional knowledge and professional skills must be taught by knowledgeable and skilled professionals themselves. The medical or aeronautical industries do not expect an academic to train a neurosurgeon or a commercial pilot; the same applies in mine ventilation.

Irrespective of what type of organisation intends to provide vocational training for ventilation professionals, it is essential to have the following three features:

- The trainers have significant ongoing and current operational ventilation experience. This means more than undertaking the occasional ventilation-related research project; it means current experience working in or leading the ventilation planning and operating processes in a mine.

- The training materials are high quality and kept up to date.

- The assessment process must be competency based. The pass/fail criteria for a neurosurgeon or a commercial airline pilot is not if he or she has sufficient head knowledge to do the job, or can do 50 per cent of the operation well, but whether they can in fact fly the plane in all circumstances or perform the surgery in all circumstances.

The authors own view as to whether the VO needs to be a qualified engineer or not is simple:

- Some of the best ventilation officers, including senior ventilation officers, I have seen on mine sites (or as consultants) have not been tertiary trained. However, these individuals have a pragmatic view of their own limitations and are willing to ask for specialist help when an issue is outside their knowledge.

- The more technical roles in ventilation (for example, roles C, D, L and generally A in Figure 2) should be engineers. It would be optional in the other roles.

- Group ventilation engineers should all be highly experienced and very competent ventilation engineers in their own right.

Training options

One key difficulty in developing and maintaining ventilation (or other technical) capability on a mine site is releasing people for training. For larger ventilation departments or those with a heavy training emphasis, this will be a significant issue. Table 1 shows some of the options that are available with their advantages and disadvantages.

TABLE 1

Training options for ventilation staff.

Option	Pros	Cons
On-site full-day	Face to face (f2f) training generally gets the best result as its easier for the trainer to pick up on body language and get the 'people chemistry' going. Full-day training allows more intense concentration and less distractions.	Must have everyone to be trained on-site at same time. Leaves a big gap in the organisation during the training, eg can't attend meetings or do other vent work. Trainer needs to get to site and may need to do so multiple times.
On-site half-day	This is also f2f training but by only running the training for a half-day each working day, there is less impact on the organisation	As for above, but there is better support for the organisation as the students are only in training for a half day each working day
Off-site but in-country	Also f2f and has the advantage that when students are away from the mine they face less distractions and cannot be asked to attend meetings or go underground to solve some problem etc Is cheaper than sending students away for training and may avoid need to obtain visas	Need suitable training facility Also leaves a gap in the organisation during the training
Online training using Microsoft Teams	Cheaper as the trainer doesn't need to get to site Means not everyone needs to be on-site at same time so long as they have good internet at their location Can be done in shorter sessions over more days which means less impact on the organisation Can use MS Teams English language captions so perhaps easier for learners to understand what the trainer is saying.	Need good internet at all locations Doesn't have quite the same interpersonal chemistry as a f2f course May have time zone issues depending on where the learners are physically located during the training sessions Generally best if each participant has his/her own headset and microphone so means more equipment required

Off-site externally assessed training at public course	Can just send one or two at a time so minimal impact on the operation. Allows students to apply what they are being taught (if run over multiple sessions) Will give a formally recognised qualification and potentially a statutory licence	May require multiple off-site sessions so extra travel expenses and time and perhaps multiple visas Students' availability needs to fit in with when the course is offered
Off-site public course	Can just send one or two students at a time so minimal impact on the operation. Compared to off-site assessed option, course is cheaper as it doesn't have the assessments	Travel time and cost but generally only one attendance is required Students' availability needs to fit in with when the course is offered
Off-site private course	Most of the issues of the on-site course Can potentially be tailored to the needs of the operation Timing can fit needs of the operation	Most of the issues of the off-site public course

A further important issue is whether the training will be assessed or not, and whether the training will provide a recognised qualification or not.

In most cases, assessing the training is useful for the students (it identifies gaps in their knowledge), their employer (it identifies who is competent) and the trainer (it identifies what type of training delivery is most effective). Competency assessment can be difficult to achieve in any meaningful manner if there is no time during the training program for the students to revise their work. However even a relatively low standard of assessment can still be useful.

The provision of a recognised qualification will be attractive to most students but the level of assessment will necessarily be much more rigorous and the assessing authority will also have its own costs that need to be charged. This will extend the cost and time of the training course. It also makes the students more desirable to other organisations and can result in poaching.

Mentoring programs and competent supervision

As noted earlier, technical skills are only one part of producing or being a competent ventilation engineer. The VO has traditionally had little line management authority, so a key aspect of the role is to be able to persuade other stakeholders to do what is needed for the ventilation system to work as intended. The job is therefore often as much about managing relationships and being an 'influencer' or 'persuader' as it is about managing technical issues. Building and maintaining relationships takes face-to-face time. Respect is earned not demanded and in a technical role, respect from operators or other technical people comes by being good communicators and demonstrably competent (ie having a proven track-record of providing useful, practical and effective advice). Such individuals generally achieve a lot more cooperation than those who do not.

A lot of these soft skills, including 'how to influence people', are best learned by observing good role models. This requires some sort of mentoring process. Ideally this would happen on a daily basis from the individual's supervisor, but mentoring can be done remotely if the two individuals are willing and able to put the time into this and if both have the commitment to working through the inevitable personality and other challenges that arise. It would certainly be desirable for all technical persons in a mine to have someone who can mentor them technically; and in addition, hopefully this mentor can also help the mentee navigate the many soft-skill issues involved in successfully persuading other stakeholders to provide the assistance that is needed to produce good ventilation outcomes.

For more senior career individuals in mine ventilation, options for mentoring include:

- The group ventilation engineer if one is available and has the necessary interpersonal skills.

- The newly established AusIMM mentoring program (AusIMM, 2022b) if it is successful.

- The individuals own professional network, eg could be a senior consultant with whom the mentee has a strong and long-standing professional relationship.

Accreditation, certification, re-certification, continuing professional development, professional memberships and the role of statutory appointments

A vocational qualification is effectively a statement that a person is <u>competent</u> ('job ready') to practice in a particular role in a particular occupation (vocation). <u>All</u> jobs requiring independent action require the incumbent to be trained and competent (even the local coffee shop barista); the more important questions are whether the person in that role should need a <u>license</u> (ie an authority or registration to practice) and/or if their role should be a <u>statutory</u> one (ie one that has specific legal responsibilities under some legislation). There are therefore three matters to consider:

- **Competency**: is the person competent ('job ready') for the role? If so, how is this assessed and by whom? A bad cup of coffee does not have the same consequences as badly repaired car brakes or a badly designed ventilation system.

- **Licencing/registration**: Does (or should) the person need a licence (or registration) to practice in that role? If so, what are the requirements for the licence or registration and who is the issuing authority? Note that a vocation may require a licence to practice (eg a doctor or nurse) but in most cases, these do not result in the person being a statutory appointment. Engineers who want to practice in Queensland need to be registered (effectively a licensing scheme), and registration is authorised and managed by a government statutory authority, the RPEQ.

- **Statutory appointment**: Should that role be a statutory appointment, ie have statutory obligations (legal responsibilities) and authorities under specific legislation? If so, what are the requirements to be appointed into the role and who sets and assesses these requirements?

Few, if any, persons would argue that the VO does not need to be competent. The more divisive question is whether the VO should be licenced and/or whether the position itself should be a statutory appointment.

It can certainly be argued that there is no need for the VO role on a mine site to be a statutory appointment. However, this argument is more a question of whether there is a need for <u>any</u> statutory appointments in a mine. Many countries do operate without any statutory appointments.

The author's own view is that in an ideal world, there would be no need for statutory appointments, but in the real world, statutory appointments do force or at least influence both the individual and their employing corporation to 'stop and think' before making potentially 'grey area' or risky decisions. Adopting statutory positions also helps to prevent some of the dangers of 'group think', which can be particularly problematic when there are very strong personalities involved on a mine site, because ultimately the statutory official (the 'ticket holder') will have a personal responsibility for discharging that particular role and that particular decision, with the individual being potentially criminally liable for breach of those duties. All statutory officials should therefore, in the author's opinion, be given assertiveness training and trained in negotiation skills. This will reduce the individual's stress levels and improve the outcomes for all involved.

There remain some significant issues regarding the system of statutory appointments generally, including for the ventilation officer role. These would include:

- The situation where the responsibilities of two or more statutory positions overlap. For example, how does the legal responsibilities of the VO interact with the legal responsibilities of the 'registered mine manager' or the 'site senior executive', both of which are statutory positions in Queensland and both of which also have responsibility for mine ventilation (Halim, 2011).

- To some extent, the tension between the statutory position holder and his or her line managers. For example, in Figure 2, the statutory position holders might be positions E to G. What happens if there is a difference of opinion between E and B?

- In both cases above, this author would consider the VO to have failed to have competently discharged his or her statutory duties if any of the following were present:

  o Ventilation design or ventilation technical advice was poor especially where this design or advice is outside the reasonable competency of the mine manager.

  o Inadequate ventilation measurements were taken or there was inadequate analysis or interpretation of those measurements. It is important that the VO is taking the correct measurements using the correct method at the correct location and correct time and then using the correct procedure to analyse and interpret the results.

  o Inadequate communication with the relevant persons who have the responsibility and authority to take the required actions. It is important that the VO actively offers advice and doesn't just passively wait to be asked his or her opinion. The VO must be sufficiently assertive to be able to get the attention of the other duty holders.

- How will the operation fill the statutory role when the VO is away from site, eg on leave, or when he or she leaves the operation.

- Who will provide accreditation for a statutory qualification? The author's view is that statutory qualifications, which are intended to result in legal appointments, should only be authorised by the regulators of the legislation that requires those appointments, although the regulator may outsource some of the aspects of that accreditation process if it sees fit, but accountability for the appointments must always remain with the regulator.

- Statutory qualifications should not be 'for life'. The trend in society is for vocational qualifications and statutory appointments to expire if they are not being used and kept up to date, and this is entirely reasonable and in fact essential to protect the integrity of the system. If an orthopaedic surgeon has not performed an operation for some years or a commercial pilot has not flown a plane for some years, should they still be allowed to practice?

- This raises the issue of re-accreditation or recertification or continuing registration. In the context of a professional qualification, re-accreditation is best achieved by a process of continuing professional education (or development). As the VO is a vocational appointment, the CPE must be more than 'head knowledge'. It must include a requirement to be 'back on the tools, ie practicing as a ventilation officer or engineer.

In supporting this, note that the 1996 Inquiry from the Queensland Moura #2 disaster stated (Windridge, 1996) (author's underlining):

> *As demonstrated repeatedly in evidence, it should not be taken for granted that a statutory certificate of competency to practise as a mine manager, undermanager or deputy carries an assurance that the person possessing it is maintaining, and where necessary developing, the original knowledge base required for the appointment.*

> *It is recommended, therefore, that the procedures for granting statutory certificates for underground coal mining and the conditions under which they are awarded, be reviewed. In particular, it is recommended that certificates not be granted for life and that a system needs to be developed and put into effect as soon as practicable that requires certificate holders to demonstrate their fitness to retain the certificate of competency on a regular basis, at intervals of not less than three and not more than five years.*

> *The process should aim to ensure that certificate holders maintain a sound knowledge base on, and keep abreast of, technical developments in coal mining and most particularly those relevant to coal mine safety*

A requirement for ongoing professional development also supports career path development as well as continuing to improve the ventilation standards in mines as new practices emerge, ie continuing professional development supports a culture of cross-pollination and continuous improvement. This requirement for ongoing professional education is already a requirement in most professional societies for AusIMM members to maintain Chartered status.

Both NSW (NSW Resources Regulator, 2022b) and Queensland (Qld Resources Safety and Health, 2022) now have quite specific ongoing requirements 'to ensure... mining professionals in safety-critical roles have the necessary skills and knowledge'. Key features include:

- Practising certificates will be valid for five years.

- Holders of the VO statutory certificate in Qld (whether coal or metal mines) must achieve 120 hours of qualifying continuing professional development over the five years, with a minimum and maximum of 24 hours and 30 hours per annum respectively. NSW is similar.

- There are opportunities for mutual recognition of practising certificates between jurisdictions.

The role of experience

Experience is the best teacher, according to one well-regarded proverb. There should be caveats on this proverb but there is no doubt that experience is very important teacher for the VO. However, in this author's experience, it is essential for individuals and management to understand that 20 years working in a mine ventilation role ('20 years' experience') could mean two very different things:

- It could mean 20 lots of one years' experience, ie in effect, one year's experience repeated 20 times, or

- It could mean 20 years of a wide variety of ventilation design and operating challenges, ie a genuine 20 years of experience.

Of course, in most cases the experience gained by an individual lies somewhere between these two, but it is important to not look at solely the 'time in the job' in terms of realistically assessing the level of experience gained by someone in these roles; the breadth and variety of the experience is pivotal.

It is incumbent on the individual, as part of their own career planning, to make sure they are getting many years of varied experience and not just repeating the same one years of experience over and over.

Soft skills and professional skills

Overall, too many technical people on mine sites, including VOs, are seen as being out of touch with the operation, or their opinions or advice are seen as being impractical or unreasonable. In most cases, this is more the fault of management (lack of support) than it is of the VO but this rejection and deprecation can create a distressing cycle in the VO producing timidity, lack of confidence, withdrawal and isolation. To avoid this, both management and the individual need to increase the individual's competence not only in the technical areas but also in the areas known as 'soft and professional skills'.

These refer to a range of essential qualities such as assertiveness training, negotiation skills, time management, personal effectiveness, team skills, communication skills, stress management, conflict management, 'emotional intelligence', natural curiosity, creativity and innovation, critical thinking, thirst for learning, problem solving and leadership skills (McCarthy, 2014; Marwein, 2021).

To repeat: it requires a range of technical _and_ soft and professional skills for a VO to be able to collaborate with others to the extent necessary to develop a ventilation plan and to then persuade or influence those necessary others to follow that plan. This recognition of the importance of soft skills is something that should be taken up in the various CPD or CPE schemes.

THE COMPETENT GROUP VENTILATION ENGINEER

The knowledge, skills and personal qualities required by a competent VO have already been discussed. However, more and more mining companies, especially the global operators, are recognising the need and the value to be added by have technical 'leads' or group engineers in business-critical technical areas. Often these business-critical areas have been identified as a result of the damage done to persons, communities or the companies own reputation (and bottom line) as a result of major fires, falls of ground, explosions or tailings dam collapses, etc. In recent times, most of the global mining companies have or are implementing the role of group ventilation engineer. This provides an exciting new career pathway for the VO who wants to become a 'mover and shaker' in the area of mine ventilation at the corporate, industry and professional level.

If the VO wants to move into a senior ventilation career role, their technical, interpersonal and leadership skills all need to move to a higher plane. For example, the skills needed and required roles to be performed by a principal ventilation engineer at one large mine site (Oyu Tolgoi) were publicly advertised by the mine owner (Rio Tinto, 2021), with key aspects summarised in Table 2. It is very interesting to note the many higher-order technical, leadership, business, communication and professional skills expected for this role.

TABLE 2

Requirements for a Principal Ventilation Engineer (Rio Tinto, 2021).

Requirements	Duties
<ul><li>An Engineering degree or equivalent mine ventilation qualifications with >10 years' underground mining experience</li><li>Proficient with Ventsim simulation software</li><li>Demonstrated experience in medium and long-term ventilation planning</li><li>Strong understanding of Occupational Hygiene Principles and Emergency Preparedness</li><li>Experienced in underground mine fire modelling, fire detection and prevention</li><li>Capable to design and evaluate ventilation infrastructure for Fixed Plant and Critical Facilities</li><li>Experience with underground mine construction and worked with EPCM</li><li>Mine heating and refrigeration/cooling experience</li><li>Competency in Risk Management and Risk Assessment.</li><li>You will be a highly motivated individual with a strong technical aptitude in the field, ideally with significant exposure to mining operations and the metals and mining industry in general</li><li>Excellent writing skills, attention to detail and an ability to problem solve</li><li>Demonstrated commercial insight, strategic thinking and business acumen;</li><li>Strong focus on Safety</li></ul>	<ul><li>Lead the program to further reduce underground major hazard risks at our assets and studies</li><li>Work with asset and studies teams to identify risks and actively engineer out hazards</li><li>Ensure the right specialist technical capabilities are developed in RTC (Rio Tinto Copper)</li><li>Work with Group Technical to enhance professional development programs</li><li>Deliver efficient short to long-term mine/plant ventilation planning services to RTC</li><li>Expand the productivity and benchmarking program across RTC, and lever to accelerate leading practice replication outcomes</li><li>Improve technical systems and processes within RTC, including data quality and utilisation</li><li>Work across asset and functional team boundaries to deliver technical optimisation opportunities</li><li>Work with Oyu Tolgoi JV team to ensure prioritisation and delivery of highest value/risk business opportunities</li><li>Support the technical work program for the Resolution and Kennecott underground projects</li><li>Support the Oyu Tolgoi team to deliver the technical program for the UG project, including mine design and production readiness</li><li>Shape and coordinate the deployment of the RTC Research and Development program, levering synergies with the CoEs (Centres of Excellence)</li><li>Work with assets/functions to utilise Strategic Production Planning (SPP) to increase long-term value</li><li>Support Ventures with M&A technical due diligence</li></ul>

TRENDS IMPACTING THE MINE VENTILATION PROFESSION AND IMPLICATIONS FOR CAREER PLANNING

There are four waves that have been impacting the mine ventilation profession over the past 20 years and these waves will continue to influence development of the profession for at least the next decade or more:

- the globalisation wave
- the changing technical role wave
- the professional standards and accreditation wave
- the technology wave.

The globalisation wave

Globalisation and especially the development of global (trans-national) mining corporations is resulting in more harmonisation of ventilation standards and good practices around the world. This will impact all mines, because even small locally-owned mines will need to lift their standards as the global miners in that country harmonise theirs. Globalisation will also mean more career opportunities for technical staff, more opportunities to experience different cultures and different management approaches to technical matters generally, a larger network of fellow ventilation professionals, more opportunities for mentoring, and more exposure to different approaches to ventilating a mine.

The changing technical role wave

In the view of this author, the nature of most technical roles is changing. There is a growing tension between increasing specialisation and increasing generalisation. This is due to at least the following factors:

- Broadening of most technical roles is due to a combination of:
  - Flattening of corporate structures which has meant the breadth of each role has often increased whilst simultaneously there is less support and mentoring from middle management, which either no longer exists or is preoccupied with other tasks.
  - More emphasis on safety, health, social and environmental impacts, as well as current organisational values such as diversity, in addition to the traditional engineering skills.
- Increased specialisation especially at the specialist level. The rapid expansion of knowledge in most technical fields means it is increasingly difficult even for specialists to remain up-to-date in their specialty area. Traditional speciality areas are fragmenting into even more specialist areas. There will be an increasing demand for providers of highly specialised niche services.
- Increased complexity and an acceleration in the development of new technologies means a greater need (and opportunity) for creative thinking, problem solving, flexibility, willingness to change and interpersonal skills. People who are resistant to change or resistant to thinking outside their or the industry's customary box will be swept away or left behind. This will particularly be an issue for older workers who are more 'invested' in the way things have always been done and are generally more reluctant to try new ideas or change their ways.

The professional standards and accreditation wave

There is still considerable resistance to adopting professional standards in mine ventilation. However, if the concept has merit, and this author believes it does, then history is on the side of professional qualifications being, initially, 'desirable' but eventually being 'required'. The issues are going to be:

- Who sets the professional or competency standards?
- Will there be a registration or licencing scheme for ventilation professionals, and if so, who will manage this scheme?

- How would the competency accreditation and/or registration schemes link into statutory accreditation, where applicable?

- What will be the role of continuing professional education in terms of ongoing registration?

- How will mutual recognition work across jurisdictions?

- Is there a future for organisations such as the MVSA and if so, what does that future look like?

The technology wave

The technology wave, as it will affect mine ventilation, has at least these four elements:

- Digital revolution, with almost all previously manual or analogue processes becoming digital.

- Big data and real-time data, so that a huge amount of monitoring information will become available in real time. This will affect the practice of mine ventilation directly, but also in ways yet to be fully understood as brainstorming comes up with ever-better ways of extracting an orebody (Glencore, undated).

- Automation and remote control as these will fundamentally change many mining processes and these new processes will need new ventilation solutions.

- Environmental and social expectations so the underground mine of the future will have a minimal surface footprint and will look and be 'clean and green', or at least in keeping with the pre-existing natural environment.

One unnecessary concern about changing technology is the view expressed to this author that there will be no need for mine ventilation professionals as 'mines are going to get rid of diesels and be electrified'. This is rather ironic as diesels only came into underground mines in the 1960s. And before diesels the industry used electric and pneumatic powered machines—and still needed mine ventilation professionals! Part of this foolish opinion relates to an equally foolish fallacy held by some mining engineers that mine airflow requirements are solely based on diesel engine power in which case, if there are no diesels, there is no need for air.

Implications for future ventilation professionals

The ventilation professional of the future, especially at a more senior level, will need:

- An increased understanding of data networks and communication and automation.

- The ability to work simultaneously in multiple teams and on multiple projects.

- To be comfortable with real-time mine ventilation monitoring and management.

- To be a very good at process engineering as mine ventilation will increasingly revolve around setting up, maintaining and managing good processes including via the VCP.

- To be a self-starter and self-directed, have excellent time management skills, be confident and comfortable at expressing their opinions (assertive), and be good at conflict resolution.

- To have a good network of technical resources and advisers that can be called on at short notice for problem solving.

- To be a good communicator and influencer and in most cases, a good mentor and mentee.

Video conferencing and virtual reality will mean that mines will become more comfortable combining on and off-site ventilation support. The author can see the time approaching where:

- The site VO could go underground with a video camera on his hard hat and conduct a real-time (from underground) conversation with an off-site specialist or start-up a 'Teams meeting' of specialists who can ask questions and provide advice or assistance with fault-finding in real-time.

- The site VO can send a drone underground on a route pre-planned in an app much like Google maps (ie the VO selects waypoints or vent stations and the drone decides the best route and time to get between those points taking into account all the various mine restrictions). The drone has sophisticated anti-collision technology and can move out of the way of mobile equipment or people. It carries a live video feed as well as takes extensive real-time wind speed, airflow, temperature, gas, DPM, and dust monitoring. It could take off automatically on its pre-programmed route (or routes) three times per day and just issue a summary or exception report to the VO or could automatically deploy in the event of certain triggers (eg reports of a fire).

The above are generally positive scenarios for the future of the VO profession. Alternative and negative scenarios could be for the role of the site VO to:

- Be subsumed by other professions such as hygienists or safety officers or other engineering professions. This is already happening to some extent due to the shortage of quality ventilation officers in part related to the shortage of mining engineering graduates. As an example, the University of Queensland has regrettably ceased offering an undergraduate mining engineering degree.

- Become increasingly de-skilled and therefore reliant on vendors to do the ventilation design for the mine. Vendors are certainly an important part of the overall mining team, but it is unrealistic to expect them to give impartial advice that is in the best interest of the mine, especially when there is no charge for that advice.

Career planning

The implications for career development and planning follow on from the various observations above. In many cases, the career VO has been too passive about their career development so it is important to note that, ultimately, it is up to the individual to take responsibility for their own career development. The employer can and should assist with career development, but the career development buck rests solidly on the individual's desk.

CONCLUSIONS

Many changes are currently impacting on the mine ventilation profession. These are interesting and even exciting times not only technically, but also in terms of the emerging recognition of the contribution of the mine ventilation professional to the operation's business success and new career opportunities and the growing status of the profession. However, there are no certainties in life and it will be up to the individual to steer his or her own career and up to industry and the various professional organisations to position this important technical role for success in the future.

REFERENCES

AusIMM, 2022a. Chartered Professionals, accessed 14 April 2022, https://www.ausimm.com/career-development/accreditation/chartered-professionals/

AusIMM, 2022b. Mentoring Program, accessed 14 April 2022, https://www.ausimm.com/career-development/mentoring/

Australasian Mine Safety Journal, 2020. Grosvenor mine explosion UPDATE, May 7, accessed 5 March 2022. https://www.amsj.com.au/grosvenor-mine-explosion-update/#:~:text=Anglo%20American%20has%20provided%20an%20update%20on%20Grosvenor,to%20their%20torso%E2%80%99s%20and%20airways%20while%20working%20underground

Brake, D J, 2019. What are the implications of the new statutory position of ventilation officer in Queensland?, *AusIMM Bulletin*, Feb, pp 60–64.

Engineers Australia, 2022. Chartered Engineer: the Measure of Excellence, accessed 14 April 2022, https://www.engineersaustralia.org.au/Chartered

Glencore, undated. Building the mine of the future, accessed 5 March 2022, https://www.glencore.ca/en/sudburyino/innovation-and-technology/building-the-mine-of-the-future

Halim, A, 2011. Ventilation Officers or Mine Managers? Who should Ultimately be Responsible for Mine Ventilation? A Comparison between Western Australia and Queensland, in *Proceedings of the Australian Mine Ventilation Conference* (The Australasian Institute of Mining and Metallurgy: Melbourne).

Hoskins, J, undated. Survey of 22 underground non coal mines, Completed for the USBM.

Loomis, I, Karmawan, K, Rachmad, D and Duckworth, I, 2006. Underground Ventilation Laboratory at PT Freeport Mining Operations, West Papua, Indonesia, *SME Annual Meeting*, 27–29 Mar 2006, St Louis.

Marks, J R, 2012. Hardrock mine ventilation-thoughts and perceptions (keynote address), *14th US Mine Ventilation Symposium*, Salt Lake City.

Marwein, K, 2021. Top 10 qualities of a great engineer—Know Here!!, accessed 5 March 2022, https://www.embibe.com/exams/top-10-qualities-of-a-great-engineer/

McCarthy, P, 2014. What makes a great engineer?, *AusIMM Bulletin*, accessed 5 March 2022, https://www.ausimm.com/bulletin/bulletin-articles/what-makes-a-great-engineer/

NSW Government, 2014. Work Health and Safety (Mines) Regulation 2014.

NSW Resources Regulator, 2021. Technical Reference Guide: Ventilation Control Plan – Guide for the development of a ventilation control plan in underground mines, CM9 reference: DOC19/110502.

NSW Resources Regulator, 2022a. 1979 Appin Colliery Explosion, accessed 28 March 2022, https://www.resourcesregulator.nsw.gov.au/safety/safety-events-and-education-programs/learning-from-disasters/learning-from-disasters/1979

NSW Resources Regulator, 2022b. Practising certificates overview, accessed 11 July 2022, https://www.resourcesregulator.nsw.gov.au/safety/certification/practising-certificates/practising-certificates-overview

Panckhurst, G, Bell, S and Henry, D, 2012. Royal Commission on the Pike River Coal Mine Tragedy, accessed 5 March 2022, https://pikeriver.royalcommission.govt.nz/Final-Report

Pike River Royal Commission, 2012. Commission's Report – Volume 1, accessed 25 March 2022, https://pikeriver.royalcommission.govt.nz/Volume-One---Recommendations

Qld Resources Safety and Health, 2022. Mining Legislation (Continuous Professional Development) Amendment Regulation 2022 – Introduction and Overview, accessed 11 July 2022, https://www.boepcs.qld.gov.au/knowledgebase/article/KA-01034/en-us

Ramjack, 2019. The Key to Improving Mine Productivity is Increasing Mine Safety, The Key to Improving Mine Productivity is Increasing Mine Safety, accessed 5 March 2022, https://ramjacktech.com/blogs/key-improving-productivity-increasing-safety/

Rio Tinto, 2021. Principal Ventilation Engineer, Underground Mining at Rio Tinto – WORK180, accessed 21 Sep 2021, https://au.work180.co/job/297186/principal-ventilation-engineer-underground-mi

Rose, H J M, 1979. Management of ventilation departments-Introduction, *2nd Mine Vent Congress*, SME, pp 17–20.

Scanimetrics, 2021. Global skills shortage impacts mining productivity, accessed 5 March 2022, https://scanimetrics.com/index.php/scanimetrics-news-menu-item/11-improving-productivity/195-global-skills-shortage-impacts-mining-productivity-2#:~:text=A%20global%20skills%20shortage%20may%20compromise%20the%20productivity,over%20the%20next%20five%20years%2C%20according%20to%20AP

Self, A, 1999. The role of the ventilation officer, *Proc Qld Mining Industry Health and Safety Conf*, pp 202–208.

Shaw, C T, 1979. A management and reward system for specialist departments in mining companies, *2nd Mine Vent Congress*, SME, pp 37–41.

Taylor, J, Coulter, J and Espitia, V, 2020. Mining Engineer Shortage Report.

WA Resources Safety, 2011. Mines Safety Bulletin 95 Ventilation standards in underground mines, Dept of Mines and Petroleum.

Wikipedia, 2022. Chartered (professional), accessed 14 April 2022, https://en.wikipedia.org/wiki/Chartered_(professional)#:~:text=A%20Chartered%20professional%20is%20a%20person%20who%20has,a%20formal%20credential%20by%20a%20relevant%20professional%20organization

Windridge, F W, 1996. Report on an accident at Moura No 2 underground mine on Sunday, 7 August 1994, Warden's Inquiry (The State of Queensland: Brisbane).

Zakharia, N, 2021. Is a mining skills shortage looming?, *Australian Mining*, 4 May.

Ventilation monitoring using fibre optics

G B French[1]

1. Product Manager Australia/New Zealand, Yokogawa Australia, Brisbane Qld 4172.
   Email: gavin.french@yokogawa.com

ABSTRACT

Post a serious heating event underground, and at the direction of the State Mining Safety Authority, a mining company needed to demonstrate the use of cutting-edge technology to mitigate future events. Yokogawa were contracted to provide a turnkey solution to help the mine achieve this objective. The solution was to use a fibre optic-based Heat Detector that will allow entry into and around the mine considering the potential volatility of the environment, also the need to redeploy as mining activities moved underground and more boreholes are added. Industrial strength fibre optic cable and a robust connector system was chosen to achieve this goal and give the customer the ability to send cable down boreholes of up to +1 km and continuously monitor the temperature of underground roadways. Being able to re-deploy the cables into different boreholes and link borehole temperature measurements gave insight into all areas on the mine. The machines deployed on-site were the Yokogawa DTSX1 heat detector incorporating the Ramen Backscatter principle. Being of a robust construction and a low power consumption device, it was easily included into the portable monitoring trailers designed by the customer and built locally.

INITIAL MEETING – SCOPE OF REQUIREMENTS

Yokogawa met with the mining company to obtain an understanding of their requirements necessary to assist in the restart the underground mining operations. After a site inspection and consultation with industry subject matter experts Yokogawa and the customer were able to discuss and prioritise the requirements to make this project a success.

1. Bring leading edge technology into the mining operation to bring real time monitoring of the ventilation system.

2. The solution needs to bring real time monitoring of the underground roadway temperature.

3. Solution needed to be transportable.

4. Needs to be flexible to allow more, or less monitoring, as required by the mining Manager.

5. Solution needed to be lowered and raised into new and existing boreholes across the entire site without breaking:

   o Cabling Solution.

6. The network created needed to be robust and resistant to damage and harsh environment:

   o Fibre Optic connector system.

7. Local interrogation of the system and adjustments made by the mine as required:

   o Existing Control System – PLC Based.

8. The system must be self-monitoring including health checks.

9. The solution needs to be able to be mounted in a portable, above ground monitoring system.

BRING LEADING EDGE TECHNOLOGY INTO THE MINING OPERATION TO REALISE REAL TIME MONITORING OF THE VENTILATION SYSTEM

The mining company that experienced the underground event was at the request of the mining authority required to employ leading edge technology to demonstrate that they are looking to proactively mitigate any future events. The mine decided to review the use of Fibre Optics as the chosen temperature measurement medium due to the ease of use as a simple light source. 'We need to know in real time the temperature of our mine that is also expandable, portable and reliable'.

THE SOLUTION NEEDS TO BRING REAL TIME MONITORING OF THE UNDERGROUND ROADWAY TEMPERATURE

All the actions of a fibre optic-based sensing system by nature occurs at the speed of light. The chosen technology has been designed to operate up to five times faster than conventional monitoring systems incorporating similar technologies. Originally designed and approved internationally for use in Tunnels and in public areas the DTS used can read, analyse, and present up to 64 000 temperature measurements across 64 km of fibre cable in a matter of seconds. The information presented to Yokogawa by the customer is the requirement for early warning should the mine temperature rise to a level that would require immediate action/investigation. Irrespective of the location of the borehole the mine requires that not only the two gas monitoring system be lowered into the underground roadway but also the temperature monitoring cable.

SOLUTION NEEDS TO BE TRANSPORTABLE

To meet this request of the customer a fibre optic system needed to meet three aspects of a transportable solution.

1. The cable needed to be able to withstand multiple handling and movements across the entire site including:

   o Coiling and uncoiling of the cable either on a drum or coiled by hand.

   o Pulling of the cable over rough terrain.

   o Extended periods in the open and exposed to high temperatures and road traffic.

2. The fibre optic terminations also needed to withstand harsh environments whilst maintaining a clean and clear contact. When the cables are not connected, they needed to have a protection cap that maintains the required IP rating and ensures the contact surface of these fibre connectors remain pristine.

3. Where the cable enters the underground roadways the end of the cable has 2 fibres joined together allowing the fibre optic signal to return up the borehole and onto the next location if one is required at the time. The spliced joint is 'Potted' to ensure it remains resistant to hits and knocks whilst installed.

In addition to the request for a transportable solution there was a need to make the entire project a 'plug n play' format. To facilitate this the fibre termination solution needed to include a main entry fibre box mounted externally on the customers Gas Monitoring Trailers. No access to the sensing unit is required simply connect the cables to the trailer and start sensing.

NEEDS TO BE FLEXIBLE TO ALLOW MORE, OR LESS MONITORING, AS REQUIRED BY THE MINING MANAGER

As mentioned before the need for a flexible solution on a plug and play format did have its challenges. To be flexible with the monitoring came down to two aspects of this monitoring system.

1. The ability of the proprietary software supplied by Yokogawa to know where the cable enters the underground roadway.

2. To disregard areas where sensing is not required. This included the following areas.

   o Above ground cable runs both between boreholes and the Gas Monitoring Trailer.

   o The cable run from the surface of the borehole down to just before the cable entered the roadway.

In essence the critical information over these long distances comes down to around 5–6 m of sensing fibre that is suspended in the roadway. The system is intuitive and can be running live as the cable is entered into the borehole. Once a change in temperature is seen on the temperature graph is the key to knowing when the cable has entered the roadway and lowering ceases.

SOLUTION NEEDS TO ALLOW FOR THE FIBRE OPTIC CABLE TO BE RAISED AND LOWERED FROM BOREHOLE TO BOREHOLE WITHOUT ANY LOSS OF SIGNAL OR CABLE BREAKAGE

Mentioned earlier the cables will be raised and lowered where it will either be simply coiled up or rewound back onto a cable drum. To achieve the longevity required by the customer Yokogawa needed to engineer a fibre optic cable that can be suspended long distances and be notch and abrasion resistant. To do this Yokogawa worked with a European manufacturer to design a 'Military Spec' cable. This cable can hold its weight and be suspended vertically well over 1000 m also provide a strong resistance to damage via a synthetic sheath and strong fibre bundle construction. Use of a Aramide Yarn centre strength member gives the cable a 2000N load bearing capability which allows such a long vertical run.

- Aramide Yarn centre member.

- Thermoplastic Polyurethane sheath.

- Short-term crush resistance 2100N.

- Temperature range whilst installed -60 to +85°C.

THE FIBRE NETWORK CREATED NEEDED TO BE ROBUST AND RESISTANT TO THE HARSH ENVIRONMENT IT WAS INSTALLED IN

Critical to any fibre optic installation is the combined fibre loss across the entire installation. Generally, where there is a connector or even a fibre splice there will be a certain level of fibre loss and these losses must be kept to a minimum to ensure peak performance. Selection of the right connector system in this case needed to not only include the impact resistance and its protection rating but also a low loss connector system. Typically, each connection point had a -0.05 Db loss with a maximum allowable of -0.3 Db loss. Lowest realised value was -0.03 Db loss and values across the entire installation ever exceeded 0.10 Db. Located within the Propriety software of Yokogawa is the ability to measure the CFL, Combined Fibre Loss, ensuring the status of each run is always know and specific points can be identified should any increase be realised.

FIG 1 – Low loss duplex connector system used in project.

LOCAL INTERROGATION OF THE FIBRE OPTIC SYSTEM AND ADJUSTMENTS TO BE MANGED AND IMPLIMENTED BY THE MINE

The Yokogawa sensing unit, (DTSX1), needed to communicate seamlessly with the existing controls system used on the mine site. MODBUS TCP was used as the output from the DTSX1 into the site PLC and this ensured the time between any temperature event and its reporting was kept to an absolute minimum. As mentioned earlier the machine will process up to 64 000 temperature measurements but given the smaller window of critical measurements Yokogawa were able to provide the mine with accurate temperature data in under 5 seconds. The Proprietary software provided allows for the mine to set their own Administrator and four only client level assess. Along with an addressable IP ensures the machine's operation and integrity of the data remain with the customer. Location for monitoring of the system was requested to be locally at the mine Administration block, this was easily achieved and implemented with their site IT Manager.

THE SYSTEM MUST BE SELF AWARE OF ITS PERFORMANCE AND BE CAPABLE OF HEALTH CHECKS

Locate within the DTSX1 is the ability of the system to monitor and report anomalies quickly and these are known as pre-alarm functions. The functionality of the machine is that when it is first connected and is given a 'Start' command it will send a preparation measurement that is used so it can identify the total length of the connected fibre, by Channel. Once known the machine will in sequence measure the temperature across each selected channel, with a maximum of four channels available. Review of the CFL is easily selected at any time also the scan time and pulse rates can be adjusted between 1 second and 1 week. The system will automatically alarm should there be any change from the original 'Preparation Measurement' and subsequent scans will show up to the anomaly giving its location as a distance (to the metre).

1500 mtrs of data across 3 boreholes

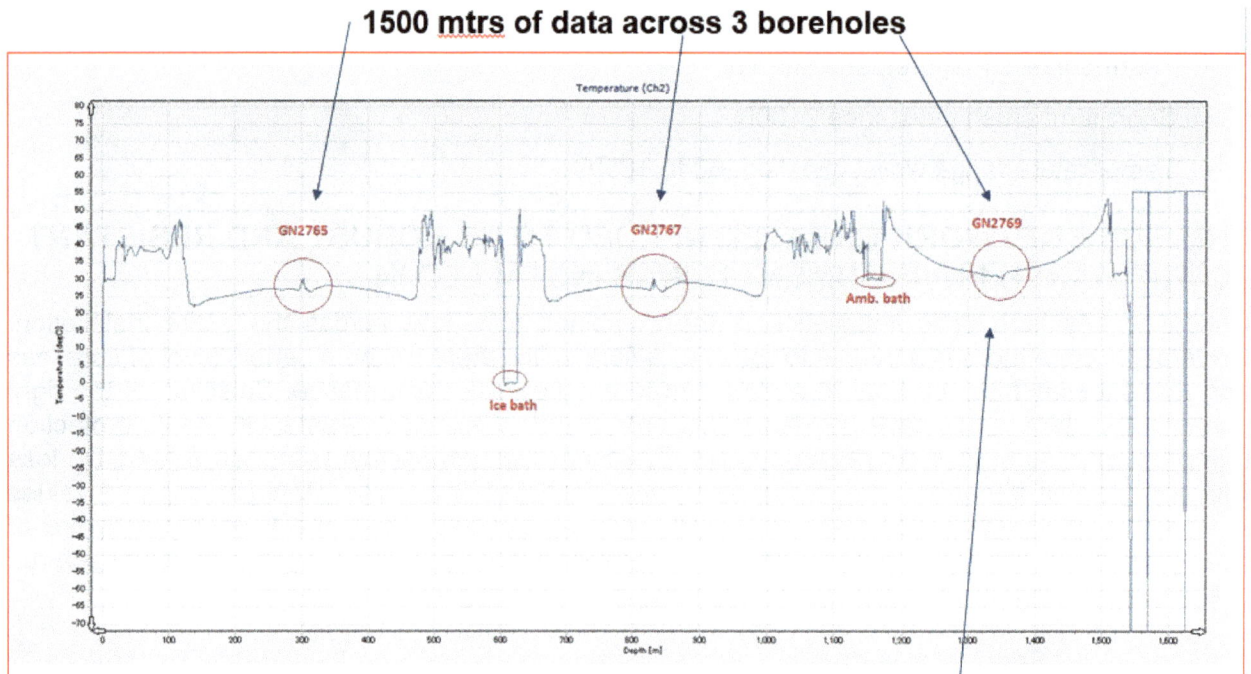

FIG 2 – Actual measurements from the DTSX1 across three boreholes.

CONCLUSIONS

Insight is the value to the customer, insight into real time and accurate reporting of the temperature underground also the ability to see temperature trending across the entire site. With these insights come early detection and reporting of events, together with portability and flexibility has given the mine an unparalleled knowledge over their mining operations underground.

ACKNOWLEDGEMENTS

Yokogawa would like to acknowledge the commissioning support from the Solutions Delivery team of Yokogawa Australia.

Mr John Vogels: Functional Expert Hybrid Systems.

Mr Ajay Trivedi: Function Expert Network and IT Systems.

Airborne contaminant monitoring strategy and technology

J H J Holtzhausen[1] and F Velge[2]

1. Principal Ventilation Consultant, OreTeck Mining Solutions, Melbourne Vic 3000.
   Email: jholtzhausen@oreteck.com
2. Managing Director, Pinssar, Brisbane Qld 4102. Email: fvelge@pinssar.com.au

ABSTRACT

Never has there been more focus on the health, safety, and well-being of our employees at the workplace. Ever evolving expectations of individuals, companies and clients are shaping regulatory reviews and advancing standards across the world in the drive to zero harm. The technology is advancing at a rapid pace, which is now allowing continuous real-time monitoring of contaminants to take place in the harshest of environments.

Many ventilation and occupational hygiene practitioners are now placing their trust in real time monitoring technologies and implementing strategies to control the hazard proactively, rather than relying on lag indicators. Accordingly, the application of monitoring technology and the correct deployment is of utmost importance and will ultimately determine the relevance of the data set being collected. Successful real time Diesel Particulate Matter (DPM) monitoring must include strategic approaches that will ensure adequate validation of site controls. This paper will explore the positioning of the DPM readers to ensure the most relevant/truly reflective sample is collected as a lead indicator. Consequences associated with poor positioning of monitors will be explored and the potential impact on the data set discussed. It is expected that the use of real time monitoring data will continue to evolve in the mining industry and could play a major role in the Zero Harm mine of the future.

INTRODUCTION

DPM (Diesel Particulate Matter) has been the topic of discussion for many years in all industries where diesel units may be applied in confined or semi-confined spaces. DPM is of particular interest in the mining, tunnelling and construction sector where large diesel combustion engines are applied to carry out work. Diesel has been the fuel source of choice as it poses less of a fire and explosive risk to the underground and confined space application. Other advantages of diesel are:

- Diesel is moderately priced.

- Diesel has a higher energy density, ie more energy can be extracted from diesel than comparable fuel sources such as gasoline.

- Diesel has a boiling point higher than water.

- Diesel engines ignite on compression and not on spark.

- Diesel engines compress at a ratio of 14:1, and up to 25:1.

- They are more rugged and reliable.

- They generate less heat that is rejected into the atmosphere etc.

While diesel, over the years, has intrenched itself into the market as an economic way of achieving work, the health impact of the units on the workforce has been questioned. In 2012 the WHO (World Health Organisation) analysed research and realised the impact of DPM on the body and declared DPM to be a Group 1 Carcinogen. They concluded that the use of diesel-powered equipment in a confined space can result in an elevated risk of workers developing cancer by between 20 and 50 per cent.

DPM monitoring and controls have been largely based on lag indicators which would see the workforce exposed to the hazard whilst awaiting monitoring results. One company have recognised a gap in the control and monitoring of DPM in the market and developed a real time monitoring solution which will assist industry to proactively monitor and understand the DPM risk at their operations. The question that remains however is 'why, where and what is the value?'. This paper is aimed at being an informative technical review to help industry understand the 'why, where and what is the value?' better.

WHY SHOULD REAL TIME MONITORING BE CONSIDERED?

The first of the questions to address is 'Why?'. Why should we as an employer or industry invest in the monitoring, control, and management of DPM exposure within the workplace. The easy answer to the question is, that it is a legal requirement widely enforced in all industries utilising diesel to conduct work, and that most governing bodies within the world legislates in some form or another that it should be controlled. The author however believes understanding the impact of DPM on the exposed person must be fully grasped to set a clear guideline of why DPM must be controlled and actively managed. The primary exposure mechanism to DPM is via inhalation, upon inhalation particles are deposited in the respiratory system. Some of the effects of exposure to DPM over time have been indicated to be:

- **Inflammation** in the lungs, involving a complex set of molecular and cellular responses resulting from exposure to exogenous stimuli such as pathogens or noxious substances, including particles or endogenous stimuli such as cytokines or danger signals.

- **Immune system response** is triggered in the body which can lead to upregulation of a different set of inflammatory genes (eg interferon-β).

- **DPM is a transport vector** for pathogens, which alter the ensuing inflammatory response as pathogens and can access sites in the lung where they would not normally be found.

- **Oxidative stress** develops when there is an imbalance between the production of ROS (Reactive Oxygen Species) and the availability of antioxidants defences.

- **Short-term exposure** may lead to headache, dizziness, and irritation of the eyes, nose and throat, severe enough to distract or disable miners and other workers.

- **Prolonged exposure may** increase the risk of cardiovascular, cardiopulmonary and respiratory diseases and lung cancer.

The health effects of exposure to DPM may not in all cases be a short-term impact on the exposed person, but in many cases lead to health effects which will:

- Impact on the quality of life as a person ages.

- Reduce the life expectancy of individuals exposed to DPM.

- Have a psychologically traumatic effect on the exposed person as well as their immediate and extended families.

- Restrict the freedom of movement due to medical needs and specialist equipment.

- May lead to regular/irregular short-/long-term hospitalisation and increased medical costs.

- Impact on the economic independence of the exposed person and their family etc.

Knowing the severity and the impact of over exposure to DPM would lead all to agree that DPM exposure must be understood, monitored, and managed, not because it is a legal requirement to do so, but because:

- We have an obligation to protect the health and safety of all under our care.

- No work is so important that it requires the sacrifice of a life.

- All employees have the right to work and earn a living, but even more importantly, to return home safely to their family and friends.

WHERE SHOULD REAL TIME MONITORING BE APPLIED?

Where to monitor has been a question that has not always been well addressed. To fully understand the 'Where' the operation must understand:

- The work methodology to be utilised at the operation.

- o In the case of mining how will the orebody be extracted.

- o In tunnelling/civils how will the excavations be completed.

- What equipment will be utilised and at what rate will the work take place.

- What is the footprint that will be created by the operation?

- Where will the high contaminant activities be taking place?

- What is the ventilation strategy and methodology to be applied?

Tunnelling

As all tunnelling projects are not the same, and must be assessed individually, it is important to fully understand the operation in terms of design, ventilation system, emitting technologies and/or the personnel interactions. This paper is therefore only able to make generic recommendations and would recommend the engagement of a professional ventilation consultant to validate any proposal and the effectiveness of the units (real time DPM monitoring units) when incorporating them into a control methodology.

It is understood that the factors to be considered are the number and overall kW of emitting equipment (fixed and mobile), number of ventilation breakaways, potential ventilation leakage points ie cross passages etc, length of excavation and applicable work zones. The monitoring units would serve best downstream from areas where diesel powered plant is operated, and with multiple units installed, the impact of DPM (Diesel Particulate Matter) across the footprint of the operation may be monitored.

Generally, for tunnelling it is recommended that the DPM readers are located at a minimum in the following locations:

- Figure 1 illustrates a positive forced ventilation system commonly associated with TBM (Tunnel Boring Machine) works. 1 × Reader at portal (Point A) positioned for TBM launch (would read the full diluted accumulation of DPM within the circuit as the quality of air at this point would be expected to be the worst representation).

- 1 × Reader at Point C in the tunnel downstream from the materials handling zone where the cut material from the TBM face is loaded onto trucks and transported to surface. The second unit will be an indication of workforce exposure close to the work zone and allow steps to be taken to mitigate exceedances prior to lag indicators received. Lag indicators such as gravimetric sampling may take three to six months for a result to be returned, in which time the workforce will continue to be over exposed.

- 1 × Reader at Point B midway between the portal and the TBM working face.

- Readers as required for cross passage excavations in the downstream position.

  - o These readers will allow the point of exceedance to be identified should an alarm level be reached.

  - o Being able to identify where the source of the exceedance is, is invaluable during the investigation and will allow the operation to act swiftly to remove employees from harm, stop contamination at the source and identify what has been the cause of the exceedance.

  - o It does happen that a vehicle with a dirty engine, poor maintenance, substandard fuels, poor operator discipline, poor ventilation controls, failure of ventilation controls etc could be to blame and being able to narrow down the potential source/zone will make the operation agile in response and give confidence to the legislator that hazards are actively managed and understood.

- As the project develops subsequent Readers may be required to give confidence to the operation that sufficient dilution of DPM is being achieved with the ventilation controls in place. The number of units are largely driven by the complexity of the project and the span. The more complex and the larger the span of the footprint, the more units should be deployed.

- Consideration for expanding the network of Readers should be guided by identifying areas where:
  - Significant diesel emission sources are present.
  - Workers spend considerable time in diesel particulate contaminated airways.
  - Teams that could help identify and establish the scope for such a project is the tunnel managers, design engineers, ventilation managers/engineers and/or OH&S consultants and personnel.

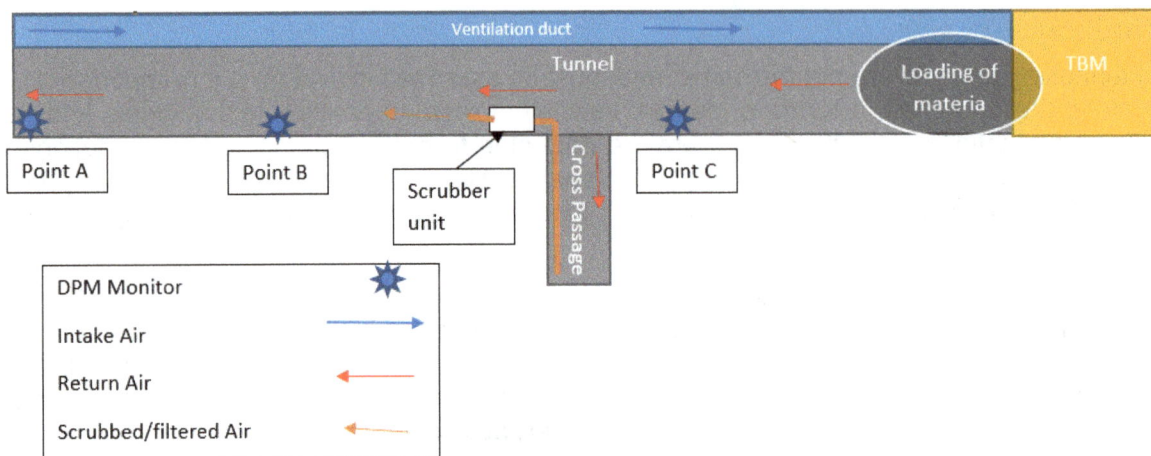

FIG 1 – Example of tunnel ventilation layout.

Mining

As is the case for tunnelling, each mining project must be assessed on its own merits fully appreciating the mine plan, work schedule, extraction rate, ventilation design and extraction methodology. In mining there are various methods applied for the extraction of the orebody which may alter the placement of the DPM monitors. This paper would only be able to make generic recommendations and would recommend the engagement of a professional ventilation consultant to validate the proposal and the effectiveness of the units when incorporating them into a control methodology.

Mining methods such as SLC (sub-level caving), BC (block caving), Narrow reef stoping, Sub-level open stoping, etc, may all have an impact on the ventilation strategy and the placement of units. As in most hard rock mechanised mines, the tramming of materials in and out of the mine is concentrated in the decline. It is important to understand the quality of air and how it is impacted on in the underground section by diesel units deployed in the operation. DPM will accumulate through the mine should adequate ventilation not be available for the dilution and removal of contaminants from the mine.

In a general hard rock mine ventilation system as illustrated in Figure 2, the following placement of DPM units would be recommended:

- Place a DPM Reader in the ramp from the portal above the first level
  - this reader will indicate the impact that diesel unit movement has on the air between the portal and the first level.
- Place a second Reader mid-way down between the bottom of the mine and the first reader
  - this reader will monitor the impact that production and tramming may have on the air quality up to that point in the mine.
- Place a third reader at the lowest point of return in the decline which will indicate the accumulation of the impact that diesel units have in the circuit up to that point.
- As the project develops subsequent Readers may be required to give confidence to the operation that sufficient dilution of DPM is being achieved with the ventilation controls in place

- o the number of units are largely driven by the complexity of the project and the span. The more complex and the larger the span of the footprint, the more units should be deployed.

- Consideration for expanding the network of Readers should be guided by identifying areas where:

  - o Significant diesel emission sources are present.

  - o Workers spend considerable time in diesel particulate contaminated airways.

  - o Team members that could help identify and establish the scope for such a project are mine managers, mining engineers, ventilation managers/engineers and/or OH&S consultants and personnel.

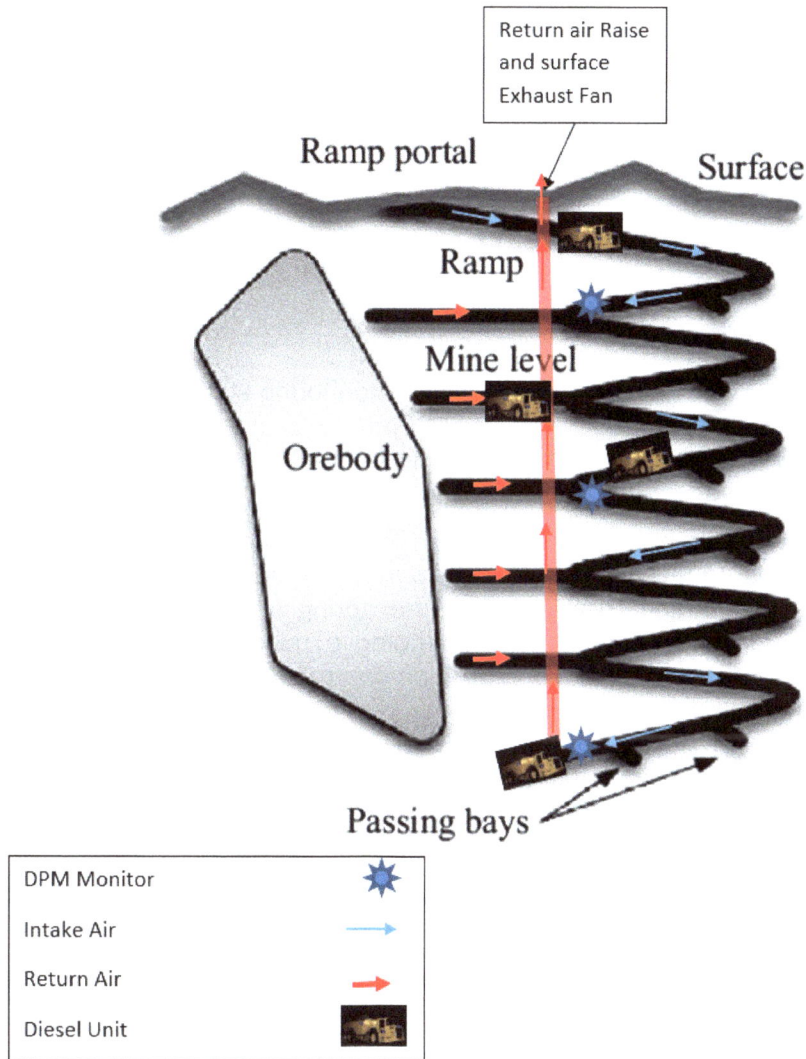

FIG 2 – Example of an underground hard rock mine.

WHAT DOES REAL TIME MONITORING LOOK LIKE?

Figure 3 is an example of what an environmental monitoring station may look like. All environmental monitoring stations will not resemble each other in function, but be site specific, considering the contaminants that must be monitored for or controlled. A vast array of sensors and technology is available on the market to capture the needs of most applications and clients. Instrumentation may be connected to a network and then report/record readings to a database or as specified by the client. Ventilation control outcomes may be programmed into the system which may allow for:

- Ventilation systems to be automatically adjusted to dilute/remove contaminants.

- Localised alarms set up and direct reading displays for underground references.

- Network alarms combined with action levels to ensure workers are removed from harm's way prior to overexposure etc.

Real Time Environmental Monitoring Station Measuring:

Dust, NO2, O2, CO, BP, Wet Bulb Temp and Airflow

FIG 3 – Environmental monitoring stations.

Statistical data analysis and trending may be done from data collected to show compliance and to use in incident investigations.

Figure 4 shows an example of the Pinssar Real Time DPM monitor that is available in the market. Controlling DPM exposure levels forms a major part of the ventilation control system methodology and why it is being implemented. Real time DPM monitoring will indicate that the operation is actively monitoring and controlling the risk of workers being exposed to exceedances. It will allow the operation to action controls or adjust the ventilation control methodology to maintain compliance.

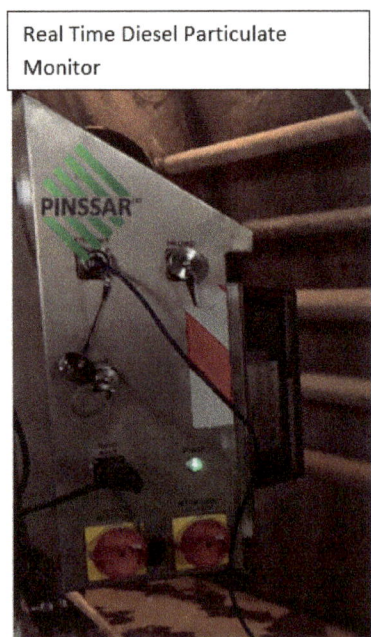

Real Time Diesel Particulate Monitor

FIG 4 – Pinssar DPM monitor.

Ensuring that maximum value is derived from a real time monitoring system requires a good quality network that is reliable, and a control room interface (Figure 5) that is easy to understand and manage. The control room will be the heart of the system where all the controls will be monitored, set-up, data collected and managed. The system should allow for an alarm function which indicates when exceedances are encountered, or when environmental conditions move outside the acceptable parameters. These parameters will be decided by the client and must be based on legal compliance requirements and occupational hygiene action level control limits.

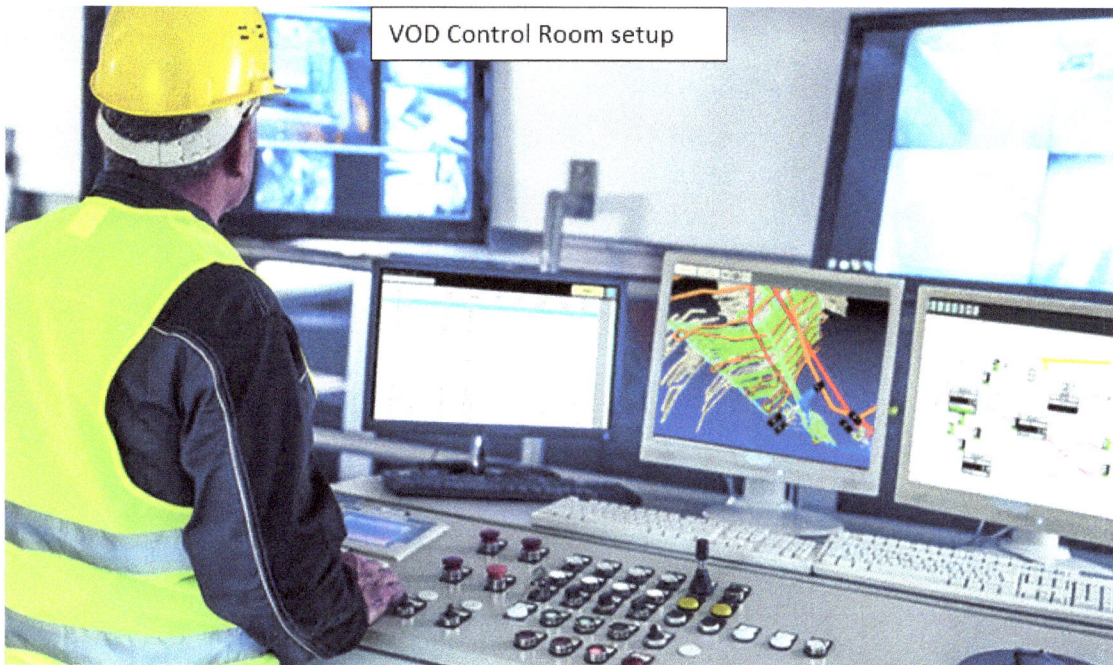

FIG 5 – VOD (Ventilation On Demand) control.

The IT back bone will form part of the interface which will be required to ensure the VOD system can function and supply the level of control required by the operation. Networking strategies and control room layouts will be important in establishing a system that will be practical, reliable and easy to use and maintain. Throughout the tendering process the ease of maintenance, accuracy and reliability must play a fundamental role in the real time monitoring system analysis and procurement process.

Positioning of the monitoring stations

It is the desire to have each unit located in the position where optimal results and consistent accuracy are achieved.

Figure 6 illustrates characteristics of airflow that must be incorporated into the monitoring plan and layout for units. Positioning units in incorrect locations will result in readings that are not quantifiable and repeatable. Installing units in positions where accurate readings may not be obtained could lead to:

- False alarms being reported when no DPM exceedance is occurring.
- No alarms being reported when DPM action levels are breached.

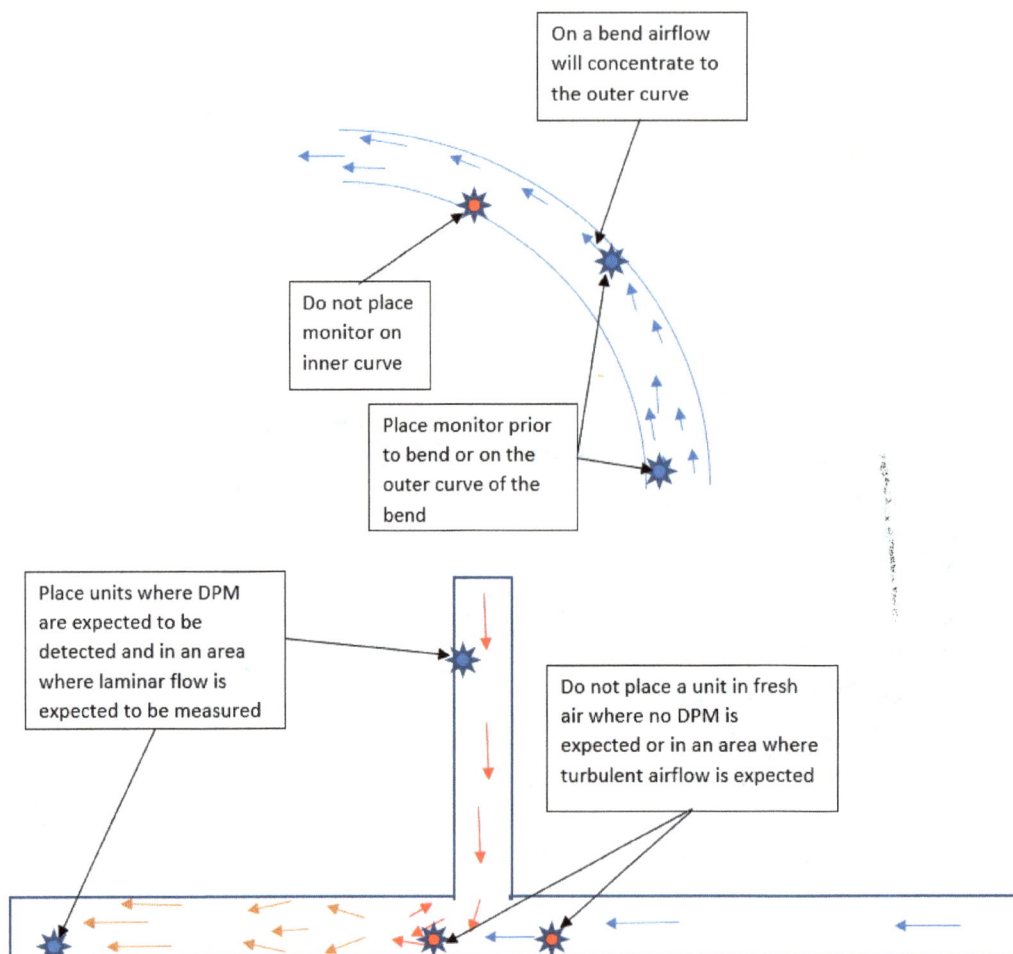

FIG 6 – Placement of monitoring stations.

Mounting heights for gas sensors are based on their density, relative to air. There are three groups:

1. Lighter than air and will be more concentrated near the ceiling/roof – hydrogen, methane (natural gas), ammonia, helium. Install at 30 cm to 90 cm from ceiling/roof/backs.

2. Similar density to air and will be diluted in air equally at all levels – carbon monoxide, nitrogen dioxide, hydrogen sulfide, oxygen, carbon dioxide.

   o For vehicle emissions, carbon monoxide monitors or carbon monoxide monitors combined with nitrogen dioxide monitors are installed at 1–2 m from the floor where the ceiling is 2–3 m high.

   o If the ceiling height is higher than three metres, eg for heavy equipment, the carbon monoxide monitors are installed at 1–1.5 m from the floor and the nitrogen dioxide monitors should be installed at 50 per cent of the ceiling height and above the vehicle height to be in the open circulation of the air.

   o If the exhaust pipes of diesel vehicles are below the vehicles, then the nitrogen dioxide monitors should be installed at 1–1.5 m from the floor. Other gases in this group are typically installed at 1–1.5 m from floor but can be installed up to 50 per cent of the ceiling height.

   o In all cases the monitors must be installed above obstructions blocking circulation of air in front of the monitors, for example, sub stations, walls, worktables, and storage racks etc.

3. Heavier than air gases will concentrate near the floor – HFCs, HCFCs, propane, butane, chlorine, most organic vapours. Install sensors at 30–90 cm from floor.

For all types of sensors you need to avoid drafts, obstacles, aerosols and silicones. Place sensors in the centre of its coverage area as much as possible. Taking all these points into consideration it would be the recommendation to place the DPM monitoring device in the 1.5–1.8 m zone above the floor level for optimal accuracy and consistence in results. It is important to evaluate and determine the most efficiently suitable location for each installation and the use of a suitably qualified expert for guidance and design of the locations where monitoring devices are placed.

CONCLUSION

The final question to touch on in this position paper is **'What is the value'**. For many organisations the value of a product must be measurable in production output, cost benefit or cost saving. To fully appreciate the value adding potential of a real time monitoring solution, it is more important to examine your companies value set and health and safety commitments. These may be similar to Pinssar Australia's:

- 'Protect every breath. Building a better world starts by building healthier workplaces' (home page on the Pinssar web page).

- Pinssar's Goal – 'The Pinssar goal is simply to help make the world a safer place and reduce the health risks when workers are exposed to deadly diesel engine exhaust emissions'.

When seeing/classifying the product as a tool to ensure the health and safety of each employee, the question should rather be – 'How much value do I place on my own health and the health of my co-workers?' When the question is personal, the answer and the figure is harder to come by. It can be said with confidence that:

- Having multiple fixed real time diesel engine exhaust emission monitors installed, validates controls, and reduces the risk associated with not identifying areas of high emittance, and or failed system controls timeously.

- The more units that are made available, the more agile the system is when monitoring, managing, and investigating DPM. With fewer units the complexity of identifying and pinpointing the source and managing the risk increases.

- Fixed instruments become constant reference points – and establish diesel emission trends which allow for data analysis and compliance verification to be confirmed.

- The data collected may be utilised to support the baseline trend and confirm regulatory compliance.

The most important benefit/value to real time monitoring, is believed to be the ability for an employer to confidently say that they are actively managing a hazard to acceptable levels and putting the welfare of the workforce and their safety before all else. Real time monitoring is useful not only for emissions control but also for:

- Re-entry after firing.

- Ventilation system control and validation.

- Emergency response management.

- Incident investigation etc.

With so many monitoring solutions being available for multiple airborne contaminants the question that must be asked is not 'Why?', but rather 'Why has this not been considered and implemented yet?'

ACKNOWLEDGEMENTS

We would like to acknowledge the team at Pinssar Australia for their collaboration in completing this paper. Their passion for continuous DPM monitoring and protecting the health and safety of workers is inspiring.

Dust and diesel characterisation in underground coalmines

N LaBranche[1] and D Cliff[2]

1. MAusIMM(CP), Research Fellow, UQ Minerals Industry Safety and Health Centre, Brisbane Qld 4072. Email: n.labranche@uq.edu.au
2. MAusIMM, Professor of OHS in Mining, UQ Minerals Industry Safety and Health Centre, Brisbane Qld 4072. Email: d.cliff@uq.edu.au

INTRODUCTION

The resurgence of Silicosis and Coal Worker's Pneumoconiosis (CWP) in coalmines has placed the spotlight back on the management of worker exposure to particulate matter across all mining sectors (Coal Workers Pneumoconiosis Select Committee, 2017).

There already exists a substantial body of knowledge on particulate matter and its impacts upon human health (National Academies of Sciences, Engineering, and Medicine, 2018). The contribution of silica to respiratory disease is well known and there is a growing concern for exposure to diesel particulate matter. However, it is also clear that there are major gaps in our understanding. In part, those gaps relate to the contribution to adverse health effects of mineralogical constituents in the dust (LaBranche *et al*, 2021a). It is also becoming evident that the size and shape of the particulate matter can affect its potential impact on human health.

The University of Queensland has developed a methodology for characterising respirable and inhalable dust samples using scanning electron microscopy. Multiple samples from numerous mining operations have already been analysed using the UQ methodology. The process has been successful in showing the variety of mineralogical components and particle size distributions present in various areas of the mines.

MATERIALS AND METHODS

While characterisation work was attempted in the 1980s, technology has changed significantly since then allowing for a much more detailed look at what is in the dust and therefore the health hazard (Lee, 1986; Mutmansky and Lee, 1984). This methodology has two key benefits including respirable dust is sampled in actual mining conditions and technology now allows for data to be obtained on individual particles, not just for the whole sample.

Sampling in actual mining conditions

Previous work often collected bulk samples from the mines, which was then crushed to analyse its components. However, a number of these studies complain about how hard it is to crush the coal to the desirable size fraction. This leads to an interesting paradox. If it is difficult to obtain certain size fractions, is the coal breaking up into those size fractions under real world conditions.

There can be several sources of dust in the mine besides the cutting of the coal seam, these may include vehicle traffic though the mine, the mining of roof or floor rock or rider seams, stone dusting activities and other activities in the mine such as shovelling. Performing sampling in actual mining conditions picks up the contributions from all these sources, which in some instances can be significant.

Samples were taken in a variety of underground locations including at the maingate and midface of the longwall, around the continuous miners, on the shuttle car, in belt roadways, and during secondary recovery activities. This characterisation technique samples the underground locations using respirable cyclone elutriators with a specialised polycarbonate filter which lends itself to microscopy analysis. Samples are normally conducted at static locations to obtain information about what various sources are generating, but personal sampling can also be performed. Figure 1 shows an example of the respirable cyclone array used and two cyclone arrays mounted to a continuous miner.

FIG 1 – (left) Respirable cyclone array on frame, (right) cyclone array mounted to continuous miner.

Data on individual particles

This characterisation technique allows for multiple parameters to be obtained for individual particles on the filter. Historic characterisation techniques only allowed for an overall particle size distribution (PSD) to be calculated and for the relative abundance of certain minerals. With data on individual particles this allows for separated PSDs to be obtained for various mineral components, ie the PSD of only the quartz in the sample.

The number of particles analysed can vary significantly depending on how much dust is on the filter and how much of the filter is analysed. Figure 2a show varied levels of particle loading on the filters. Sampling is often only performed for an hour or two in order not to overload the filter. A filter becomes overloaded when there is not sufficient space between the particles to differentiate them under the microscope. Overloading is normally easy to spot by checking the overall PSD curve. To date up to 758 000 particles have been analysed on a single filter. But to save time and money, smaller punches from the filter can be taken for a representative sample as shown in Figure 2b.

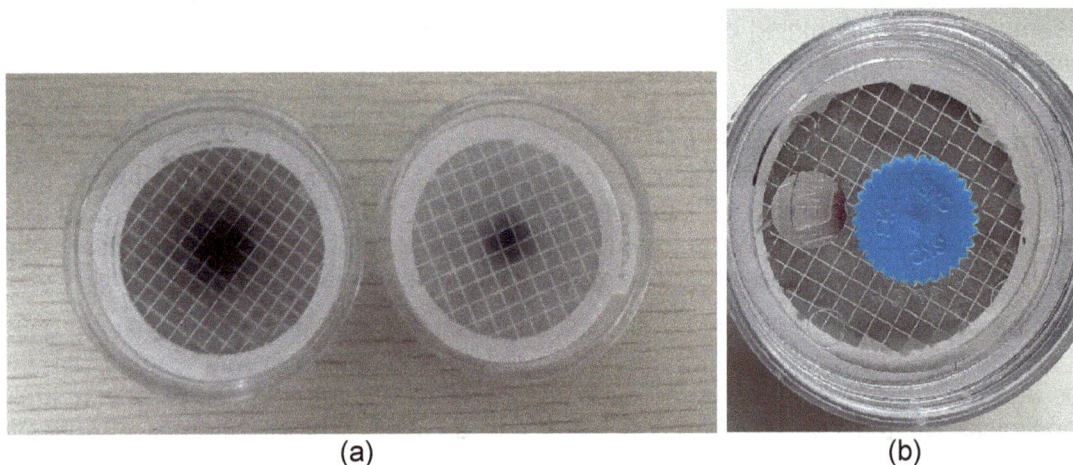

(a)   (b)

FIG 2 – (a) Varied levels of particle loading on filter, (b) hole punched out of filter for analysis.

RESULTS

A total of eight underground coalmines have been analysed so far. The filters from Mines 1–5 were analysed by Emily Sarver *et al* at Virginia Tech using scanning electron microscopy with energy dispersive X-ray SEM-EDX (Johann-Essex *et al*, 2017; Johann-Essex, Keles and Sarver, 2017; LaBranche *et al*, 2021b; Sarver, Keles, and Rezaee, 2019a, 2019b). The filters for Mines 6–8 were analysed on the Mineral Liberation analyser at the Julius Kruttschnitt Mineral Research Centre, which is part of the Sustainable Minerals Institute at The University of Queensland (LaBranche *et al*, 2022).

This data is also showing respirable particles are more complex than initially thought. The particles look to be agglomerating at this small size producing multiple mineralogies in one particles, not just particles of single mineralogies like was initially thought.

Figure 3 shows an example of the data obtained for particles using the MLA. The legend below shows the false colour applied to the various particles. This figure shows the variety of shapes and mineralogies present in the samples, sorted by descending particle size.

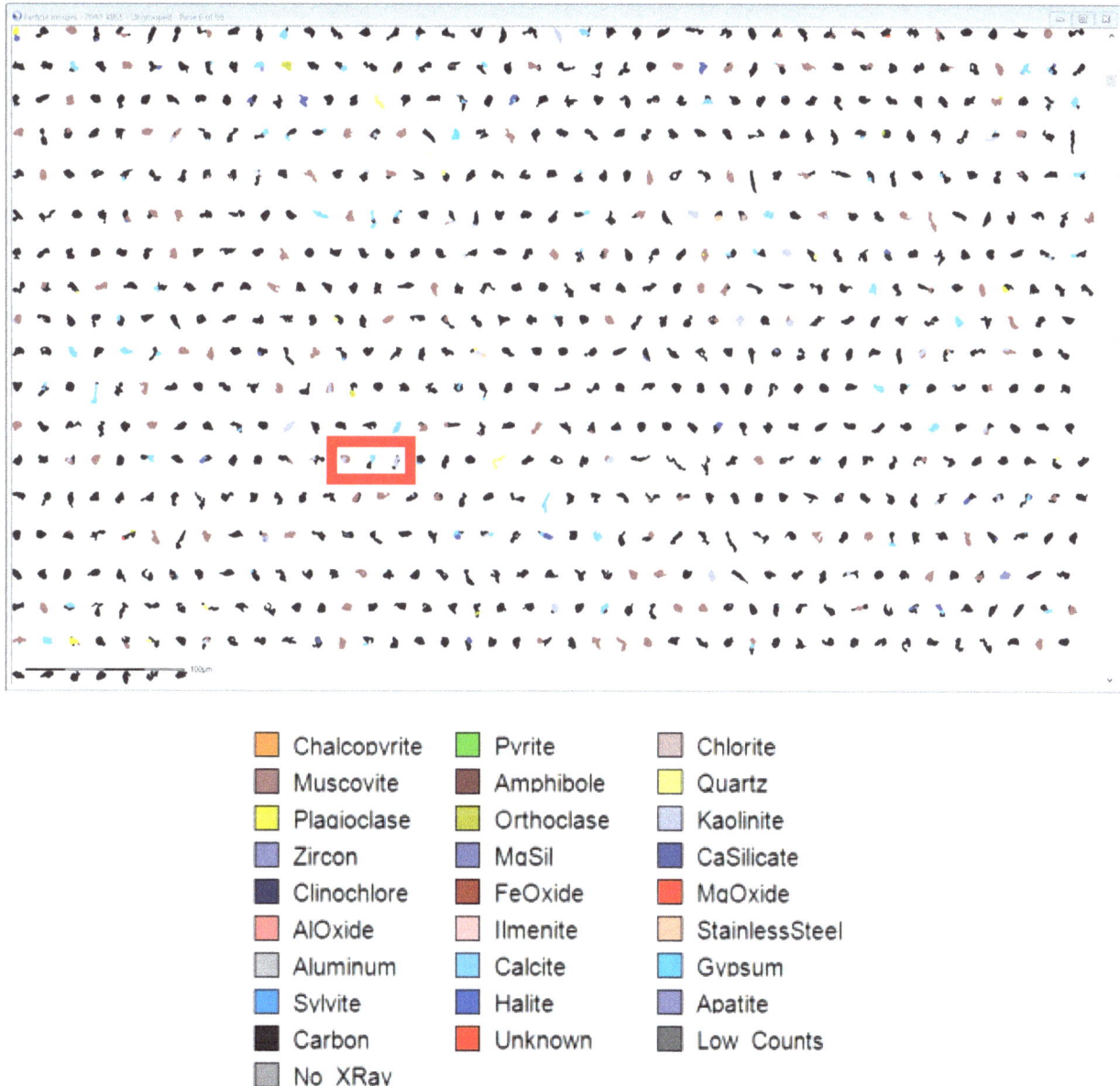

Chalcopyrite	Pyrite	Chlorite
Muscovite	Amphibole	Quartz
Plagioclase	Orthoclase	Kaolinite
Zircon	MgSil	CaSilicate
Clinochlore	FeOxide	MgOxide
AlOxide	Ilmenite	StainlessSteel
Aluminum	Calcite	Gypsum
Sylvite	Halite	Apatite
Carbon	Unknown	Low Counts
No XRay		

FIG 3 – False colour particle images generated by the Minerals Liberation Analyser.

Figure 4 shows the three individual particles inside the red box in Figure 3. As can be seen here, the MLA is able to indicate which portions of the particle are which mineralogy and map them. The number next to the mineralogy is the area of the particle that the mineralogy represents. Densities are applied to the various mineralogies to be able to calculate a weight for the individual mineralogies as well.

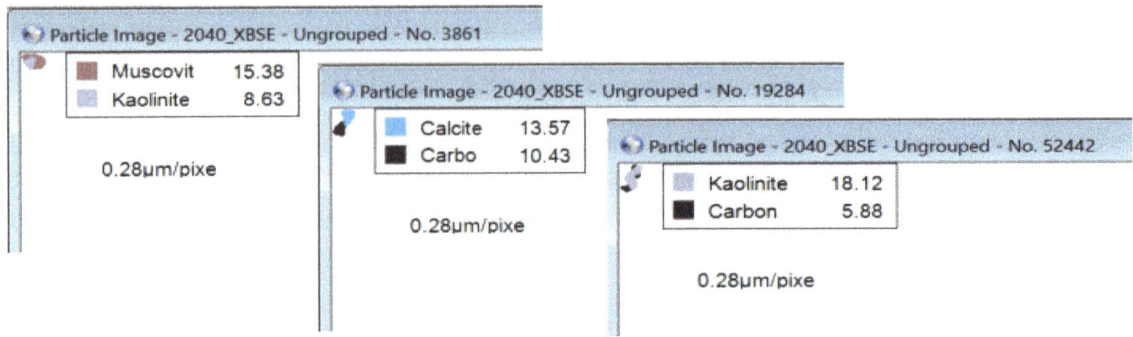

FIG 4 – Individual particle images from the MLA.

Diesel particulate matter

The energy dispersive spectra (EDS) methodology used is unable to differentiate the carbon fraction between the coal and diesel particulate matter (DPM) present (LaBranche *et al,* 2021b). The MLA imaging will show the shape of the particles which can provide some indications of which particles are agglomerated clusters of DPM but will not give definitive information on what is coal and what is DPM. Figure 5 is an example of these outputs. Here the particles have been filtered for those that are at least 50 per cent carbon by weight. Many of these particles were found to also contain chlorite.

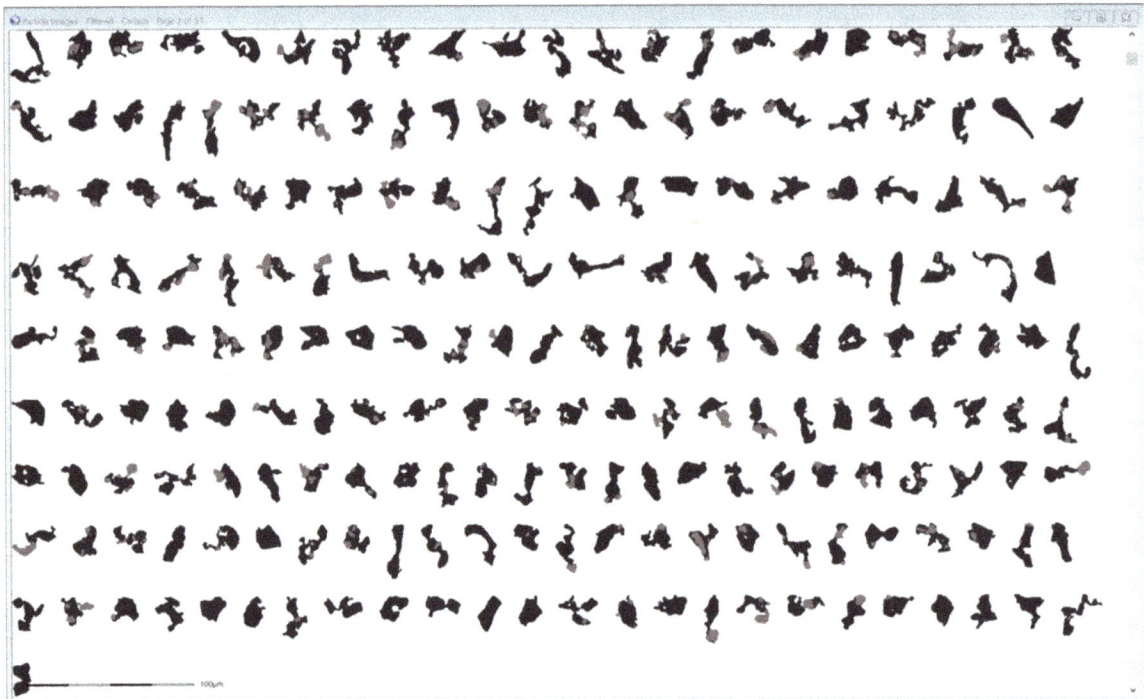

FIG 5 – Example of particles >50 Wt per cent carbon; black is carbon, light brown is chlorite.

DPM is most easily identified optically using the scanning electron microscope (SEM). Figure 7 and Figure 8 show two different SEM images from Australian underground coalmines at different levels of magnification. The dark black dots are the holes in the filter media which are 0.4 microns in size.

The DPM consists of long changes of solid carbon cores surrounded by absorbed hydrocarbons and other smaller particles as shown in the schematic representation in Figure 6. These wispy particles can be seen across the filters in both SEM images below.

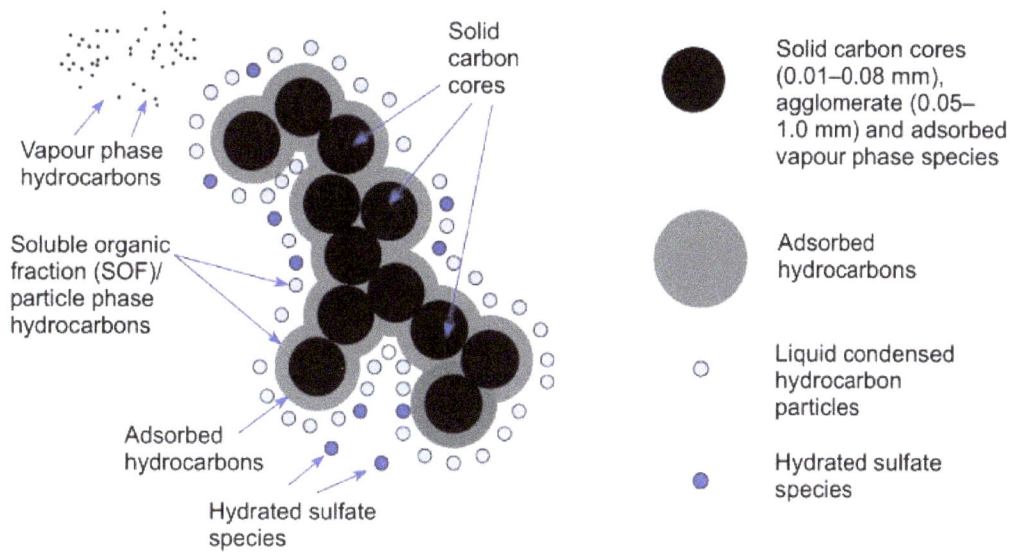

FIG 6 – Schematic representation of diesel particulate matter (Twigg and Phillips, 2009).

| 3/7/2019 | HV | WD | mag ◻ | det | spot | ⊢————— 5 µm ————— |
| 7:57:14 AM | 15.00 kV | 12.4 mm | 20 000 x | BSED | 5.5 | |

FIG 7 – Filter image at 20 000× magnification.

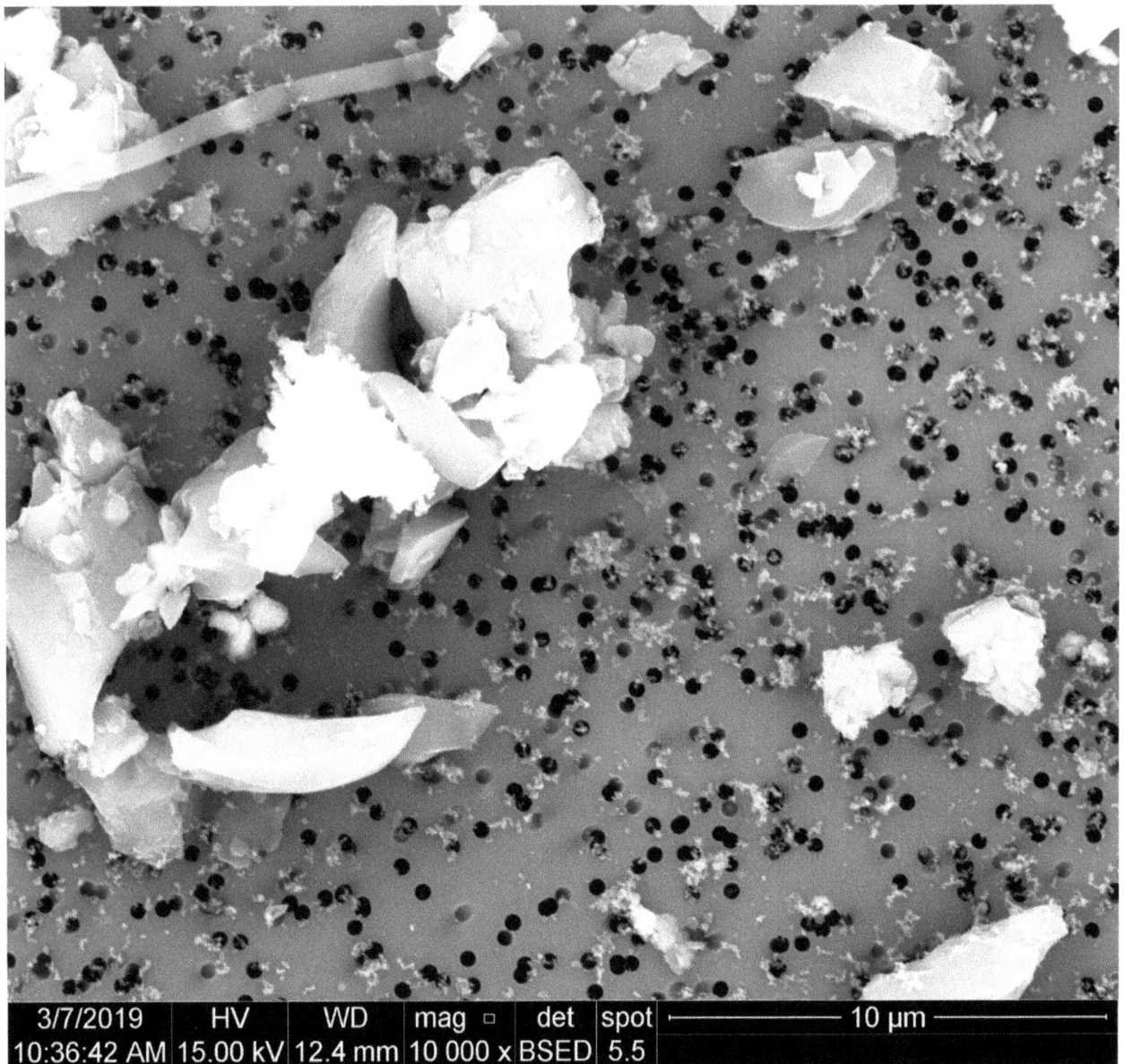

FIG 8 – Different filter image at 10 000× magnification.

Figure 8 also shows a long fibre in the upper left hand corner. Elongate mineral particles such as these can be particularly harmful because their shape allows them to penetrate further into the lungs.

Some of the DPM particles observed on the filters are up to two microns in length. Personal sampling for DPM sampling in Queensland is done with cyclone elutriators that have a maximum particle size cut-off of 1 micron. It is not currently known whether the agglomeration to these larger sizes is happening in the mine, in the cyclone or on the filter. Further research needs to be done to verify that the former two agglomerations are not taking place and allowing oversize DPM particles to be excluded.

CONCLUSIONS

There is much that characterisation data can tell us about the size, shape and chemical composition of the dust, which can further be related to the health hazard. A database of exposure characterisation is needed for all high dust hazard industries.

New techniques need to be developed to better be able to differentiate the carbon fraction between coal dust and diesel particulate matter. We do not currently know the PSD of the DPM that is being emitted in underground mines.

ACKNOWLEDGEMENTS

The authors would like to acknowledge the support of Emily Sarver *et al* at Virginia Tech for the analysis of filters from Mines 1–5 and the optical SEM images and the Sustainable Minerals Institute including Elaine Wightman and Kellie Teale as well as the School of Earth and Environmental Sciences including Kelly Johnstone and Michelle Elmes.

REFERENCES

Coal Workers Pneumoconiosis Select Committee, 2017. *Black Lung White Lies: Inquiry into the re-identification of Coal Workers' Pneumoconiosis in Queensland.* https://www.parliament.qld.gov.au/Documents/TableOffice/Tabled Papers/2017/5517T815.pdf

Johann-Essex, V, Keles, C and Sarver, E, 2017. A Computer-Controlled SEM-EDX Routine for Characterizing Respirable Coal Mine Dust, *Minerals, 7*(1), doi:10.3390/min7010015

Johann-Essex, V, Keles, C, Rezaee, M, Scaggs-Witte, M and Sarver, E, 2017. Respirable coal mine dust characteristics in samples collected in central and northern Appalachia, *International Journal of Coal Geology,* 182:85–93, doi:10.1016/j.coal.2017.09.010

LaBranche, N, Cliff, D, Johnstone, K and Bofinger, C, 2021a, 10–12 February. Respirable Coal Dust and Silica Exposure Standards in Coal Mining: Science or Black Magic? Paper presented at the Resource Operators Conference 2021, University of Southern Queensland.

LaBranche, N, Keles, C, Sarver, E, Johnstone, K and Cliff, D, 2021b. Characterization of Particulates from Australian Underground Coal Mines, *Minerals, 11*(5), doi:https://doi.org/10.3390/min11050447

LaBranche, N, Teale, K, Wightman, E, Johnstone, K and Cliff, D, 2022. Characterization Analysis of Airborne Particulates from Australian Underground Coal Mines using the Mineral Liberation Analyser, *Minerals.*

Lee, C, 1986. *Statistical Analysis of the Size and Elemental Composition of Airborne Coal Mine Dust,* PhD, The Pennsylvania State University.

Mutmansky, J M and Lee, C, 1984. An Analysis of the Size and Elemental Composition of Airborne Coal Mine Dust, Paper presented at the Coal Mine Dust Conference, West Virginia University.

National Academies of Sciences, Engineering, and Medicine, 2018. *Monitoring and Sampling Approaches to Assess Underground Coal Mine Dust Exposures,* Washington, DC: The National Academies Press, https://doi.org/10.17226/25111.

Sarver, E, Keles, C and Rezaee, M, 2019a. Beyond conventional metrics: Comprehensive characterization of respirable coal mine dust, *International Journal of Coal Geology,* 207:84–95, doi:10.1016/j.coal.2019.03.015

Sarver, E, Keles, C and Rezaee, M, 2019b. Characteristics of respirable dust in eight appalachian coal mines: A dataset including particle size and mineralogy distributions, and metal and trace element mass concentrations, *Data Brief,* 25:104032, doi:10.1016/j.dib.2019.104032

Twigg, B M V and Phillips, P R, 2009. Cleaning the Air We Breathe – Controlling Diesel Particulate Emissions from Passenger Cars, *Platinum Metals Review, 53*(1):27–34, doi:10.1595/147106709x390977

Inclusion of Safescape Laddertube to enhance emergency egress in ventilation shafts

B Murphy[1] and S Durkin[2]

1. Business Development Officer, Safescape, Bendigo Vic 3555. Email: bronwyn@safescape.com
2. FAusIMM, Managing Director, Safescape, Bendigo Vic 3555. Email: steve@safescape.com

ABSTRACT

Safescape Laddertube provides an improved system for emergency egress in ventilation shafts. The addition of Laddertube in a rise has very little impact on ventilation flow, studies have indicated that the difference in flow velocities are difficult to measure due to the minimal impact. Due to the size and open-ended nature of Laddertube, the air velocity inside and out of the Laddertube is mostly linear. Laddertube can be used either permanently or during an emergency as a secondary ventilation duct. One project involved a 385 m ladder installed in a surface return air raise that could be made to bypass the primary circuit and supply fresh air to a sea container fresh air base at the bottom of the raise.

Laddertube has been designed to be installed to any feasible length and can be fitted with a climb assist system ensuring enhanced speed and reduced fatigue during ascent. The climb assist used in association with Laddertube is a Tractelift system which applies a weight relief of up to 36 kg off the climber's body weight when using the ladder. This system has successfully been tested for ladders up to 600 m in length and is currently in use for a ladder of 185 m length.

The use of refuge chambers in underground coalmines has long been a contentious issue due to the concern that while it may provide temporary shelter, the risk of being exposed to high heat or the inability for rescuers to access them due to fire or explosive atmosphere means that they may be effectively trapped. The additional application for coal environments includes Laddertube installed alongside a refuge chamber concealed behind a blast proof door ensuring a secondary means of emergency egress. In certain operations designs may allow for this chamber to be provided with positive pressure fresh air.

INTRODUCTION

Safescape Laddertube was developed as a secondary means of egress specifically for the mining industry. Secondary means of egress, or escapeways, are integral to mining operations as without them various regulatory bodies will not allow production activities to begin. Escapeways are used as the fastest and shortest path to escape a mine emergency when access to a refuge chamber is not possible or practical. As a mine develops, escapeways are installed in various locations to allow evacuation of personnel away from the emergency or potential danger. Events that can trigger an evacuation range from rockfalls, heavy vehicle and light vehicle accidents and the greatest risk, an underground fire. Depending on the nature of the event the scale of the evacuation could be localised, or in the case of a spreading fire, an entire underground workforce may need to be evacuated.

In large scale evacuation events, ventilation systems may influence the escape route of miners. If suitable to the mine design, escapeways should be considered to be installed in fresh air intakes, however this cannot be assumed to always be the safest escape route as in an intense underground fire, the ventilation flow may reverse.

Laddertube is a fully enclosed polyethylene escapeway, providing the miner with added protection, includes a fall arrest system and has been designed with mines rescue techniques in mind. When installed in a ventilation shaft, Laddertube has been studied to have very little impact on flow within the shaft.

LADDERTUBE DESIGN

In 2010, after many years of concept development, Laddertube was produced (design details as shown in Figure 1). Laddertube was designed to be an alternative to traditional escapeways with a focus on increased safety features and durability.

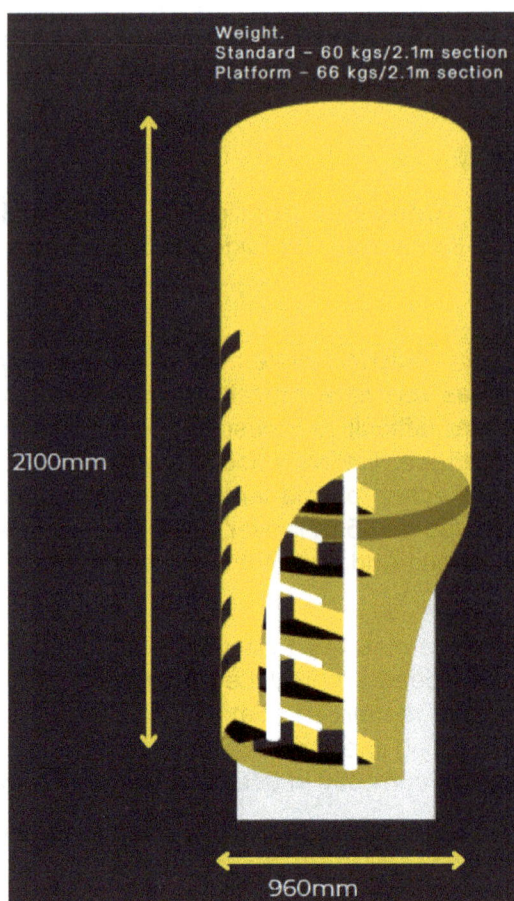

Weight.
Standard – 60 kgs/2.1m section
Platform – 66 kgs/2.1m section

2100mm

960mm

FIG 1 – Safescape Laddertube design.

Laddertube is made from polyethylene plastic which will not rust, corrode or degrade over time, resulting reduced maintenance. It is light weight, easy to install and suitable for blasted alimak and raise bore rises in varying diameters. The enclosed nature offers some degree of protection from fretting material, and for more serious ground condition concerns Laddertube can be installed with Perlite or Poly Liners (Deadman *et al*, 2016).

Polyethylene plastic requires significant energy to ignite and is similar to many other plastic products commonly used in mining operations, such as poly pipes. Laddertube has undergone testing for fire resistance and has been observed in real fire emergencies where the ladder was in close proximity to the fire source and showed no signs of igniting or sustaining damage. Standard Laddertube is suitable for almost all underground mine conditions, however if increased fire resistance is required to satisfy specific environments or local regulations Laddertube can be produced using fire retardant powder. The fire retardant Laddertube achieves a V02 rating under the UL94 flame retardancy criteria.

The design of Laddertube complies with the Western Australia Mines Safety and Inspection Regulations 1995 and the design and testing criteria of AS 1657 and MSHA requirements. Mines rescue techniques were kept in mind when considering the straight line ladder system with landing gate style platforms, ergonomic step design and smooth internal handrails. The smooth interior of Laddertube allows for mines rescue personnel equipped with breathing apparatus to slide through the ladder, preventing any catch points.

LADDERTUBE IMPACTS ON VENTILATION FLOW

Laddertube can be installed to any practical length and in various diameter and rise completion methods. To take advantage of mine designs and reduce the economic impact of drilling additional escapeway rises, it has been common practice to incorporate escapeways into the ventilation network.

Incorporating escapeways into a mine ventilation network has the potential to significantly impact the mine's ventilation flow. Modern mine ventilation systems created by engineers are now typically highly detailed and frequently updated, laying out nearly every pathway within a mine plan (Stewart *et al*, 2017). Modifications to the ventilation system need to be recorded in the relevant simulation program and evaluated to determine the effects (Widzyk-Capehart and Watson, 2001). The design, installation method and length of the escapeway are factors that could contribute to changes in air movement.

Emergency scenarios should also be considered when looking to install an escapeway in a vent rise. In the event of an underground fire or explosion, smoke and gas will travel through the ventilation system causing visibility, and air quality issues. Multiple escapeways should be considered in all areas of the mine should an emergency occur.

In order to assess if Laddertube would have an impact on vent flow two tests were completed, one where an escapeway was installed within a large mine with high volume ventilation, the other a smaller Laddertube installed in a mine with a known small circuit vent system. When trying to measure the Laddertube vent impact in the large mine, the problem arose that a pressure difference between the top and bottom of the raise changed more when a truck was moving in the decline, or through other operations while the vent officer was moving from the top to the bottom of the raise to take pressure measurements. It was deemed impossible to measure the subtle pressure differential between the before installation and after installation scenarios

The second test was conducted on a 9 m Laddertube installed in a blasted rise at the Colorado School of Mines Edgar Mine, shown in Figure 2, which has a vent circuit with a known volume of air and steady pressure. Measurements were taken in the rise before removing an existing timber ladderway, with the rise empty and again with the Laddertube installed. The results indicated a negative difference in vent flow between the timber ladder in the rise and an empty rise, but no difference between an empty rise and when Laddertube was installed.

FIG 2 – Safescape Laddertube installed at Colorado School of Mines Edgar Mine.

As Laddertube is a physical structure mounted within the rise, there must be an impact on ventilation but we've been unable to define the K factor. The reason for this is the impact is so small as to not be measurable in an operational mine.

When Laddertube is installed in a vent rise, to maintain the vent size and flow, the ladder has to be open at both ends otherwise it will effectively reduce the area of the raise by 0.72 m². If the ladder remains open, the velocity will essentially be the same inside the Laddertube as within the rise. If the air velocity is so high as to make climbing the escapeway difficult, a ventilation curtain can be used to reduce the velocity within the ladder as shown in Figure 3. Alternatively, velocity in the rise could be controlled by modifying the fan speed, which might be an appropriate option in the event of an emergency.

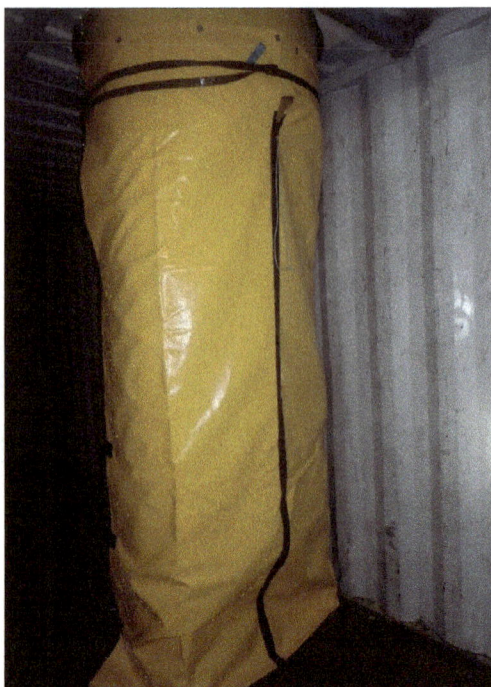

FIG 3 – Safescape Laddertube ventilation curtain.

LADDERTUBE INSTALLED IN VENTILATION RISES

Laddertube is able to be installed in various diameter and completion style rises and is not length dependent, thus making it desirable for a ventilation installation. Since 2010, Safescape Laddertube has been installed in various fresh air vents and return airways.

One particular ventilation project stands out from the norm, in 2012 Safescape was consulted by Cracow mine site in relation to two long rises that required an escapeway solution that was practical and economical. The site had an existing winder escapeway which was proving to be costly when it came to yearly maintenance checks, training operational personnel and for required replacement items. Instead of replacing the existing system, it was deemed more time and cost-effective to install Laddertube with a climb assist in a new exhaust air rise. The first rise was an 186 m escapeway installed in a 3.5 m diameter return air way, shown in Figure 4. The second escapeway was installed within the large 4.5 m diameter 384 m return airway connected to the surface, shown in Figure 5.

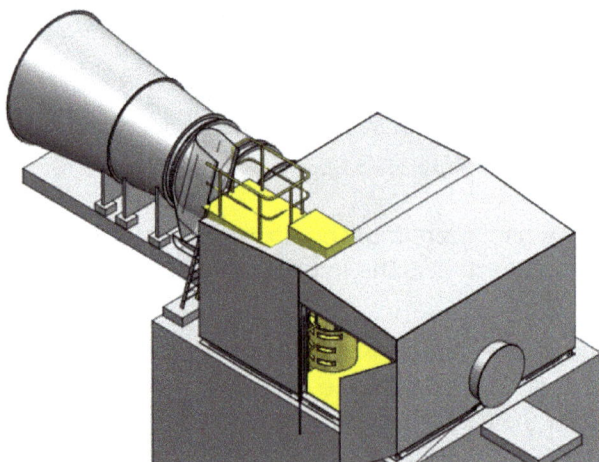

FIG 4 – 186 m Safescape Laddertube installed at Cracow Mine Site return airway.

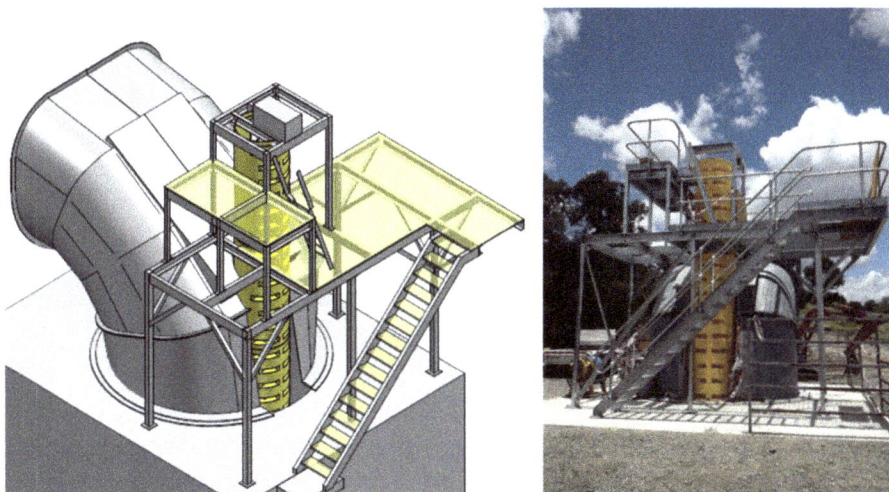

FIG 5 – 384 m Safescape Laddertube installed at Cracow Mine Site return airway.

The 184 m escapeway was installed successfully. The top of the rise was completed with a ventilation support room surrounding the Laddertube. A ventilation curtain was added to the escapeway to reduce airflow within the Laddertube for comfort without the need to stop the fan.

With the 385 m rise the Laddertube was sealed at the point it exited the evasé, creating a short circuit path in the primary vent circuit allowing fresh air to be delivered to the bottom of the rise. Installation of vent curtains at the top and bottom of the Laddertube reduced leakage to acceptable limits. The full area of the Laddertube is unavailable for primary ventilation in normal use. The rise area was reduced by 4.5 per cent. This did not pose an issue to site as the rise was drilled larger than the vent flow requirement for the operation. The temporary short-circuit feature was ideal as should the escapeway be required in an emergency situation the climbers will have access to fresh air without changing the fan speed.

There are several arguments for installation of long-term infrastructure within fresh air intake development. Return air pathways are ultimately exposed to potentially harmful emissions created during mining operations, including blast fumes, dust and humidity. One of the primary benefits of Laddertube by design, is that the presence of potentially corrosive elements does not result in structural deterioration, when compared to metal or wooden structures.

While the Laddertube itself does not deteriorate, the experience of climbing in return air rises, compared to fresh air rises can be dramatically different. In relatively wet mines a layer of greasy dust is deposited and increases over time. In dry mines, the dust can build-up to a point the drive is coated in the material and the dust can be easily suspended, making the environment unpleasant and difficult for working in the area. Any mine that executes independent firing activities provide a special section of the tag board for personnel working within return airways, with access approval required from the shift boss. These tag boards are often positioned deep within the mine in fresh development areas, yet return airways can be accessed from many places some distance from the tag boards, it presents a risk of personnel either intentionally or inadvertently entering a return airway and being exposed to blast fumes. In an emergency, access to escapeways installed in return airways is not required.

While there are a number of benefits of installing Laddertube in fresh air intakes, providing underground personnel with an improved working environment is the primary benefit.

LADDERTUBE USE IN MINE RESCUE SCENARIOS – FIRE SIMULATION CASE STUDY

The Eastern Goldfields Chamber of Minerals and Energy Mines Rescue Committee (MRC) is the organisation body for the annual Underground Mines Emergency Response Competition held in the Kalgoorlie region of Western Australia. It is also the only event that stages an engineered, fuel controlled live fire event which exposes competing teams to live fire in the underground environment. The focus of the event is team safety, preservation of life if safe to rescue, observation and reporting of mine conditions in the fire affected area and to control the fire by indirect methods.

The competition in 2018 placed the event at the base of the host mine decline on the level where the primary ventilation exhausted through the stope and a level by level longhole developed Return Air Way (RAW). The fire was fuelled by liquid petroleum gas released from liquid withdrawal cylinders, plumbed into a purpose-built gas regulator fitted with direction flow control, flash back arrestors. A compressed nitrogen purge acts as an emergency extinguishment. Over the burn duration of a scenario up to 384 000 BTU's of energy are converted into heated gases, carbon and water vapour. This released energy changes the established ventilation pathways in the area, requiring the event planners to contain the ventilation changes to a localised area and also to anticipate how the repeat generation of gases over multiple scenarios could be safely exhausted (Byrne, 2022).

The fire event was placed in a loading bay ahead of the RAW, with an open stope further ahead and a stockpile opposite the loading bay where a Laddertube was installed as the termination of the escapeway system, shown in Figure 6. The Laddertube had a passive downcast vent flow due to the top level being sheltered off the decline and the bottom landing ahead of the RAW on the fire level. Pyrotechnic smoke was used to reduce the team's visibility as they turned in from the decline to the fire level, the smoke was laid between the RAW and the loading bay. To ensure the smoke stayed in the intended area and to prevent any combustion gases exhausting into the smoke a parachute was installed ahead of the loading bay and stockpile. The scenario required the competing team to leave a fresh air base above the fire level and proceed under self-contained regenerative breathing apparatus to the fire level.

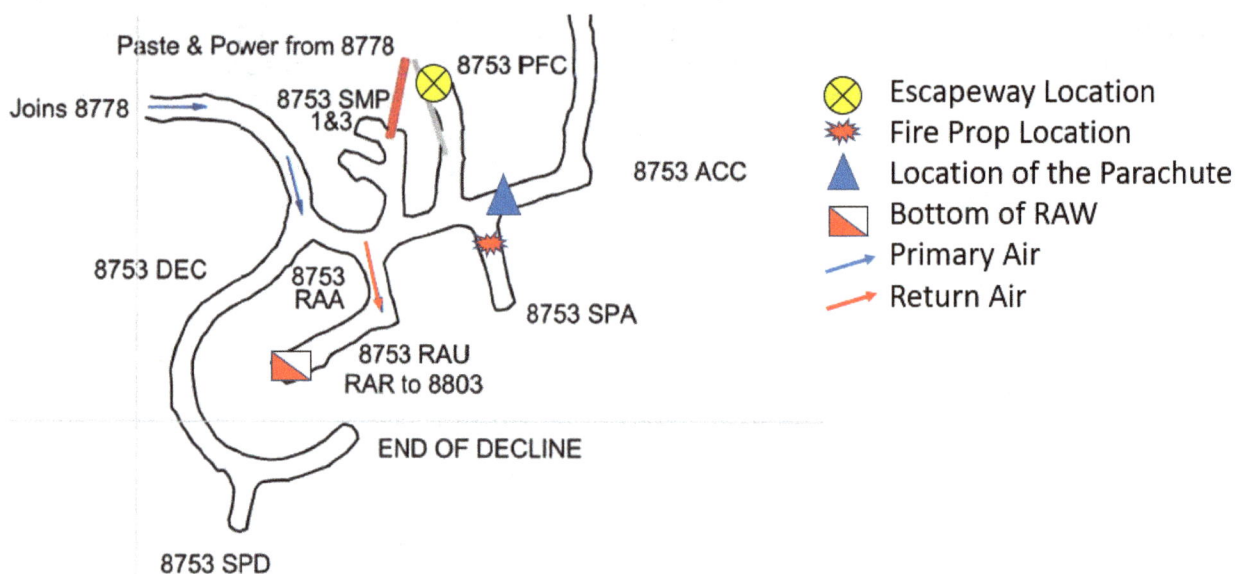

FIG 6 – Mines rescue fire simulation location map.

Prior to the test burn the event management team recognised that due to the higher profile of the loading area and stockpile, to the access drive, the heated smoke and gases would rise up the escapeway, removing the ladder as a safe means of egress and the polluted flow would contaminate the decline of the fresh air base. This assessment was validated by the use of pyrotechnic ventilation testing smoke. The event could not be moved to another location leaving the management with two options:

1. Remove the parachute at the stope to increase the ventilation flow past the stockpile and the fire loading bay, but risking accumulation of pollution in the stope. The hazard of miasmic accretion as the gases cooled and it's behaviour could not be predicted to mitigated this option.

2. Increase the velocity and volume of air cast down the escapeway to create a fresh air station at to the base landing of the Laddertube. Increasing the flow down the escapeway and creating a stopping between the landing of the Laddertube and the live fire area would overbalance the system.

Option 2 was the preferred option and the characteristics of enclosed Laddertube were critical to the success of the process. The profile of Laddertube at the top level allowed for the fashioning of a volute to direct air via a pneumatic fan down the escapeway conduit. The aperture at the bottom landing of the Laddertube increased the velocity of air into the space behind the stopping at ground level so that the fresh air side was overbalanced to the fire. This process allowed the gas controller to be in fresh air and event spectators to be able watch the scenario through a viewing slot in the stopping area.

LONG VENTILATION ESCAPEWAYS AND CLIMB ASSIST SYSTEMS

Escapeways are necessary secondary pathways for escape should an emergency event underground occur. In most underground emergency scenarios, escapeways are not used for miners to escape directly but for mine rescue teams to access miners that are located in a refuge chamber.

As stated by Stewart *et al* (2017), ladders are used at a heavy travel time at metabolic cost due to physical exertion, confined space and limited access space for multiple personnel. This paper also suggests that an estimated travel time using an escapeway for a person of average fitness would be approximately 300–400 m vertically per hour including rests. Safescape has been installing Laddertube for over 10 years and would estimate the travel time for an for a person of average fitness to ascend an escapeway to be 100–200 m vertically per hour. With advancement of new mine development and processing technologies, mines are able to go deeper and target varying ore grades significantly influencing the mine life. As a result escapeways are required to connect extensive development levels and can extend over a vertical length of 1.5 km.

Historically, being a miner was always a very physically strenuous job, with limited access to transport, hands on manual labour and exhaustive scheduling. The industry has changed with fly-in – fly-out operations, vehicle access and machine assistance which increases sedentary jobs on a mine site. The aerobic fitness levels of miners has significantly fallen as a result of the more inactive lifestyle. Brake (1999) indicated that a guiding principle when developing an emergency escapeway system was that the escapeways on-site must be capable of being traversed by mine workers who meet the minimum physical fitness requirement of the mine.

With gaining interest in installing long escapeways in vent rises, or otherwise, Safescape recognised the necessity of adding a climbing aid to Laddertube. When climbing a long ladder or escapeway, most people are negatively impacted with added stress to the upper body and lower legs due to the length of the ladder and carrying the added weight of mining equipment including harnesses and self-rescuers. With prolonged exposure, the climber may be prone to injury, even with frequent rests. When escapeways are in use, time is usually of the essence in the emergency situation, a solution that would decrease climbing time, decrease physical stress and reduce the potential for injury would significantly improve the use of mine escapeways.

In 2013, Safescape installed an Extended Ladder Climb Assist with an Easy Climb Controller and motor at the top of the Laddertube that controls an endless round reinforced polymer belt mounted to either side of the product (Figure 7). The climber connected to the belt using a rope grab attached to their harness. This system provides weight relief for the climber and could be set-up to 55 kg according to climber weight. This results in a corresponding reduction of weight that the climber's legs and arms must support during the climb. The Extended Ladder Climb Assist system consisted of a series of remote receivers spaced at intervals throughout the ladder and a signal cable interconnecting the remotes to the controller mounted near the motor. A display is provided at each receiver to display approximate position of the climber within the Laddertube.

This system achieved the desired results of alleviating physical stress on the climber and increased the speed of the ascent. One of the primary issues with this system was that the receivers occasionally would not connect, causing the belt to stop and require maintenance.

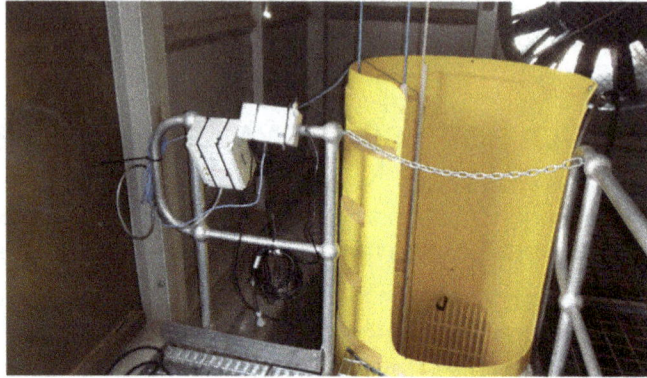

FIG 7 – Extended Ladder Climb Assist with an Easy Climb Controller.

Looking toward future installations it was clear that the climb assist system could be simplified so that it required less maintenance and installation requirements. The Tractelift system by Tractel was selected as another climbing aid to trial for use with Laddertube. The Tractelift Climb Assist System (CAS) is a climbing aid which applies a weight relief of up to 36 kg off the climber body weight when ascending ladders. This system can be installed in a similar fashion to the Extended Ladder Climb Assist with a motor at the top of the Laddertube and continuous braided core plastic cable to which the climber attaches as shown in Figure 8. It doesn't however, have the receiver system required for belt movement, personnel position and weight tracking capability. The CAS motor is controlled by a time delay relay and switch, which is activated each revolution of the driver pulley. To start the motor, the belt is pulled upwards which will initiate the power and rotation of the belt. The motor will keep running as long as the signal is repeated with each pulley rotation. When a climber stops climbing or resists the pull of the belt, the signal is not repeated within the determined time lag and the motor will stop.

FIG 8 – Tractelift system installed in Safescape Laddertube.

During the initial mine test of a 20.9 m escapeway, the mines rescue team volunteered to participate in the trial of the system. Many of the rescue team had never climbed a Laddertube before which was beneficial to simulate a miner utilising the system in an emergency for the first time. The heart rate of each climber was taken prior to climbing the ladder without the CAS, after the climb at the top of the ladder then again before and after using the CAS system. Heart rates were taken manually by the rescue team. The time for ladder ascent both with, and without, the CAS was recorded. The climbers rested for ~2–5 minutes between climbs and the final climber rested 10 minutes between the trial climbs. The results of the mines rescue trial are presented in Table 1.

TABLE 1

Climb Assist trial results from Mines Rescue Personnel.

Climber	Resting heartrate	Heartrate without CAS	Heartrate with CAS	Climb time without CAS	Climb time with CAS	Notes
		Mines Rescue Testing of CAS				
1	108 bp min	114 bp min	108 bp min	2:17	1:17	
2	93 bp min	126 bp min	96 bp min	2:13	1:10	
3	105 bp min	117 pb min	111 bp min	2:50	1:42	
4	107 bp min	120 bp min	110 bp min	4:44	2:35	With BioPack (15 kg)
5	90 bp min	102 bp min	87 bp min	2:44	1:14	With BioPack (15 kg) 10 min rest between climbs

The CAS impact on user heart rate and climbing time indicates that the system alleviates fatigue and allows personnel to ascend the ladder quicker in the case of an emergency. This is extremely important in long ladders where climbers may face upwards of 100 m of climbing length under high pressure and stressful conditions. The improved climbing rate and increased endurance would allow climbers to get out of harm's way faster and less fatigued.

In late 2021 Safescape installed a 180.6 m, 80 degree angled escapeway equipped with a CAS at a gold mine in New South Wales, Australia. A number of modifications were made to the Tractelift system to comply with the mine site standards. These modifications included, moving the emergency stop to the column of the motor unit for easier access for climbers or mines rescue personnel, changing electrical wiring to steel braided wire cables and hardwiring the system into the mine electrical grid.

This Laddertube has been climbed utilising CAS by a number of mine site personnel, including members of the mines rescue team, mine engineers and a superintendent. Overall the feedback was positive, indicating that the weight relief was felt and appreciated during the ascent. The system however is not without fault, as most climbers indicated that the weight relief seemed less effective at the bottom of the ladder and increased in the top half of the climb. It was also noted that closer to the bottom of the ladder it was more difficult to start and stop the system.

The Tractelift CAS system is supplied with an indicated maximum installation length of 100 m and designed for vertical ladders. For the system to be successfully integrated with Safescape Laddertube it needs to be able to support longer escapeways, up to 400 m in extreme cases, and ladders installed at angles from 60–90 degrees. The concerns with starting/stopping the motor and weight relief variance, may be a function of the length of the climb assist cable and potential stretching of this during use and over time.

Based on the CAS limitations and required modifications, Safescape is now developing a system internally to address these issues and improve the system so that it can better suit installation in the mining environment and within Laddertube.

LADDERTUBE INSTALLATION IN COALMINES

As summarised by Gillies and Wu (2004), mine fires remain among the most serious hazards in underground mining. The threat that fire presents depends on the nature and amount of flammable material, the ventilation system arrangement, the duration of the fire, the extent of the spread of combustion products, the ignition location and, very importantly, the time of occurrence (Conti, 2001).

The response to fire by mining personnel will depend upon all of these factors. In the case of an emergency evacuation in a coalmine, common protocol involves the use of staggered caches of oxygen self-rescuers along access route, usually the intake entries are dedicated as the primary escapeways. As miners travel the egress pathway self-rescuers are used and replaced at the designated checkpoints. If the evacuation is due to fire or an explosion in the mine, likely the air will be hot and visibility will be low, and a chance of further fire or explosion exists. Interconnection of drives or leakage through stoppings due to mine design may further pose a risk of air contamination in the escapeway.

Unlike underground hard rock mines, the use of refuge chambers in coalmines has raised concerns that while it may provide temporary safety for the miners, the risk of exposure to high heat or entrapment from later fire or explosions may be too high. A solution to this scenario is proposed, by using a refuge chamber off the main escape route, accessed through an explosion proof door at intervals determined by the distance miners can travel using their oxygen self-rescuers. In the isolated area, fresh air would be supplied directly from surface. Within this area, separate to the refuge chamber a Laddertube escapeway would be installed connecting to surface or a secondary development heading allowing for isolated emergency egress as shown in Figure 9. At the top of the ladder if exiting at surface, it would be important to have a self-contained structure which could be opened from the inside and houses a system for ventilating the chamber. It would be recommended that the refuge chamber have independently supplied fresh air as the isolated area may be temporarily contaminated as miners travel through the sealed access door to the escapeway.

By providing two systems of escape in the mine and temporary refuge, risk of fire exposure and entrapment would be minimised.

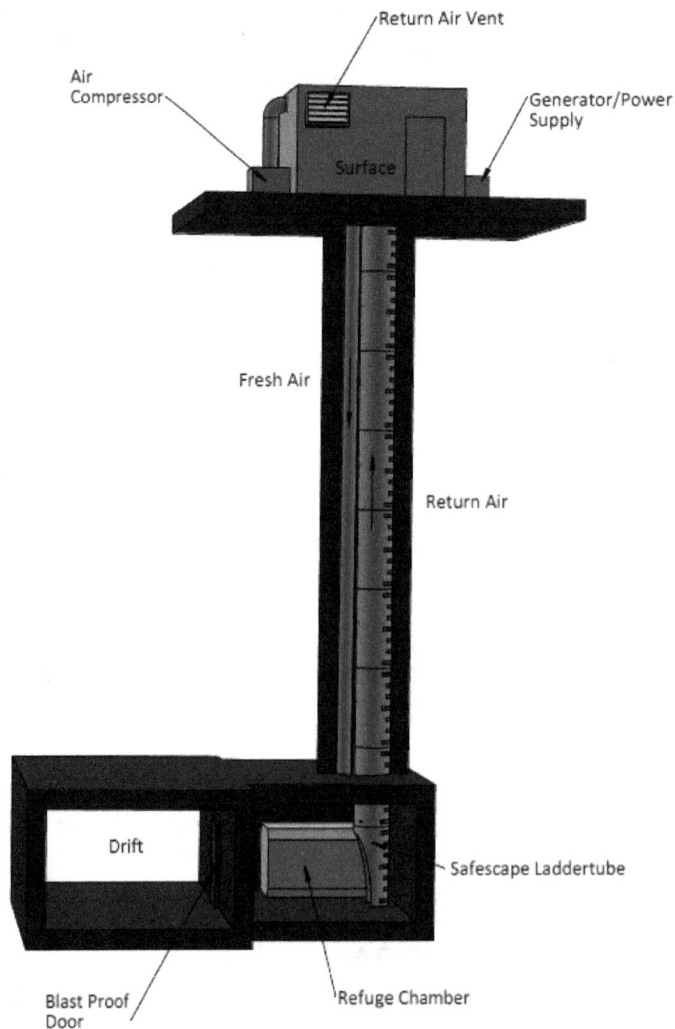

FIG 9 – Safescape Laddertube alternative egress solution in coalmines.

Laddertube is made from standard polyethylene which is not suitable for the coal environment due to its static insulative properties. To overcome this issue, Laddertube can be made using a carbon nanotube based polymer, this material also offers an increase in fire retardancy (Figure 10).

FIG 10 – Safescape Laddertube – carbon nanotube based polymer.

CONCLUSION

Safescape Laddertube has been shown to have minimal impact on ventilation flow, thus making it an ideal solution for incorporation into vent rises as part of the mines emergency egress plan. Ventilation engineers would not be required to account for the installation of Laddertube in their simulation programs unless the ladder is capped to prevent air flow such as in a return air rise. It also offers the capability to be made to bypass the primary air circuit and supply fresh air to mine personnel.

Laddertube can be installed to any practical length and within various rise developments which is beneficial for installation in fresh air or return air rises. As vent rises can dramatically vary in length, with some rises exceeding 200 m, it would be advantageous to have a climbing aid installed within a ventilation escapeway. Providing an easy to use system and potential climbing relief for mining personnel during an emergency evacuation should be considered a priority when developing an secondary egress plan.

Safescape have presented a potential solution to possible entrapment and prolonged hazard exposure due to isolated refuge chambers in coalmines. Installing Laddertube alongside refuge chambers within an evacuation route allows miners to access multiple escape routes out of the mine.

REFERENCES

Brake, R, 1999. Entrapment and escape from metal mines: A Case Study, *Qld Mining Ind Occ Health and Safety Conference*, Qld Mining Council, pp 136–146.

Byrne, T, 2022. Maintaining a Positive Pressure Environment in Laddertube to Exclude Fire Gases from an Escapeway, A Case Study, *Underground Mines Emergency Response Competition 2018*.

Conti, R S, 2001. Responders to underground mine fires, *Proceedings of the 32nd Annual Conference of the Institute on Mining Health, Safety and Research*, University of Utah.

Deadman, A, Durkin, S and Lawrence, V, 2016. Escapeway Solutions, paper 218, *ISMS 2016, 3rd International Symposium on Mine Safety*.

Gillies, S and Wu, H W, 2004. Case studies from simulating mine fires in coal mines and their effects on mine ventilation systems, *Proceedings of the 2004 Coal Operators' Conference*, Mining Engineering, University of Wollongong, 18–19 February 2019.

Stewart, C M, Aminossadati, S M and Kizil, M S, 2017. Emergency egress pathway prediction using ventilation models, *Australian Mine Ventilation Conference*, Brisbane, Australia, 28–30 August 2017.

Widzyk-Capehart, E and Watson, B, 2001. Agnew Gold Mine expansion mine ventilation evaluation using Ventsim, *Proceedings of the 7th International Mine Ventilation Congress,* 17–22 June 2001.

Oxidative combustion law of water-soaked coal under oxygen-lean condition

H Y Niu[1], Y C Bu[2], S P Li[3], Y X Yang[4] and T Qiu[5]

1. Professor, School of Civil and Resource Engineering, University of Science and Technology Beijing, Beijing 100080, China. Email: niuhuiyong@163.com
2. School of Civil and Resource Engineering, University of Science and Technology Beijing, Beijing 100080, China. Email: buyunchuan1994@126.com
3. School of Civil and Resource Engineering, University of Science and Technology Beijing, Beijing 100080, China. Email: 971938861@qq.com
4. School of Civil and Resource Engineering, University of Science and Technology Beijing, Beijing 100080, China. Email: 734699789@qq.com
5. Beijing University of Chemical Technology, College of Mechanical and Electrical Engineering, Beijing 100080, China. Email: 1449395109@qq.com

ABSTRACT

The coal left in the goaf is easily submerged by groundwater damage. During the mining process of the underlying coal seam, it is necessary to drain the water in the overlying goaf. The coal sample is prone to spontaneous combustion after being soaked in water and air-dried. To explore the effect of micro-functional group transformation of water-immersed coal samples on the spontaneous combustion characteristics of coal samples during the low-temperature oxidation process. Using the experimental method of infrared spectroscopy, the migration law of special functional groups under different oxygen concentrations of water-soaked coal samples was analysed. By means of thermal analysis, the endothermic and exothermic properties of the functional groups in the water-soaked coal samples were studied. The results show that: the higher the oxygen concentration, the more oxygen-containing functional groups in the water-soaked coal sample, compared with the 18 per cent oxygen concentration, the 14 per cent oxygen concentration has a small increase, and the chain length is shortened. The temperature increase together with the oxygen concentration content determines the functional group content, the pyrolysis process produces some functional groups. The lower the oxygen concentration, endothermic rate decreases and endothermic temperature increases, the higher the ignition temperature of coal oxy-combustion, the lower the total heat release, the higher the burnout temperature, and the faster the oxygen-absorbing weight gain and the conversion of functional groups. Below 14 per cent oxygen concentration, the oxidative combustion process will be greatly reduced. The results provide a theoretical basis for preventing long-term water-soaked coal body oxidative spontaneous combustion in shallow buried coal seams.

INTRODUCTION

As the main fossil energy, coal is widely used in the world, and still plays an irreplaceable role in the fields of coal chemical industry and power generation (Green *et al*, 2012; Xie *et al*, 2019; Chen *et al*, 2015; Saikia *et al*, 2015). However, in the actual coal mining process, the coal samples left in the goaf are prone to uncontrollable spontaneous combustion, which will lead to a series of problems such as serious waste of coal resources, worker casualties and environmental pollution (Li *et al*, 2018a, 2018b; Stracher and Taylor, 2011). In the north-western region of my country, most of the coal seams are shallowly buried and close to each other. After coal mining, the gobs can easily flow into the groundwater due to the cracks between the rocks. When mining the underlying coal seam, it is necessary to drain water from the overlying goaf, which will cause the phenomenon of water-gas replacement. Over time, the water-soaked coal in the upper gob is gradually air-dried, creating conditions for further oxidation. The increase of air leakage and the change of coal-oxygen contact amount will cause the slow accumulation of heat, which will lead to spontaneous combustion of coal.

Some scholars have done research on the effect of moisture on the spontaneous combustion of coal (Zhao *et al*, 2015; Xu *et al*, 2013; Beamish and Hamilton, 2005), in which water immersion is the main way for external moisture to affect the spontaneous combustion of coal. Zhao *et al* (2021) analysed the low-temperature oxidation characteristics and microscopic characteristics of water-immersed coal samples. The research showed that water immersion would increase the content of functional groups, promote the release of indicator gases, and promote the coal oxidation reaction. Lu *et al* (2021) analysed the apparent morphology and oxidative combustion process of the coal samples immersed in water for seven years. After immersion in water, the concentration of cations that inhibit spontaneous combustion decreased, the pore size and specific surface area increased, the molecular side chains were shortened, and the oxygen-containing functional groups were increased, which promoted oxidative spontaneous combustion. Yang *et al* (2017) studied the composition of dissolved substances in water immersed coal and its oxidative combustion law, and concluded that the higher the degree of coal metamorphism, the less the total dissolved substances, and the more developed pores of the coal sample after water immersion, which is more prone to spontaneous combustion. Zhai *et al* (2019) believes that changes in the number of functional groups and characteristic temperature of water-immersed coal are the main reasons for the enhancement of coal spontaneous combustion strength. Wang *et al* (2003) analysed the oxygen consumption of coal samples and the generation rate of carbon and oxygen gas at different moisture contents. The higher the moisture content, the greater the oxygen consumption. In the oxidation process, water reacts with carbonyl groups to form carboxyl groups. McCutcheon and Zhao (2013, 2015) analysed that there is an optimal moisture content in the coal sample to ensure that the internal structure of the coal molecule is fully contacted with oxygen, and more oxygen-containing functional groups are generated during the oxidation process, which promotes the release of the index gas and improves the oxidation rate of the coal sample. In our previous studies (Bu *et al*, 2020; Xu *et al*, 2021), the micropore structure and the transfer of active sites were mainly changed after immersion of the oxidised coal, and the micro-reaction mechanism for processing coal samples was constructed, resulting in different degrees of coal spontaneous combustion.

In addition, some scholars have studied the oxidative combustion characteristics of coal samples under different air leakage conditions. Zhao *et al* (2022) believes that 10 per cent oxygen concentration is the key critical concentration to inhibit and promote various active functional groups, which in turn affects the gas release characteristics. The main functional groups associated with the characteristic parameters of spontaneous combustion are CH_2 and CH_3. Zhou *et al* (2019) pointed out that below a certain oxygen concentration, the participation of free radicals in the reaction slows down and the CO gas release decreases. The increase in temperature has little effect on the change trend of its free radical concentration. Shi *et al* (2021) found that increasing the volume fraction of oxygen and the heating rate can promote the coal-oxygen reaction and reduce the hysteresis effect during the heating process. Zhou *et al* (2010) studied the generation rules of CO, CO_2, C_2H_4 gas under the conditions of 21 per cent, 17 per cent, 13 per cent, 9 per cent, and 5.8 per cent oxygen concentration, and concluded that the generation of oxidation products showed a 'hysteresis effect' with the decrease of oxygen concentration.

To sum up, most scholars have mainly analysed the influence of micro-macro changes of coal with different water contents on combustion, and a few scholars have analysed the oxidation process of coal under low oxygen concentration. The research on water-soaked coal samples under different air leakage conditions is rarely mentioned. In the goaf, the different oxygen supply due to changes in geological fissures has a non-negligible impact on the oxidative combustion of water-soaked coal. In this paper, the water-soaked coal is used as the research object, and the infrared spectrum changes of coal samples under different oxygen concentrations and the changes of thermal weight loss parameters are analysed, and the oxygen-lean combustion characteristics of the water-soaked coal are obtained. Provide theoretical guidance for the prevention and control of spontaneous combustion of water-soaked coal in goafs with large water content.

EXPERIMENTS AND METHODS

Preparation of coal samples

The coal sample used in this experiment was long-flame coal from a mining area in Xinjiang, and it was sealed, stored and transported to the laboratory. After the surface oxide layer was stripped, the coal sample was crushed and sieved into four parts with 40–80 mesh (0.18–0.38 mm). The coal samples were placed in a vacuum drying oven (30°C, relative pressure -0.08 MPa) for 72 hr. After taking it out, put it in a programmed heating box, set an initial temperature of 40°C, a gas flow of 50 mL/min, and maintain a constant temperature for 1 hr at oxygen concentrations of 3 per cent,

5 per cent, 10 per cent, 15 per cent, and 21 per cent, respectively. The treated coal samples are placed in a sealed bag.

Experimental system

Infrared spectroscopy

The infrared spectrum analysis experiment used a TENSOR-37 Fourier transform infrared spectrometer (Germany, Bruker Spectroscopy Instrument Co., Ltd.), with dry KBr as he diluent. The coal samples treated with different oxygen concentrations were placed in a crucible for grinding, and then placed in a vacuum drying oven for 12 hr. The infrared spectral information of the samples was collected from 400 ~ 4000 cm^{-1}, the sample was scanned 32 times with a resolution of 4 cm^{-1}.

Thermogravimetric experiment

The thermal analysis experiment used a STA6300 synchronous thermal analyser. The initial temperature is set to 30°C, the heating rate is 10°C/min, the oxygen concentration is 21 per cent, 18 per cent, 14 per cent, 10 per cent, the termination temperature is 800°C, and the coal sample mass is 20±1 mg. The experimental data is extracted to obtain TG – DSC curves were analysed.

EXPERIMENTAL RESULTS AND ANALYSIS

FTIR analysis

The content of active functional groups in coal is closely related to the speed of oxidative combustion of coal, and the degree of branch chain rupture and the mutual transformation of functional groups in treated coal samples determine the content of functional groups in coal samples. A large number of functional groups are distributed on the surface of coal, such as alkyl groups, hydroxyl groups, carbonyl groups, carboxyl groups, etc. During the oxidation process, the functional groups will release heat along with the cleavage of bonds, which affects the spontaneous combustion of coal. The change of oxygen supply has a great influence on the further conversion of functional groups. Using infrared spectroscopy, the changes in the functional groups of water-soaked coal treated with different oxygen concentrations can be quantitatively determined, and the curves are shown in Figure 1.

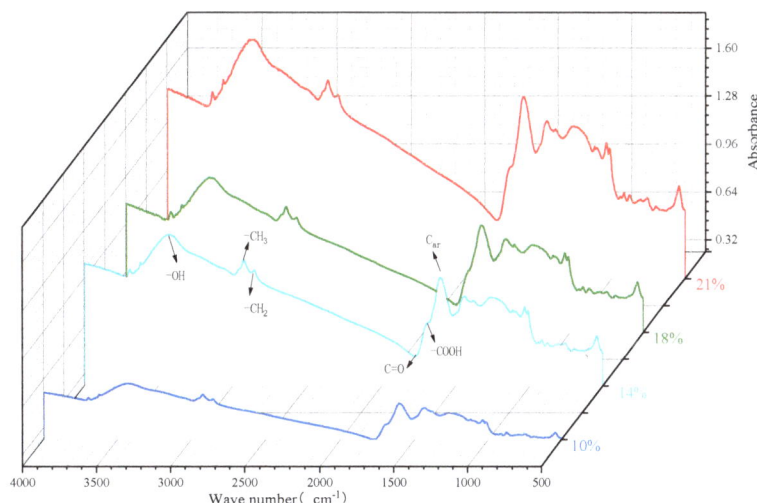

FIG 1 – Infrared spectrum of coal sample.

Combined with the previous research results (Zhou *et al*, 2017; Parsa *et al*, 2018, 2019), the fitting curves at wavenumbers 2800 cm^{-1}–3000 cm^{-1} and 1500 cm^{-1}–1750 cm^{-1} are shown in Figure 2, and the corresponding functional group peak positions are found. The peak positions of the corresponding functional groups are, methyl (2854 cm^{-1}), methylene (2922 cm^{-1}), hydroxyl (2851 cm^{-1}), carbonyl (1651 cm^{-1}), carboxyl (1685 cm^{-1}), aromatic carbon (1615 cm^{-1}), all positions have been marked in the figure.

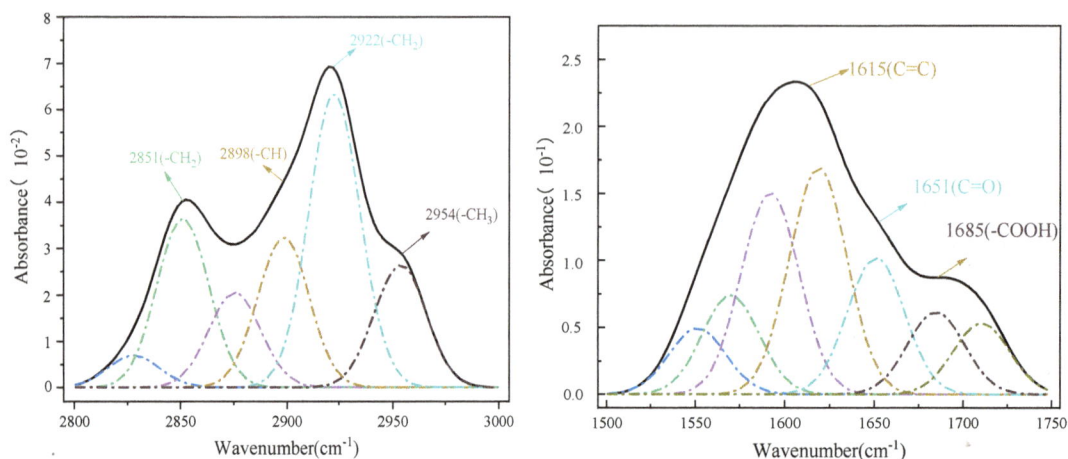

FIG 2 – Infrared fitted spectrum.

Figure 3 shows the changes of functional groups of coal samples treated with different oxygen concentrations. As the oxygen concentration of the treated coal samples decreased, the hydroxyl group gradually decreased, and the decrease of 21 per cent from the 18 per cent oxygen concentration treatment to the 14 per cent oxygen concentration treatment. The change trend of oxygen-containing functional groups (carbonyl, carboxyl) is similar, showing a decreasing trend as a whole, with a small increase under the treatment of 18 per cent to 14 per cent oxygen concentration, the alkane length first increases, then decreases and then increases, and is smaller at 14 per cent oxygen concentration.

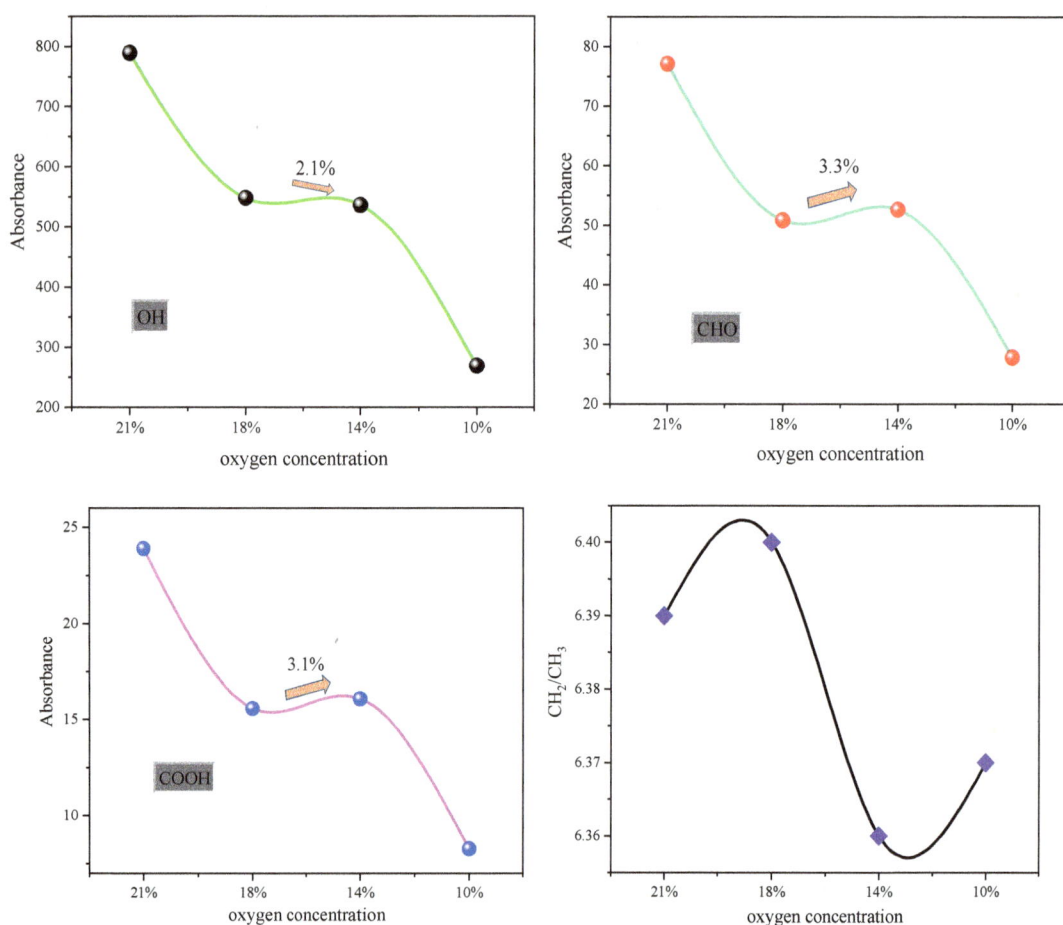

FIG 3 – Changes in each functional group.

The changes of functional groups in the water-leached coal treated with different oxygen concentrations were greatly affected by the oxygen concentration, indicating that oxygen is an important intermediate substance to promote the generation of functional groups in coal samples.

When the oxygen concentration is high, with the cleavage of the side chain group and the violent oxidation, more oxygen-containing functional groups are generated, and the reactivity of coal molecules is high. Compared with the 18 per cent oxygen concentration, there is a slight increase in the oxygen-containing functional groups of the coal sample with 14 per cent oxygen concentration. During the pyrolysis process, the coal sample also has functional group transformation, and the molecular groups with side chain scission at low temperature generate oxygen-containing functional groups.

Lean oxygen combustion ignition mechanism

The thermogravimetric curve can reflect the quality change of the coal sample during the pyrolysis process, and the oxidative combustion process of the water-soaked coal under different oxygen concentrations can be judged by the quality change. The TG curve of the coal sample is shown in Figure 4. The pyrolysis process of coal samples is a special chemical reaction process, in which there is no lack of mutual transformation and cracking of various microscopic functional groups, in which the temperature change is the dominant factor. According to previous studies (Kizgut and Yilmaz, 2004; Qi *et al*, 2017; Xu *et al*, 2020), coal samples were divided into several special temperature points for analysis. They are the critical temperature (T_1), the weight loss temperature (T_2), the oxygen absorption weight gain temperature (T_3), the ignition point temperature (T_4), the maximum weight loss temperature (T_5), and the burnout temperature (T_6). The change trend is listed in Table 1.

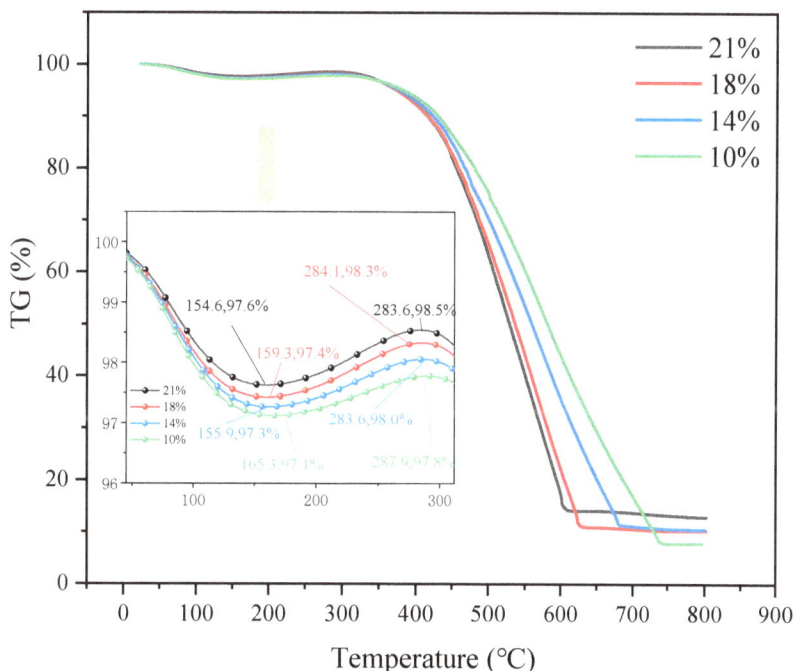

FIG 4 – Thermogravimetric characteristic temperature points of coal samples.

TABLE 1

Characteristic temperature of coal sample.

Raw coal	21%	18%	14%	10%
T$_1$	83.8	83.5	83.42	83.4
DTG	61.4	65.4	67.3	67.5
T$_2$	154.6	159.3	155.9	165.3
TG	97.6	97.4	97.3	97.1
T$_3$	283.6	284.1	283.6	287.9
TG	98.5	98.3	98.0	97.8
T$_4$	412.2	417.7	439.5	445
T$_5$	468.8	505.4	538.9	558.8
DTG	1054.4	944.9	751.8	649.9
T$_6$	610.1	625.3	683.8	740.8
TG	14.3	11.7	11.3	8.1

According to the characteristic temperature points, the coal sample pyrolysis process is divided into five stages. They are the first stage (T$_1$-T$_2$) and the initial oxidation stage. This stage is mainly the evaporation of water on the surface of coal particles and the desorption process of gas, and the mass loss is small. The second stage (T$_2$-T$_3$), the oxygen absorption and weight gain stage, with the increase of temperature, a series of functional group reactions occur inside the dried coal particles, and strong chemical adsorption occurs, and the mass gradually increases. The third stage (T$_3$-T$_4$), the rapid oxidation stage, the chemical bond breaking is strengthened, the gas is precipitated in large quantities, the oxygen is consumed in large quantities, and the mass reduction is accelerated until the coal sample catches fire. In the fourth stage (T$_4$-T$_6$), the violent combustion stage, the organic molecules speed up the reaction, and the oxygen supply is not enough to support the intensification of the combustion until it burns out. Figure 5 shows the characteristic temperature points of thermogravimetric loss at different stages and the stage changes under 21 per cent oxygen concentration.

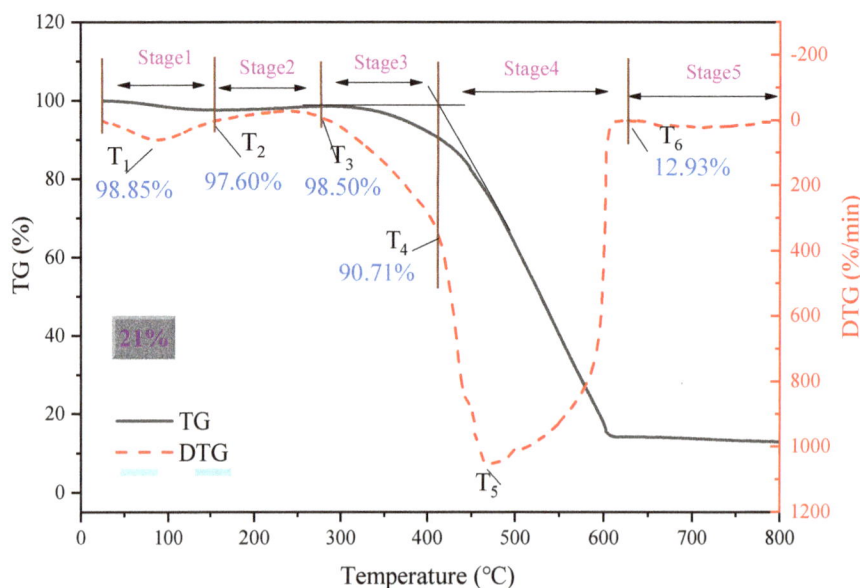

FIG 5 – Thermogravimetric characteristic temperature points of coal samples.

During pyrolysis, coal gradually loses mass along with the breaking of chemical bonds, such as bonds between aliphatic chains and aromatic hydrocarbons (Bu *et al*, 2022; Tian *et al*, 2018). The temperature T_1 tends to decrease with the decrease of oxygen concentration, and the variation of DTG increases. Lower oxygen concentration is helpful for low-temperature pyrolysis of water-soaked coal samples. Under low-temperature oxidation, the conversion of functional groups requires less oxygen. In addition, combined with infrared spectroscopy experiments, the content of various functional groups increased under high oxygen concentration (21 per cent), including hydrophilic functional groups (Han *et al*, 2016; Li *et al*, 2019), which delayed the process of water evaporation to a certain extent. At this time, the promoting effect of oxygen is not obvious, and the speed of water evaporation is mainly affected by temperature.

The physical structure of coal undergoes obvious changes after immersion in water (Song *et al*, 2018; Zhong *et al*, 2019). The relatively dry coal pore structure is exposed to the oxygen environment, which further promotes the process of oxygen absorption and weight gain. The maximum weight gain temperature difference (T_1-T_2) at 21 per cent oxygen concentration is larger, and the weight gain is more. The oxygen demand for oxygen in the weight-enhancing stage is relatively high, and the coal-oxygen composite reaction and chemical conversion are accelerated, thereby promoting the formation of oxygen-containing functional groups in the coal and enhancing the third-stage high-temperature oxidative cracking process.

The ignition temperature is an important indicator to judge the difficulty of burning coal samples. As the oxygen concentration decreases, the ignition temperature gradually increases. Compared with the 21 per cent oxygen concentration, the 18 per cent oxygen concentration increased by 1.3 per cent, the 14 per cent oxygen concentration increased by 6.2 per cent, the 10 per cent oxygen concentration increased by 7.3 per cent, and the 14 per cent oxygen concentration was the inflection point of increase. Each functional group is decomposed and broken by heat, and the quality decline is accelerated. The change of oxygen supply mainly affects this process. As the oxygen concentration decreases, the burnout temperature increases gradually, the pyrolysis process is delayed, the residual amount decreases, and the pyrolysis becomes more thorough.

Thermal effect of coal combustion

To further analyse the heat release characteristics of water-soaked coal under different oxygen concentrations, the extracted DSC comparison curves are shown in Figure 6. The curves of coal samples with different oxygen supply concentrations have similar trends, which are divided into two processes: endothermic process and exothermic process, the exothermic period lasts for a long time. It is mainly reflected in the growth period (T_b-366°C), the plateau period (366°C-406°C), the rapid increase period (406°C-T_c), and the drop period.

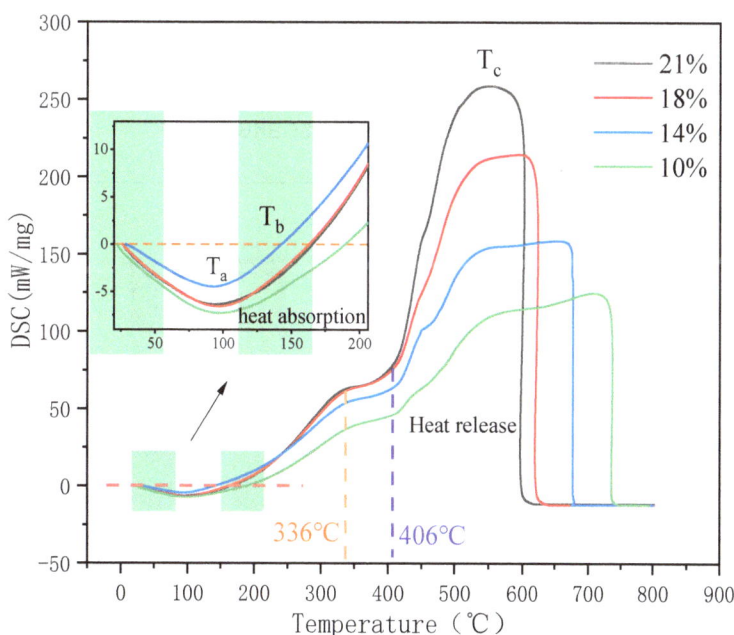

FIG 6 – DSC changes of coal samples.

Scholars usually use thermal characteristic parameters to express the judgment index of the oxidative combustion process of coal samples (Zhu *et al*, 2020). According to the DSC curve, several characteristic temperature point parameters are obtained as shown in Table 2. T_a is the maximum endothermic rate temperature, T_b is the initial exothermic temperature, and T_c is the maximum exothermic rate temperature. In general, when the oxygen concentration is low, T_a and T_b increase, while the two temperature points of 14 per cent oxygen concentration are small. Combined with the conclusion that the oxygen-containing functional group increases and the chain length decreases, the possible reason is that there is a parallel pyrolysis reaction and oxidation reaction in the low-temperature oxidation process. At 14 per cent oxygen concentration, the cleavage of its functional groups is relatively accelerated, the heat release increases, the actual heat absorption decreases, and the exothermic temperature is reached faster. T_c increases with decreasing oxygen concentration, and the maximum heat release decreases gradually, it shows that the oxygen concentration plays a key role in the fracture of macromolecular structure in coal during high temperature pyrolysis.

TABLE 2

Variation of thermal parameters of coal samples.

Raw coal	T_a	Maximum endothermic power (mW/mg)	T_b	T_c	Maximum exothermic power (mW/mg)
21%	93.8	-6.34	165.5	553.1	258.8
18%	97.7	-6.57	163.3	593.2	214.30
14%	92.8	-4.51	142.9	651.9	158.41
10%	96.8	-7.28	189.6	705.4	124.76

Integrating the DSC curve to obtain the endothermic and exothermic heat of the coal sample oxidative combustion process is shown in Figure 7. There is a small amount of moisture in the pores of the soaked coal, which absorbs a lot of heat during low-temperature oxidation and evaporation (Deng *et al*, 2016; Fry *et al*, 2009). At 10 per cent oxygen concentration, the heat absorption of the coal sample increases significantly, indicating that the low-temperature oxidation process is hindered, the chemical bond rupture is delayed, and the heat release is reduced. Oxygen-containing functional groups accumulate and distribute at high oxygen concentration, and water molecules are adsorbed on the coal pore surface under the action of capillary force, and the heat absorption is relatively large. The heat release is mainly affected by the concentration of oxygen. Compared with the 21 per cent oxygen concentration, the 14 per cent oxygen concentration has a larger reduction in the heat release (reduced by 5 per cent). The 14 per cent oxygen concentration is an important inflection point to reduce the further oxidation of water-soaked coal.

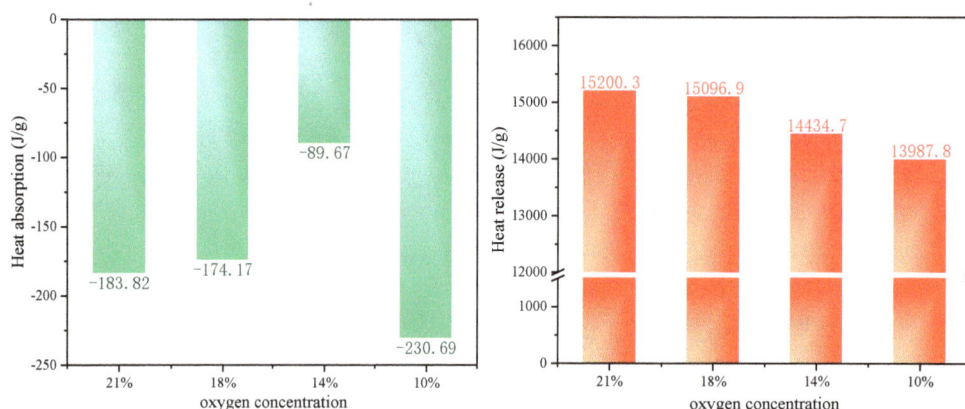

FIG 7 – Changes in heat absorption and release of coal samples.

CONCLUSIONS

With the decrease of oxygen concentration, the main functional groups showed a decreasing trend. At 14 per cent oxygen concentration, the oxygen-containing functional groups increased slightly, the aliphatic hydrocarbon functional groups were cleaved faster, and the side chain length decreased. During the low temperature oxidation process of coal samples, the occurrence of functional groups is affected by both oxygen concentration and temperature, and more functional groups are produced during the pyrolysis process.

Lower oxygen concentration accelerates the reduction of the endothermic temperature and the increase of the endothermic rate. The oxygen-absorbing weight gain process requires a large amount of oxygen molecules to participate in the transformation, and the oxygen concentration determines the weight gain change. The lower the oxygen concentration, the higher the burnout temperature of the coal sample, and the lower the total heat release. The pyrolysis process with 14 per cent oxygen concentration will make the ignition temperature increase rapidly and the heat release decrease rapidly.

ACKNOWLEDGEMENTS

The authors wish to acknowledge gratefully the financial support of the research funding provided by the National Natural Science Foundation of China (Nos. 52174163, 51874131 & 51474106).

REFERENCES

Beamish, B and Hamilton, G, 2005. Effect of moisture content on the R 70, self-heating rate of Callide coal, *Int J Coal Geol*, 64:133–138.

Bu, Y, Niu, H, Wang, H, Wang, H Li, S, Yang, Y, Qiu, T and Wang, G, 2022. Study on pore structure change and lean oxygen re-ignition characteristics of high-temperature oxidized water-immersed coal, *Fuel*, 323:124346.

Bu, Y, Xu, Y, Chen, M and Liu, Z, 2020. Effect mechanism of initial oxidation temperature on secondary oxidation characteristics for soaked long-flame coal, *J. Saf. Sci. Technol*, 16:64–69.

Chen, J, Lin, M, Cai, J, Yin, H, Song, X and Li, A, 2015. Thermal characteristics and kinetics of refining and chemicals wastewater, lignite and their blends during combustion, *Energ Convers Manag*, 100:201–211.

Deng, J, Zhao, J, Zhang, Y, Huang, A, Liu, X, Zhai, X and Wang, C, 2016. Thermal analysis of spontaneous combustion behavior of partially oxidized coal, *Process Safety and Environmental Protection*, 104:218–224.

Fry, R, Day, S and Sakurovs, R, 2009. Moisture-induced swelling of coal, *Int J Coal Prep Utilization*, 29:298–316.

Green, U, Aizenshtat, Z, Metzger, L and Cohen, H, 2012. Field and laboratory simulation study of hot spots in stockpiled bituminous coal, *Energy Fuel*, 26:7230–7235.

Han, Y, Bai, Z, Liao, J, Bai, J, Dai, X, Li, X, Xu, J and Li, W, 2016. Effects of phenolic hydroxyl and carboxyl groups on the concentration of different forms of water in brown coal and their dewatering energy, *Fuel Process Technol*, 154:7–18.

Kizgut, S and Yilmaz, S, 2004. Characterization and non-isothermal decomposition kinetics of some Turkish bituminous coals by thermal analysis, *Fuel Process Technol*, 85:103–111.

Li, J, Li, Z, Wang, C, Yang, Y and Zhang, X, 2018a. Experimental study on the inhibitory effect of ethylenediaminetetraacetic acid (EDTA) on coal spontaneous combustion, *Fuel Process Technol*, 178:312–321.

Li, J, Li, Z, Yang, Y, Kong, B and Wang, C, 2018b. Laboratory study on the inhibitory effect of free radical scavenger on coal spontaneous combustion, *Fuel Process Technol*, 171:350–360.

Li, J, Li, Z, Yang, Y, Niu, J and Meng, Q, 2019. Insight into the chemical reaction process of coal self-heating after N2 drying, *Fuel*, 255:115780.

Lu, W, Li, J, Li, J, He, Q, Hao, W and Li, Z, 2021. Oxidative kinetic characteristics of dried soaked coal and its related spontaneous combustion mechanism, *Fuel*, 305:121626.

McCutcheon, A and Wilson, M, 2013. Low-temperature oxidation of bituminous coal and the influence of moisture, *Energy Fuels*, 17:929–33.

Parsa, M and Chaffee, A, 2018. The effect of densification with NaOH on brown coal thermal oxidation behaviour and structure, *Fuel*, 216:548–558.

Parsa, M and Chaffee, A, 2019. The effect of densification with alkali hydroxides on brown coal self-heating behaviour and physico-chemical properties, *Fuel*, 240:299–308.

Qi, X, Li, Q, Zhang, H and Xin, H, 2017. Thermodynamic characteristics of coal reaction under low oxygen concentration conditions, *J Energy Inst*, 90:544–555.

Saikia, B, Khound, K and Baruah, B, 2015. Extractive de-sulfurization and de-ashing of high sulfur coals by oxidation with ionic liquids, *Energ Convers Manag*, 81:298–305.

Shi, X, Zhang, Y, Chen, X, Zhang, Y and Ma, T, 2021. Numerical study on the oxidation reaction characteristics of coal under temperature-programmed conditions, *Fuel Process Technol*, 213:106671.

Song, S, Qin, B, Xin, H, Qin, X and Chen, K, 2018. Exploring effect of water immersion on the structure and low-temperature oxidation of coal: a case study of Shendong long flame coal, China, *Fuel*, 234:732–737.

Stracher, G and Taylor, T, 2011. Chapter 6 – The effects of global coal fires, *Coal Peat Fires A Global Persp*, pp 101–114.

Tian, B, Qiao, Y, Lin, X, Jiang, Y, Xu, L, Ma, X and Tian, Y, 2018. Correlation between bond structures and volatile composition of Jining bituminous coal during fast pyrolysis, *Fuel Process Technol*, 179:99–107.

Wang, H, Dlugogorski, B and Kennedy, E, 2003. Role of inherent water in low-temperature oxidation of coal, *Combust Sci Technol*, 175:253–270.

Xie, J, Ni, G, Xie, H, Li, S, Sun, Q and Dong, K, 2019. The effect of adding surfactant to the treating acid on the chemical properties of an acid-treated coal, *Powder Technol,* 356:263–272.

Xu, T, Wang, D and He, Q, 2013. The study of the critical moisture content at which coal has the most high tendency to spontaneous combustion, *Coal Prep,* 33:117–127.

Xu, Y, Bu, Y and Wang, L, 2021. Re-ignition characteristics of the long-flame coal affected by high-temperature oxidization & water immersion, *J Clean Prod*, 315:128064.

Xu, Y, Bu, Y, Chen, M, Yu, M and Wang, L, 2020. Characteristics for the Oxygen-lean Combustion and Residual Thermodynamics in Coalfield-Fire Zones within Axial Pressure, *Acs Omega,* 35:22502–22512.

Yang, Y, Li, Z, Si, L, Gu, F, Zhou, Y, Qi, Q and Sun, X, 2017. Study Governing the Impact of Long-Term Water Immersion on Coal Spontaneous Ignition, *Arab J Sci Eng*, 42:1359–1369.

Zhai, X, Wang, B, Wang, K and Obracaj, D, 2019. Study on the Influence of Water Immersion on the Characteristic Parameters of Spontaneous Combustion Oxidation of Low-Rank Bituminous Coal, *Combust Sci Technol*, 191:1101–1122.

Zhao, H, Yu, J, Liu, J and Tahmasebi, A, 2015. Experimental study on the self-heating characteristics of Indonesian lignite during low temperature oxidation, *Fuel,* 150:55–63.

Zhao, J, Wang, W, Fu, P, Wang, J and Gao, F, 2021. Evaluation of the spontaneous combustion of soaked coal based on a temperature-programmed test system and in-situ FTIR, *Fuel,* 294:120583.

Zhao, J, Zhang, Y, Song, J, Zhang, T, Ming, H, Lu, S, Deng, J and Shu, C, 2022. Microstructure of coal spontaneous combustion in low-oxygen atmospheres at characteristic temperatures, *Fuel,* 309:122132.

Zhong, X, Kan, L, Xin, H, Qin, B, Dou, G. 2019. Thermal effects and active group differentiation of low-rank coal during low-temperature oxidation under vacuum drying after water immersion, *Fuel*, 236:1204–212.

Zhou, B, Yang, S, Wang, C, Hu, X, Song, W, Cai, J, Xu, Q and Sang, N, 2019. The characterization of free radical reaction in coal low-temperature oxidation with different oxygen concentration, *Fuel,* 262:116524.

Zhou, C, Zhang, Y, Wang, J, Xue, S, Wu, J and Chang, L, 2017. Study on the relationship between microscopic functional group and coal mass changes during low-temperature oxidation of coal, *Int J Coal Geol*, 171:212–222.

Zhou, F, Shao, H and Li, J, 2010. Experimental Research on Combustion Product Formation During Coal Spontaneous Combustion Under Reduced Oxygen Concentrations, *J China U Min Techno*, 39:808–811.

Zhu, H, Zhao, H, Wei, H, Wang, W, Wang, H, Li, K, Lu, X and Tan, B, 2020. Investigation into the thermal behavior and FTIR micro-characteristics of re-oxidation coal, *Combust Flame*, 216:354–368.

Characterisation of respirable dust generated from full scale laboratory concrete cutting tests with conical picks at three stages of wear

S Slouka[1], J Brune[2], J Rostami[3], C Tsai[4] and E Sidrow[5]

1. Research Assistant, Colorado School of Mines, Golden CO 80401. Email: sslouka@mines.edu
2. Research Professor, Colorado School of Mines, Golden CO 80401. Email: jbrune@mines.edu
3. Research Professor, Colorado School of Mines, Golden CO 80401. Email: rostami@mines.edu
4. Associate Professor, University of California Berkely, Los Angeles CA 90095.
   Email: candacetsai@ucla.edu
5. Research Assistant, University of British Columbia, Vancouver Canada V6T 1Z4.
   Email: evan.sidrow@stat.ubc.ca

ABSTRACT

Airborne rock dust poses serious long-term health complications to workers in the underground mining and civil environment where rock excavation is utilised. During drilling, airborne rock particles are immediately released into the breathable environment where additionally, the settled dust can be reintroduced into the air further down in production. The purpose of this study is to build a characterisation suite of dust particles such that underground operations can implement more appropriate dust suppression and mitigation techniques in accordance to pick wear. Therefore, concrete is cut to create a preliminary dust characterisation baseline, where future cutting tests will be performed on limestone and sandstone rocks.

In this study, three conical pick wears (new, moderately worn, and fully worn picks) are used to generate dust from a concrete block and characteristics are statistically compared between the dusts. Airborne respirable particles are collected during cutting and deposited fine material, or particles left on the surface of the block, are collected after cutting. The dusts are then analysed through standardised methods, field emission scanning electron microscope image capture with image analyses, and laser diffraction methods. Findings reveal that the worn pick generated the highest concentration of respirable dust, all picks generated dust containing quartz, the three picks generally generated respirable dusts with similar particle shapes (in terms of roundness, roughness, and aspect ratio), and the worn pick generated the largest deposited fine material particles.

INTRODUCTION

Airborne dust particles less than 100 µm in aerodynamic diameter pose a major respiratory health concern for occupational workers in underground mining and civil environments. Exposure to silica containing particles less than 4 µm in aerodynamic diameter, or respirable particles, can cause irreversible diseases such as coal workers pneumoconiosis (Department of Health and Human Services, 2011), silicosis (Ross and Murray, 2004) and other lung diseases (Pope III *et al*, 2002; Stansbury, 2018). Although the US Mine Safety and Health Administration (MSHA) recognised and addressed these major issues in the 1970s, there has been an increase in lung disease cases in the United States since the 1990s with a continued rise in numbers in recent years (Blackley *et al*, 2016; Lee *et al*, 2014; Zosky *et al*, 2016; Doney *et al*, 2019; Hall *et al*, 2019). Even with MSHA regulations such as Lowering Miners' Exposure to Respirable Coal Mine Dust Including Continuous Personal Dust Monitors established in 2014 (MSHA, 2014) which changed the permissible respirable dust concentration limit from 2 to 1.5 mg/m^3 and requires continuous personal dust monitors to be worn by mine operators, modern mine workers are still contracting irreversible lung diseases.

It is not understood why there is an increase in modern miner's lung diseases. Therefore, there are many research projects investigating other reasons aside from the known fact that dust concentration impacts human health (MSHA, 1977; Centers of Disease Control and Prevention, 1974). Other investigations include analysing nano-particle presence, particle size presence, mineralisation analysis, and how particle shape affects deposition into the human lung (Fan and Liu, 2021; Sarver, Keles and Rezaee, 2019; Labranche *et al*, 2021; Zellnitz *et al*, 2019). There is also limited understanding of how dust characteristics change as picks wear. Picks generate dust at the tip because they strike the surface to break rock and will be changed over time as they wear down (Plis, Wingquist and Roepke, 1988; Roepke, Lindroth and Myren, 1976), but it is uncertain if the pick wear drastically alters dust characteristics. The only evidence shows that dust concentration changes with the geometry and wear of pick tips, but there are no further investigations (Plis, Wingquist and Roepke, 1988; Roepke and Hanson, 1983; Hanson and Roepke, 1979; Fowell and Ochei, 1984; Zhou *et al*, 2020; Qayyum, 2003).

Therefore, this research focuses on the concentration, mineral presence, particle shapes, and particle size distributions for respirable dust generated from three different conical picks during full scale cutting of a concrete block. A conical pick is used for experimentation because they are commonly used in hard rock cutting machines for excavation (Talbo and Sevigné, 1986; Roxborough, King and Pedroncelli, 1981; Su and Akkaş, 2020; Shao, 2016). A concrete block is used because it provides a reasonably homogenous baseline sample without joint sets and known composition to compare to future limestone and sandstone rock dust characterisation studies.

With limited research on metal and non-metal dust, the intent of the research is to continue building a preliminary baseline dust characterisation suite such that future underground operations can make improved decisions in dust suppression and mitigation strategies in relation to the wear of their picks. The research findings presented are built off preliminary experiments conducted with two conical picks on an igneous rock block (Slouka, Rostami and Brune, 2021) to continue the understanding of how pick wear influences dust characteristics.

METHODS

Sample preparation and full scale testing

A concrete block was used for experimentation with a Linear Cutting Machine (LCM) at the Earth Mechanics Institute (EMI) at the Colorado School of Mines (CSM). Concrete is used because it is drilled on worksites and has a known composition with a reasonably homogenous media. The sample did not contain any joint sets or discontinuities in structure, which provided the most uniform and homogeneous sample possible for consistent cutting tests compared to other future rock samples to be tested. The concrete was a well-mixed, uniform small aggregate mixture. Additionally, the sample was tested in the rock mechanics lab at CSM where the rock strength properties are as follows: 34.2 MPa unconfined compressive strength, 3.1 MPa Brazilian tensile strength, 1.13 Cerchar abraisivity index, and 6.0 kN/mm punch penetration energy slope index.

The most homogeneous concrete on the market available was poured into a metal rock box that was then placed onto the LCM sled. The sled moves the box linearly such that the pick contacts the concrete's surface at a constant speed of 250 mm/sec (10 inch/sec) and in a linear manner to simulate rock cutting in a full scale testing process. The length of each cut line is about 1.1 m (3.5 feet) and therefore, each cut line takes 4.2 seconds. Multiple lines of cuts are made across the concrete surface as seen in Figure 1 at 3.81 cm (1.5 inch) spacings, which is a typical cutting spacings for a machine cutting a similar rock with a USC of 34 MPa. The multiple cut lines across the surface of the block create one pass across the concrete surface where dust is generated and collected. Therefore, each dust collection sample is dust generated while cutting lines for one pass left to right at the desired spacing.

The penetration, or depth, of the pick into the rock sample was 0.51 cm (0.2 inches) for the experiments. These parameters were used because they are representative of industry spacings and penetrations used for the specific rock type and are determined by optimising the specific energy and normal forces when cutting various hard rocks picks (Bilgin *et al*, 2006).

With this, the LCM ensures constant cutting speed during the cuts with a linear variable displacement transducer (LVDT) sensor. The sensor measures the location of the rock box (sample displacement) in real time during testing and the speed is validated after each cut. Additionally, the penetration down into the rock sample is fixed by inserting steel plates between the cross frame and the machine's main structure. For a 5 mm (0.2 inch) penetration depth, a 5 mm plate would be used to assure the accuracy of the measured penetration, while it allows for stiffness of the cutting unit, since the vertical load is spread over a large area provided by the metal sheet. For other penetration values other than 5 mm, multiple of the available 5 mm sheets, or inserts with appropriate thickness (ie 2.5, 10, 12.5 mm), are used.

FIG 1 – The Linear Cutting Machine with the first set of conditioning cut lines on the surface of the concrete sample. The pick is encompassed by the dust curtain.

Dust collection set-up

Two collection methods were used to obtain representative samples of dust generated from cutting the concrete block. The first method uses nylon 10 mm Dorr-Oliver cyclones with a 50 percent cut point of 4 μm aerodynamic diameter particles to collect the airborne respirable particles. The second method uses a vacuum (shop-vac) to collect the particles left behind along the line of cut (crushed zone under the bit) on the surface of the rock after each cut. The particles collected with the vacuum are the fine materials that would be transported into downstream operations if the cuts were performed in a real mine. Fines were vacuumed up after each cut and sieved through the number 200 mesh, where particles passing the 200 mesh are analysed.

The Dorr-Oliver cyclones remove larger airborne particles and deposit the respirable particles onto 37 mm diameter polycarbonate (PC) filters and 37 mm diameter PVC filters. Four cyclones inside the dust curtain were equidistant and evenly spaced around the pick to collect dust with the various filters. Polycarbonate filters are used for these experiments because the substrate is best suited to analyse particles in a scanning electron microscope (Johann-Essex, Keles and Sarver, 2017; Sellaro, Sarver and Baxter, 2015; Sellaro, 2014). PVC filters are used to obtain the concentration and mineralogy because the material has a stable weight and high collection efficiency (Vaughan, Milligan and Ogden, 1989; Soo *et al*, 2016; Bogen *et al*, 2011; Lowry and Tillery, 1979). The cyclones are connected with Tygon® tubing to air pumps running at 1.7 L/min, the recommended flow rate to use for maximising collection efficiency (NIOSH, 1998). Due to humidity changing the collection efficiency of cyclones, tests are conducted within a range that does not affect collection efficiency, which was between 20 per cent and 30 per cent humidity (Chen and Huang, 1999; Vinson *et al*, 1984; Volkwein and Thimons, 2001; Marple and Rubow, 1984).

The cyclones and pumps are integrated to an automated dust collection system as seen in Figure 2. This system eliminates human error in sampling dust particles during full scale cutting tests and additionally clears out dust from prior cuts to flush out the system with fresh air. With a laser measuring the movement of the concrete box, the data interface uses the laser readings to open and close the electric ball valves, as well as turn the flushing system on and off. The automated dust collection system ensures consistency and therefore, more reliable comparability, between dusts generated with the various picks. Lastly, a rotameter is used to verify pump flow rates before and after cutting and a real-time concentration monitor is used to ensure the flush system removes all particles in the dust curtain before the next cut begins.

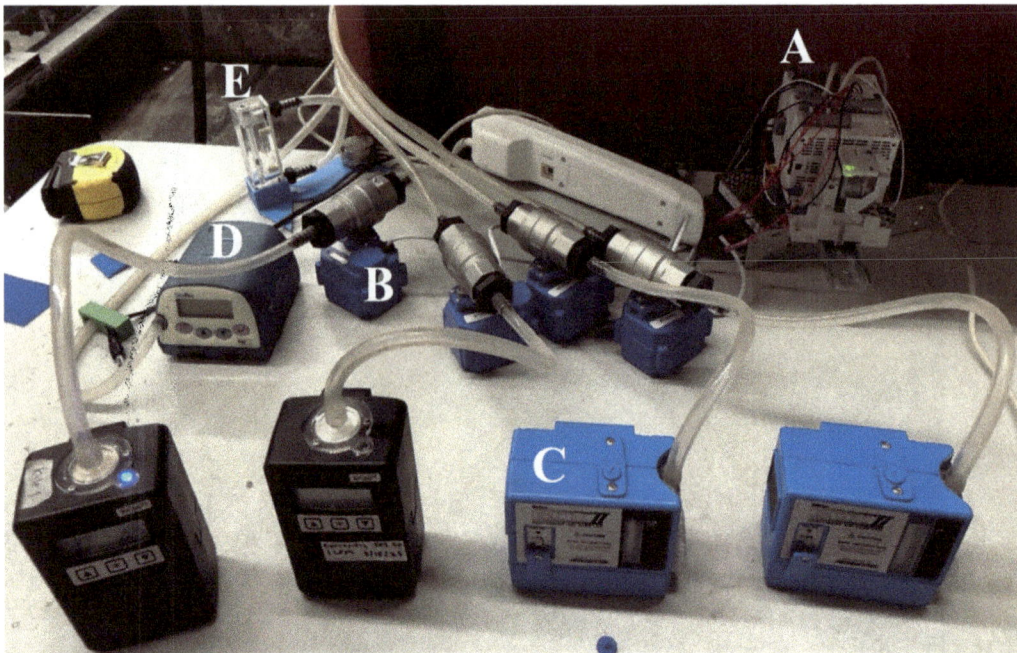

FIG 2 – Dust collection system: (a) Data acquisition interface and power unit that controls when dust is collected or flushed out of the system; (b) Electric ball valves to control the on/off air flow-through the cyclones; (c) Pumps running at 1.7 L/min; (d) Real-time dust concentration monitor; (e) Rotameter currently placed in-line before testing to ensure pump is running at 1.7 L/min.

Lastly, a curtain is installed around the LCM pick and saddle to confine the dust generated from cutting and to keep other unrepresentative lab dusts from being collected. Figure 1 shows the dust curtain around the pick with the fringed bottom to smoothly run along the surface of the concrete block. The cyclones are inside the dust curtain and are piped out to the pumps with Tygon® tubing.

Pick wear measurement and wear quantification

Conical picks are used for cutting the concrete where a new, moderately worn (assumed mid-life), and worn (assumed end-life) pick are tested. As seen in Figure 3, a circle is superimposed at each of the pick tips, where the new pick tip has a diameter of 3.58 mm (0.141 in), the moderately worn pick has a diameter of 7.52 mm (0.296 in), and the worn pick tip has a diameter of 10.36 mm (0.408 in). Diameter of the tip is used to quantify pick wear because the angle that makes the tip will stay the same throughout cutting and only the tip itself will become more blunt over time. With this, the moderately worn and fully worn picks were generated by artificially wearing down a duplicate new pick tip by hand with a Dremel and a lathe.

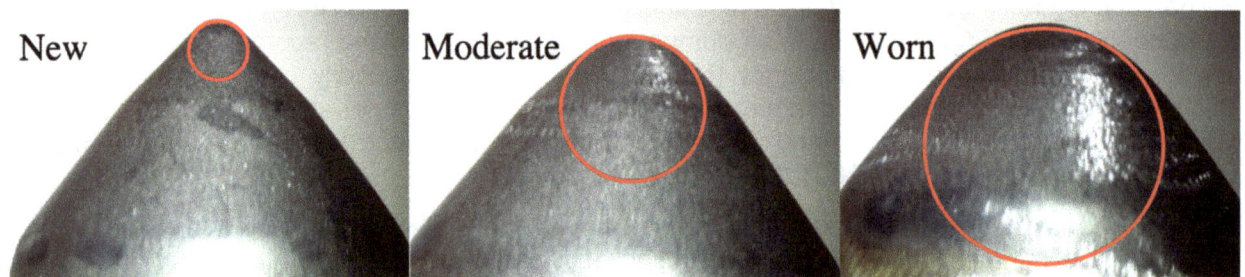

FIG 3 – Comparison of a new, moderately worn, and worn pick tip with 0.7× magnification. A circle is superimposed to show the increase in pick tip radius with increase of wear.

Particle characterisation

The airborne respirable particles less than 10 µm in diameter are characterised in terms of concentration, mineral content, particle shape, and particle size distribution. Fines materials, which were particles left behind on the surface on the sample and sieved to obtain particles smaller than number 200 mesh, are characterised in terms of particle shape and particle size distribution. Figure 4 shows the purpose of each instrument used for dust collection with the characterisation output obtained for the various particles collected. With limited rock sample to cut, one dust sample with one duplicate for each pick wear was obtained and results are the average of the two samples.

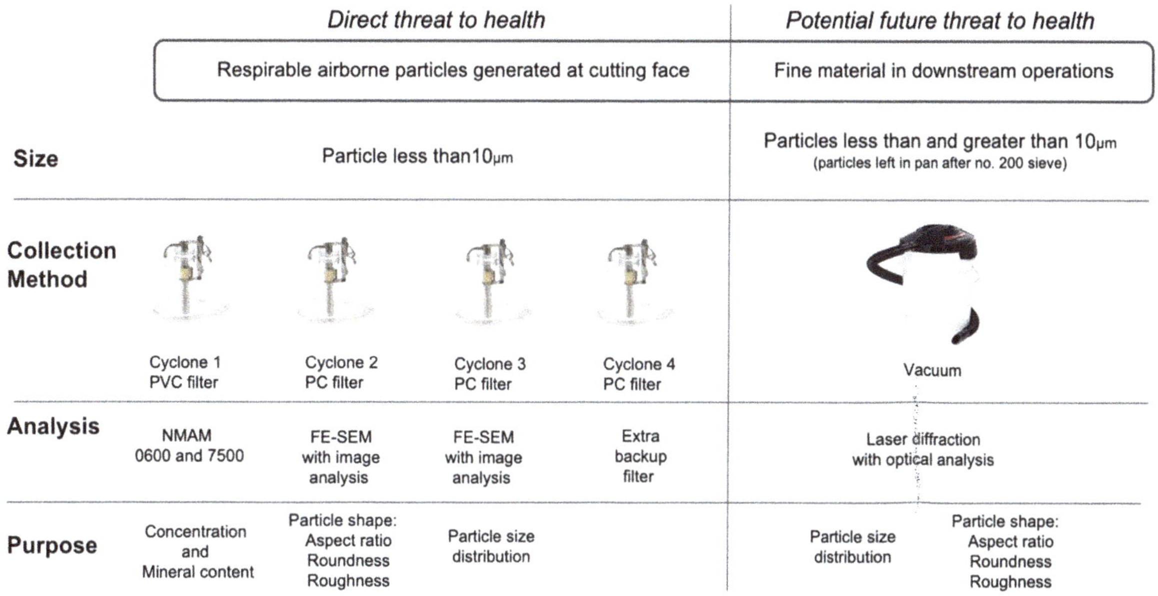

FIG 4 – Representation of the instruments used, analyses performed, and characterisation output obtained for the various particles collected.

TEST RESULTS

NIOSH Manual of Analytical Methods

The NIOSH Manual of Analytic Methods (NMAM) analyses are performed on the dust collected by a professional third party. The NMAM 0600 standard provides the dust concentration normalised to the duration of collection time and the NMAM 7500 standard provides the micrograms of cristobalite, quartz, and tridymite detected in the respective collected samples. It is critical to perform these tests on the samples because these are the standardised tests in the United States performed in industry that are used to regulate and mitigate dust exposures.

The results obtained from experiments are presented in Table 1 and are an average of two duplicate tests. As seen in the results from the NMAM 0600 standard, the concentration of the dust from the new to worn pick in µg/min increases with increase in pick wear. With this, the results from the NMAM 7500 standard reveal that all picks generated dust that contained silica containing minerals, mainly quartz with traces of cristobalite.

TABLE 1

Results from NMAM 0600 and 7500 lab tests.

Analysis	Result	New Pick	Moderately Worn Pick	Worn Pick
NMAM 0600	Concentration	µg/L	µg/L	µg/L
		150	210	270
NMAM 7500	Cristobalite	µg/sample	µg/sample	µg/sample
		20	14	17
	Quartz	320	290	360
	Tridymite	0	0	0

Particle shapes

To determine the particle shapes, a field emission scanning electron microscope (FE-SEM) and image analysis program is used to capture and process images from the PC filter surfaces. A Tescan FE-SEM at a voltage of 15 kV with back scattering electron (BSE) detection is used to obtain images. Then, Clemex Vision PE® software processes the images, detects particles, and performs calculations of roundness, aspect ratio, and roughness of particles. The Clemex Vision PE® software is programmed to detect the particles with binarisation by grey thresholding (Lane, Martin and Pirard, 2008; Chopard *et al*, 2019) to distinguish the difference between particles and the background. The process moving from an FE-SEM image to grey thresholding, to highlighting individual particles is shown in Figure 5. Other commands are also added to the program, such as bridging and object transfer, which separates particles that are clumped together and removes particles intersecting the edge of the image during detection, respectively.

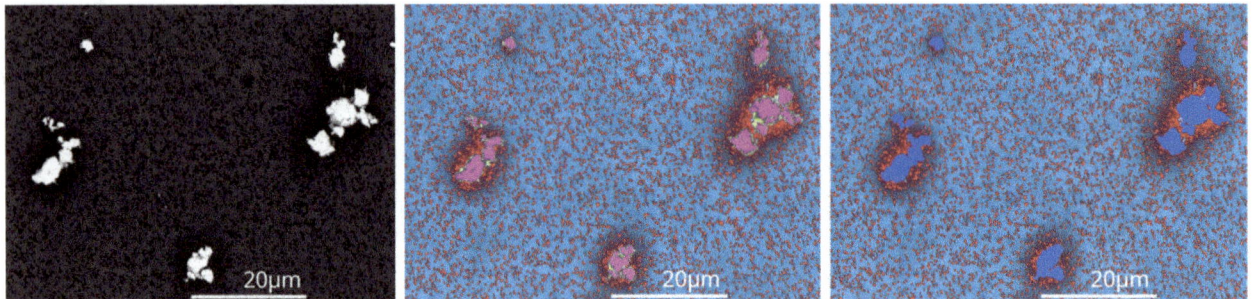

FIG 5 – Raw FE-SEM image (left), the image under grey thresholding (middle), and the program identifying and highlighting individual particles in blue (right) that are then analysed for shape features.

Particle roundness is calculated to determine how close to a perfect circle the particle of interest is via Equation 1. Particle aspect ratio is calculated to determine the elongation of particles via Equation 2. Particle roughness is calculated to determine the smoothness of the particle perimeter via Equation 3. Convex perimeter is used to determine the particle roughness, where the convex perimeter is the perimeter of the object if a rubber band were placed around the particle as shown in Equation 4. With this, *length* is the longest measurement across an object, *width* is the shortest distance measured across an object. The graphical data is presented in Figure 6 as histograms to visualise the number of particles with specific values for the roundness, aspect ratios, and roughness particle counts.

$$\text{Roundness} = \frac{4\,(\text{area})}{\pi\,(\text{length})^2} \tag{1}$$

$$\text{Aspect Ratio} = \frac{\text{length}}{\text{width}} \tag{2}$$

$$\text{Roughness} = \frac{\text{ConvexPerimeter}}{\text{Perimeter}} \tag{3}$$

$$\text{ConvexPerimeter} = \sum \text{ferets}\left[2\tan\left(\frac{\pi}{2(\text{number of ferets})}\right)\right] \tag{4}$$

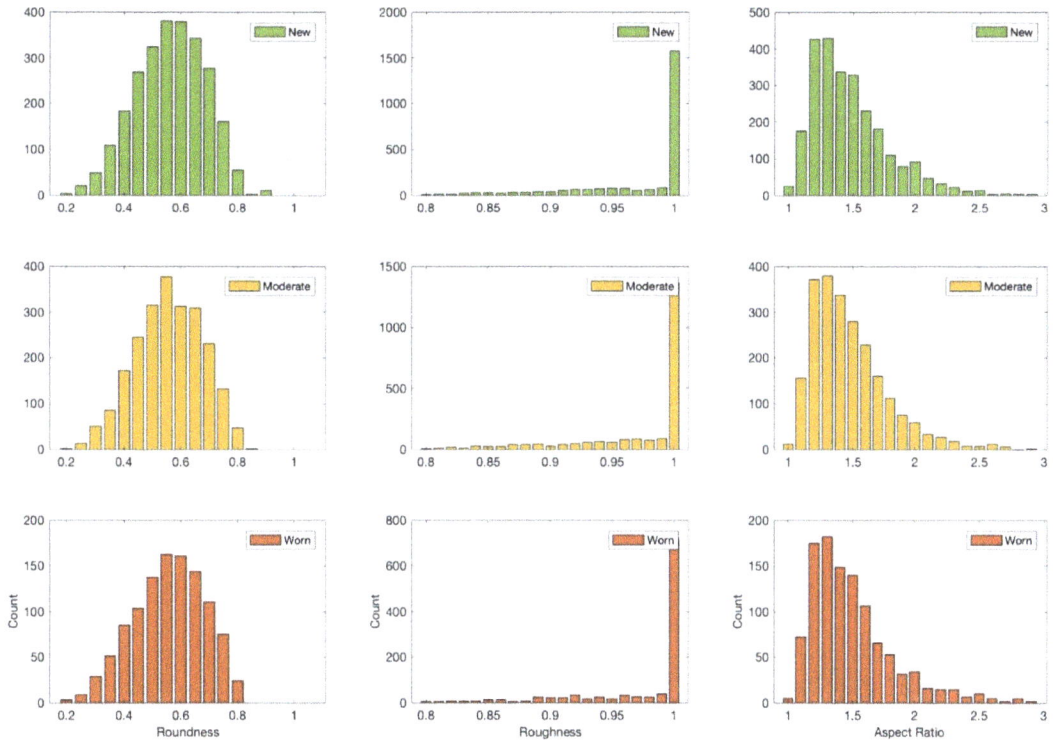

FIG 6 – Respirable particle counts of the roundness in the left column, roughness in the middle column, and aspect ratio for dust particles deposited on PC filters.

A camera simultaneously enabled during the laser diffraction analysis was used to collect particle shape characteristics for the fines material. Results of the counts of particles for respective shape measurements are provided in Figure 7.

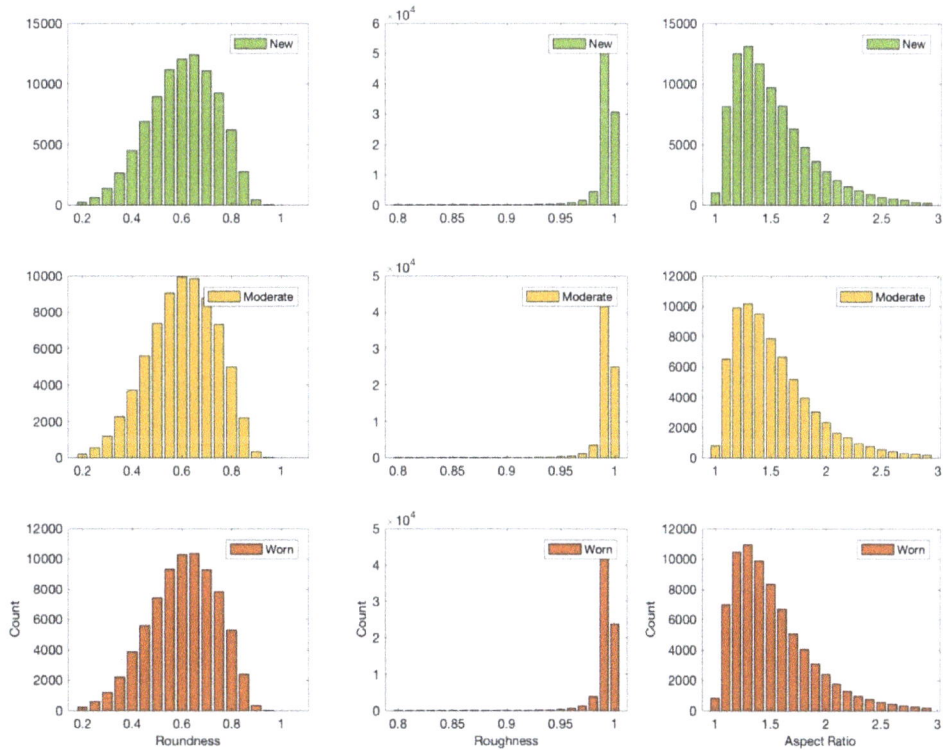

FIG 7 – Fine material particle counts of the roundness in the left column, roughness in the middle column, and aspect ratio for dust particles left on the surface of the block.

Particle size distributions

FE-SEM analysis is used to determine the particle sizes for the respirable particles and laser diffraction is used to determine the size distribution of the fines material. The size distributions for respirable particles are shown in Figure 8 and the size distributions for the fines material is shown in Figure 9. The aerodynamic diameter of the particles are calculated from the physical diameters using the Cunningham correction factor and slip correction factor.

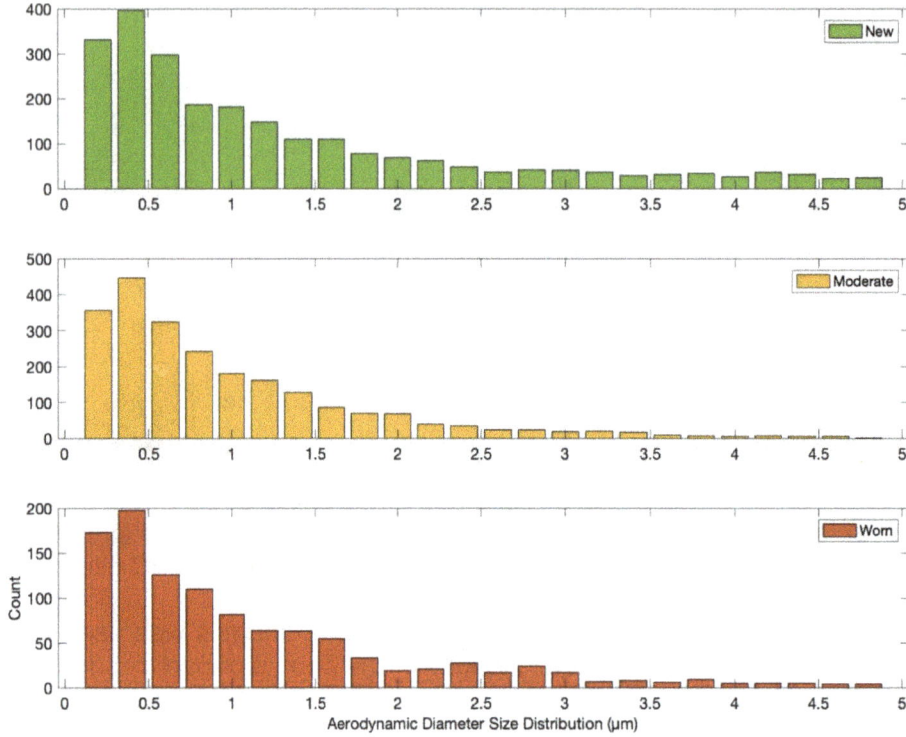

FIG 8 – Size distributions of respirable particles obtained with FE-SEM and image analysis.

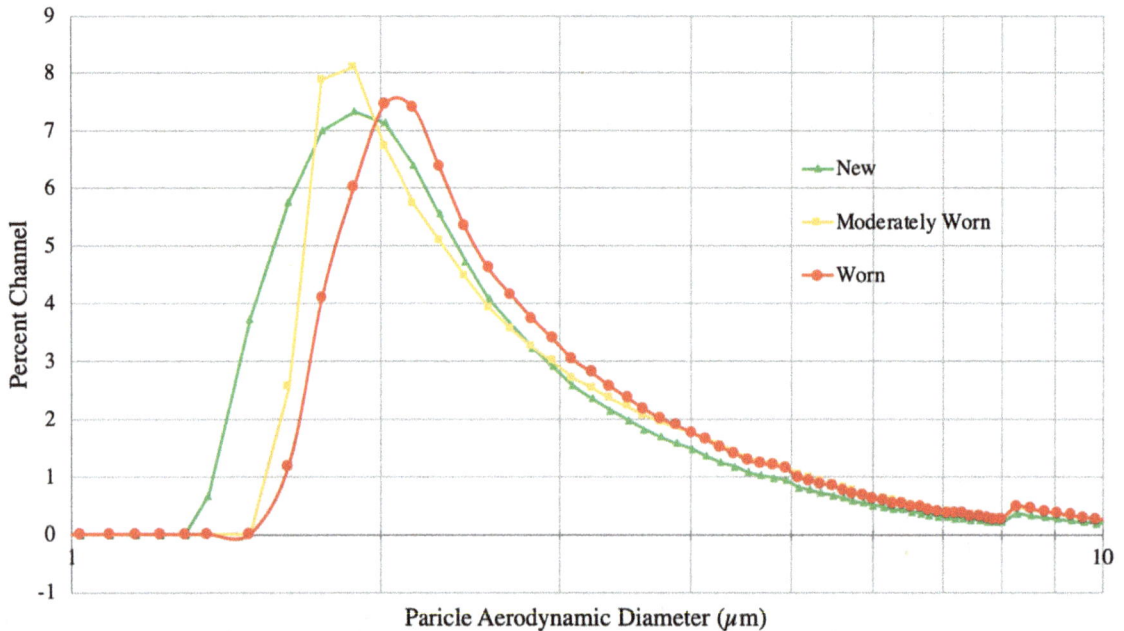

FIG 9 – Size distributions of fines material particles obtained with laser diffraction.

DISCUSSION AND ANLAYSIS

NMAM standards

The results obtained by the NMAM 0600 reveal concentrations of respirable dust particles increasing (from 150 µg/L to 210 µg/L to 270 µg/L) as the wear of pick increases from new to worn. With this, the generation of more dust particles from a worn pick compared to a new pick confirms findings in other experiments where a duller pick tip generates more dust (Plis, Wingquist and Roepke, 1988). Therefore, it is recommended that more caution is taken when cutting with picks that are more worn in comparison to new picks because it is expected that more dust is generated with worn picks. This could mean increasing the ventilation flows or incorporating water suppression systems.

The NMAM 7500 standard revealed that all picks generated dust containing hazardous minerals. Quartz and a fraction of cristobalite were detected in all the samples, which are both minerals containing silica. However, the quantitative differences in mineral concentrations could result from the nature of the grains in the concrete block. Therefore, it could be the concrete mixture composition that caused the difference in amount of quartz and cristobalite instead of the wear of the pick. It is inconclusive whether the change in amount of quartz between picks is due the concrete or due to the changing pick tips. Further research needs to confirm the findings presented in this paper with additional cutting tests.

Particle shapes

The results obtained on the roundness, aspect ratio, and roughness measures of respirable particles and fines material particles show that there is little to no statistical differences between the shapes generated by various picks. The Kolmogorov-Smirnov statistical test (KS test) is performed between each of the data sets to test the hypothesis that the distributions of particle shapes generated from the picks are similar between picks. A significance level of 0.05 is used in the statistical analysis meaning that p-values below 0.05 provide strong statistical evidence to reject the hypothesis test.

Visually observing Figure 6 reveals that the particle shapes look similar between pick wears. The KS test confirms and provides evidence that respirable dust generated by the new, moderately worn, and fully worn picks share similar roundness and aspect ratio measures, as all the p-values are 0.174 or larger. However, the statistical test suggests that the worn pick generates dust with different roughness measures compared to the new and moderately worn picks. The p-value obtained for the KS test was 0.021 between the new and worn pick, and 0.008 between the moderately worn and worn picks for roughness measures. Aside from these values, there is an overall strong statistical evidence that all the picks generate respirable dust particles of similar shapes.

With this, although a few of the nine total p-values calculated from KS tests for the fines material data sets are below the 0.05 threshold for shape measurements, the graphs in Figure 7 provide strong visual evidence that the fines material particle shapes generated are all similar between the three pick wears. Overall, there is strong statistical and graphical evidence that the fines material dust particle roundness, aspect ratio, and roughness shapes generated from all the picks are similar.

Particle size distribution

The results obtained for the size distributions of the respirable particles show that all the picks generated similar sized dust particles. The histogram representations of data exhibit similar distributions where additionally, the highest count of particle size detected from every pick was within the 0.25 µm–0.5 µm bin.

The results for the size distributions of the fines material particles show that as the pick wear increases, the generated particle sizes also increase. As seen in Figure 9, each pick generates particles with a similar single modal peak distribution curve. However, there is a clear shift in the curves where the new pick generates smaller particles and the worn pick generates the largest particles, with the moderately worn pick generating particle sizes between the two. Additionally, the new pick generated particles that are as small as 1.35 µm in aerodynamic diameter and the worn pick generated particles that are only as small as 1.62 µm in aerodynamic diameter.

Smaller particles sizes generated from the new pick could possibly be attributed to the smaller surface area point force that occurs during rock cutting with newer picks (Dogruoz and Bolukbasi, 2014; Evans, 1965; Hurt and Evans, 1981; Dogruoz *et al*, 2016). With this, the single, smaller point of contact could lead to further crushing of the rock in this zone. However, there is some uncertainty in the shift of particle size distribution in these experiments. There was only one duplicate dust collection performed for each pick wear in the full-scale cutting tests (for an average of two tests results), and while the general trend is intuitive, the quantitative results and conclusions drawn need to be confirmed with repeated tests.

CONCLUSIONS

Full scale cutting tests of a concrete sample with a new, moderately worn, and worn conical pick are used to generate dust for analysis. Characterisation of the respirable airborne dust from all three wear conditions resulted in obtaining and comparing data for concentration, mineral presence, particle shape, and particle size distribution. Additionally, characterisation of the fines material generated from all three wear conditions resulted in analysing particle shape and particle size distribution.

The worn pick generated the most overall respirable dust and all three picks generated dust containing quartz with traces of cristobalite. In general, all the picks generated respirable and fine particles with statistically similar particle shapes in terms of roundness, aspect ratio, and roughness measures. The particle size distributions have statistical evidence that the respirable particles generated from all the picks are also similar in size. However, as the pick wear increases, the picks generate fines material particles that slightly increase in size.

All picks generate dust containing quartz minerals and the concentration of dust increases with the increase in pick wear. Therefore, when drilling concrete with conical picks, appropriate engineering dust mitigation measures should be considered to protect workers with all levels of pick wear. Additionally, dust suppression measures will possibly need to be increased as drilling continues with the pick wear increasing.

Confident recommendations on bit management and dust suppression protocol cannot yet be drawn from the results obtained in these experiments. Future tests need to be performed on other material to confirm these preliminary trends and comparisons drawn from the results. Additionally, further understanding on the effects of particle shapes and sizes on the human respiratory system is necessary because until these complex interactions are well accepted, it is still unclear which pick generates the most toxic dust to miners. With further full scale testing and advances in particulate to lung interactions, the dust characterisations obtained from these experiments will eventually aid in engineering dust mitigation measures to protect miners and limit the number of respiratory diseases in the underground working environment.

ACKNOWLEDGEMENTS

The research leading to these results received funding from NIOSH/CDC under contract Agreement 75D30119C05413 (Improving Health and Safety of Mining Operations Through Development of the Smart Bit Concept for Automation of Mechanical Rock Excavation Units and Dust Mitigation).

All authors certify that they have no affiliations with or involvement in any organisation or entity with any financial interest or non-financial interest in the subject matter or materials discussed in this manuscript.

On behalf of all authors, the corresponding author states that there is no conflict of interest.

Authors would also like to acknowledge various individuals for their guidance and contributions, which includes Muthu Vinayak Thyagarajan, Brent Duncan, Bruce Yoshioka, Joe Chen, Susann Stolze, Michelle Reiher, Brian Asbury, and Tom Lillis.

REFERENCES

Bilgin, N, Demircin, M A, Copur, H, Balci, C, Tuncdemir, H and Akcin, N, 2006. Dominant rock properties affecting the performance of conical picks and the comparison of some experimental and theoretical results, *International Journal of Rock Mechanics and Mining Sciences*, 43(1):139–156, https://doi.org/10.1016/j.ijrmms.2005.04.009

Blackley, D J, Crum, J B, Halldin, C N, Storey, E and Laney, A S, 2016. Resurgence of progressive massive fibrosis in coal miners — Eastern Kentucky, 2016, *Morbidity and Mortality Weekly Report*, 65(49):1385–1389, https://doi.org/10.15585/mmwr.mm6549a1

Bogen, K T, Brorby, G, Berman, D W, Sheehan, P and Floyd, M, 2011. Measuring Mixed Cellulose Ester (MCE) Filter Mass Under Variable Humidity Conditions, *The Annals of Occupational Hygiene,* 55(5):485–494, https://doi.org/10.1093/annhyg/mer003

Centers of Disease Control and Prevention, 1974. *Recommendations for an occupational exposure standard for crystalline silica,* pp 1–23, https://www.cdc.gov/niosh/pdfs/75-120a.pdf

Chen, C C and Huang, S H, 1999. Shift of aerosol penetration in respirable cyclone samplers, *American Industrial Hygiene Association Journal*, 60(6):720–729, https://doi.org/10.1080/00028899908984494

Chopard, A, Marion, P, Royer, J J, Taza, R, Bouzahzah, H and Benzaazoua, M, 2019. Automated sulfides quantification by multispectral optical microscopy, *Minerals Engineering*, 131(October 2018):38–50, https://doi.org/10.1016/j.mineng.2018.11.005

Department of Health and Human Services, 2011. Coal Mine Dust Exposures and Associated Health Outcomes: A Review of Information Published Since 1995. *Current Intelligence Bulletin*, 64.

Dogruoz, C and Bolukbasi, N, 2014. Effect of cutting tool blunting on the performances of various mechanical excavators used in low- and medium-strength rocks, *Bulletin of Engineering Geology and the Environment*, 73(3):781–789, https://doi.org/10.1007/s10064-013-0551-y

Dogruoz, C, Bolukbasi, N, Rostami, J and Acar, C, 2016. An Experimental Study of Cutting Performances of Worn Picks, *Rock Mechanics and Rock Engineering*, 49(1):213–224, https://doi.org/10.1007/s00603-015-0734-x

Doney, B C, Blackley, D, Hale, J M, Halldin, C, Kurth, L, Syamlal, G and Laney, A S, 2019. Respirable coal mine dust in underground mines, United States, 1982-2017, *American Journal of Industrial Medicine*, 62(6):478–485, https://doi.org/10.1002/ajim.22974

Evans, I, 1965. The force required to cut coal with blunt wedges, *International Journal of Rock Mechanics and Mining Sciences And*, 2(1):1–12, https://doi.org/10.1016/0148-9062(65)90018-5

Fan, L and Liu, S, 2021. Respirable nano-particulate generations and their pathogenesis in mining workplaces: a review, *International Journal of Coal Science and Technology*, 8(2):179–198, https://doi.org/10.1007/s40789-021-00412-w

Fowell, R J and Ochei, N N, 1984. A comparison of dust make and energy requirements for rock cutting tools, *International Journal of Mining Engineering*, 2(1):73–83, https://doi.org/10.1007/BF00880859

Hall, N B, Blackley, D J, Halldin, C N and Laney, A S, 2019. Current Review of Pneumoconiosis Among US Coal Miners, *Current Environmental Health Reports*, 6(3):1–11, https://doi.org/10.1007/s40572-019-00237-5.Current

Hanson, B D and Roepke, W W, 1979. Effect of Symmetric Bit Wear and Attack Angle on Airborne Respirable Dust and Energy Consumption, *Report of Investigations - United States, Bureau of Mines, 8395.*

Hurt, K G and Evans, I, 1981. Point Attack Tools: an Evaluation of Function and Use for Rock Cutting, *Mining Engineer*, 140(234):673–683.

Johann-Essex, V, Keles, C and Sarver, E, 2017. A computer-controlled SEM-EDX routine for characterizing respirable coal mine dust, *Minerals*, 7(1):14–16, https://doi.org/10.3390/min7010015

Labranche, N, Keles, C, Sarver, E, Johnstone, K and Cliff, D, 2021. Characterization of particulates from Australian underground coal mines, *Minerals*, 11(5):1–10, https://doi.org/10.3390/min11050447

Lane, G R, Martin, C and Pirard, E, 2008. Techniques and applications for predictive metallurgy and ore characterization using optical image analysis, *Minerals Engineering*, 21(7):568–577, https://doi.org/10.1016/j.mineng.2007.11.009

Lee, K, Lee, E J, Lee, S Y, In, K H, Kim, H K and Kang, M S, 2014. Resurgence of a Debilitating and Entirely Preventable Respiratory Disease among Working Coal Miners, *American Journal of Respiratory and Critical Care Medicine*, 190(6):708–709.

Lowry, P L and Tillery, M I, 1979. Filter Weight Stability Evaluation, *United States,* https://doi.org/10.2172/5794614

Marple, V A and Rubow, K L, 1984. Respirable dust measurement, *Research Report,* Minnesota University Particle Lab.

MSHA, 1977. *Federal Mine Safety & Health Act of 1977 (Mine Act).*

MSHA, 2014. Lowering Miners' Exposure to Respirable Coal Mine Dust, Including Continuous Personal Dust Monitors, Federal Register, *Research in International Economics by Federal Agencies*, 79(84):24814–24994, https://doi.org/10.7312/schw92626-009

NIOSH, 1998. NIOSH manual of analytical methods (NMAM), In *NIOSH Manual of Analytical Methods, 4th Edition,* (3):1–6.

Plis, M, Wingquist, C and Roepke, W, 1988. Preliminary Evaluation of the Relationship of Bit Wear to Cutting Distance, Forces and Dust Using Selected Commercial and Experimental Coal- and Rock-Cutting Tools, *US Bureau of Mines: Report of Investigations, 9193.*

Pope III, C A, Burnett, R T, Thun, M J, Calle, E E, Krewski, D and Thurston, G D, 2002. Lung Cancer, Cardiopulmonary Mortality and Long-term Exposure to Fine Particulate Air Pollution, *The Journal of the American Medical Association*, 287(9):1132–1141, http://jama.jamanetwork.com/article.aspx?doi=10.1001/jama.287.9.1132

Qayyum, R A, 2003. *Effects of bit geometry in multiple bit-rock interaction*, PhD thesis, West Virginia University, USA.

Roepke, W, Lindroth, D P and Myren, T A, 1976. Reduction of dust and energy during coal cutting using point-attack bits, *Bureau of Mines, 8185*.

Roepke, W W and Hanson, B D, 1983. Effect of asymmetric wear of point attack bits on coal-cutting parameters and primary dust generation, *US Department of the Interior, Bureau of Mines, 8761*.

Ross, M H and Murray, J, 2004. Occupational respiratory disease in mining, *Occupational Medicine*, 54(5):304–310, https://doi.org/10.1093/occmed/kqh073

Roxborough, F F, King, P and Pedroncelli, E J, 1981. Tests on the Cutting Performance of a Continuous Miner, *Journal of The South African Institute of Mining and Metallurgy*, 81(1):9–25, https://doi.org/10.1016/0148-9062(81)90237-0

Sarver, E, Keles, C and Rezaee, M, 2019. Characteristics of respirable dust in eight Appalachian coal mines: A dataset including particle size and mineralogy distributions and metal and trace element mass concentrations, *Data in Brief*, 25:104032, https://doi.org/10.1016/j.dib.2019.104032

Sellaro, R M, 2014. Development and Demonstration of a Standard Methodology for Respirable Coal Mine Dust Characterization Using Sem-Edx, MSc thesis, Virginia Polytechnic Institute and State University.

Sellaro, R, Sarver, E and Baxter, D, 2015. A Standard Characterization Methodology for Respirable Coal Mine Dust Using SEM-EDX, *Resources*, 4(4):939–957, https://doi.org/10.3390/resources4040939

Shao, W, 2016. A study of rock cutting with point attack picks, The University of Queensland, School of Mechanical and Mining Engineering, Australia.

Slouka, S, Rostami, J and Brune, J, 2021. Characterization of respirable dust samples generated from picks at differing stages of wear, *Mine Ventilation*, pp 198–207, https://doi.org/10.1201/9781003188476-20

Soo, J C, Monaghan, K, Lee, T, Kashon, M and Harper, M, 2016. Air sampling filtration media: Collection efficiency for respirable size-selective sampling, *Aerosol Science and Technology*, 50(1):76–87, https://doi.org/10.1080/02786826.2015.1128525

Stansbury, R C, 2018. Progressive massive fibrosis and coal mine dust lung disease: The continued resurgence of a preventable disease, *Annals of the American Thoracic Society*, 15(12):1394–1396, https://doi.org/10.1513/AnnalsATS.201809-598ED

Su, O and Akkaş, M, 2020. Assessment of pick wear based on the field performance of two transverse type roadheaders: a case study from Amasra coalfield, *Bulletin of Engineering Geology and the Environment*, 79(5):2499–2512, https://doi.org/10.1007/s10064-019-01712-x

Talbo, H and Sevigné, C, 1986. *Longwall mining system U.S. Patent No. 4,382,633.*

Vaughan, N P, Milligan, B D and Ogden, T L, 1989. Filter weighing reproducibility and the gravimetric detection limit, *Annals of Occupational Hygiene*, 33(3):331–337, https://doi.org/10.1093/annhyg/33.3.331

Vinson, R P, Williams, K L, Schnakenberg, G H J and Jayaraman, N 1984. *The Effect of Water Vapor and Water Droplets on the RAM-1 (Preliminary Results)* (Vol. 8898).

Volkwein, J C and Thimons, E D, 2001. New tools to monitor personal exposure to respirable coal mine dust, *Proceedings of the 7th International Mine Ventilation Congress,* Cracow, Poland, pp 143–150.

Zellnitz, S, Zellnitz, L, Müller, M T, Meindl, C, Schröttner, H and Fröhlich, E 2019. Impact of drug particle shape on permeability and cellular uptake in the lung, *European Journal of Pharmaceutical Sciences*, 139(September):105065, https://doi.org/10.1016/j.ejps.2019.105065

Zhou, W, Wang, H, Wang, D, Zhang, K, Du, Y and Yang, H, 2020. The effect of geometries and cutting parameters of conical pick on the characteristics of dust generation: Experimental investigation and theoretical exploration, *Fuel Processing Technology*, 198(October 2019):106243, https://doi.org/10.1016/j.fuproc.2019.106243

Zosky, G R, Hoy, R F, Silverstone, E J, Brims, F J, Miles, S, Johnson, A R, Gibson, P G and Yates, D H, 2016. Coal workers' pneumoconiosis: An Australian perspective, *Medical Journal of Australia*, 204(11):414-418.e2, https://doi.org/10.5694/mja16.00357

Active explosion barrier conformance to the Australian Recognised Standard 21 – a total mine solution

A Späth[1] and B Belle[2,3]

1. Explospot Systems Australia Pty Ltd, Woodridge Qld 4114.
   Email: arend.spaeth@explospot.com
2. Anglo American Coal Australia, Brisbane City Qld 4000.
   Email: bharath.belle@angloamerican.com
3. School of Minerals and Energy Resources Engineering, University of NSW, Sydney NSW 2000, Australia; University of Queensland, Brisbane Qld 4000, Australia.; Department of Mining Engineering, University of Pretoria, Hatfield 0028, South Africa.

ABSTRACT

Active explosion barriers are used to contain and suppress a methane gas and/or coal dust explosion propagation, including localised mine fires. Early research indicated that these active explosion barriers could be used on a mechanised cutting machine, making it the only technology that could suppress an explosion during its propagation phase at the cutting face. Thereby not only adding protection during mining operations but also reducing the development of toxic gases, fatal temperature increase and fatal overpressure build-up. During their evolution over the last two decades, further advantages of Active explosion barriers were proven in mine operations in countries such as South Africa and China. Characteristics such as effectivity, efficiency, mobility and space restrictions allow these systems to be used in various locations within a mine operation, including highwall operations. This improved technology contribution to safer coal mining was recognised by mine operators, workers and regulators. The Queensland's Recognised Standard 21 (Resources Safety and Health Queensland (RSHQ), 2021) stipulated active explosion barriers as a technology to ensure the safety within their underground coalmines. The mobility and flexible implementation of the active explosion barriers allow a feasible and practical solution for all critical locations for explosion and fire management in an underground coalmine. The conformance to the Recognised Standard can be achieved all underground coal mining operations such as Bord and Pillar, and Longwall mining. This paper describes the conformance, implementation and use of Active explosion barriers for Longwall mining and development as well as Bord and Pillar mining in Queensland's underground coalmines.

INTRODUCTION

The latest version of the Recognised Standard 21 was introduced in December 2021 (Resources Safety and Health Queensland (RSHQ), 2021). The purpose of a recognised standard is that a standard may be made for safety and health, stating ways to achieve an acceptable level of risk to persons arising out of coal mining operations (RSHQ, 2021). The application framework of the Recognised Standard 21 is to achieve an acceptable level of risk in relation to the hazard of methane gas and coal dust explosions (RSHQ, 2021). The standard is a technical guidance in implementing the section 303A Explosion Barriers (Queensland Government, 2021) of the Queensland Coal Mining Safety and Health Regulation (CMSHR) of 2017. According to the Recognised Standard 21 published in December 2021, the section 303A stipulates:

> 1. The underground mine manager for an underground mine must ensure—
>
>> (a) explosion barriers are installed and maintained in the part of each of the following roadways within an ERZ1 in a part of the mine where coal is being extracted—
>>
>>> (i) a return roadway;
>>>
>>> (ii) a single entry drive;
>>>
>>> (iii) a roadway in which a coal conveyor is installed; or
>>
>> (b) explosion barriers that are active barriers, devices or systems are installed and maintained on plant within an ERZ1 in a part of the mine where coal is being

extracted in a way that would prevent the propagation of a coal dust explosion to a place mentioned in paragraph (a).

The mining industry currently recognises two types of explosion barriers. These two types of barriers are called 'active' and 'passive' explosion barriers. The passive explosion barrier is a non-electrical device that requires external, physical forces to activate or trigger the suppression process. In case of coal dust explosions, the preceding pressure wave is considered as the external force that will trigger the suppression sequence. The active explosion barrier is an electromechanical device (or group of devices) that commonly use optical detectors to detect either or both methane gas and coal dust explosions. Due to its ability to detect a methane gas explosion at a very small distance and its compact build, active explosion barriers can be used on other plants close or at the coal cutting face (Späth, Belle and Philips, 2017). The same devices can however also be used when being placed at a far distance from the production face (Späth and Belle, 2020). As active explosion barriers are equally effective at both a close proximity to the ignition source as well as at a far distance from the ignition source, a mine can use the devices at either a location mentioned in section 303A (1)(a) (Queensland Government, 2021) or (b) CMSHR 2017. The placement of the active explosion barrier can therefore be determined by the following factors stipulated in Table 1.

TABLE 1
Guideline to an Active Explosion Barrier placement.

Requirement	Category
Conformance to the Recognised Standard 21	Mandatory
Acceptable Risk	Assessment
Operational Practicality	Application Specific

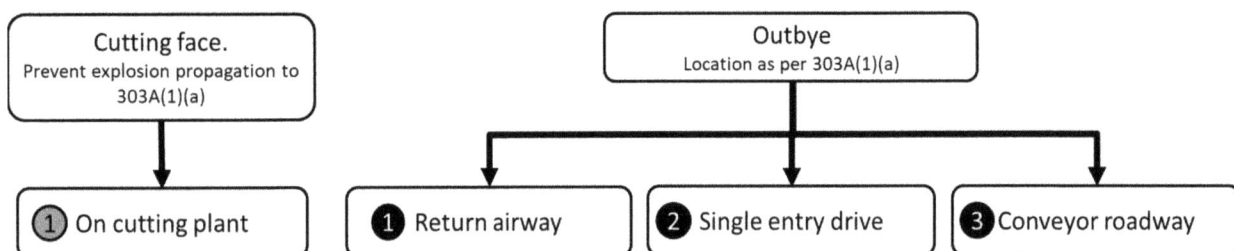

FIG 1 – 303A(1) Explosion Barrier (Queensland Government, 2021); active explosion barrier placement.

CONFORMANCE

The Recognised Standard 21 requires conformance to design parameters and to the validation and testing of an explosion barrier (RSHQ, 2021). While the Recognised Standard 21 sets a general installation bracket for active explosion barriers, the design and installation parameters of an active explosion barrier is limited to the specification of the detectors, required amount of suppression material and placement and spray characteristic of the dispersion nozzles. The sum of these parameters will determine any specific location within the framework set in Recognised Standard 21.

Installation

When considering the use of active explosion barriers, the guidance by a risk assessment and the definition of the 'acceptable risk' is of importance. The Recognised Standard 21 allows two approaches when using active explosion barriers. The first approach being the installation of an active explosion barrier on a plant at the coal cutting face. This approach results in the highest degree of safety to the mine. The reasons for this being, that any ignition developed explosion being suppressed at its ignition source (cutting face) will result in minimal harm to personnel and machinery. The temperature rise, pressure build-up and toxic gas accumulation are prevented when suppressing an explosion during its propagation phase. In case of a longwall this can be reached by integrating an active explosion barrier into eg a shearer. Figure 2 is a schematic of the resulting risk area when installing an active explosion barrier in a shearer.

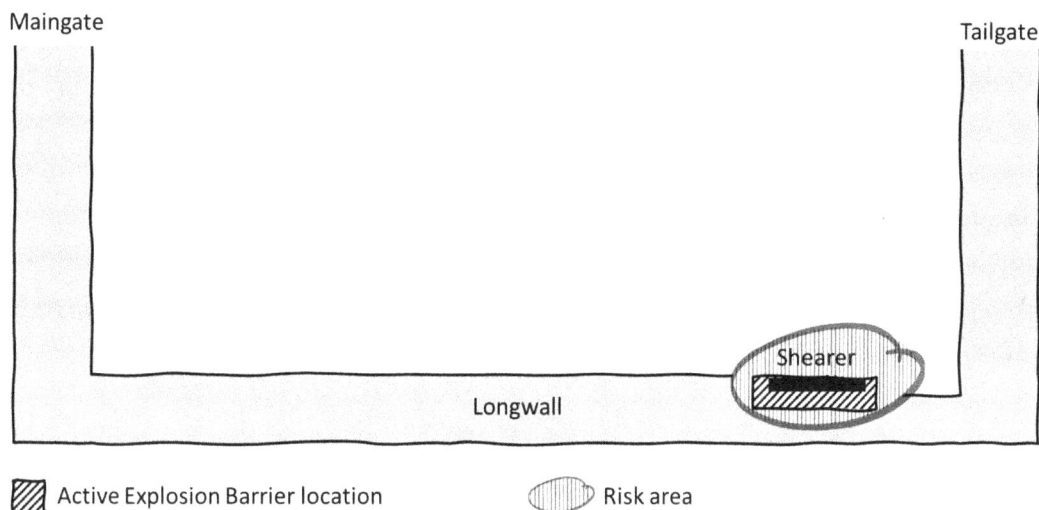

Active Explosion Barrier location   Risk area

FIG 2 – Shearer; Machine mounted active explosion barrier.

Due to the space restrictions and the practical implementation of an active explosion barrier on a longwall shearer has not yet matured and requires additional research or field evaluations. Current retrofit concepts do not fulfill the minimum requirements to ensure an optimal function of the active explosion barrier. One example being the shearer-mounted barriers, that do not provide possibilities to apply suppression material to the outbye sides of both maingate and tailgate cutter drums.

Unlike the shearer-mounted barriers, machine mounted active explosion barriers used in room and pillar and development operations have been proven to function effectively for many years with bord and pillar installations as early as 1999 in South Africa (Späth and Belle, 2020). Figure 3 illustrates the risk area if an active explosion barrier that is installed on a continuous miner or road header in a development section of a mine. Figure 4 and Figure 5 illustrate the active explosion barrier design and implementation into a continuous miner.

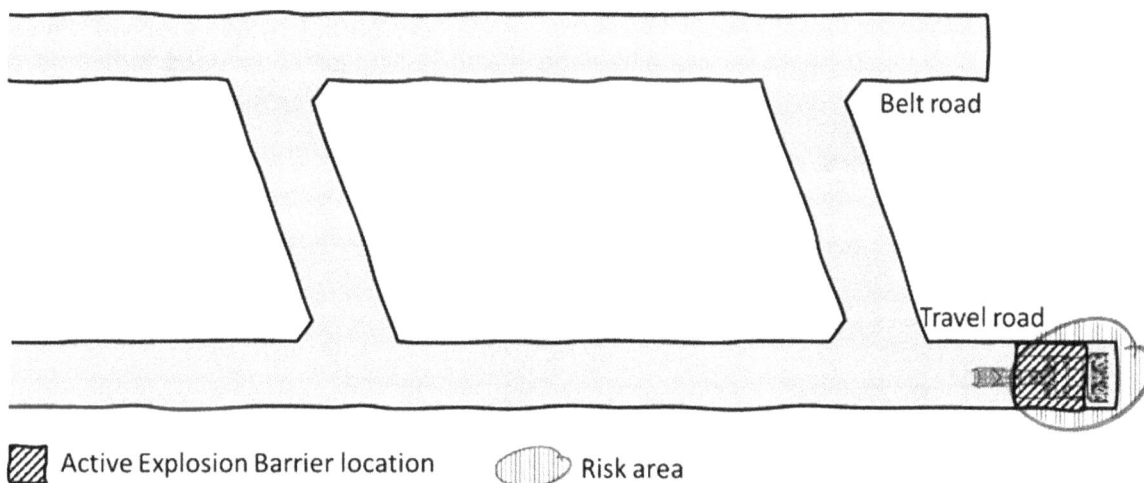

Active Explosion Barrier location   Risk area

FIG 3 – Continuous miner; machine mounted active explosion barrier.

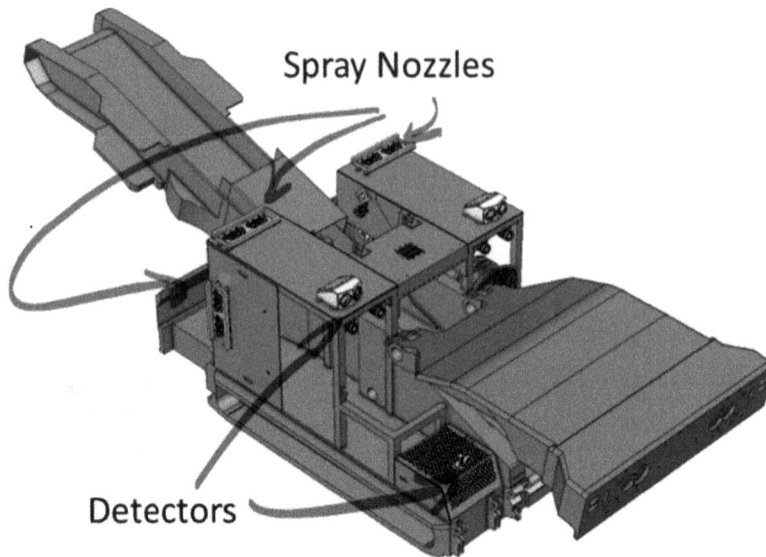

FIG 4 – Continuous miner; Active explosion barrier design.

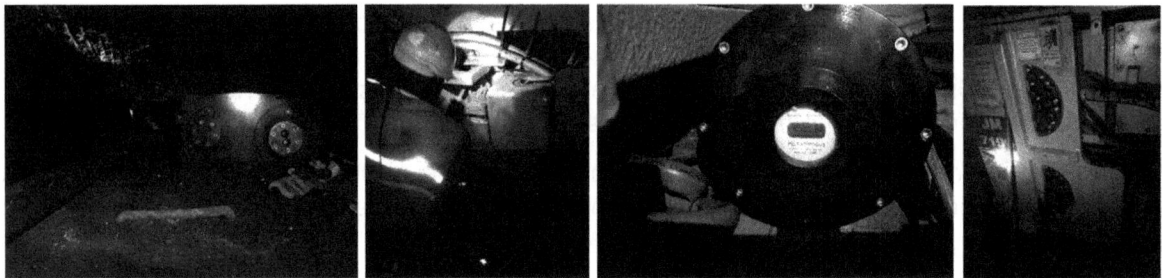

FIG 5 – Continuous miner integrated active explosion barrier: (left to right) Detector, Suppression Canister, System Controller, Spray Nozzles.

A similar risk area to that of a continuous miner-mounted active barrier can be reached if an active explosion barrier is installed on a continuous haulage system. Using the continuous haulage systems as a plant, conforms to the Recognised Standard 21 (RSHQ, 2021) and CMSHR section 303A (1)(b) (Queensland Government, 2021), if the barrier is located on that haulage system when being located adjoining the continuous miner and within the travel road. This active explosion barrier application will prevent an ignition propagating to a place, as required as required in section 303A(1)(b) (Queensland Government, 2021). Figure 6 is a schematic representation of the risk area when having an active explosion barrier installed on a continuous haulage system.

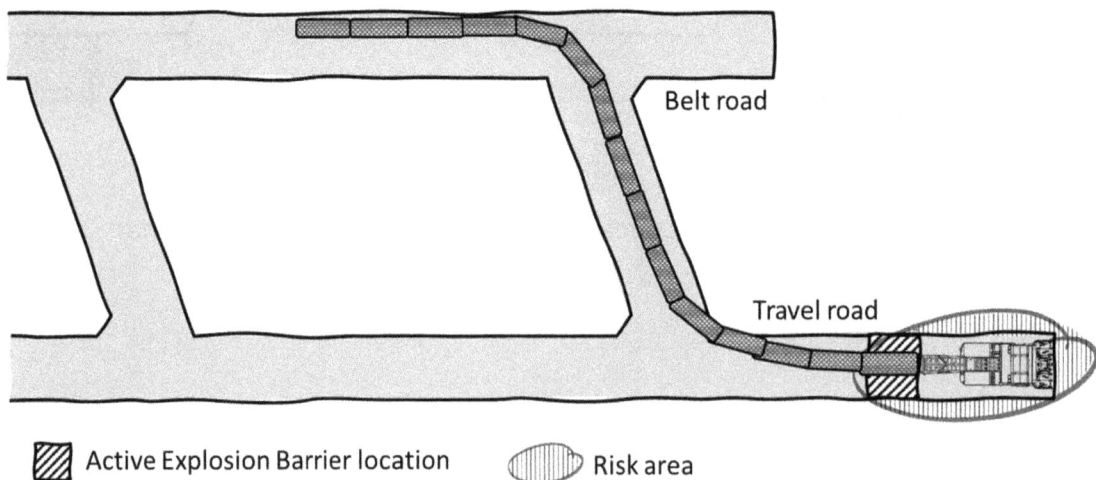

Active Explosion Barrier location   Risk area

FIG 6 – Continuous Haulage System; machine mounted active explosion barrier.

Although a machine mounted active explosion barrier at the cutting face results in the smallest possible risk area in theory, it is not always possible to effectively install an active explosion barrier on any machine. Should it not be possible to place an active explosion barrier on a plant located at the cutting face, then the section 303A (1)(a) Explosion Barriers (Queensland Government, 2021) CMSHR 2017 can be applied. Thereby the active explosion barrier should be installed in an outbye position of the production face. In case of a Longwall mining operation, the Recognised Standard 21 (RSHQ, 2021) requires the active explosion barrier to be located outbye of the production face in both the return airway and entry drive. The active explosion barrier in the single entry drive should be located within 150 m outbye of the production face (area labelled 'A' in Figure 7). Another active explosion barrier should be located in the return airway area labelled 'B' in Figure 7.

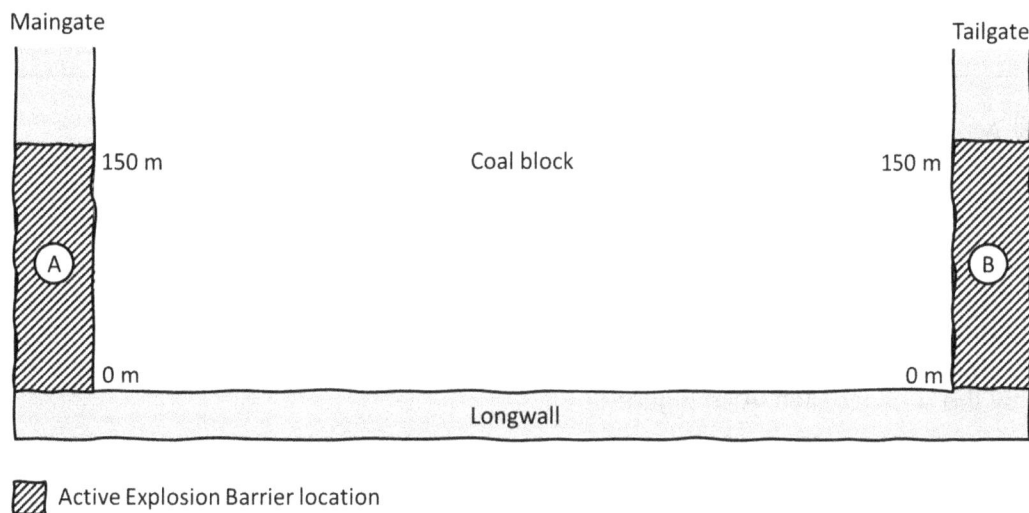

Active Explosion Barrier location

FIG 7 – Longwall outbye active explosion barrier location.

The Beam Stage Loader (BSL) is a plant to integrate the active explosion barrier in the single entry. The total length of a BSL can vary but the active explosion barrier can commonly be installed within 30 m outbye of the production face. By integrating the active explosion barrier into the BSL, the Recognised Standard 21 (RSHQ, 2021)conformance can be ensured at all times. The BSL will, by design, remain within a fixed distance to the production face. The active explosion barrier will therefore remain within the same distance to the production face and the limits of the Recognised Standard 21 at all times.

The return airway of a longwall will under production circumstances not have any fixed plants related to the production. This environment requires a standalone active explosion barrier to contain an explosion in that application. The Recognised Standard requires an active explosion barrier to be located within 150 m outbye of the intersection with the longwall. Current solution of standalone active explosion barriers is mounted on sleds, wheeled, tracked or mounted to a monorail. The solution to be selected for any return airway can be based on the degree of automation in the return airway or use of common equipment within the mine.

Placing an active explosion barrier further outbye of the production face also applies to room and pillar or development operations. The Recognised Standard 21 (RSHQ, 2021) requires the active explosion barrier to be located 150 outbye of the last open cut through of a return airway. Plants that are suited to carry an active explosion barrier within the roadway are feeder breakers or boot ends. The reasons to use these plants are their placement within the roadway, the close proximity to the last open cross through. An active explosion barrier installed on these plants conforms with the Recognised Standard 21 due their placement within 150 m of the last cut through in a return airway and/or conveyor roadway. This location is labelled 'A' in Figure 8. The feeder breaker or boot end will typically be located within the first few metres outbye of the last cut through. As the cutting face progresses, these two plants will follow the cutting face.

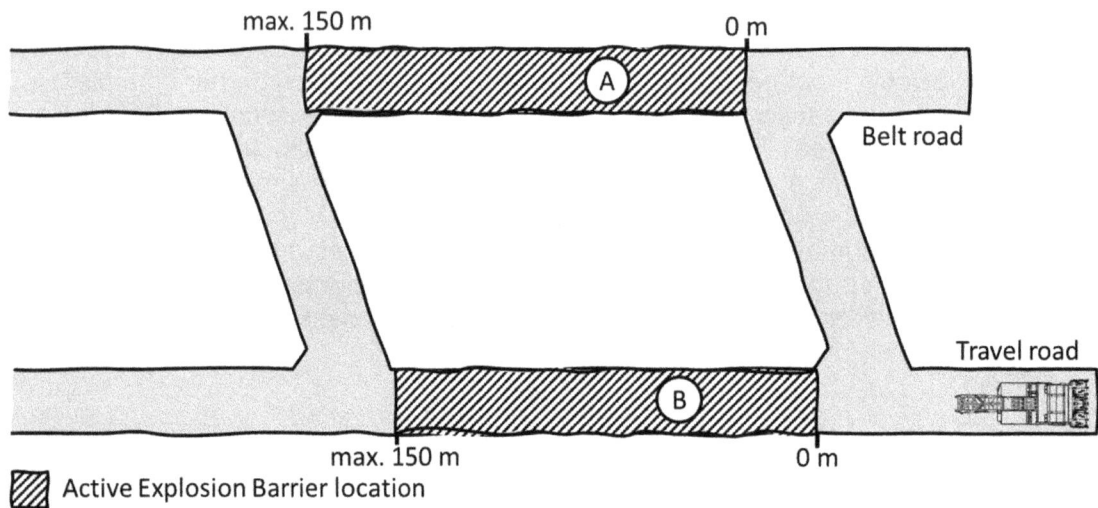

FIG 8 – Development and room and pillar operation; Outbye active explosion barrier placement.

In case the feeder breaker or boot end are not used to install an active explosion barrier along the conveyor roadway, a standalone active explosion barrier can be used. These barriers are designed to be mobile and maintain the appropriate distance to the last cut through. The standalone active explosion barriers are required to maintain within a distance of 150 m of the last cut through as indicated by the area labelled 'A' in Figure 8.

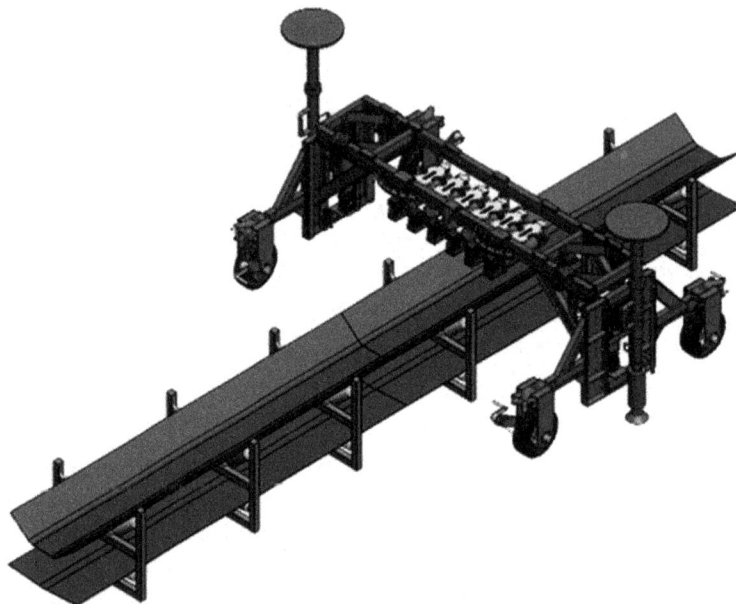

FIG 9 – Standalone; Wheeled active explosion barrier.

If the active barriers are not located at the cutting face, then further active barriers will be required in any single entry drive. The barriers are required to be located within 150 m of the production face (Figure 8 labelled 'B').

Validation and testing

The Recognised Standard 21 (RSHQ, 2021) requires the following system validation to allow conformance:

- Proven method to suppress coal dust explosion.

- Proven system to suppress a coal dust explosion.

- Evidence of testing conducted by an independent, recognised testing facility.

- Design of an explosion barrier must be based on peer-reviewed research.
- Tested at a large-scale explosion facility.

Explosion barriers have been researched for many years independent of passive or active methodology. In 1995 du Plessis, Brandt and Vassard began researching active explosion barriers in South Africa. In 1996, Phillips concludes his report for the Safety in Mines Research Advisory Committee of South Africa on methods to reduce the risk of explosions, in that active explosion barrier could be used to mitigate the risk of explosions (Phillips, 1996). Genc (2000) established in his research done in South Africa that active explosion suppression barriers are capable to mitigate the risk of methane gas explosions. Zou and Panawalage (2001) describes the different barrier principals available in the United States, Germany and UK, noting in his paper that active explosion barriers have '...remarkable advantages over passive barriers...' in regard to both function and usability. These researchers based their recommendations and results on active explosion barriers designed in a way as to utilise a sensor that detects an explosion, a controlling unit that processes the events, the suppression material and a device that effectively disperses the suppression material in the roadway. These systems would stop and suppress a methane gas and/or coal dust explosion by dispersing the material into the approaching explosion. This being done either by using stored pressure (high or low pressure) or by create an instantaneous pressure to disperse the suppression material into the roadway (Zou and Panawalage, 2001). Besides for research purposes, the South African 'Explospot' system is the only active barrier system that could be industrialised and used effectively under operational environments (Späth et al, 2019). The design of the active barrier system being based on the protocol for machine mounted active explosion barriers (du Plessis, 2001) and confirmed by research and monitoring by forums such as the Safety in Mines Research Advisory Committee of South Africa (Moolman, 2005). The proven operational effectiveness of these systems was confirmed in South Africa by the Department of Minerals and Resources (DMR) making them mandatory on mechanised plants used to cut through burnt coal (DMR, 2003). The Explospot active explosion barrier is reported to be continuously implemented in various coalmines on different plants in South Africa since 2000 (Explospot Systems Pty Ltd, 2021). In 2016 further testing was done in a coalmine in South Africa, confirming the effectiveness of standalone active explosion barriers in conveyor roadways (Späth, Belle and Philips, 2017). These tests conducted in an underground operation environment were proved to be effective that the Department of Minerals and Energy (DME) added these systems as part of the updated Guidelines for the Compilation of Mandatory Code of Practice the Prevention of Flammable Gas and Coal Dust Explosion in Collieries (DME, 2020).

Quality assurance

The quality assurance for any safety related system is essential. This is also required by the Recognised Standard 21. Active explosion barriers are electromechanical devices that ensure their function by both electronic and mechanical principals. Due to all components being electronically driven and used in an underground coal mining environment, the active explosion barrier components need to be certified to IEC 60079 and ANZ 60079. The manufacturer's quality management system is required to fulfill the current ISO/IEC 80079–34 standards (ISO, 2018). These standards and certifications are based on ISO 9001 but refer to special requirements needed to fulfill critical safety requirements and detailed component traceability. The active barrier components are inspected individually and the manufacturer regularly audited by a third party. The documentary evidence is published and can be provided on request by the supplier.

The function of an active explosion barrier can be tested generically and documented using high-speed video footage. All parameters of the barrier can be monitored using telemetry data logging. The function and reaction and changed parameters can therefore be documented transparently.

CONCLUSIONS

Active explosion barriers are a proven method of mitigating the risk of methane gas and coal dust explosions and equally to manage localised underground mine fires on conveyor belts and transfer points. These systems have undergone rigorous testing in large scale testing facilities in South Africa, America, China and Europe. These systems have been proven to function and suppress explosions in operational coalmine for many years. The flexibility of these system enables a mine to use the same components in various application within their mine. The active explosion barrier is always placed in the most inbye location within the guideline of the Recognised Standard 21. Where possible even in such a way as to suppress a methane gas explosion prior it propagating to a coal dust explosion. The mobility and flexibility of these systems ensure the highest possible degree of safety. The ability of an active explosion barrier to detect a methane-initiated explosion enables the system also to be used to detect a coal oxidation or friction related coalmine fires on surface, underground or coal processing plants. This is of particular advantage when installing these systems along the roadway or any haulage plant. As these barriers can cope with variable logic, they can also be used to detect fires on conveyors and active applicable suppression sequences for these events.

ACKNOWLEDGEMENTS

The authors would like to acknowledge the significant contribution made by Professor Emeritus Huw Philips to the research related to frictional ignition, methane and coal dust explosions and active explosion barriers and safer mining in general. He was a 'health and safety mining production pioneer'. His dedication during his lifetime to teaching, academics and industrial research made a notable difference in the mining industry worldwide. Professor Huw Philips passed on 26 January 2022. He was a mentor, a colleague and personal friend to both authors and will be dearly missed.

REFERENCES

Department of Minerals and Energy (DME), 2020. Guidelines for the Compilation of Mandatory Code of Practice the Prevention of Flammable Gas and Coal Dust Explosion in Collieries, South Africa.

du Plessis, J J L, Brandt, M P and Vassard, P S, 1995. Assessment of explosion barriers, COL 010, SIMRAC Research Report, South Africa.

du Plessis, J J L, 2001. Evaluation protocol for active machine-mounted explosion suppression systems for continuous miners, CSIR Miningtek Report No. EC01–0149, pp 1–8 [C8].

Explospot Systems Pty Ltd, 2021. Archives, South Africa.

Genc, B, 2000. *Coal mine explosion suppression using active and passive barrier systems*, MSc dissertation, University of the Witwatersrand Johannesburg.

International Organization for Standardization, 2018. ISO/IEC 80079–34:2018, Explosive atmospheres — Part 34: Application of quality management systems for Ex Product manufacture.

Moolman, C J, 2005. Development and testing of on-board systems for single- and double-pass continuous miners in high-seam mining conditions. Final Report. Safety in Mines Research Advisory Committee, South Africa.

Phillips, H R, 1996. Identify methods to reduce the risk of explosions and fires caused by frictional ignition hazards, SIMRAC Report COL 226, pp 1–50 [B47].

Queensland Government, 2021. Coal Mining Safety and Health Regulation (CMSHR) 303A Explosion Barriers, Australia.

Resources Safety and Health Queensland (RSHQ), 2021. Recognised Standards 21, Underground Explosion Barriers.

Späth, A, Belle, B and Phillips, H, 2017. Introducing a new age of highly effective, automatic explosion suppression barriers, in *Proceedings of the Australian Mine Ventilation Conference*, pp 197–202 (The Australasian Institute of Mining and Metallurgy: Melbourne).

Späth, A, Silva, J, Steeves, L and Perera, I E, 2019. Active Explosion Barrier Systems for North American Underground Coal Mines, *North American Mine Ventilation Symposium*, Canada.

Späth, A and Belle, B, 2020. Active Roadway Explosion Barrier evolution in coal mining, in *Proceedings of the Resource Operators Conference*, pp 119–131.

Zou, D H and Panawalage, S, 2001. Passive and Triggered Barriers in Underground Coal Mines – A literature review of recent research, Dalhousie University, Halifax, Nova Scotia.

Ventilation considerations and modelling of lithium battery fires in underground mining

C M Stewart[1]

1. MAusIMM, Director, Minware, Cleveland Qld 4163. Email: craig@minware.com.au

ABSTRACT

Lithium battery electric powered vehicles and machines (BEVs) continue to gain traction in underground mining use, with predictions they will dominate the underground light and heavy vehicle fleets for some mines within the next decade. There are clear advantages for ventilation systems through reduced heat output and the elimination of harmful exhaust gases and diesel particulates. However, the risks and effects of BEV fire emergencies for underground personnel and ventilation systems are still largely in the early stages of analysis and understanding.

This paper conducts an industry review of a selection of lithium battery fire incidents and research. A research paper review on lihthium fire behaviour, intensity and toxicity will be analysed and comparisons made with equivalent diesel machinery fires for context. Guidelines for mine fire modelling parameters will be offered to assist in predicting the behaviour, heat and emissions from BEV fires, and recommendations and conclusions will be drawn on the specific hazards of lithium battery fires compared to conventional diesel powered machines.

INTRODUCTION

Lithium battery powered mobile mining equipment, often referred to as battery electric vehicles (BEVs) have advanced rapidly in recent years and are expected to replace at least a portion of diesel powered internal combustion engine vehicles (ICEVs) in many mines in the future. The benefits to underground mines include reduced heat, potentially reduced airflow requirements, zero noxious gas emissions, lower noise levels and potentially lower operating costs (Sandvik, 2021b). However, the hazards of underground lithium battery fires have yet to be fully evaluated, with toxic emissions, heat release behaviour and emergency response protocols still being assessed, hindered by a lack of real-world examples of actual battery fire emergencies or full-scale underground research.

The heat generated from underground fires can cause considerable disruption to ventilation system airflow and direction, and the spread of toxic or explosive gases and smoke can endanger the lives of any personnel caught in the fire region. Computer fire modelling has progressed rapidly in recent decades with the development of software to model the effects of large fires on underground ventilation systems. Numerous case studies exist of both actual emergencies and research fires to validate and gain an understanding of the behaviour, intensity and gases released from fires in confined mine spaces. In particular, full scale diesel machine fires have been studied intensively by researchers such as Hansen and Ingason (2013) to establish fire behaviour, heat release rates and gas release.

The risk of BEV fires is acknowledged due to the now extensive history of surface consumer EV vehicles and associated fires, as well as a number of notable underground BEV fires and the investigations which have now been published (Figure 1). What has not been fully assessed is whether a BEV fire is more hazardous than an equivalent ICEV diesel machine fire.

FIG 1 – Photo of Sudbury/Onaping BEV fire, Ontario (courtesy Glencore).

To effectively predict outcomes and impacts of fire on mine ventilation systems, modelling of underground fires requires consideration of the following data.

- The products burned and the relative heat of combustions.
- The burning behaviour of the products and associated heat release rates.
- The oxygen consumption and yields of noxious gases.

UNDERGROUND LITHIUM BATTERY FIRES

Canada is acknowledged as an early adopter of underground lithium battery equipment, the technology stimulated by government funded research and subsidies into reduced carbon emissions for underground mines. Several recent reports of underground lithium battery fires have been published, and the list is unfortunately expected to grow as underground use of BEVs becomes more prevalent.

In 2020, a fire occurred at the Onaping Depth BEV fire (Glencore, 2020) an electrical short circuit through unfused shunt modules on the boom lift BEV battery started a fire that mostly destroyed the vehicle. The fire temporarily trapped two technicians behind the vehicle who later managed to escape as mine rescue attempted to suppress the fire. Thick black smoke and burning equipment temperatures of nearly 400°C were observed three hours after the initial report. The fire was declared extinguished more than five hours later.

In 2019, a fire occurred at a charging/parking bay in a Southern Districts Ontario mine (Ontario Mine Rescue, 2021) destroying three adjacent parked BEV transports. The fire started from a faulty charging cable and spread to the adjacent BEVs, consuming and destroying the vehicles and associated lighting and electrical equipment as shown in Figure 2.

FIG 2 – BEV before and after – underground fire southern districts (Ontario Mine Rescue, 2021).

In 2021, at a mine in Sudbury, Ontario, a BEV battery was removed from underground service due to a fault resulting from a failed battery coolant system. When back on the surface, the battery caught fire while being transported by truck to the supplier. The battery is believed to have ignited from a thermal runaway event resulting from initial damage caused by overheating underground.

In 2018, in the Red Lake District of Ontario, a fire started at a tool battery charging bay resulted in 30 batteries being destroyed. The Mine Rescue team attended and extinguished the batteries approximately one hour later. The fire was thought to have started from a faulty charger.

There is insufficient data at this early stage to suggest BEV fires are more common or more dangerous than diesel machine fires. Indeed, statistics from consumer internal combustion engine vehicles (ICEVs) versus electric vehicles (EVs) in collision accidents suggest the opposite (Winton, 2022). No fatalities directly related to lithium battery fires have been recorded to date in underground mines, although any incidents no doubt have potential.

LITHIUM BATTERY TYPES

Lithium ion batteries (LIBs) are available in a variety of chemistries, the more common chemistries in transport applications including iron phosphate (LFP), nickel cobalt manganese (NMC), nickel cobalt aluminium oxide (NCA) and lithium titanate (LTO).

Most consumer road electric vehicles (EVs) currently use higher energy density NCM and NCA chemistries, whereas LFP batteries are proving the current preferred choice for underground BEV's and LTO batteries are becoming more prominent in surface mining electric equipment. Battery technology preference is driven by a combination of factors including cost, specific energy capacity (Figure 3), robustness and safety, speed of recharging and life cycles of charging. The energy capacity contributes to the size and weight of the battery, with lower energy capacity contributing to larger and heavier battery sizes.

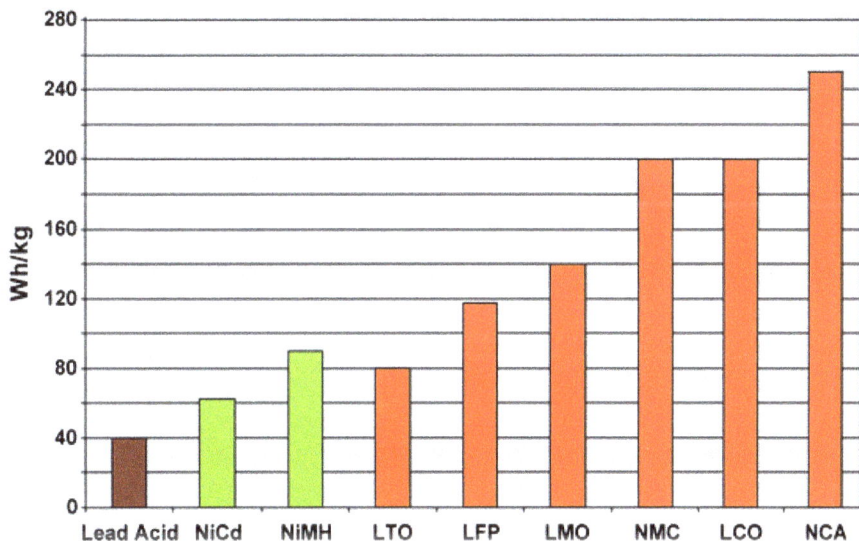

FIG 3 – Specific energy capacity of battery types (Battery University, 2021).

Mining technology has leaned towards batteries with a greater inherent safety design, better thermal stability, lower cost, and improved life cycle charging (ideally LFP and LTO) despite having lower specific energy capacity (therefore requiring larger, heavier batteries).

COMBUSTION OF LITHIUM BATTERIES

In all LIBs, the batteries consist of an anode, cathode, and electrolyte with separators between (Figure 4a). The combustion sources of the batteries can be broken into four categories. The relative masses of these potential combustible sources and other components are shown in Figure 4b.

1. The cathode (for an LFP battery, lithium iron phosphate), although this contributes to a relatively low heat of combustion.

2. The anode, commonly a graphite material or a less combustible substance such as lithium titanate.

3. A variety of types of carbonate electrolyte solvents (Wang *et al*, 2005). Under heat, the electrolyte can evaporate, decompose, and produce flammable gases such as hydrogen, and methane, as well as toxic gases such as hydrogen fluoride.

4. The (typically) plastic case and internal separators. While typically being made of fire retardant material, the products will eventually burn under intense heat.

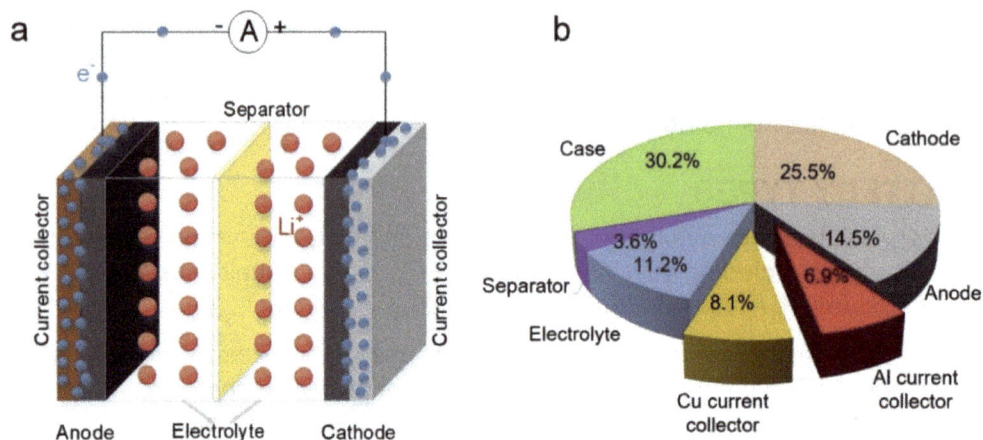

FIG 4 – Mass composition of typical lithium ion battery.

Battery packs are typically assembled with a large number of smaller battery cells in a variety of physical forms, including cylindrical cells, prismatic or pouch packs.

Ignition of lithium batteries can be initiated from several sources:

- External non-electrical fires or heat that spreads into the battery packs.

- Electrical fires from short circuits, bridged fuses, faulty switches etc.

- Internal battery thermal runaway and combustion from cell failure.

Thermal runaway can occure due to the degradation of lithium cells, failing internal conductors or from physical damage causing the uncontrolled release of kinetic (electrical) energy. It can also occur from overheating due to a lack or failure of internal cooling. If sufficient heat is generated (typically >120–150°C) the reaction becomes self-sustaining and a thermal runaway effect can spread to surrounding battery cells resulting in an eventual fire. Due to higher contained kinetic energy, thermal runaway is more likely with batteries that have a high state of charge (SOC).

Over-heating of lithium cells results in the venting of flammable gas and fire that may flare or jet from the battery pack if ignition occurs. Suppression of lithium battery fires is performed by cooling with water or foam. Dry powder and inert gas extinguishers are not generally effective on lithium fires as the thermal runaway event and venting of explosive gases continues even if the flame is suppressed.

Sun *et al* (2020) reviewed a range of laboratory full-scale EV battery fire tests and noted that the total fire heat release of a LIB was around seven times the rated electrical capacity of the battery, with the (SOC) causing only minor variation in total heat released.

Other researchers achieved even higher energy combustion values compared to electrical capacity in laboratory conditions, recording 12× for Chen *et al* (2018) and 14× for Larsson *et al* (2014), the difference likely due to the more complete combustion achieved for small batteries in externally heated (gas-fired) combustion environments. Given most mining machinery is surrounded by other combustible items such as tyres and hydraulic oils, the higher 14× value seems prudent to use until further validation for underground fires can be achieved.

Full scale large battery tests indicated the total heat of combusted products averaged 28 MJ/kg for both 50 per cent and 100 per cent SOC LFP cells (Wang *et al*, 2017). The total amount of battery mass combusted however was only 26 per cent, so it can be argued the total heat released from the entire battery mass is only approximately 7 MJ/kg. Wang also noted that batteries with full SOC exhibited more volatile combustion behaviour with more frequent gas flaring (jetting) and venting.

While a lithium LFP battery fire may only have 15 per cent (7 MJ/kg) of the thermal fire energy per mass unit compared to the equivalent diesel fuel source (47 MJ/kg), the mass of the batteries carried on a BEV is likely to be 5× to 10× more than diesel fuel. The thermal heat released in the event of a battery fire may therefore be similar or greater to a diesel fire depending on manufacturer battery size preferences. The research also suggests that while battery mass is an indicator of potential heat release, experimental data more commonly uses the rated maximum stored electrical energy as a more reliable indicator of heat released when considering different battery chemistries.

The potential total heat release of a LFP battery (assuming a factor of 14× the rated electrical capacity) can therefore be considered in the order of:

$$THR = \frac{14 \times 3600 \times C_b}{1000} \tag{1}$$

Where:

THR   is total heat release MJ

C_b is battery capacity in kWh

For LFP Lithium batteries, the THR for LFP batteries can be approximated to:

$$THR = 50.4 \times C_b \ (For\ LFP\ batteries\ only) \tag{1a}$$

For reference, an equivalent diesel fuel tank heat release can be calculated by:

$$THR = H_c \times \rho_d \times C_d \tag{2}$$

Where:

H_c is the heat of combustion of diesel fuel in MJ/kg (= 47MJ/kg typically)

C_d is capacity diesel fuel in litres

ρ_d is density of diesel fuel in kg/litre (= 0.84 kg/L typically)

For a diesel fuel tank of a given capacity at nominal fuel density, this can be approximated to:

$$THR = 39.5 \times C_d \ (for\ diesel\ fuel) \tag{2a}$$

Calculating combustion heat output for BEV versus ICEV (an example)

The primary battery module of a Sandvik LH518B Loader (353 kWh capacity) is reported to weigh 9920 kg (Sandvik, 2021b). The module weight however cannot be assumed to be the battery weight as a significant portion is contributed from the heavily protected battery compartment, tray frames, counterweights, and support infrastructure. The actual LFP battery mass is not published but could be expected to be in the range of 3900 kg assuming a modest energy density conversion of 11 kg/kWh (Long, 2019). The arrangement of the battery module is shown in Figure 5.

FIG 5 – The detachable battery module of an LH518B loader (Sandvik, 2021a).

For the LH518B Loader, the battery thermal energy of combustion can be estimated using Equation 1a:

$$THC = 50.4 \times 353 = 17{,}791 \; MJ$$

The thermal energy of combustion THC of an equivalent capacity diesel Sandvik TH517i Loader carrying a 100 per cent fuel tank load of 580L (Sandvik, 2021c) can be estimated using Equation 2a:

$$THC = 39.5 \times 580 = 22{,}910 \; MJ$$

The combustion heat output of either machine's fuel source is therefore similar (disregarding other combustible sources such as tyres and hydraulics). However, if (for example) the fuel level and electrical (SOC) were reduced by 50 per cent, the BEV loader would have a greater thermal combustion output than the diesel loader (a battery combustion heat output is similar for most SOCs).

It can be concluded the total heat output of equivalent BEV powered machines (assuming other combustible components such as tyres and hydraulics are equivalent) is similar or even higher in some circumstances.

PRODUCTS OF COMBUSTION

Products of LIB combustion are not dissimilar to a diesel fire and primarily include carbon dioxide and carbon monoxide. However, several additional dangerous gases are also released including hydrogen, methane and hydrocarbon volatiles (which can flare or explode if allowed to accumulate) as well as hydrogen fluoride and phosphate gases, which are extremely toxic to humans.

There are several measures of considering gases released in fires. The most convenient for fire modelling is the *yield* rate of gas, specified as kilograms gas released per kilogram of fuel mass burned (kg/kg). Hydrocarbon fires such as diesel for example release carbon dioxide (3 kg/kg fuel) and carbon monoxide (0.01 to 0.2 kg/kg fuel, depending on the equivalence ratio of fuel to oxygen).

Laboratory testing of lithium fires (Larsson *et al*, 2017) suggests hydrogen fluoride gas is yielded at the rate of 20 g to 200 g gas per kWh of lithium battery. A level of 30 ppm of HF is considered immediately dangerous to life or health (IDLH) and the lethal 10 min toxicity value (AEGL-3) is 170 ppm. For comparison, carbon monoxide has a 1200 ppm IDLH and 1700 ppm AEGL-3 (NRC, 2003). The toxicity of HF gas from a lithium battery fire may therefore be of an order of magnitude or more than the equivalent carbon monoxide released.

Full scale and experimental laboratory results from various sources (Larsson *et al*, 2017; Wang *et al*, 2017; Quintiere *et al*, 2016) reported a large and variable range of gas yield values, making fire gas production estimation difficult. A broad approximation of experimental values for LFP batteries has been converted to a yield value against an equivalent battery mass and electrical energy capacity in Table 1.

TABLE 1

Typical gas yields from LFP batteries.

Gas types	Mass yield LFP		Energy yield LFP	
	Low kg/kg	High kg/kg	Low kg/kWh	High kg/kWh
O_2	-0.3975	-0.3375	-3.8813	-4.5713
CO_2	0.4500	0.5300	5.1750	6.0950
CO	0.0038	0.0133	0.0431	0.1524
NO	0.0021	0.0021	0.0242	0.0242
SO_2	0.0023	0.0023	0.0265	0.0265
HCL	0.0003	0.0003	0.0035	0.0035
HF	0.0017	0.0174	0.0200	0.2000

An interesting observation for carbon dioxide and carbon monoxide is that the yield of CO increases with the battery state of charge (SOC), ranging from a 1:120 part CO/CO_2 ratio for a discharged battery to a 1:40 ratio for a fully charged battery. Another noteworthy observation is that toxic hydrogen fluoride emissions *increase* with *decreasing* SOC, an opposite trend to carbon monoxide.

A theoretical example of gas toxicity for a BEV fire

At an assumed HF yield between the minimum and maximum values in Table 1 for a 354 kWh Sandvik BH550b truck battery (Sandvik, 2021b), the release of toxic gases from the LIB would be:

$$Best\ Case\ Hydrogen\ Fluoride\ Mass\ Release\ BEV\ =\ 354\ \times 0.02 = 7\ kg$$

$$Worst\ Case\ Hydrogen\ Fluoride\ Mass\ Release\ BEV\ =\ 354\ \times 0.20 = 71\ kg$$

$$Carbon\ Monoxide\ Mass\ Battery\ =\ 354\ \times 0.152 = 54\ kg$$

To compare the toxicity of diesel fire carbon monoxide and lithium battery fire hydrogen fluoride, a worst-case diesel fuel-rich yield of 0.2 kg/kg (Pitts, 1994) of a fully fuelled Sandvik TH551i loader with 840 L diesel produces the following. The tyre mass is assumed common in both cases:

$$Carbon\ Monoxide\ Mass\ Diesel\ =\ 840\ \times 0.84\ \times 0.2 = 141\ kg$$

$$Carbon\ Monoxide\ Mass\ Tyres\ =\ 3800\ \times 0.2 = 760\ kg$$

The IDLH level of Hydrogen Fluoride is 30 ppm (NRC, 2003) and carbon Monoxide is 1200 ppm, indicating hydrogen fluoride is 40× more toxic per unit concentration than the equivalent carbon monoxide. For a full BEV fire combined with burning tyres the carbon monoxide released is:

760 kg tyres + 54 kg battery = 814 kg total CO, equivalent in toxicity to 20.4 kg of HF

When compared to best and worst-case HF releases, it can therefore be argued that HF production for a full BEV fire including tyres could have the potential to be anywhere from 35 per cent to 350 per cent as toxic as the carbon monoxide released during the same fire. Full scale consumer EV experiments by Sun *et al* (2020) measuring HF gas released support the upper limit values of 350 per cent.

If the battery fire burned alone (for example in a charging bay) without other hydrocarbon combustibles, the HF gas may be between 4× to 40× more toxic than the carbon monoxide released from a battery-only fire.

HEAT RELEASE RATE

The final consideration for modelling underground fires is the heat release rate, or the rate at which energy is released into the surrounding environment. Large scale testing of BEV fires demonstrates typical time squared (T-Square) curve behaviour, where the fire initially grows rapidly, before peaking and then decaying over the remaining period as shown in Figure 6. The experimental work by Larsson *et al* (2017) on smaller scale battery packs shows a clear correlation of increased HRR (more rapid burning) with battery SOC above 50 per cent while lower SOC's resulted in lower and longer HRR curves. Higher SOC's also appear to increase short-term HF production although the total HF production over time is lower.

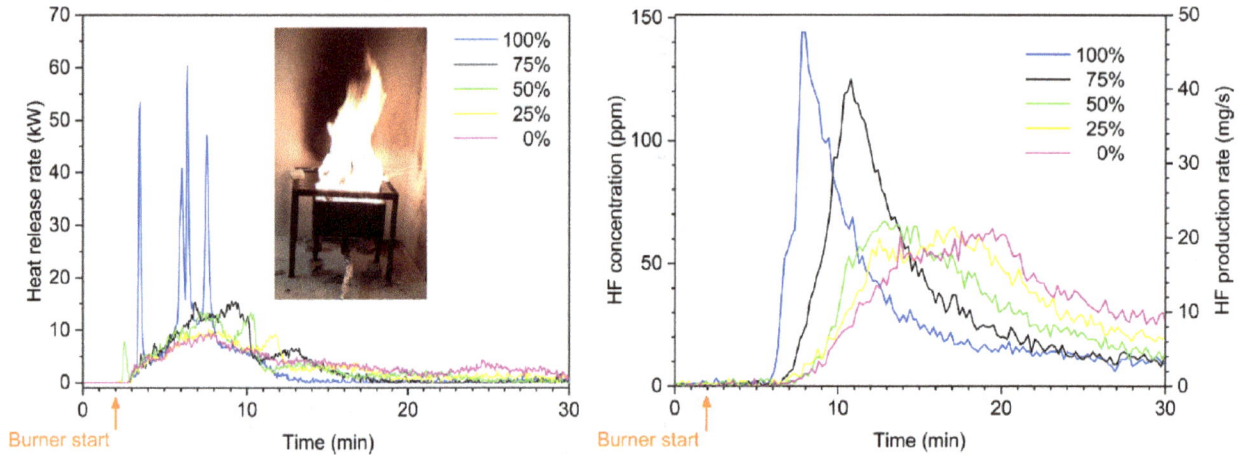

FIG 6 – HRR and HF production of lithium battery fires (Larsson *et al*, 2017).

The battery of a BEV fire is likely to be only part of the whole fire combustion output, with hydraulics, tyres and plastics on BEVs contributing to the remaining bulk of the fire. When a battery fire is overlaid with a hydrocarbon fire on a BEV, full scale vehicle tests on surface consumer vehicles by Sun *et al* (2020) showed very little difference in total burn time or combustion heat output compared with internal combustion vehicles. This assumption is supported by full scale experiments observed on consumer grade electric vehicles (EVs), plug in hybrids (PHEVs) and internal combustion engine vehicles (ICEs) by Sun *et al* (2020) as shown in Figure 7. The only minor difference observed is that diesel pool fires developed from ruptured fuel tanks may create shorter, higher peaks in initial HRR compared to a slightly slower and lower HRR from LIB fires.

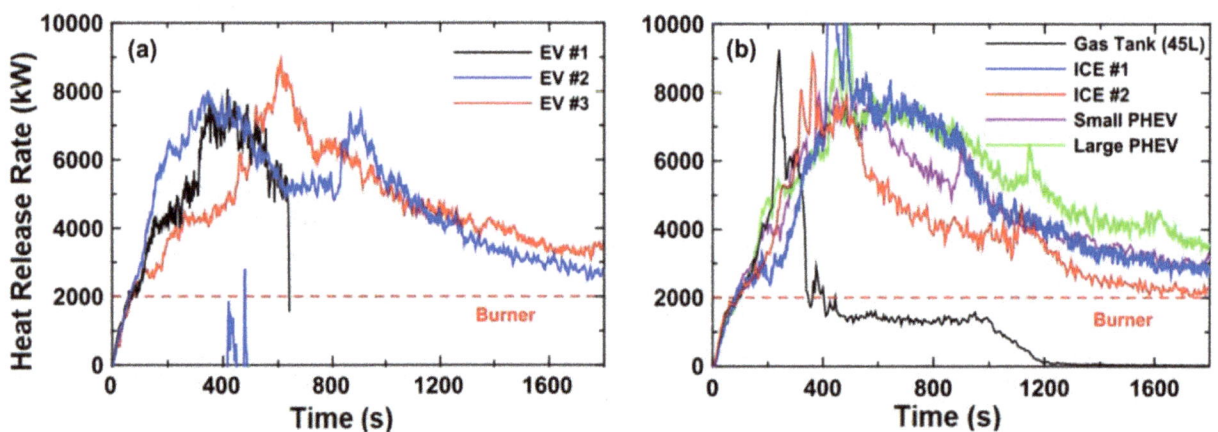

FIG 7 – Comparison of HRR for EVs and PHEVs/ICEs (Sun *et al*, 2020).

MODELLING OF BEV FIRES

The simulation of underground BEV fires is not necessarily more complex than for diesel fires. The diesel portion of the fire combustible mass can simply be replaced with the battery portion using appropriate heat and gas yield factors derived from the mass (kg) or energy capacity (kWh) of the battery. Caution should be taken if using the quoted manufacturer battery module mass, as the figure may be substantially larger than the true battery mass due to the included module casing. Using battery capacity yield factors instead of weight-based factors may produce more reliable results.

For full BEV equipment fires, a simplified HRR curve similar in heat output and combustion time to a diesel machine can be used. While a complex HRR curve representing both the BEV battery and other components could be constructed, there is not enough experimental data on underground BEVs to support any assumptions at this time. Total burn time is likely to be consistent with diesel machines as the tyres are still likely the longest burning items. Both HF and CO gas are likely to be released in dangerous concentrations, depending on fire size.

Where BEV batteries are modelled individually (for example in a charging bay environment), the total burn time should reflect the size and capacity of the battery, growing linearly with the kWh rating. The research for large lithium battery fires (up to 100 kWh) did not observe any fires taking longer than one hour to reach maximum HRR although it is conceivable larger underground batteries could take longer due to the larger battery packs used. Special consideration should be given to the production of hydrogen fluoride gas, which is likely to be of an order of magnitude more toxic than the carbon monoxide released in a pure battery fire scenario.

CONCLUSIONS

Underground BEV fires are likely to have similar heat release rates (HRR), combustion times and total heat of combustion (THC) to equally sized diesel machines if the whole machine is engulfed. While the release of carbon monoxide may be slightly less, the release of hydrogen fluoride gas may match or exceed the toxicity of carbon monoxide by several times and should be considered in any entrapment or rescue situation. CO scrubber-type rescuers (not oxygen generating re-breather types) or CO filters on refuge stations are unlikely to be effective against HF. The levels of soot (smoke) released and subsequent loss of visibility are also likely to be slightly greater for a BEV fire.

Modelling of BEV fires can be performed using similar assumptions to diesel machines, but with lithium battery yield assumptions used instead (summarised in this paper for heat and gas released). No significant changes are considered necessary for total burn times and HRR, and as either hydrogen fluoride or carbon monoxide released could be equally as toxic, only one of the toxic gases may need to be modelled. The presence of either gas in the simulation should be warning enough of the danger to life or health.

For lithium battery fires only (without BEV involvement), standard T-Squared HRR curves can also be used. While total heat release over time is similar at any battery SOC, higher peak HRR will be reached sooner with batteries with high SOCs. Larger heavy vehicle underground batteries may take one hour or more to reach peak HRR and double this time to decay depending on SOC and chemistry however further experimental study is required to validate this. For a lithium battery fire alone, hydrogen fluoride gas production is likely to be an order of magnitude more toxic than the carbon monoxide produced and should therefore be considered separately during modelling and emergency response.

Other hazards such as jetting, flaring and explosive gas release from high SOC batteries are unpredictable and unlikely to be able to be accurately modelled. These hazards need to be part of a broader emergency risk and management plan for lithium batteries, as does lithium battery fire protection, firefighting, escape and emergency procedures. As with diesel equipment fires, while modelling of lithium battery fires and equipment can be helpful, it only forms part of a complete emergency management plan. Assuming everything has already been done to prevent an LIB fire, ventilation design must accommodate the hazards of fires, particularly to control or isolate personnel from the release and spread of highly toxic hydrogen fluoride gas. For example, a dedicated exhaust ventilation option for battery charging, storage and changeout is highly recommended, as well as more generic strategies such as careful placement of refuge bay and escape ways.

REFERENCES

Battery University, 2021. *BU-205: Types of Lithium-ion*, Available: https://batteryuniversity.com/article/bu-205-types-of-lithium-ion.

Chen, M, Liu, J, Dongxu, O, Cao, S, Wang, Z and Wang, J, 2018. A Simplified Analysis To Predict The Fire Hazard Of Primary Lithium Battery, *Applied Sciences*, 8:2329.

Glencore, 2020. Incident Summary – Onaping Depth BEV Fire, Glencore Internal Document Public Release, https://www.workplacesafetynorth.ca/sites/default/files/uploads/Glencore_INO_BEV-Incident-Summary-Extended-2021-01-20.pdf.

Hansen, R and Ingason, H, 2013. Heat Release Rate Measurements Of Burning Mining Vehicles In An Underground Mine, *Fire Safety Journal*, 61:12.

Quintiere, J G, Crowley, S B, Walters, R N, Lyon, R E and Blake, D, 2016. Fire Hazards of Lithium Batteries, In: *Transportation*, Virginia: National Technical Information Services NTIS.

Larsson, F, Andersson, P, Blomqvist, P and Mellander, B-E, 2017. Toxic Fluoride Gas Emissions From Lithium-Ion Battery Fires, *Scientific reports*, 7:1–13.

Larsson, F, Andersson, P, Blomqvist, P, Lorén, A and Mellander, B-E, 2014. Characteristics of lithium-ion batteries during fire tests, *Journal of Power Sources*, 271:414–420.

Long, T R and Misera, A M, 2019. Sprinkler Protection Guidance for Lithium-Ion Based Energy Storage Systems, *Fire Protection Research Foundation*, Final Report.

National Research Council (NRC), 2003. *Acute Exposure Guideline Levels for Selected Airborne Chemicals*, Volume 3, 8.

Ontario Mine Rescue, 2021. BEV Emergency Response Incident Review and Best Practices, *Virtual Symposium: Battery Electric Vehicle Safety in Mines Jan 2021*, https://www.workplacesafetynorth.ca/sites/default/files/uploads/Ontario-Mine-Rescue-BEV-Emergency-Response-2020-01-20.pdf.

Pitts, W M, 1994. *Global Equivalence Ratio Concept And The Prediction Of Carbon Monoxide Formation In Enclosure Fires* (Nist Monograph 179).

Sandvik, 2021a. Sandvik lh518b Battery Electric Loader, Sandvik Publication.

Sandvik, 2021b. Sandvik th550b Battery Electric Truck, Sandvik Publication.

Sandvik, 2021c. Toro™ lh517i Safer, Stronger, Smarter, Sandvik Publication.

Sun, P, Bisschop, R, Niu, H and Huang, X, 2020. A Review Of Battery Fires In Electric Vehicles, *Fire technology*.

Wang, Q, Huang, P, Ping, P, Du, Y, Li, K and Sun, J, 2017. Combustion Behavior Of Lithium Iron Phosphate Battery Induced By External Heat Radiation, *Journal of Loss Prevention in the Process Industries*, 49:961–969.

Wang, Q, Sun, J and Chu, G, 2005. Lithium Ion Battery Fire And Explosion, *Fire Safety Science*, 8:375–382.

Winton, N, 2022. Electric Car Fire Risks Look Exaggerated, But More Data Required For Definitive Verdict, *Forbes Article*, https://www.forbes.com/sites/neilwinton/2022/03/02/electric-car-fire-risks-look-exaggerated-but-more-data-required-for-definitive-verdict/?sh=1986fe322327

Simulation study on crucial parameters of long-compressive and short-suction ventilation in large section roadway excavation of LongWangGou coalmine

H Wen[1], W Mi[2], S Fan[2], Y Xu[3] and X Cheng[2]

1. School of Safety Science and Engineering, Xi'an University of Science and Technology, Xi'an 710054, China. Email: wenhxust@163.com
2. School of Safety Science and Engineering, Xi'an University of Science and Technology, Xi'an 710054, China.
3. School of Resources and Safety Engineering, Central South University, Changsha 410083, China.

ABSTRACT

LongWangGou coalmine auxiliary inclined shaft heading face has a large section, fast heading speed, and large dust production. The problems of high concentration and wide distribution of dust in the roadway air are very prominent. To improve the efficiency of roadway ventilation and dust removal, long-compression and short-suction (after this referred to as LCSS) are adopted. The dust concentration and airflow field distribution in the roadway are numerically simulated by FLUENT. The change of the position between the pressure duct and the exhaust duct and the different pressure-extraction ratios affect the distribution of airflow velocity, the formation of vortex zone, and the distribution of dust concentration in the roadway. When L_y = 25 m, L_c = 5.0 m, W = 1.2, a wind curtain is formed at 5 m in front of the heading face to prevent dust from spreading deeper into the roadway.

INTRODUCTION

Currently, the comprehensive heading machine is primarily used in mine roadway excavation, which has the advantages of good roadway forming effect and high heading efficiency. But in the process of heading will produce large amounts of dust, dust removal effect is poorer. This problem has been widely concerned by scholars. Based on this, some scholars believe that the air volume and the length of the pressure duct are the key factors affecting the dust distribution in the roadway, and use numerical simulation technology to study this problem, and find that the dust concentration distribution on the heading face is not uniform, especially the dust concentration on the side of the extraction duct is significantly higher than the average (Geng et al, 2018). The dust deposition in the roadway is mainly affected by the airflow pattern and the working state of the conveyor belt. Given this phenomenon, the Euler-Euler method is established on the numerical calculation platform of Gas-solid two-phase flow in the roadway to visually simulate the movement characteristics of dust particle diffusion, settlement, and aggregation under the action of complex airflow, and the temporal and spatial variation of dust distribution in the hybrid ventilation system. The k-ε turbulence model and intermediate grid simulate the dust movement caused by the airflow around the conveyor belt in the coalmine roadway (Stovern et al, 2014; Wang et al, 2015; Torno et al, 2020). Therefore, to find the best ventilation mode, scholars used numerical simulation technology to compare the distribution and diffusion of dust in roadways under various ventilation modes, including long-compressive and short-suction (LCSS). The results show that the long-compression and short suction ventilation mode has a better dust suppression effect. Still, this ventilation mode is easy to form a continuous vortex field in the return airflow, which is not conducive to the effective control of dust concentration (Hua et al, 2018; Kurnia et al, 2014; Shi et al, 2008; Du et al, 2010). According to the current research results, the formation of an air curtain can effectively limit the movement of dust in space (Xiao et al, 2019; Li et al, 2021; Liu et al, 2010), and the proper position of the air duct and the outlet angle of air curtain will affect the formation of air curtain and the effect of dust suppression (Li et al, 2014; Xing et al, 2005; Liu et al, 2009; Tan et al, 2015). In the actual test of the underground coalmine, the effective diffusion coefficient of dust was evaluated by adding tracer gas into the airflow of the roadway. The technology was developed to form a triangular air curtain using a high-speed air field generated by two fan outlets to reduce dust concentrations within the operator range of machinery from 2000 mg·m⁻³ to 300 mg·m⁻³ (Cai et al, 2019, 2020; Chekan et al, 2006; Widiatmojo et al, 2015). In addition, some scholars established the gas-liquid-dust three-phase coupling numerical model according to the specific conditions of underground coalmines, and developed equipment such as external jet dust removal of road header and wet dust collecting network. Through various means and methods, improved the ventilation efficiency, reduced the dust concentration, and limited the spread of dust (Tan *et al*, 2014; Wang *et al*, 2018; Nie *et al*, 2017a, 2017b; Hua *et al*, 2018; Cheng *et al*, 2012; Chen *et al*, 2018; Yu *et al*, 2018; Yin *et al*, 2019; Moloney *et al*, 1997).

Most of the roadways studied by scholars are small and medium sections, and the maximum is not more than 18 m². The research on ventilation and dust removal of roadways with section areas of 20 m² ~ 30 m² is rarely reported. At the same time, in the ventilation and dust removal design of roadway, the empirical equation (Ly ≤ (4–5)S$^{1/2}$, Lc ≤ 1.5S$^{1/2}$) is only applicable to medium and small section roadway. Therefore, when long-compressive and short-suction (LCSS) ventilation is used in large cross-section roadways, how to determine the relative position of the pressure pipe and the extraction pipe, and how to determine the reasonable pressure-extraction ratio are rarely reported. On this basis, this study takes the ventilation of heading face of auxiliary inclined shaft roadway in LongWangGou coalmine as the background, the cross-sectional area of heading face is 25 m², adopts long-compressive and short-suction ventilation, and the air volume is 1500 m³·min^{-1}. By comparing L_y (the distance between the outlet of the pressure pipe and the heading face), L_c (the distance between the inlet of the suction pipe and the heading face) and W (the ratio of the suction pressure of the pressure pipe to the suction pipe), the dust concentration distribution and airflow field change of the horizontal cross-section at Z = 1.5 m (height of breathing zone), horizontal cross-section at Z = 2.5 m (height of driver's position), and the tail of the road header are obtained. By comparing with the actual measurement results, the best combination of ventilation and dust removal in this LCSS ventilation mode is obtained. The actual measurement results show that the simulation results are accurate and reliable. It is hoped to provide a reference for ventilation and dust removal of large section roadway under long-compressive and short-suction (LCSS) ventilation mode.

MODEL ESTABLISHMENT

Mathematical model

Model assumptions

Before the establishment of the model, make the following two assumptions:

1. Consider the wind as an incompressible fluid at a constant temperature.

2. Roadway heading face is the only source of roadway dust, and dust is regarded as spherical particles, without considering the interaction between dust particles.

Therefore, the discrete phase model is selected to study the migration law of dust. The movement state of dust particles in the roadway is affected by the airflow in the roadway, which is shown in Equation 1:

$$\frac{du_p}{dt} = F_D\left(u - u_p\right) + \frac{g_n\left(\rho_p - \rho\right)}{\rho_p} + F_s \tag{1}$$

Where:

u	is the fluid phase velocity, m·s^{-1}
u_p	is the particle velocity, m·s^{-1}
μ	is the fluid dynamic viscosity, pa·s
ρ	is the fluid density, kg·m^{-3}
ρ_p	is the particle density, m·s^{-2}
g_n	is the acceleration of gravity, m·s^{-2}
F_D	is the lifting force of unit mass dust, kg·m^{-3}

Governing equation

The dust particles in this simulation follow the law of conservation of mass and the law of conservation of momentum, as shown in Equations 2 and 3. The mathematical expression of the stress state of dust particles is shown in Equation 4.

$$F_D = \frac{18u}{\rho_p d_p^2} \times \frac{C_D R_e}{24} \tag{2}$$

$$\frac{\partial \rho}{\partial t} + \frac{\partial (\rho u_i)}{\partial x_i} = S_m \tag{3}$$

$$\frac{\partial (p u_i)}{\partial t} + \frac{\partial (p u_i u_j)}{\partial x_j} = -\frac{\partial p}{\partial x_i} - \frac{\partial \tau_{ij}}{\partial x_j} + \rho g_i + F_i \tag{4}$$

Where:

d_p	is the particle diameter, m
C_D	is the drag coefficient
R_e	is the relative Reynolds number
P	is the static pressure
τ_{ij}	is the stress tensor
ρg_i	is the gravitational volume force in the I direction
F_i	is the external volume force

The expression of stress tensor (τ_{ij}) is shown in Equation 5.

$$\tau_{ij} = \left[u \left(\frac{\partial u_i}{\partial x_j} + \frac{\partial u_j}{\partial x_i} \right) \right] - \frac{2}{3} u \frac{\partial ul}{\partial xl} \delta_{ij} \tag{5}$$

Physical model

Model establishment

Establish the model according to the actual size. Among them, the establishment of the coordinate system, the section size of the auxiliary inclined shaft of LongWangGou, and the parameters of the wind duct are shown in Figure 1. This study simulated the dust concentration distribution at different heights in the Z direction under various combinations of L_y from 15 m to 30 m (spacing of 2.5 m) and L_c from 2.5 m to 7.5 m (spacing of 0.5 m or 1 m). The TGrid type is used to divide the roadway grid. The unit is tetrahedral, and locally includes hexahedral, vertebral, and wedge units. Hex/Cooper type is used to separate the wind tube grid, and the unit theme is hexahedral. The number of grids has a significant influence on the calculation results. Therefore, the average dust concentration on the measuring line Y = 15 m (tail of the road header) and Z = 1.5 m (breathing zone) was compared. At the same time, taking into account the time required for calculation, an independence test was conducted for the five grid partitions of 84525, 96427, 119629, 135074, 161743, as shown in Figure 2. Considering the calculation accuracy and calculation time, the 119629 grid method is adopted.

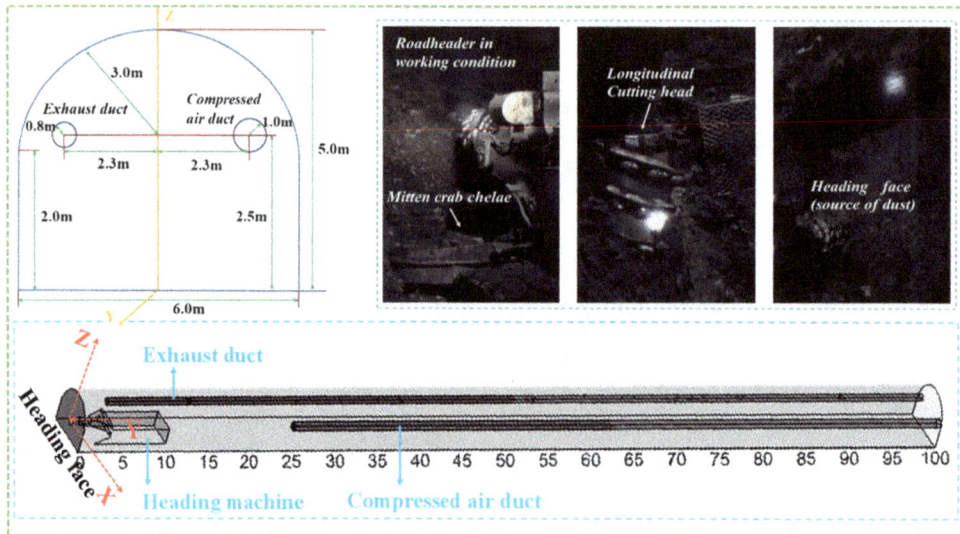

FIG 1 – Three-dimensional models of the roadway under LCSS ventilation mode.

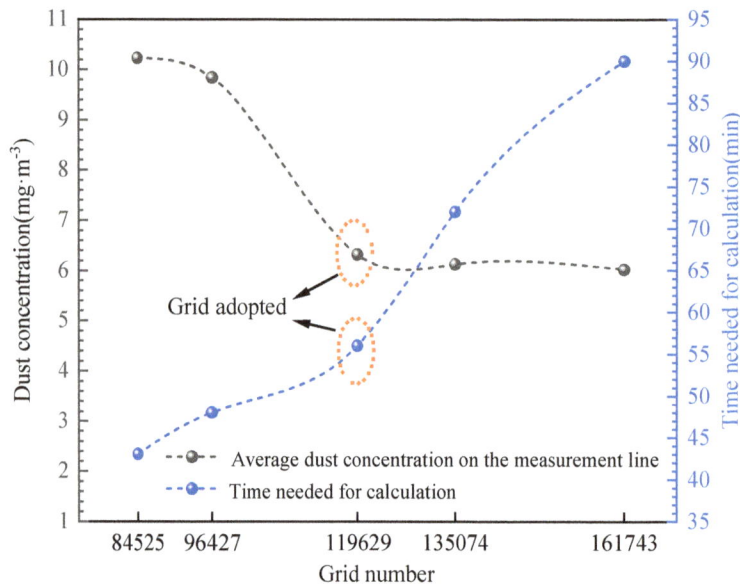

FIG 2 – Grid independent analysis.

Parameter setting

The field measurement shows that the roadway air volume is 1500 m³·min⁻¹, the dust concentration near the heading face is 450 mg·m⁻³, the wind speed is 0.3 m·s⁻¹, and the dust yield of the heading face is 0.003 kg·s⁻¹. There is a lot of turbulence in the roadway airflow field, so the standard turbulence model is adopted. The discrete phase model (DPM) was used, and close the energy equation. Specific parameter settings are shown in Table 1.

TABLE 1

Model parameter setting.

Injection source parameters	Parameter setting	Computer model	Model setting
Type of jet source	Surface jet	Solver	Uncoupled solution method
Mass flow rate (kg·s⁻¹)	0.003	Time	Steady-state
Particle size distribution	Rosin-Rammler	Turbulence model	Standard k – ε model
Minimum particle diameter (m)	3.61×10^{-7}	Discrete phase model	Open
Maximum particle diameter (m)	2.478×10^{-4}	Energy equation	Close
Median diameter (m)	4×10^{-5}	Phase to phase coupling frequency/Step-1	10
Distribution index	1.34	Time step (s)	0.01
Random orbit model	Open	Calculation steps	10 000
		Resistance characteristics	Spherical particles

DETERMINATION OF CRUCIAL VENTILATION PARAMETERS

For the distribution and migration law of dust concentration in the roadway under LCSS ventilation mode, it is discussed from three crucial parameters: the determination of the best position of the compressed air duct, the determination of the best position of the exhaust air duct, and the pressure-extraction ratio (expressed by *W* in this paper). The appropriate parameter values are selected by comparing of ventilation and dust removal effects.

Determine the optimal value of L_y

Fix the position of the exhaust pipe and move the air pressure pipe to make the distance between it and the heading face be 15 m ~ 30 m, with an interval of 2.5 m for each movement. The dust diffusion results are shown in Figure 3. The horizontal section of the height of the breathing belt, the horizontal section of the height of the driver's position, and the longitudinal section of X = 0 all show that the dust diffusion range decreases first and then increases with the change of L_y, and reaches the minimum at 25 m.

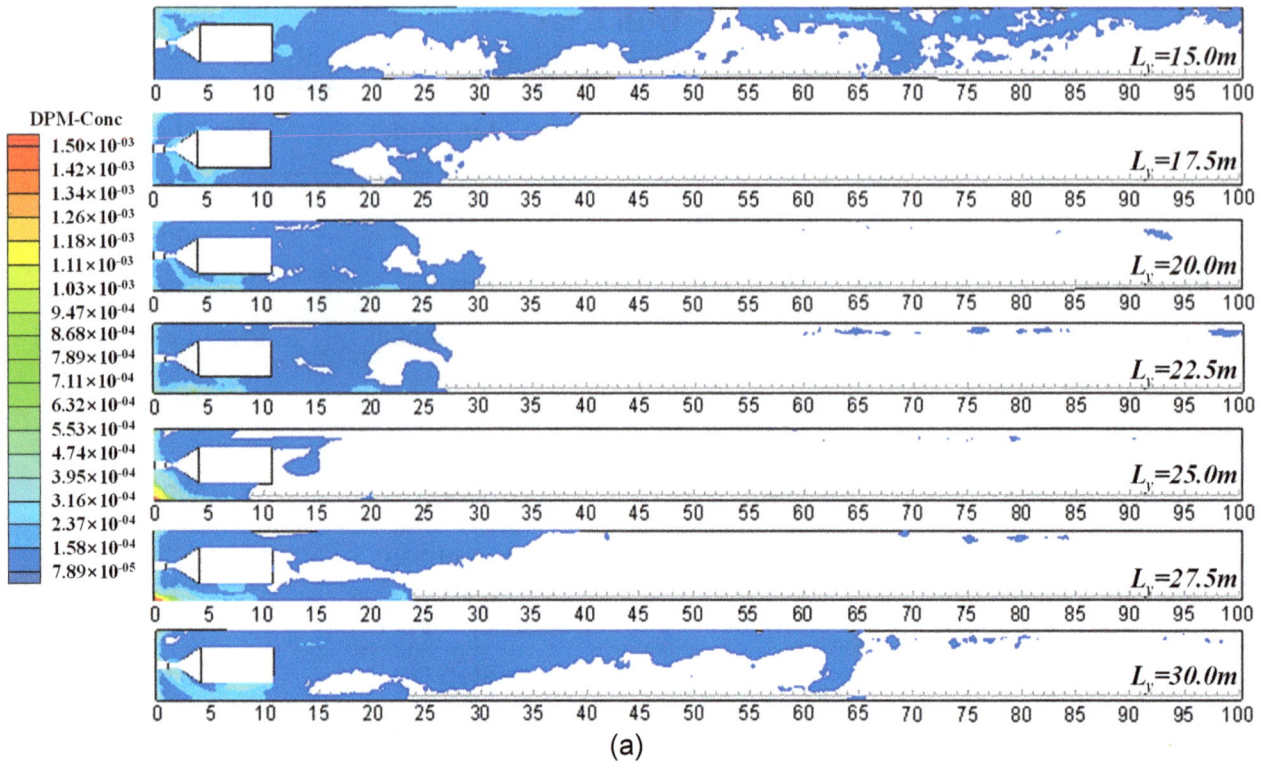

(a)

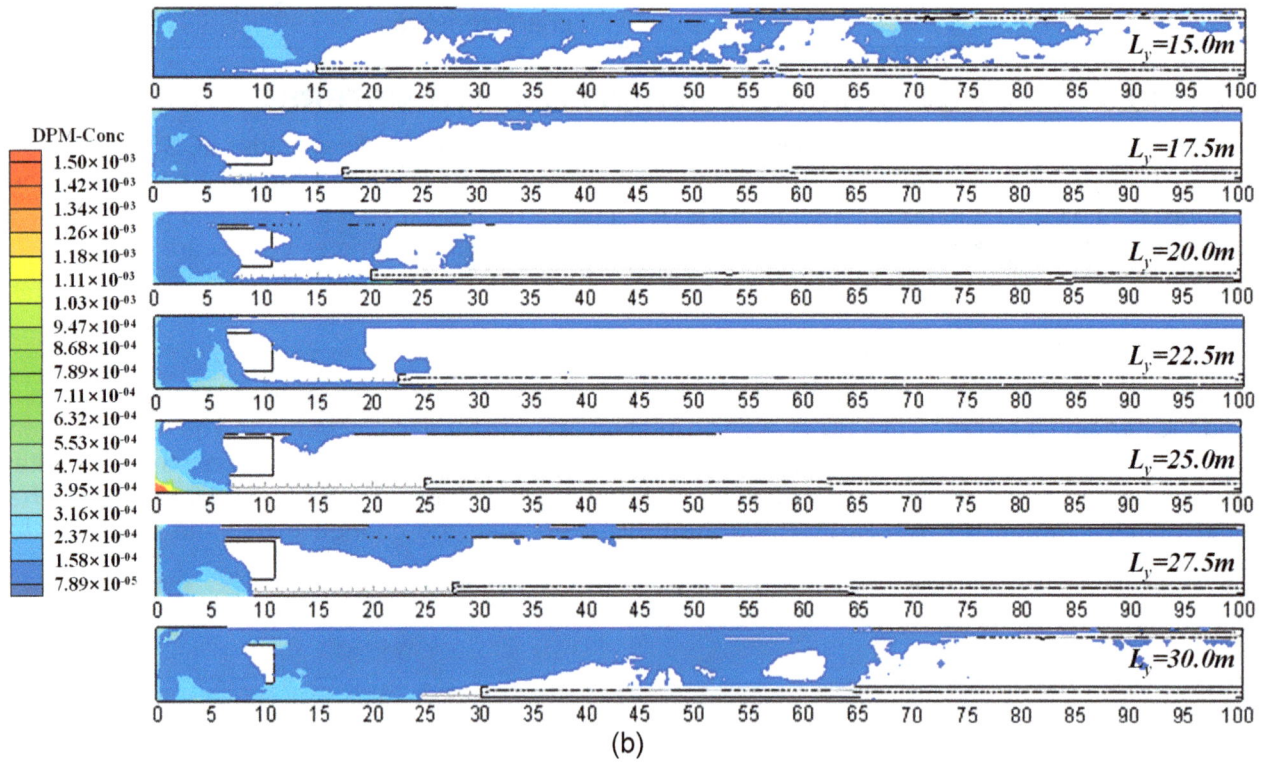

(b)

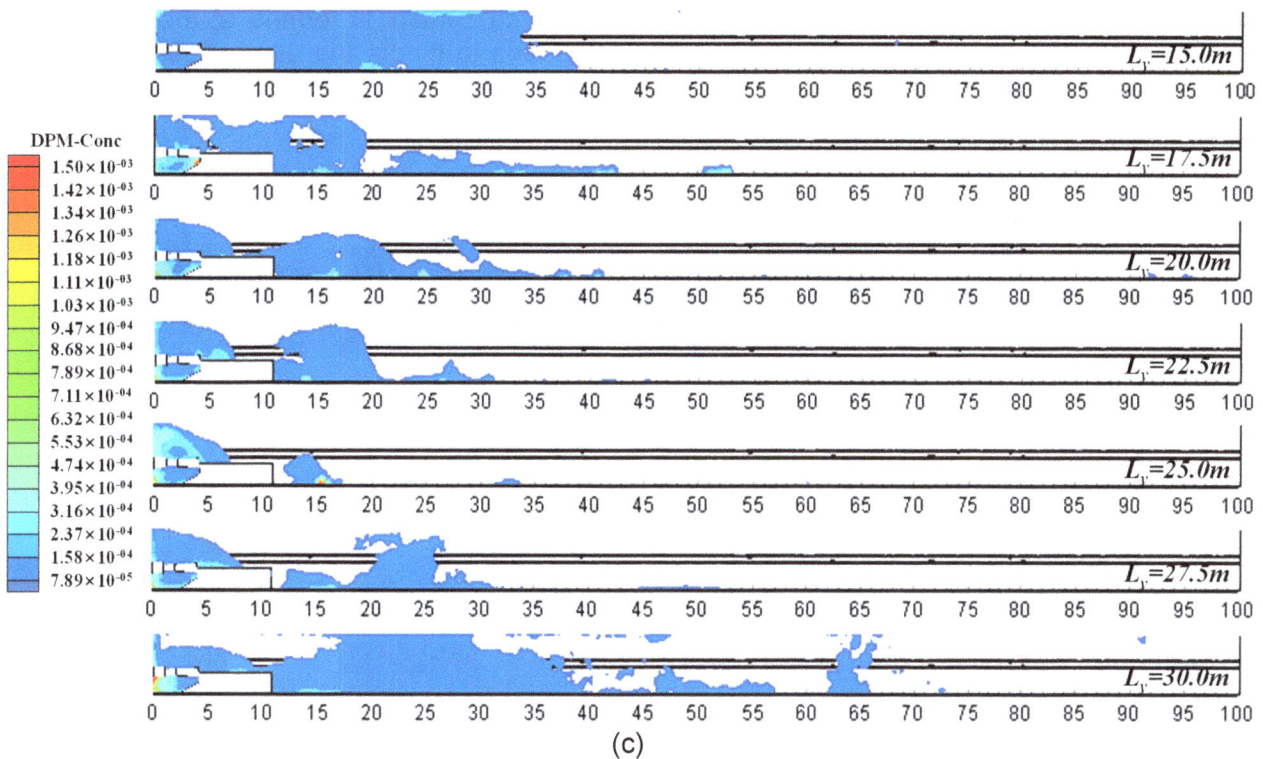

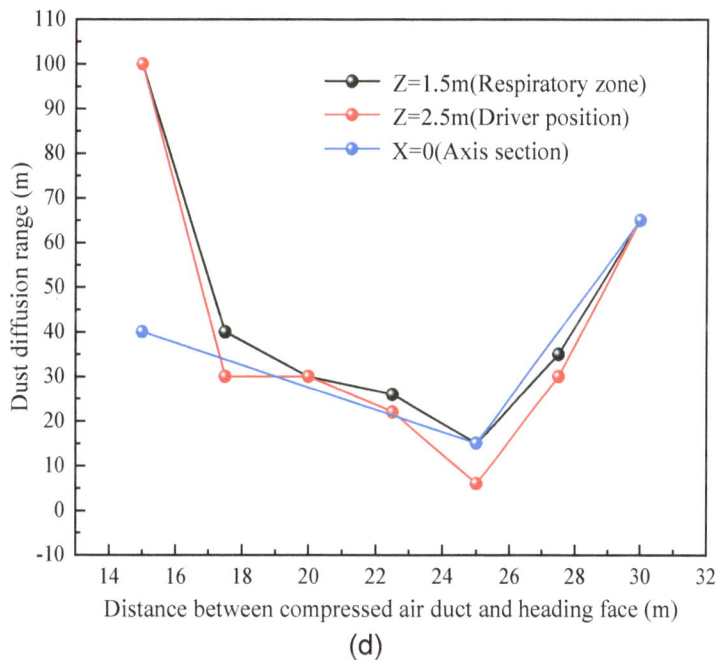

FIG 3 – Dust distribution at different locations under different L_y conditions: (a) Z = 1.5 m (Respiratory zone); (b) Z = 2.5 m (Driver's position); (c) X = 0 (Longitudinal section along roadway axis); (d) Variation of dust diffusion distance with different L_y.

The range of 25 m ~ 30 m away from the heading face is a region with intensive personnel activities. It can also be seen from (c) in Figure 3 that the dust concentration in this region is the lowest when L_y = 25 m, and the dust diffusion range is the smallest.

As seen from Figure 4, the dust concentration distribution at the driver's position is mainly concentrated in the space between both sides of the road header and the vertical wall below the roadway arch waist, and the concentration in the upper space is very small. With the increase of L_y, the airflow velocity entering the position of the driver decreases, and the concentration increases due to dust accumulation. Among them, when L_y = 25 m, in the height plane Z = 2.5 m, the jet flow from the pressure duct is sucked to the side of the exhaust duct under pressure difference. Due to the turning of the jet flow from the pressure duct, a 'wind curtain' is formed at the side of the compressed air duct and 5 m away from the heading face to hinder the dust diffusion, the dust concentration is low, and the driver's working position is also within the effective range of airflow. The 'wind curtain' is shown in Figure 5.

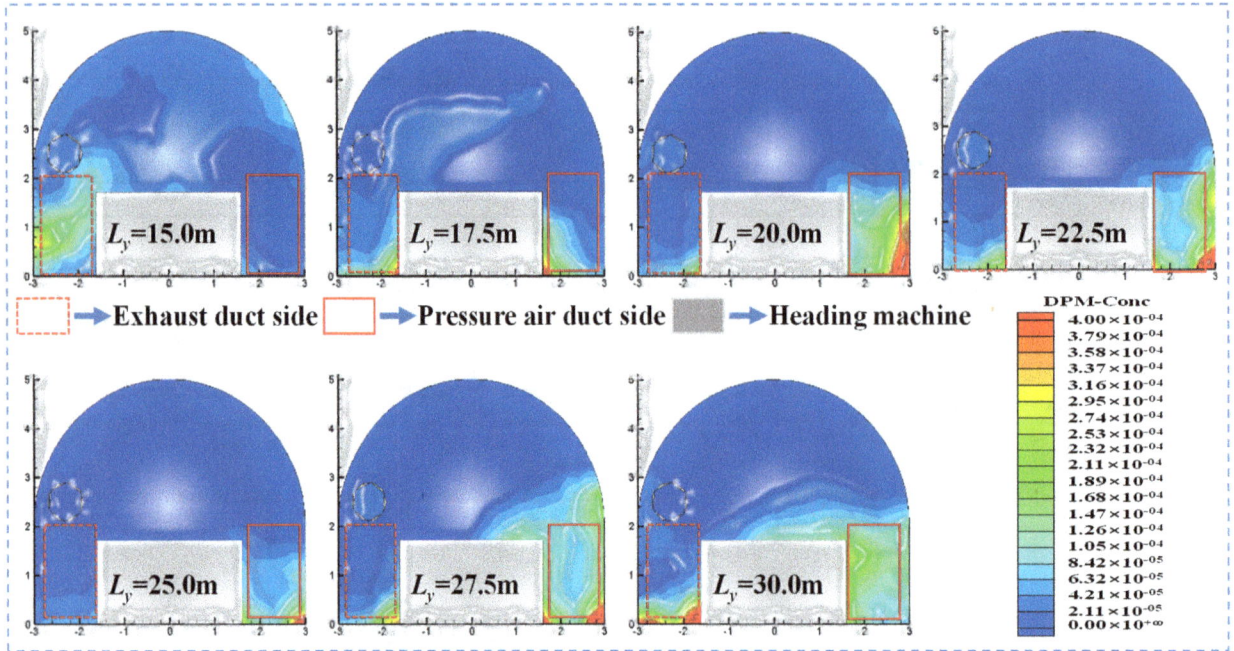

FIG 4 – Dust concentration distribution at the driver's position under different L_y conditions.

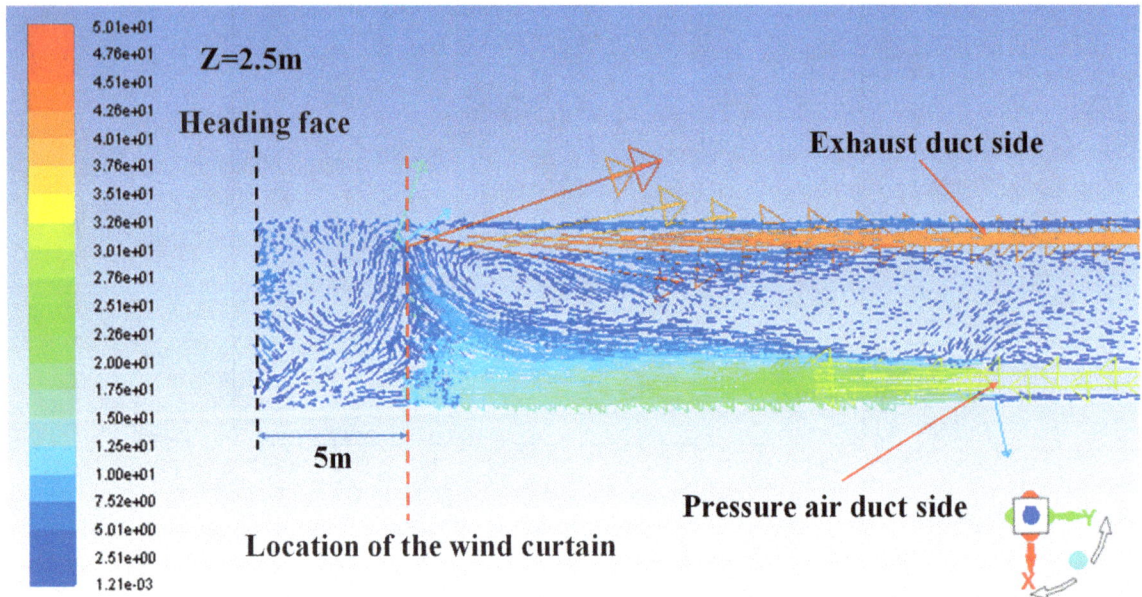

FIG 5 – Diagram of wind curtain.

As shown in $(a_1) \sim (a_7)$ of Figure 6, at the working position of the road header driver, the airflow flows from the pressure air duct side to the exhaust air duct side, forming vortex, convergence, or diversion at different locations, and the airflow wind speed decreases with the increase of L_y. At the same time, the plane where the pressure air duct outlet is located is intercepted, and the airflow field distribution of the plane is drawn, as shown in $(b_1) \sim (b_7)$ in Figure 6. Because the air velocity near the air outlet of the compressed air duct is as high as 31.85 m·s⁻¹, causing the air pressure in the area where the high-speed airflow reaches to become smaller, but the speed of the airflow near the air inlet of the exhaust pipe is 4 m·s⁻¹ ~ 8 m·s⁻¹. The air pressure in this area is high. Under pressure difference, the airflow in the plane flow field flows from the side of the exhaust pipe to the side of the pressure air pipe. On the premise of Z = 2 m, when L_y = 15 m ~ 17.5 m, most sections of the roadway are eddy current areas. The specific analysis is shown in Table 2.

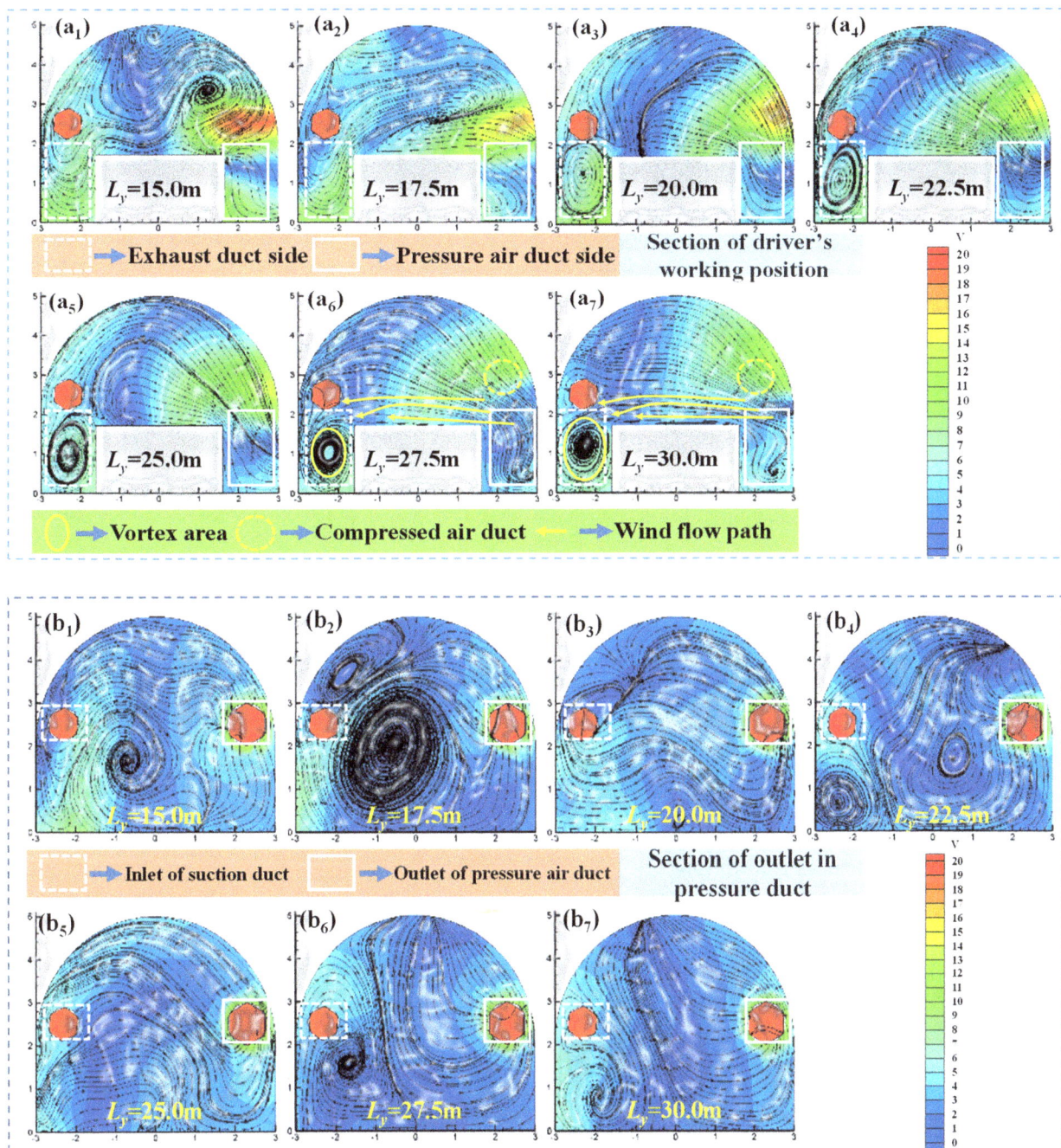

FIG 6 – Plane flow field of driver's position and compressed air duct outlet under different L_y.

TABLE 2

Wind field distribution of driver position section and pressure duct outlet section under different L_y.

		Z = 2.5 m
(a₁) ~ (a₇)	L_y = 15 m	The outlet wind speed of the pressure duct is 19 m·s⁻¹, and the driver's position is located in the jet region of the pressure duct.
	L_y = 17.5 m	The wind speed of the driver's position plane is 15 m·s⁻¹, and the wind speed above and below the extraction duct is 4 m·s⁻¹ and 10 m·s⁻¹.
	L_y = 20 m	The wind speed of the driver's position plane is 15 m·s⁻¹, and the eddy current is located between the roadway wall below the exhaust duct and the road header. The airflow speed in this area is 15 m·s⁻¹, and the airflow merges in the middle of the plane.
	L_y = 22.5 m ~ 25 m	The airflow of the pressure duct flows to the side of the exhaust duct along the vault of the roadway, and a vortex is formed under the exhaust duct.
	L_y = 27.5 m ~ 30 m	The airflow of the pressure duct flows under the suction duct along the outer surface of the road header and forms eddy currents.
(b₁) ~ (b₇)	L_y = 15 m ~ 17.5 m	Eddy currents occupy an ample space in the roadway.
	L_y = 20 m ~ 22.5 m	The range of eddies becomes smaller.
	L_y = 25 m	The airflow of the pressure duct is divided into upper and lower parts.
	L_y = 27.5 m ~ 30 m	A vortex is formed near the extraction duct, and the airflow flows to the vortex area and the inlet of the extraction duct.

Figure 7 shows the distribution of airflow field in the roadway under different conditions in the section along the axial direction of the roadway at this height when Z = 2.5 m. With the increase of L_y, the vortex formed under pressure difference is farther from the heading face. The specific changes in eddy currents at each stage are shown in Table 3.

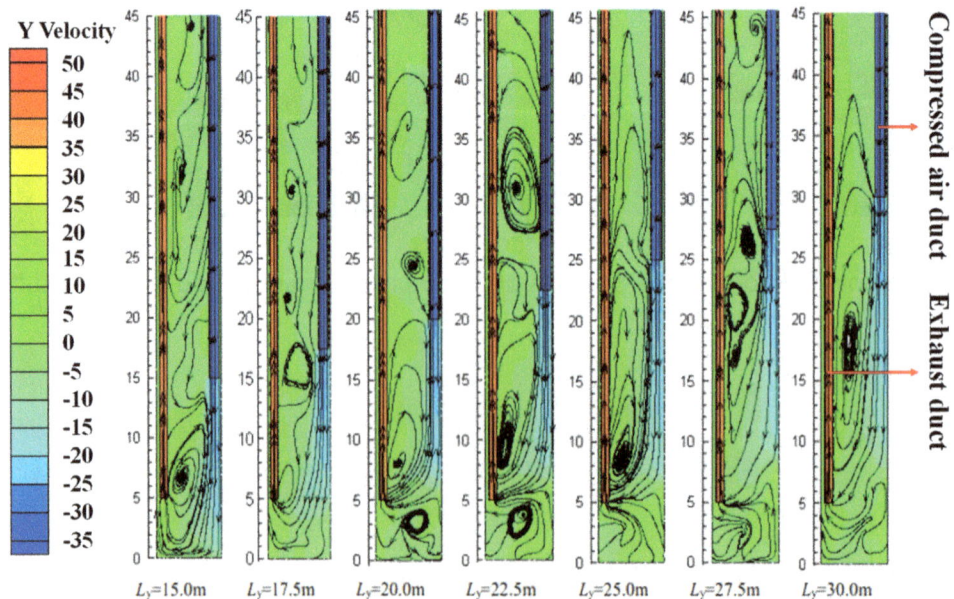

FIG 7 – Cross-section of roadway airflow field under different L_y conditions when Z = 2.5 m.

TABLE 3
The distribution of horizontal wind field at Z = 2.5 m height under different L_y.

	Z = 2.5 m
L_y = 15 m ~ 17.5 m	The airflow at the outlet of the pressure duct reaches the heading face at the speed of 5 m·s⁻¹ ~ 10 m·s⁻¹, and takes away the dust from the heading face, resulting in a lot of dust suspended in the vortex area and unable to be dispersed by the airflow.
L_y = 20 m ~ 22.5 m	The end of the effective influence range of the pressure duct outlet jet is just to the inlet of the extraction duct. The fresh airflow from the pressure duct is pumped directly from the extraction duct and forms a vortex zone near the heading face. The vortex area is wrapped with a large amount of dust, which is not conducive to dust removal.
L_y = 25 m	The airflow at the outlet of the pressure duct forms a vortex at 7 m away from the heading face, and a small amount of airflow is drawn into the suction duct along the outer edge of the vortex under the suction of the suction duct air. The eddy current acts as a 'wind curtain', limiting most of the dust within 5 m from the heading face. At this time, the jet effect of the pressure duct is poor, and the negative pressure suction effect of the suction duct is dominant.
L_y = 27.5 m ~ 30 m	The airflow coming from the pressure duct forms an eddy current at 18 m from the heading face. This position is precisely located in the activity area of the staff. The dust concentration in the area is much higher than 10 mg·m⁻³ due to the entrainment of eddy current, which cannot meet the dust removal requirements.

To sum up, in this LCSS ventilation mode test, the best distance between the outlet of the compressed air duct and the heading face is 25 m.

Determine the optimal value of L_c

When Z = 1.5 m, as shown in Figure 8a that with the increase of L_c, the dust diffusion range in the roadway first decreases from large to small, and then increases from small to large. When Z = 2.5 m, as shown Figure 8b, the dust distribution range in the roadway takes the tail of the road header as the dividing line. As can be seen from Figure 8c, The accumulation law of dust with L_c is similar to that of dust with L_y. In different cases, the diffusion distance of dust in the roadway is represented by the histogram length of Figure 8d. The detailed description of the picture is shown in Table 4.

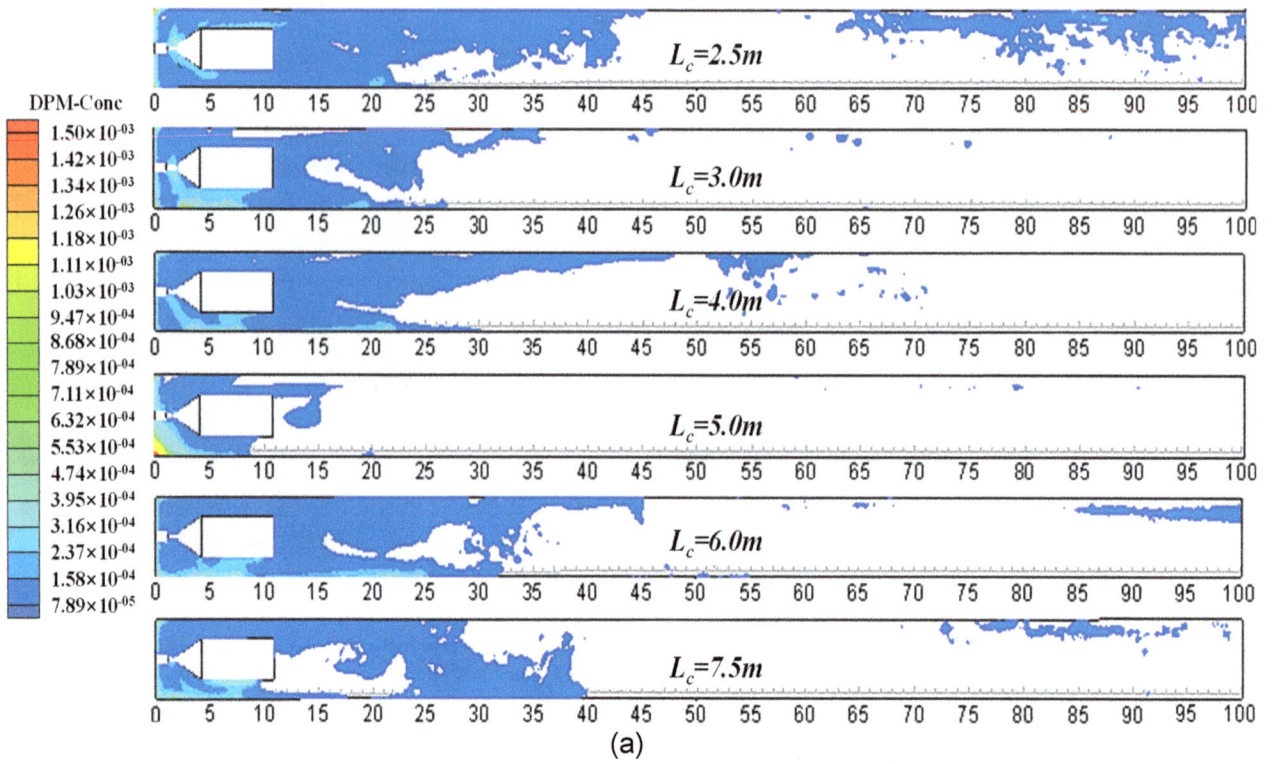

(a)

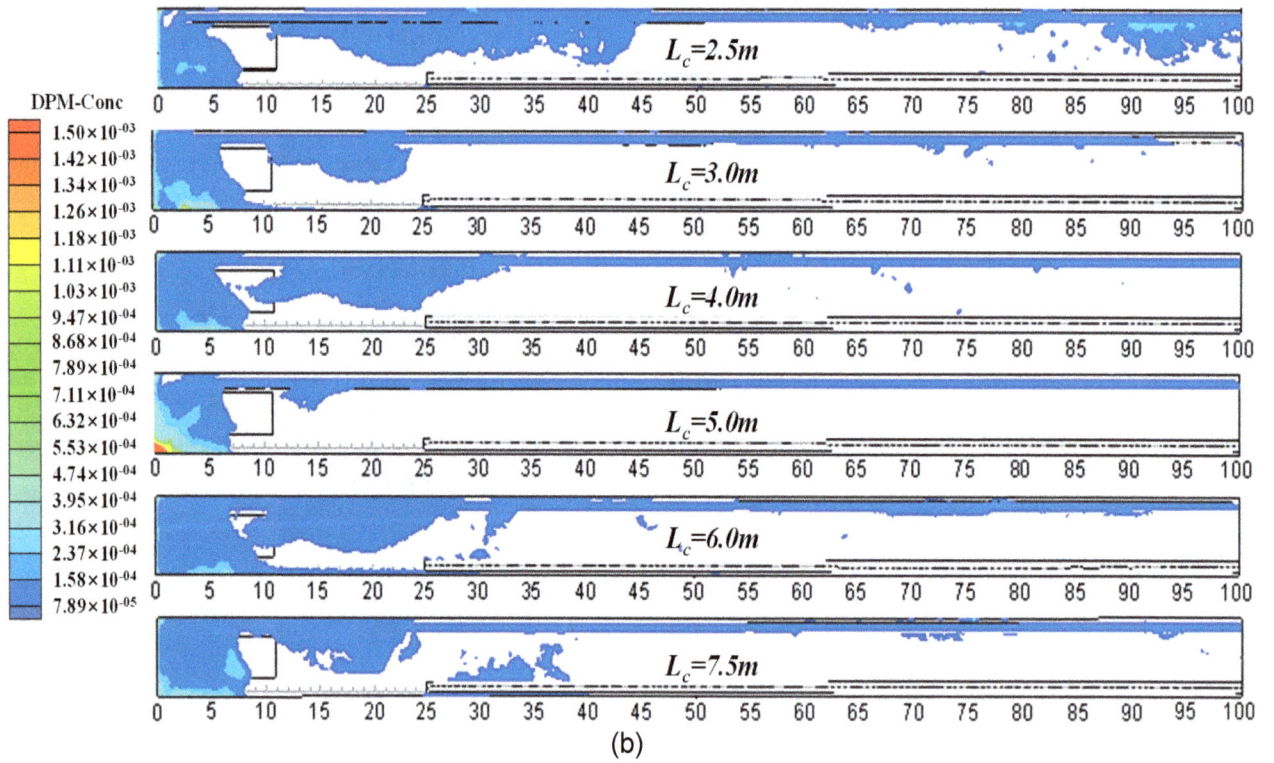

(b)

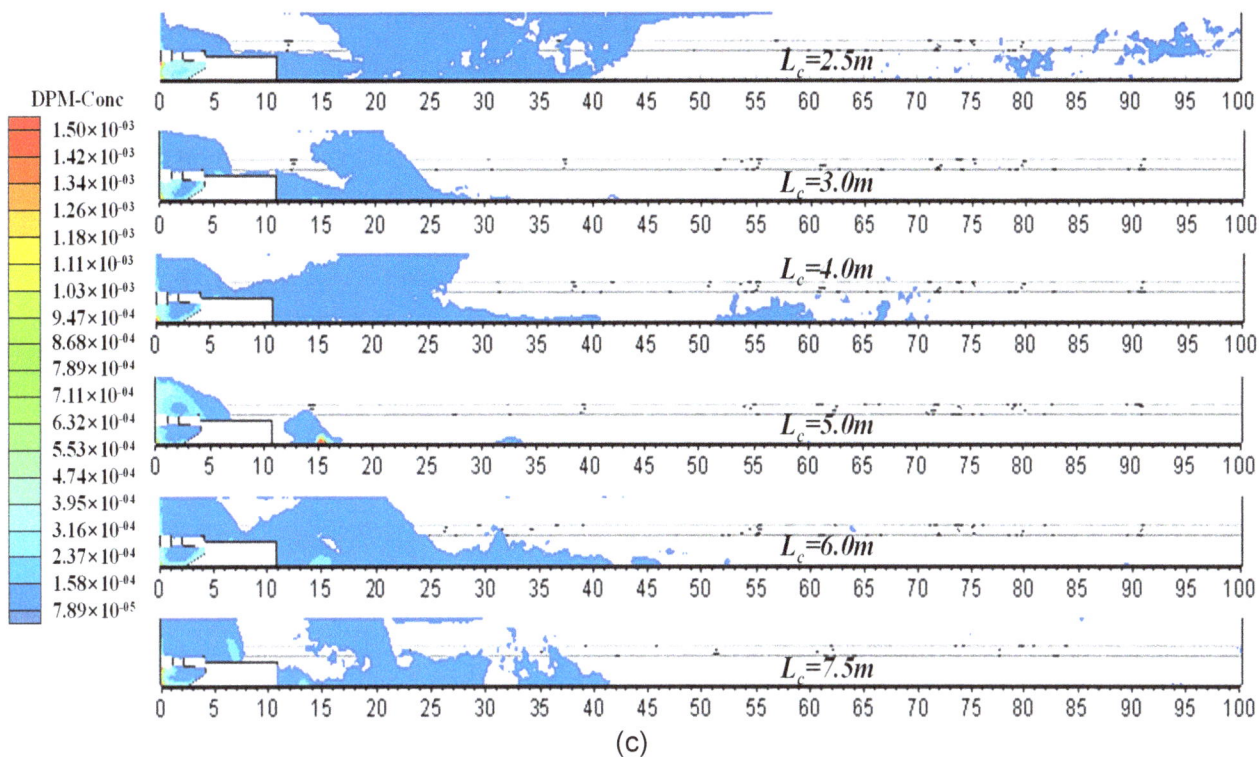

(c)

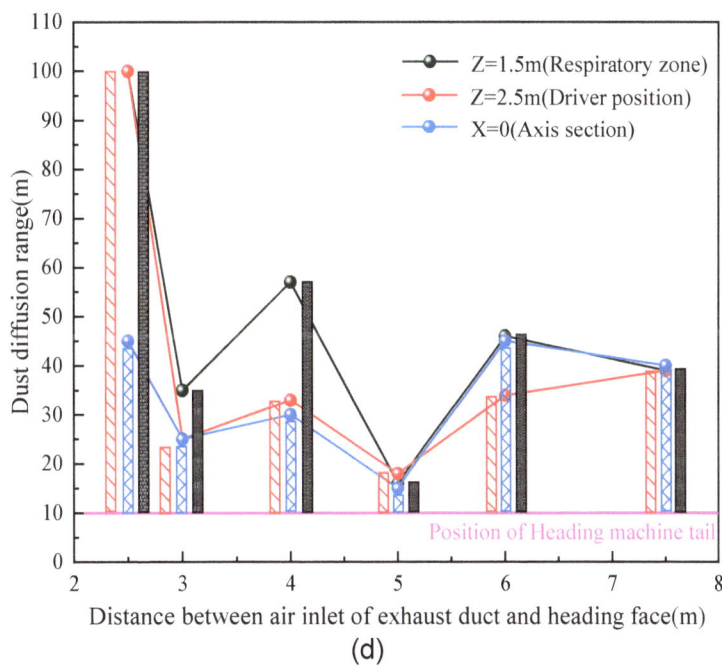

(d)

FIG 8 – Dust concentration at different locations under different L_c conditions: (a) Z = 1.5 m (Respiratory zone); (b) Z = 2.5 m (Driver's position); (c) X = 0 (Longitudinal section along roadway axis); (d) Variation of dust diffusion distance with different L_c.

TABLE 4

Dust diffusion distribution in cross-section plane at different locations under different L_c.

Z = 1.5 m	L_c = 2.5 m ~ 4 m L_c = 6 m ~ 7.5 m	There is dust distribution within the range of 0 m ~ 40 m and 63 m ~ 100 m from the heading face, and the dust concentration between the road header and the roadway wall is relatively high at the side of the pressure duct.
	L_c = 5 m	Dust is distributed within the range of 0 m ~ 15 m away from the heading face, and the dust concentration is higher at the corner of the heading face. There is no human activity in this area, which does not influence the heading work.
Z = 2.5 m	L_c = 2.5 m ~ 7.5 m	The dust diffused throughout the roadway. It is widely distributed in the range of 0 m ~ 45 m away from the heading face. In the range of 0 m ~ 7 m, dust is covered, and the dust concentration in the corner of the side of the pressure duct is higher. A range of 15 m ~ 20 m from the heading face is the active area of staff. Dust distribution in this area is smaller only when L_c = 5 m.
X = 0	L_c = 2.5 m ~ 7.5 m	The influence of roadway dust is concentrated in 0 m ~ 40 m from the heading face, and the higher dust concentration is at the bottom of the heading face. For the personnel activity area at the tail of the road header, only when L_c = 5 m is the most minor affected by dust.

It can be seen from the analysis that when L_c = 2.5 m ~ 4 m, the negative pressure range is small, and the effect is poor due to the close distance between the inlet of the extraction duct and the heading face. In addition, the entrainment effect of eddy current makes the dust cannot be effectively discharged through the extraction duct, thus forming accumulation in the range close to the heading face. When L_c = 5 m, under the dual effect of the jet flow of the pressure duct and the negative pressure of the extraction duct, dust accumulation was formed at the corner of the heading face, but no dust accumulation was formed in the staff activity area at the tail of the road header. When L_c = 6 m ~ 7.5 m, the distance between the inlet of the extraction duct and the outlet of the pressure duct is relatively close, forming a solid eddy current effect near the tail of the road header, wrapping dust and causing considerable area accumulation.

Figure 9 shows the dust concentration distribution in roadway sections at different L_c. When L_c = 2.5 m, dust accumulated on both sides of the road header near the floor area, the average dust concentration was 310 mg·m⁻³. When L_c = 3 m ~ 4 m, the average dust concentration in the roadway is 200 mg·m⁻³. When L_c = 5 m, the average dust concentration is 20 mg·m⁻³. When L_c ≥ 6 m, the average dust concentration is 150 mg·m⁻³.

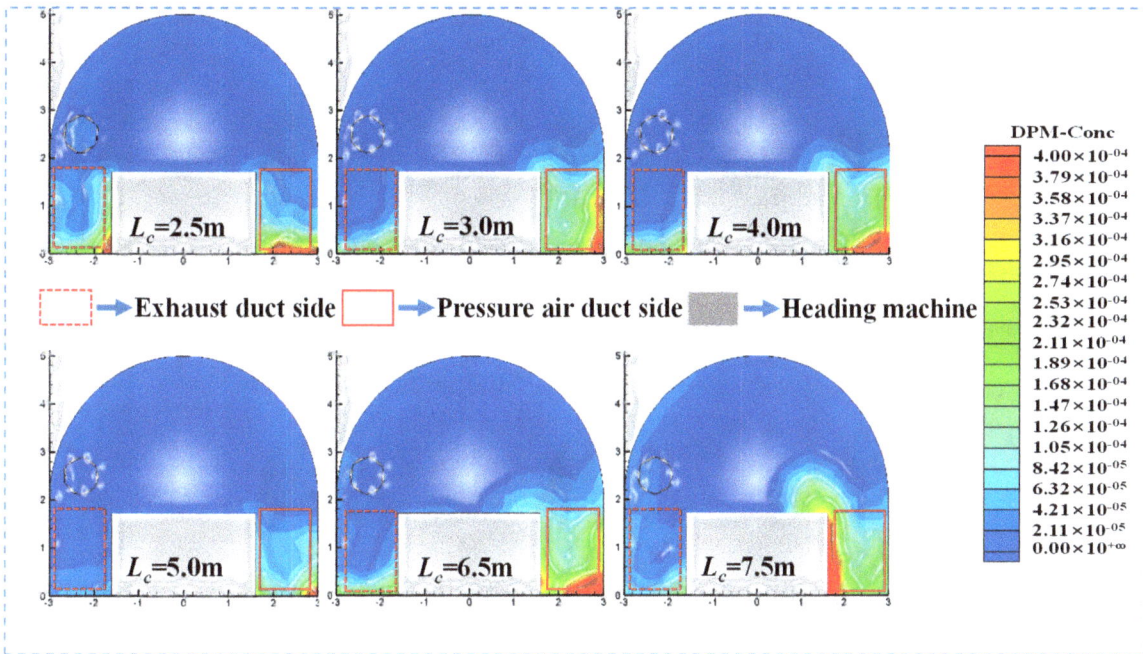

FIG 9 – Dust concentration distribution at the driver's position under different L_c conditions.

Figure 10 shows the wind field distribution of the roadway section (c_1) ~ (c_6) at the driver's position and the roadway section (d_1) ~ (d_6) at the inlet of the extraction duct at different L_c. The rainbow column in the figure represents the average wind speed in m·s^{-1}. The details are shown in Table 5.

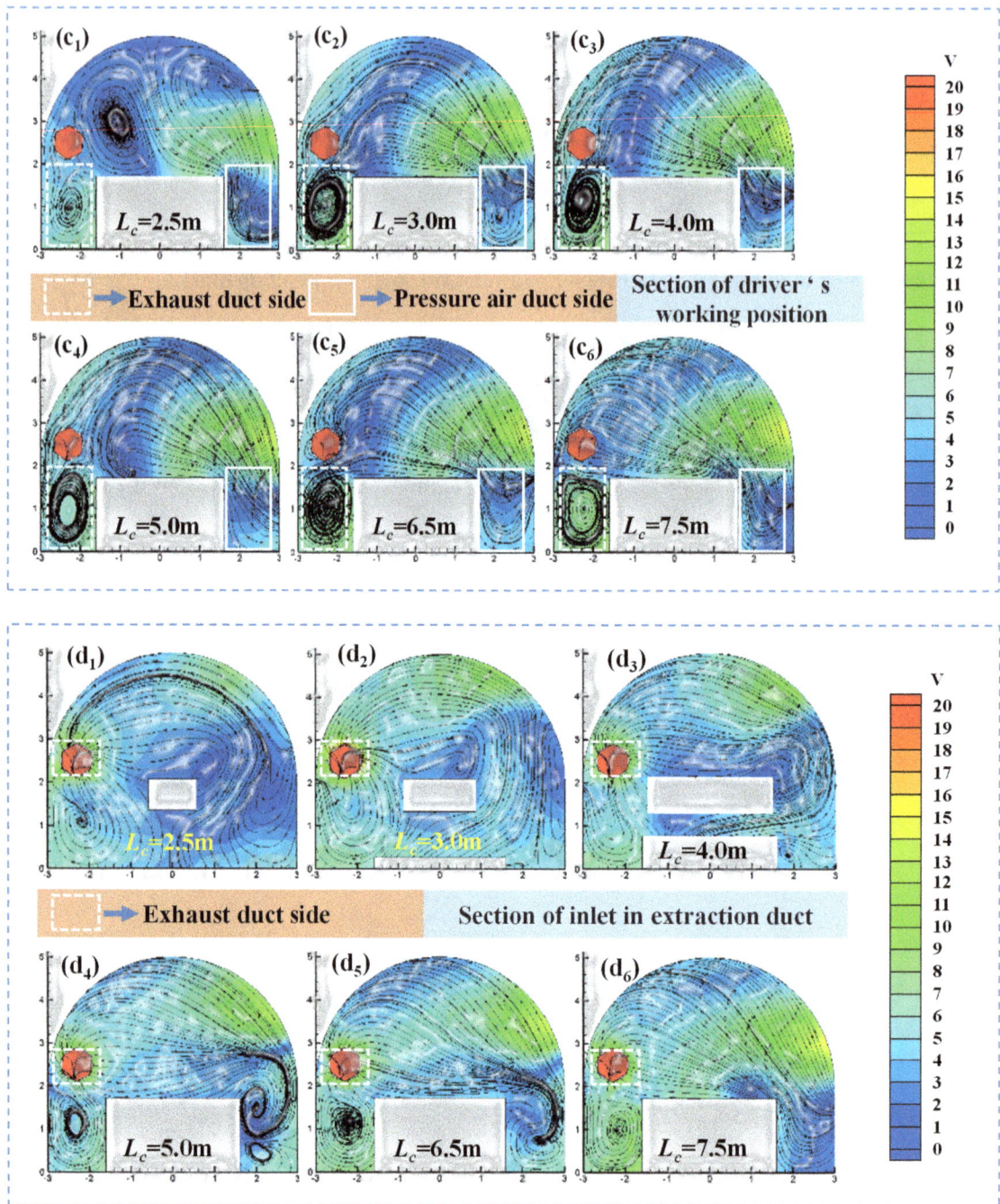

FIG 10 – The plane flow field of driver's position and air inlet of exhaust pipe under different L_c.

TABLE 5
Distribution of airflow velocity in roadway section at different locations with different L_c.

(c_1) ~ (c_6)	L_c = 2.5 m	The wind speed in the vortex region formed below the extraction duct is 5 m·s⁻¹, and the wind speed in the vortex region with a diameter of 2 m at X = -1 m is 2 m·s⁻¹.
	L_c = 3 m ~ 4 m	Part of the airflow flows to the extraction duct along the surface of the road header and is inhaled by a vortex below the exhaust duct. The wind speed in the vortex region below the extraction tube is about 12 m·s⁻¹. The other part of the wind flow along the roadway vault flows to extract the pipe. The middle of the roadway wind speed is 3 m·s⁻¹.
	L_c = 5 m	The jet flow of the pressure duct flows to the extraction duct along the roadway vault and the surface of the road header. Part of the airflow is sucked into the vortex below the extraction duct, and the wind speed of the vortex is 6 m·s⁻¹.
	L_c = 6 m ~ 7.5 m	The wind trajectory is the same as L_c = 5 m, and the airflow velocity at the vortex increases to 9 m·s⁻¹.
(d_1) ~ (d_6)	L_c = 2.5 m	The height of the pressure jet is 4.5 m. The airflow flows along the vault of the roadway to the inlet of the extraction duct, forming a vortex with a diameter of 2 m and a wind speed of 4 m·s⁻¹ below the inlet of the extraction duct.
	L_c = 3 m ~ 4 m	The height of the pressure jet is 4 m. The airflow flows along the roadway vault to the inlet of the extraction duct, and the wind speed near the roadway vault is 7 m·s⁻¹. The airflow forms a long and narrow vortex near the roadway wall below the extraction duct.
	L_c = 5 m ~ 6.5 m	The height of the pressure jet becomes 3 m, and a half vortex is formed on the side of the pressure duct. The wind speed in the vortex area is 3 m·s⁻¹, and the wind speed on the roadway vault is about 12 m·s⁻¹.
	L_c = 7.5 m	The height of the pressure jet becomes 2.5 m, and the wind flow flows from the pressure duct to the outlet duct. The wind speed in the area below the pressure duct is 2 m·s⁻¹.

The detailed description of Figure 11 is shown in Table 6.

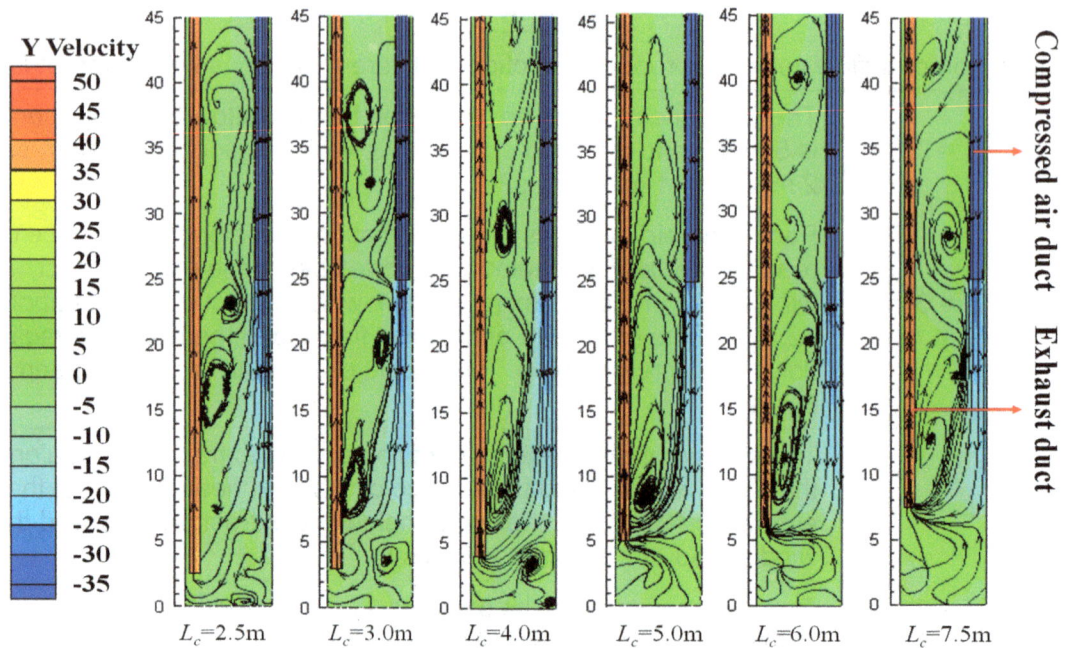

FIG 11 – Cross-section of roadway airflow field under different L_c conditions when Z = 2.5 m.

TABLE 6

Wind field distribution at different L_c on the horizontal section of Z = 2.5 m.

	Z = 2.5 m
L_c = 2.5 m	The jet of the pressure duct forms a vortex at 16 m from the heading face. Since the inlet of the extraction duct is too close to the heading face, some dust has been diffused outside the effective range of negative pressure before being sucked into the extraction duct. At the same time, the entrainment of eddy current further promotes dust diffusion.
L_c = 3 m ~ 4 m	The eddy current is formed at 7 m from the heading face. At this time, the negative pressure of the extraction pipe plays a leading role in roadway dust reduction.
L_c = 5 m	The high-pressure airflow forms an eddy current at 9 m away from the heading face, which plays the role of 'wind curtain'. The dust diffused from the heading face is blocked by the unit at the front end of the inlet of the extraction duct. The inlet of the exhaust pipe is within the optimal extraction range, and most dust is discharged through the exhaust pipe, so the dust removal effect is the best.
L_c = 6 m	The jet of the pressure duct forms a vortex of 9 m in diameter at 11 m from the heading face. The pressure airflow has little effect on the front of the outlet of the exhaust pipe, and the dust-containing airflow is driven by the pressure airflow.
L_c = 7.5 m	The distance between the inlet of the extraction duct and the outlet of the pressure duct is too close, which is not conducive to ventilation and dustfall.

Thus, the optimal distance between the exhaust pipe inlet and the heading face of the roadway is 5 m.

In summary, in this LCSS ventilation mode test, the best distance between the pressure duct outlet and the heading face is 25 m by numerical simulation, and the best distance between the exhaust pipe inlet and the heading face is 5 m (L_y = 25 m, L_c = 5 m).

Determine the optimal value of the pressure-extraction ratio (W)

Pressure-extraction ratio is a crucial factor affecting the change of dust concentration and airflow field in the roadway. Figure 12 shows the dust distribution on the horizontal section of the breathing zone (Z = 1.5 m) when the pressure-extraction ratio is different.

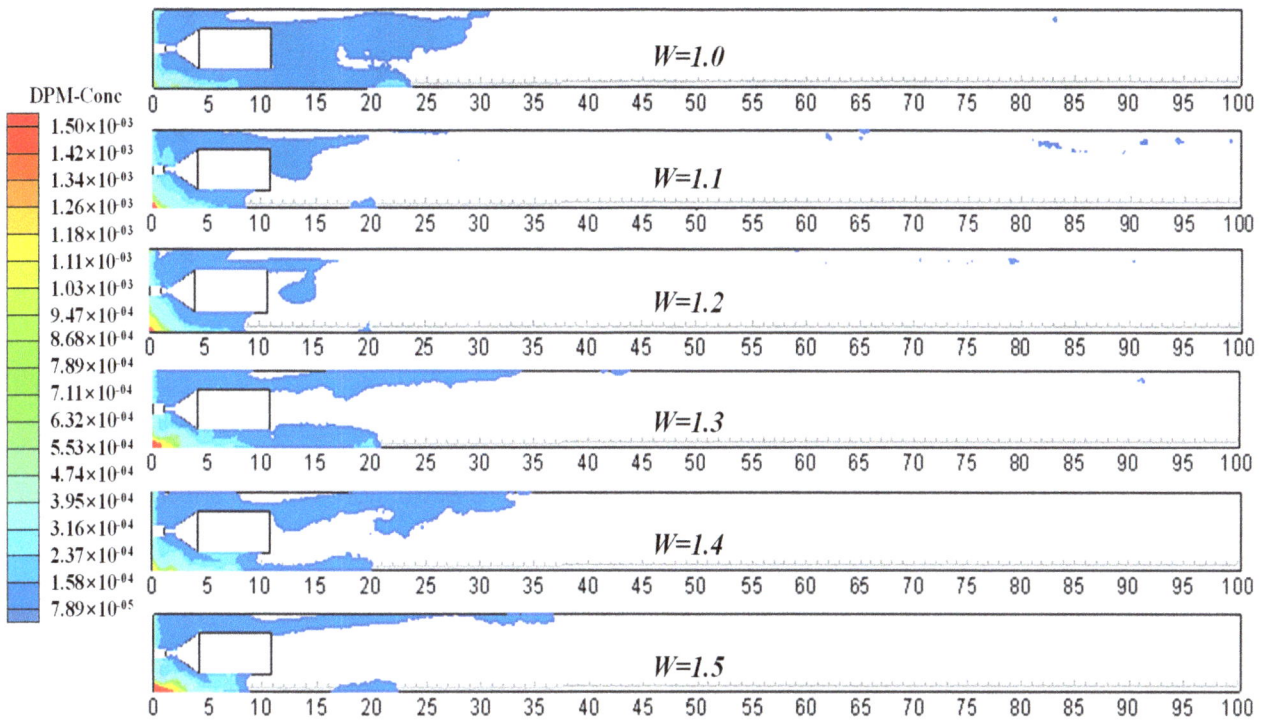

FIG 12 – Roadway dust distribution under different pressure-extraction ratios when Z = 1.5 m.

As shown in Figure 12, with the increase of *W*, the dust diffusion range first decreases and then increases. The content of personnel activities in the above-known roadway excavation process is 15 m ~ 20 m behind the tail of the road header.

It can be seen from the diagram that no matter which value is taken as the pressure-extraction ratio, dust will accumulate in the corner of the heading face. When W = 1.0, the dust diffusion distance is 30 m. When W = 1.1 ~ 1.2, the diffusion distance is 15 m. When W = 1.3 ~ 1.5, the diffusion distance is 35 m. The dust generated in the LCSS ventilation mode forms a 'strip' distribution near the wall on both sides of the roadway. The dust concentration in Figure 12 is 78 mg·m^{-3} to 157 mg·m^{-3}, and the dust concentration in the middle of the roadway is lower than 10 mg·m^{-3}. When W ≥ 1.2, the dust concentration in the driver's position is greater than 10 mg·m^{-3}, and the dust diffusion distance in the respiratory belt is too long, which is not conducive to roadway ventilation and dust removal.

Figure 13 shows the variation of airflow field in the axial section along the roadway under different pressure-extraction ratios. As shown in the figure, the wind speed at the pressure pipe side is 15 m·s^{-1}, while the wind speed at the extraction pipe side is 5 m·s^{-1}. The high-speed airflow from the outlet of the pressure duct forms eddy currents on the side of the exhaust duct. The size and position of eddy currents change with the increase of the pressure-extraction ratio. There are eddy zones of different sizes in 5 m from the heading face. The vortex zone draws in dust from the heading face to prevent its massive spread. However, when the dust is sucked into the vortex zone, it is not conducive to dust extraction and emission.

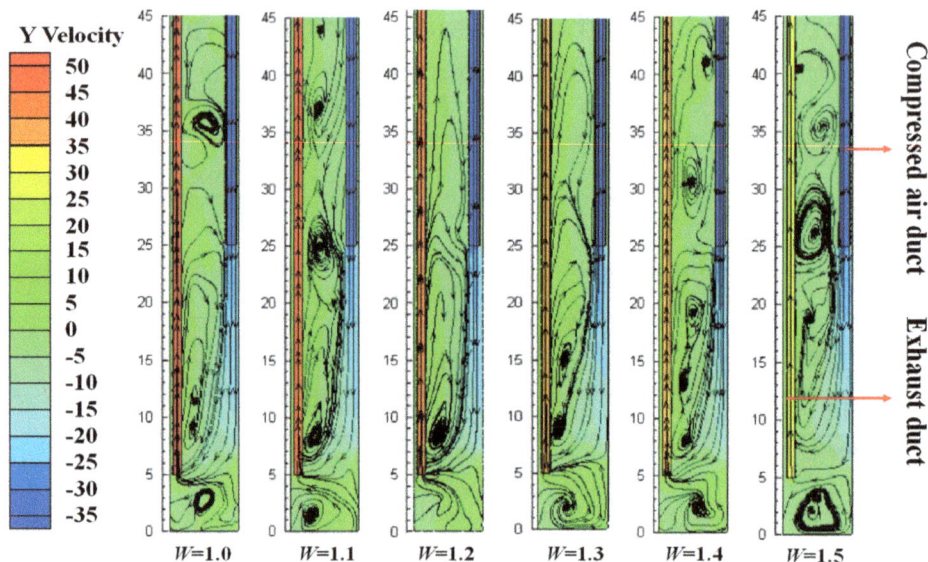

FIG 13 – Distribution of airflow field in the roadway under different pressure-extraction ratio.

The vortex formation in the roadway changes with the pressure-extraction ratio, as shown in Table 7.

TABLE 7

Formation position and size of the vortex on the side of a ventilation tube in the roadway with different W.

W	Distance from heading face (m)	Maximum diameter (m)	Remark
1.0	8	17	Eddy current along the roadway axis.
	7	13	
1.1	25	5	–
	32	3	
1.2	8	20	No eddy current in 5 m range of heading face.
1.3 ~ 1.4	≤5	-	Formation of eddy current, inhalation of dust.
	5 ~ 25	-	Not conducive to dust control and drainage.
1.5	7 ~ 40	-	Large span and many vortex centres.

Under different pressure-extraction ratios, the maximum diffusion distance of dust in the roadway is 37 m, as shown in Figure 12. Given this, the variation rule of dust concentration and wind speed in the roadway is shown in Figure 14. As can be seen from the figure, with the increase of distance from the heading face, the dust concentration in the roadway becomes lower and lower. In addition, the dust concentration within 5 m distance from the heading face decreases significantly, and the dust concentration within 5 m ~ 10 m distance decreases slightly. The dust concentration increases somewhat in the range of 10 m ~ 15 m. After 15 m, the dust concentration begins to decline slowly until it is 30 m away from the heading face and the dust concentration remains below 10 mg·m⁻³. At the same time, with the increase of distance, the wind speed in the roadway increases first and then decreases. The wind speed increased sharply within 5 m of the excavation face. In the range of 5 m to 10 m, the wind speed decreases rapidly, and its value varies from 4 m·s⁻¹ to 4.6 m·s⁻¹. In the field of 10 m ~ 20 m, the wind speed decreases, but the decreasing rate drops. At 20 m away from the heading face, the wind speed reduces to 3.8 m·s⁻¹ ~ 4 m·s⁻¹. In the range of 20 m ~ 30 m, the wind speed decreases considerably, and at 30 m, the wind speed drops to 1 m·s⁻¹. In the span of 30 m ~ 45 m, the wind speed decreases slowly, and in the position of 45 m, the wind speed decreases to 0.3 m·s⁻¹ ~ 0.8 m·s⁻¹.

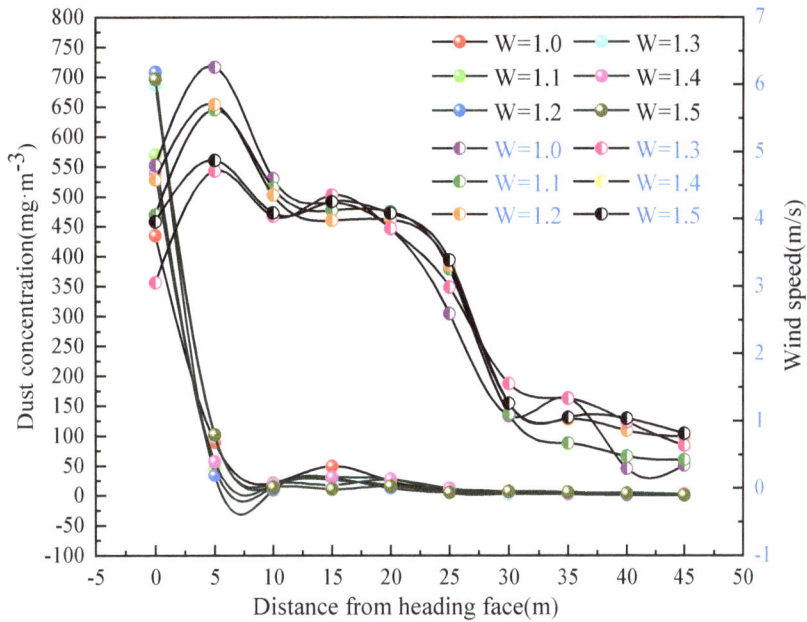

FIG 14 – Variation curve of roadway dust concentration and wind speed with the distance from the heading face.

When W = 1.2, the wind speed of the heading face is 4.6 m·s⁻¹, and the dust concentration is 707 mg·m⁻³. The wind speed 5 m away from the heading face is 5.7 m·s⁻¹, and the dust concentration decreased to 3.9 mg·m⁻³. The wind speed is 3.9 m·s⁻¹, and the dust concentration is 30.27 mg·m⁻³ at 15 m away from the heading face. After 25 m from the heading face, the dust concentration decreases by 4.6 mg·m⁻³, and the wind speed decreases rapidly.

As seen from Figures 15 and 16, the dust concentration in the roadway decreases with the increase of the distance from the heading face. The dust concentration 10 m away from the heading face is 10 mg·m⁻³, and the dust concentration on the exhaust duct side is significantly lower than that on the compressed air side. As can be seen from Figure 15, on the side of the compressed air duct, the dust concentration of working flour reaches the lowest when W = 1.0, which is 674 mg·m⁻³. After the distance from the excavation face increases to 1 m ~ 2 m, the dust concentration increases slightly, and the dust concentration decreases continuously after the length exceeds 2 m. When W = 1.1 ~ 1.2, the dust concentration continues to decline, and the decline rate is first large and then trim. W = 1.3 ~ 1.5, the dust concentration decreased first and then increased slightly, and continued to decline again after the distance reached 2 m. Figure 16 shows that the dust concentration decreases the most at a distance of 1 m from the heading face on the side of the exhaust pipe. When W = 1.2, the dust concentration at the heading face reached 384 mg·m⁻³, and the dust concentration at the working position of the road header driver was lower than 10 mg·m⁻³.

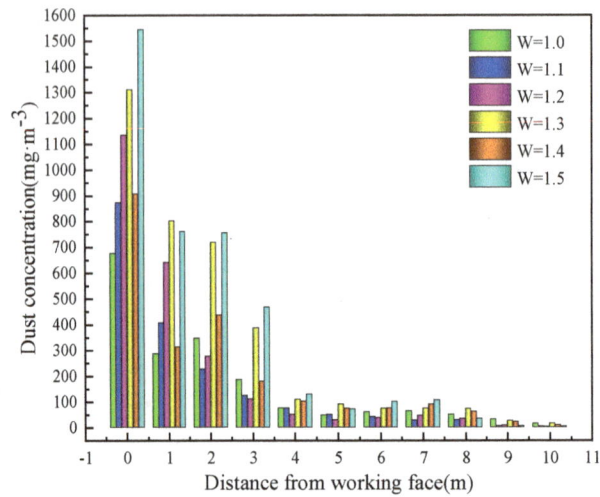

FIG 15 – Variation of dust concentration on the side of the pressure air duct in Roadway with distance.

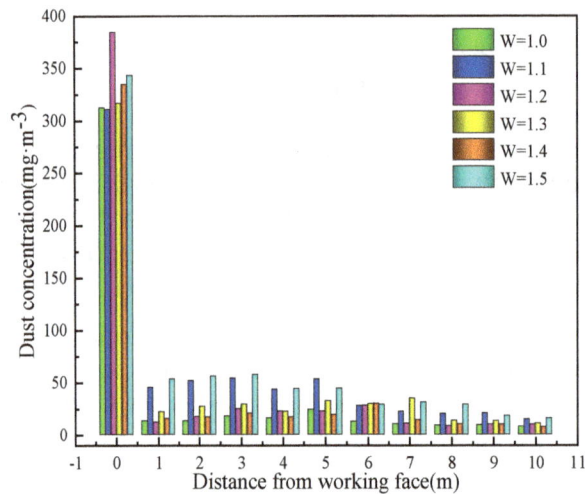

FIG 16 – Variation of dust concentration on the side of exhaust duct in Roadway with distance.

Figures 17 and 18 describe the variation trend of airflow velocity on the pressure air duct side and the suction air duct side with the increase of the distance from the heading face under the conditions of different pressure-extraction ratios. Within the range of 5 m away from the heading face, the airflow velocity at the pressure air duct side is more significant than that at the suction air duct side. The maximum airflow velocity is found in area 10 m ~ 15 m away from the heading face. After that, the airflow velocity in the roadway continues to decrease with the continuous increase of distance.

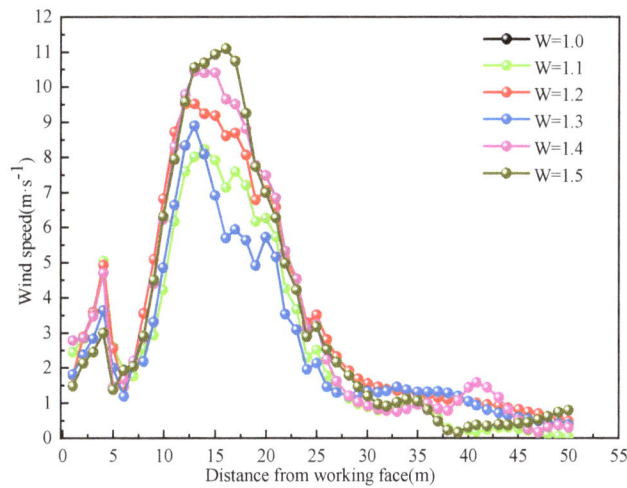

FIG 17 – Variation of airflow velocity with distance on the side of the pressure air duct in Roadway.

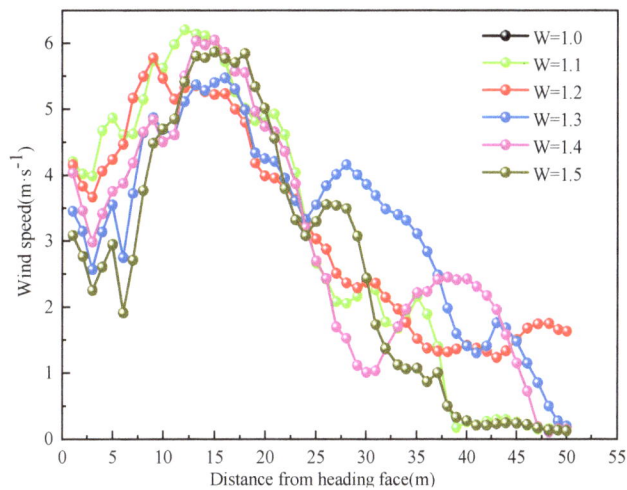

FIG 18 – Variation of airflow velocity with distance on the side of the pressure air duct in Roadway.

On the side of the pressure pipe, as shown in Figure 17, the wind speed variation trend is the same under different pressure-extraction ratios. In the range of 5 m ~ 7 m from the heading face, the wind speed increases at first and then decreases rapidly. At distances greater than 7 m, the wind speed on the roadway rises again until it peaks at about 15 m and then continues to decline. On the side of the exhaust pipe, as shown in Figure 18, the wind speed variation trend is not affected by the change in the compression-extraction ratio. The wind speed on the road rises and then falls, reaching its peak 11 m away from the heading face. Compared with Figure 17, the airflow field at the side of the exhaust pipe is more chaotic and unstable. When $W = 1.2$, the wind speed change in the above two figures is relatively smooth, representing the axial wind speed change rule in the roadway.

Therefore, it can be determined that in this LCSS ventilation mode, when the ratio of inlet air pressure to negative exhaust pressure is $W = 1.2$, the effect of roadway ventilation and dust removal is the best. The optimal combination of crucial parameters in this study is $L_y = 5$ m, $L_c = 5$ m, and $W = 1.2$.

Comparison between measured data and numerical simulation results

Along the direction of the -Y axis, in the range of 0 m ~ 50 m from the heading face, the interval is 5 m (except the 10 m position). The results of numerical simulation and actual measurement of roadway dust concentration are shown in the histogram in Figure 19. It can be seen from the diagram that the dust concentration of the simulated and measured roadway decreases with the increase of the distance from the heading face. The actual measured concentration value is higher than the simulated value, which is due to the airflow at the entrance of the inclined shaft carrying a small amount of dust into the roadway.

In addition, nine representative locations were selected for error analysis, and standard deviation (SE) was selected for evaluation, as shown in the illustration in Figure 19. As shown in the figure, the error bar at the suction inlet is the largest, where the standard deviation is 25.45584. The reason is that the airflow at the inlet of the inclined shaft contains a certain concentration of dust, and the error rod range at other positions is tiny. In the Z direction, there are seven groups of measured data height in the respiratory belt (Z = 3.5 m), the standard deviation is 0.42426 ~ 5.33866, the standard deviation of the tail of the road header (Z = 1.5 m) is 1.22329, and the standard deviation of the driver position (Z = 2.5 m) is 0.50205.

The above comparison shows that the discreteness of the simulation results and the measured results is small, which verifies that the error of the simulation results is within the allowable range and the simulation results are reliable.

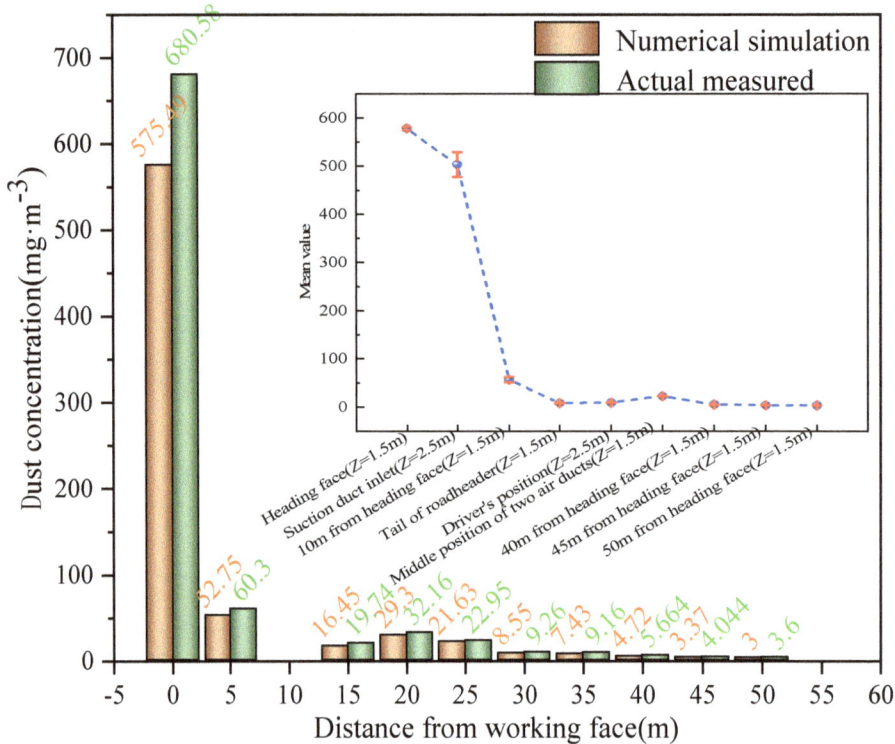

FIG 19 – Comparison between numerical simulation results and measured data.

CONCLUSIONS

This study used Gambit to establish the ventilation model of LCSS. Fluent was used to compare the roadway dust transport law and airflow field distribution under different values of L_y, L_c, and W. LCSS ventilation mode has the characteristics of large airflow velocity, large jet area, strong dust removal capacity, forming a 'wind curtain' to inhibit dust diffusion at the outlet of the compressed air duct. The ventilation mode of LCSS is adopted in the heading face, which solves the problems of extensive dust range and high dust concentration in the heading face. Test shows that the change of the relative position of the air outlet of the compressed air duct and the air inlet of the exhaust air duct in the roadway will affect the shape of the flow field, the distribution of wind speed, the distribution of eddy current and the disturbance degree of airflow in the reflux area, to change the distribution of dust in the roadway.

In this experiment, under LCSS ventilation mode, the distance between the outlet of compressed air duct and roadway excavation face is L_y = 27 m, the distance between the inlet of exhaust air duct and roadway excavation face is L_c = 5 m, The pressure-extraction ratio is W = 1.2, a 'wind curtain' will be formed in front of the activity area of the road header driver and operators to effectively prevent the dust from spreading to the depth of the roadway, and a large amount of dust will be discharged through the exhaust pipe. The numerical simulation and measurement results show that the exhaust pipe in LCSS ventilation plays a leading role. The above combination of pressure air duct and exhaust air duct effectively weakens the diffusion and settlement of dust, minimises the pollution area in the roadway, and is most conducive to ventilation and dust removal.

ACKNOWLEDGEMENTS

Thanks to Prof Xiao Junfeng, Dr Xu Feng, Prof Lu Ping, Dr Shen Xue and Dr Chen Zhongwei from Anhui Jianzhu University. Your academic views and research results have important guiding significance for this paper. Thanks to Ding Xiaomin, senior mining engineer, Manager of Xinbai Coalmine Company, Huating Coal Industry, Gansu Province, and Dr Xu Yu, School of Resources and Safety Engineering, Central South University, for your constructive guidance on numerical simulation of this manuscript.

REFERENCES

Cai, P, Nie, W, Chen, D, Yang, S and Liu, Z, 2019. Effect of air flowrate on pollutant dispersion pattern of coal dust particles at fully mechanized mining face based on numerical simulation, *Fuel*, 239:623–635, https://doi.org/10.1016/j.fuel .2018.11.030

Cai, P, Nie, W, Liu, Z, Xiu, Z, Peng, H, Du, T and Yang, B, 2020. Study on the air curtain dust control technology with a dust purifying fan for fully mechanized mining face, *Powder Technology,* 374:507–521, https://doi.org/10.1016/j. powtec.2020.06.082

Chekan, G J, Colinet, J E and Grau, R H, 2006. P Impact of fan type for reducing respirable dust at an underground limestone crushing facility, in *Proceedings of the 11th US/North American Mine Ventilation Symposium*, Penn State Univ, https://www.cdc.gov/niosh/mining/UserFiles/works/pdfs/ioftf.pdf, Accessed 1 March 2022.

Chen, D, Nie, W, Cai, P and Liu, Z, 2018. The diffusion of dust in a fully-mechanized mining face with a mining height of 7 m and the application of wet dust-collecting nets, *Journal of Cleaner Production*, 205:463–476, https://doi.org/10. 1016/j.jclepro.2018.09.009

Cheng, W, Nie, W, Zhou, G, Yu, Y, Ma, Y and Xue, J, 2012. Research and practice on fluctuation water injection technology at low permeability coal seam, *Safety science*, 50(4):851–856, https://doi.org/10.1016/j.ssci.2011.08.021

Du, C, Wang, H, Jiang, Z, He, Z and Hu, G, 2010. Numerical simulation of dust distribution in fully mechanized excavation face with long compressive short suction ventilation, *Journal of Beijing University of science and technology,* 32(08):957–962, https://doi.org/10.13374/j.issn1001–053x.2010.08.027

Geng, F, Luo, G, Wang, Y, Peng, Z, Hu, S, Zhang, T and Chai, H, 2018. Dust dispersion in a coal roadway driven by a hybrid ventilation system: A numerical study, *Process Safety and Environmental Protection*, 113:388–400, https://doi.org/10.1016/j.psep.2017.11.010

Hua, Y, Nie, W, Cai, P, Liu, Y, Peng, H and Liu, Q, 2018a. Pattern characterization concerning spatial and temporal evolution of dust pollution associated with two typical ventilation methods at fully mechanized excavation faces in rock tunnels, *Powder Technology,* 334:117–131, https://doi.org/10.1016/j.powtec.2018.04.059

Hua, Y, Nie, W, Wei, W, Liu, Q, Liu, Y and Peng, H, 2018b. Research on multi-radial swirling flow for optimal control of dust dispersion and pollution at a fully mechanized tunneling face, *Tunneling and Underground Space Technology*, 79:293–303, https://doi.org/10.1016/j.tust.2018.05.018

Kurnia, J C, Sasmito, A P and Mujumdar, A S, 2014. Dust dispersion and management in underground mining faces, *International Journal of Mining Science and Technology,* 24:39–44, https://doi.org/10.1016/j.ijmst.2013.12.007

Li, X, Zhao, X, Jiang, Y, Zhang, M, Wang, L, Liu, Y, Xiao, D, Xu, X, Li, Z and Wang, Y, 2021. Air curtain dust-collecting technology: Influence factors for air curtain performance, *Journal of Wind Engineering and Industrial Aerodynamics*, 218:104780, https://doi.org/10.1016/j.jweia.2021.104780

Li, Y, Li, Z and Gao, L, 2014. The layout of the air duct in mechanized excavation face based on wind flow and dust distribution, *Acta coal Sinica*, 39(S1):130–135, https://doi.org/10.13225/j.cnki.jccs.2013.0788

Liu, R, Li, X, Shi, S, Wang, P and Wang, W, 2009. Study on the influence of outlet angle of air curtain at full-mechanized work face on its dust-isolating effectiveness, *China Safety Science Journal (CSSJ)*, 19(12):128–134; p 205, https://doi.org/10.16265/j.cnki.issn1003–3033.2009.12.011

Liu, R, Shi, S, Li, X and Wang, P, 2010. Influence of Outlet Velocity of Air Curtain on Effectiveness of Dust-isolating, *2010 4th International Conference on Bioinformatics and Biomedical Engineering (ICBBE 2010)*, https://doi.org/10.1109/ ICBBE.2010.5516578

Moloney, K, Lowndes, I, Stokes, M and Hargrave, G, 1997. Studies on alternative methods of ventilation using Computational Fluid Dynamics (CFD), scale and full-scale gallery tests, in *Proceedings 6th International Mine Ventilation Congress*.

Nie, W, Liu, Y, Wang, H, Wei, W, Peng, H, Cai, P, Hua, Y and Jin, H, 2017b. The development and testing of a novel external-spraying injection dedusting device for the heading machine in a fully-mechanized excavation face, *Process Safety and Environmental Protection,* 109:716–731, https://doi.org/10.1016/j.psep.2017.06.002

Nie, W, Wei, W, Ma, X, Liu, Y, Peng, H and Liu, Q, 2017a. The effects of ventilation parameters on the migration behaviors of head-on dust in the heading face, *Tunneling and Underground Space Technology,* 70:400–408, https://doi.org/10.1016/j.tust.2017.09.017

Shi, X, Jiang, Z, Zhou, S and Cai, W, 2008. Experimental study on dust distribution regularity of fully mechanized mining face, *Journal of China Coal Society,* 33:1117–1121, https://kns.cnki.net/kcms/detail/detail.aspx?FileName=MTXB 200810008&DbName=CJFQ2008

Stovern, M, Felix, O, Csavina, J, Rine, K P, Russell, M R, Jones, R M, King, M, Betterton, E A and Saez, A E, 2014. Simulation of windblown dust transport from a mine tailings impoundment using a computational fluid dynamics model, *Aeolian Res,* 14:75–83, https://doi.org/10.1016/j.aeolia.2014.02.008

Tan, C, Jiang, Z, Chen, J and Wang, P, 2014. Numerical simulation of influencing factors on dust movement during coal cutting at fully mechanized working faces, *Journal of University of Science and Technology Beijing,* 36(06):716–721, https://doi.org/10.13374/j.issn1001–053x.2014.06.002

Tan, C, Jiang, Z, Wang, M and Chen, Y, 2015. Similarity experiment on multi-source dust diffusion law is fully mechanized caving face, *J China Coal Soc,* 40(01):122–127, https://doi.org/10.13225/j.cnki.jccs.2014.0105

Torno, S, Toraño, J and Álvarez-Fernández, I, 2020. Simultaneous evaluation of wind flow and dust emissions from conveyor belts using computational fluid dynamics (CFD) modeling and experimental measurements, *Powder Technology,* 373:310–322, https://doi.org/10.1016/j.powtec.2020.06.061

Wang, H, Nie, W, Cheng, W, Liu, Q and Jin, H, 2018. Effects of air volume ratio parameters on air curtain dust suppression in a rock tunnel's fully-mechanized working face, *Advanced Powder Technology,* 29(2):230–244, https://doi.org/10.1016/j.apt.2017.11.007

Wang, Y, Luo, G, Geng, F, Li, Y and Li, Y, 2015. Numerical study on dust movement and dust distribution for hybrid ventilation system in a laneway of coalmine, *Journal of Loss Prevention in the Process Industries,* 36:146–157, https://doi.org/10.1016/j.jlp.2015.06.003

Widiatmojo, A, Sasaki, K, Sugai, Y, Suzuki, Y, Tanaka, H, Uchida, K and Matsumoto, H, 2015. Assessment of air dispersion characteristic in underground mine ventilation: Field measurement and numerical evaluation, *Process Safety and Environmental Protection,* 93:173–181, https://doi.org/10.1016/j.psep.2014.04.001

Xiao, D, Li, X, Yan, W and Fang, Z, 2019. Experimental investigation and numerical simulation of small-volume transverse-flow air curtain performances, *Powder Technology,* 352:262–272, https://doi.org/10.1016/j.powtec.2019.04.063

Xing, D, 2005. Discussion on the reasonable position of suction and pressure air duct for driving mixed ventilation, *Mining Safety and Environmental Protection,* 04:25–27, https://kns.cnki.net/kcms/detail/detail.aspx?FileName=ENER 200504010&DbName=CJFQ2005

Yin, S, Nie, W, Liu, Q and Hua, Y, 2019. Transient CFD modeling of space-time evolution of dust pollutants and air-curtain generator position during tunnelling, *Journal of Cleaner Production,* 239:117924, https://doi.org/10.1016/j.jclepro.2019.117924

Yu, H, Cheng, W, Peng, H and Xie, Y, 2018. An investigation of the nozzle's atomization dust suppression rules in a fully-mechanized excavation face based on the airflow-droplet-dust three-phase coupling model, *Advanced Powder Technology,* 29(4):941–956, https://doi.org/10.1016/j.apt.2018.01.012

Experimental study on the mechanism of gas-coal dust synergistic control based on compound reagents enhanced water spray technology

B Zhao[1], S Li[2], H Lin[3], H Shuang[4], M Yan[5], D Yan[6], Y Cheng[7] and E Yang[8]

1. Doctor, College of Safety Science and Engineering, Xi'an University of Science and Technology, Xi'an, Shaanxi 710054, China. Email: zhao0220bo@163.com
2. Professor, College of Safety Science and Engineering, Xi'an University of Science and Technology, Xi'an, Shaanxi 710054, China. Email: lisg@xust.edu.cn
3. Professor, College of Safety Science and Engineering, Xi'an University of Science and Technology, Xi'an, Shaanxi 710054, China. Email: lhaifei@xust.edu.cn
4. Professor, College of Safety Science and Engineering, Xi'an University of Science and Technology, Xi'an, Shaanxi 710054, China. Email:shuanghaiqing@163.com
5. Professor, College of Safety Science and Engineering, Xi'an University of Science and Technology, Xi'an, Shaanxi 710054, China. Email: minyan1230@xust.edu.cn
6. Doctor, College of Safety Science and Engineering, Xi'an University of Science and Technology, Xi'an, Shaanxi 710054, China. Email: yan_dongjie@qq.com
7. Master, College of Safety Science and Engineering, Xi'an University of Science and Technology, Xi'an, Shaanxi 710054, China. Email: 937351045@qq.com
8. Doctor, College of Safety Science and Engineering, Xi'an University of Science and Technology, Xi'an, Shaanxi 710054, China. Email: yangerhao@hotmail.com

ABSTRACT

Gas and coal dust, as potential harmful elements (PHEs) produced during mining, are considered two serious threats to the safe mining and the health of miners because of its explosive and pathogenic properties. To solve this problem, compound reagents enhanced water spraying technology was used to control the gas and coal dust in excavation working face in this study, and spray testing platform and wind tunnel were used to investigated the suppression effect for gas and coal dust. In addition, sink test, and agglomeration experiments were also used to analyse its suppression mechanism. Testing results show that the dust reduction efficiencies of compound reagents enhanced water spraying technology for four different particle sizes were 80 per cent, 91.43 per cent, 94.74 per cent and 94.46 per cent, and improved by 14 per cent, 10.86 per cent, 8.42 per cent and 4.25 per cent compared to water spraying. Moreover, the gas suppression efficiency also can be increased more than 20 per cent. Sink test, and agglomeration experiments results show that compound reagent improves the spreading efficiency of the liquid on the surface of the coal by changing its wettability, and this process can not only slow down the desorption rate of gas by forming a water-sealing effect in the coal surface indicated, but also promote the agglomeration of fine-grained coal dust.

INTRODUCTION

Coal is the main energy source in China, and its proportion of primary energy consumption in 2021 was 56 per cent. Coal dust and gas, as potentially harmful elements generated during the extraction and mining process, are considered as a serious threat to the underground working environment, the health of coalminers and safe production (Li *et al*, 2021; Zhao *et al*, 2022). In recent years, with the improvement of mining mechanisation and intelligence, the degree of coal fragmentation had increased significantly, which leaded to an increase of coal dust produced during mining (Zhou *et al*, 2017). Moreover, it has been shown that the higher the degree of coal fragmentation and the smaller the particle size, the faster the gas desorption rate (Liu and Liu, 2015), which was easily leaded to gas concentration exceeded in working face. High concentration of gas and dust mixture not only pollute the underground working environment, but also easily cause gas explosion and secondary explosion of dust, which seriously affect the safety production and health of underground workers (Prostański, 2018; Xu *et al*, 2018; Zhou and Qin, 2021). Therefore, it is important to achieve green and efficient mining and protect workers' health by carrying out coal dust-gas synergistic prevention and control in excavation face.

At present, many scholars have conducted a lot of research on gas control mechanism and technology. Li *et al* (2022) studied the effect of non-ionic surfactant APG0810 on the modification of coal structure and gas desorption inhibition effect, and proposed to control the gas concentration in working face by water injection in coal seam inhibiting gas desorption. Lin *et al* (2019) and Liu (2019) studied the wettability of surfactant and salt compound solution on coal, and studied the effect of compound solution on gas desorption process of coal by using HCA high pressure gas adsorption and desorption device. Zhang (2021) investigated the effect of surfactant-assisted spraying technology on gas desorption pattern of coal by self-developed multi-factor influenced coal body gas desorption spraying experimental platform, and the influencing factors were studied in combination with this test platform. Chen *et al* (2013) used the adsorption-water injection-desorption test platform to comparatively study the effects of water injection and surfactant injection solution on gas desorption from coal; and concluded that surfactant can effectively reduce the surface tension of the solution and make it spread rapidly on the surface of coal bodies, thus slowing down the rate of gas desorption.

Because surfactant has well wetting effect on the coal, it can effectively improve the capture efficiency of coal dust by liquid droplets. Therefore, surfactants are also commonly used as additives to improve dust reduction efficiency. Wang, X, *et al* (2019a, 2019b, 2019c) analysed the wetting mechanism of coal body with four different types of surfactants using contact angle and surface tension experiments; and the results concluded that anionic and nonionic surfactants were more effective than cationic and amphoteric surfactants for wetting coal. Li *et al* (2016) developed compounding solution by different anionic surfactants and analysed the effect of the compounded solutions on the improvement of coal wetting properties. Wang, P, *et al* (2019a, 2019b) used water spraying technology to control coal dust at working face under coalmine and studied the effect of different types of nozzles and atomisation methods on the dust reduction effect. In order to further improve the effectiveness of dust spraying, some scholars began to develop high-efficiency, multi-kinetic dust suppressants. Wang *et al* (2021) used Bacillus subtilis as a medium to synthesize BDS (biological dust suppressant) by bio-culture method and optimised its synthesis method by response surface optimisation method, which improved the dust suppression properties of the solution to some extent. Zhou *et al* (2020) and Yan *et al* (2021) used molecular modification chemistry to develop more efficient dust suppressants and investigated the dust suppression effect and mechanism by macroscopic experiments and microscopic simulations. Zhao *et al* (2022) developed a dust suppressant with efficient wetting and agglomeration properties by using a wetting agent compounded with an agglomeration agent, and investigated the mechanism of the effect of the compound reagent on coal dust wetting-agglomeration using surface tension, sink experiments, spraying experiments and electron microscopy scanning.

A lot of research has been carried out on gas control and coal dust control, and good control effectiveness have been achieved. The water spray technology can not only effectively control diffusion of coal dust during the driving and mining process, but also slow down gas desorption rate by spreading the solution on coal surface, thus reducing the gas concentration at the working face, which also can effectively reduce technology costs and optimise staff allocation. However, there is less research on coal dust-gas synergistic control method by water spray technology, and the mechanism of water spray on coal dust-gas synergistic control was not found. Therefore, the water spray gas desorption suppression test and water spray dust reduction test were selected to analyse gas and dust suppression effects under spray conditions. Synergistic suppression mechanism of gas and coal dust was obtained through wetting experiment and agglomeration experiment. The research results provide a certain theoretical basis for the application of coal dust-gas synergistic control technology.

EXPERIMENTAL SET-UP

Samples preparation

Coal sample selected

Coal samples were selected from the third mining area of Shaanxi Chenghe Shanyang Coal Mine Co. Firstly, the coal samples were crushed by crusher and 500 g coal samples with different diameter: 0.1~0.5 cm, 0.5~1 cm, 1~1.5 cm and 1.5~2 cm were sieved by standard sieves for water spray inhibition gas desorption experiment. In addition, 100 g coal samples with smaller diameter: 100~150 mesh; 150~200 mesh; 200~250 mesh; 250~300 mesh were sieved by small aperture standard sieve for wetting, agglomeration and spray dust reduction test. The industrial analysis of the coal samples was shown in Table 1.

TABLE 1

Industrial analysis indexes of experimental coal samples.

Coal rank	M_{ad}(%)	A_d(%)	V_d(%)	FC_{ad} (%)
Bituminous coal	0.58	11.02	12.15	76.38

Compound solution

The results obtained by authors and their team had shown that 0.05 wt% XTG+0.05 wt% SDBS compound solution not only has high wetting characteristics of coal dust, but also has good agglomeration characteristics of coal dust (Zhao *et al*, 2022). Therefore, 0.05 wt% XTG solution was selected to compound with 0.05 wt% SDBS for next experiment.

Firstly, weigh 499.5 g of deionised water in a beaker and place the beaker in a stirrer at constant temperature and start stirring; then weigh 0.25 g of xanthan gum and slowly add it to the deionised water and stir until all dissolved; finally weigh 0.25 g of SDBS and slowly add it to the XTG solution and stir until all dissolved.

Coal dust wettability experiment

The dust reduced effect of solution on coal body can be evaluated by direct and indirect methods (Chen *et al*, 2019). Indirect methods were mainly determined by contact angle and sinking experiments; since the contact angle is the contact between the surface and the droplet during the determination, it does not directly reflect the wetting effect of the solution on the coal dust (Xu *et al*, 2019), so the sinking method was used to determine its wettability in this study.

Firstly, measured 100 mL of the compound reagent in the sink test tube, weighed the prepared coal powder mass 0.1 g by electronic scale and slowly added it to the surface of the solution and recording the final sinking time of coal dust until all coal dust settled. For the same solution, three parallel tests were taken and the average value was taken as the final settling time.

Coal dust agglomeration experiment

Coal powder was dried by vacuum drying oven at 60°C for 6 hrs. Weigh four portions of 4 g of coal powder in the reaction dish and label it with the corresponding spraying reagent. The prepared reagent solution was quantitatively sprayed onto the coal dust surface by spraying device. Placed them in a cool and ventilated place natural air dry to constant weight. According to the experimental requirements, a high-definition camera was used to record the agglomeration pattern of coal dust after spraying and obtained the corresponding agglomeration pictures

Water spraying inhibition gas desorption test

Test bench introduction

Multi-factor influenced coal body gas desorption spraying experimental platform developed By Xi'an University of Science and Technology was used to investigate the effect of spraying on gas desorption inhibition. The test stand mainly consists of the model body, air flow system, solution spraying system, gas adsorption system and data acquisition system, which were shown in Figures 1 and 2.

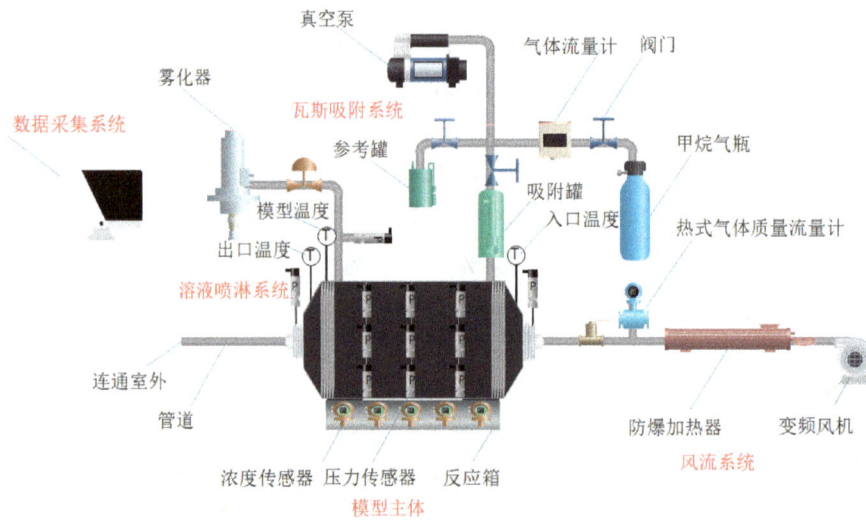

FIG 1 – Experimental system.

FIG 2 – Multi-factor influenced coal body gas desorption spraying experimental platform.

Experimental steps:

1. Coal sample preparation. Firstly, 100 g of the prepared coal sample was weighed and dried under vacuum at 60°C for 6 hr in a vacuum drying oven, ready for the experiment.

2. Negative pressure degassing. Prepared coal sample was placed into adsorption tank. Close the gas inflating valve, open the degassing valve, and finally open the vacuum pump and degas for 3 hrs under constant temperature.

3. Gas adsorption. After degassing was completed, close the degassing valve, open the gas filling valve and fill with 0.8 MPa gas, adsorption to saturation at constant temperature.

4. Gas desorption inhibition experiment by spraying. Firstly, open the data acquisition system, set the save path and start data acquisition; then open the spray system and ventilation system; after the air flow in the experimental chamber became stable, open the valve at the bottom of the adsorption tank and inject the gas-containing coal body into the chamber, real-time monitoring of gas concentration through gas concentration sensing in the chamber.

5. End of the experiment. When the gas concentration stabilised, stop the experiment and save the data.

Water spraying dust reduction test

Test bench introduction

Spray dust reduction wind tunnel simulation experimental platform developed by Xi'an University of Science and Technology was used to investigate the dust reduction efficiency under water spray. The experimental platform mainly consists of a simulated tunnel, dust generation system, dust monitoring system, spray system, exhaust gas treatment system, ventilation system, and spray booster system, which were shown in Figure 3.

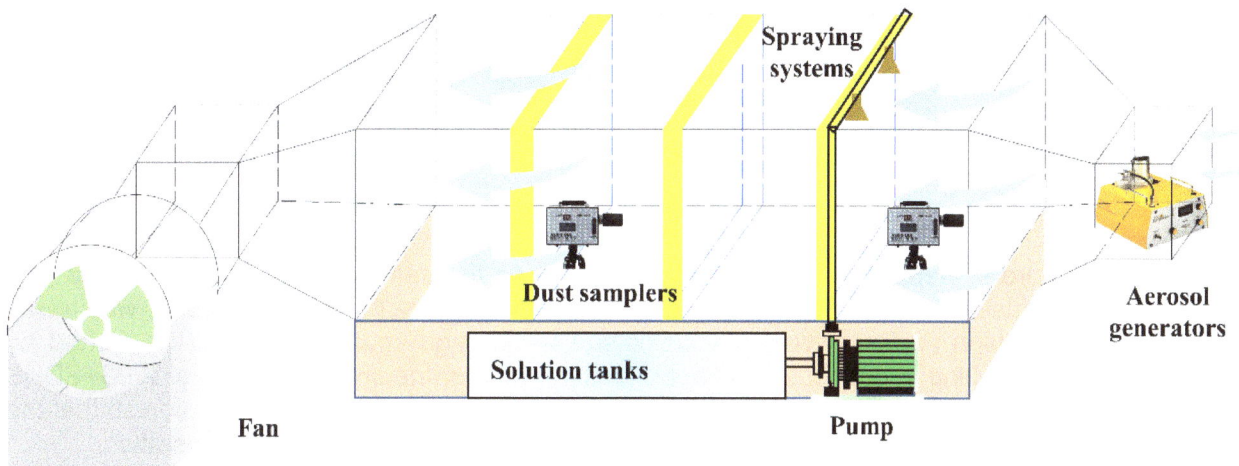

FIG 3 – Spray dust reduction wind tunnel simulation experimental platform.

Experimental steps:

1. Coal sample preparation. Firstly, 20 g of prepared coal samples with different particle sizes were weighed and dried under vacuum at 60°C for 6 hrs in a vacuum drying oven to prepare for the experiment.

2. Firstly, weigh 14 985 g of deionised water in the solution bucket, place the solution bucket in the stirrer and start stirring, then weigh 7.5 g each of xanthan gum and SDBS reagent, slowly add to deionised water and stir until all dissolved.

3. Dust sampling. Put the dust sampling filter paper in the vacuum drying oven for drying, and then put it in the analytical balance to weigh the mass of the filter paper without sampling. Then the filter paper was installed in the dust sampler and the dust sampler was placed in the designated location in the simulated tunnel ready for sampling.

4. Starting the dust reduction experiment. Firstly, turned on the fan and adjusted the wind speed to 0.5 m/s; after the wind flow stabilised, turned on the dust sampler, dust aerosol generator and spraying system at the same time to start the spraying dust reduction test.

5. Results analysis. After sampling, took out the sampling filter paper and weigh its mass after drying. Dust concentration was calculated based on the sampled gas flow rate and sampling time.

RESULTS AND DISCUSSION

Wettability results analysis

The wetting characteristics of the compound solution and deionised water on coal dust were investigated by sinking experiments and sinking results were shown in Figure 4.

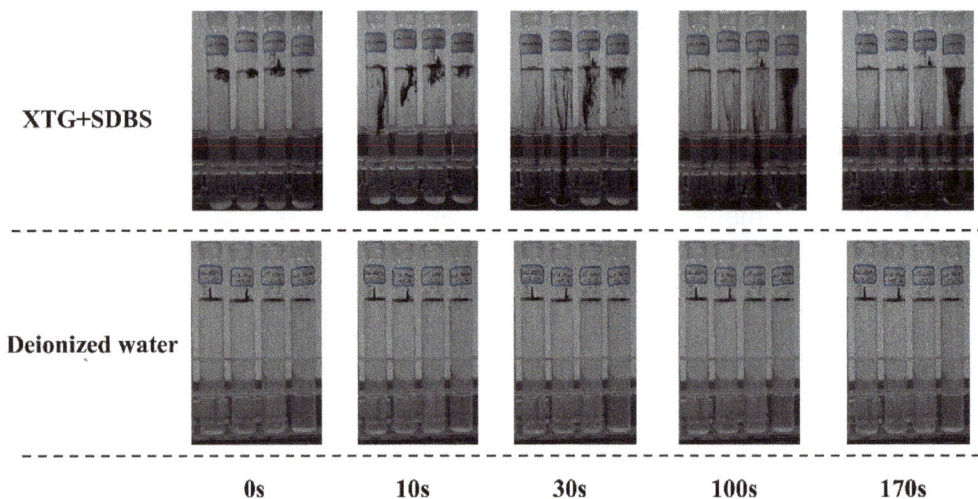

FIG 4 – Sinking results.

As shown in Figure 4, wettability had a significant effect on the settling time of coal dust. No sink occurred of coal dust with four different particle sizes on the surface of deionised water, which was mainly due to the high surface tension of deionised water, and coal dust cannot be wetted by deionised water, resulting in coal dust was not effectively sink on deionised water. For XTG+SDBS complex reagent, the surfactant can effectively reduce the surface tension of the solution so that the aqueous solution can wet the coal dust more efficiently thus causing the coal dust to settle.

The particle size of coal dust had a certain influence on the settling time. As shown in Figure 5. there was no sink of coal dust with four different particle size on the surface of deionised water within 200 s. The settling time on the surface of XTG+SDBS composite reagent gradually increases with the decrease of coal dust particle size. It took longest time for coal dust with particle size 250~300 mesh to settle completely on the surface of XTG+SDBS compound solution and sinking time was 164.78 s.

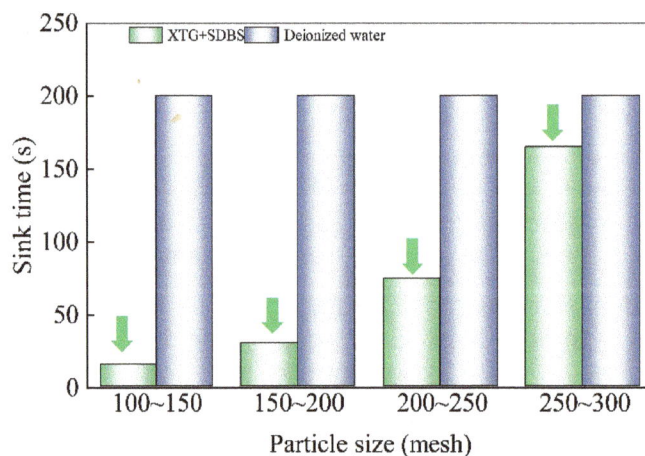

FIG 5 – Sink time with different solution and particle size.

The sinking rate of coal dust with different particle sizes in compound solution was analysed and the results were shown in Figure 3. When the particle size of coal dust is 120~140 mesh, the settling rate can reach 6.62 mg/s. As the particle size increases, the settling rate gradually decreases, and when the particle size of coal dust is reduced to 250–300 mesh, the settling rate decreases to 0.61 mg/s. The relationship between particle size and sedimentation rate can be described by power function, which was shown in Figure 6.

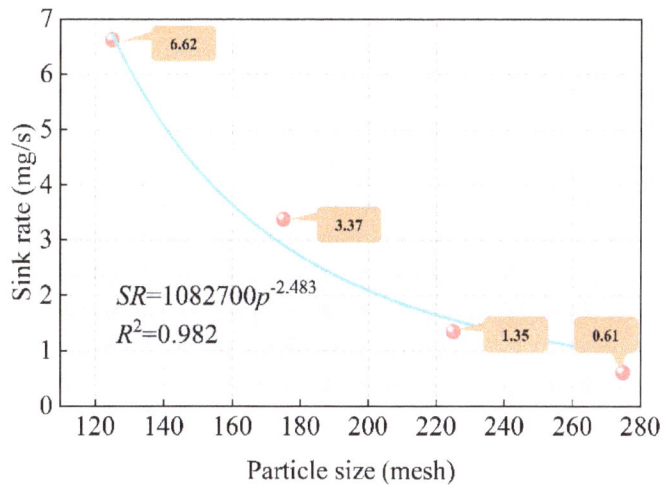

FIG 6 – The relationship between particle size and sedimentation rate.

Analysis of coal dust agglomeration characteristics

The spraying device was used to study agglomeration characteristics for coal dust with different particle size, and high-definition camera was used to record agglomeration characteristics of coal dust after spraying compound reagent and deionised water respectively, which was shown in Figure 7.

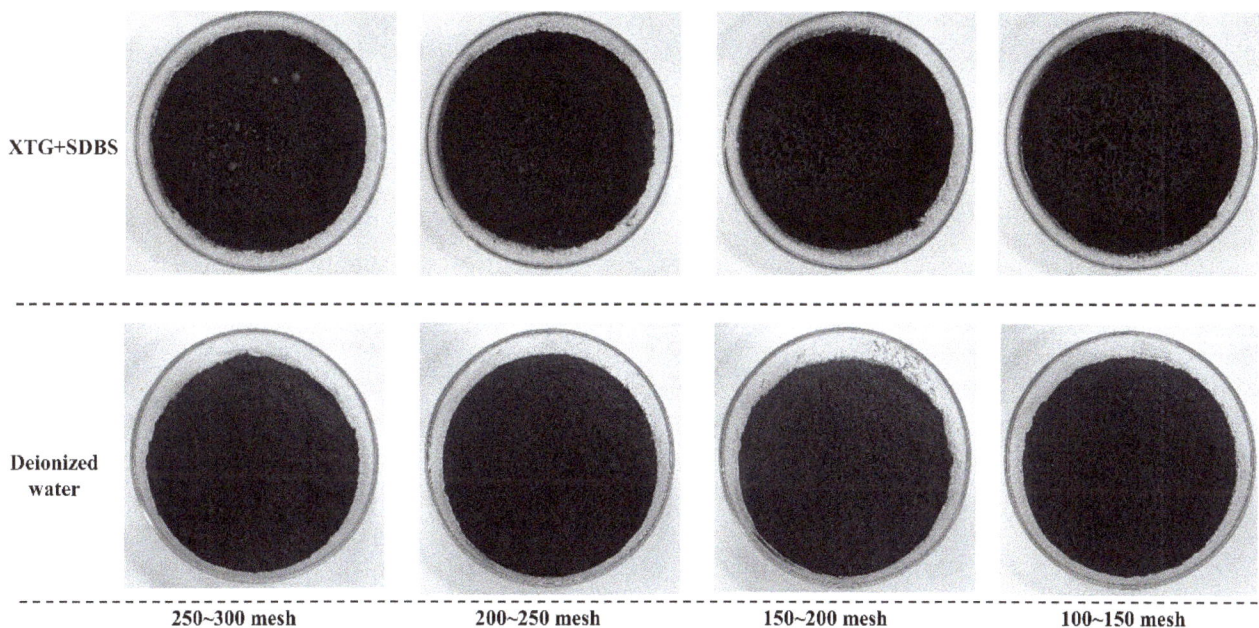

FIG 7 – Agglomeration characteristic results.

As shown in Figure 4, the deionised water spray-treated coal samples showed smaller pores and the surface was too soft to form larger agglomerates. Compared with the deionised water treated coal samples, the XTG+SDBS compound solution spray-treated coal samples showed large agglomerates with structure like mesh. This agglomerate can effectively gather fine particles of coal dust to form larger particles, which can not only prevent secondary dust, but also accelerate the settling of suspended coal dust.

The agglomeration effect of XTG+SDBS compound solution and deionised water on coal dust with different particle sizes was also different. As shown in Figure 4a, when particle size was 250~300 mesh, most of the coal samples treated with compound solution were more dispersed, and some of coal dust can agglomerate into water bead-like agglomerates. When the particle size increased to 150~200 mesh, the agglomerates in middle area begun to appear irregularly reticulated, which was better than that of teardrop agglomerates. When the particle size increased to 100~150 mesh, agglomerate area further increased, the irregular mesh structure became more obvious, and the agglomeration effect increased significantly. Coal dust treated by deionised water spray shown a looser and softer agglomerates, showing different regular hollow structures in middle, which could not form larger agglomerates and had poor agglomeration effect.

Inhibition effect of gas desorption

Multi-factor influenced coal body gas desorption spraying experimental platform was used to investigate the inhibition effect under different spraying solution and particle size, and the results were shown in Figure 8.

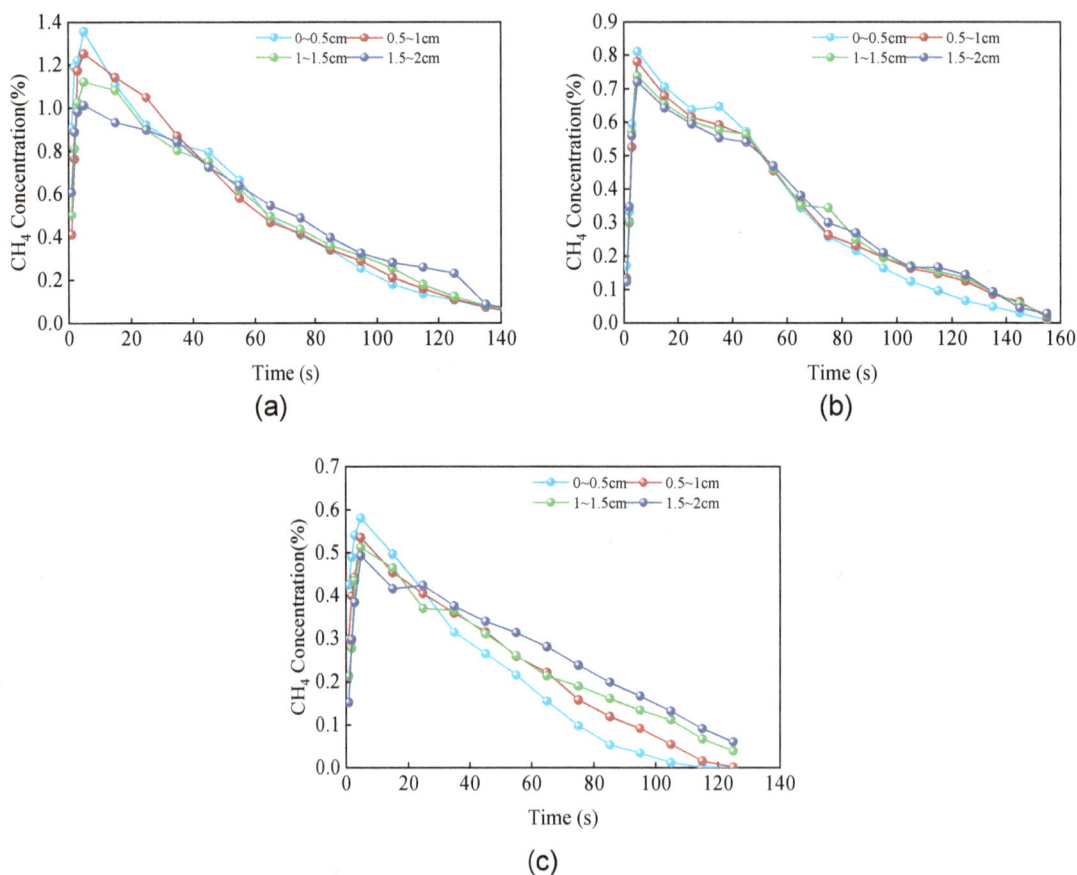

FIG 8 – Inhibition effect of gas desorption: (a) Without water; (b) Spraying deionised water; (c) Spraying compound solution.

From Figure 8, it can be seen that the gas concentration in experimental chamber increased rapidly in first 5 s when no spraying measures, water spraying measures and XTG+SDBS compound reagent spraying measures were taken. Due to adsorption pressure reduce to atmospheric pressure, the pressure gradient increases, resulting in rapid desorption of gas and reaching a peak at 5 s. With the gas gradually desorbed, the gas desorption rate gradually decreased, and the gas concentration in experimental chamber gradually decreased after 5 s under the influence of ventilation.

Compared the gas suppression effect with no spraying measures, water spraying measures and XTG+SDBS compound solution spraying measures, it can be seen that the maximum gas concentration in the experimental chamber was 0.93~1.11 per cent without any measurements; When water spraying measures were taken, the maximum gas concentration inside the experimental chamber decreased slightly, with the maximum gas concentration ranging from 0.72 to 0.81 per cent. When XTG+SDBS compound solution spray measures were taken, the gas concentration in the experimental chamber was rapidly reduced to 0.49 per cent~0.58 per cent, which had a significant inhibitory effect on gas desorption.

Inhibition effect of gas desorption for coal samples with different particle size by spray was different. As shown in Figure 8a, without any measurements, the smaller particle size was, the greater gas desorption rate and desorption amount would be. Therefore, as the particle size decrease, the maximum gas concentration in chamber gradually increase. Because the small size coal body has higher gas desorption rate, the gas desorption amount of large coal particle size is higher than that of small coal particle size after 40 s. The gas concentration of the large-size coal sample in experimental chamber is gradually larger than that of the small-size coal sample after 40 s.

Spray dust reduction analysis

Spray dust reduction wind tunnel simulation experimental platform was used to simulate dust reduction efficiency with no spraying measures, water spraying measures and XTG+SDBS compound reagent spraying measures under 0.5 m/s wind velocity. The results were shown in Figure 9.

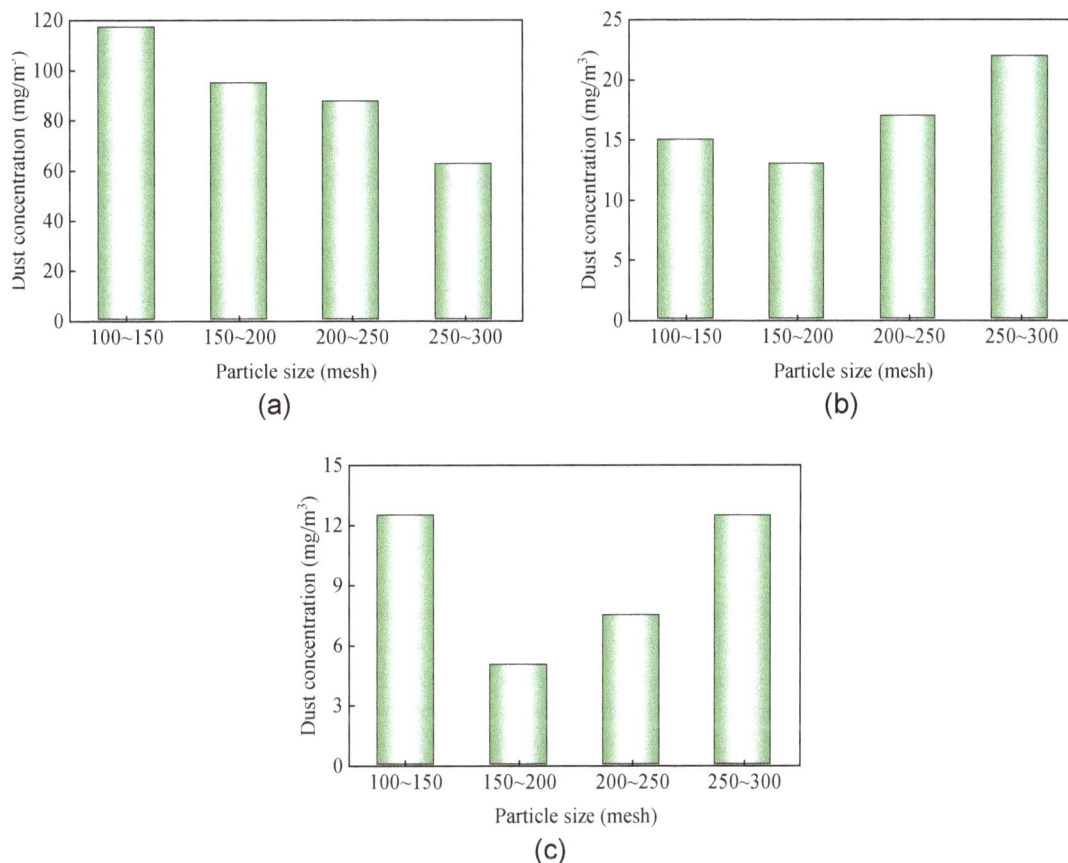

(a)

(b)

(c)

FIG 9 – Spraying dust reduction effect: (a) Without spraying measures; (b) Water spraying measures; (c) XTG+SDBS compound reagent spraying measures.

Seen in Figure 9a, coal dust with different particle sizes were generated by dust generators. Without spraying measures, the dust concentration in the simulated tunnel decreased as the particle size decrease. 100~150 mesh dust particles are not easily carried by the wind flow under 0.5 m/s wind velocity. They exist longer in the simulated roadway, and the mass expelled with the wind flow is small. Therefore, the smaller the particle size was, the lower dust concentration would be.

As shown in Figure 9b, When water spraying measures were taken, the dust concentration of 250~300 mesh, 200~250 mesh, 150~200 mesh and 100~150 mesh were reduced from 62.5 mg/m^3, 87.5 mg/m^3, 95 mg/m^3 and 117.5 mg/m^3 to 22.5 mg/m^3, 17 mg/m^3, 13 mg/m^3 and 11.5 mg/m^3 respectively, and the dust reduction efficiency were 64 per cent, 80.57 per cent, 86.32 per cent and 90.21 per cent.

As shown in Figure 9b, when XTG+SDBS compound solution measures were taken, the dust concentration of 250~300 mesh, 200~250 mesh, 150~200 mesh and 100~150 mesh were reduced from 62.5 mg/m^3, 87.5 mg/m^3, 95 mg/m^3 and 117.5 mg/m^3 to 12.5 mg/m^3, 7.5 mg/m^3, 5 mg/m^3 and 6.5 mg/m^3, respectively, and the dust reduction efficiency were 80 per cent, 91.43 per cent, 94.74 per cent, 94.46 per cent.

CONCLUSION

1. Agglomeration and wetting effects of XTG+SDBS compound solution on coal samples with different particle sizes are different. As the particle size decreases, the time, dust particles full sink on the surface of the solution, gradually increases, and its wettability was worse. The relationship between wetting rate and particle size can be described by $SR=1082700P^{-2.483}$. With the decrease of particle size, wetting rate decreased sharply first and became stable after that. XTG+SDBS compound solution can make coal dust agglomerate into mesh structure and form large agglomerates, which have a better suppression effect on coal dust. Moreover, The agglomeration effect gradually increases with the increase of coal particle size.

2. XTG+SDBS compound solution spraying can effectively inhibit gas desorption, and inhibition effect is not the same for different particle size coal samples. The gas concentration in experimental chamber increased to the peak rapidly, and then decreased slowly. As the particle size decrease, the peak concentration gradually increase. Before 40 s, the smaller the particle size of coal sample is, the higher the gas concentration will be. After 40 s, The gas concentration of 100~150 mesh coal sample is gradually higher than that of small particle size. In the particle size range of 100~300 mesh, the gas suppression effect becomes more obvious with the increase of particle size.

3. XTG+SDBS compound solution spraying can effectively suppress dust, and suppression effect is not the same for different particle size. Compared coal dust concentration under four different particle size with no spraying measures, water spraying measures and XTG+SDBS compound reagent spraying measures. The inhibition efficiency of XTG+SDBS compound solution spray on 250~300 mesh, 200~250 mesh, 150~200 mesh and 100~150 mesh coal dust were 80 per cent, 91.43 per cent, 94.74 per cent and 94.46 per cent respectively. Compared with water spray, the suppression efficiency was improved by 14 per cent, 10.86 per cent, 8.42 per cent and 4.25 per cent respectively.

ACKNOWLEDGEMENTS

This study was funded by the National Natural Science Foundation Key Project of China (51734007), the National Natural Science Foundation of China (52074217, 52174207), the Natural Science Foundation of Shaanxi Province (2020JC-48). Thanks to Prof Li Shugang and Prof Lin Haifei for their guidance on this study. In addition, Thanks for the guidance and assistance for this research from the Mine Ventilation Conference 2022 organising team.

REFERENCE

Chen, S, Jin, L, Chen, X and Sun, C, 2013. Analysis on influence of water infusion by adding surfactant upon gas desorption, *Mining safety and environmental protection*, 40:12–14.

Chen, Y, Xu, G, Huang, J, Eksteen, J, Liu, X and Zhao, Z, 2019. Characterization of coal particles wettability in surfactant solution by using four laboratory static tests, *Colloids and Surfaces A: Physicochemical and Engineering Aspects*, 567:304–312, https://doi.org/10.1016/j.colsurfa.2019.01.068

Li, J, Zhou, F and Liu, H, 2016. The Selection and Application of a Compound Wetting Agent to the Coal Seam Water Infusion for Dust Control, *International Journal of Coal Preparation and Utilization*, 36:192–206, https://doi.org/10.1080/19392699.2015.1088529

Li, S, Yan, D, Yan, M, Bai, Y and Yue, M, 2022. Effect of alkyl glycoside on coal structure modification and methane desorption characteristics, *Journal of China coal society*, 47:286–296, https://doi.org/10.13225/j.cnki.jccs.yg21.1887

Li, S, Zhao, B, Lin, H, Shuang, H, Kong, X and Yang, E, 2021. Review and prospects of surfactant-enhanced spray dust suppression: Mechanisms and effectiveness, *Process Safety and Environmental Protection*, 154:410–424, https://doi.org/10.1016/j.psep.2021.08.037

Lin, H, Tian, J, Liu, D, Yan, M and Ding, Y, 2019. Inhibitory effect of SDBS and CaCl₂ compound solution on gas desorption of coal, *China safety science journal*, 29:149–155, https://doi.org/10.16265/j.cnki.issn1003-3033.2019.11.024

Liu, D, 2019. Experimental study on the effect of surfactant on the gas desorption of coal, MA thesis, Xi'an University of Science and Technology, https://kns.cnki.net/KCMS/detail/detail.aspx?dbname=CMFD202001&filename=1019618995.n.

Liu, Y and Liu, M, 2015. Effect of particle size on difference of gas desorption and diffusion between soft coal and hard coal, *Journal of China coal society*, 40:579–587, https://doi.org/10.13225/j.cnki.jccs.2014.0380

Prostański, D, 2018. Development of research work in the air-water spraying area for reduction of methane and coal dust explosion hazard as well as for dust control in the Polish mining industry, *IOP Conf Ser: Mater Sci Eng*, 427:012026, https://doi.org/10.1088/1757–899X/427/1/012026

Wang, H, He, S, Zhang, Q and Zhao, X, 2021. Experimental study on synthesis of biological dust suppressant by microbial fermentation, *Journal of China coal society*, 46:477–488, https://doi.org/10.13225/j.cnki.jccs.XR20.1962

Wang, P, Shi, Y, Zhang, L and Li, Y, 2019a. Effect of structural parameters on atomization characteristics and dust reduction performance of internal-mixing air-assisted atomizer nozzle, *Process Safety and Environmental Protection*, 128:316–328, https://doi.org/10.1016/j.psep.2019.06.014

Wang, P, Zhang, K and Liu, R, 2019b. Influence of air supply pressure on atomization characteristics and dust-suppression efficiency of internal-mixing air-assisted atomizing nozzle, *Powder Technology*, 355:393–407, https://doi.org/10.1016/j.powtec.2019.07.040

Wang, X, Yuan, S and Jiang, B, 2019a. Wetting Process and Adsorption Mechanism of Surfactant Solutions on Coal Dust Surface, *Journal of Chemistry*, 2019:1–9, https://doi.org/10.1155/2019/9085310

Wang, X, Yuan, S and Jiang, B, 2019b. Experimental investigation of the wetting ability of surfactants to coals dust based on physical chemistry characteristics of the different coal samples, *Advanced Powder Technology*, 30:1696–1708, https://doi.org/10.1016/j.apt.2019.05.021

Wang, X, Yuan, S, Li, X and Jiang, B, 2019c. Synergistic effect of surfactant compounding on improving dust suppression in a coal mine in Erdos, China, *Powder Technology*, 344:561–569, https://doi.org/10.1016/j.powtec.2018.12.061

Xu, C, Wang, D, Wang, H, Ma, L, Zhu, X, Zhu, Y, Zhang, Y and Liu, F, 2019. Experimental investigation of coal dust wetting ability of anionic surfactants with different structures, *Process Safety and Environmental Protection*, 121:69–76, https://doi.org/10.1016/j.psep.2018.10.010

Xu, G, Chen, Y, Eksteen, J and Xu, J, 2018. Surfactant-aided coal dust suppression: A review of evaluation methods and influencing factors, *Science of The Total Environment,* 639:1060–1076, https://doi.org/10.1016/j.scitotenv.2018.05.182

Yan, J, Nie, W, Xiu, Z, Yuan, M, Zhou, W, Bao, Q, Peng, H, Niu, W and Yu, F, 2021. Development and characterization of a dust suppression spray agent based on an adhesive NaAlg−gln−poly/polysaccharide polymer, *Science of The Total Environment*, 785:147192, https://doi.org/10.1016/j.scitotenv.2021.147192

Zhang, K, 2021. Experimental study and application of the effect of external leaching solution on coal gas desorption, MA thesis, Xi'an University of Science and Technology, doi: 10.27397/d.cnki.gxaku.2021.000891

Zhao, B, Li, S, Lin, H, Cheng, Y, Kong, X and Ding, Y, 2022. Experimental study on the influence of surfactants in compound solution on the wetting-agglomeration properties of bituminous coal dust, *Powder Technology*, 395:766–775, https://doi.org/10.1016/j.powtec.2021.10.026

Zhou, G, Ding, J, Ma, Y, Li, S and Zhang, M, 2020. Synthesis and performance characterization of a novel wetting cementing agent for dust control during conveyor transport in coal mines, *Powder Technology*, 360:165–176, https://doi.org/10.1016/j.powtec.2019.10.003

Zhou, Q and Qin, B, 2021. Coal dust suppression based on water mediums: A review of technologies and influencing factors, *Fuel,* 302:121196, https://doi.org/10.1016/j.fuel.2021.121196

Zhou, Q, Qin, B, Ma, D and Jiang, N, 2017. Novel technology for synergetic dust suppression using surfactant-magnetized water in underground coal mines, *Process Safety and Environmental Protection*, 109:631–638, https://doi.org/10.1016/j.psep.2017.05.013

Operations

A review of the GAG as an effective firefighting tool

D Chalmers[1], G Si[2] and F Soleimani[3]

1. Senior Lecturer, MERE, UNSW Sydney NSW 2052. Email: d.chalmers@unsw.edu.au
2. Senior Lecturer, MERE, UNSW Sydney NSW 2052. Email: g.si@unsw.edu.au
3. Casual Academic, MERE, UNSW Sydney NSW 2052. Email: f.soleimani@unsw.edu.au

ABSTRACT

After the sustained use of the GAG (Górniczy Agregat Gaśniczy) engine at a Queensland coalmine, it has raised questions as to its effectiveness as a remote firefighting tool.

All firefighting apparatus have advantages and limitations, however the deployment of any apparatus must be based on its likelihood to succeed and effectively extinguish the fire or bring it under control.

A recent study conducted for ACARP by researchers at UNSW has highlighted that there is more to be learned about the GAG. As coalmines are mining at greater depths and fan pressures are much greater than they were when the GAG was introduced, deploying the GAG may be a total waste of resources and may place operators at greater risk.

When the GAG unit is positioned on the surface of the mine, the pressure it generates has to overcome buoyancy of the exhaust gases and the frictional resistance of the borehole and part of the mine circuit. Buoyancy in deeper holes can lead to the GAG pressure relief system opening, venting out the products to the atmosphere without delivering any product to the mine workings. Additionally, frictional resistance is a function of distance. Deeper holes will have a greater frictional resistance than shallow holes. Delivering sufficient products from the unit will be limited by the available pressure that has to overcome the buoyancy, and increased mine resistance that consumes any residual pressure.

By documenting the limitations of this unit it will enable better decisions to combat a fire or spontaneous combustion event.

INTRODUCTION

Inertisation of goaf areas is one of the most effective mechanisms to combat the rise of spontaneous combustion in that it excludes oxygen from the reactive material. Other techniques such as pressure limitation and pressure balancing can assist in prevention of an uncontrolled heating. Inactive goaves can be sealed and pressure balanced, however cracking through to the surface can open leakage paths giving rise to oxygen ingress. Active goaves are more problematic in that there is a balance between maintaining low oxygen levels in the goaf areas and controlling the methane/gas levels on the working face. Fluctuations in barometric pressure provide a further level of complexity causing contraction and expansion of the goaf gas volume.

This inertisation of the atmosphere around a fire or heating can be achieved by several techniques. These techniques include self inertisation by sealing or by providing additional inert gas from an external source.

One external source is The GAG-3A Jet engine. The GAG (Górniczy Agregat Gaśniczy) translates from Polish as Mine Fire Extinguisher. This study examines the application of docking stations to aid in the deployment of the GAG-3A to modern underground coalmines.

The Polish Mines Rescue service brought the GAG unit to Australia for ACARP Project C6019 and operational support was provided by SIMTARS.

The GAG-3A is described as an inert gas generator that utilises a small aviation jet engine with a single stage afterburner. The purpose of the afterburner in this instance was to provide a more complete combustion. Normally, a jet engine has a thrust nozzle to provide the motive force to propel the aircraft. In the case of the GAG-3A this was removed and Bell *et al* (1997) stated that there was no discernible thrust generated.

As the exhaust gases from the jet are around 480°C, a series of water cooling ducts 8 m in length form the exhaust delivery system. This adds considerable water vapour to the inert gas bringing the open circuit capacity to approximately 20 m³/s and 2 per cent oxygen. It also contains water as a fine mist. 'The presence of this water vapour may improve the cooling and quenching capacity of the gases, although the exact effect is not yet fully understood' Bell *et al* (1997).

PREVIOUS WORK

Several trials were undertaken to demonstrate the applicability of the GAG as an effective inertisation tool. The findings showed that the products from the GAG differentially accumulated to the higher side of the mine workings. It was recommended that 'each usage of the GAG would be dependent on-site-specific factors and may not be suitable for every mine' (Bell *et al,* 1997).

The deployment of the GAG at Loveridge Mine (2003) and Pinnacle Mine (2003–2004) showed that the GAG was susceptible to high backpressures due to buoyancy effects, barometric fluctuations and mine resistance. Figure 1 shows the leakage of GAG product from the mine at the point of delivery on the Sugar Run Slope, Loveridge Mine.

FIG 1 – Escaping GAG product (from Mucho *et al*, 2005).

Mucho *et al* (2005) report that the GAG backpressure reached 2367 Pa.

The Pinnacle Mine deployment reported less leakage issues however, they reported similar periods to the Loveridge Mine where the backpressure varied in a similar fashion. The report suggested that it took approximately seven days to inert the mine, however seven days after the commencement of the GAG a pressure increase occurred on the transducers emplaced to detect underground explosions indicating that there had been a possible explosion underground.

The GAG was deployed at Newlands (Southern) Colliery as part of a trial to validate the Ventgraph simulation program. A GAG port was fitted to a shaft capping and an 80 m³/sec fan was set-up on the old belt drift. This fan enabled the ventilation of the outbye part of the mine.

The GAG was started about 4:30 pm 15 December 2005 and ran until mechanical failure at 2:40 am 16 December 2005.

During the injection process using the GAG, the volume was measured at 6.5 m³/s. Underground gas readings were taken using a tube bundle system and it was observed that although the shaft bottom area inertised, the gas did not migrate throughout the mine as predicted.

The main issues during the operation were the backpressure from the buoyancy of the gas in the shaft and the cooling of the gas reducing the quantity (Stone, D.) as reported by Tomlinson (2006).

Kestrel Mine was at the end of its economic life and in 2013 it was thought that it would be an opportunity to use the GAG to inert the underground mine atmosphere.

The planned inertisation circuit was the entire 1 km length of the cross measure conveyor drift on a 1:4 decline from the surface to 3CT at pit bottom; continuing along the belt roadway with approximately 2 km of conveyor belt, towards the LW311 Block; then returning to the up-cast shaft. Two trials were conducted with the first running a low ventilation rate of 30 m³/s and the second with the fans off, relying on the GAG to inject the inert gas into the workings.

The GAG circuit was segregated so that the GAG flow could be analysed. And 10 m³/s flow rate was established in this circuit on day one of the trial. With the additional air flowing in a parallel M & M drift, so that observation could be made by personnel from QMRS. An extensive monitoring system was set-up monitoring temperature and gases.

Several key points can be drawn from this exercise.

- As the run progressed the pressures inside the drift increased and the leakage of GAG product increased proportionately.

- The temperature measurement showed a significant trend in that only the first 200 m of the drift approached the GAG output temperature of 89°C. Beyond 232.6 m shows that the atmosphere had not been heated as much by the GAG exhaust. The authors referred to a significant cooling event that was attributed to a broken compressed air line.

- The velocity of the GAG product as measured by the temperature indicates that the GAG product slowed with depth. This is consistent with the buoyancy of the GAG product producing leakage, and the loss of volume due to shedding the moisture from the atmosphere.

- Damage to infrastructure due to heat and rib spalling occurred with the deployment of the GAG.

The planned inertisation circuit was the entire 1 km length of the cross measure conveyor drift on a 1:4 decline from the surface to 3CT at pit bottom; continuing along the belt roadway with approximately 2 km of conveyor belt, towards the LW311 Block; then returning to the up-cast shaft. Two trials were conducted with the first running a low ventilation rate of 30 m³/s and the second with the fans off, relying on the GAG to inject the inert gas into the workings.

The GAG circuit was segregated so that the GAG flow could be analysed. And 10 m³/s flow rate was established in this circuit on day one of the trial. With the additional air flowing in a parallel M & M drift, so that observation could be made by personnel from QMRS. An extensive monitoring system was set-up monitoring temperature and gases.

Figure 2 shows the progression of hot gases down the drift. It is clear from this graph that the velocity of hot gases slows progressively as the depth increases. It is unclear from this report, the mechanisms involved in this process. There would be some cooling occurring, water dropping out of the GAG product, leakage due to rising buoyancy and stratification of the GAG product.

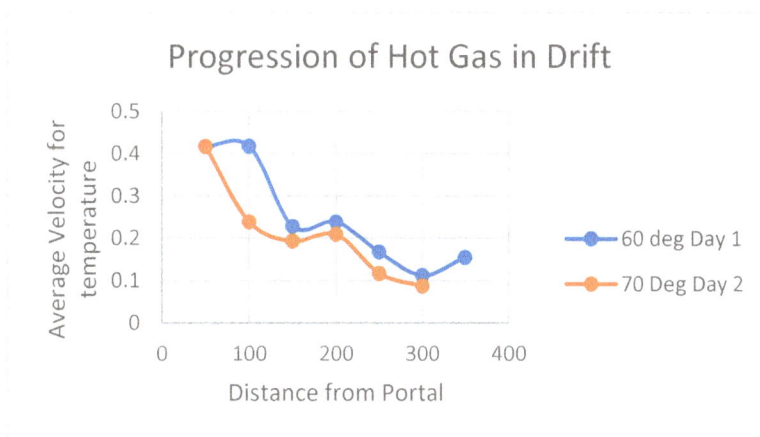

FIG 2 – Temperature–velocity relationship (after Watkinson *et al*, 2014).

The influence of each of these would affect the arrival time for the GAG products. However, the velocity of a 20 m³/s body of gas in a drift should be constant. Similarly the velocity of dry gas should also be constant if the water had dropped out of the product.

The gas readings also show that diffusion of the hot gases is a significant factor in the behaviours exhibited in the drift. The gas readings taken show that the lower oxygen concentrations occurred in the lower (below waist height) portion of the roadway.

Watkinson *et al* (2014) suggested that measuring the progression of the temperature front would equate to the speed of the GAG product moving down the drift. They also state that the cross-sectional area of the conveyor drift is 16.3 m²/s.

TABLE 1

Expected velocities for GAG product.

Volume m³/s	Velocity m/s
25	1.53
20	1.23
7	0.43

The velocity of the Temperature drops from 0.4 m/s to less than 0.1 m/s over the 300 m length of conveyor drift. This supports the concept that the GAG product is stalled in the drift and that the rise in temperature is due to conduction/convection heating up the interface between the GAG product and the atmosphere in the drift. While considerable stratification would be taking place the hot gases would be moving into the cooler atmosphere faster than the conveyor atmosphere moving into the GAG product layer. This would account for the presence of higher concentrations of inerts in the floor region as the cooler gases descend.

RESEARCH PROJECT

The literature search raised several questions about its performance and its ability to inertise portions of a mine. It was clear that on face value that the GAG product was of a similar density to methane due to its temperature and its moisture content. As such the buoyancy of this gas would make it difficult to send down a borehole, shaft or inclined drift as it would provide a substantial backpressure on the GAG unit.

The frictional resistance of the mine would also contribute to the back pressure on the GAG unit however this would diminish as the airways filled with the less dense GAG product.

The ambient air temperatures would also affect the buoyancy pressure, cooler days/nights providing greater buoyancy than warm to hot days and nights.

The questions that the project sought to answer included:

- What was the maximum depth that the GAG could push product into a mine through various shafts/boreholes ranging from 1 m–5.5 m?

- Are the current docking stations on Mine A and Mine B able to provide GAG product into each mine to control and extinguish a modest fire underground?

- What would be the conditions required if the fans for the mine were stopped and the GAG provided the pressure to push the GAG product underground?

Methodology

Two validated coalmine ventilation models were provided and a third model was constructed in Ventsim® as a sanity check and familiarisation model. Ventsim® was chosen as it was the simulation program that was used by these mines and a GAG simulator tool as one of the available features.

The plan was to provide a fire simulation on or near the tailgate of the longwall and attach the GAG simulator to the nearest Shaft or Docking Station on the model. Monitor the oxygen levels along the pathways to the fire and measure the time taken to reduce the oxygen level below 4 per cent to indicate that the GAG product had reached the fire.

The plan was then to provide modified scenarios to provide a good statistical base for the project conclusions.

Results

The first simulation run highlighted a problem with the GAG simulator and with considerable help and advice from Ventsim support it was concluded that the GAG Simulator did not perform as the GAG would on the mine. The program uses the available information for the GAG, namely gas concentration, moisture content, temperature and volume. On discovering this the research team searched through all available documents to find pressure-quantity performance curves (PQ).

This was a key piece of information that would allow for appropriate simulations to be undertaken. Throughout the project it became clear that this performance curve did not exist and that a change of tack was needed to ensure that the project would create meaningful results.

The PQ curve was developed on the assumptions that the open circuit capacity of the GAG was 25 m^3/s and that the stall pressure was 2.4 kPa.

This curve was applied to the GAG simulator and models were then tested using this assumed curve. Each simulation showed that the GAG product would not enter the mine and that the stall pressure was exceeded in each case.

Once it was realised that the GAG could not deliver to any of the three models without going into stall the simulations were run again without the assumed fan curve to determine the pressures that would need to be applied to the docking points to force the GAG product into the mine.

Table 2 shows the pressures required at the docking station for Mine A with four different flow rates and mine fans running and a second set of simulations with the mine fans off.

TABLE 2

Summary of eight simulations for Mine A.

	GAG Flowrate (m3/s)	Other fans status	Required pressure to deliver GAG (kPa)	O2%	Temperature at the monitoring point (°C)
1	14	on	38.3	18.6	47.1
2	16	on	45	18.4	48.2
3	18	on	51.3	18.1	49.1
4	20	on	57.5	17.9	49.8
5	22	on	63.3	17.6	50.4
6	14	off	37.3	2.9	81.9
7	16	off	43.9	2.9	81.2
8	18	off	50.3	2.9	80.5
9	20	off	56.4	2.9	79.8
10	22	off	62.3	2.9	79

Similarly, Table 3 shows the results of Mine B simulations with the same criteria. The overall pressures are lower due to the mine geometry. The minimum pressure required was 10 times the stall pressure of the GAG.

In both mines the size of the fire was relatively small and the GAG was deployed relatively quickly on the docking stations.

TABLE 3

Summary of eight simulations for Mine B.

Fan Flowrate (m3/s)	Other fans status	Required pressure to deliver GAG (kPa)	O2%	Dry bulb Temperature at the monitoring point (°C)
14	on	25.6	18.3	50.9
16	on	30.6	18	52.6
18	on	35.7	17.6	54
20	on	40.5	17.3	55.2
22	on	45.3	17	56.2
14	off	25.4	2.9	83
16	off	30.5	2.9	82.6
18	off	35.5	2.9	82.1
20	off	40.4	2.9	81.7
22	off	45.2	2.9	81.1

DISCUSSION

For a fluid to flow it is bound by the basic laws of physics including Bernoulli's, Atkinsons and thermodynamic equations. These laws govern the direction of flow and the interaction of them provide the underpinning knowledge enabling the simulation of ventilation models.

Since the GAG output is hot at approximately 89°C natural ventilation pressure (buoyancy) is substantial. Figure 3 shows the relationship between the maximum pressure of the GAG (2400 Pa) and the available pressure depending on the depth of the borehole/shaft and the ambient temperature of the air in the borehole/shaft. The hotter the air is the deeper the GAG can penetrate due to the lower temperature differential.

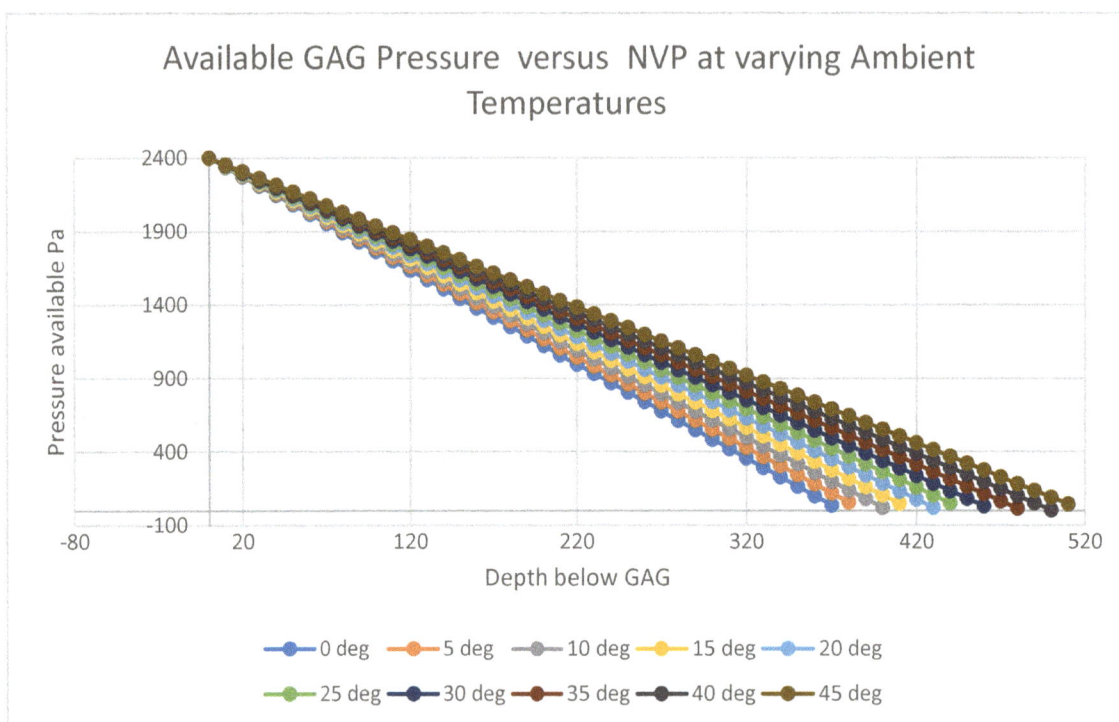

FIG 3 – Effect of ambient temperature on GAG buoyancy.

At 350 m depth and an ambient temperature of 25°C the available pressure would be about 500 Pa. This assumes that the shaft is fully charged with GAG product.

Without the main fans running the 500 Pa has to overcome the frictional resistance of the airways between the GAG and the open exhaust of the mine. The resistance of the mine then determines the quantity that flows.

This means that there are 1900 pascals pressure on the docking station and surrounds facilitating leakage. This leakage will be hot, moist and laden with dissolved acidic gases, especially CO_2. As it seeps out it will dissolve cementitious materials, natural and man-made. It will expand and corrode steel. All these actions will increase the leakage and reduce the delivery of the GAG product into the mine.

At 89°C the GAG product will heat up the docking station and cowling faster than the concrete, causing the cowling to buckle and move increasing leakage and the concrete to spall and fret.

Pipework in the airway will expand as it is heated and will raise its temperature from 25°C to 89°C, it will lengthen by 83 mm/100 m. This lengthening would buckle the pipeline and cause it to break. A compressed air line would then discharge its contents into the inertisation area. The compressed air will be cold and would displace the GAG product. There would be similar issues with fixed guides and dolly car rails.

Since the GAG is unable to generate the high pressures that are required to deliver the GAG product, there have to be other mechanisms in action to get the GAG product underground. In the incidents where the GAG has delivered product underground to control a fire/spontaneous combustion incident, the intensity of the event has generated sufficient heat to provide a draw of the product into the mine.

Where the GAG was deployed to inert mines where no fire was present, the GAG had difficulty in providing the desired result. Mines reported significant damage to roof/ribs and infrastructure due to the heat, moisture and the corrosive nature of the moisture entrained on the product stream.

In all cases the amount of GAG product delivered was far less than the stated 20–25 m^3/s output. While 20 m^3/s may leave the GAG much of the water will precipitate/condense out of the product stream and considerable leakage of product during deployment as the backpressure rises further limits the available GAG product.

If the GAG product makes it into the mine workings it is likely to stratify in the mine workings and will work its way up dip due to its buoyancy. The potential for two airway masses to be drawn to the fire zone is considerable giving rise to pockets of higher oxygen concentration mixes and pockets of inert gases. The ACARP Report C23006 (Watkinson *et al*, 2014) documented the behaviour of the GAG at Kestrel North. This report documented that the rate of the GAG product slowed with depth. Stratification was evident in the distribution of the GAG products in the roadways. It also documented the rupture of a compressed air line in the conveyor drift and this was a direct result of thermal expansion. This could also occur with methane drainage ranges once they have been exposed to hot GAG product.

With a fire with the intensity to generate enough pressure to bring the GAG product underground would also be lowering the pressure on all of the mine workings that are connected to the fire. An open longwall goaf would be expanding as the underground pressure falls and would bring the contents of that goaf into the adjacent airways. This would give rise to the potential for a gas explosion or several explosions as time progresses.

CONCLUSIONS

While the GAG is reported to produce 20–25 m^3/s of inert product, the reality is that this is far less with the loss of water and cooling that may take place.

The performance curves relating to pressure and quantity do not exist in Australia and how the GAG performs cannot be modelled.

The use of Ventsim models to predict the effectiveness of the GAG in creating an inert atmosphere cannot give reliable information due to the absence of the performance curve to model GAG behaviour.

Much of the pressure, if not all the pressure generated by the GAG is wasted due to the buoyancy of the GAG product.

Once the GAG product leaves the shaft and enters the mine workings its behaviour between the shaft and the fire is unknown.

The heat and water-based fluid of the GAG product does considerable damage to the infrastructure in the mine. Pipework, gas drainage ranges, roof supports and rails will lengthen when heated. Similarly, roof materials/ribs and cementitious grouts will be eroded by the corrosive nature of the fluids, leading to falls of ground and increased leakage.

If the GAG can be deployed as close as physically possible so that the exhaust impinges on the fire then the moisture in the GAG product would be useful in fighting the fire.

RECOMMENDATIONS

This paper has summarised the shortcomings and some of the issues related to the operation of the GAG. While it is seen that the GAG has considerable potential to provide inert gases to assist in the control of fire and flammable atmospheres, in its current form it has some significant drawbacks. In order that the GAG becomes a more effective tool it is recommended that:

1. The Pressure–Quantity characteristic for the GAG be developed as a matter of priority.

2. An investigation on cooling the GAG product so that the buoyancy of the GAG product no longer hampers the delivery of inert material underground. This would also remove the hazards created by thermal expansion.

3. An investigation into removing the moisture from the GAG exhaust before it is delivered underground to reduce the development of corrosive materials and increasing the density of the GAG product.

ACKNOWLEDGEMENTS

The authors would like to thank ACARP for providing funding for the research presented in this paper under ACARP Project C20113. Also, Thanks to the mines who supplied the ventilation models for use in this project and the support from Howden – Ventsim for their assistance and contribution.

REFERENCES

Bell, S, Humphreys, D, Harrison, P, Hester, C and Torlach, J, 1997. *Practical Demonstration & Evaluation of Jet Engine Inertisation Techniques*. ACARP C6019.

Mucho, T P, Houlison, I R, Smith, A C and Trevits, M A, 2005. Coal mine inertisation by remote application, in the *Proceedings of the 2005 US National Coal Show*, pp 1–14.

Tomlinson, A, 2006. Newlands uses GAG on Southern, *Australian Longwall Magazine*, March.

Watkinson, M, Liddell, K, Muller, S, Gido, M and Nissen, J, 2014. *A practical research study into the environmental and physical impacts during an underground mine inertisation with the GAG Jet*. ACARP C23006.

Technology

Are Battery Electric Vehicles (BEVs) the answer to the challenge of complying with the European Union (EU)'s new TLV-TWA for diesel exhaust contaminants?

A Halim[1]

1. MAusIMM(CP), Senior Lecturer, Mining and Rock Engineering, Luleå University of Technology, Luleå SE-971 87, Sweden. Email: adrianus.halim@ltu.se

ABSTRACT

The European Union (EU) have recently issued new Threshold Limit Value-Time Weighted Average (TLV-TWA) for diesel exhaust contaminants, which are CO, NO_2 and DPM. The new limits for CO and NO_2 were issued in 2017 whilst the one for DPM was issued in 2019. However, in order to allow EU mines to formulate strategies to comply with these new limits, the EU granted a transition period in which mines could use the old limits until the new limits are enforced. The new limit for CO and NO_2 will be enforced from 21 August 2023 whilst the one for DPM will be enforced from 21 February 2026. The new limit for NO_2 and DPM is very low, 0.5 ppm for NO_2 and 0.05 mg/m^3 of Elemental Carbon (EC) for DPM. Combination of these two limits would be the lowest among all mining jurisdictions worldwide. It poses a huge compliance challenge to EU mines if they keep using diesel machines. Therefore, a study and a field trial were carried out by the author in 2019–2021 to investigate whether replacing diesel machines with Battery Electric Vehicles (BEVs) would be the best answer to this challenge. This paper outlines results of this study and field trial.

INTRODUCTION

One of the main objectives of underground mine ventilation is to dilute atmospheric contaminants produced by mining activities to a safe concentration level. Traditionally, the safe concentration level of a contaminant is defined as the concentration level below the limit prescribed in the local Occupational Health & Safety (OH&S) regulation. The limit that has been traditionally used in mining OH&S regulations worldwide is Threshold Limit Value-Time Weighted Average (TLV-TWA), which is defined as 'maximum contaminant concentration where personnel may be repeatedly exposed without adverse effect for a normal 8 working hours per day or 40 working hours per week'. Since many mines worldwide use working shifts longer than 8 hours per day, the actual limit for this situation has to be adjusted according to the length of the shift. The common method used to do this adjustment is Brief and Scala method, as shown in Appendix 1. Using this method, the limits for shifts longer than 8 hours are calculated as lower than the TLV-TWAs. For example, the limits for 12-hour working shift, which is used by many mines worldwide, are calculated as half of the TLW-TWAs. This makes sense since long duration exposure is more hazardous than short duration exposure, therefore warrants lower exposure limits.

Interestingly, TLV-TWA for the same contaminants varies across mining OH&S regulations worldwide, as shown in Table 1. The European Union (EU) recently issued new TLV-TWA for diesel exhaust contaminants, which are CO, NO_2 and DPM (EU Commission 2017, 2019). The new limits for CO and NO_2 were issued in 2017 whilst the one for DPM was issued in 2019. The new limit for NO_2 (0.5 ppm) is half of the previous limit (1 ppm), whilst the one for DPM (0.05 mg/m^3 of Elemental Carbon – EC) is the first time the DPM exposure limit is prescribed in the EU's OH&S legislation. As shown in Table 1, it is the lowest TLV-TWA for DPM among mining OH&S regulations worldwide.

Member states of the EU such as Sweden and Finland subsequently have to incorporate these new limits in their OH&S regulation. Since the reduction of NO_2's TLV-TWA is huge and the new DPM's TLV-TWA is the lowest in the world, EU mines cannot immediately comply with these new limits. Therefore, in order to allow EU mines to formulate strategies to comply with these new limits, the EU granted a transition period in which the mines could use the old limits until the new limits are enforced (Halim, Bolsöy and Klemo, 2020). The new limit for CO and NO_2 will be enforced from 21 August 2023 whilst the one for DPM will be enforced from 21 February 2026. Note that Sweden and Finland have incorporated NO_2's new TLV-TWA in their OH&S regulation, as shown in Table 1. However, their mines are still using the old TLV-TWA of 1 ppm until 21 August 2023.

TABLE 1

TLV-TWA of diesel exhaust contaminants in various mining OH&S regulations worldwide.

Contaminant/jurisdiction	NO$_2$ (ppm)	CO (ppm)	DPM (mg/m^3)
The European Union (EU)	0.5*	20	0.05 of Elemental Carbon (EC)**
Sweden	0.5***	20	n/a****
Finland	0.5***	20	n/a****
Australia	3	30	0.1 of Elemental Carbon (EC)
Canada (Ontario)	3	25	0.4 of Total Carbon (TC)
USA	5	50	0.16 of Total Carbon (TC)
South Africa	3	30	n/a

* Set in 2017 but will be enforced from 21 August 2023.

** Set in 2019 but will be enforced from 21 February 2026.

*** Set in 2018 but will be enforced from 21 August 2023.

****Currently not available but will be 0.05 after 21 August 2026 when the EU limit is enforced.

Sources: EU Commission (2017, 2019), Arbetsmiljöverket (2018), Sosiaali – ja terveysministeriön (2018), SafeWork Australia (2015, 2019), Government of Ontario (2020), MSHA (2018), Stanton, du Plessis and Belle (2014).

The combination of the new TLV-TWA for NO$_2$ and DPM would be the lowest among mining OH&S regulations worldwide. It poses a huge compliance challenge for EU mines if they keep using the existing diesel machines. In this circumstance, the most obvious solution to comply with these new limits is significantly upgrading the capacity of the existing primary and secondary mine ventilation systems. The upgrade of each system is very costly and subsequently increases the power cost of each system by a significant margin because fan power is proportional to the cube of airflow quantity. In long-term, it would severely impact the mine cash flow since ventilation power cost could comprise up to 40 per cent of the total mine electricity costs (Halim and Kerai, 2013). Moreover, it could cause some delays in production and development since ventilation in some parts of the mine must be cut-off during the upgrading.

The second option is employing remote controlled machines that are operated from the surface. This reduces workers' exposure to diesel exhaust emissions. However, this technology has currently been applied successfully only on drill rigs and load-haul-dumps (LHDs). Trucks are still required to be operated manually, means that their drivers could still be exposed to diesel exhaust emissions, even if the trucks are fitted with a filtered air conditioning system in their cabin. Moreover, this option is also expensive, which, in many cases, is not affordable to small and medium-sized mines.

The third option is controlling or doing something on the source of the contaminants, which is diesel machines. This option could avoid the requirement to upgrade the existing primary and secondary ventilation systems and it could be done without causing delays in production and development. It also negates the requirement to employ remote controlled machines. Moreover, controlling the source of the contaminants is the most effective solution to comply with the exposure limits and offers the best long-term solution to this challenge. This option is also costly, but since it is the most effective solution and it offers the best long-term solution, it has the best value of money spent among these three options.

Therefore, EU mines are faced with the following two questions:

1. Will all diesel machines have to be fitted with Euro VI compliant engines? (Euro VI is the cleanest emission standard for heavy machine diesel engines sold within the EU and the European Economic Area – EEA).

2. Will all diesel machines have to be replaced with electric ones, in this case Battery Electric Vehicles (BEVs)?

As an effort to assist these mines to answer these two questions, the author carried out a study and a field trial involving BEVs, in cooperation with two EU major mining companies and an EU Original

Equipment Manufacturer (OEM) during 2019–2021. This paper outlines results of this study and field trial.

FITTING ALL DIESEL MACHINES WITH EURO VI COMPLIANT ENGINES

The author, in cooperation with Luossavaara-Kiirunavaara Aktiebolag (LKAB), a Swedish state-owned iron ore mining company, carried out a study about developing strategies to reduce ventilation and heating costs in Konsuln iron ore mine in Northern Sweden. Results of this study have been published by Gyamfi, Halim and Martikainen (2022). One part of this study is investigating whether the existing ventilation system could cater the increase of the existing mine diesel fleet. This increase would be required because the mine was planning to increase its output from 0.8 million tonnes per annum (Mtpa) to a rate between 1.8 and 3 Mtpa. The investigation assumed a production rate of 3 Mtpa because it is the most challenging design case among all possible production rates between 1.8 and 3 Mtpa. The new fleet consists of four Epiroc ST18 LHDs and 12 Scania R500 trucks, all of them are fitted with Euro VI compliant engines. The investigation was done using diesel simulation in VentSim Design software and was based on 8-hour and 12-hour working shifts. This is because the mine is currently using 8-hour shifts but it may use 12-hour shifts in the future. Therefore, the following exposure limits were used in this investigation:

CO: 20 ppm for 8-hour shifts and 10 ppm for 12-hour shifts

NO_2: 0.5 ppm for 8-hour shifts and 0.25 ppm for 12-hour shifts

DPM: 0.05 mg/m^3 of EC for 8-hour shifts and 0.025 mg/m^3 of EC for 12-hour shifts

All machines were assumed to run at 45 per cent of their engine rated power in the simulation, based on LKAB's experience. The investigation found that the existing ventilation system would be able to cater the new diesel fleet because the maximum concentration of all diesel exhaust contaminants was predicted to be below the above exposure limits, as shown in Table 2. This indicates that Euro VI compliant engines are so clean and fitting all diesel machines with these engines would make compliance to the new limits possible. It must be noted that the CO concentration was predicted to be very close to the limit for 12-hour shifts, as shown in Table 2. It must also be noted that the contaminant emission rates of the Euro VI compliant engines used as input data of the simulation are for brand new engines. These values can change for the worse when the engines are poorly maintained and this often happens in reality. The low contaminant emission rates depend heavily on the good operation of the aftertreatment devices. Poor maintenance of these devices can increase the contaminant emission rates (Hines, 2019). In the investigated scenario, poor maintenance would cause an increase of CO emissions from the new fleet and its concentration may exceed the limit for 12-hour shifts as well as the limit for 8-hour shifts. The emissions of NO_2 and DPM would also increase and their concentration may exceed their limit for both shifts.

TABLE 2

Highest predicted values of diesel exhaust contaminants concentration for Konsuln mine's new diesel fleet fitted with Euro VI compliant engines (Gyamfi, Halim and Martikainen, 2022).

Contaminant	Highest value
CO	9.1 ppm
NO_2	0.11 ppm
DPM	0.016 mg/m^3 of EC

Moreover, having contaminant concentrations hovering close to their exposure limit is a poor practice because they could easily exceed the limit when something goes wrong, such as poor maintenance of engine aftertreatment. The principle of As Low As Reasonably Achievable (ALARA) or As Low As Reasonably Practicable (ALARP) should be used in controlling air quality, means that the contaminant concentrations should be kept much lower than their exposure limit. In the investigated scenario, this means that the existing ventilation system must be significantly upgraded, which requires high capital expenditures (CAPEX) and will results in significantly higher ventilation operating costs. More details of this investigation are in Gyamfi, Halim and Martikainen (2022).

REPLACING ALL DIESEL MACHINES WITH BATTERY ELECTRIC VEHICLES

Electric vehicles (EVs) do not produce toxic gases and DPM, hence replacing diesel machines with EVs should eliminate these contaminants in the areas where mobile equipment are operating. Among all types of EVs, BEVs are the most suitable for all mining operations because they are the most flexible. Cable-trailed EVs are inflexible because of their cable. It is impractical to move them to another level because this requires relocating transformers and electric cables. Trolley-assisted EVs require expensive fixed overhead trolley lines, thus limiting their range. The road surface must be very level to allow the pantograph to stay in contact with the overhead wires. This means the roads require more maintenance than the normal underground mine roads, increasing the mine maintenance costs. Moreover, the system cannot be installed in mines that have poor ground condition and plenty of groundwater ingress since stable openings and dry condition are required to prevent any damage to the overhead wires.

The author, in cooperation with Agnico Eagle Finland Oy, a Finnish subsidiary of Agnico Eagle Mines Ltd of Canada and Epiroc AB, a well-known Swedish OEM, carried out a field trial of Epiroc's BEVs at Kittilä mine during 2019–2020. The field trial was carried out as a part of the EU's project Sustainable Intelligent Mining System (SIMS) (https://www.simsmining.eu/project/battery-driven-equipment/). The tested BEVs were MT42 Battery truck, ST14 Battery LHD and Boomer E2 Battery jumbo drill rig. Details of the trial are described in Halim *et al* (2022).

Air quality and air temperatures were measured in the trial area of MT42 Battery and ST14 Battery (level S150). The results have been published by Halim *et al* (2022). Gases measurements were done using Dräger X-am 5600 monitors hung on the footwall drive wall. An example of these measurement results is shown in Figure 1. It clearly shows that gases were only detected when diesel machines were running in the trial area and no gases were detected when only BEVs were running in the trial area. The other measurements also detected no gases when only BEVs were running in the trial area.

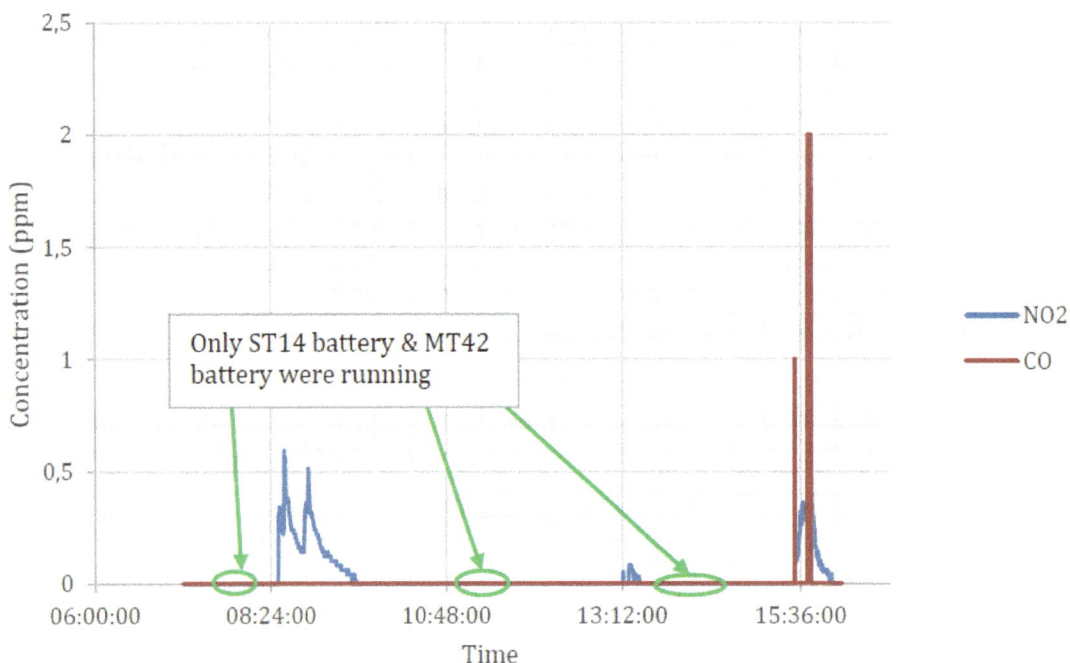

FIG 1 – An example of gases measurement results in the trial area of MT42 Battery and ST14 Battery (Halim *et al*, 2022).

Table 3 shows results of DPM measurements in the same trial area. DPM was sampled using a personal sampling unit hung on the footwall drive wall. This unit included an SKC AirChek TOUCH 220–5000TC pump that draws air into a filter via an SKC GS-1 cyclone fitted with a DPM filter cassette. The samples were then sent to a laboratory in the USA where the EC concentration was determined using NIOSH 5040 method. It took about three months to get the results from the laboratory. Due to the delay of the start of the field trial and the time consumed to get the results from the laboratory, only two DPM samplings could be done during the trial: one when only diesel machines were running and another when only BEVs were running. Currently, there are no laboratories in Finland or in other EU countries that can analyse DPM samples using NIOSH 5040 method, so this is why the results had to be sent to an American laboratory.

The amount of EC in the sample that was taken when only BEVs were running was below the detection limit, so the laboratory could only estimate its concentration, placing it at less than 0.002 mg/m^3 This concentration is very low and can be considered as virtually non-existent.

TABLE 3

DPM measurement results in the trial area of MT42 Battery and ST14 Battery (Halim *et al*, 2022).

Only BEVs running on level S150	Only diesel machines running on level S150
Estimated to be less than 0.002 mg/m^3 of EC (virtually non-existent)	0.046 mg/m^3 of EC

More details of these measurements and other measurements and observations during the field trial can be read in Halim *et al* (2022).

Results of gases and DPM measurements show that replacing diesel machines with BEVs will eliminate gases and DPM from the areas where mobile equipment are operating. Even though gases are also produced by blasting, workers are mostly exposed to the ones emitted by diesel machines because no personnel are allowed to enter a newly blasted area until the blasting fumes is cleared. Therefore, EU mines will easily comply with the new exposure limits when BEVs are used instead of diesel machines, irrespective of the length of the shifts.

DISCUSSION

The investigation outlined earlier in this paper indicates that fitting all diesel machines with Euro VI compliant engines would be problematic since high maintenance standard and increasing airflow quantity will be required to keep exhaust contaminants concentration far below the new limits. The measurements of air quality taken during the field trial of BEVs at Kittilä mine clearly demonstrate that gases and DPM are not going to be present when BEVs are running instead of diesel machines. This result was expected because BEVs do not emit gases and DPM. Based on the study and the field trial, replacing diesel machines with BEVs would definitely answer the challenge to comply with EU's new exposure limits.

However, there are remaining concerns that must be addressed before BEVs replace diesel machines. These concerns were mentioned by Kittilä mine workers and management of several mines that are planning to use BEVs. They are machine productivity, fire safety, battery swapping, user-friendliness of the machines, potential electrical-related accidents, changes in mine management, changes in mine design, changes in procedures and power management (Halim *et al*, 2022). These concerns are understandable because BEVs are relatively new and many of their aspects are still not well understood. An ongoing EU funded project, Next Generation Carbon Neutral Pilots for Smart Intelligent Mining Systems (NEXGEN SIMS, www.nexgensims.eu) is an effort to address some of these concerns. The project is the successor of SIMS project and the author is heavily involved in it. More research and development (R&D) and field trials should be carried out to make BEVs viable mining machines in the near future.

Despite ongoing considerable efforts to reduce diesel exhaust emissions such as improving emission standards (Euro stages and Tier standards) and improving aftertreatment devices, the compliance issue is not going to disappear if mines keep using diesel machines. This is because these efforts are akin to treating the symptoms rather than removing the cause. The author believes that removing the cause, which is the use of diesel machines, is the best option to address the compliance issue since history has shown that OH&S regulations becomes more stringent along with time (TLV-TWAs will be further reduced in the future).

CONCLUSIONS

The study and the field trial outlined above strongly indicates that BEVs would be the better answer to the challenge of complying with the EU's new TLV-TWA for diesel exhaust contaminants than fitting the existing diesel machines with Euro VI compliant engines. However, since the usage of BEVs in underground mining is relatively new, many aspects are still not well understood; consequently, there are still many concerns among mine personnel. More studies and field trials must be done to address these concerns and to make BEVs as viable mining machines in the near future.

ACKNOWLEDGEMENTS

For investigating the option of fitting all diesel machines with Euro VI compliant engines, the author acknowledges great support from co-investigators from LKAB: Mr. Seth Gyamfi and Dr. Anu Martikainen and other personnel from LKAB and EOL Vent Mining AB whose names are stated in Gyamfi, Halim and Martikainen (2022).

For carrying out BEVs field trial at Kittilä mine, the author is very grateful for the financial support from the EU's Horizon 2020 program (Grant Agreement no. 730302). The author also acknowledges great support from the following personnel during the field trial: Dr. Joel Lööw and Prof. Jan Johansson from Luleå University of Technology, Mr. Jan Gustafsson from Epiroc AB, Mr. Andre van Wageningen from Agnico Eagle Finland Oy and Prof. Karoly (Charles) Kocsis from University of Utah, as well as other personnel from Kittilä mine and Epiroc AB whose names are stated in Halim *et al* (2022).

REFERENCES

Arbetsmiljöverket, 2018. AFS 2018:1 Hygieniska gränsvärden (in Swedish) [online]. Available from: https://www.av.se/globalassets/filer/publikationer/foreskrifter/hygieniska-gransvarden-afs-2018–1.pdf

European Union (EU) Commission, 2017. *Commission Directive (EU) 2017/164 of 31 January 2017 establishing a fourth list of indicative occupational exposure limit values pursuant to Council Directive 98/24/EC and amending Commission Directives 91/322/EEC, 2000/39/EC and 2009/161/EU (Text with EEA relevance)* [online]. Available from: https://eur-lex.europa.eu/legal-content/EN/TXT/PDF/?uri=CELEX:32017L0164&from=EN

European Union (EU) Commission, 2019. *Directive (EU) 2019/130 of The European Parliament and of The Council of 16 January 2019 amending Directive 2004/37/EC on the protection of workers from the risks related to exposure to carcinogens or mutagens at work* [online]. Available from: https://eur-lex.europa.eu/legal-content/EN/TXT/PDF/?uri=CELEX:32019L0130&from=FR

Government of Ontario, 2020. *Current Occupational Exposure Limits for Ontario Workplaces Required under Regulation 833* [online]. Available from: https://www.labour.gov.on.ca/english/hs/pubs/oel_table.php#qr

Gyamfi, S, Halim, A and Martikainen, A, 2022. Development of strategies to reduce ventilation and heating costs in a Swedish sublevel caving mine – a unique case of LKAB's Konsuln mine, *Mining, Metallurgy & Exploration,* 39(2): 221–238. Available from: https://doi.org/10.1007/s42461–021–00483-y

Halim, A and Kerai, M, 2013. Ventilation Requirement for 'Electric' Underground Hard Rock Mines – A Conceptual Study, in *Proceedings of The Australian Mine Ventilation Conference 2013,* pp 215–220 (The Australasian Institute of Mining and Metallurgy: Melbourne).

Halim, A, Bolsöy, T and Klemo, S, 2020. An overview of the Nordic mine ventilation system, *CIM Journal,* 11(2):111–119.

Halim, A, Lööw, J, Johansson, J, Gustafsson, J, van Wageningen, A and Kocsis, K, 2022. Improvement of working conditions and opinions of mine workers when Battery Electric Vehicles (BEVs) are used instead of diesel machines – Results of field trial at the Kittilä mine, Finland, *Mining, Metallurgy & Exploration,* 39(2):203–219. Available from: https://doi.org/10.1007/s42461–021–00506–8

Hines, J, 2019. *The Role of Emissions Based Maintenance to Reduce Diesel Exhaust Emissions, Worker Exposure and Fuel Consumption,* PhD thesis, The University of Wollongong, Wollongong, Australia. Available from: https://ro.uow.edu.au/cgi/viewcontent.cgi?article=1857&context=theses1

Mine Safety and Health Administration (MSHA), 2018. *30 Code of Federal Regulations (CFR)* [online]. Available from: https://www.govinfo.gov/content/pkg/CFR-2018-title30-vol1/pdf/CFR-2018-title30-vol1.pdf

SafeWork Australia, 2015. *Guide to Managing Risks of Exposure to Diesel Exhaust in the Workplace* [online]. Available from: https://www.safeworkaustralia.gov.au/system/files/documents/1702/guidance-managing-risks-exposure-diesel-exhaust-in-the-workplace.pdf

SafeWork Australia, 2019. *Workplace Exposure Standards for Airborne Contaminants* [online]. Available from: https://www.safeworkaustralia.gov.au/system/files/documents/1912/workplace-exposure-standards-airborne-contaminants.pdf

Sosiaali – ja terveysministeriön, 2018. Haitallisiksi tunnetut pitoisuudet (HTP) arvot 2018 (in Finnish) [online]. Available from: https://julkaisut.valtioneuvosto.fi/bitstream/handle/10024/160967/STM_09_2018_HTParvot_2018_web.pdf?sequence=1&isAllowed=y

Stanton, D W, du Plessis, J J L and Belle, B K, 2014. Occupational Exposure Limits for Airborne Pollutants, in *Ventilation and Occupational Environment Engineering in Mines* (J J L du Plessis, ed.), pp 237–254 (Mine Ventilation Society of South Africa: Johannesburg).

APPENDIX 1

Adjustment of concentration limit for non-8-hour working shifts using Brief and Scala method

Brief and Scala method to calculate concentration limit for working period with duration other than 8 hours is as follows:

$$\text{Adjusted limit for } h \text{ working hours per day} = \frac{8 \times (24 - h) \times TLV\ TWA}{16 \times h}$$

Using this method, the gases' and DPM's limit for 12-hour shifts is calculated as half of their TLV-TWA.

Development of mine intelligent ventilation technology in western China

S G Li[1], J N Gao[2], F L Wu[3], X T Chang[4] and Z G Yan[5]

1. Professor, College of Safety Science and Engineering, Xi'an University of Science and Technology, Xi'an 710054, China. Email: lisg@xust.edu.cn
2. Lecturer, College of Safety Science and Engineering, Xi'an University of Science and Technology, Xi'an 710054, China. Email: 943157956@qq.com
3. Professor, College of Safety Science and Engineering, Xi'an University of Science and Technology, Xi'an 710054, China. Email: 15038537@qq.com
4. Professor, College of Safety Science and Engineering, Xi'an University of Science and Technology, Xi'an 710054, China. Email: changxt@xust.edu.cn
5. Associate Professor, College of Safety Science and Engineering, Xi'an University of Science and Technology, Xi'an 710054, China. Email: 393826629@qq.com

ABSTRACT

Mine intelligent ventilation is an important guarantee to realise intelligent mining. It is the integration of the Internet of Things, big data, artificial intelligence, internet, cloud computing with ventilation technology, forming a complete ventilation system with intelligent perception, intelligent decision-making, intelligent emergency response and airflow control. In order to fully grasp the progress and trends of mine intelligent ventilation research in western China. The paper based on literature analysis and field application research, comprehensively expounds the main research and practical results of mine intelligent ventilation in western China, mainly analyses the development status of key technologies such as accurate and rapid determination of mine ventilation parameters, intelligent analysis and decision-making of ventilation and intelligent regulation of ventilation equipment, and introduces the field test or application of mine intelligent ventilation technology in western China. It lays the foundation reference for related research and improving the intelligent level of mine ventilation.

INTRODUCTION

With the integration and penetration of technologies such as big data, artificial intelligence, internet and cloud computing, and China's continuous exploration of the process of industrialisation of mines, China's intelligent construction of mines is in the ascendant. By the end of 2020, taking coalmines as an example, China has built 494 intelligent mining faces, and the mining, excavation, electromechanical and transportation systems of coalmines have fully transformed from mechanisation and automation to intelligence. With the construction of mine intelligence in China, as an important part of mine production, the intelligent construction of mine ventilation system is crucial to ensure the normal operation of mine intelligent mining.

The so-called mine intelligent ventilation refers to the operation of real-time on-demand air supply in normal periods and emergency air flow control in disaster periods by an intelligent control system according to changes in temperature and humidity, toxic and harmful gases and dust concentrations in roadways. Many scholars and enterprises have made extensive explorations in the construction of mine intelligent ventilation. In terms of theoretical research, the calculation of natural air distribution, and the on-demand optimisation and control of air flow models and algorithms as the theoretical core of mine ventilation, have been gradually improved, which has laid a theoretical foundation for the realisation of mine ventilation intelligence; in terms of key technologies, the main focus is on accurate and rapid determination of ventilation parameters, optimal arrangement of sensors, intelligent diagnosis of ventilation faults, real-time calculation of ventilation network and other technologies; in terms of equipment system research and development, it is mainly reflected in ventilation parameter measurement and monitoring equipment, intelligent ventilation control platform, intelligent variable frequency ventilator, and remote intelligent control wind damper and windows, local intelligent ventilation systems, disaster emergency wind control equipment, etc. In the past three or four years, China has also successively carried out research on the goals, connotations, logical structure, key technology systems and application scenarios of mine ventilation intelligent construction, aiming to strengthen top-level design, unify construction standards, clarify implementation paths, and improve mine ventilation intelligence level. For this reason, the author intends to start from the key technologies to realise the intelligentisation of mine ventilation, and introduce the development status of intelligent ventilation technology in mines in western China, so as to provide reference for related research.

MINE INTELLIGENT VENTILATION TECHNOLOGY SYSTEM

The mine intelligent ventilation system is a deep integration of geographic information, communication, mobile internet, Internet of Things, big data, artificial intelligence, cloud computing, automatic control, surveying and monitoring, intelligent equipment etc with mine ventilation technology and equipment, integrating a comprehensive and autonomous perception of mine ventilation and climate environment parameters, the operation status of ventilation facilities and equipment, the real-time and efficient interconnection of ventilation perception systems with ventilation networks, ventilation facilities and equipment, disaster prevention systems, and major production systems, ventilation air volume adjustment plans, emergency air flow control strategies for disasters, and precise coordinated control of ventilation facilities and equipment to achieve intelligent linkage control of the entire process of mine ventilation. Accordingly, the mine intelligent ventilation technology system mainly includes intelligent sensing technology, intelligent analysis and decision-making technology, and intelligent control technology. The overall technical architecture of mine intelligent ventilation is shown in Figure 1.

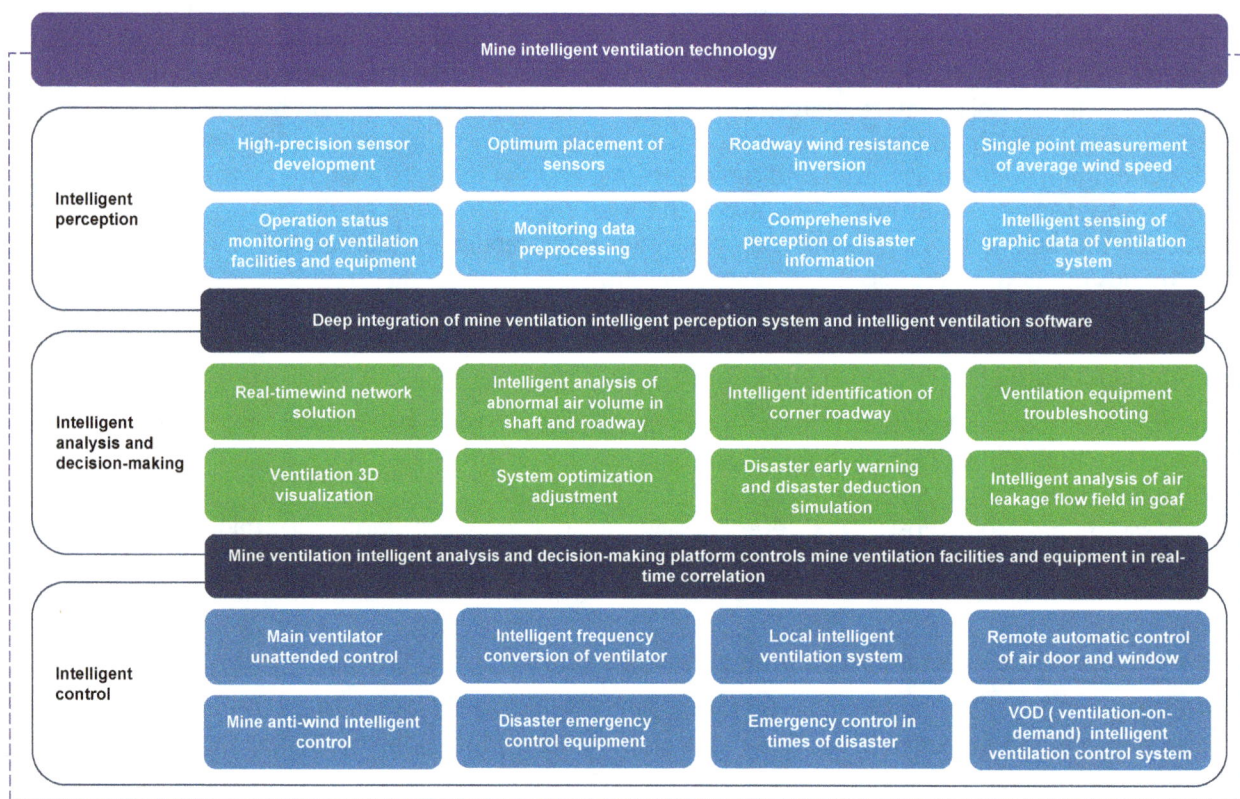

FIG 1 – Mine intelligent ventilation technology system.

1. Intelligent perception technology: develop full-scale high-precision wind speed, wind pressure, wind quality and other types of sensors, study the relationship between the wind speed at a single point in the mine roadway and the average wind speed in the section, determine the best placement of sensors in the monitoring section, and optimise perception point distribution in non-blind area for network ventilation parameters of the overall wind speed, develop noise processing algorithms for sensor monitoring data, and establish an accurate inversion model for the resistance coefficient of the entire wind network to achieve rapid and accurate automatic measurement of mine ventilation parameters and online accurate monitoring of ventilation state parameters in the entire mine, comprehensive perception of mine disaster information, online monitoring of the operating status of ventilation facilities and equipment, to ensure the accuracy of mine ventilation network calculation and intelligent wind control.

2. Intelligent analysis and decision-making technology: analyse in real-time and mining in-depth mine ventilation monitoring data, establish a real-time calculation model of mine ventilation network, a ventilation anomaly identification model, and a ventilation equipment fault diagnosis model, to realise the rapid construction of mine ventilation network model and mine ventilation network solution in real-time, 3D dynamic visualisation display of ventilation system, dynamic monitoring and early warning of ventilation network, rapid determination of optimal adjustment scheme of ventilation system, evolution simulation of ventilation disaster, rapid calculation and early warning of the influence range of ventilation disaster, rapid decision-making of emergency wind flow control scheme in disaster, personnel safety rapid formulation of escape routes and intelligent identification of dangerous areas in gobs provide important technical support for on-demand air supply during normal periods of mine ventilation systems and emergency air flow control during disasters.

3. Intelligent control technology: develop remote intelligent adjustment dampers and windows, develop local intelligent ventilation systems and main ventilation intelligent control systems, research linkage control technology for mine disaster areas, develop 'ventilation on demand' (VOD) intelligent ventilation technology, and realise remote visualisation and precise interconnection control of ventilation facilities and equipment, unattended main ventilator, intelligent frequency conversion control of ventilator, intelligent control of local ventilation, linkage isolation and wind control of disaster area, laying a hardware foundation for the accurate implementation of mine ventilation control decision-making scheme.

RESEARCH PROGRESS IN INTELLIGENT PERCEPTION TECHNOLOGY OF MINE VENTILATION PARAMETERS

Whether it is online accurate perception of the operating status of the wind network, real-time accurate calculation and intelligent accurate control, accurate ventilation parameters are required as the basis. Therefore, the development of accurate and rapid testing technology and equipment for ventilation parameters is of great significance for mine intelligent ventilation.

At present, the mine wind speed sensors are mostly thermal, differential pressure and vortex street sensors, with low measurement accuracy and insensitive response, which cannot meet the requirements of intelligent ventilation for accurate and rapid wind speed testing. Ultrasonic velocity measurement technology has been introduced into China's mine wind measurement operation due to its advantages of wide test range and high precision, China developed ultrasonic high-precision wind measurement devices (as shown in Figure 2), it can test the wind speed of 0.1~15 m/s with a resolution of 0.01 m/s and an accuracy of ±0.1 m/s (Zhang, 2016, 2019a; Zhang, Yao and Zhao, 2020).

FIG 2 – Ultrasonic accurate wind measuring device.

The inhomogeneity of the wind speed distribution in the roadways indicates that the wind speed at the roadway point tested by the sensor cannot represent the average wind speed. The works (Ji, 1997; Lu, Zhang and Fan, 2015; Zhou *et al*, 2015; Song *et al*, 2016; Zhao *et al*, 2014; Zhang, 2019b) studied the roadway wind speed field and the relationship between point wind speed and average wind speed by analysis, laboratory similar model test, numerical simulation and other methods. Zhang (2018) used numerical simulation method to obtain the distribution area (green area) of the wind speed range of wind measuring station section $[0.97V_a, 1.03V_a]$ (the error of the average wind speed V_a of the test section of the wind speed sensor is less than 3 per cent), as shown in Figure 3.

The works (Wang, 2022; Wang *et al*, 2016) developed a fully automatic mine wind measurement system (as shown in Figure 4) and a multi-point mobile wind measurement device (as shown in Figure 5), using high-precision wind speed sensor to test the wind speed of multiple points in the wind measurement section of the mine roadway and get the average value, so as to realise the automatic and unmanned accurate wind measurement of the whole section of mine roadway. Shanxi Tiandi Wangpo Coal Industry Co., Ltd. has put into use a multi-point mobile wind measurement device, and the wind measurement accuracy has reached more than 95 per cent; Shaanxi Coal North Mining Co., Ltd. Zhangjiamao Coal Mine has equipped with an automatic mine wind measurement system, which takes only two minutes to complete the mine wind measurement.

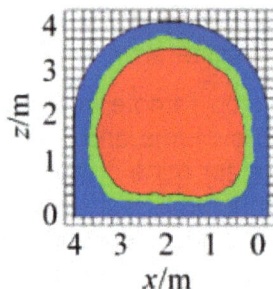

FIG 3 – Distribution area of air velocity range [0.97V_a, 1.03V_a] on section of wind measuring station.

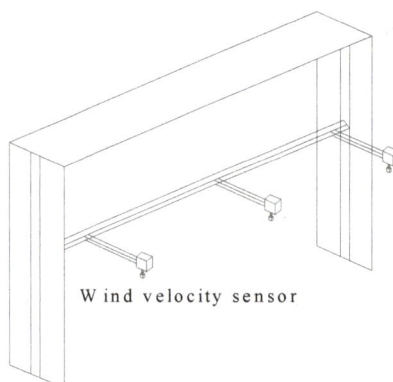

FIG 4 – Gantry type automatic wind measurement system.

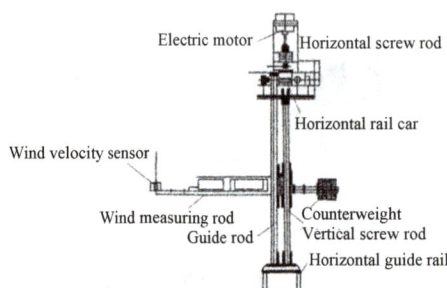

FIG 5 – Multi-point mobile wind measurement device.

The real-time and accurate measurement of wind pressure is an important means to monitor the distribution of ventilation resistance in mines, and to identify abnormal wind flow conditions and ventilation structure failures. It is also of great significance to the real-time calculation of mine ventilation network and the intelligent regulation of wind flow. Commonly used mine ventilation parameter detection instruments have low measurement accuracy, resulting in low reliability of resistance measurement results for large-section and small-resistance roadways; wind pressure sensors mostly use capacitance or semiconductor pressure-sensing principles, which have low measurement accuracy, insensitive response, and cannot meet the requirements of mine intelligent ventilation for real-time and accurate monitoring of wind pressure. In recent years, China adopted the American MENSOR silicon resonant principle, developed the CPD120 smart wind pressure gauge, the absolute pressure resolution is accurate to 1 Pa, and has functions such as automatic storage of monitoring data and wireless transmission, and has developed a high-precision pressure sensor with automatic compensation with a measurement accuracy of 1 Pa, supplemented by ventilation resistance measurement and verification, which realises real-time online accurate perception of mine ventilation resistance, the technical principle is shown in Figure 6 (Zhou *et al*, 2020), which solves the time-consuming and labour-intensive problem of manual resistance measurement.

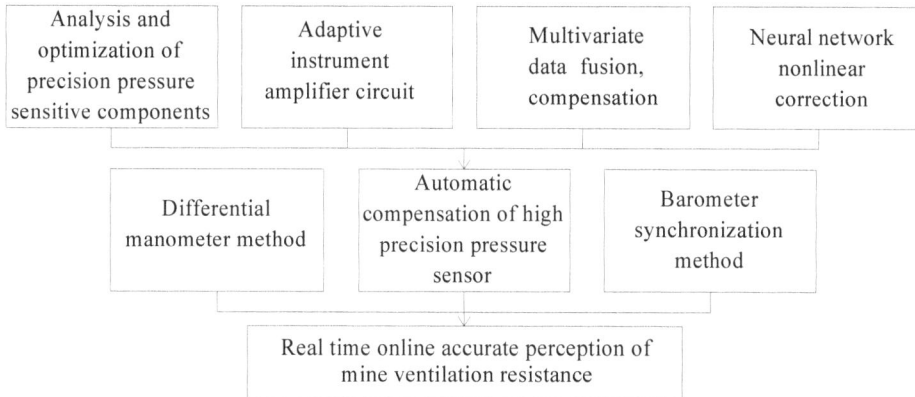

FIG 6 – Real time online accurate perception of mine ventilation resistance.

Whether it is optimisation of ventilation system, system transformation, real-time calculation of wind network or intelligent control of wind flow, the wind resistance of each branch in the wind network needs to be known. At present, the wind resistance obtained by the calculation of ventilation resistance and the Atkinson formula has a large error, which cannot meet the requirements of mine intelligent ventilation. Chen and Zhao (1994) studied the measurement adjustment method of mine ventilation network, and improved the resistance measurement accuracy. Wu, Zhao and Lei (2021) given a friction and wind resistance calculation method that can characterise the three-dimensional surface roughness of the roadway based on fractal theory, and combines the three-dimensional laser scanning technology to obtain roadway friction and wind resistance quickly and accurately. Compared with the time-consuming and laborious work of measuring ventilation resistance, the wind measurement operation is relatively easy. The works (Zhou, Lyu and Liu, 2004; Huang *et al*, 2016) studied the wind resistance measurement technology, its technical principle: take the air volume of the roadway and the working point of the main ventilator as the known quantity, changing the wind resistance of one or several branches in the wind network, the air volume distribution of the wind network under different working conditions is obtained, and then the number of equations for solving the wind resistance is increased, and then the roadway wind resistance is inversely calculated.

$$\sum_{j=1}^{n} c_{ij}^{k} dig(q_{j}^{k} \left| q_{j}^{k} \right|) r_{j}^{k} = \sum_{j=1}^{n} c_{ij}^{k} h_{j}^{k}$$

(1)

Where:

c_{ij}^{k}	is the circuit matrix of the *k*-th air regulation
i	is the circuit number
j	is the branch number
q_{j}^{k}	is the branch air volume in the *k*-th air regulation, m³/s
r_{j}^{k}	is the branch wind resistance in the *k*-th air regulation, N·s²/m⁸
h_{j}^{k}	is the ventilator pressure in the *k*-th air regulation, Pa.

This technology has reliable accuracy in solving wind resistance, and greatly reduces the workload of measurement.

Artificial neural networks, support vector machines and other artificial intelligence technologies have superior nonlinear processing capabilities and are suitable for the prediction of ventilation resistance coefficients. Wei, Sun and Deng *et al* (2018) used artificial neural networks, support vector machines and other methods to establish a roadway friction resistance coefficient prediction model; Deng (2014) established a mathematical model of ventilation resistance coefficient inversion based on the principle of least squares, and applied intelligent optimisation algorithms such as genetic algorithm and particle swarm algorithm to the ventilation resistance coefficient inversion problem.

The optimisation problem of sensor layout to meet the needs of accurate perception of ventilation parameters of the whole wind network and to optimise the economy is the focus of mine intelligent ventilation research. Liu, Ma and Yang (2017) used variable fuzzy set theory to construct a variable fuzzy optimal model for wind speed sensors, accordingly, calculates relative wind speed sensors, which obtain the reasonable weight of the wind speed sensor installed on each branch, use the breadth-first search algorithm to obtain the minimum spanning tree of the ventilation network, and determine the basic loop of the ventilation network combined with the root-finding method, so as to quickly and accurately determine the specific loop branch where the wind speed sensor is installed, the optimal location scheme of the wind speed sensor is obtained, and the real-time dynamic monitoring of the air volume of the ventilation network without blind spots is realised; Li *et al* (2021b) used the minimum spanning tree algorithm to solve the minimum spanning tree of the ventilation network graph with weights assigned to the branches, and obtains the network residual branch combination with the largest weight in the ventilation network graph, determine the monitoring distribution point as the largest residual branch combination, and calculate the air volume of each branch of the ventilation system by monitoring the residual branch air volume metre; Li *et al* (2021a) used the directed passage matrix to determine the coverage of the branch, and uses the branch that covers the most directed passages as the sensor branch to monitor the air volume change of the entire ventilation network with the least number of sensors.

CONSTRUCTION PROGRESS IN INTELLIGENT ANALYSIS AND DECISION-MAKING PLATFORM OF MINE VENTILATION

The construction of mine ventilation intelligent analysis and decision-making platform is mainly reflected in the development of mine ventilation network calculation software, and the rapid diagnosis of ventilation faults. Since the 1970s, Chinese scientific and technical personnel have invested a lot of research in mine ventilation network analysis, ventilation visualisation and ventilation simulation, and have developed a large number of software with functions such as ventilation network diagram drawing and wind network calculation. The ventilation system analysis software includes CFIRE, mine ventilation visualisation system, and mine ventilation auxiliary design system developed by Professor Chang Xintan's team; Mine Ventilation Simulation System (MVSS) developed by Professor Liu Jian's team; Mine Ventilation Management System (MVMS) developed by Professor Wang Deming, etc. These software have been widely used in the actual mine ventilation design and daily management. After nearly 50 years of development, China's software for static analysis of ventilation systems has become more mature, and the main functions of the software have also become more complete.

In recent years, based on mine ventilation theory, geographic information, communication, cloud computing, internet, sensors, Internet of Things and other technologies, Chinese researchers have used ventilation parameter monitoring sensors with full wind network coverage to optimise site selection, monitoring data noise reduction and pre-processing, and multiple data information fusion and redundancy analysis and other methods to analyse in real-time and in-depth mining for mine ventilation monitoring data, have established a real-time calculation model of mine ventilation network, the mine ventilation network calculation software and ventilation monitoring system are integrated. Developed a series of mine ventilation intelligent auxiliary decision-making systems that integrate the functions of real-time calculation of mine ventilation network, formulation of wind flow control schemes in daily and disaster periods, and online ventilation monitoring and early warning, and realise the show of the mine ventilation system diagram, ventilation parameter changes, real-time calculation results of ventilation network, ventilation abnormal diagnosis and early warning through the form of two-dimensional and three-dimensional dynamic visualisation. Fang and Ma (2016) proposed an integrated algorithm of ventilation monitoring and ventilation simulation, through the method of sharing the mine monitoring system database, realises the dynamic solution of the wind network and the online real-time simulation of the ventilation system based on the internet technology; Luo *et al* (2019) established a data-driven dynamic calculation model of the ventilation network, optimised the sensor arrangement process for monitoring ventilation parameters, and developed an online monitoring system for mine ventilation, realised the real-time monitoring of mine ventilation simulation, intelligent online solution and 3D dynamic online. Xi'an University of Science and Technology has developed a visual mine ventilation network solution model based on the mine ventilation network parallel solution model, the interface is shown in Figure 7, which realises fast and accurate calculation of large-scale ventilation networks; an intelligent ventilation software system has been developed, and the interface is shown in the Figure 8, the system realises the functions of safety tracking, monitoring and early warning, daily ventilation management, medium and long-term analysis and design, disaster deduction and emergency ventilation disposal; Based on cloud computing technology, a three-dimensional intelligent real-time dynamic solution system for mine ventilation network is developed, it has realised the functions of whole wind network monitoring, real-time calculation of wind network, three-dimensional ventilation simulation, abnormal air volume analysis, demonstration of disaster avoidance route, and formulation of ventilation plan. The three-dimensional display of the ventilation system is shown in Figure 9.

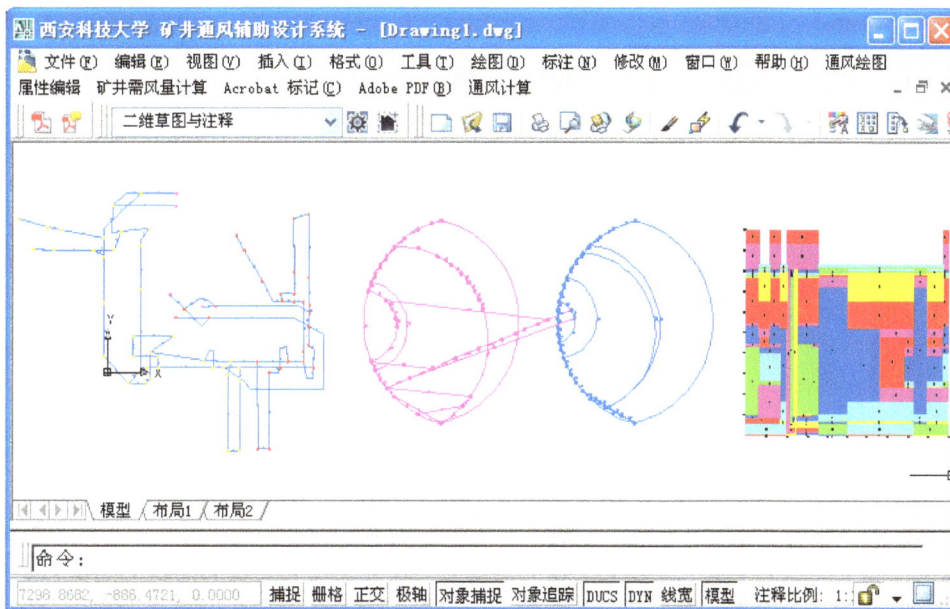

FIG 7 – Mine ventilation network solution software.

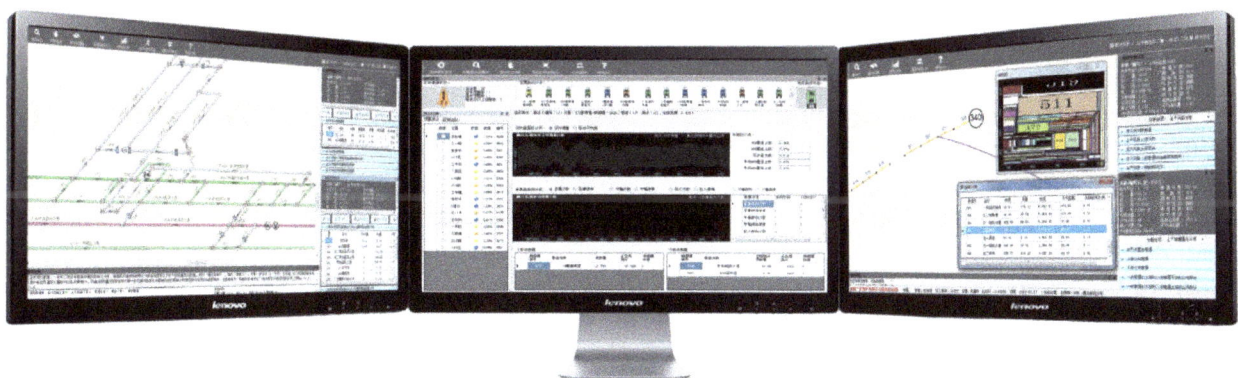

FIG 8 – Intelligent ventilation software system (Fan, Li and Yan, 2020).

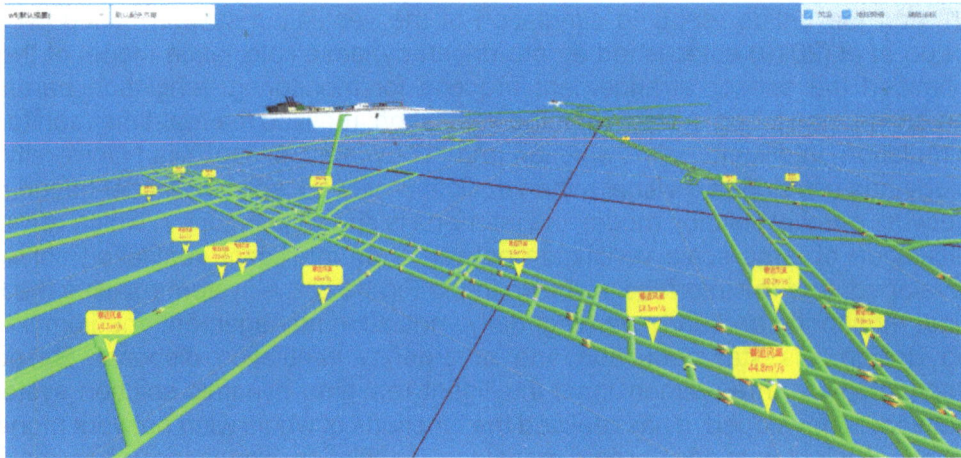

FIG 9 – 3D display of mine ventilation system.

The intelligent diagnosis of mine ventilation faults mainly includes abnormal analysis of the air volume of the roadway, optimised layout of monitoring points, identification of the angle-connected roadway, and fault diagnosis of ventilation equipment.

From the perspective of the entire mine ventilation network, the reasons for the abnormal air volume in the roadway (the collapse of the roadway, the unclosed or damaged air door, the extension and scrapping of the roadway, and the emptying of the coal bunker) can all be attributed to the change of the branch wind resistance. The mine wind network is an overall structure, and the change of the wind resistance of any branch will cause the change of the air volume of the relevant branch or even all branches in the whole network. The sensitivity of the branch air volume to the change of other branch wind resistance is defined as sensitivity.

$$d_{ij} = \lim_{\Delta R_{ij} \to 0} \Delta Q_i / \Delta R_j = \partial Q_i / \partial R_j \qquad (2)$$

Where:

$\triangle Q_i$ is the change amount corresponding to the air volume of any branch i when the wind resistance R_j of branch j in the wind network changes $\triangle R_j$.

The works (Wang, Wu and Wang, 2008; Wu, 2011a, 2011b; Jiang, 2011) applied the sensitivity theory to the research of air volume regulation, outlier analysis of air volume, identification of corner-connected roadways, and analysis of wind network stability. Xi'an University of Science and Technology developed a sensitivity calculation program and integrated it into the ventilation network calculation software, and realised the calculation and view of the sensitivity of complex networks under Auto CAD.

Research on automatic and accurate identification of corner-joined roadways. Zhao, Liu and Yang (2001) proposed a seven-tuple algorithm, the automatic identification program of the angle-connected roadway is used to realise the accurate and rapid identification of the angle-connected roadway, and the air leakage and seepage area in the gob is analysed as an air channel in the wind network; Li *et al* (2012) proposed the node location method, developed automatic identification software of corner-connected roadways; Zhao *et al* (2009) studied the mathematical model of corner-connected structure recognition based on parallel computing, and realised automatic identification of corner-connected structures in super-large wind networks; Yan *et al* (2018) proposed an equivalent attribute graph model, and established a high-risk area identification algorithm based on the subgraph isomorphism method. While analysing and comparing the topology of the wind network, the attributes of the roadway are included in the judgment. Through the relationship between roadway attributes and risks, structures that do not have production risks are eliminated, and finally the high-risk areas of corner connection in the complex wind network are automatically identified.

In the research of intelligent analysis technology of air leakage flow field in gob. The works (Wu *et al*, 2016, 2017) studied the Jacobian matrix characteristics of mine ventilation network and its parallel solution technology, realised the rapid solution of large-scale ventilation network, and provided technical support for the use of network solution method to study gob flow field; The works (Wu, Luo and Chang, 2019; Wu, Gao and Chang, 2020; Wu and Luo, 2020) studied the ventilation problem of the coupling of mine ventilation network and gob flow field, and proposed a numerical simulation method for coupling the one-dimensional mine ventilation network and the two/three-dimensional gob flow field–the finite flow tube method (The coupling network of the one-dimensional mine ventilation network and the two/three-dimensional gob flow field is shown in Figure 10), developed a visualised mine ventilation network and gob air leakage integrated simulation software (i-MVS), and realised the partition of the finite element mesh of the gob, ventilation network modelling, and dynamic generation of gob flow pipes, a coupling model of mine ventilation network and gob flow field based on the finite-flow pipe method is established (the simulation results of the coupling of mine ventilation network and gob flow field are shown in Figure 11), developed the software can determine the areas with gas explosion and coal spontaneous combustion hazards in gob and mine ventilation network (the distribution of gas concentration mine ventilation network and gob is shown in Figure 12), and can also analyse the risk analysis in various mines ventilation scheme.

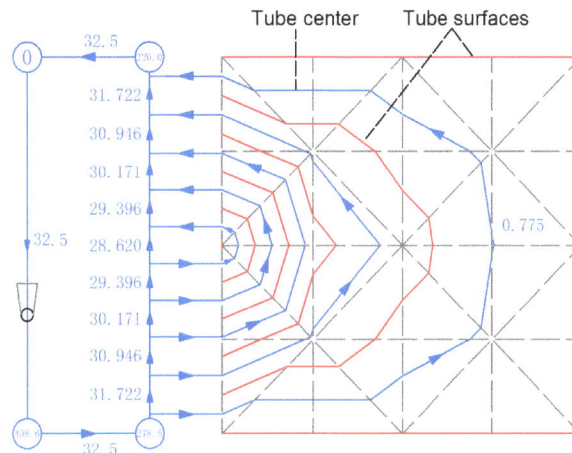

FIG 10 – The coupling network (blue colour) of the one-dimensional mine ventilation network and the 2D/3D gob flow field.

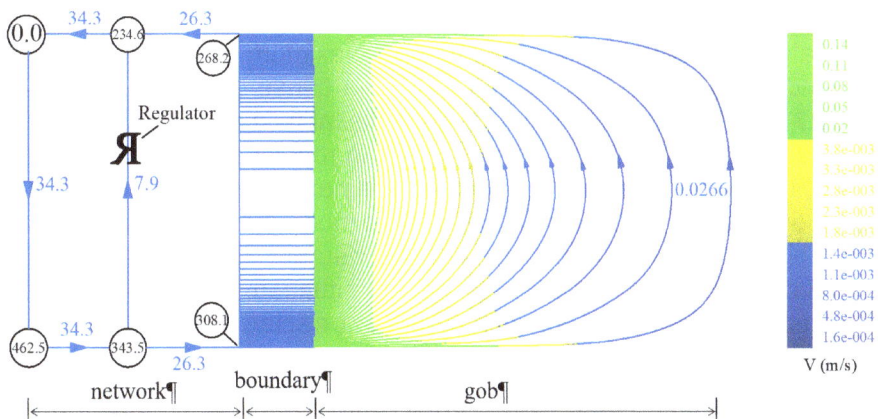

FIG 11 – The coupling simulation results of mine ventilation network and gob flow field.

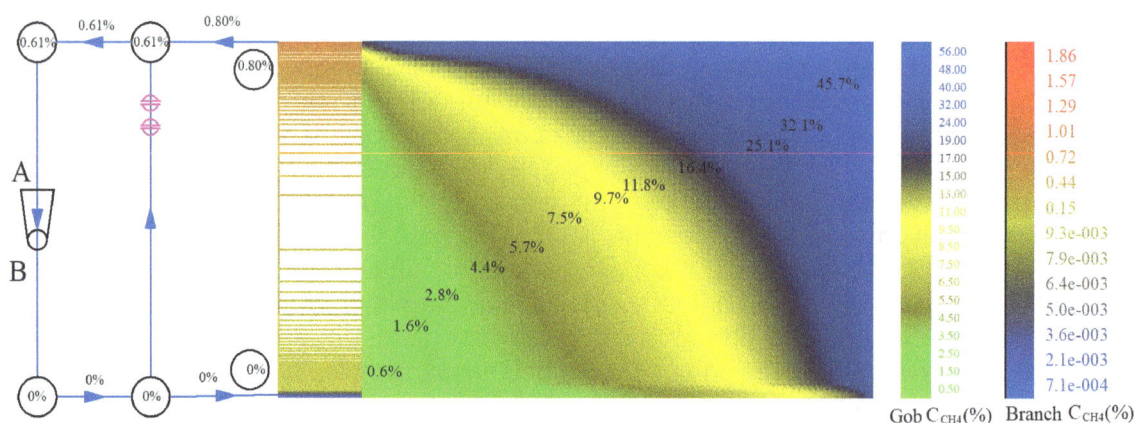

FIG 12 – Methane concentration in the ventilation system.

With the development of modern digital signal processing, computer, virtual instrument, artificial intelligence and other technologies, the fault diagnosis technology of mine ventilator has entered the field of intelligence. China has applied artificial intelligence technology and related theories to develop intelligent diagnosis technology for mine ventilator faults, and has achieved a series of fruitful results. The works (Jing, Leng and Li, 2004; Jing and Hua, 2007) used wavelet packet channel energy decomposition technology to extract feature vectors of different frequency bands reflecting different working states of ventilators as fault samples of BP neural network and support vector machine, taking the trained network and classifier as the fault intelligent classifier to realise the intelligent diagnosis of ventilator faults. Aiming at the problem of missed diagnosis of mine main ventilator fault diagnosis relying on a single signal source; The works (Fu, Li and Zhu, 2007; Lei and Ren, 2007; Ling and Huang, 2008) studied that the multi-sensor information fusion technology combined with BP neural network and D-S evidence theory is studied to diagnose the fault of mine ventilator, and realised the automatic and accurate diagnosis of ventilator faults; Li and Li (2010) applied the holographic spectrum technology of multi-sensor information fusion theory to the fault diagnosis of mine ventilators, and combined with virtual instrument technology, developed a virtual instrument for ventilator fault diagnosis, and realised the accurate diagnosis of ventilator faults; Cui and Cheng (2021) used fuzzy reasoning theory to convert the constructed main ventilator fault rule base into a fuzzy matrix of fault relations, and uses the maximum membership function to calculate the highest confidence level, based on which to find the cause of the ventilator failure; Jing *et al* (2021) used Unity3D, 3dsMax, SciFEA etc to build a digital twin model of the ventilator, and uses PREspective to communicate with the ventilator's PLC in real time, maps the ventilator's operating status to the digital twin model in real time, and combines expert knowledge and machine learning, historical data etc to establish a predictive fault diagnosis model for ventilators, and use the improved whale optimisation algorithm to optimise probabilistic neural network to conduct predictive fault diagnosis for ventilators. Compared with the actual situation, if the diagnosis is wrong, the parameters in the predictive fault diagnosis model will be corrected to achieve real-time and accurate diagnosis of ventilator faults.

RESEARCH AND DEVELOPMENT PROGRESS IN INTELLIGENT REGULATION EQUIPMENT OF MINE VENTILATION

The main function of mine ventilation facilities is to control the air flow, the dampers and windows are distributed throughout the underground, it takes a lot of manpower to manually adjust the air volume. More importantly, when a disaster occurs in the underground, timely control of the ventilation system can effectively prevent the spread of the disaster. In order to realise the intelligent regulation of ventilation system and improve the technical level of mine emergency and disaster relief, China has carried out research and development work on remote automatic control of dampers and remote automatic adjustment of windshields.

The remote automatic control damper is an intelligent ventilation facility for mine air flow. At present, automatic dampers have been widely used in Chinese mines. This type of damper sends and receives signals by means of mechanical triggering, ultrasonic wave, radar microwave, infrared induction, electromagnetic induction, pulsed light source photosensitive etc, and controls the mechanical transmission mechanism through relay, microcontroller or PLC, the automatic opening and closing of the air door is driven by means of compressed air, hydraulic pressure or motor, which solves various problems of manual opening and closing, the pressure structure around the door frame solves the problem of the opening and closing failure of the air door due to the deformation of the roadway. At the same time, it has voice sound and light alarm prompt automatic air door lock, intelligent video surveillance technology to identify vehicles and pedestrians, infrared detection to prevent people from trapping cars and other functions (Wu *et al*, 2010; Shao *et al*, 2008; Sun *et al*, 2021; Fu, Wei and Pan, 2022; Zhang *et al*, 2020).

The actual automatic damper is shown in Figure 13, and the automatic damper video monitoring function interface is shown in Figure 14. For the emergency rescue of underground fire accidents, on the basis of the automatic damper, China has used communication technology to design the ground central station and damper controller substation, and developed the remote automatic damper system, which realised the purpose of emergency smoke and gas evacuation, following this workflow: monitors the disaster situation according to the sensor → the signal is transmitted to the ground to alarm → the wind control and disaster relief decision-making → the ground control the opening and closing of the underground automatic damper → the status feedback of the wind control facilities → the ground display wind control effect (Zhu *et al*, 2000; Wang *et al*, 2021). The interface of the intelligent remote control system on the ground of ventilation facilities is shown in Figure 15.

FIG 13 – Automatic damper of Yubei Caojiatan Mining Company.

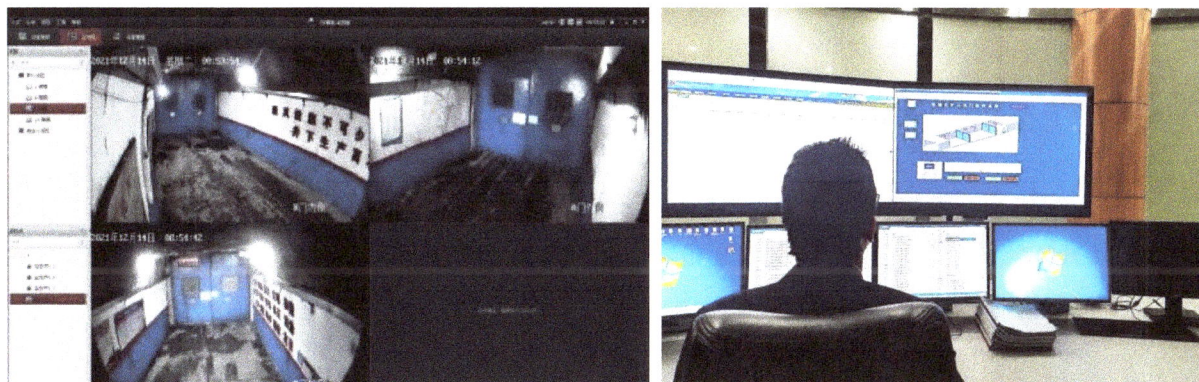

FIG 14 – Remote monitoring system of automatic damper in Yubei Xiaobaodang Mining Company.

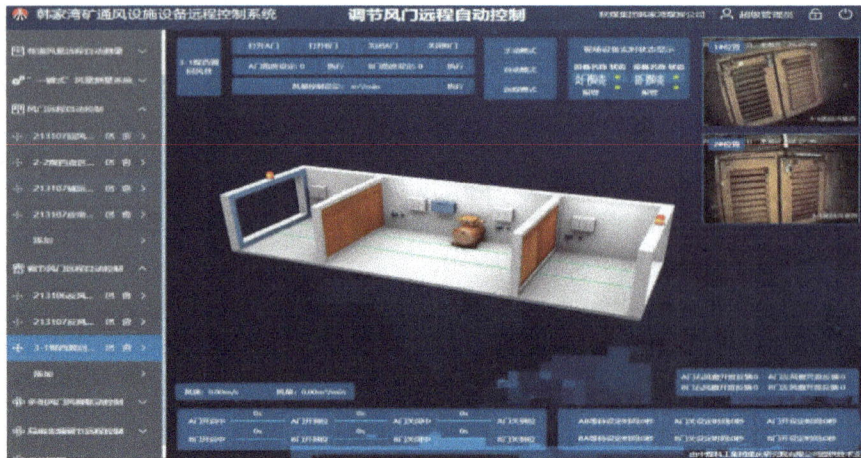

FIG 15 – Ground intelligent remote control system for ventilation facilities in Hanjiawan Coal Company of Northern Shaanxi Co., Ltd. (Bai *et al,* 2021).

The remote automatic adjustment of air windows is an intelligent ventilation facility for quantitative adjustment of mine air volume. By integrating automation, sensors and Internet of Things technologies, China has developed high-precision, high-sensitivity intelligent air doors and windows, the quantitative relationship model between wind resistance increment and wind window adjustment parameters is constructed, forming the intelligent remote control technology of air volume, so as to realise the intelligent and precise control of the mine ventilation system. Shandong Lion King Software Co., Ltd. has designed a multi-layer push-pull type, louver type, shutter type and other structures of intelligent control damper and damper, as shown in Figure 16. The design principle is to use a stepper motor to accurately control the movement of the windshield through the synchronous belt, the material level opening sensor feeds back the opening of the wind window in real time, the controller accepts the remote control command to drive the actuator, and feeds back the collected sensor and wind window status data to the remote system in real time, realising the remote automatic closed-loop precision control of the wind window. This type of intelligent control air door and window has several control modes such as remote automatic adjustment, local remote control adjustment, automatic and manual switching adjustment etc, which can meet the emergency response under different ventilation conditions (Lu and Yin, 2020); Zhou *et al* (2020) based on the adjustment of the characteristic curve of the wind and window, integrating the automatic control technology, the structural characteristics of the adjusting wind window and the network structure characteristics of the branch, a PID adjustment model of the branch air volume is constructed, and the continuous and precise control of the adjusting wind window is realised.

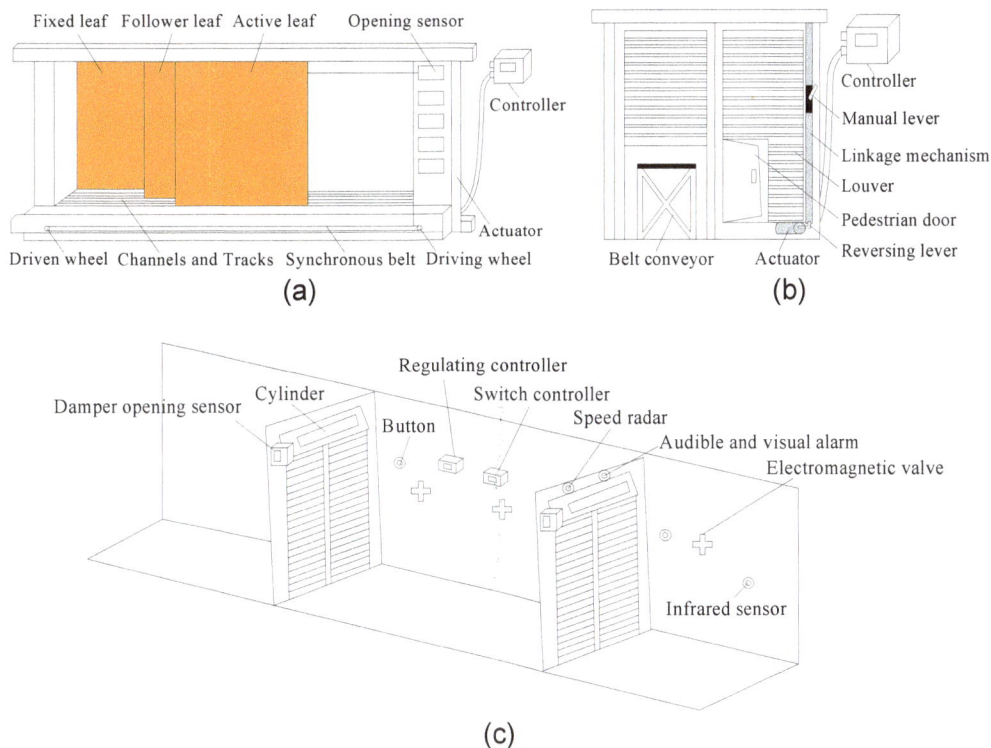

FIG 16 – Mine ventilation intelligent regulation facility: (a) Multi-layer push-pull adjustable window; (b) Louvered regulating door; (c) Intelligent control damper.

The ventilation rate of the traditional tunnelling face is determined by quantitative calculation in advance according to factors such as temperature, gas emission, carbon dioxide emission, and the maximum number of workers, and after takes its maximum value, According to this, the quantitative air supply of the local ventilator belongs to the ventilation mode of one wind blowing, It is impossible to intelligently control the air flow according to the changes of environmental parameters, resulting in waste of energy consumption caused by excessive air volume or gas exceeding the limit caused by insufficient air volume. In recent years, integrating communication, sensors, Internet of Things, automation, frequency conversion control and other technologies, China has developed an intelligent control system for local ventilators as shown in Figure 17. The system can intelligently sense environmental parameters through sensors and feed back to the intelligent control system in real time to adjust the frequency and speed of the local ventilators, thereby changing the output air volume and air pressure of the ventilators, so as to realise on-demand air supply, automatic gas discharge, automatic switching of the main and standby ventilators, and remote unmanned start-up, thereby ensuring the safety, efficiency and energy saving of local ventilation (Cui, 2016; Rui, 2021; Liu, 2017; Cheng, 2020). Zhang (2008) combined the pulsating air supply technology with the frequency conversion speed regulation technology, and developed an intelligent ventilation control system for the excavation face, the system analyses and determines the gas concentration collected by the gas sensor, changes the ventilator speed to adjust the air volume, air pressure, and the switching cycle and switching ratio of the local ventilator, thereby realising intelligent control of the ventilation of the excavation face; Ping An Electrical Group has developed an intelligent local ventilation system, which consists of local ventilator, mine flow channel inverter, intelligent control switch etc systematically adopts a control system composed of PLC and inverter, used the change of gas concentration as the main reference to adjust the ventilator speed, so as to realise the automatic adjustment of air volume and control of gas discharge (Chen, Xiao and Hu, 2009; Wang *et al*, 2012).

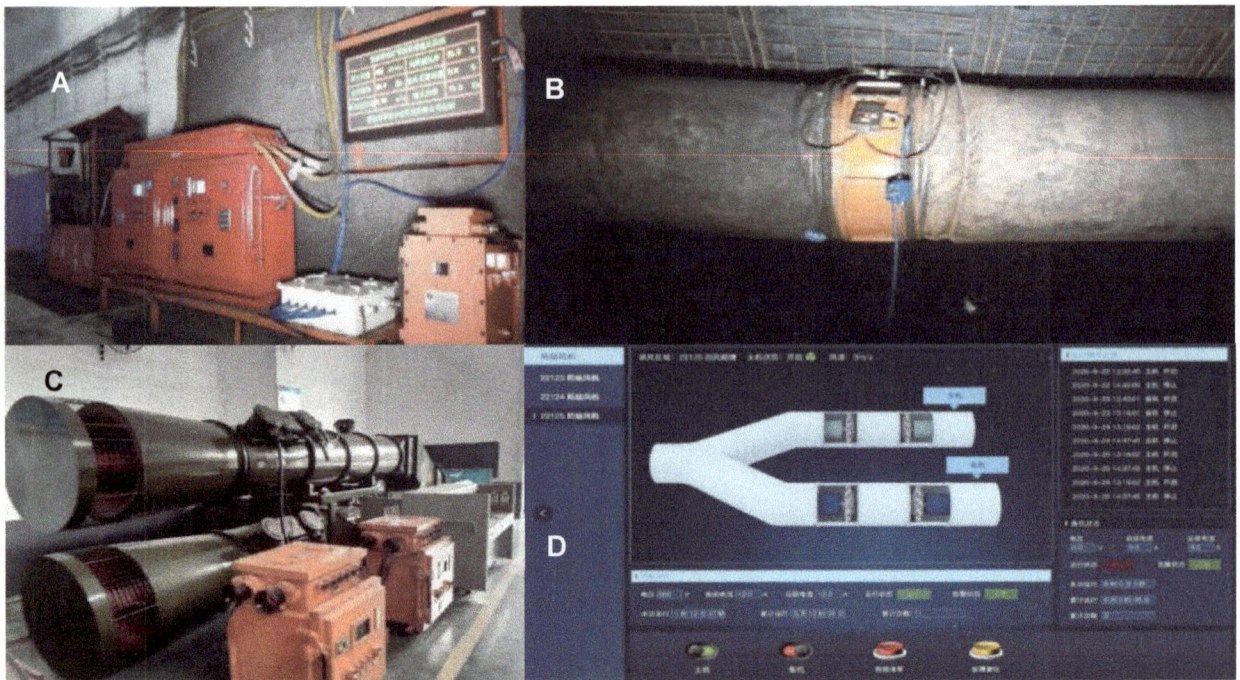

FIG 17 – Mine intelligent local ventilation system: (a) Frequency converter; (b) Wind speed sensor; (c) Mine intelligent local ventilator; (d) Intelligent regulation and control system.

CONCLUSIONS

With the in-depth integration of big data, artificial intelligence, Internet of Things, cloud computing and other high-tech technologies and mine ventilation technology and equipment, Intelligent integrated management and control of mine ventilation characterised by 'comprehensive and autonomous perception of ventilation parameters, deep and efficient integration of monitoring data, intelligent analysis and decision-making of ventilation status, and precise coordinated control of ventilation facilities/equipment' has become an important mode of mine ventilation intelligent upgrade in western China. In recent years, mines in western China have firmly grasped the development opportunities brought by the new generation of information technology, and have achieved a series of important achievements in intelligent ventilation technology, software and equipment. However, judging from the current development status of mine intelligent ventilation technology in western China, the intelligent construction of mine ventilation is still in its infancy, and it is necessary to continue to explore and tackle key technologies, so as to promote the development of mine ventilation engineering from experience to high-tech, and to improve the development level of China's mines in intelligence, automation, and less labour in western China.

ACKNOWLEDGEMENTS

The authors sincerely thank the scholars who provided references, whose papers provided important references for this paper. This work is supported by the National Natural Science Foundation of China [grant numbers 51974232].

REFERENCES

Bai, M B, Yao, Y H, Luo, G and Gao, J W, 2021. Research and application of intelligent ventilation technology in Hanjiawan Coal Mine, *Mining Safety and Environmental Protection,* 48(5):87–91 (in Chinese).

Chen, C H X, Xiao, W L and Hu, X M, 2009. Introduction of intelligent local ventilation system, *Jiangxi Coal Science & Technology,* (3):109–112 (in Chinese).

Chen, K Y and Zhao, Y H, 1994. The correlated adjustment of condition equations for resistance measurement of mine ventilation network, *Journal of China University of Mining and Technology,* 23(1):80–89 (in Chinese).

Cheng, L H, 2020. Application research on intelligent control system of ventilation in heading face, *Shandong Coal Science and Technology,* (11):131–133 (in Chinese).

Cui, B W, 2016. Application of intelligent frequency conversion technology in mine ventilation system, *Inner Mongolia Coal Economy,* (8):3–4 (in Chinese).

Cui, H F and Cheng, F L, 2021. Fault diagnosis of mine main ventilator based on improved fuzzy reasoning theory, *Coal Technology,* 40(10):112–115 (in Chinese).

Deng, L J, 2014. *Study on mine ventilation resistance coefficient inversion*, PhD thesis, Liaoning University of Engineering and Technology, Fuxin (in Chinese).

Fan, J D, Li, C H and Yan, Z H G, 2020. Overall architecture and core scenario of a smart coal mine incorporating 5G technology ecology, *Journal of China Coal Society,* 45(6):1949–1958 (in Chinese).

Fang, B and Ma, H, 2016. Mine ventilation network application monitoring database and application dynamic solver, *Journal of Liaoning Technical University (Natural Science),* 35(12):1439–1442 (in Chinese).

Fu, S H, Li, H T and Zhu, Q, 2007. A fault early warning and software development of the main ventilator in mine, *Journal of Beijing University of Technology,* 33(8):809–812 (in Chinese).

Fu, Z H, Wei, Q and Pan, G H, 2022. A damper system based on AI image intelligent recognition, *China Patent* CN215595636U.

Huang, J X, Yang, Z H M, Zhang, L and Su, X Q, 2016. Measuring airflow to evaluating resistance based on nonlinear optimization and independent partition, *Coal Technology,* 35(3):165–167 (in Chinese).

Ji, C H S, 1997. Analysis of Wolonin's basic theory of mine ventilation, *Nonferrous Metals Engineering*, (2):1–5 (in Chinese).

Jiang, S H, 2011. Theory and application of mine ventilation network sensitivity, *Safety in Coal Mines,* 42(1):96–99 (in Chinese).

Jing, H X, Huang, Y R, Xu, S H Y and Tang, C H L, 2021. Research on the predictive fault diagnosis of mine ventilator based on digital twin and probabilistic neural network, *Industry and Mine Automation,* 47(11):53–60 (in Chinese).

Jing, S H X Hua, W, 2007. The mine ventilator fault diagnosis based on wavelet packet and support vector machine, *Journal of China Coal Society,* 32(1):98–102 (in Chinese).

Jing, S H X, Leng, J F and Li, Z H, 2004. Study on the mine ventilator fault diagnosis based on wavelet packet and neural network, *Journal of China Coal Society*, 29(6):736–739 (in Chinese).

Lei, M and Ren, Z H, 2007. Build the mine main ventilator fault diagnosis model based on multi-sensor information fusion technology, *Industry and Mine Automation,* 10(5):62–64 (in Chinese).

Li, B R, Wang, W, Chen, F M Liu, N, 2021a. Optimal arrangement of wind speed sensor based on directed path matrix method, *Industry and Mine Automation*, 47(5):52–57 (in Chinese).

Li, M and Li, Y, 2010. Virtual instrument of fault diagnosis for mine ventilator based on holographic spectrum, *Coal Science and Technology,* 38(3):93–96, p 108 (in Chinese).

Li, W, Zhang, L, Wang, H F and Guo J H, 2012. Distinguishment on diagonal air way based on improved node location method, *Coal Science and Technology,* 40(11):77–79 (in Chinese).

Li, Y J, Wu, J K, Li, Y H Yao, P Y, 2021b. Optimization on monitoring layout of mine ventilation network based on the principle of minimum spanning tree, *Mining Research and Development,* 41(7):172–175 (in Chinese).

Ling, L Y and Huang, Y R, 2008. Fault diagnosis of mine ventilator base on multi sensor information integration, *Coal Science and Technology,* 36(6):72–74 (in Chinese).

Liu, P, 2017. The research and application of intelligent mine local ventilation equipment, *Coal Mine Modernization,* (5):87–88 (in Chinese).

Liu, Y X, Ma, H and Yang, H R, 2017. Mine wind sensor variable fuzzy optimization model, *Journal of Liaoning Technical University (Natural Science),* 36(10):1031–1035 (in Chinese).

Lu, G L, Zhang, M H and Fan, C Q, 2015. Analysis of fixed-point wind speed measurement method based on CFD, *Coal Technology,* 34(5):156–157 (in Chinese).

Lu, X M and Yin, H, 2020. The intelligent theory and technology of mine ventilation, *Journal of China Coal Society,* 45(6):2236–2247 (in Chinese).

Luo, G, Zou, Y H, Ning, X L Jin, N Q J, 2019. Research and application of online monitoring technology for mine ventilation network, *Mining Safety and Environmental Protection,* 46(5):47–50, p 55 (in Chinese).

Rui, G X, 2021. Study on intelligent reconstruction scheme of ventilation system in Chahasu coal mine, *Journal of North China Institute of Science and Technology,* 18(8):40–44 (in Chinese).

Shao, H, Jiang, S H G, Qin, J H, Wu, Z H Y and Wang, L Y, 2008. Program design on auto air-door base on PLC control, *Coal Science and Technology,* 36(4):66 (in Chinese).

Song, Y, Liu, J, Li, X B, Liu, Y H, Tian, R X and Zhao, C S, 2016. Experiment and numerical simulation of average wind speed distribution law of airflow in mine tunnel, *China Safety Science Journal*, 26(6):146–151 (in Chinese).

Sun, Y J, Dong, K W, Yun, X, Zhou, Y, Wang, W W, Cheng, X Z H and Zhou, J S, 2021. An intelligent collaborative linkage control method for downhole dampers based on video detection, *China Patent* CN112308032A.

Wang, E, Zhang, L, Li, W and Sang C, 2016. Key technology of multipoint mobile wind-measured device, *Safety in Coal Mines,* 47(6):97–99, p 103 (in Chinese).

Wang, G F, 2022. New technological progress of coal mine intelligence and its problems, *Coal Science and Technology,* 50(1):1–27 (in Chinese).

Wang, H G, Wu, F L and Wang, Y, 2008. The airflow abnormal value analysis of mine ventilation on network based on the sensitivity, *Safety in Coal Mines,* (9):86–88 (in Chinese).

Wang, K, Cai, W Y, Gao, S W, Chen, X Y and Zhang, Y C H, 2021. Research on the linkage reliability of mine fire wind and smoke flow emergency control system, *Journal of China University of Mining and Technology,* 50(4):744–754 (in Chinese).

Wang, W C, Qiao, W, Li, G and Tian, C H, 2012. Application of mine intelligent partial ventilation system in Huhewusu coal mine, *Safety in Coal Mines,* 43(6):114–116 (in Chinese).

Wei, N, Sun, Y S H N, Deng, L J, Huang, D and Guo, X, 2018. Influence factors analysis and prediction on mine ventilation resistance coefficient based on SVM, *Journal of Safety Science and Technology,* 14(4):39–44 (in Chinese).

Wu, B, Zhao, C H G and Lei, B W, 2021. Characterization and calculation method of friction resistance based on fractal theory of roadway rough surface, *Journal of China University of Mining and Technology,* 50(4):633–640 (in Chinese).

Wu, F L and Luo, Y, 2020. An innovative finite tube method for coupling of mine ventilation network and gob flow field: Methodology and Application in Risk Analysis, *Mining, Metallurgy and Exploration,* 37(5):1517–1530.

Wu, F L, 2011a. Sensibility calculation and stability analysis on a complicated mine ventilation network, *Science and Technology Review,* 29(19):62–65 (in Chinese).

Wu, F L, 2011b. Using sensitivity to optimize mine air regulation point and parameter, *Mining Safety and Environmental Protection,* 38(5):1–3, p 7 (in Chinese).

Wu, F L, Gao, J N, Chang, X T and Li, L Q, 2016. Symmetry property of Jacobian matrix of mine ventilation network and its parallel calculation model, *Journal of China Coal Society,* 41(6):1454–1459 (in Chinese).

Wu, F L, Gao, Y C H and Chang, X T, 2020. Boundary coupling model and solution technology of mine ventilation network and gob flow field, *Journal of Central South University (Science and Technology),* 51(8):2333–2342 (in Chinese).

Wu, F L, He, X C H, Chang, X T, Ma, L and Li, C H, 2017. Research on simulation technology of surface air leakage of shallow-buried goaf based on network calculation, *Industry and Mine Automation,* 43(12):64–69 (in Chinese).

Wu, F L, Luo, Y and Chang, X T, 2019. Coupling simulation model between mine ventilation network and gob flow field, *Journal of the Southern African Institute of Mining and Metallurgy,* 119(10):783–792.

Wu, H W, Zhang, Y M, Wu, Z H Y, Jiang, S H G, Wang, L Y and Wang, J, 2010. Development, analysis and comparison of technologies of mine-used automatic air doors, *Industry and Mine Automation,* 36(1):61–65 (in Chinese).

Yan, Z H G, Chang, X T, Fan, J D, Wang, Y P and Zhao, P X, 2018. Recognization on high risk region of mine ventilation system in coal mine based on subgraph isomorphism, *Coal Science and Technology,* 46(11):63–68 (in Chinese).

Zhang, G J, 2008. Design and application of intelligent ventilation control system to mine roadway heading face, *Coal Science and Technology,* 36(5):76–79 (in Chinese).

Zhang, J L, Wang, S H, Li, Z H, Liu, B S and Zhang L M, 2020. Human-vehicle identification automatic air door control system and control method, *China Patent* CN108561171B.

Zhang, L, 2018. Optimized study on location to measure average air velocity with air velocity sensor in wind measuring station of underground mine, *Coal Science and Technology,* 46(3):96–102 (in Chinese).

Zhang, Q H, 2016. Development and prospect of mine ventilation technology and equipment, *Coal Science and Technology,* 44(6):146–151 (in Chinese).

Zhang, Q H, Yao, Y H and Zhao, J Y, 2020. Status of mine ventilation technology in China and prospects for intelligent development, *Coal Science and Technology,* 48(2):97–103 (in Chinese).

Zhang, S L, 2019a. Development status and trend of mine ventilation resistance measurement technology, *Safety in Coal Mines,* 50(6):188–191 (in Chinese).

Zhang, S L, 2019b. Study on measurement and change law of wind speed in cross-section of coal mine ventilation roadway, *Mining Safety and Environmental Protection,* 46(4):17–20 (in Chinese).

Zhao, D, Huang, F J, Chen, S, Wang, D and Wang, D W, 2014. Relationship between point air velocity and average air velocity in circular roadway, *Journal of Liaoning Technical University (Natural Science),* 33(12):1654–1659 (in Chinese).

Zhao, D, Liu, J, Pan, J T and Ma, H, 2009. Analysis of identifying diagonal structure of ventilation network based on parallel computing, *Journal of China coal society,* 34(9):1208–1211 (in Chinese).

Zhao, Q L, Liu, J and Yang, C H X, 2001. A new approach to automatic identification of diagonal branches in a complicated mine ventilation network, *Journal of Safety and Environment,* 1(6):19–21 (in Chinese).

Zhou, F B, Wei, L J, Xia, T Q, Wang, K, Wu, X Z H and Wang, Y M, 2020. Principle, key technology and preliminary realization of mine intelligent ventilation, *Journal of China Coal Society,* 45(6):2225–2235 (in Chinese).

Zhou, L H, Lyu J and Liu, X J, 2004. The principle and implementation of calculating resistance through surveying airflow quantity, *Journal of Xi'an University of Science and Technology,* (2):148–150, p 165 (in Chinese).

Zhou, X H, Meng, L, Li, C Y, Feng, C C, Zhou, Y B and Shi, M J, 2015. Experimental study on determination and correction method of wind speed in circular pipe, *Journal of Liaoning Technical University (Natural Science),* 34(5):156–157 (in Chinese).

Zhu, H Q, Liu, X S H, Zhang, W K and Zhang, F Y, 2000. Research and application of auto air flow control system at initial mine fire accident, *Coal Science and Technology,* (2):26–28 (in Chinese).

Comparison of heat, noise and ore handling capacity of battery-electric versus diesel LHD

C McGuire[1], D Witow[2], M Mayhew[3] and K Bowness[4]

1. Mine Ventilation Engineer, Hatch Ltd, Mississauga Ontario, Canada.
   Email: chris.mcguire@hatch.com
2. Mine Ventilation Systems Lead, Hatch Ltd, Sudbury Ontario, Canada.
   Email: darryl.witow@hatch.com
3. Founder, Mayhew Performance Ltd, Sudbury Ontario, Canada.
   Email: mike@mayhewperformance.com
4. Sudbury Integrated Nickel Operations, Sudbury Ontario, Canada.
   Email: keith.bowness@glencore.ca

ABSTRACT

As mobile equipment electrification in underground mining continues to gain popularity, control of heat and humidity in the workplace is becoming a key driver in sizing of ventilation and mine air cooling systems. For design and sizing of ventilation and cooling systems to be accurate, and for the mine mobile equipment fleet to be sized appropriately to meet production goals, vehicle productivity metrics, heat modelling, and energy balances for electrified fleets need to consider operation through a realistic duty cycle that is reflective of various tasks performed during a shift.

The authors conducted a real-world comparison of two 8 cubic yard (14 metric tonne) class Load-Haul-Dump (LHD) machines – one diesel and one battery-electric vehicle (BEV) – performing identical muck-haul-dump operations on consecutive days at an underground nickel/copper hard rock mine in Ontario, Canada.

The battery-electric LHD averaged 8.9 per cent faster cycle time than the diesel vehicle (gross, without accounting for charging time), and was able to transport 25 per cent more material per cycle due to a larger bucket capacity. When including the charging time, average net cycle times were equivalent for the two vehicles over the 200 cycles. The BEV carried more payload over the eight hour period due to the increased capacity per cycle.

The testing identified an approximately five-fold reduction in heat output from the BEV in comparison to the diesel LHD on a basis of heat output per tonne of rated bucket capacity. Heat output from the fast charger was found to be approximately 4 per cent of the charging rate, with negligible temperature change observed in the main drift. External noise levels in the mine were found to be 14 dB lower for the BEV during transit and 11 dB lower during bucket loading. Fuel/energy costs for BEV electricity input were approximately four times lower per operating hour than diesel fuel cost based on typical 2021 electrical and diesel unit rates for a grid-connected mine in Ontario, Canada.

INTRODUCTION

Electrification of mobile equipment fleets in underground mining operations remains a key strategy to economically mining increasingly challenging orebodies, and as part of the solution to achieve net zero carbon emissions in the mining industry. Where ventilation and/or refrigeration requirements are constraints, battery-electric vehicles (BEVs) offer significant opportunities due to their reduced heat and humidity output and the reduction, or in many jurisdictions absence, of regulated minimum ventilation rates.

Ventilation planning for a BEV-based mine, in the absence of regulated volume requirements for BEVs, is increasingly dependent on defining airflows that maintain safe workplace temperatures. Therefore, understanding of electrified mobile equipment fleet heat generation is of emerging importance. As the global fleet of BEVs in operation continues to increase, additional opportunities to survey and quantify heat output will allow for effective planning and efficient use of ventilation capital, plus subsequent operating costs, for new mines and expansion projects. Direct comparison of a BEV against an existing diesel vehicle allows for clear, concise understanding of the potential improvements.

In addition to reduced heat generation and elimination of diesel particulate matter (DPM) emissions, BEVs are purported to be significantly quieter than diesel vehicles due to the replacement of the internal combustion engine with variable-speed electric motors. While this benefit is widely stated (eg multiple times in Global Mining Guidelines Group, 2022), the authors were unable to find published data on a literature review. Thus, the testing also included installation of external microphones on the drift walls to capture sound levels during loading and tramming of material, and personal monitors were fitted to the LHD operators to quantify impacts within the cab of an operating vehicle.

TEST DESCRIPTION

Testing was completed at the Nickel Rim South Mine, an underground hard rock mine operated by Glencore's Sudbury Integrated Nickel Operations in Ontario, Canada. The LHDs were operating in a transfer mucking application – moving material from an orepass to a rockbreaker grizzly station approximately 270 m away across level ground. The charging bay for the BEV unit was located within the haulage route in an unused storage area. The test route and location of the charging bay is shown in Figure 1. Five key locations for data collection are also shown; these are defined further in Table 1.

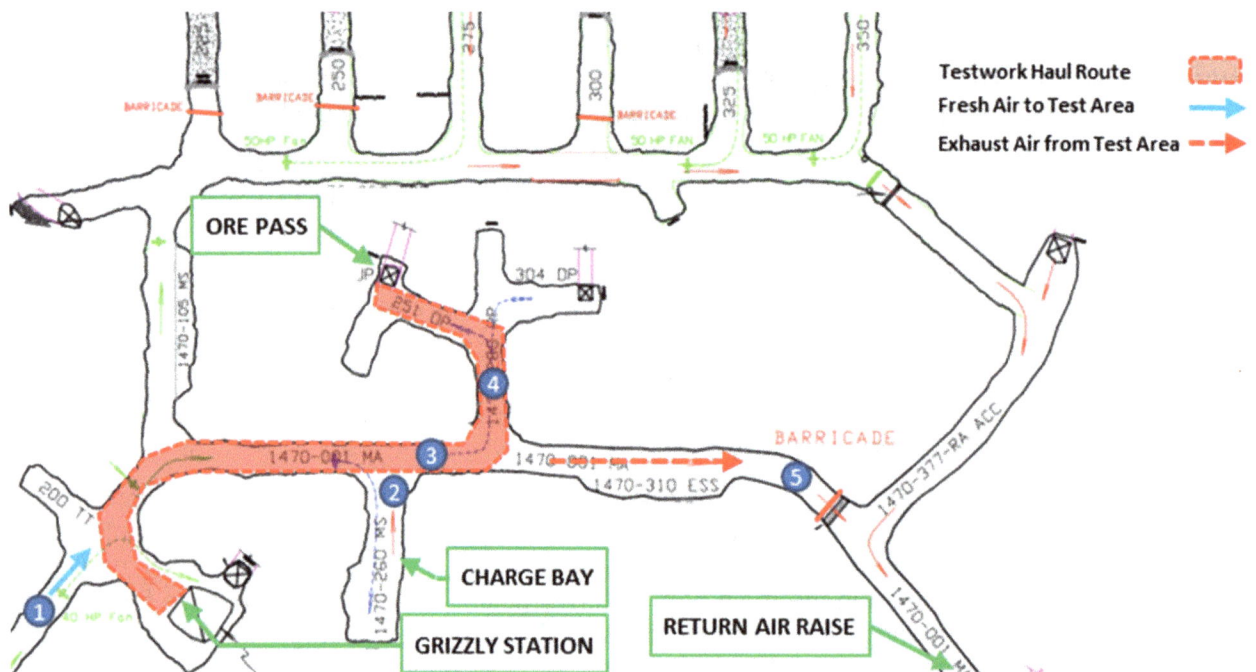

FIG 1 – Plan view of test location including tramming route, charge bay location, and measurement locations (denoted 1 through 5).

Ventilation in the testing location includes flow-through ventilation from the mine's internal ramp, through the testing area to an automated flow control regulator at the return air raise. Secondary ventilation fans direct air into the grizzly station, the charge bay, and the orepass access from the main ventilation stream in the tramming drift.

Continuous air monitoring was established at key locations within the test area in order to capture the temperature and humidity throughout. Additionally, noise sensors were installed in the tram drift and the orepass access to allow monitoring of the pass-by noise exposure and the bucket loading noise exposure, respectively. Spot measurements of the rock surface/skin temperature and equipment temperature were taken using an infrared thermometer from a safe location within the charge bay as the vehicles passed by in the tram route. Referring to the measurement locations shown in Figure 1, the measured parameters at each location are described in Table 1.

TABLE 1

List of test parameters and locations.

ID	Location description	Parameters measured	Remarks
1	Level Inlet	Temperature, relative humidity, barometric pressure	Continuous monitoring data logger installed
2	Charge Bay Outlet	Temperature, relative humidity, barometric pressure	Continuous measurement during charging only
3	Tramming Drift	Noise (LHD travel), rock skin temperature, vehicle temperature	Continuous noise monitoring, spot temperature measurements
4	Orepass Outlet	Noise (LHD bucket loading), temperature, relative humidity, barometric pressure	Continuous monitoring logger and continuous noise monitoring installed
5	RAR Regulator	Air flow, temperature, relative humidity, barometric pressure	Continuously monitored by mine's existing air quality monitoring station
N/A	LHD Cabin	Operator noise exposure	Continuous personal monitor worn

In addition to the continuous logging of level flow reported from the return air raise (RAR) air monitoring stations, a series of field measurements were conducted with handheld thermal anemometer traverses throughout the test area and pitot tube traverses of secondary ducts to corroborate the logging data. The typical ventilation profile in the test location during the test period was as follows:

- Total flow in the level access (Location 1) of 68 m³/s
- Flow in the primary tramming route (Location 3) of 22 m³/s
- Ducted secondary flow to the grizzly station of 21 m³/s
- Ducted secondary flow to the charge bay of 17 m³/s
- Ducted secondary flow to the orepass of 28 m³/s.

Some recirculation was observed in the orepass ventilation fan which was consistent throughout the testing. Logging instruments were left in place for 18 hours prior to executing the tests, allowing for baseline heat loads to be observed. The local heat impacts from this recirculation in this secondary fan is accounted for in the analysis.

The vehicles involved in the study were 8 cubic yard, 14 metric tonne class articulated LHDs with identical frame sizes. The diesel unit, fitted with an aftermarket bucket, was rated for 11.9 t capacity, while the BEV unit was rated for a volume associated with 14.9 t. While this discrepancy does not allow for comparison on an equivalent basis, the study outcomes related to heat and humidity are normalised to be on the basis per tonne of rated haulage capacity in an attempt to make the comparison as direct as practically possible with the vehicles involved. It is also important to note that load measurements were not available at the time of the test, so comparison of fill factors and actual quantity of material moved are not included.

RESULTS

Vehicle performance overview

Vehicle productivity over the test work period is shown in Figure 2.

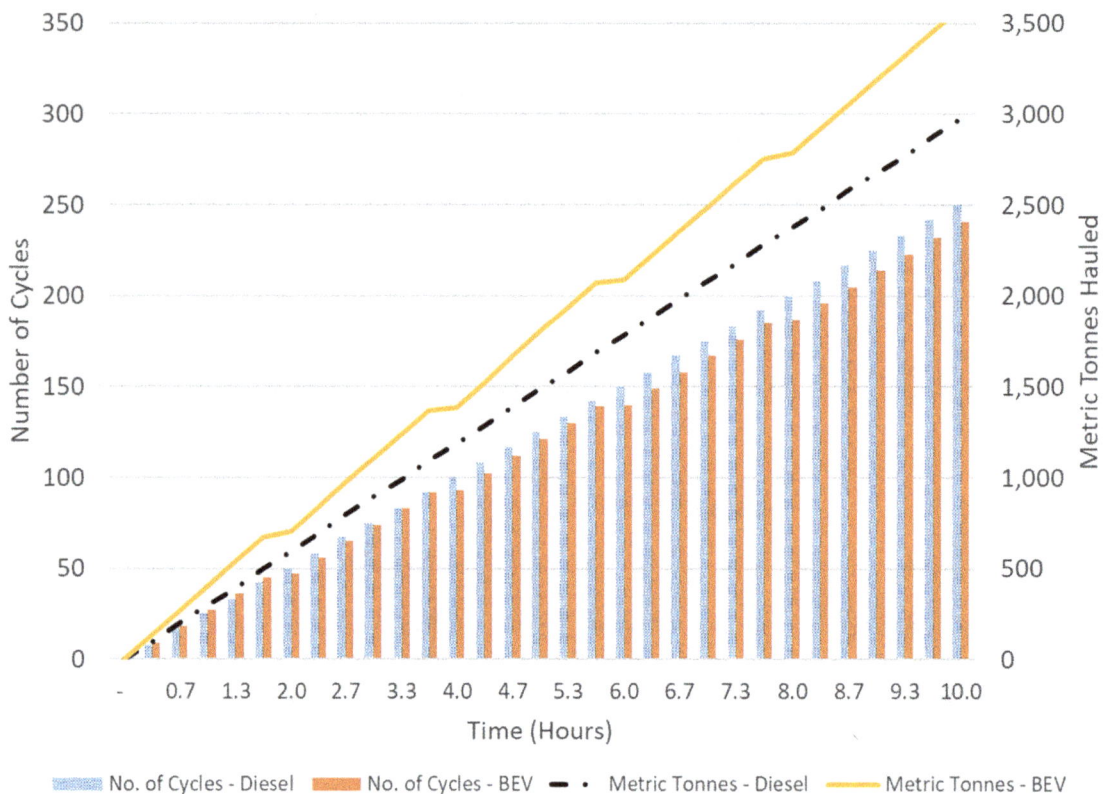

FIG 2 – Vehicle productivity comparison.

The above chart demonstrates that the BEV completes more cycles and moves more tonnes than its diesel counterpart even with the required stoppage to charge the battery.

For the diesel LHD, 169 data points were captured over two shifts with an average cycle time of 144 seconds. In comparison, 87 data points were captured for the BEV with an average cycle time of 132 seconds. The BEV averaged 8.9 per cent faster cycle time, on a gross basis without accounting for the charging time. On a net basis, the average cycle times were equivalent after 200 cycles, but the BEV equipment carried 4 t more per cycle and, as a result, more tonnes over the eight hour period.

Temperature profiles

Psychrometric properties were logged in the orepass return air and can be used to compare the workplace temperature rise during the mucking cycles with the diesel and battery electric LHDs. The workplace temperature is reported as the wet bulb globe temperature (WBGT), calculated as the sum of 70 per cent of the wet bulb temperature and 30 per cent of the dry bulb temperature, given that no solar radiation is present in the underground environment. The WBGT is used as a metric of workplace heat stress by the American Congress of Governmental Industrial Hygienists and incorporates both the air temperature and humidity and therefore gives an understanding of the true workplace temperature 'felt' by a worker as their body regulates its internal temperature. WBGT profiles for the orepass outlet during the morning and afternoon test periods are shown in Figure 3.

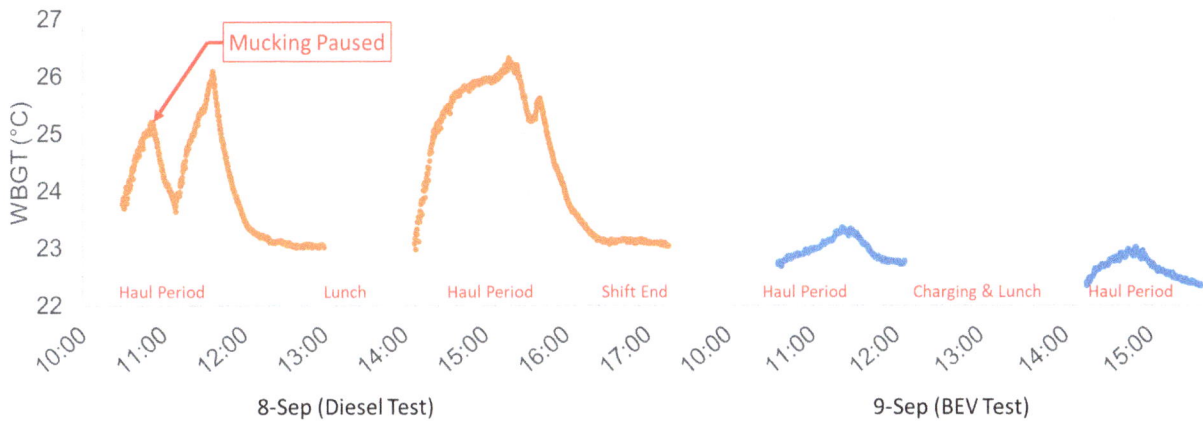

FIG 3 – Temperature profiles in the orepass access during mucking with diesel LHD (left) and BEV LHD (right).

As seen in Figure 3, the orepass exhaust air temperature continuously increased while the LHD was operating and then slowly decreased between runs and while the vehicle was not actively hauling (ie vehicle paused, on lunch break, or charging for the battery electric case).

The plot clearly demonstrates the potential advantages from a ventilation and cooling perspective with the battery electric compared to the diesel LHD. The observed temperature rise was approximately 3.1°C to 3.3°C WBGT for the diesel LHD and approximately 0.7°C for the BEV. This translates to an observed temperature rise which is 4.4 to 4.7 times greater in the diesel case than the BEV case. The smaller increase in temperature for the BEV case is attributed to its greater efficiency compared to the diesel LHD, for the electric motor relative to the internal combustion engine plus any reductions in braking heat associated with regenerative braking present on the BEV.

It is important to note that this test reflects the observed temperature rise in this specific scenario only and is not a generalised value that can be applied accurately in every scenario. The observed temperature rise is application-specific and depends on a variety of parameters (eg inlet air conditions, air flow quantity, vehicle duty cycle etc); however, it is still useful to provide a general indication of the expected impact on workplace temperatures between the two vehicle types in workplaces supplied with auxiliary ventilation.

Heat output comparison

Load-haul-dump cycle

A key aspect for examination as part of the test work was the difference in heat output for the different vehicle types. A simplified theoretical heat release is typically estimated for the vehicles based on work output and powertrain efficiencies. Specifically, diesel combustion engines are assumed to be around 33 per cent efficient, while electric motors are significantly more efficient at around 95 per cent. The useful power output (minus any work done against gravity) plus the inefficiency in its delivery are modelled as the combined heat release from the vehicle. BEVs, being approximately three times more efficient than those with diesel engines, are therefore expected to produce approximately one third the heat (for identical cycle power requirements on level ground) in this simplified approach.

Limitations of this heat quantification methodology are that it relies on the assumed engine/motor efficiency (which could vary) and does not account for other differences in the BEV construction and operation that impact heat output (eg lack of torque converters and the potential benefits of regenerative braking). Therefore, it is important to undertake field measurements to provide more real-world data on heat release and performance. Such field measurements were performed as part of this test work analysis. These measurements were used to perform a heat balance on 1470L to quantify the heat release from each LHD and the difference between the diesel and battery electric LHD operation. The approach focused on determining the total heat release from the LHD by measuring the heat reporting to the following areas:

- ventilation air stream

- LHD chassis, power train and tires
- surrounding strata.

Heat added to the air stream was calculated based on the difference in enthalpy between the level exhaust air and intake air. This difference was calculated for each data point and summed over the duration of the respective operating period to quantify the heat release rate. This exercise was performed for periods with and without LHD operation. This enabled a baseline testing area background heat measurement. This value accounts for constant heat loads across the level such as auxiliary fans and the electrical substation so they can be subtracted to isolate the LHD heat.

To support the analysis, intake air psychrometric properties were taken from the data logger located at Location 1 (refer to Figure 1). Exhaust air psychrometric properties were taken using a thermal anemometer (with logging capabilities) placed at Location 5. Airflow data was taken from the local Air Monitoring Station (AMS) at Location 5 when available, however this AMS experienced an extended communications failure throughout the diesel trial and for much of the day shift BEV trial as well. As a result of this instrument failure, manual velocity traverses were taken periodically with a thermal anemometer, averaged, and applied to the diesel LHD analysis.

The second component of the heat analysis focused on heat produced by the LHD that reported to the vehicle itself. The average temperature rise of the vehicle (ie the vehicle body) was captured via thermometer readings during the test work period. Fluid temperatures in the diesel LHD were monitored using on-board sensors. Data was not available for applicable fluid temperatures in the battery electric LHD and had to be assumed for the analysis based on previous test work data. In all cases, the temperature rises were converted into the stored heat energy based on the material masses and heat capacities.

Heat transferred to the surrounding rock (ie tunnel floor, walls and roof) was the last component of the analysis. Strata heat is dependent on a variety of factors, both for the rock (ie type, specific heat capacity and thermal diffusivity) and the air (ie flow and temperature). Strata heat was quantified for the purposes of this analysis using theoretical equations for heat flux at the rock surface based on measured surface temperatures.

The total heat emitted from each LHD was then equated as the sum of the three reported heat locations as discussed above, with the background heat load baseline first subtracted from the air stream value. The results are presented under LHD Heat Output in Table 2.

TABLE 2

Diesel versus battery-electric LHD – summary comparison of energy and heat output.

Parameter	Units	Diesel LHD	Battery LHD	Ratio of diesel/battery
Energy input				
Average power input	kW	413	98	4.2
Average power per metric tonne of muck capacity	**kW/t**	**34.7**	**6.5**	**5.3**
LHD heat output				
Heat to air stream	*kW*	*340*	*72*	*4.7*
Heat to vehicle	*kW*	*35*	*17*	*2.1*
Heat to strata	*kW*	*7*	*4*	*1.8*
Total heat output	kW	382	93	4.1
Total heat output per metric tonne of muck capacity	**kW/t**	**32.1**	**6.2**	**5.2**

This total heat output was then compared to the estimated LHD energy input calculated during the haul cycle (ie tramming, loading and dumping). For the battery electric LHD, this was equivalent to the total measured machine energy usage as measured and reported by the on-board systems, net of any regeneration from the regenerative braking system. For the diesel LHD, the measured fuel consumption was converted to an average power output using the typical energy density of diesel fuel. The results are summarised under Energy Input in Table 2.

The heat balance shows a heat output slightly less than the measured total energy input for both vehicles. This could be due to limited precision in physical measurements which could be further refined in future testing. With 92 per cent of the heat input accounted for in the diesel case and 95 per cent in the BEV case, the authors consider the sampling efforts strongly supported by the energy balance.

One conclusion that can be drawn from the heat balance is that the largest portion of the LHD heat reported to the air stream for both the diesel and battery electric cases. For this application specifically, there were minimal strata heat impacts due to the relatively similar temperature (and therefore small driving force for heat transfer) between the air and rock. Although there is some uncertainty in the quantification of the heat reporting to the vehicle, the air stream is still observed to be the largest receptor of the LHD heat.

The total heat output was revised to be based on rated bucket capacity (11.9 t for the diesel LHD and 14.9 t for the BEV) in an effort to form a valid comparison between the operating cases. With this adjustment, the ratio of heat for the diesel to battery case is approximately 5.3, which exceeds the theoretical value of about 3.0 based on the difference in drivetrain efficiencies. Potential reasons for this improvement could include:

- The lack of torque converters on the battery electric LHD. Specifically, during high-torque stall conditions (eg when crowding into the muck pile), the electric motors can produce high torque with no mechanical 'slip' at low speeds, while the diesel engine runs at high speed with elevated slip at the torque converter that is released as heat.

- Differences in operation of the vehicle's hydraulic pump. The BEV's onboard battery management system runs the hydraulic pump 'on demand' which limits bypass when hydraulic power is not needed. Diesel equipment, on the other hand, typically have the hydraulics mechanically linked to the engine, leading to operation of the pump at a higher duty and running more frequently, with the excess pump power lost as heat.

- Benefits from regenerative braking, which enables the BEV to perform the same hauling cycle with a lower net energy consumption. Although the elevation change for this test location was minimal, some regeneration benefits were observed on the vehicle telemetry (reduction of 5 kW on average from the total energy input) during machine deceleration. Applications with more opportunity for regenerative braking (eg larger elevation changes, especially with hauling loaded down-ramp), will help further differentiate workplace heat savings for BEVs.

It should be noted that the results presented here are application – and equipment-specific and cannot be used as a generalised diesel to battery LHD heat comparison for all scenarios.

Charge cycle

The test work analysis also sought to quantify the heat impacts from the battery electric charging cycle. The LHD in use for the test work included plug-in fast charging capabilities with dual chargers installed in the dead-ended charge bay excavation. The LHD remained parked inside the charge bay during charging – no battery swap was used. Psychrometric properties of the air were measured at the charge bay inlet and outlet during three charge cycles, with the air flow (provided by an auxiliary air supply duct) remaining constant throughout. The results are summarised in Table 3 and are based on a representative full-charge cycle from the day shift.

TABLE 3

Summary of charging heat measurements.

Parameter	Units	Value
Change in state of charge	%	91
Energy input to battery	kWh	178
Average heat to air measured	kW	19–24
Total energy input to air during charge cycle	kWh	5.7–7.4
Energy losses from charger to battery	%	3.2–4.2
Temperature rise – charge bay	°C WBGT	0.7 (Max)
Temperature rise – main drift	°C WBGT	Negligible

The observed heat output to the air is consistent with previously reported range of 15.1 to 45.3 kW (average of 20.6 kW) from an operating mine (Ross and Mayhew, 2019). The percentage loss from the charger is on the low end of estimated range of 5–10 per cent indicated in Recommended Practices for Battery Electric Vehicles in Underground Mining (Global Mining Guidelines Group, 2022).

Although a temperature rise was observed in the charge bay during the charging cycle, the heat impacts seen in the main drift and the flow-through ventilation circuit were negligible due to much higher airflows diluting the temperature increase.

Noise generation

External noise

A total of 25 pass-by events we recorded by the microphone stationed in the tramming drift (refer to Location 3 in Figure 1). The sound level of a pass by was characterised by a five second A-weighted equivalent sound pressure level (LAeq$_{5\text{-sec}}$). The Leq$_{5\text{-sec}}$ was evaluated by averaging samples taken within a five second time window centred about the peak level. Figure 4a and 4b shows a typical sound pressure sampling data set for the diesel and BEV LHD pass by, respectively. The 5-second time window used to evaluate the LAeq$_{5\text{-sec}}$ is highlighted.

Similarly, 25 bucket loading events recorded for each of the diesel vehicle (DV) and BEV (refer to Location 4 in Figure 1). The loading sound level was characterised by averaging over 30-second time windows (LAeq$_{30\text{-sec}}$). Noise levels from loading of the diesel LHD and BEV LHD are shown in Figure 4c and 4d, respectively.

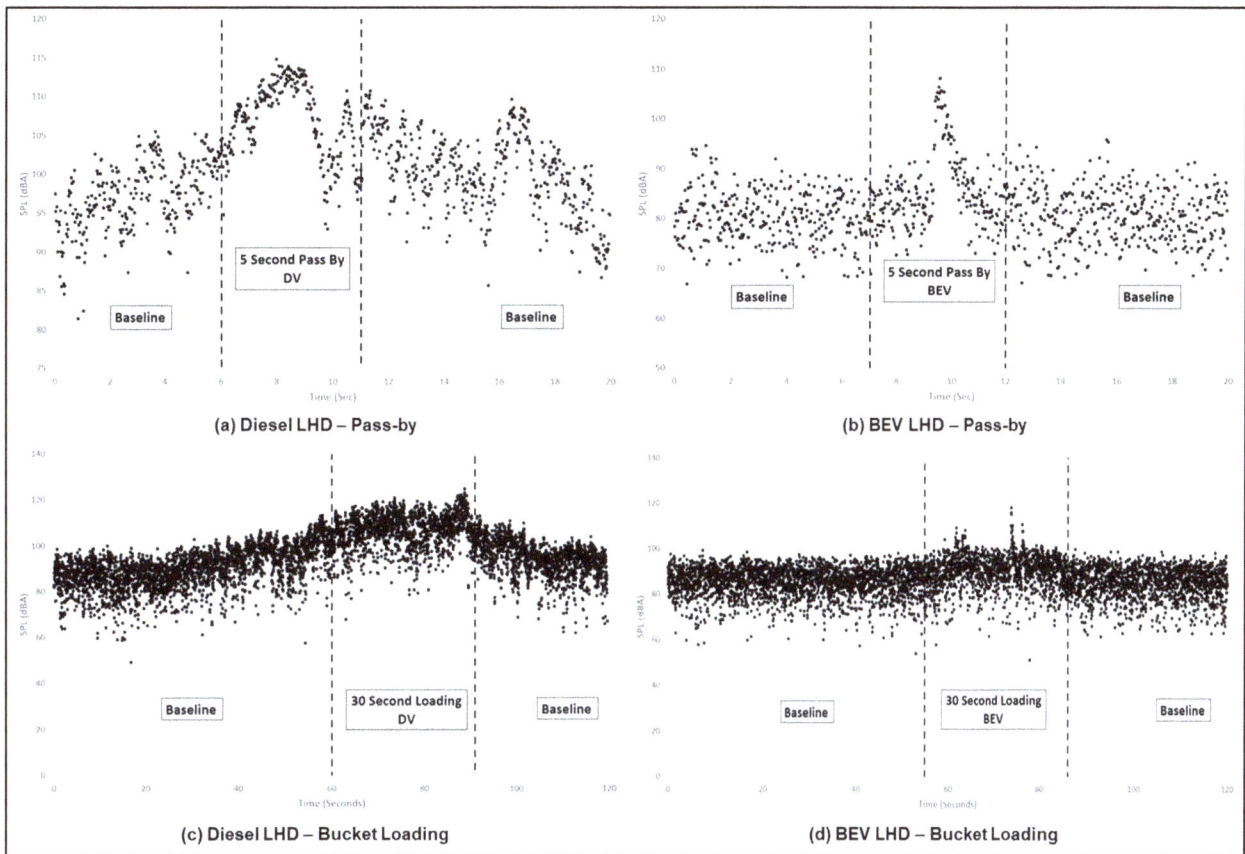

FIG 4 – Noise profile comparison during tramming and bucket loading.

The average pass by Leq$_{5-sec}$ and loading noise Leq$_{30-sec}$ levels across all 25 events for are presented in Figure 5. External noise levels were found to be approximately 14 dB lower for the BEV during transit and 11 dBA lower during bucket loading.

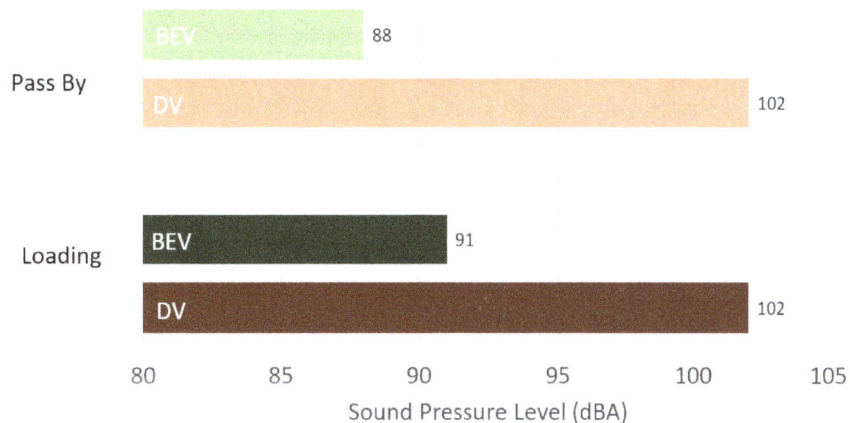

FIG 5 – Comparison of average external noise levels.

The average unweighted spectra of pass-bys and loadings are presented in Figure 6. The spectral plots show that the diesel vehicle is louder than the BEV across all bands above 63 Hz. The diesel vehicle also exhibits a tonal characteristic at 125 Hz during pass-bys, whereas the BEV does not exhibit tonality as its spectrum remains flat across all bands. Though tonality does not change the overall noise level and occupational exposure limits, it gives the noise source a distinctive sound. A distinctive sound is often a source of annoyance, but in some cases can be essential such as for reversing alarms and warning sirens.

The 125 Hz level is likely attributed to the engine exhaust of the diesel vehicle (DV).

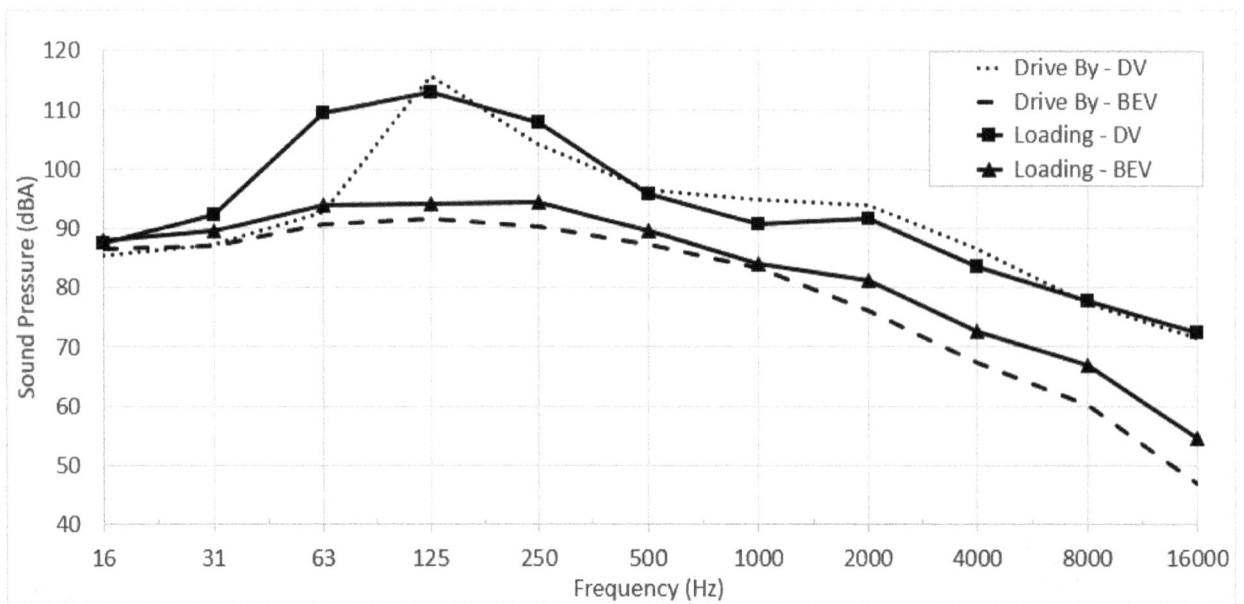

FIG 6 – External noise spectral plots.

Internal cab noise

Noise levels in the cab were measured over one full day shift and one full night shift for each of the two vehicles. Measurements were taken while vehicles were in operation and paused during breaks in testing including operator lunch breaks and charging stops.

Figure 7 compares the in-cab noise levels over a two-hour period for each vehicle. Internal cab noise levels were generally higher for the diesel vehicle, with the minimum five-minute average levels ($LAeq_{5-min}$) being 81 dBA for the BEV and 86 dBA for the diesel vehicle.

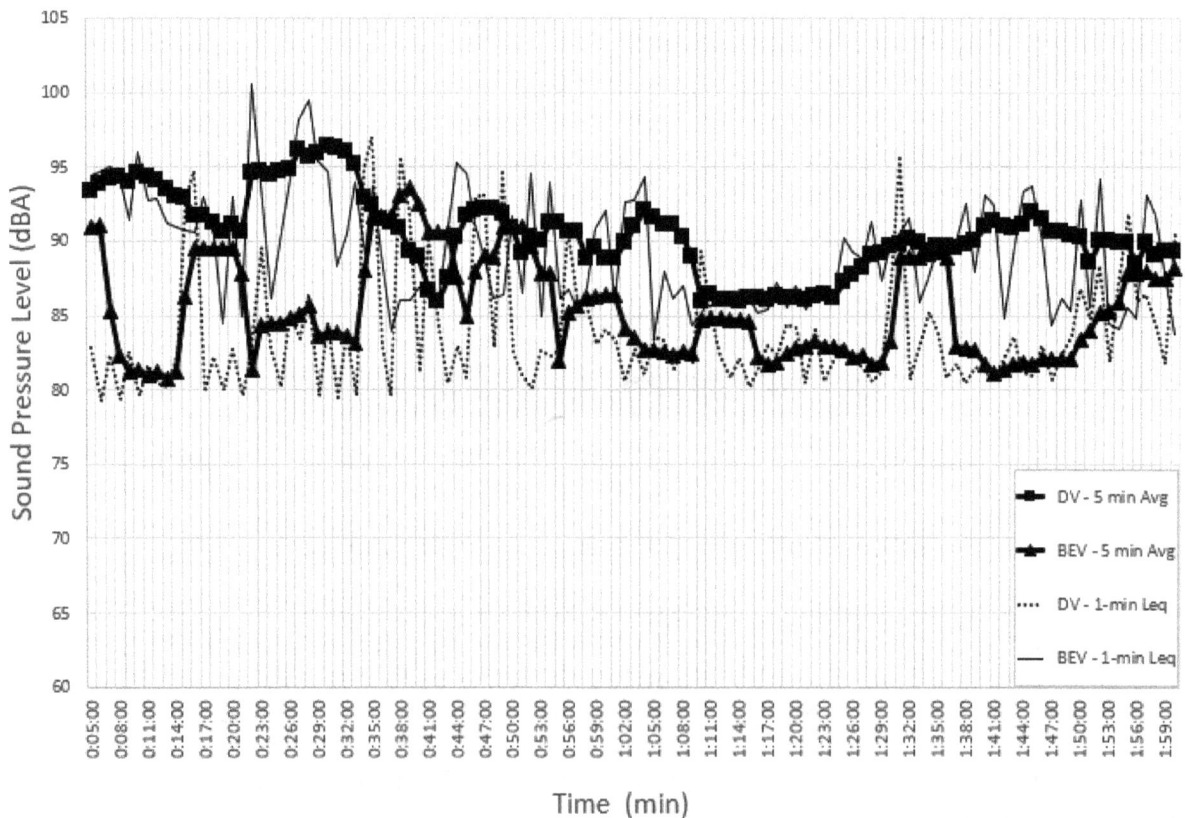

FIG 7 – Two-hour sample of in-cab noise levels.

Noise levels within the cab are a combination of vehicle movement and powertrain noise as well as other sources such as radio communications and vehicle systems such as backup alarms. Figure 8 compares the average in-cab noise exposure level between vehicles as well as work shifts. Operators within the diesel vehicle cab experienced noise levels 3–4 dB higher than those within the BEV for a given shift. The exposure on night shifts tended to be lower than the day shift for both vehicle types. The lower internal cab noise at night is attributed to less frequent radio communication which is typical for reduced mine-wide activity overnight.

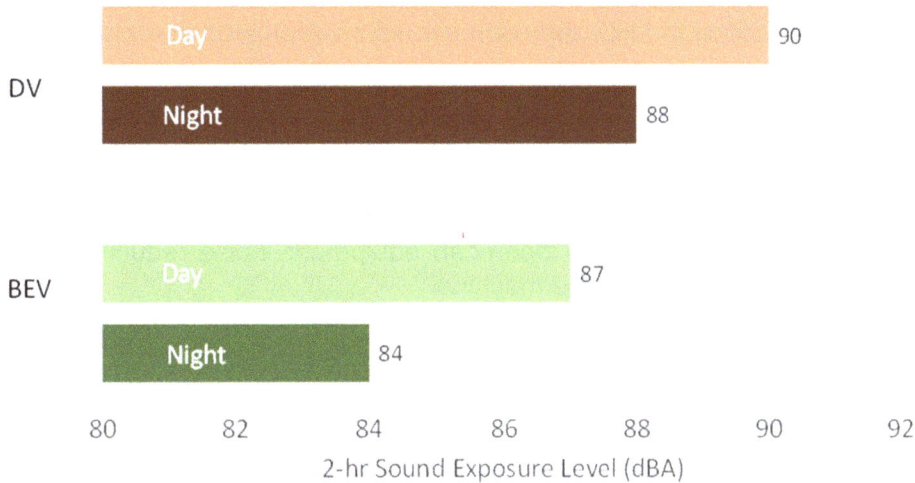

FIG 8 – Comparison of average in-cab noise levels.

Energy and fuel cost of operation

Total energy input from Table 2 can be used to estimate the fuel and energy cost required to move a tonne of rock (based on rated bucket capacity) for comparison. A cost estimate is presented in Table 4, using representative unit costs from typical mining operations in Ontario, Canada as of 2021. It can be seen that the input cost per tonne of rated capacity is approximately five times lower for the BEV as compared to the diesel LHD, or four times lower on the basis of input cost per vehicle operating hour.

TABLE 4

Fuel and energy cost comparison.

Parameter	BEV	Diesel vehicle
Average power input	98 kW	413 kW
Energy/fuel consumption per metric tonne of rated capacity	0.24 kWh/t	0.14 L/t
Unit cost	$0.09 / kWh	$0.75 / L
Energy/fuel cost per tonne of rated capacity	$0.02/t	$0.10/t
Energy/fuel cost per operating hour	$8.82/hr	$31.13/hr

CONCLUSIONS

Testing of a battery-electric LHD in transfer mucking application identified increased haulage productivity, in addition to reductions in heat, humidity and noise in comparison with the equivalent diesel vehicle. Heat output was approximately five times less for the BEV than the diesel LHD in this application, based on the same quantity of rock hauled. This is a more significant reduction than expected when considering drivetrain efficiency differences alone and is in line with previous test work which found as much as eight-fold reduction in heat output (McGuire *et al*, 2019).

Despite the fact that this was a repeated tramming path with minimal grade, regenerative braking effects were found to reduce energy input by 5 per cent, and therefore correspondingly reduce heat output to the air stream by limiting excess fuel consumption or mechanical braking waste heat.

Charging heat was found to be approximately 20 kW, which in this case had negligible impact outside of the charge bay heading itself. Representing a fraction of the LHD heat, fast charging infrastructure and on-shift charging can be deployed by some operations without detrimental impacts on downstream workplaces. Proper planning is still recommended, as multiple chargers located in close proximity and operating simultaneously could result in a marked increase in dry bulb temperature in the ventilation stream.

Noise level improvement in the mine was substantial, with 11 to 14 dB reduction observed. This will improve ability for personnel to safely communicate in proximity to large BEVs. Safety risk associated with the lack of advance notice of LHD approach should be evaluated and planned for accordingly in future mining operations.

Operator noise exposure in the enclosed vehicle cabin was also observed to reduce by approximately 3 dBA in the BEV. The smaller magnitude of reduction indicates that the operator's cab is reasonably well insulated from external noise, and that contribution from inside the cab (eg radio communication) are notable contributors to an operator's exposure over the course of a working shift. For operations that deploy open-cab equipment, noise reductions for the operator could be closer to that seen external to the machine.

ACKNOWLEDGEMENTS

Test work performed in collaboration with Hatch, Mayhew Performance Ltd, and Caterpillar.

Data collected and reproduced with permission from Sudbury Integrated Nickel Operations and Nickel Rim South Mine.

REFERENCES

Global Mining Guidelines Group, 2022. Recommended Best Practices for Battery Electric Vehicles in Underground Mining — Version 3 (GMG07-EM-2022).

McGuire, C, Macdonald, K, Armburger, J and Baumann, J, 2019. Direct comparison of heat produced by battery vs. diesel LHD, in *Proceedings of the 17th North American Mine Ventilation Symposium* (A Masideh, A Sasmito, F Hassani and J Stachulak, eds), pp 403–413 (Canadian Institute of Mining, Metallurgy and Petroleum: Montreal).

Ross, S and Mayhew, M, 2019. Ventilation Strategy Related to Battery Powered Equipment at Macassa Mine, in *Proceedings of the 17th North American Mine Ventilation Symposium* (A Masideh, A Sasmito, F Hassani and J Stachulak, eds), pp 387–393 (Canadian Institute of Mining, Metallurgy and Petroleum: Montreal).

A new gas emission prediction model based on Australian longwall mining conditions

Q Qu[1], R Balusu[2] and B Belle[3,4]

1. Senior Research Engineer, CSIRO, Brisbane Qld 4069. Email: Qingdong.qu@csiro.au
2. Senior Principal Research Engineer, CSIRO, Brisbane Qld 4069. Email: Rao.balusu@csiro.au
3. Group Ventilation and Gas Manager, Anglo American Metallurgical Coal, Brisbane Qld 4000. Email: bharath.belle@angloamerican.com
4. Adjunct Professor at School of Minerals and Energy Resources Engineering, The University of NSW, Sydney, Australia; The the University of Queensland, Brisbane, Australia; and University of Pretoria, South Africa.

ABSTRACT

Reliable prediction of longwall gas emission sources and rate plays a critical role in mine ventilation and methane drainage capacity planning. Over the years, Australian longwall mines have been using the historic Winter method and the Flugge method in modelling specific gas emissions (SGE). The predicted results often experience high discrepancies from the operational reality. To improve gas emission prediction, post-mining gas contents were measured in several mines in Qld and NSW by safely drilling core holes into the longwall goaf. The results revealed many new insights into gas emission characteristics from different coal seams and the drawbacks of the European methods in their use in high-production gassy Australian longwall mines. Based on measurement data, a new, time-dependent gas emission prediction model has been developed. Considering several factors such as caving, desorption and production rate, the new model can predict more accurate and meaningful gas emissions to meet today's needs in mine gas emission management. A case study is provided to demonstrate the application of the new SGE model.

INTRODUCTION

Methane emitted from coal mining acts as a potent greenhouse gas, in addition to its threat to mining safety and productivity. In large-scale longwall mining, extraction of coal induces relaxation and fracturing of a large area of strata above and below the mining seam (Guo *et al*, 2012), which in turn causes a substantial amount of methane to be released from adjacent coal seams and gas-bearing strata. In Australian gassy longwalls, it is reported that the specific gas emission (SGE) rate, which refers to the total gas emission rate per tonne of coal produced, can be up to 20–45 m^3/t (Meyer, 2006; Qu *et al*, 2016; Belle, 2017).

Gas emission prediction plays a critical role in mine ventilation and methane capture planning. Conventionally, this work is performed by using empirical techniques. A range of methods has been developed in different coal mining countries based on their specific mining conditions (Curl, 1978; Yu, 1990; Noack, 1998). Curl (1978) conducted a detailed review of the European methods, including the Gunther, Jeger, INIEX, Winter, Flugge, Schultz, Lidin, Barbara Mine and Mining Research and Development Establishment (MRDE) methods. These methods are similar in principle, all predicting a gas emission zone and the degree or percentage of gas emission from coal seams at different distances from the worked seam. When used in a specific mining condition, these methods can predict substantially different gas emissions.

Australian coalmines have been long using historic European methods to predict gas emissions (Lunarzewski and Battino, 1983; Battino *et al*, 1988; Meyer, 2006; Black, 2017; Belle, 2017). Although various efforts were made to improve gas emission predictions in Australian coalmines (Hargraves, 1986; Jensen *et al*, 1992; Lunarzewski, 1998; Ashelford, 2003; Guo *et al*, 2009; Booth *et al*, 2017), no alternate and improved empirical methods have been developed. The Flugge method (Flugge, 1971) and the Winter method (Winter, 1975) are still commonly used in Australian longwall mines. Figure 1 illustrates and compares the two methods. Their prediction results, however, often show high discrepancies from reality. For instance, Belle (2017) reported that the errors in using the Winter model could be up to 50 per cent overestimation for some panels to underestimation by 290 per cent using operational data.

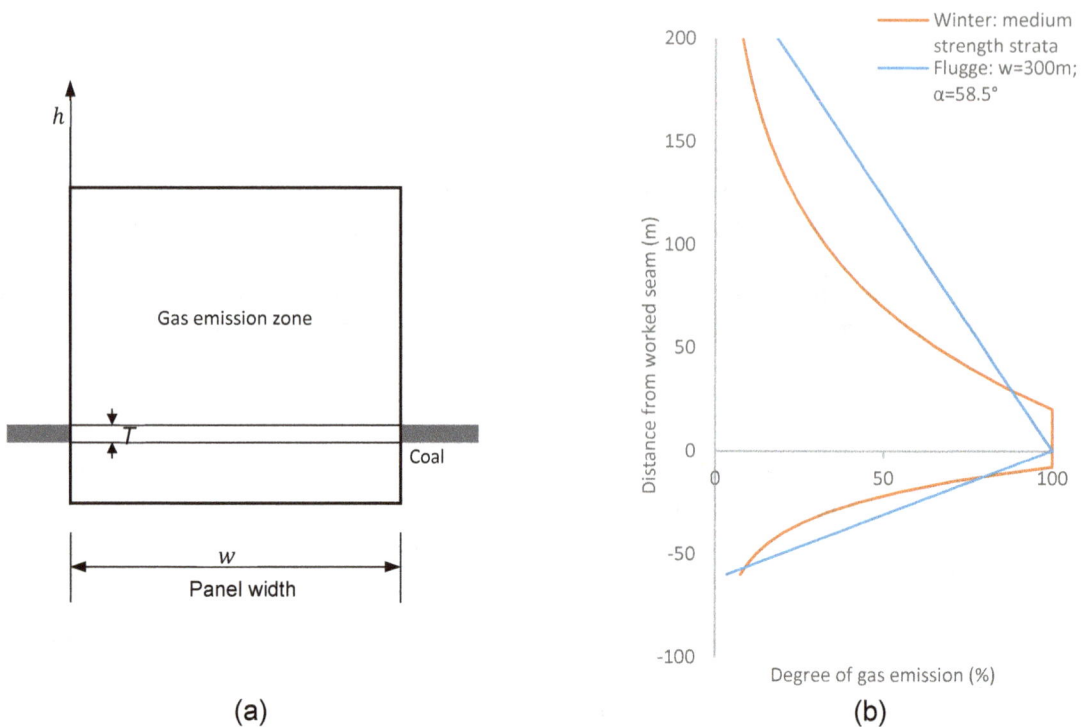

FIG 1 – The historic Flugge and Winter models: (a) the assumed gas emission zone; (b) gas emission degree with varying distance from the working seam.

In the past decade, significant efforts to measure gas release from different coal seams were made at various longwall mines in Australia by safely drilling core holes into the goaf. These measurements enabled, for the first time in Australia, to verify the extent of the gas emission zone and the proportion of gas released from different coal seams. Based on these measurement data, a new SGE prediction method that accounts for production rate has been developed. This paper briefly introduces the development and the application of the new CSIRO SGE model.

POST-MINING GAS CONTENT MEASUREMENTS

Measurements of post-mining gas contents in different seams by safely drilling core holes into the longwall goaves were carried out at four longwall mines, including three (Mine A, B, C) located in the Bowen Basin, Qld and one (Mine D) in the Hunter coalfield, NSW. The geological and mining conditions differ in these two regions, and the details and the gas content testing data can be found in Qu *et al* (2019a). The measured degree of gas emission from coal seams at different distances is shown in Figure 2. The Winter model and the Flugge model are also shown in the Figure 2 for comparison.

The key observations from the measurement results include:

- In general, the measured degree of gas emission decreases with the increasing distance from the worked seam. The Winter model and the Flugge model roughly capture the trend but match loosely in their magnitude.

- In the Qld mines, the Winter model significantly underestimates gas emissions from coal seams 75–150 m above the mining seam. The linear Flugge model predicts an average line of measured data for these coal seams by overestimating and underestimating alternatively but can overestimate 20 per cent from a thick coal seam located between 175 m and 225 m.

- In the NSW mine, the Winter model significantly overestimates gas emission proportions for both roof and floor seams. In contrast, the Winder model predicts approximately for the roof seams at one testing point but also overestimates gas emissions from floor coal seams. It is noted that Mine D was a multi-seam mine with a supercritical longwall width (410 m). This special mining condition has a significant influence on the calculation of the gas emission degree and the prediction using the Flugge model (Qu *et al*, 2022).

- The vertical extent of the gas emission zone in the roof strata varies at different mines but appears to be approximately 45 times the mining thickness in all the studied mines. The vertical extent in the floor strata is limited to below 60 m.

- The gas emission zone appears to be more laterally extensive than the mined area as generally assumed in the historic models.

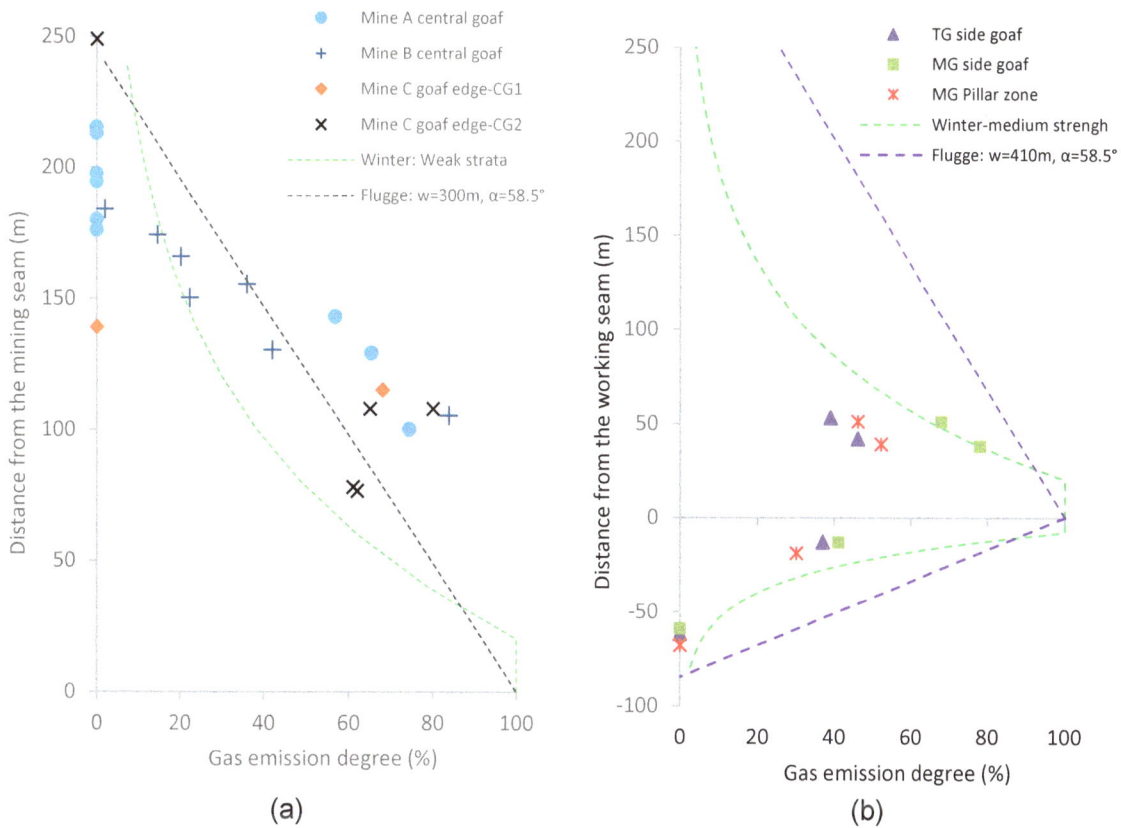

FIG 2 – Measured gas emission degrees in different coal seams and their comparison to the Winter and Flugge models: (a) three mines in Qld (panel width 300 m); (b) a mine in NSW (panel width 410 m).

Depending on the location and size of the seam gas reservoirs, an erroneous prediction in gas emission degree can result in substantial overestimation or underestimation of gas drainage capacity as learned from the operational data. For example, using the Flugge model to predict gas emissions from floor coal may result in substantial overestimation, whilst using the Winter model to predict gas emissions from a group of coal seams between 75 and 150 m may result in significant underestimation.

The drawback of the European models demonstrated from these measurements is likely due to the differences in the geomechanical, mining and gas drainage conditions between Australian and European mines.

A NEW TIME-DEPENDENT GAS EMISSION PREDICTION METHOD

To improve gas emission prediction accuracy, a new gas emission prediction model that can better represent the measured gas emission degrees is required. Further, to meet the requirements of longwall production planning, the new method should equally consider the production rate. Based on the measurement data, a time-dependent gas emission prediction method has been developed and briefly described in this section. The full details can be found in Qu *et al* (2019a).

Linking gas emissions with the caving model

The measured data shows that the variation of gas emission degree in the overlying coal seams correlates better with multipliers of extraction thickness (T) than distance, as shown in Figure 3. The variation trend exhibits a three-zone characteristic:

- Zone I (0 – 25 T): the degree of gas emission is high (80–90 per cent) and reduces gradually with increasing distance.

- Zone II (25 – 45 T): the degree of gas emission reduces quickly from about 80 per cent to zero.

- Zone III (>45 T): no gas is released from this zone.

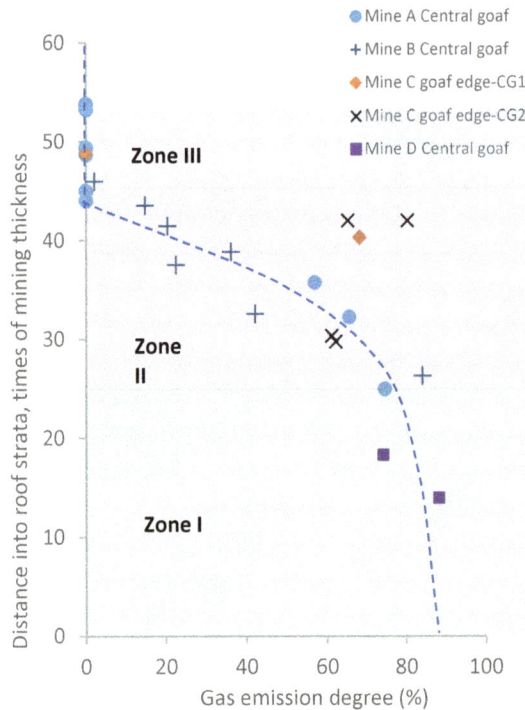

FIG 3 – Correlation of gas emission degrees with the multipliers of the extraction thickness.

The three-zone division of gas emission patterns in the roof strata conforms with the caving zone models with respect to deformation characteristics and gas migration patterns (Qu *et al*, 2015). Zone I in Figure 3 reflects the connective fracture zone where gas released from coal seams can flow into the goaf almost unimpededly. Zone II corresponds to the de-stressed dilation zone where horizontal permeability is well developed, and vertical gas migration is hampered and depends on strata permeability. Zone III conforms with the constrained zone where the deformation of the strata is constrained, and no mining-induced fractures are developed.

The close correlation of gas emission patterns with the caving models suggests that the caving model parameters shall be accounted for in quantifying gas emissions. This is because the extent of the deformation zones, and hence the gas emission zone, differ in different mines.

Despite other factors (eg longwall width and strata strength) that may affect the heights of the deformation zones, the extent of the gas emission zone may be estimated based on the mining thickness, as that is often used in typical caving models. For example, the caving models in Bai and Kendorski (1995) and Mark (2007) suggest that the fractured zone can extend up to 24–30 T, and the dilated zone can extend up to 60 T. The best match with the measurement results at various mines in this study suggests the two heights of the fractured and dilated zones are 25 T and 50 T, respectively. These two numbers are used as default values in the development of the new method. Their values, however, could be easily adjusted according to operational data.

For the floor coal seams, a similar three-zone also exists, but the depths of the fractured zone and the dilated zone appear to be independent of mining thickness (Qu *et al*, 2019b). In this CSIRO SGE model, the default values for the depths of the two zones are set as 15 m and 60 m, respectively.

Taking time factor into consideration to account for production rate

Airey (1968) conducted experiments on gas emissions from broken coal, from which he developed an empirical formula to describe the cumulative volume of gas emission with time:

$$Q_t = Q(1 - \exp(-(\frac{t}{t_0})^n)) \tag{1}$$

Where, Q_t = total gas released between time zero and time t; Q = the total desorbable gas content; t = the time after the start of desorption; t_0 = a time constant; and n = an index.

The fraction of gas emissions over time t can then be expressed as:

$$F_t = \frac{Q_t}{Q} = 1 - \exp(-(\frac{t}{t_0})^n) \tag{2}$$

Equation 2 enables a time-dependent estimation of the gas emission degree by assigning values to t_0 and n. For the parameter n, a value of 0.33 is generally used for bituminous coal. The parameter t_0 depends on the coal particle sizes that make up the coal fragment. For solid coal containing a crack structure (eg coal seams in the fractured zone), t_0 is a function of permeability (Airey, 1968).

Figure 4 compares a gas desorption testing curve of a coal core sample from a Qld mine with Airey's formula. The coal-core sample can be regarded as a representation of the de-stressed coal seams in the fractured or dilated zone. The excellent match demonstrates the suitability of using Airey's formula to estimate the degree of gas emission when considering time dependence.

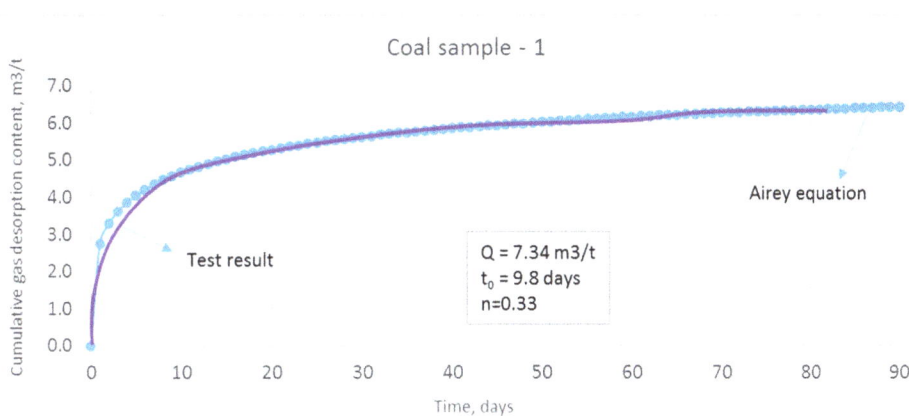

FIG 4 – Comparison of tested cumulative gas desorption content and the Airey equation.

Using numerical modelling as an approach, a function of t_0 with distance under prescribed heights of the fractured zone and the dilated zone is developed (Figure 5a). The resultant gas emission degree (Figure 5b) matches reasonably well with the measured gas emission degrees in different mines.

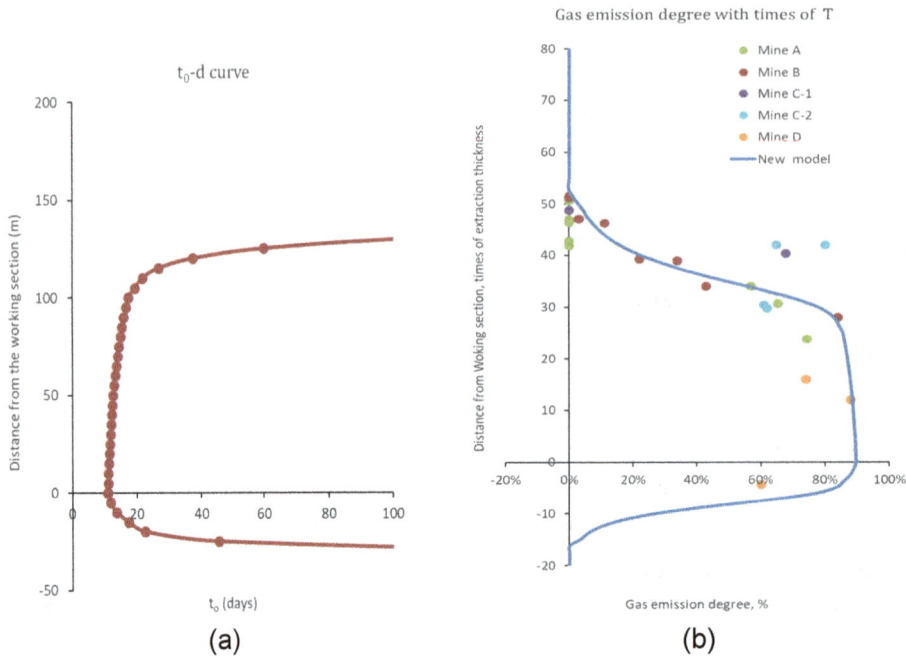

FIG 5 – (a) the developed function of t_0; and (b) the resultant gas emission proportion using Equation 2 in comparison with the measured gas emission degrees.

A SPREADSHEET TOOL FOR PRACTICAL USE

Based on the time-dependent gas emission prediction method described above, a CSIRO SGE spreadsheet tool was further developed to facilitate the uptake of the new method.

Principles

The incorporation of the time-dependent gas emission degree method (Equation 2) is realised by discretising the goaf into various unit zones, as shown in Figure 6. Each unit zone has a uniform length equal to the average daily retreat rate (eg 10 m per day in the figure). As such, the goaf is converted to various unit zones with different ages of gas emission. For example, the first 10 m has a gas emission age of one day, and the next unit zone has a gas emission age of two days.

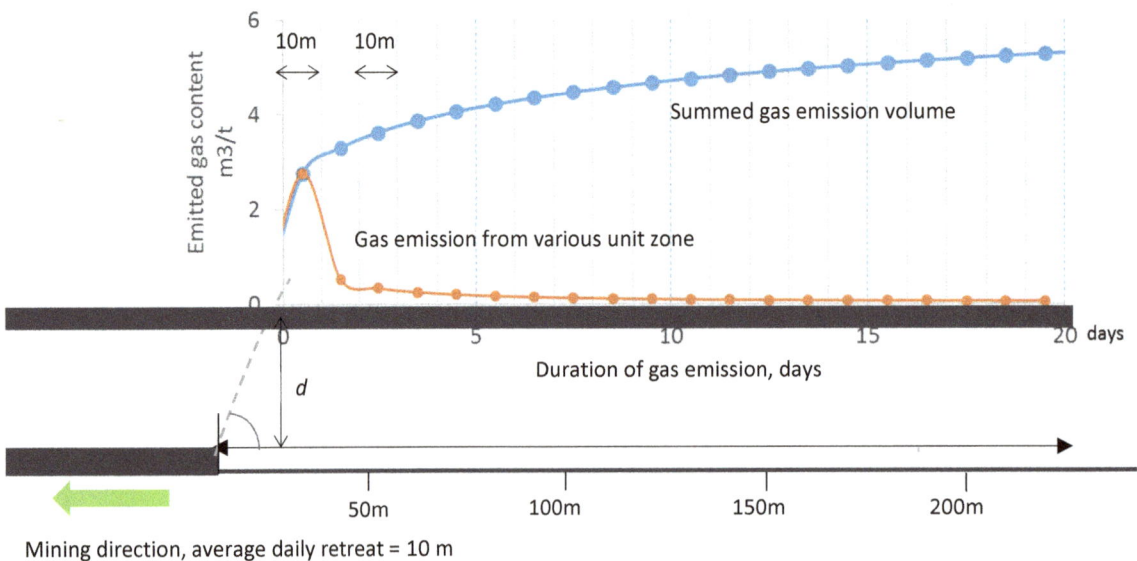

FIG 6 – Converting the length of goaf into a gas-emission-time zone to calculate cumulative proportion of gas emission.

With t_0 determined, the gas emission degree for the seam can be calculated in excel by:

$$\eta = 1 - \exp[-\left(\frac{\left(L - h \cdot \cot(\alpha)\right)/\bar{v}}{t_0}\right)^{0.33}]$$ (3)

Where $\bar{v}$ = average daily retreat rate, m/day; α = front angle of break; L is the length of the goaf.

With η calculated, the SGE component from a coal seam can be obtained by Equation 4.

$$q = \frac{c \cdot V}{M \cdot \bar{v}} = \frac{c \cdot Q_i \cdot \eta \cdot m \cdot \bar{v}}{M \cdot \bar{v}} = c \cdot Q_i \cdot \eta \cdot \frac{m}{M}$$ (4)

Where:

c is the total coefficients of other factors affecting gas emission (discussed in the following subsection)

Q_i is the initial gas content of the coal seam

η is the gas emission degree of the coal seam

m is the unit mass per metre, $m = \gamma \cdot t \cdot W \cdot 1$

M is the unit mass per metre of the working seam

Other factors considered in the spreadsheet tool

In addition to gas emission from adjacent coal seams, the reported longwall gas emission rate is affected by many other factors. In the new model, the factors listed in Table 1 are considered. It is noted that these factors are preliminarily quantified based on the authors' existing knowledge and available operational data. They may be subject to change in accordance with new data and understandings.

TABLE 1

Factors considered in the new SGE model.

Factors	Description and required input parameters and calculations
Rate of production	The rate of production reflects the longwall retreat rate, which further determines the age of the goaf district. A higher production rate generally results in lower SGE at a specific longwall.
Gas drainage intensity	A coefficient is quantified based on the spacing of gas drainage boreholes.
Background rib emissions	A value under site-specific conditions is required to be input into the spreadsheet tool. The value will be automatically added to the total gas emission rate. Where no data is available, the default value is set as 300 L/s.
Adjacent goaf leakages	A value under site-specific conditions is required to be input into the spreadsheet tool. Where no data is available, the default value is set as 300 L/s.
Gas-in-place variation	The total available gas-in-place can vary significantly along the length of the longwall. A coefficient is required to be calculated based on specific conditions.
Surround excavation situations	The first longwall of an area generally has a higher gas emission rate than the longwall with one side excavated. The effect is accounted through incorporating a coefficient for the lateral extent.

The spreadsheet tool

Figure 7 shows a snapshot of the SGE prediction summary page of the spreadsheet tool. The prediction results are automatically calculated and updated as the input parameters vary. This automatic calculation enables an easy assessment and refinements of input parameters to reconcile with actual data.

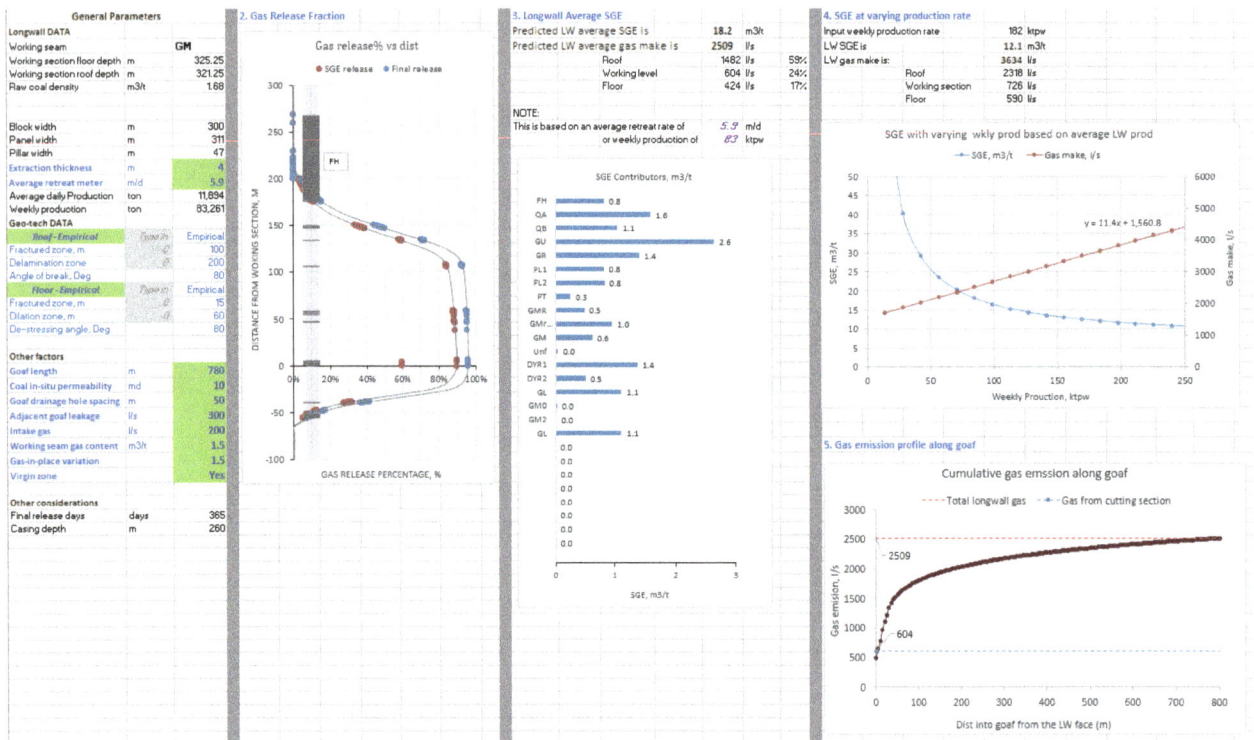

FIG 7 – A snapshot of the dashboard page of the gas emission prediction spreadsheet tool showing key prediction results and input parameters.

The prediction can quantify five aspects of gas emission from a longwall panel:

- The percentage of gas emission in each coal seam.

- SGE and its components from each coal seams.

- The absolute gas emission rate and its breakdown into the roof, floor and the working level.

- SGE variations with a weekly production rate.

- Cumulative gas emission rate profile into the longwall goaf.

Advancements in the new gas emission prediction method are apparent with respect to its fundamentals, the multiplicity of predictions, accuracy, and factors taken into consideration. As the new model incorporates many more parameters, it is unavoidably more complex than conventional models. However, with the development of the spreadsheet tool, the use of the new CSIRO model is relatively straightforward.

A CASE STUDY

The study longwall, LWA2 (Figure 8), had a width of 300 m and an average mining thickness of 4.0 m. Gas and production data in the first 780 m were used for comparison. It took 133 days for the longwall to retreat the first 780 m, resulting in an average retreat rate of 5.9 m/day and an average daily production rate of 12 489 t. This results in a ROM coal density of 1.68 t/m^3.

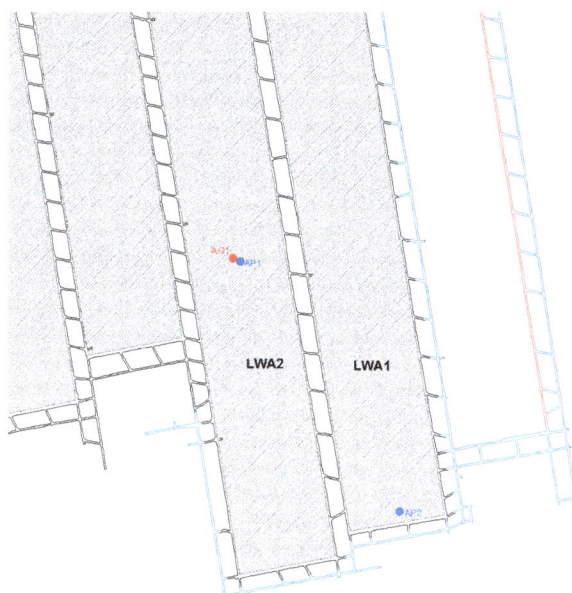

FIG 8 – Layout of Mine A.

On the first two days of the longwall commissioning to start excavation, the return had a gas make of 595–450 L/s. A gas make of 200 L/s and 300 L/s were therefore assumed for intake gas and adjacent goaf leakage, respectively.

The working seam gas content was set as 1.5 m³/t as suggested by gas compliance measurement at AP2 (1.95 m³/t) and AP1 (0.97 m³/t) area. Such a significant difference in the working seam gas content also highlights the difficulty in predicting gas emissions from the working seam.

Table 2 compares the desorbable gas-in-place of coal seams situated in the roof gas emission zones. The available gas-in-place is much higher in the inbye section of the longwall (AP1) than at AP2, which is based on the prediction. Within the gas emission zone excluding the FH seam, the gas-in-place density is 35.1 m³/m² at AP1 and 72.4 m³/m² at AP2. Based on the results, the gas-in-place coefficient is estimated as 1.5.

TABLE 2

Analysis of gas in place at different locations.

		AP1							AP2				
Coal	Distance	Thick	Density	Qm	CH4%	Available in place	Coal	Distance	Thick	Density	Qm	CH4%	Available in place
FH	178.0	39.0	2.0	1.4	0.94	102.3	FH	192.0	44.1	1.8	3.8	0.87	257.4
QA	143.5	1.5	1.6	5.0	0.99	12.3	QA	147.1	1.5	1.7	3.5	0.99	9.4
QB	130.0	0.8	1.6	4.4	1.00	5.3	QB	131.4	0.8	1.6	4.6	1.00	5.6
GU	101.6	0.8	1.5	4.9	0.99	5.8	GU	108.1	1.6	1.8	3.0	0.99	8.5
GR	54.6	0.8	1.5	2.2	0.98	2.5	GR	54.0	2.3	1.4	7.0	0.99	21.9
PL	50.5	2.6	1.5	2.4	0.98	9.1	PL	50.8	3.0	1.5	5.9	0.99	27.0
Total gas in place						137.4							329.9
Total gas excluding FH						**35.1**							**72.4**
Gas in place variation ratio								**1.53**					

The first 740 m from the longwall start-up line was in the virgin zone, with no excavations present on the two sides of the longwall. The 'virgin area' scenario was selected for the impact of surrounding excavations.

The detailed coal log at AP2 was copied into the spreadsheet tool, resulting in a data table of 480 rows. Assignment of gas content to each coal slice is conducted based on the measured gas content at AP2 and variations of ash content in each coal slice.

The operation data of LWA2 was used to compare with the model predictions. The average longwall gas emission in the field in the first 780 m of LWA2 was 2692 L/s, of which 1540 L/s was from goaf drainage, and 1152 L/s was reported in the return roadway. This results in an average SGE of 18.6 m³/t. The prediction (Figure 7) is close to the actual value.

Figure 9 shows the SGE with varying production rates in both weekly and seven-day rolling average. As can be seen, the predicted results are in close agreement with the actual data in terms of both value and trend, and the agreement is especially good when the daily production rate exceeds 8000 t. This demonstrates the excellent practical capability of this spreadsheet tool in predicting gas emissions for different production targets. This is important in assisting the planning and design of the goaf drainage capacity for given production targets.

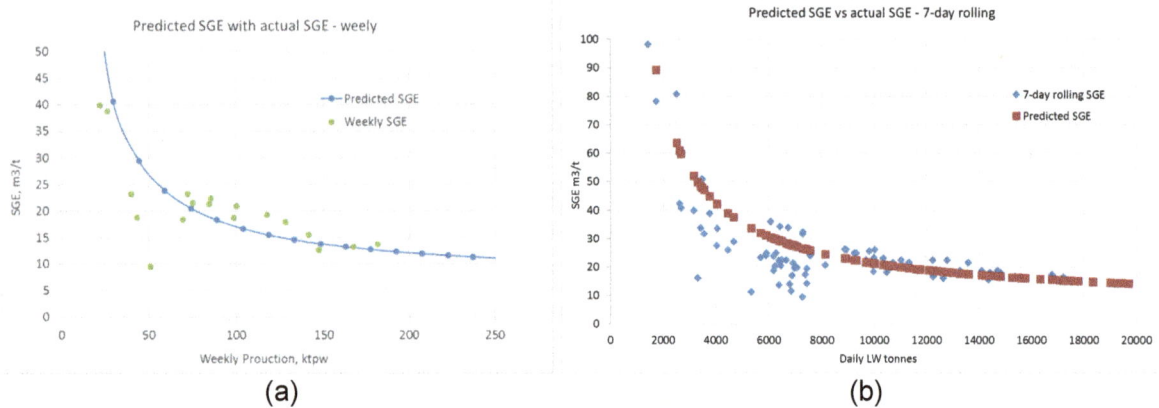

FIG 9 – Predicted SGE with varying production rate using the spreadsheet tool in comparison to the actual data at LWA2 of Mine A: (a) based on weekly production rates; (b) based on seven-day rolling daily production rate.

CONCLUSIONS

Gas emission prediction plays a critical role in mine ventilation and gas drainage capacity planning, to meet coal production targets and safety standards. The Winter and Flugge models, developed in West Germany in the 1970s, are the two common methods used in Australian coalmines. However, in practice, the predictions are loosely correlated with actual gas emissions.

Post-mining gas content measurements were conducted at four different longwall mines in Australia. The results show that the empirical models have drawbacks in predicting gas emissions for roof seams located 75 m above and also for the floor coal seams. The measurements also suggest that the gas emission zone increases with the mining thickness in its vertical extent, and is more laterally extensive than the mined area in a high permeability environment.

Based on the measurement data, a new, time-dependent gas emission prediction model was developed by incorporating Airey's equation that describes gas emission processes from broken coal. The model also takes into account the geo-mechanical caving zones. The results of this model match reasonably well with field measurement data at the project study mines.

A CSIRO spreadsheet tool was developed for the practical use of this new method. Other factors affecting longwall gas emissions are also quantified and accounted for in the spreadsheet tool. These factors include production rate, size of the goaf, gas-in-place variations, background rib emissions, leakage from neighbouring goaves and gas-drainage intensity.

A case study demonstrates that, with appropriate input parameters, predictions from the new method are very close to the actual observations for SGE, as well as its variation with production rate. Such a prediction provides support to facilitate mine planning and ventilation management on a weekly basis, during which mining parameters, production rate and gas-in-place may all vary.

It is worth noting that a robust method requires a long period of continuous improvement and refinement with practical experience and data. Regular and appropriate use of the spreadsheet tool and the method is encouraged at various mines where the reservoir, mining and operational parameters may differ significantly. While the newly developed CSIRO SGE model provides increased confidence in the prediction of longwall gas emissions, operations are encouraged to incorporate flexible and contingency gas drainage capacity to manage mine safety.

ACKNOWLEDGEMENTS

The authors are highly grateful to the Australian Coal Association Research Program for funding the development of the new SGE prediction method through ACARP Project C25065. Special thanks are due to the industry monitors of this project for their suggestions, monitoring and management of the project. The measurement of pre – and post-mining gas contents at Mine D in the Hunter Coalfield was supported by the Australian Government through the 'Coal Mining Abatement Support Package Program (CMATSP)'. We would like to thank the management and staff of all the study mines for their tremendous support. We also wish to express our gratitude to Dr Roy Moreby, UK, and our CSIRO colleagues Dr Hua Guo and Dr Andy Wilkins for their contributions to the project.

REFERENCES

Airey, E M, 1968. Gas emission from broken coal, An experimental and theoretical investigation, *Int J Rock Mechanics and Mining Sciences*, 5:475–494.

Ashelford, D, 2003. Longwall 'Pore Pressure' Gas Emission Model, in N Aziz and B Kininmonth (eds.), *Proceedings of the 2003 Coal Operators' Conference*, Mining Engineering, University of Wollongong.

Bai, M and Kendorski, F S, 1995. Chinese and North American high extraction underground coal mining strata behaviour and water protection experience and guidelines, *14th Conference on Ground Control in Mining*, pp 209–217.

Battino, S, Lunarzewski, L and Truong, D, 1988. Towards a reliable gas emission prediction method for Australian longwall mining, *Proceedings of 21st century Higher Production Coal Mining System Symposium*, The AusIMM Illawarra Branch, pp 315–322.

Belle, B, 2017. Optimal goaf hole spacing in high production gassy Australian longwall mines – operational experiences, In *Proceedings of The Australian Mine Ventilation Conference*, pp 1–12 (The Australasian Institute of Mining and Metallurgy: Melbourne).

Black, D J, 2017. Gas reservoir and emission modelling to evaluate gas drainage to control tailgate gas concentrations and fugitive emissions, In *Proceedings of the 2017 Australian Mine Ventilation Conference* (The Australasian Institute of Mining and Metallurgy: Melbourne).

Booth, P, Brown, H, Nemcik, J and Ren, T, 2017. Spatial context in the calculation of gas emissions for underground coal mines, *International Journal of Mining Science and Technology*, 27:787–794.

Curl, S J, 1978. Methane prediction in coal mines, IEA Coal Research, Report Number ICTIS/TR 04 Dec 1978, 79 p.

Flugge, G, 1971. The application of the Trogg theory to the zone of gas emission (Die Anwendung der Trog theorie auf den Raum der Zusatsausgasung), *Glueckauf – Forschungshefte*, 32:122–129 (in German).

Guo, H, Adhikary, D P and Craig, M S, 2009. Simulation of mine water inflow and gas emission during longwall mining, *Rock Mech Rock Eng*, 42:25–51.

Guo, H, Yuan, L, Shen, B, Qu, Q and Xue, J, 2012. Mining-induced strata stress changes, fractures and gas flow dynamics in multi-seam longwall mining, *Int J Rock Mech Min Sci*, 54:129–139.

Hargraves, A J, 1986. Seam gas and seam gas drainage, In C H Martin (ed), *Australasian Coal Mining Practices*, Monograph Series No 12, pp 393–411 (The Australasian Institute of Mining and Metallurgy: Melbourne).

Jensen, B, Gillies, A D S, Anderson, J M and Jones, N, 1992. Review of methane emission and prediction research in longwall coal mines, *Proceedings Aus Inst Min Met*, 1:11–17.

Lunarzewski, L and Battino, S, 1983. Prediction of required ventilation levels for longwall mining in Australian gassy coal mines, *Symposium on Ventilation of Coal Mines*, Wollongong.

Lunarzewski, L W, 1998. Gas emission prediction and recovery in underground coal mines, *Int J Coal Geol*, 35(1):117–145.

Mark, C, 2007. Multiple-seam mining in the United States: background, In *Proceedings of the New Technology for Ground Control in Multiple-seam Mining*, NIOSH, Pittsburgh Research Laboratory, pp 3–14.

Meyer, T, 2006. Surface Goaf Hole Drainage Trials at Illawarra Coal, In N Aziz and B Kininmonth (eds.), *Proceedings of the 2006 Coal Operators' Conference*, University of Wollongong.

Noack, K, 1998. Control of gas emissions in underground coal mines, *International Journal of Coal Geology*, 35(1998):57–82.

Qu, Q, Guo, H and Balusu, R, 2022. Methane emissions and dynamics from adjacent coal seams in a high permeability multi-seam mining environment, *Int J Coal Geol*, 253:103969.

Qu, Q, Guo, H and Loney, M, 2016. Analysis of longwall goaf gas drainage trials with surface directional boreholes, *Int J Coal Geol*, 156:59–73.

Qu, Q, Rao, B, Wilkins, A and Moreby, R, 2019a. Specific Gas Emission Patterns from Different Coal Seams, ACARP Project C25065 Final Report.

Qu, Q, Wilkins, A, Qin, J and Khanal, M, 2019b. Floor Seam Gas Emission Characteristics and Optimal Gas Drainage Strategies in Longwall Mining, ACARP Project C26050 Final Report.

Qu, Q, Xu, J, Wu, R, Qin, W and Hu, G, 2015. Three-zone characterisation of coupled strata and gas behaviour in multi-seam mining, *International Journal of Rock Mechanics and Mining Sciences*, 78:91–98.

Winter, K, 1975. Extent of gas emissions in zones influenced by extraction, *16th International Conference on Coal Mine Safety Research*, US Bureau of Mines, Washington. 17 p.

Yu, Q, 1990. *Coal Mine Gas Prevention*, China University of Mining and Technology Press, Xuzhou (In Chinese).

Reducing ventilation change downtime and increasing operational uptime

J A Rowland[1]

1. Managing Director, Dallas Mining Services Pty Ltd, Wollongong NSW 2500.
   Email: jr@dallasmining.com.au

ABSTRACT

When executing a ventilation change the mine ventilation circuit is in a state of flux throughout the entire change period with discrete mine areas both under ventilated and/or over ventilated during the change process. Whilst routine risk reduction strategies such as removing electrical power and people from the mine are commonly deployed it is wise to complete the change in the shortest practical time frame. This not only minimises the chance of unacceptable concentrations of contaminants accumulating around the mine but reduces the impedance on production activities which are routinely suspended during these change periods.

This paper firstly highlights the importance of the maintenance of an accurate ventilation model and discusses issues which should be considered to assist to maintain in it a suitable state of health. Further it provides some tips on how to monitor that your model is in an acceptable state of tune so that it can be confidently deployed in the change process. Finally, it details the sequential process how a finely tuned and accurate ventilation model can be utilised to carry out major ventilation changes according to appliance pressure adjustments only and without the need to constantly measure and check resultant branch flows throughout the whole change process. Also discussed are some simple practical tips which will help ensure that the ventilation change process will go according to the devised plan on the day and in the most appropriate order including checking the efficiency and accuracy of the change, after it is completed.

INTRODUCTION

The author has been utilising this ventilation change method since starting as a consultant Ventilation Officer in Queensland's Bowen Basin at around the turn of this century. It certainly isn't a ground-breaking technique but surprisingly one that the author only rarely sees in the field in all the NSW and Queensland underground coalmines that he has consulted to, over the past 20 odd years. The author has frequently explained the method to both university undergraduates and Ventilation Officers around the country over that time, but it seems to have gained little traction, which is perplexing because it is blindingly simple and reliable, as long as the mine retains an accurate and well validated ventilation model. Hopefully this paper will arm site staff with sufficient knowledge to gain the confidence to execute changes in this manner as it has the ability to significantly reduce ventilation change duration thereby increasing safety and increasing operational uptime as a result. The method relies totally on a well validated ventilation model of the mine being available immediately prior to the ventilation change planning process phase beginning. As such the author has expanded on this in detail herein given its pivotal importance in that change process. It further relies on the personnel planning the change ensuring that a risk-based approach is utilised on developing the actual change steps so that critical splits are not left unventilated for extended periods. The model change sequence generated in the Ventsim software is ideal for this. The order of the change steps is critical to ensure that splits that cannot afford to be under ventilated for long periods are treated with the respect they deserve and thus supplied with the appropriate ventilation quantities in the shortest practicable time frame. The longer durations of reduced quantities during the change process must be shouldered by the more benign splits whilst the total circuit is in such a dynamic state of flux and the ventilation change sequence model set is ideal for identifying this. Further, the Ventsim software can graph key flows on any chosen branch which is a handy tool to rapidly see the flow variations at a particular location across each stage. Finally, it is imperative to ensure that the ventilation change process itself can be carried out successfully on the day without any glitches that might negatively affect the outcome or the execution of the change process itself.

THE NEED FOR A VENTILATION MODEL

The practical change process using a ventilation model obviously requires the maintenance of an accurate ventilation model with which to plan the intended circuit adjustments. Planning a ventilation change with a grossly inaccurate model is likely fraught with more danger than the Ventilation Officer using just his knowledge of first principles and his experience and familiarity with the mine to do an 'ad hoc' ventilation change on the run, as it were. Such changes were almost 'custom and practice' before the advent of modern ventilation modelling software packages 30 years ago when the author was in charge of ventilation duties, as an add on to an operational role. Despite these packages now being both readily available and universally utilised in virtually all underground operations the regulators haven't bothered to identify any minimum criteria the model should comply with, when utilised for such work.

Bizarrely the word 'model', as it relates to ventilation, doesn't exist in the Queensland Coal Mining Safety and Health Regulation 2017 current as of 1 January 2022. That's said the Queensland regulators rightly demand and actively encourage its use even though the existing regulations don't attempt to manage such use, in any way whatsoever.

The only reference to the ventilation modelling process in the NSW 'Work Health and Safety (Mines) Regulation 2014 under the Work Health and Safety (Mines) Act 2013 states as follows: *Section 61 Modelling to take place before changes to ventilation system:*

- *The mine operator of an underground mine must ensure that before any significant change is made to the ventilation system for the mine, modelling of that change is carried out.*

Summarily there is insufficient legislation in both states to ensure ventilation changes are always performed using the best available quality tool and the best industry practices.

Arguably the maintenance and proper use of a ventilation model could be likened in some way to the maintenance and use of a mine gas monitoring system. Imagine if there were no prescriptive demands on gas monitoring systems and mine site personnel just randomly decided on issues such as monitor types, monitor locations, calibration standards, frequency of calibration etc. It doesn't even seem logical that such a situation could ever exist when related to gas monitoring systems in the 21st century. How is it that there are no legislated guidelines or controls on ventilation model retainment, maintenance, accuracy, and performance of such a pivotally important and critical tool?

Thankfully, the lack of legislation from the regulator can be offset to some degree by random procedural inclusions sometimes contained in mine site Ventilation Control System Management Plans. Some of these plans specify when models need to be calibrated along with detailed ventilation change processes as examples. That said there is a very broad range of varying requirements contained within these plans, along with corresponding actions to manage such requirements. As a point of note the author has never seen a management plan which specifies a minimum accuracy range that the modelling tool might need to meet, before being deemed appropriate to be utilised in the ventilation planning/change process.

WHAT SORT OF MODEL ACCURACY MIGHT BE SUITABLE?

Most Ventilation Officers seem to agree that an accuracy band of somewhere around ±10 per cent would be a suitable accuracy range for the model to normally lie within on the flows, differential pressures and absolute roadways pressures, when they are determined. The author agrees that such a goal would be suitable, but it is very difficult to set a hard and fast rule for when the model would be considered 'in compliance' or 'out of compliance' because there will always be values that lie well outside such a range, for a number of reasons.

Examples include:

- If magnitudes are very small, then errors will be far greater from a % perspective. Consider a 10 Pa measurement across a vent bag underground that shows a value of 20 Pa in the model. A 10 Pa error is almost indiscriminate but it represents a 100 per cent error between the measured and modelled value, and it's a similar situation for very small flows.

- Consider four insets leading from the main returns to an upcast shaft. The total flow measured might correlate well with the total shaft flow in the model, but it is not unusual if the individual flows vary widely, due to varying shock losses in the individual branches at shaft bottom. Whilst this can be fixed it is of little importance with overall flow being the most important criteria. As a result, it may be prudent to exclude such discrete flows in the correlation process.

Notwithstanding these issues the author believes that it would be wise to establish discrete pass/fail criteria at individual mine sites so that ventilation model tuning can be done in a timely manner. Such accuracy criteria when utilised, and it is done at a number of mine sites, allows the Ventilation Officer to be forewarned of looming inaccuracies well before the model slips completely out of specification. Conversely, when such correlation accuracy is not monitored it often comes as a shock to the Ventilation Officer when the model accuracy is closely scrutinised, and it is found to be well out of specification. As such it is prudent to graph model performance against measured data regularly so that model health can be constantly monitored, and a number of sites do this on a monthly basis.

Figure 1 shows the real measured survey flow data compared to the mine supplied model values gleaned from the same locations. You will note that less than 50 per cent of the supplied model values lie within the ±10 per cent band the author has added to the graph. In the authors opinion, this model would be considered out of calibration from a flow perspective. Figure 2 shows the rebuilt model values compared to the measured flows which shows the model recalibrated satisfactorily from a flow perspective. Figure 3 shows the measured survey pressure data compared to the supplied model values gleaned from the same locations. You will note that all measured pressure values lie either on or outside the ±10 per cent band the author has utilised. In the authors opinion this model is considered badly out of calibration from a pressure perspective. Figure 4 shows the rebuilt model values compared to the measured pressures at the same locations which shows the model recalibrated satisfactorily, from a pressure loss perspective.

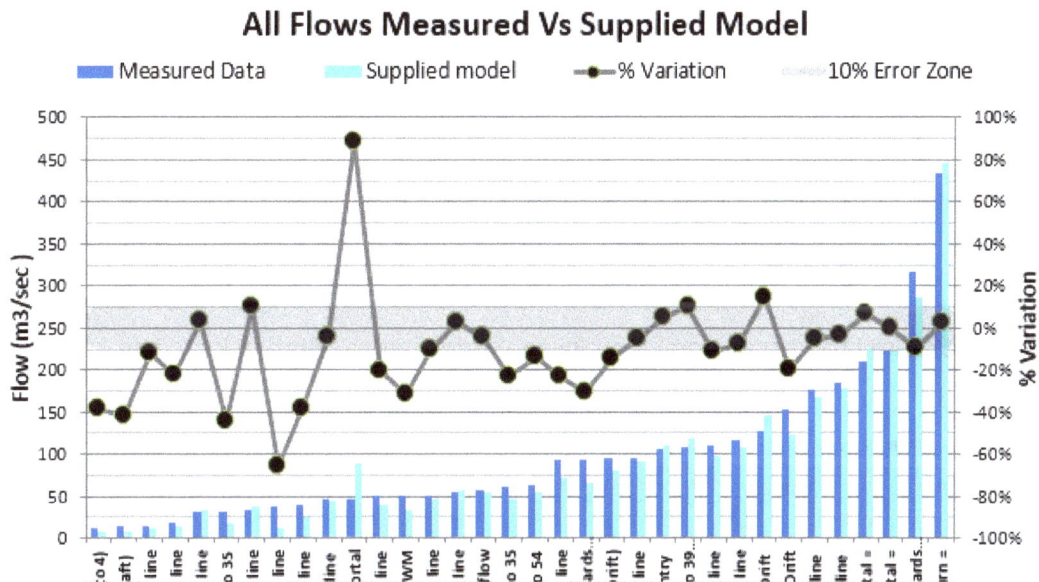

FIG 1 – Measured flows versus supplied model values (Station locations at 'Y' mine).

All Flows Measured Vs Rebuilt Model

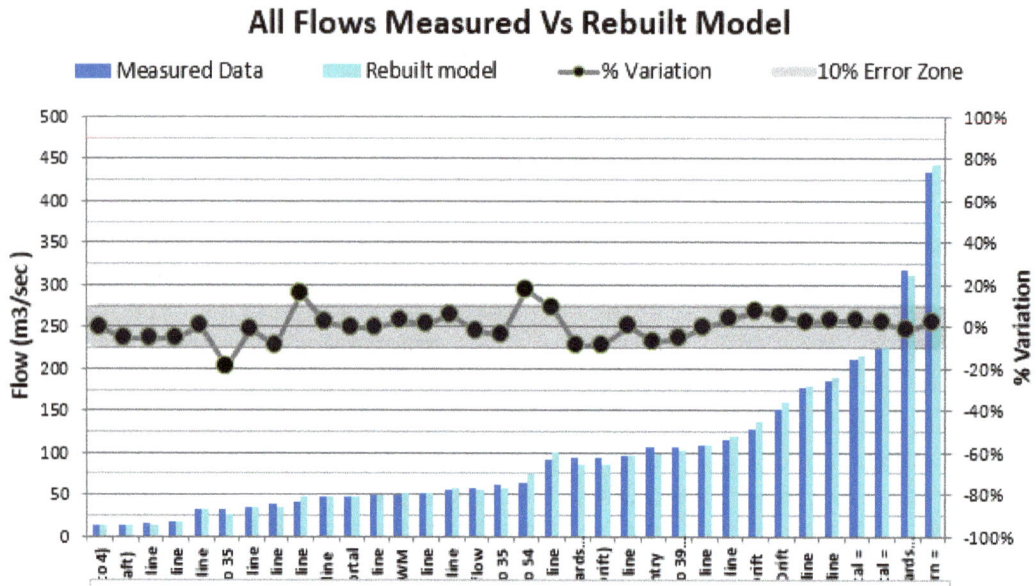

FIG 2 – Measured flows versus rebuilt model values (Station locations at 'Y' mine).

All Pressures Measured Vs Supplied Model

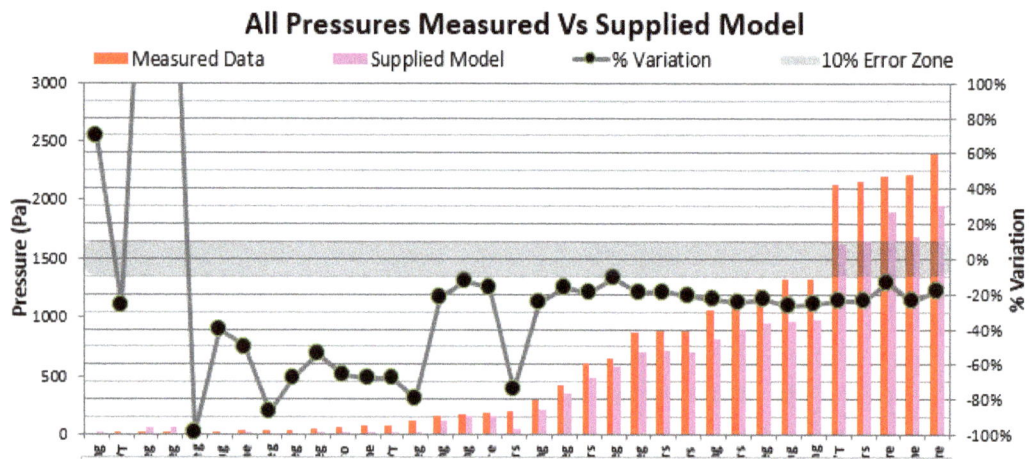

FIG 3 – Measured pressures versus supplied model values (Station locations at 'Y' mine).

All Pressures Measured Vs Rebuilt Model

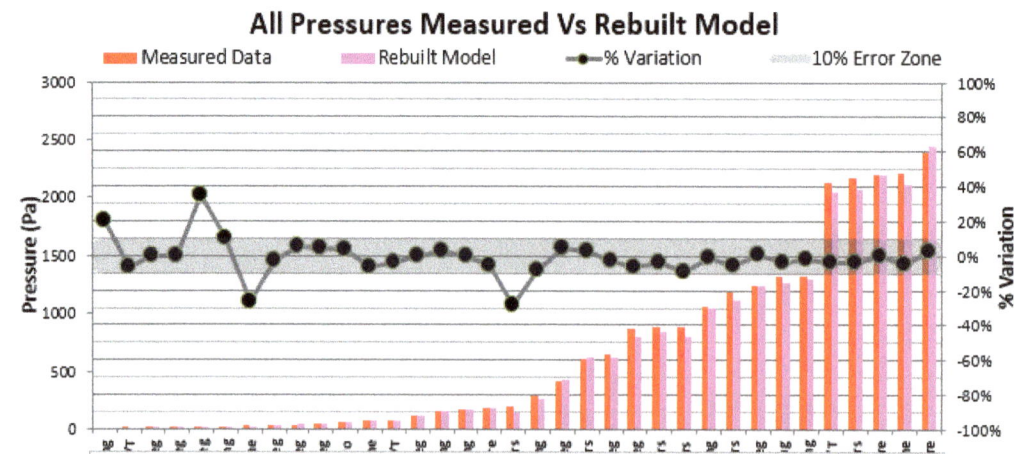

FIG 4 – Measured pressures versus rebuilt model values (Station locations at 'Y' mine).

As you can understand ventilation surveys aren't an exact art and it is very difficult, from the authors experience, to recalibrate much better than that shown in Figures 2 and/or 4, regardless of the time spent tuning it. Results are really mine specific depending on the complexity of the mine, the standard of appliances, local variations in leakage magnitudes, underground housekeeping, and the like. The author believes each mine should decide at what point their model either passes or fails and rectify where necessary to ensure ventilation changes go close to 'as planned'.

VALIDATING AND MAINTAINING AN ACCURATE VENTILATION MODEL

The word validate means to 'to recognise, establish, or illustrate the worthiness' of something. To do this then one must assume that the data set that is used as the adjustment criteria, is as accurate as practicably possible.

This point cannot be overstressed because if the underground survey does not supply an accurate pressure/quantity data set then, by adjusting the model accordingly, it is feasible to destroy the thing that you are actually trying to fix!!

The survey should be done with the utmost care because it is the ultimate accuracy of these results that will decide the future value of the model, as a useful management tool. In the authors opinion it is simply not satisfactory to have multiple underground officials collecting data over a period of weeks in a dynamic mine and then adjusting the ventilation model in line with that collected data. Much more commitment to quality and accuracy is required than that. Ideally all the quantity and pressure data should be collected over the shortest time frame possible ie sequential days if required and by the same person using the same stations and same determination techniques. Anything less and it should come as no surprise seeing the model slip rapidly out of calibration.

Summarily it is of prime importance to treat every ventilation survey as importantly as the survey you would pay a consultant to do, and this is absolutely crucial if the model is adjusted every time a routine survey is carried out. The author has seen models literally destroyed in a matter of weeks by errant updates on-site, using poorly collected and/or drip-fed data collected over extended time frames.

There are some simple tips in maintaining a ventilation model so that firstly it doesn't slip out of tune and secondly when it does slip out of tune the model owner will be well aware of it losing tune in time to arrest the problem.

SHOW YOUR CIRCUIT REGULATORS THE RESPECT THEY DESERVE

In all underground operations the branch regulators are the valves that allow you to control your mine wide flows. Usually, all splits bar one are artificially regulated (although sometimes two or more individual splits may be unregulated, depending on the resistance and flow requirements in those splits.) Out of all the quantity and pressure readings that are collected during the underground survey the only ones that allow you to determine the actual branch resistances are the branches that you regularly and repeatedly measure the flows and the pressures across. Good guess! You've got it!! They are called the regulators!! Why is then that the only circuit branches that have had their resistance values scientifically determined by $R=P/Q^2$ are usually the first one's Ventilation Officers are willing to destroy, in an effort to force the model to mimic underground flows?

The learning is: NEVER fabricate new resistances for your regulators just because the flows across the regulator/s in the model don't replicate the flows in the mine. If you change the resistance of the regulator so that the flow across it emulates the in-pit flow it simply hides the error in the pressure drop across it. (If you reduce the resistance to increase the model flow the regulator pressure drops and if you increase the resistance to lower the model flow the regulator pressure increases.) What is always consistent in this scenario is that the flows will look reasonable, and the erroneous pressures are basically hidden from view. The author has been to mines where 500+ Pa has been noted across a regulator in the model only to find the device fully open underground. Because of this almost endemic phenomenon the most common description the author receives when enquiring on the health of a ventilation model is 'the flows are pretty good, but the pressures are a worry'. If the flows are correct and the pressures are erroneous, the branch resistances are also erroneous. It really is that simple.

At one unnamed Queensland operation called Mine 'X' the author rebuilt the model a few years ago to a very fine state of tune with most pressures and flows within around a ±2 per cent band. As always, during this rebuild, every regulator resistance (17 in total) was assigned its resistance according to the pressure and flow measured across it. In under two years 14 of those 17 regulators had been assigned approximate 'orifice' areas that ensured model regulator flows were in line with measured underground flows. The 'orifice' feature is somewhat useful for modelling when a series of ventilation doors are intentionally left open during a ventilation change but it was certainly not designed to replace measured pressures and quantities across regulators. Worse still is the 'louvered regulator open % recommendation' function which is probably the most misused function ever added to the Ventsim software, in the authors opinion. What is wrong with measuring the pressure drop across a regulator and the flow running through it and determining the actual resistance and utilising that on the branch in the software? During a normal mine survey these are the only key resistances that are physically measurable so why would the ventilation practitioner just ignore such a determination and replace it with something that simply hides the model errors?

In Figure 5 you will note validity of the flows at this aforementioned operation after those two years had passed. You will note the flow validity isn't awe inspiring and could arguably be regarded as unsatisfactory. They were only even this good though because the regulator resistances were aborted to provide such a result.

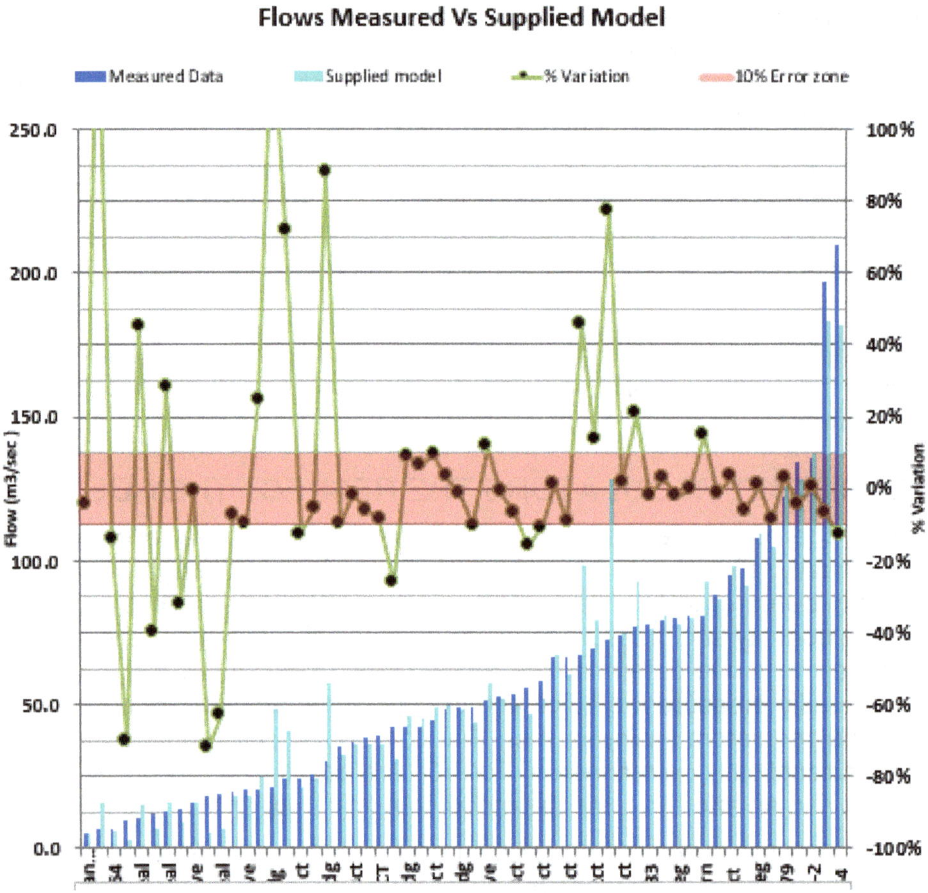

FIG 5 – 'X' Mine measured flows versus supplied model values (Station locations at 'X' mine).

In Figure 6 you will note the very poor validity of the pressures at this same operation after those two years had passed. It is unlikely the site was aware of the gravity of these pressure errors due to the not unusual focus on ensuring mine wide flows resemble ventilation model flows, at the expense of these pressures.

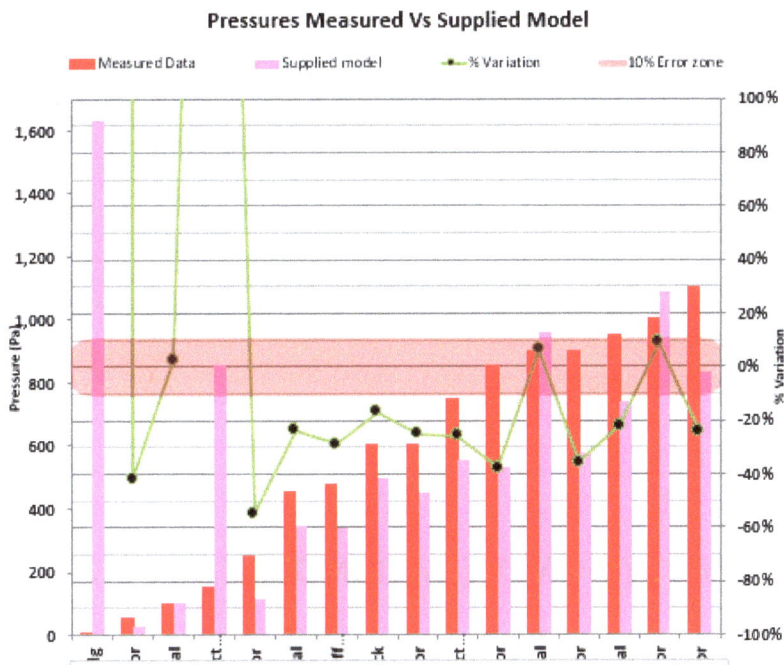

Pressures Measured Vs Supplied Model

FIG 6 – 'X' Mine measured pressures versus supplied model values (Station locations at 'X' mine).

Such a phenomenon is easily the most common problem the author has noted when he has found models well out of tune over the last 20 years. Hardly a mine has been exempt from this during the authors consultancy career. If Ventilation Officers can't resist the temptation to alter the regulator resistances to change flows, then it may be prudent to at least keep a separate model that contains the correct regulator resistances so that validity, or lack thereof, can be identified in a timelier manner.

Summarily Ventilation Officers must resist the temptation to alter the only resistances that they have actually measured. In line with Scott Morrisons quote regarding pandemic panic buying, 'stop it, it's not sensible, it's not helpful, don't do it!

WHY WOULD YOU WANT TO EXECUTE THE CHANGE BY ONLY ADJUSTING PRESSURES ACROSS REGULATORS?

The answer is quite simply speed. If you plan a ventilation change using flow alterations across the regulators with no knowledge of the expected change in pressure across the regulators you will need to alter the regulator numerous times, possibly up to four or five times which involves laboriously checking the branch flows with an anemometer. You get the idea: adjust regulator a bit then measure two or three anemometer flows then adjust regulator a bit more and measure two or three anemometer flows and so on. It's laborious work and extremely slow. The author has seen cases where excessive pressures on regulators has denied access across the regulator from the downstream adjustment side to the upstream flow measurement side which has resulted in walking long distances to get the quantity reading. If this needs to be done multiple times it's not hard to convince people it is easier, faster, and much more accurate to simply change the resistance of the regulator according to the pressure across it. This can literally be done in minutes or less then you can jump in the truck and drive off to execute the next sequential change step.

By way of explanation, you will note in Table 1 the author has predetermined the % regulator adjustments according to % variation required sequentially at each step in the change. You will note the change simply involves either completely closing and/or opening appliances and altering regulators by a predetermined % when you get to that step in the change process.

TABLE 1

Sequential ventilation change steps showing % regulator adjustments.

Ventilation change to put new L/W block on line	Regulator Pressures (Pascals)		
	Before Change	After Change	% Change
Step 1. Remove stopp face & bleeder, close M/G and T/G to bleeder	N/A	N/A	N/A
Step 2. Set M/G homo reg R =0.38878	Fully Open	890	N/A
Step 3. Close T/G doors	Fully Open	255	N/A
Step 4. Open L/W reg	Fully Closed	Fully Open	N/A
Step 5. Open flank reg R= 0.40903	311	265	-15%
Step 6. Close Mains A Hdg reg R= 0.94907	727	758	4%
Step 7. Close Mains D Hdg reg R= 1.03234	744	773	4%
Step 8. Open new M/G 3 reg R= 0.32791	782	549	-30%
Step 9. wind fan speed up to 71.5%			

PLANNING AND EXECUTING VENTILATION CHANGES USING PRESSURES ALONE

This is the crux of this paper and from hereon it is assumed that the person planning such a change is using a well validated ventilation model in the order of at least ±10 per cent on the flows and the pressures generally. The following steps summarise the overall process the author has utilised to make such changes basically throughout his career. The ultimate desire, which requires little to no explanation, is obviously to minimise the actual duration of the ventilation change to enable the revised circuit to be handed back to the mine operator as soon as practicably possible, under its new configuration and with close to expected circuits flows throughout the mine.

Survey and model update prior to the change

It is highly recommended that the final mine wide survey and model tweak prior to carrying out the actual ventilation change is done before and as close as practicable to the planned change date. Further, it is preferable that no other minor ventilation changes are planned in the period between doing this survey and executing the planned major change.

After the survey is completed and the numbers are crunched the model must be updated to the finest tune practicable to a level within the 10 per cent error band, as previously discussed.

This final model which represents the circuit layout at the time of the planned change is then used as the base model for the ventilation change planning process.

Mine design prior to the change

For this example, the author has built a very simple longwall mine model of the fictional 'Southfork' mine detailed in Figure 7 that will be utilised to explain the change process. The model shows the remnant flows around the mine in the various locations prior to executing a ventilation change to put the new L/W online.

You will note the following flows and circuit design:

- The mine has a flank return established around L/W 1 to be used to ventilate inbye the operating L/W's that has a flow of 30 m³/s.

- M/G 1 (which is soon to be converted into L/W 2 T/G) has 35 m³/s flowing up to feed the flank return and L/W faceline at this time.

- M/G 2 is still operating in the development panel design with 61 m³/s outbye the last open C/T with the new L/W face and L/W bleeder roads virtually shutdown with closed regulator/stoppings in both driveages.

- There is 30 m³/s at both Mains A Hdg and E Hdg faces.

- M/G 3 panel is idle with only 10 m³/s but awaiting start-up after the ventilation change is carried out.

- The main fan is flowing 208 m³/s at a total pressure of 3.005 kPa and spinning at 57.5 per cent of full speed.

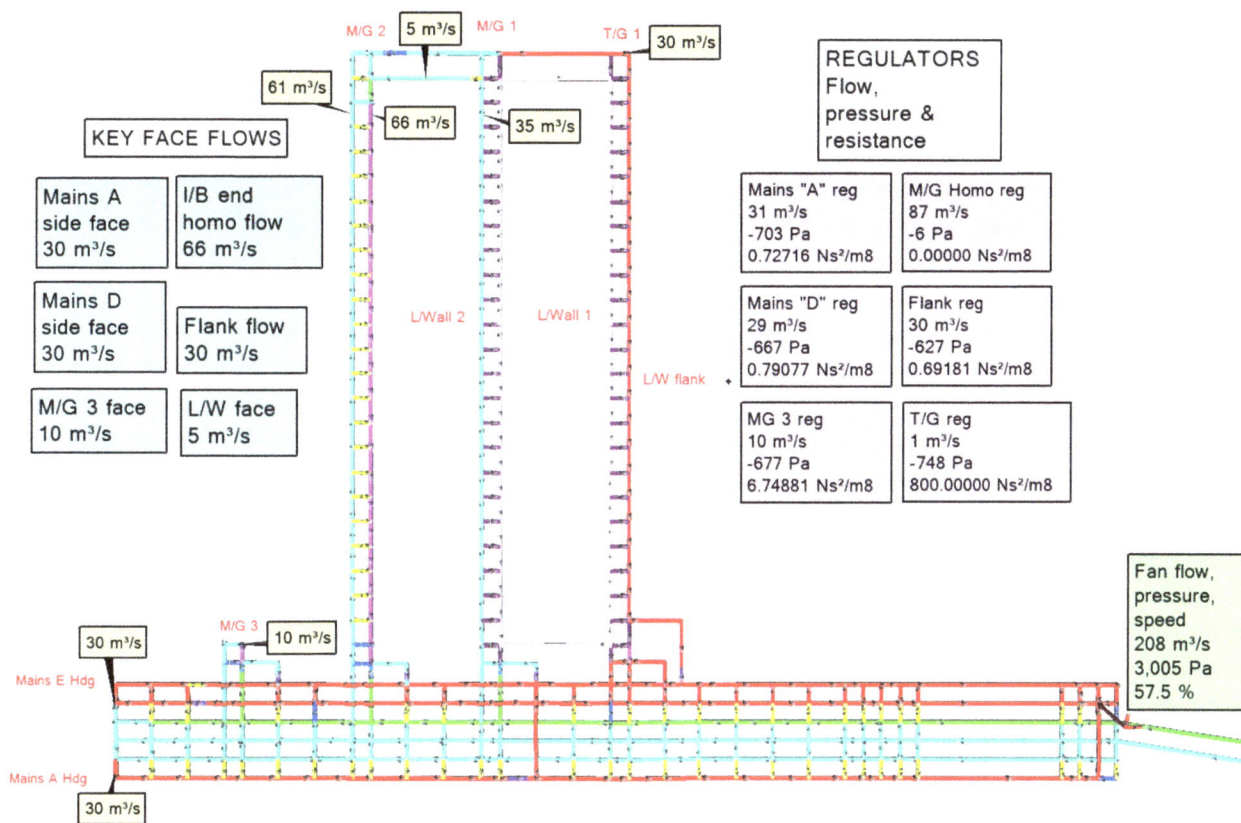

FIG 7 – Minewide flows and design just prior to enacting ventilation change.

This base model is used as the template for the change.

How the mine needs to be ventilated after the change

The author uses 'stages' in the Ventsim program to save making numerous individual change models and to allow rapid comparison and analysis of the circuit performance at each stage of that change process. The first step is to copy the base model from above into the next stage as a detached copy. This model is then adjusted, in no particular order, until all the flows are satisfied at each location. It doesn't matter what order the changes are done in as long as the final model depicts the circuit as you wish it to be after the ventilation change is completed, as shown in Figure 8. This is then saved in this example as '*Longwall running at end of change*'. You will note the following flows and circuit design:

- The mine still has a flow of 30 m³/s in the flank return established around L/W 1.

- The L/W face flow has been set to 60 m³/s.

- The L/W is employing a homotropal design with the L/W belt remaining on return and you will note a flow of 10 m³/s heading outbye from the last open C/T and 101 m³/s in the adjacent intake which is feeding the homotropal belt, the L/W face and the flank return.

- There is still 30 m³/s of air required at both Mains A Hdg and E Hdg faces.

- M/G 3 panel is ready for production with 50 m³/s allocated to that panel.

- The main fan is flowing 261 m³/s at a total pressure of 4.564 kPa and spinning at 71.5 per cent of full speed.

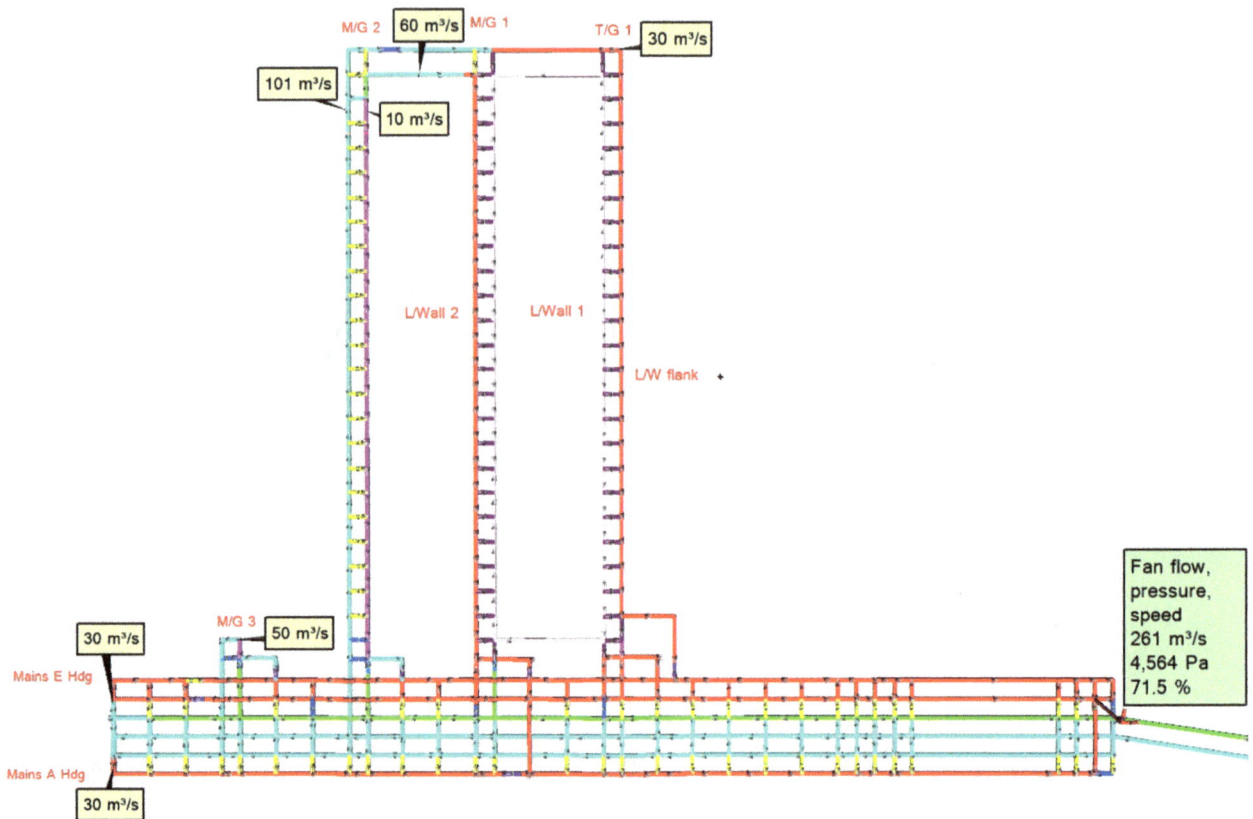

FIG 8 – Planned mine wide flows at the completion of the ventilation change.

The change order process

The model depicted above is how the mine is intended to be ventilated but so far, no regard has been paid to the order of the change process underground. It is always wise, if possible, to make changes in the same localities at the same time but not if that risks the integrity of the ventilation circuit by leaving vulnerable areas inadequately ventilated for too long and conversely benign areas overventilated for too long. This is where considerable thought and trial and error needs to be exercised to choose the most appropriate change sequence so that any perilous contaminants that might accumulate be it gas, DPM, heat, dust etc are minimised. A risk-based analysis with the appropriate people should be utilised to ensure the change step order is the most appropriate and issues such as demarcation or dropping electrical power off areas and the like is considered.

Critical Note: It is imperative that the sequential order of the change is carried out in exactly the same order underground as that detailed in the planned change list in the model. If not, the expected regulator pressures at any point in time will be completely erroneous and the change will fail as expected. It is vitally important to understand that.

In Figure 9 you will note the stages decided upon by the author to demonstrate this simple ventilation change. At the top of the list is the previous L/W sealed which is the starting point. This stage is then copied down as a detached copy and relabelled in line with the intended Step 1 and the actions detailed in Step 1 are then made to the model.

FIG 9 – Planned stage list of sequential underground adjustments.

The change process from start to finish relies on removing or erecting complete appliances such as removing stoppings or closing doors by way of example and discrete adjustment of chosen regulators in the correct order.

You will note the author has added both regulator and key flow message box summaries to the model as well as regulator locations where relevant, to enhance the understanding of circuit changes.

Expected circuit performance pursuant to each planned change

Step 1

After executing the initial Step 1 the expected model performance at that time is shown in Figure 10.

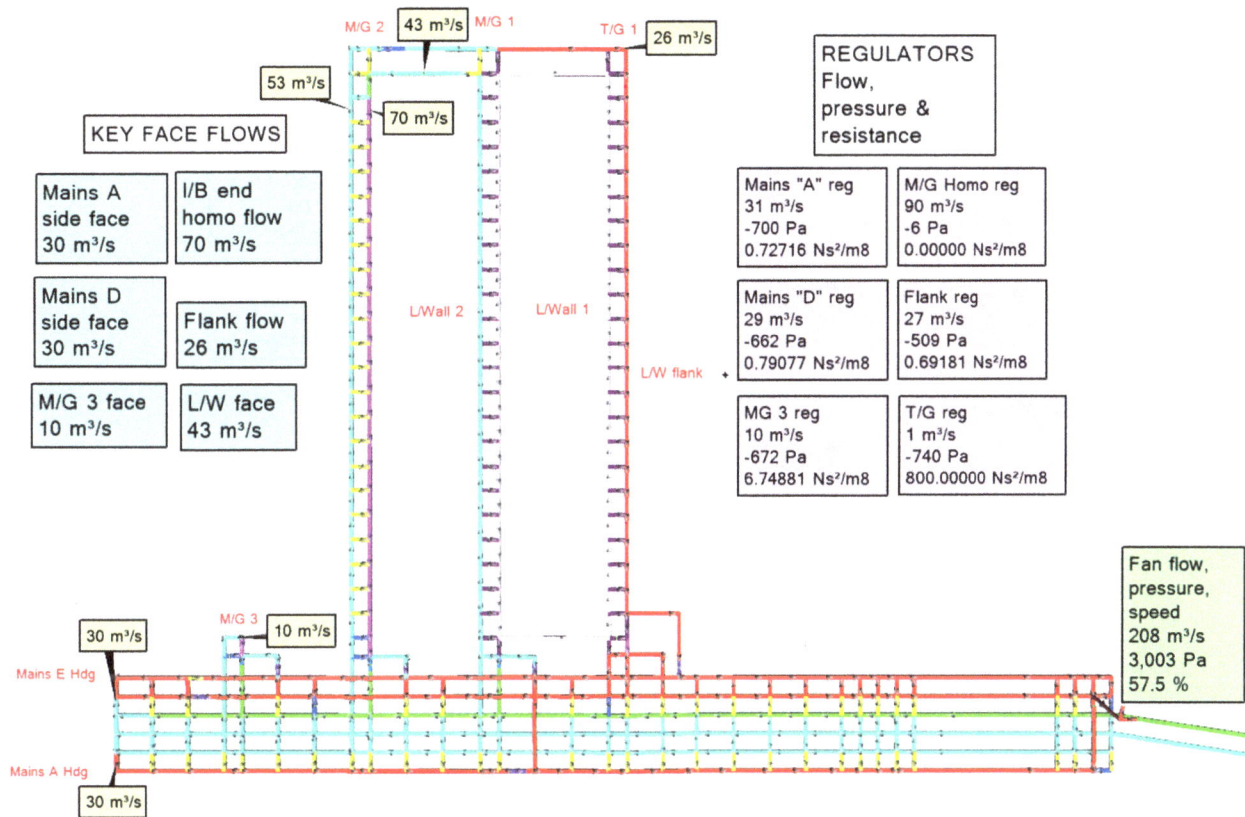

FIG 10 – Expected circuit performance after Step 1.

This involves removing stoppings on the faceline and the rear bleeder behind the face then closing the M/G and T/G side of the face between the face and the bleeder. L/W face ventilation direction is from tail to main and there is still ample flow in all panel areas. This is how the circuit is expected to perform after that work is carried out underground.

Step 2

The Step 1 model is then copied down as a detached copy and relabelled as Step 2. This step involves changing the resistance of the homotropal regulator to what it needs to be at the end of the completed ventilation change. The completed model at the bottom of the list 'Longwall running at end of change' contains the exact regulator resistances of every regulator in the mine as at the end of the change process. In this case the regulator resistance from only the M/G homotropal regulator from model 'Longwall running at end of change' is copied into the regulator branch in the Step 2 model which is then run (and resaved) just as if you had done that step underground. In Figure 11 you will note the expected mine wide flows after Step 2 is executed, according to the modelling and you will also note the expected pressure of 890 Pa across the homotropal M/G regulator, as noted in the data box.

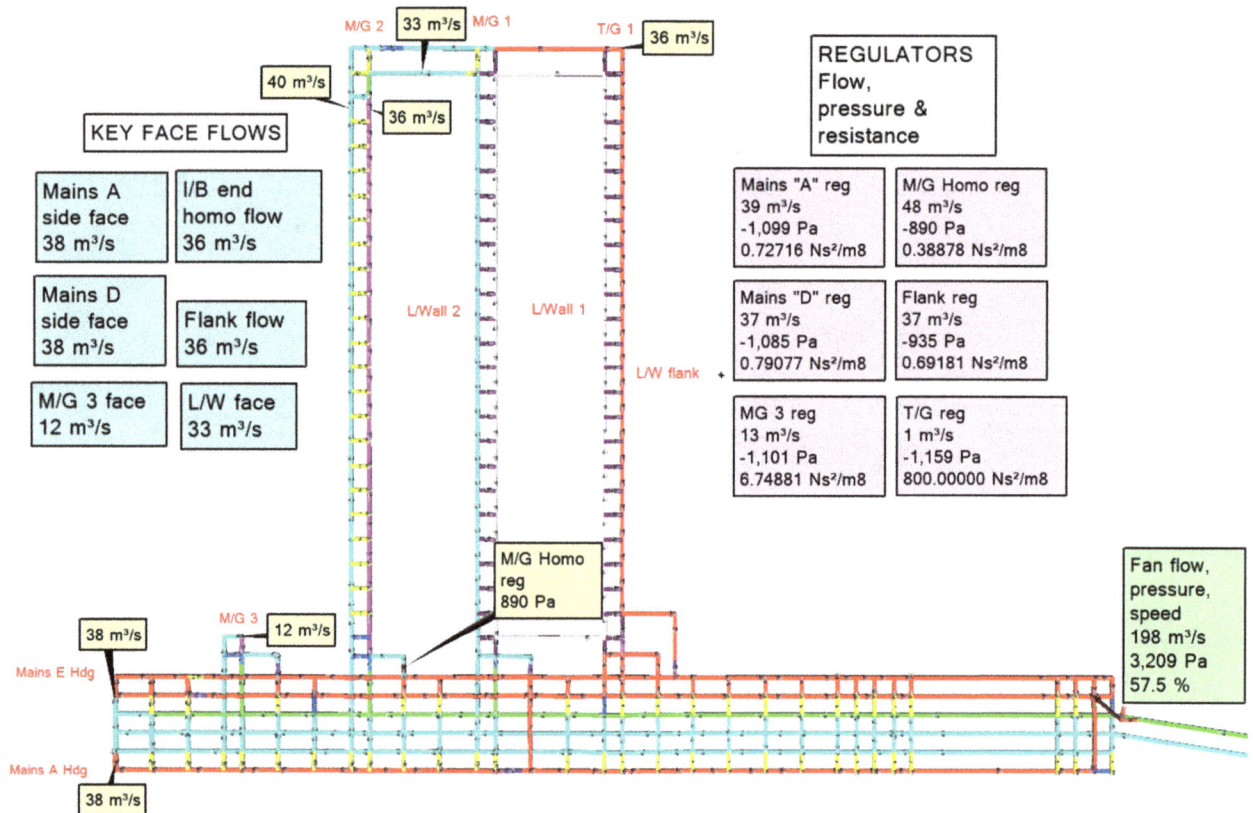

FIG 11 – Expected circuit performance after Step 2.

Pursuant to altering this regulator in Step 2 you will note mains flows and the flank bleeder flow will increase slightly along with a reduction in the homotropal flow and the L/W T/G flow. L/W face ventilation direction is still from tail to main and there is still ample flow in all panel areas. This is how the circuit is expected to perform after that work is carried out underground, at the completion of Step 2.

Step 3

The Step 2 model is then copied down as a detached copy in the stage list and relabelled as Step 3. Step 3 involves closing the T/G entry doors so this door branch in the model is assigned a value typical for a good set of double machine doors which, in this case, is 400 Ns^2/m^8. The model shows an expected temporary stagnation of the L/W T/G flow as you will note in Figure 12 with only 2 m^3/s available on the L/W face flowing from tailgate to maingate. The temporary stagnation only affects the L/W until the T/G regulator around the corner can be opened, which is the immediate next step.

Of more interest is the expected pressure across the now closed T/G doors which is only 255 Pa because, at this time, the L/W T/G regulator remains tightly closed. This is absolutely invaluable data because you can determine exactly how much load will be required to close the final door and thus determine whether say two men or a machine are required to safely close the door. To illustrate that lets assume the last door to be closed has a total surface area of 7.5 m^2. If a pressure of 255 Pa is applied to a door of that area it would require a maximum force of around 195 kg to hold it open, then safely and slowly close it. For the record if the T/G regulator was opened first then the T/G doors were closed there would be 780 Pa on the doors requiring 600 kg of force to safely close them. This highlights the crucial importance of the ordered sequence of the change. The software is invaluable in assisting to determine such order.

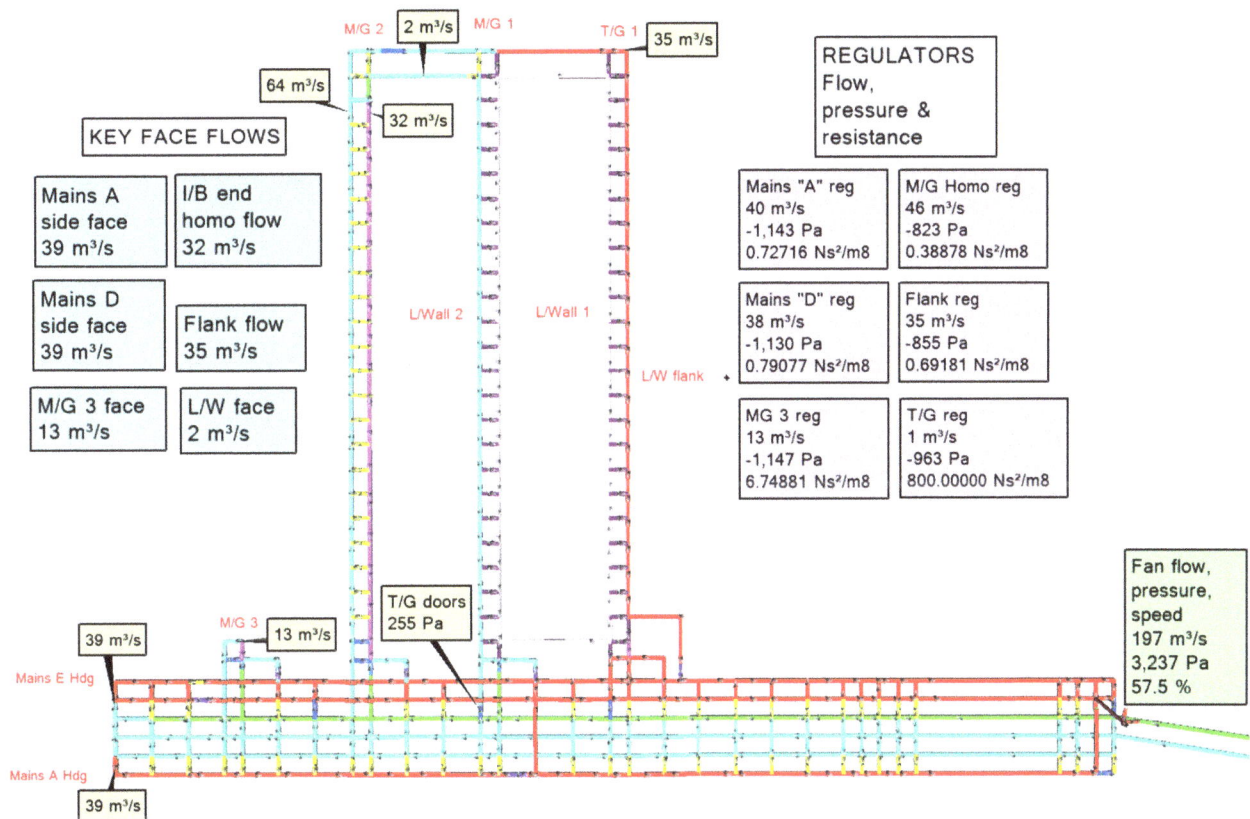

FIG 12 – Expected circuit performance after Step 3.

Step 4

The Step 3 model is then copied down as a detached copy in the stage list and relabelled as Step 4. Step 4 involves removing the resistance in the model on the L/W T/G regulator branch and then running the software (and saving).

You will note the expected flow conditions in Figure 13 after the T/G regulator is opened. You will further note that flank return flow is slightly down and the homotropal belt flow, L/W face and the mains face flows are very close to final flow quantities desired after the change is completed.

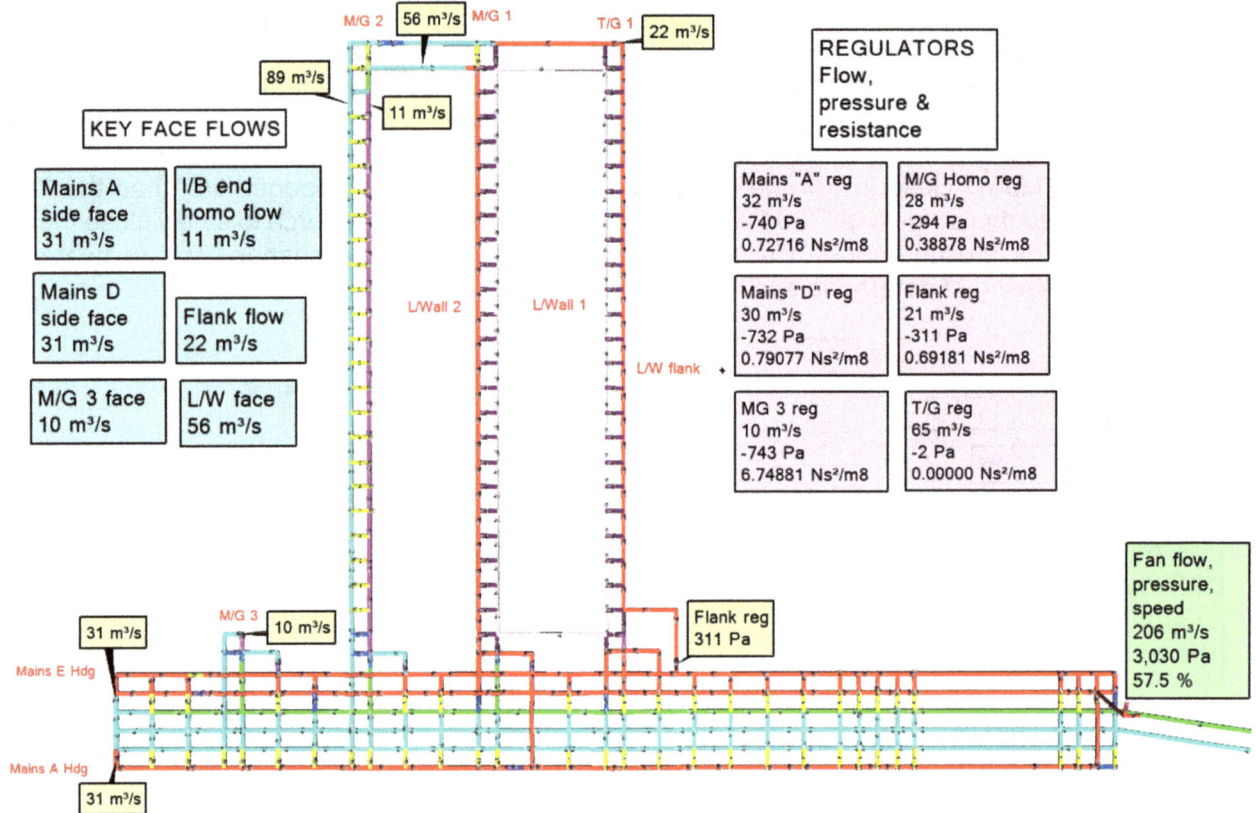

FIG 13 – Expected circuit performance after Step 4.

Step 5

The Step 4 model is then copied down as a detached copy in the stage list and relabelled as Step 5. This step involves a minor opening of the flank regulator situated adjacent to the E Hdg main return O/B the L/W T/G. The regulator resistance for the flank regulator in the final model 'L/W running end of change' is copied from that model into the regulator branch in the Step 5 model which is then run (and saved).

According to the modelling, the pressure across the flank regulator by reducing this resistance will go from 311 to 265 Pa at this time. This represents a reduction in the regulator pressure of 15 per cent as you will note in Table 1.

Expected results after executing Step 5 are shown in Figure 14.

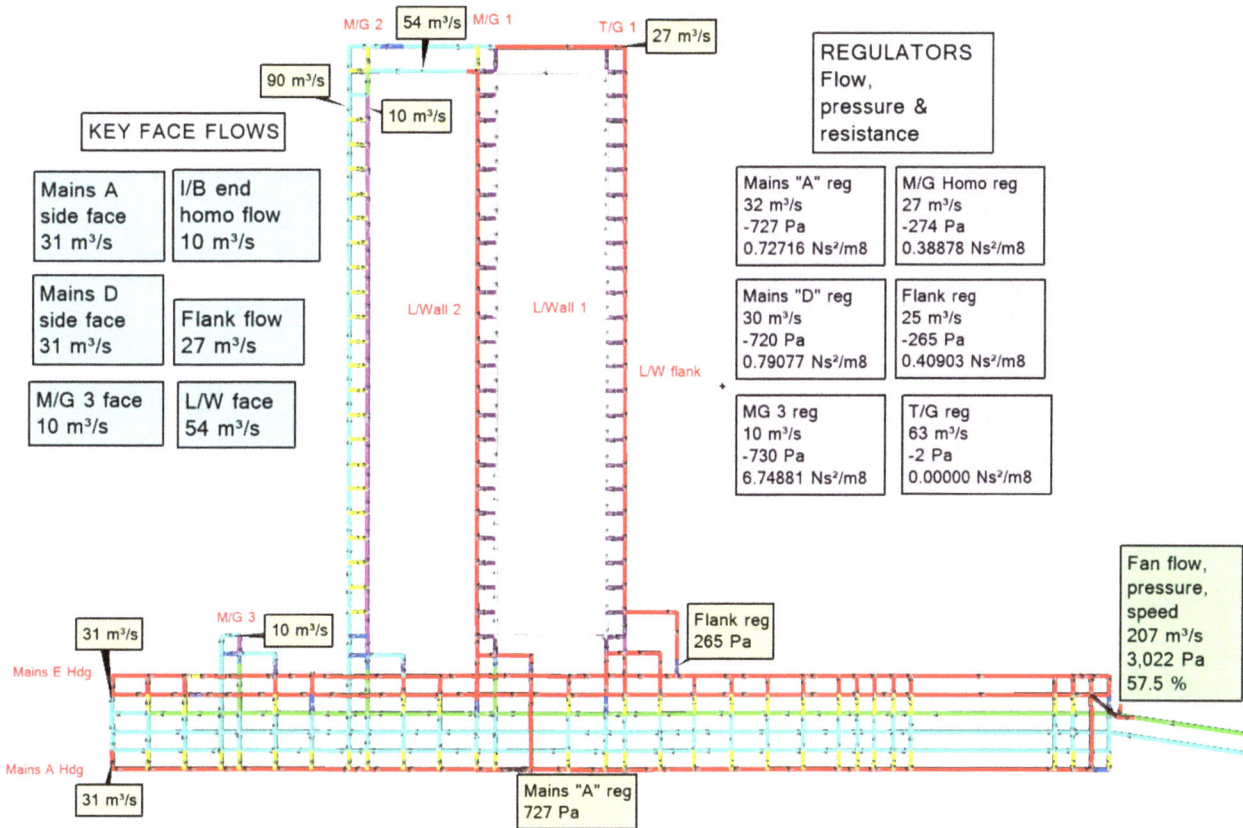

FIG 14 – Expected circuit performance after Step 5.

Step 6

The Step 5 model is then copied down as a detached copy in the stage list and relabelled as Step 6. This step involves a very minor closure of the Mains A Hdg regulator. The A Hdg Mains regulator resistance in the final model 'L/W running end of change' is copied from that model into the Mains regulator branch in the Step 6 model which is then run (and saved).

According to the modelling the pressure across the Mains A Hdg regulator will increase from 727 to 758 Pa, according to the modelled results. This represents an increase in the regulator pressure of just 4 per cent as you will note in Table 1. Expected results after executing Step 6 are shown in Figure 15.

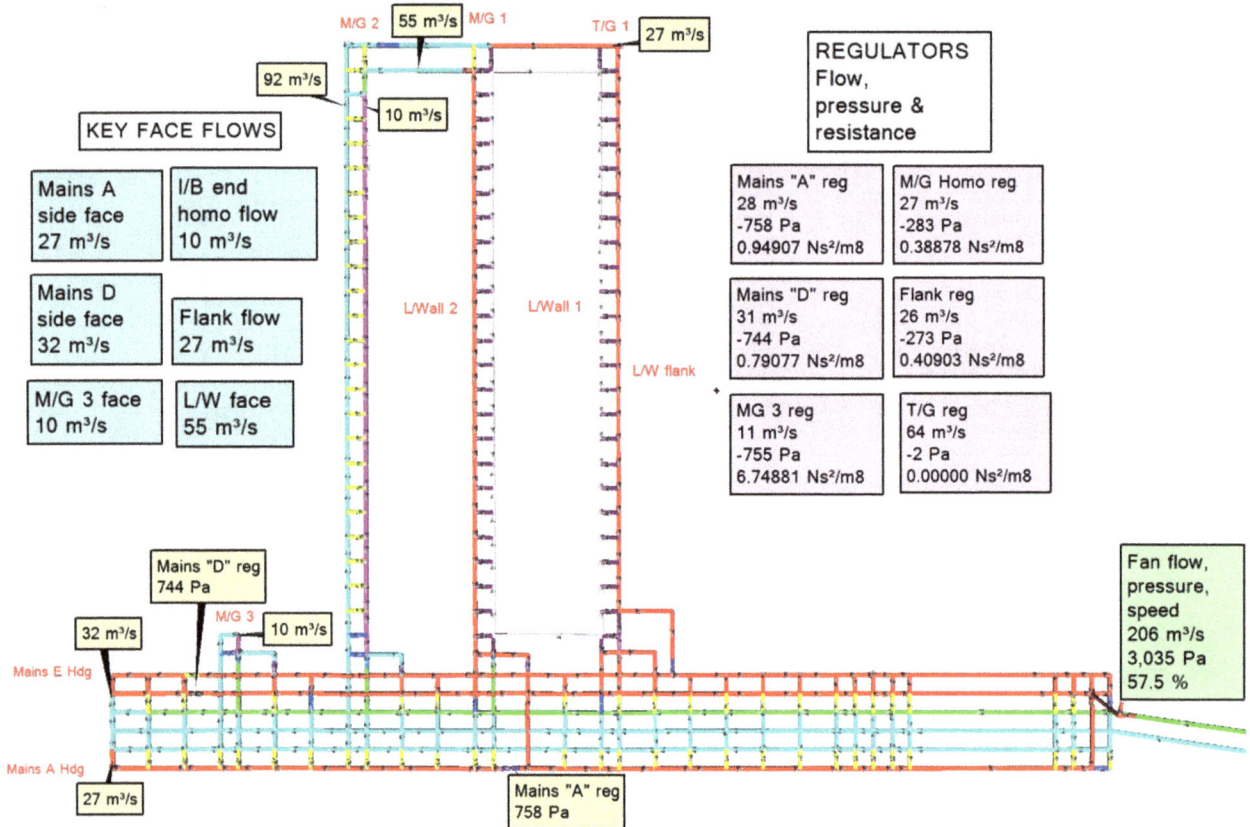

FIG 15 – Expected circuit performance after Step 6.

Step 7

The Step 6 model is then copied down as a detached copy in the stage list and relabelled as Step 7. This step involves a very minor closure of the other Mains regulator in D Hdg. The D Hdg Mains regulator resistance in the final model 'L/W running end of change' is copied from that model into that Mains regulator branch in the Step 7 model which is then run (and saved).

According to the modelling the pressure across the Mains D Hdg regulator will increase from 744 to 773 Pa, according to the modelled results. This represents another very minor increase in the regulator pressure of just 4 per cent as you will again note in Table 1. Expected results after executing Step 7 are shown in Figure 16.

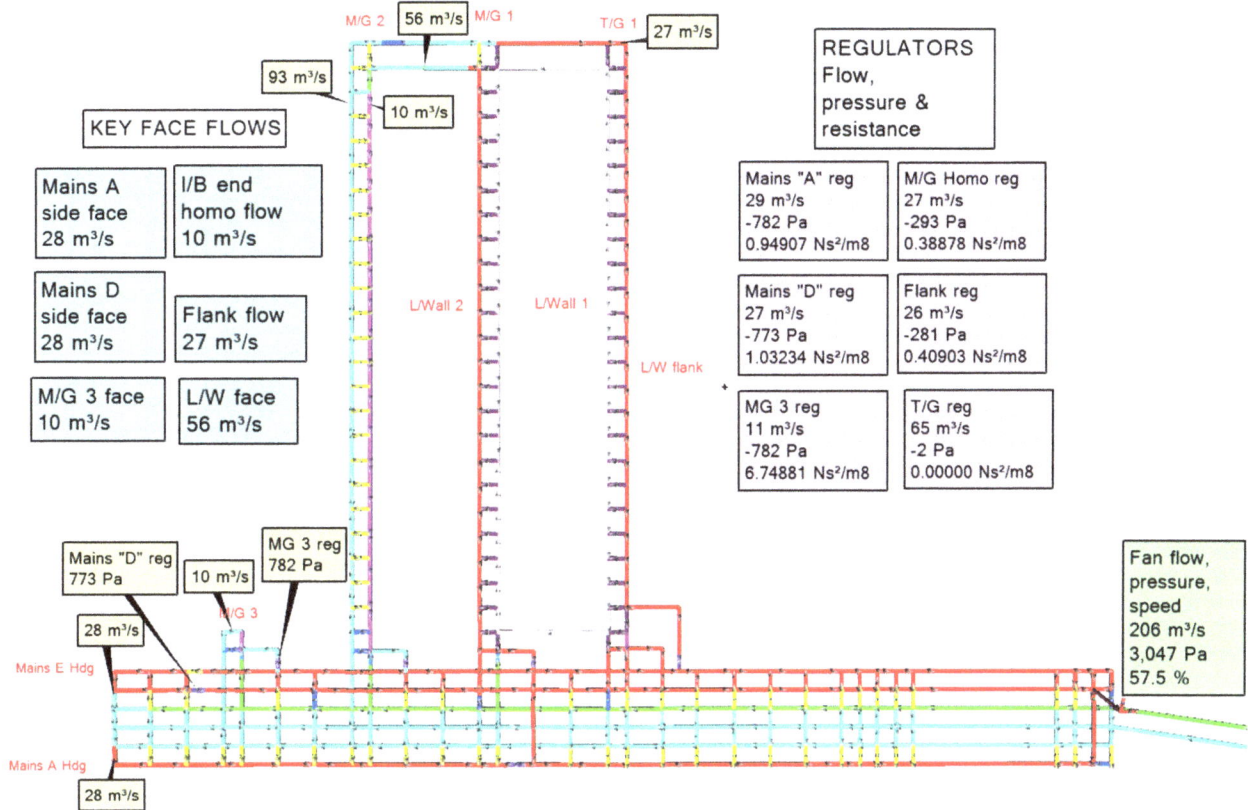

FIG 16 – Expected circuit performance after Step 7.

Step 8

The Step 7 model is then copied down as a detached copy in the stage list and relabelled as Step 8. This step involves a substantial opening of the M/G 3 panel regulator to ready the place for production. The M/G 3 regulator resistance in the final model 'L/W running end of change' is copied from that model into the M/G branch in the Step 8 model which is then run (and saved).

According to the modelling the pressure across the M/G 3 regulator when opened will reduce the regulator pressure by 30 per cent from 782 to 549 Pa, according to the modelled results as noted in Table 1. This represents a large reduction in pressure and a corresponding increase inflow in this split. As would be expected this will temporarily rob all other splits of some flow until the final change of a fan speed increase is enacted. Expected results after executing Step 8 are shown in Figure 17.

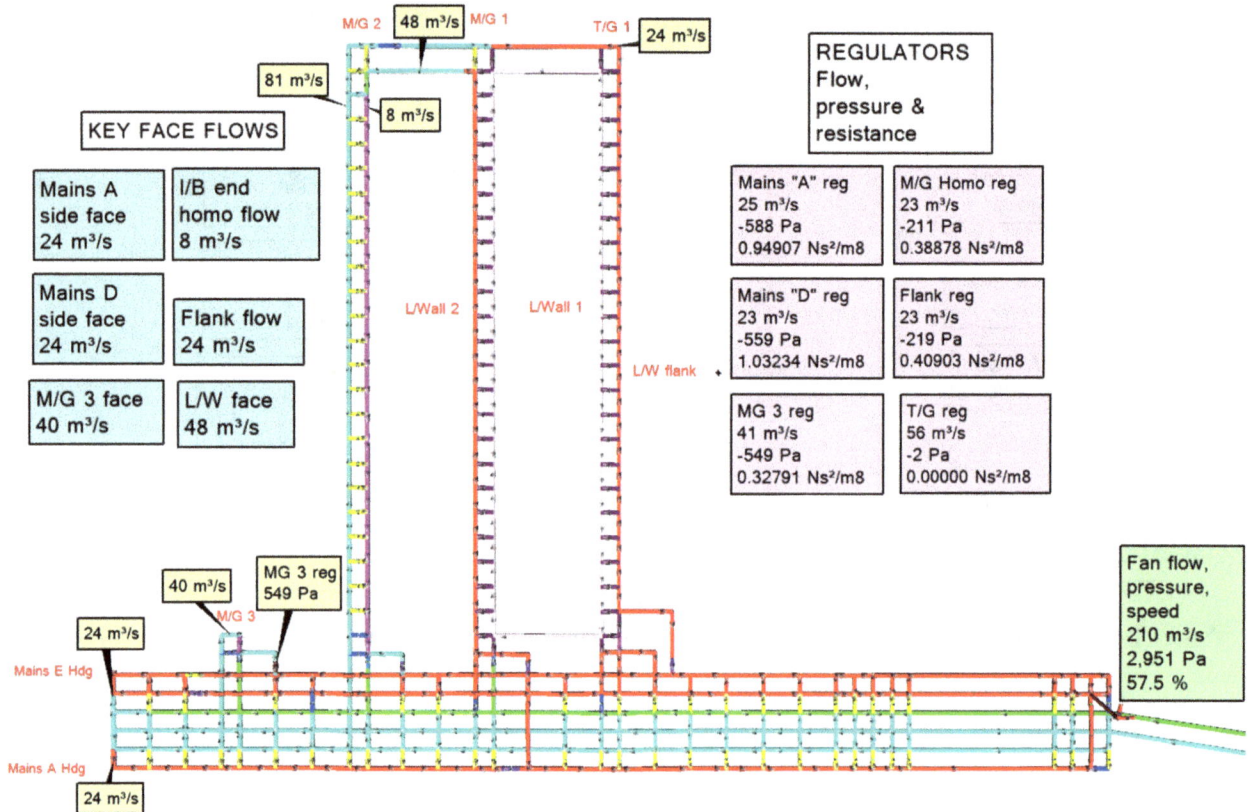

FIG 17 – Expected circuit performance after Step 8.

Step 9

The Step 8 model is then copied down as a detached copy in the stage list and relabelled as Step 9. This step simply involves increasing the fan speed to satisfy all required panel flows. The fan speed detailed in the final model 'L/W running end of change' is altered in the Step 9 model and the model is then run (and saved).

The expected performance of the model at the completion of the ventilation change is shown in Figure 18. This is now a completely independently built model to the one entitled 'L/W running end of change' but the model performance should be, and is, identical.

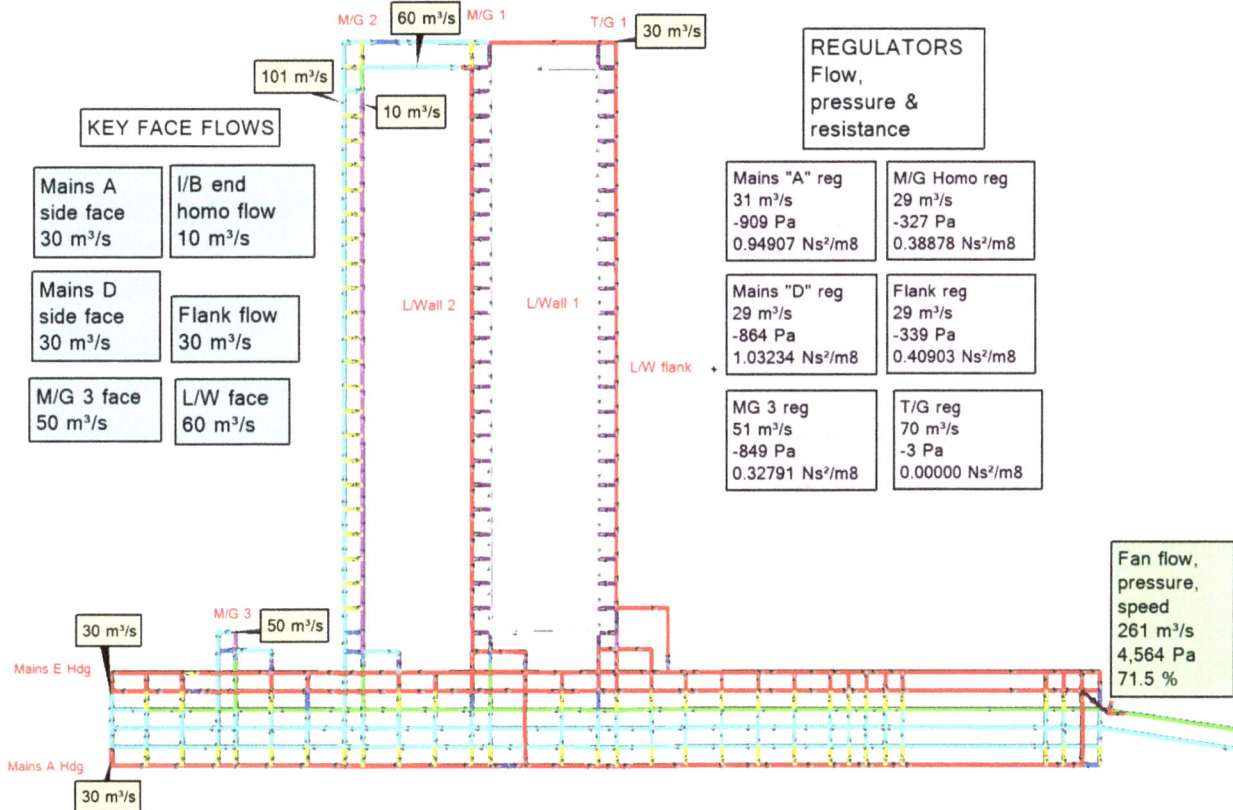

FIG 18 – Expected circuit performance after Step 9.

If you refer back to Figure 8 detailing the performance of the 'L/W running end of change' model you will note that performance is identical regarding pressures and flows. This fact ensures that all resistances have been adjusted accordingly throughout the process and the fan % change is appropriate at the end of the job, to achieve the desired results.

Stage graphs

To compliment the data shown in the individual model circuit performance summaries above the Ventilation Officer can rapidly graph any criteria across all the sequential stages of the ventilation change. The example illustrated in Figure 19 you will note the flows across the L/W face throughout the change process. In the authors Ventsim version '5.4.7.2' the process to get the stage graph is as follows:

- Select branch of interest which in this case is the L/W face branch.

- Open the 'edit box' but please note function is not available in the old style edit box but just the new one.

- Click on 'tools', 'open stage graph', 'get graph' and the graph will be generated before your eyes. You can then view any criteria you like, that is relevant to the selected branch across all the stages.

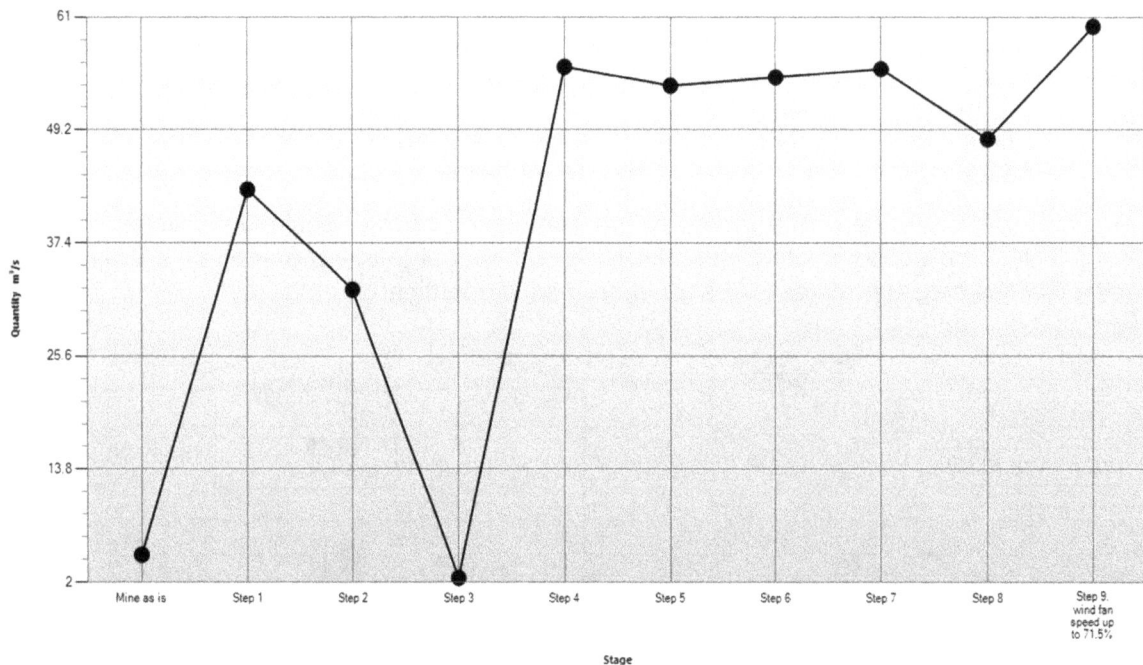

FIG 19 – Example of Stage Graph showing L/W face flow at each change stage.

Physically carrying out the ventilation change

As detailed earlier it is <u>absolutely critical</u> the physical change is done in identically the same order and manner as the planned change otherwise the regulator pressures that need to adjusted will be nonsensical and erroneous.

It is vitally important to ensure that everything is ready to go for the ventilation change. The last thing you need is a simple obstacle that may greatly lengthen the time frame of the change or worse still force you to abort the ventilation change because the change steps haven't been checked underground. The author believes it is imperative to organise a transport, if possible, the day before the planned change and visit all sites to ensure all preparatory works have been satisfactorily completed. Issues that should be checked include, but are by no means limited to:

- Ensure that all regulators that need to be adjusted are serviceable and operable. It is common for regulators, especially louvre style ones that haven't been operated for a long period of time to seize up in the linkages. Further, most louvre types are operated with a lever that is sometimes detached from the regulator itself so ensure that the lever is on-site and everything is operable.

- Ensure that all the regulators can be finely enough adjusted so that the actual intended pressure can be put across the regulator at such locations. It may be advisable to site some brattice at the regulator sites to assist with allocating the correct pressure if required.

- Very importantly, you must ensure that accurate pressures can be measured across all regulators. It is little to no use trying to measure the pressure across a regulator with a manometer if one manometer branch is in the air stream and the air is travelling at say 30 or 40 m/s through the regulator. Given the huge change in the static pressure which will result because of the elevated velocity pressure such a scenario would likely destroy the success of the ventilation change. The author recommends that during this check phase it is wise to run fish tank or tube bundle tube through the regulator and leave it tied in the rib in still air and away from any direct velocity. Don't rely on tube that someone ran ten years ago without checking its integrity. The author has seen multiple times where tubes have been damaged due to the vibration of the air rubbing through the tube over extended periods. Whilst this sounds alarmist you simply cannot afford to make an error when setting your regulator pressures. They are absolutely crucial to the change process.

- Where doors need to be closed that have been open for extended periods close them one at a time to make sure they fully close to the frame. Any number of issues may stop them closing such as floor heave or slack, seized hinges etc so it is paramount to check they are all serviceable.

- Where stoppings need to be removed visit the site to ensure nothing has been left in the C/T that might prevent you from accessing it. You may need a machine to push it over so ensure both the machine will be available and access to the stopping face itself will be maintained then demarcate the area as required.

- Whilst the regulations in Queensland demand rated appliances are utilised it is not practicable and arguably impossible to exercise a complex ventilation change without using temporary ventilation appliances during the actual change process. Where temporary stopping frames such as the brattice/mesh/acro prop style below have been erected, which is normally a necessity during ventilation changes, ensure issues such as those labelled on Figure 20 are all addressed.

- Ensure you can access return doors, crossover conveyors etc. It is advisable to travel the route as you would during the actual ventilation change to eliminate unnecessary surprises on the day of the change.

- Finally, and importantly, it is a fact that when you get to the regulator there is virtually no chance whatsoever that the pressure on the regulator will be as the model predicted it to be. That said if you have an accurate model and you've planned the change correctly it should be close. The author specifies the % change that should be made across the regulator for this very reason. If the change process predicts the regulator pressure should be say 360 Pa and needs to be reduced by 10 per cent and when you get to the regulator and find 400 Pa across it, then the author would reduce that by 10 per cent, and keep moving.

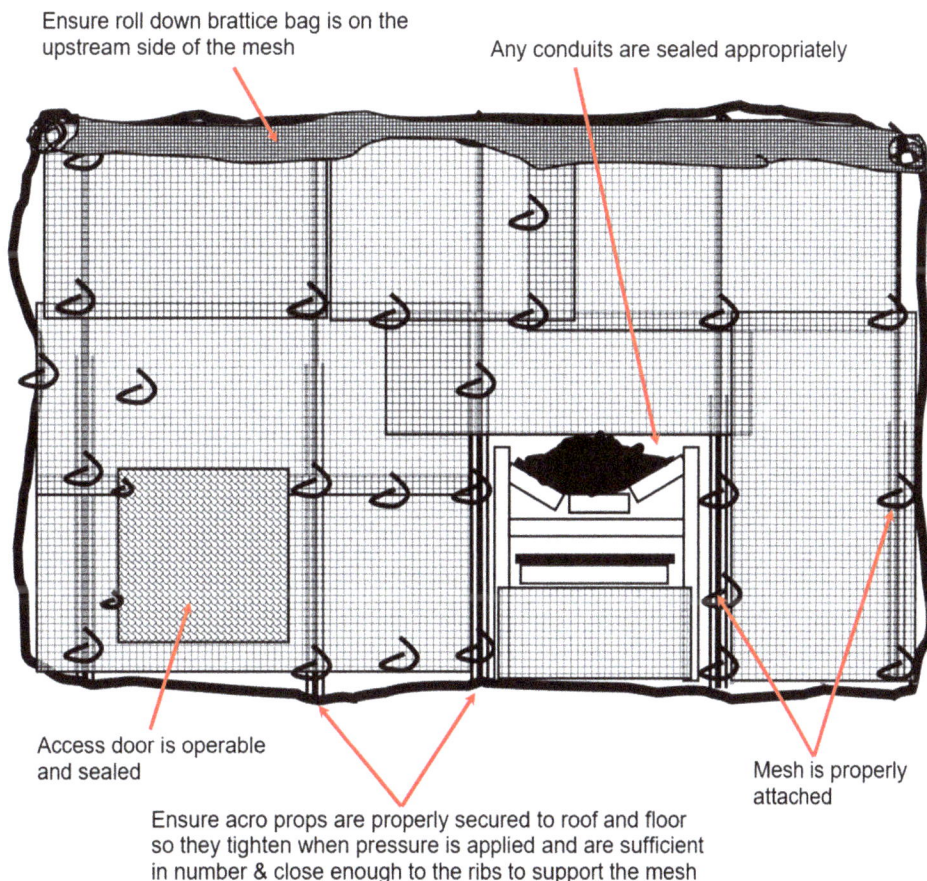

FIG 20 – Check temporary appliances are properly constructed.

Post ventilation change

Pursuant to completing the change there is generally a statutory requirement to remeasure flows at various locations specified in the legislation. Regardless of this requirement it is strongly advisable for the Ventilation Officer to resurvey the mine so that the new circuit can be accurately reflected on the model and the model tuned in line with the new circuit configuration. The measurements taken after the survey will never perfectly replicate the expected results because measuring velocities, areas and pressures in an operating coalmine is anything but an exact science. It is extremely useful and arguably mandatory to ascertain how well the planned execution of the change that was planned in the office mirrored the actual change result underground. Without analysing this it is impossible to quantify how well the model has performed, in delivering its result.

CONCLUSIONS

During any ventilation change in an underground mine the circuit is in a state of flux given the dynamic nature of the change itself. Parts of the mine are over ventilated at times and others are obviously under ventilated at times throughout the duration of the change. Whilst somewhat dependant on the criticality of the contaminants involved, it is still difficult to argue that the time taken to affect the change and restabilise the ventilation circuit in its new design shouldn't be reduced as much as possible. This is especially the case if one considers a gassy coalmine given the risk to the personal safety of the people employed if things don't go as planned. There are numerous examples that are well documented both in Australia and all over the world where ventilation changes have resulted in highly dangerous conditions being established underground, some of which have culminated in catastrophic disasters with the ensuing loss of life.

Thankfully most have not resulted in such a catastrophic conclusion but all these incidents may have if all the holes in the Swiss cheese causation model had been aligned on the day.

The author is well aware of a number of incidents that have resulted in explosive mixtures of methane being transported around the workings and exiting the mine through operating main fan sets without incident, but indeed with obvious catastrophic potential.

One of the tools in our armoury is the utilisation of ventilation modelling software with which to plan such changes so that risks during and after the change are minimised as far as reasonably practicable. There is little use using such software however if it is not well calibrated and thus can't be relied upon to deliver the expected result. It is entirely unacceptable for the police to book someone for speeding without a calibrated radar detector or fine someone for driving under the influence of alcohol without a properly calibrated breathalyser unit. Why would the proper use of and maintenance of a ventilation simulation software program be considered any differently?

Disappointingly, in our underground mines there is no statutory requirement to maintain a calibrated ventilation model and many the author has audited are in a very poor state of calibration when the model criteria is compared to the actual ventilation circuit performance underground. It is certainly in the interest of the Ventilation Officer to ensure that his model is kept in a calibrated state and his employer needs to provide the resources to enable him to do this. It is simply not acceptable to expect the tool to answer the questions asked of it if management isn't willing to commit the resources to maintain it adequately and consistently.

An accurately maintained ventilation model in tandem with the appropriate mine site ventilation knowledge and acumen from the Ventilation Officer is something that all mine sites should aspire to. If the model is properly maintained and provides consistently accurate results the ventilation officer will quickly learn just how valuable a tool it can be. Once the VO has faith in the performance of the model during such changes, he will likely nurture it more and continue to improve and rely on its performance under such duties.

It is an extremely satisfying experience, as all Ventilation Officers will attest, to execute a ventilation change which requires either no or minimal subsequent adjustments to circuit splits at the end of this change process.

If the change can be executed expeditiously and accurately then the circuit can be made available for the end users in the shortest practicable time frame thereby reducing downtime, improving safety, and increasing the mine operational uptime. That has to be a win, win, win.

The results of supplementary sampling to evaluate suitability of the instrumentation for monitoring of DPM concentration

S Sabanov[1], A Zeinulla[2], N Magauiya[3], N Kuzembayev[4] and M Torkmahalleh[5]

1. Associate Professor, School of Mining and Geosciences, Nazarbayev University, Nur-Sultan, 010000, Kazakhstan. Email: sergei.sabanov@nu.edu.kz
2. Undergraduate student, School of Mining and Geosciences, Nazarbayev University, Nur-Sultan, 010000, Kazakhstan. Email: aibyn.zeinulla@nu.edu.kz
3. Masters student, School of Mining and Geosciences, Nazarbayev University, Nur-Sultan, 010000, Kazakhstan. Email: nursultan.magauiya@nu.edu.kz
4. Masters student, School of Mining and Geosciences, Nazarbayev University, Nur-Sultan, 010000, Kazakhstan. Email: nursultan.kuzembayev@nu.edu.kz
5. Assistant Professor, School of Engineering, Nazarbayev University, Nur-Sultan, 010000, Kazakhstan. Email: mehdi.torkmahalleh@nu.edu.kz

ABSTRACT

Supplementary sampling of particulate mass concentration, the number concentration and the average particle diameter of nanoparticles using a TSI Dust-Trak DRX Aerosol Monitor, Naneos Partector 2 has been conducted. The aim of this study was to evaluate suitability of the instrument for long-term monitoring of diesel particulate matter (DPM) at conditions of polymetallic mines. Sampling was conducted in an area of the active ore heading, requiring the use of a diesel-powered face haul-age loader R1700 and a mine truck Caterpillar AD-45. The sampling was done for 126 minutes at a distance of 30 metres from the mine face. Results of this field work can be used to expand the sampling program to other mines. Expected outcomes of this study are to develop cost-effective DPM control strategies, improve mine ventilation, and protect mineworkers' health.

INTRODUCTION

Diesel engines are the main sources of hazardous gases (CO, CO_2, NO_x, SO_2, and hydrocarbons (HC)) and submicron particles known as diesel particulate matter (DPMs) in underground mines (Kurnia *et al*, 2014). All sources of DPM have the same physical characteristics, with the major elements being elemental carbon (EC) and organic carbon (OC), as well as numerous gases and aerosols created by incomplete combustion (Chang *et al*, 2019). However, in untreated exhaust, the connection between EC and OC fractions depends on engine running circumstances, engine type, fuel type, and several other factors (Bugarski *et al*, 2012). The total carbon (TC) value is obtained by summing the EC and OC numbers together, and it generally accounts for 80 per cent of the DPM (Kimbal *et al*, 2012). Since 20 May 2008, US Federal M/NM Mine Regulations have set a limit of 160 µg/m³ of total carbon (TC) for an eight-hour time weight average (TWA). Where TC is the sum of OC (organic carbon) and EC (elemental carbon) concentrations (OSHA-MSHA, 2013). Other surrogates for measuring DPM concentrations, such as black carbon (BC), PM1, and, most recently, LDSA, have been studied (Black, 2019; Volkwein *et al*, 2017; Huynh *et al*, 2019).

The amount of airflow in an underground mine is directly proportional to the diesel equipment's engine power and the unit airflow need (Halim, 2017; Kocsis and Hardcastle, 2002). Ventilation rate, diesel emission rate, exhaust flow direction, and drift face shape can all influence DPM dispersion (Zheng *et al*, 2015). It has been demonstrated that DPM residence duration in the air is influenced by its concentration and size; also, DPM effective density is substantially inversely governed by its size (Keskinen and Rönkkö, 2010; Wichmann, 2008; Maricq and Xu, 2004). The US Mine Safety and Health Administration (MSHA) released a particulate index (PI) for each permitted engine as a guideline for evaluating the optimal amount of ventilation required to manage DPM emissions. It has also created a guideline for observing control methods for DPM based on three factors: high power (more than 150 hp), high emissions (higher than 0.30 g/hp-hr), and high usage (greater than six hours each shift) (Kittelson, 1998; Haney and Saseen, 2000; Mischler and Colinet, 2009; Haney, 2012).

NIOSH 5040 method is an established technique to measure DPM concentration in underground mines. The NIOSH 5040 technique is an analytical method that measures EC and OC components of a DPM sample collected on a quartz fibre filter (Birch, 2002). However, NIOSH 5040, samples must be sent to a laboratory to be analysed with results coming back weeks later. Moreover, NIOSH 5040 only results in the average DPM concentration for an entire shift (160 µg/m³). The time and cost involved in the process make it a less practical method for mines that are looking to monitor the effectiveness of DPM controls. A real time method for measuring elemental carbon would give a mine the capability to evaluate where the highest concentrations of DPM are in the mine (Janisko and Noll, 2010; Noll and Janisko, 2007). Noll and Janisko (2013) examined a portable real-time DPM monitoring instrument FLIR Airtec which showed its effectiveness as results were equivalent to NIOSH 5040 method. Bugarski and Stachulak (2014) presented DPM threshold limit values (TLV) over countries (Table 1). The suitability of the TSI DustTrak DRX for DPM measurement was verified in underground conditions of polymetallic mines (Sabanov et al, 2021). The aim of this study was to evaluate suitability of the instrumentation used to monitor concentrations of DPM at a polymetallic metal mine operating in Kazakhstan. Sampling was carried out in an active ore heading region, necessitating the deployment of a diesel-powered face haulage loader R1700 and a Caterpillar AD-45 mine truck. Results of this field work will be used to expand the sampling program to other mines. Expected outcomes of this study are to develop cost-effective DPM control strategies, improve mine ventilation, and protect mineworkers' health.

TABLE 1

DPM threshold limit values (TLV) over countries (Bugarski and Stachulak, 2014).

Country	Threshold value	Surrogate
USA	160 µg/m³	TC
Canada		
Ontario	400 µg/m³	TC (or 1.3*EC)
Quebec	600 µg/m³	TC
Germany		
Tunnelling and non-coal mining	300 µg/m³	EC
Other than tunnelling and non-coal mining	100 µg/m³	EC
Other than tunnelling and non-coal mining when OC>EC	150 µg/m³	EC
Australia	100 µg/m³	EC

The Lung Deposited Surface Area Concentration (LDSA) is a different way of describing particle toxicity. The challenge with current mass-based exposure limits is that they fail to account for ultrafine particles (less than 100 nanometres), which can penetrate deep into human airways and lodge in lung alveoli. This is especially concerning because most diesel engine particle emissions (measured by number) fall into the ultrafine range (Braakhuis et al, 2014). In distinct metropolitan contexts, LDSA concentrations, size distributions, and height profiles have been measured in ambient conditions (Kuuluvainen et al, 2016; Kuula et al, 2020; Tran et al, 2020). LDSA concentrations have been connected to emissions from combustion sources. Afshar-Mohajer et al (2020) used a variety of measures, including LDSA, to investigate the variability of aerosol concentrations in different processing zones of a taconite mine. Huynh et al (2019) used numerous particle concentration measures, including LDSA, to investigate ambient fine particle concentrations in six taconite mines. LDSA concentrations in their investigation ranged from 50 to 300 µm²/cm³ depending on the processing region.

METHODOLOGY

Study site description

The experimental part of the research program was performed in the metalliferous mine operating in the Eastern Kazakhstan. Diesel exhaust sampling was done at the # level +335 m level which had

450 tons/day of total material moved. The mine is operating 365 days a year and operates in two 11-hour shifts each day, with two mid-shift blasting breaks. The mine employs a push-pull ventilation system with one primary exhaust fans and several auxiliary forced fan to give fresh air. The sampling station was placed in the Drift #30 while ore mucking, and haulage was performed in the Drift #31. Figure 1 shows location of the sampling station with haulage and drift lengths obtained from the map provided by the running company. The height of the Drift #30 was 5.2 m and width 5.5 m (Figure 1).

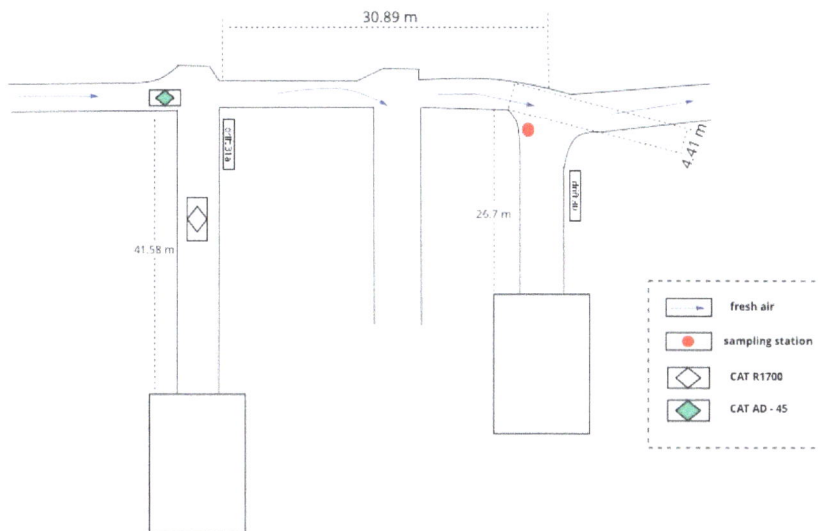

FIG 1 – Layout of the sampling station.

Sampling station was placed approximately one metre inside the Drift #30 due to safety concerns. Figure 2 indicates the location of the TSI DustTrak and Naneos Partector 2. The positions of the monitors correspond to the recommendations ie DustTrak was placed 90 cm above the floor and Naneos Partector 2 was placed in average mineworker's face level (160 cm). This arrangement of the monitors ensures effective collection of exhaust samples as due to buoyance DPM mainly accumulates at the roof region, which could possibly result in the misleading data (Zheng *et al*, 2017).

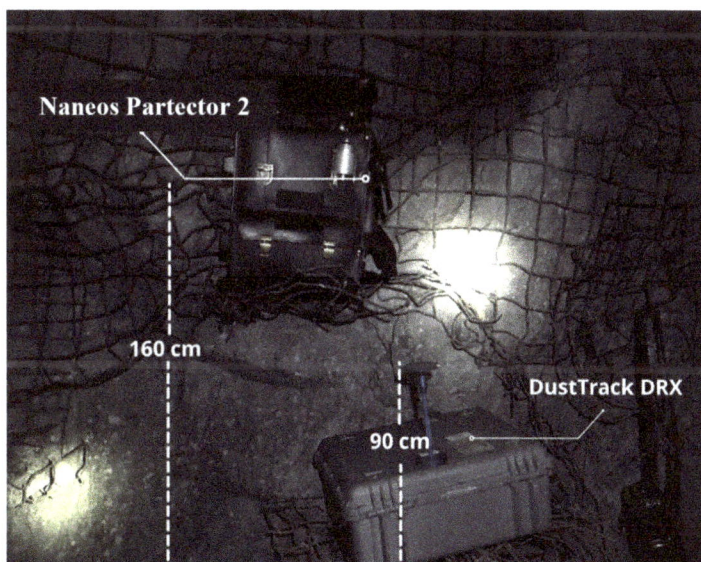

FIG 2 – Location of monitors.

Measurements were conducted in an area of the active ore heading, requiring the use of diesel-powered equipment: the loader Caterpillar R1700 and the truck Caterpillar AD-45. Caterpillar R1700 has an engine model Cat C11 ACERT with the engine power 241 kW, and Caterpillar AD-45 has engine model Cat C18 ACERT with the engine power 438 kW. The sampling was done for 126 minutes approximately 30 m away from the equipment in the downstream area. Mine environment conditions obtained from the anemometer are shown in Table 2.

TABLE 2

Mine environment conditions obtained from the anemometer.

Environmental conditions	Values
Mean temperature	29°C
Mean humidity	53%
Mean pressure	963 hPa
Airflow velocity	1.1 m/s

Sampling equipment

TSI DustTrak DRX's PM1 mass concentration results can be interpreted into TC. PM1 concentration is commonly thought to be used as a DPM level since it is the size range that encompasses practically all DPM (Bertolatti *et al*, 2011). The following linear connection was observed (Stephenson *et al*, 2006) based on an experiment using the NIOSH 5040 DPM sampling method and a TSI DustTrak real-time aerosol monitor:

$$y = 0.2316x + 32.699 \tag{1}$$

where:

y = TC by NIOSH 5040 method in µg/m^3.

x = DustTrak™ mass concentration in µg/m^3.

DPM mass concentrations were measured using a TSI DustTrak DRX Aerosol Monitor Model 8533EP. The TSI DustTrak DRX Aerosol Monitor detects both the mass and size fraction of aerosols at the same time. It contains a light-scattering laser photometer that enables for real-time monitoring of aerosols. DustTrak's logging interval for calculating average particle mass concentration was set to one minute and zero-check was conducted before the sampling.

The Naneos Partector 2 is the tiny multimeric nanoparticle detector which depicts lung deposited surface area (LDSA) and diameter of the particles. The Partector 2 employs twin non-contact detection stages to measure not only lung-deposited surface area, but also particle number concentration and average particle diameter. Given that the dispersion of diesel particles and mine dust varies from 3 to 500 nm and 1 to 20 nm, respectively (Bugarski *et al*, 2012; Haney and Saseen, 2000), the Partector 2 was used to compare emissions with the DustTrak. The collection tap was changed prior to the sampling and all necessary calibrations were done beforehand. Airflow and temperature measurements were taken by the VelociCalc® Air Velocity Meter.

RESULTS AND DISCUSSION

TSI DustTrak DRX results on particles concentration is summarised in Table 3.

TABLE 3

PM1 concentrations (TSI DustTrak DRX).

PM1 concentrations	µg/m^3
Average	3480
Minimum	160
Maximum	4460

It can be observed from Figure 3 that for the first 2500 seconds PM1 is not exceeding the obtained average 3480 µg/m^3 (red line) concentration. During this time approximately three load-haul-dump cycles were performed. After 2500 seconds PM1 values exceed the average value and mainly locates above it. This happened because the airflow velocity dropped to 0.9 m/s.

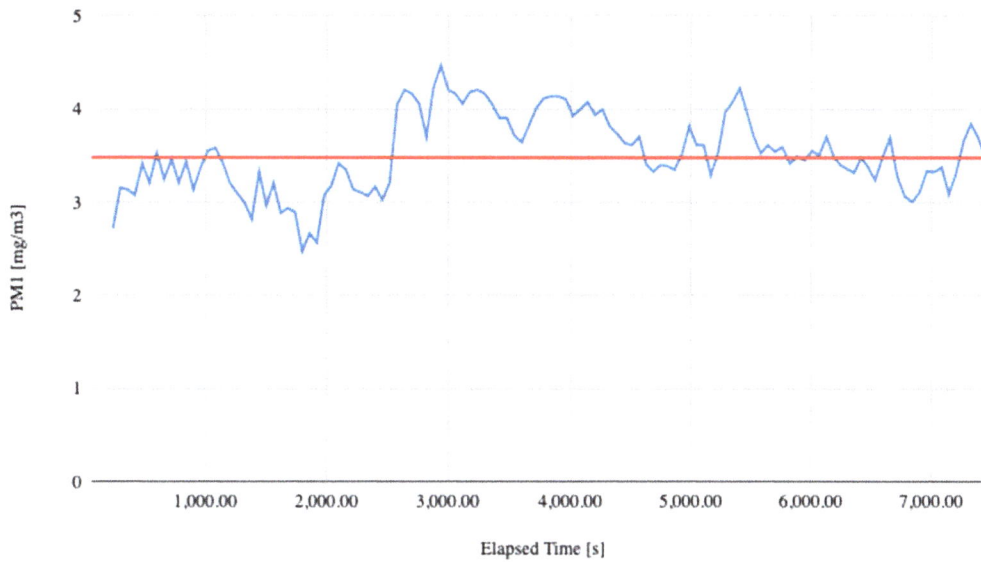

FIG 3 – TSI DustTrak PM1 concentrations versus time.

Using the Equation 1 the DustTrak mass concentration values calculated to TC concentration made 850.25 μg/m^3 in accordance with the NIOSH 5040 method.

Figure 4 depicts the distribution of particle's diameter gathered by Naneos Partector 2 and shows an average diameter of 119 nm.

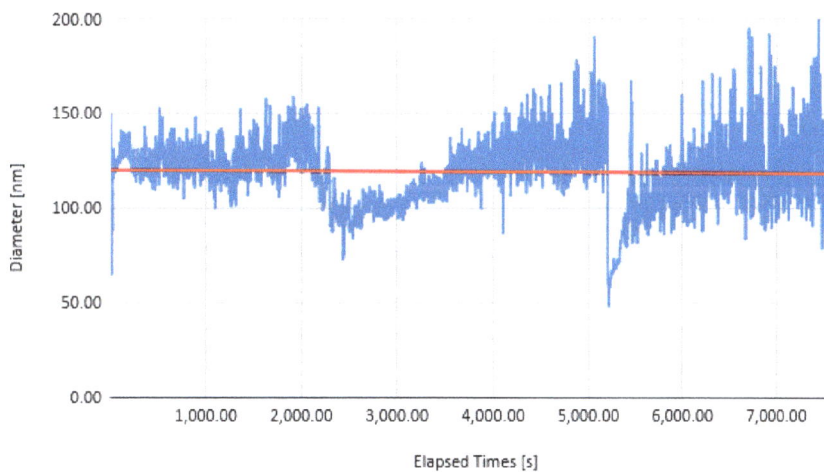

FIG 4 – Naneos Partector 2 particles diameter versus time.

Contaminants accumulation with an average value of 1284 μm^2/cm^3 indicated with the red line in Figure 5. Comparing to the research of Huynh *et al* (2019) and Afshar-Mohajer *et al* (2020) this concentration is quite high.

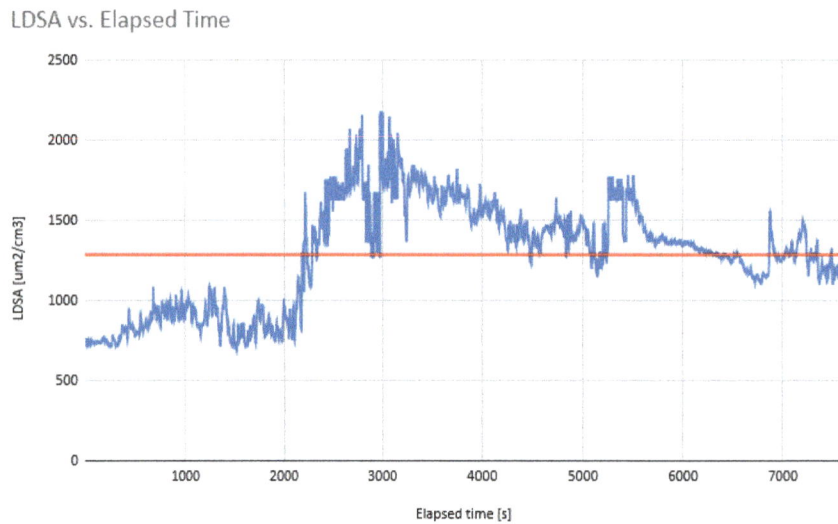

FIG 5 – Naneos Partector 2 LDSA concentration versus time.

Previous studies conducted by Sabanov *et al* (2021) using the TSI DustTrak DRX during 30 minutes in the underground metalliferous mine PM1 distribution shown in Figure 6 was received. The test was conducted near the Caterpillar R2900 during active load-haul cycles with the air velocity of 1.1 m/s. The movement of the Caterpillar R2900 loader powered by a C15 engine to and away from the measuring station caused the variability in concentrations at peak spikes in the range of 300–450 µg/m^3. However, in contrast with the experimental values obtained during this sampling program, PM1 concentrations are differ significantly. This could be due to the presence of the additional Caterpillar AD-45, the instrument arrangement and lower ventilation rate.

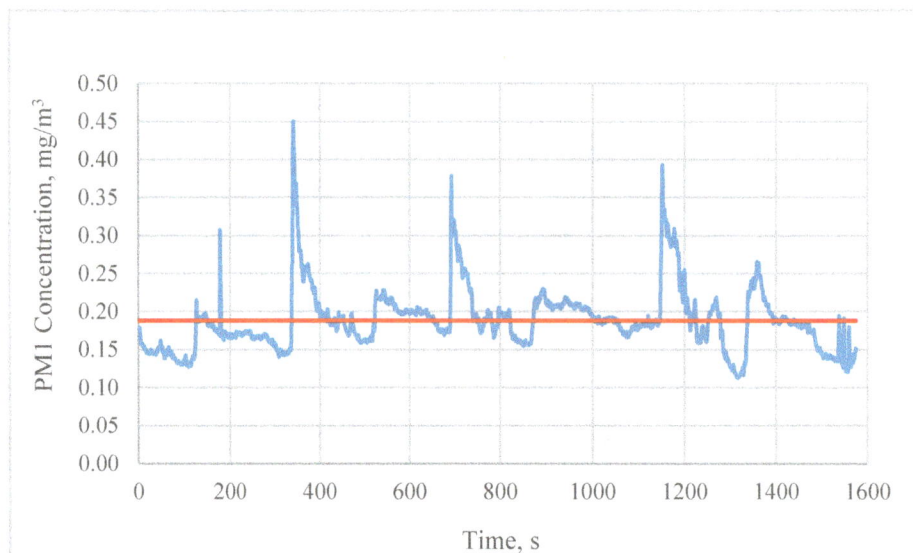

FIG 6 – TSI DustTrak PM1 concentration versus time (Sabanov *et al*, 2021).

This study presents a comparison of two different real time instruments and demonstrated a conversion factor between a mass based concentration and a particle surface area concentration (Figure 7).

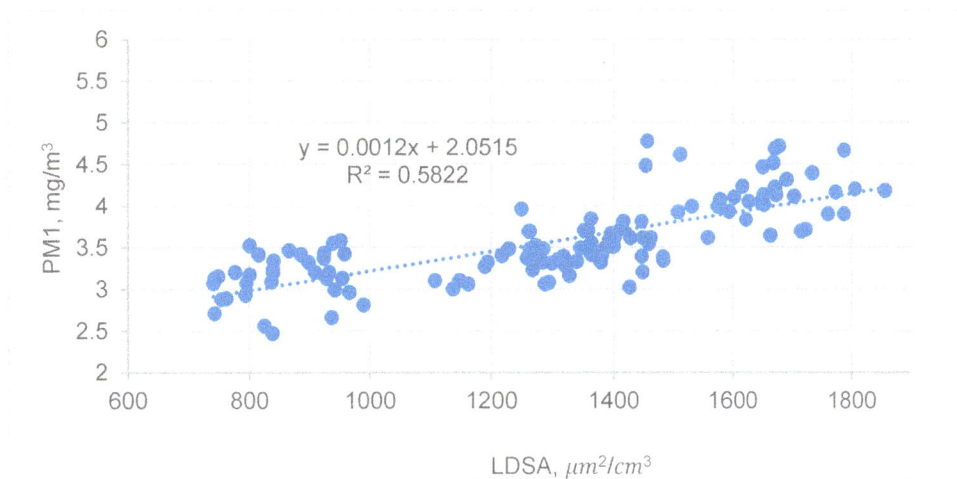

FIG 7 – A conversion between PM1 and LDSA.

The present study provided a useful equation to understand LDSA through PM1 concentrations.

CONCLUSION

Supplementary sampling of particulate mass concentrations, the surface area concentrations and the average particle diameter of nanoparticles using the TSI Dust-Trak DRX Aerosol Monitor and Naneos Partector 2 has been conducted. The aim of this study was to evaluate suitability of the instruments for long-term monitoring of diesel particulate matter (DPM) at conditions of polymetallic mines. Sampling was conducted in an area of the active ore heading, requiring the use of a diesel-powered face haul-age loader R1700 and a mine truck Caterpillar AD-45. The sampling was done for 126 minutes at a distance of 30 m from the working equipment and the average concentration of 850 μg/m^3 of TC was determined. LDSA readings from Naneos Partector 2 indicated concentration of 1284 μm^2/cm^3 and the distribution of particles shows an average diameter of 119 nm. These findings suggest that using the TSI DustTrak DRX Aerosol Monitor to undertake more comprehensive continuous monitoring of DPM concentration is a reasonable procedure to apply. Expanding the sampling program with the addition of LDSA as a DPM comparison parameter could also benefit in understanding of reading DPM. Long-term monitoring can help us understand how DPM acts in a specific location or after a ventilation change. Because of the short truck operation times, sampling was done only for a limited duration of time. Due to safety reasons, sample sites were positioned roughly one metre inside neighbouring drifts, potentially affecting the results. The findings from these field measurements will be utilised to expand the sampling program to other areas of the mine. The program's goals are to create cost-effective DPM control measures, improve mine ventilation, and preserve the health of mineworkers. This study provides useful information regarding particle surface area. Particle surface area could be a novel measure to understand the particle toxicity as the surface contact between the particles and human cells are critical. The present study provided a useful equation to understand LDSA through PM1 concentrations.

ACKNOWLEDGEMENTS

This study was supported by Nazarbayev University Grant Program: FY2020-CRP-1 Collaborative Research Project # 091019CRP2104.

REFERENCES

Afshar-Mohajer, N, Foos, R, Volckens, J and Ramachandran, G, 2020. Variability of aerosol mass and number concentrations during taconite mining operations, *Applied Occupational and Environmental Hygiene*, 17:1–14, https://doi.org/10.1080/15459624.2019.1688823

Bertolatti, D, Rumchev, K and Mullins, B, 2011. Assessment of diesel particulate matter exposure among underground mine workers, *WIT Transactions on Biomedicine and Health*, 15.

Birch, M E, 2002. Occupational Monitoring of Particulate Diesel Exhaust by NIOSH Method 5040, *Applied Occupational and Environmental Hygiene*, 17(6):400–405, doi:10.1080/10473220290035390

Black, S, Wilkinson, S, van den Berg, L and Manns, K, 2019. Tracer gas study of nano diesel particulate matter behaviour in secondary ventilation practices, *Mine Ventilation Journal*, pp 166–178.

Braakhuis, H M, Park, M V, Gosens, I, De Jong, W H and Cassee, F R, 2014. Physicochemical characteristics of nanomaterials that affect pulmonary inflammation, *Particle and Fibre Toxicology*, 11(1):18.

Bugarski, A D, Janisko, S J, Cauda, E G, Mischler, S E and Noll, J D, 2012. Controlling Exposure to Diesel Emissions in Underground Mines, pp. 280–326.

Bugarski, A D and Stachulak, J, 2014. Diesel Aerosols and Gases in Underground Mines, Workshop material on *The Australian Mine Ventilation Conference*, Sydney Australia.

Chang, P, Xu, G, Zhou, F, Mullins, B, Abishek, S and Chalmers, D, 2019. Minimizing DPM pollution in an underground mine by optimizing auxiliary ventilation systems using CFD, *Tunnelling and Underground Space Technology*, 87(Nov2018):112–121.

Halim, A, 2017. Ventilation requirements for diesel equipment in underground mines – Are we using the correct values?, in *Proceedings 16th North American Mine Ventilation Symposium*, Soc Mining, Metall Explor, pp 1–7.

Haney, R A and Saseen, G P, 2000. Estimation of diesel particulate concentrations in underground mines, *Mining Engineer*, 52:60–64.

Haney, R A, 2012. Ventilation requirements for modern diesel engines, in *Proceedings of 14th North American Mine Ventilation Symposium*, pp 249–256.

Huynh, T, Ramachandran, G, Quick, H, Hwang, J, Raynor, P C, Alexander, B H and Mandel, J H, 2019. Ambient Fine Aerosol Concentrations in Multiple Metrics in Taconite Mining Operations, *Annals of Work Exposures and Health*, 63(1):77–90, https://doi.org/10.1093/annweh/wxy086

Janisko, S and Noll, J D, 2010. Field Evaluation of Diesel Particulate Matter Using Portable Elemental Carbon Monitors, in *Proceedings of 13th United States/North American Mine Ventilation Symposium*.

Keskinen, J and Rönkkö, T, 2010. Can real-world diesel exhaust particle size distribution be reproduced in the laboratory? A critical review, *J Air Waste Manag Assoc,* 60:1245–1255.

Kimbal, K C, Pahler, L, Larson, R and Vanderslice, J, 2012. Monitoring diesel particulate matter and calculating diesel particulate densities using grimm model 1.109 real-time aerosol monitors in underground mines, *Journal of Occupational and Environmental Hygiene*, 9(6):353–361.

Kittelson, D B, 1998. Engines and nanoparticles: A review, *J Aerosol Sci,* 29:575–588.

Kocsis, C K and Hardcastle, S, 2002. Ventilation System Operating Cost Comparison between a Conventional and an Automated Underground Metal Mine, in *Society for Mining, Metallurgy and Exploration*.

Kurnia, J C, Sasmito, A P, Wong, W Y and Mujumdar, A S, 2014. Prediction and innovative control strategies for oxygen and hazardous gases from diesel emission in underground mines, *Science of the Total Environment*, 481(1):317–334.

Kuula, J, Kuuluvainen, H, Niemi, J V, Saukko, E, Portin, H, Kousa, A, Aurela, M, Rönkkö, T and Timonen, H, 2020. Long-term sensor measurements of lung deposited surface area of particulate matter emitted from local vehicular and residential wood combustion sources, *Aerosol Sci Technol,* 54:190–202. https://doi.org/10.1080/02786826. 2019.1668909

Kuuluvainen, H, Rönkkö, T, Järvinen, A, Saari, S, Karjalainen, P, Lähde, T, Pirjola, L, Niemi, J V, Hillamo, R and Keskinen, J, 2016. Lung deposited surface area size distributions of particulate matter in different urban areas, *Atmos Environ,* 136:105–113, https://doi.org/10.1016/j.atm osenv.2016.04.019

Maricq, M M and Xu, N, 2004. The effective density and fractal dimension of soot particles from premixed flames and motor vehicle exhaust, *Journal of Aerosol Science,* 35(10):1251–1274, doi: 10.1016/J.JAEROSCI.2004.05.002

Mischler, S E and Colinet, J F, 2009. Controlling and Monitoring Diesel Emissions in The United States, in *Proceedings of the Ninth International Mine Ventilation Congress* (D C Panigrahi, ed), 2:879–888.

OSHA-MSHA, 2013. Diesel Exhaust/Diesel Particulate Matter, December 2013, https://www.osha.gov/Publications/OSHA -3590.pdf

Noll, J and Janisko, S, 2007. Using laser absorption techniques to monitor diesel particulate matter exposure in underground stone mines, *Smart Biomedical and Physiological Sensor Technology*, vol 6759.

Noll, J and Janisko, S, 2013. Evaluation of a Wearable Monitor for Measuring Real-Time Diesel Particulate Matter Concentrations in Several Underground Mines, *Journal of Occupational and Environmental Hygiene*, 10(12):716–722, DOI: 10.1080/15459624.2013.821575

Sabanov, S, Zeinulla, A, Brune, J and Amouei Torkmahalleh, M, 2021. The results of the evaluations of the instrumentation used to monitor concentrations of DPM, in *Proceedings of 18th North American Mine Ventilation Symposium*, pp 133–138.

Stephenson, D J, Spear, T and Lutte, M, 2006. Comparison of sampling methods to measure exposure to diesel particulate matter in an underground metal mine, *Mining Engineering*, 58(8):39–42; 44–45.

Tran, P T M, Ngoh, J R and Balasubramanian, R, 2020. Assessment of the integrated personal exposure to particulate emissions in urban micro-environments: A pilot study, *Aerosol Air Qual Res*, 20:341–357, https://doi.org/10.4209/ aaqr.2019.04.0201

Volkwein, J C, Barrett, C, Sarver, E and Hansen, A D A, 2017. Application of an environmental 'black carbon' particulate sensor for continuous measurement of DPM in three underground mines, *Mine Ventilation Journal*, pp 143–149.

Wichmann, H-E, 2008. Diesel Exhaust Particles, *Inhal Toxicol*, 19:241–244.

Zheng, Y, Thiruvengadam, M, Lan, H and Tien C, J, 2015. Simulation of DPM distribution in a long single entry with buoyancy effect, *International Journal of Mining Science and Technology*, 25(1):47–52, https://doi.org/10.1016/j.ijmst.2014.11.004.

Zheng, Y, Lan, H, Thiruvengadam, M, Tien C, J and Li, Y, 2017. Effect of single dead end entry inclination on DPM plume dispersion, *International Journal of Mining Science and Technology*, 27(3):401–406.

Application of foam as a ventilation control for spontaneous combustion management in a longwall goaf

O Salisbury[1], G Linde[2], G Parker[3], B Williams[4], J Killerby-Smith[5], T Mackey[6], N Buchanan[7], B Beamish[8] and M Brady[9]

1. Technical Services Manager, Whitehaven Coal, Narrabri NSW 2390.
   Email: osalisbury@whitehavencoal.com.au
2. General Manager, Whitehaven Coal, Narrabri NSW 2390.
   Email: glinde@whitehavencoal.com.au
3. Operations Manager, Whitehaven Coal, Narrabri NSW 2390.
   Email: gparker@whitehavencoal.com.au
4. Mining Engineering Manager, Whitehaven Coal, Narrabri NSW 2390.
   Email: bwilliams@whitehavencoal.com.au
5. Senior Mining Engineer, Whitehaven Coal, Narrabri NSW 2390.
   Email: jkillerby@whitehavencoal.com.au
6. Ventilation Officer, Whitehaven Coal, Narrabri NSW 2390.
   Email: tmackey@whitehavencoal.com.au
7. Mining Engineer, Whitehaven Coal, Narrabri NSW 2390.
   Email: nbuchanan@whitehavencoal.com.au
8. Managing Director, B3 Mining Services Pty Ltd, Darra Qld 4076.
   Email: basil@b3 miningservices.com
9. Director, Joncris Sentinel Services, Rockhampton Qld 4700. Email: michael@joncris.com.au

ABSTRACT

Ingress of air into the longwall goaf environment needs to be managed appropriately to reduce the risk of creating a spontaneous combustion event. The maingate goaf edge is an area of particular concern in this respect, as air often penetrates for some distance into the goaf along this region inbye of the longwall face. Since the flow rate in this part of the goaf is also quite low, an opportunity exists to disrupt the air pathway by implementing a barrier to the airflow. Previous attempts using plugs made from an assortment of materials have resulted in varying degrees of success. More recently, Narrabri Coal Operations (NCO) has introduced the use of a chemical foam that is mixed with nitrogen to produce a standing barrier with the consistency of shaving cream. The implementation of this practice has consistently proven to be very effective for reducing the oxygen concentration along the maingate goaf edge by systematically retreating the foaming unit with the longwall. This paper presents how the foam is applied in the underground goaf environment to prevent oxygen ingress and uses results from gas monitoring to confirm the foam effectiveness.

INTRODUCTION

The occurrence of spontaneous heating development within a longwall goaf is closely associated with the airflow characteristics behind the face. Air leakage from the longwall to the goaf due to the existence of pressure differentials can create an ideal environment for coal oxidation to occur with minimal heat loss. A 'critical velocity zone' exists at some distance in from the goaf fringe and often this may occur along the maingate (MG) goaf edge. The preventive practice of proactive goaf inertisation using nitrogen has been adopted by some mines to overcome this issue and modelling has been used to help identify the best location for the nitrogen injection. Other mines have used physical barriers to the airflow path along the MG side of the goaf with varying degrees of success. This paper presents the mine site experience with using a more versatile chemical foam that is mixed with nitrogen to produce a flexible intact barrier to the airflow path along the MG side of the goaf.

MINE BACKGROUND

Narrabri Coal Operations (NCO) is located in the Gunnedah Basin 28 km south of Narrabri as shown in Figure 1. NCO exports a high energy thermal and mid volatile PCI coal with low ash, low sulfur and low phosphorus. Longwall operations commenced in 2012. General operating parameters are 9–10 m thick seam, coal strength 20–30 MPa, depth of cover 200–420 m, conglomerate unit 20–

30 m thick above the seam, 400 m wide longwall with three heading gate roads, and ventilation quantities of 450 m³/s at 4 kPa.

FIG 1 – Narrabri Mine location.

NCO coal has gas contents up to 14 m^3/t with Carbon Dioxide (CO_2) being the predominant seam gas, The coal also has a high propensity to spontaneous combustion, with R_{70} values in the range of 4–8°C/h. During the initial six years of longwall mining no spontaneous combustion events were detected. Since September 2018 NCO has successfully dealt with four spontaneous combustion events (Salisbury, Linde and Beamish, 2022). The third event was effectively brought under control through the injection of Nitrogen (N_2) foam (Coal Foam EXT-ECO) from underground to the hotspot site. This enabled longer term remediation of the site to be actioned in a staged manner that removed the risk to the mine and personnel. The consistency of the N_2 foam is similar to shaving cream as can be seen in Figure 2. The successful implementation of the N_2 foam prompted the mine to explore more sequential use of the Coal Foam – GOAF FOAM product, as attempts at using physical barriers along the MG side of the goaf for preventing oxygen ingress was not achieving the desired result.

FIG 2 – Photograph of N_2 foam from a pipeline feed showing stand-up capability.

APPLICATION OF NITROGEN FOAM TO PREVENT OXYGEN INGRESS

Prior to the 8 April 2021 the normal goaf ventilation was operating in longwall 109 with nitrogen injection at 16CT as shown in Figure 3. This ventilation mode still enabled oxygen penetration some considerable distance back into the goaf from the longwall face. On the 8 April 2021, a ventilation change occurred that placed MG109 A and B headings inbye of the longwall on return pressure, which reversed the pressure differential on the MG seals aimed at reducing air ingress into the goaf. At the same time the nitrogen injection was advanced to 13CT as shown in Figure 4. However, it can be seen that even with this change the oxygen front is still able to be measured for some distance back from the face along the MG seal side of the goaf.

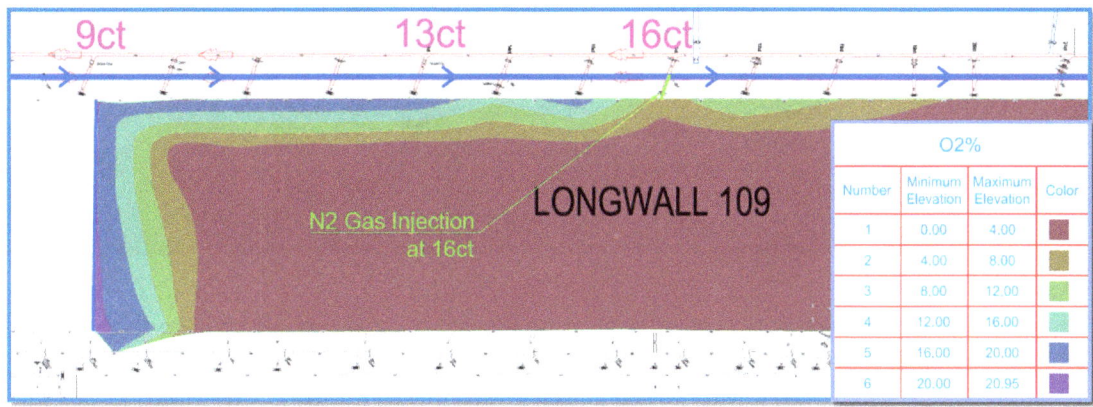

FIG 3 – Traditional inbye longwall ventilation.

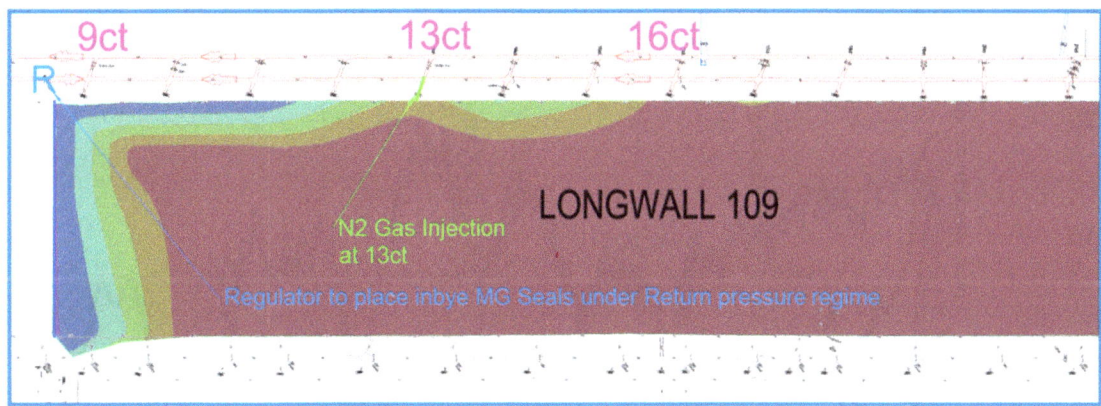

FIG 4 – Goaf seals on return pressure.

On the 27 April 2021, less than 1 ppm Ethylene (C_2H_4) was detected at MG109 14CT seal 400 m behind the longwall face. During the previous couple of weeks longwall retreat rates had been very slow due to mining encountering two unexpected faults running parallel to the longwall face. In line with the mine Trigger Action Response Plan (TARP) an Incident Management Team (IMT) was formed and event management commenced with:

- Involvement of external experts.

- Mobilisation of Mine shield, additional diesel N_2 generator and N_2 foam unit.

- Drilling of boreholes into the goaf.

- Longwall ventilation quantity reduced.

- N_2 foam injection at outbye seal site to reduce air flow along the MG rib line. As foaming continued, foam eventually migrated to the MG corner of the longwall and needed to be dispersed.

The ability to exclude air from the goaf was the key element in controlling this event. This was a combination of the ventilation circuit put in place prior to the event and the use of N_2 foam (Coal Foam – GOAF FOAM) in outbye seals. The GOAF FOAM was used in this particular application to provide a lasting foam barrier rather than the fire-fighting Coal Foam EXT-ECO FOAM. Subsequently, as the longwall retreated the foam injection point was relocated outbye to successive cut-throughs to maintain the integrity of the foam barrier behind the longwall face. The impact of the N_2 foam injection is shown by the dramatic decrease in Oxygen (O_2) levels at all of the MG seals (Figure 5). The time sequence of foam injection shown in Figures 6 and 7 for 12CT highlights that the impact of the foam injection is not immediate and it can take five to six days for the foam barrier to build to a sufficient level to resist the oxygen ingress into the goaf.

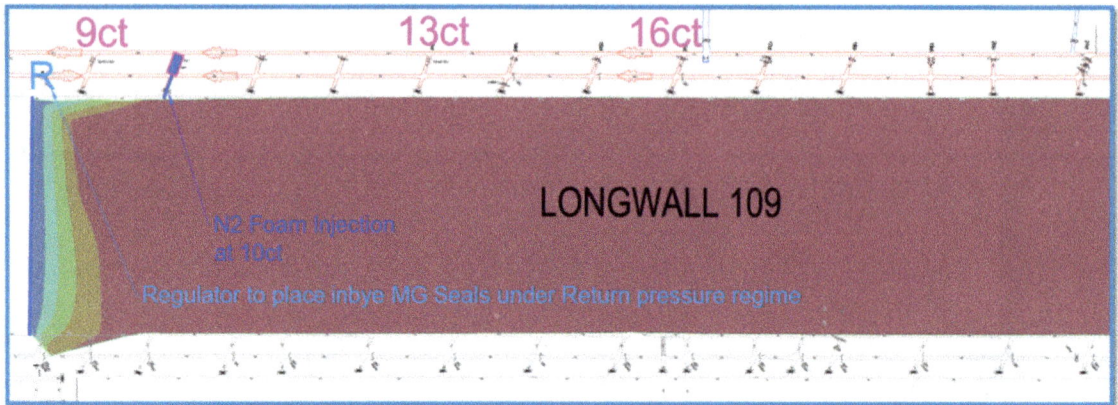

FIG 5 – Goaf seals on return pressure and N_2 foam injection.

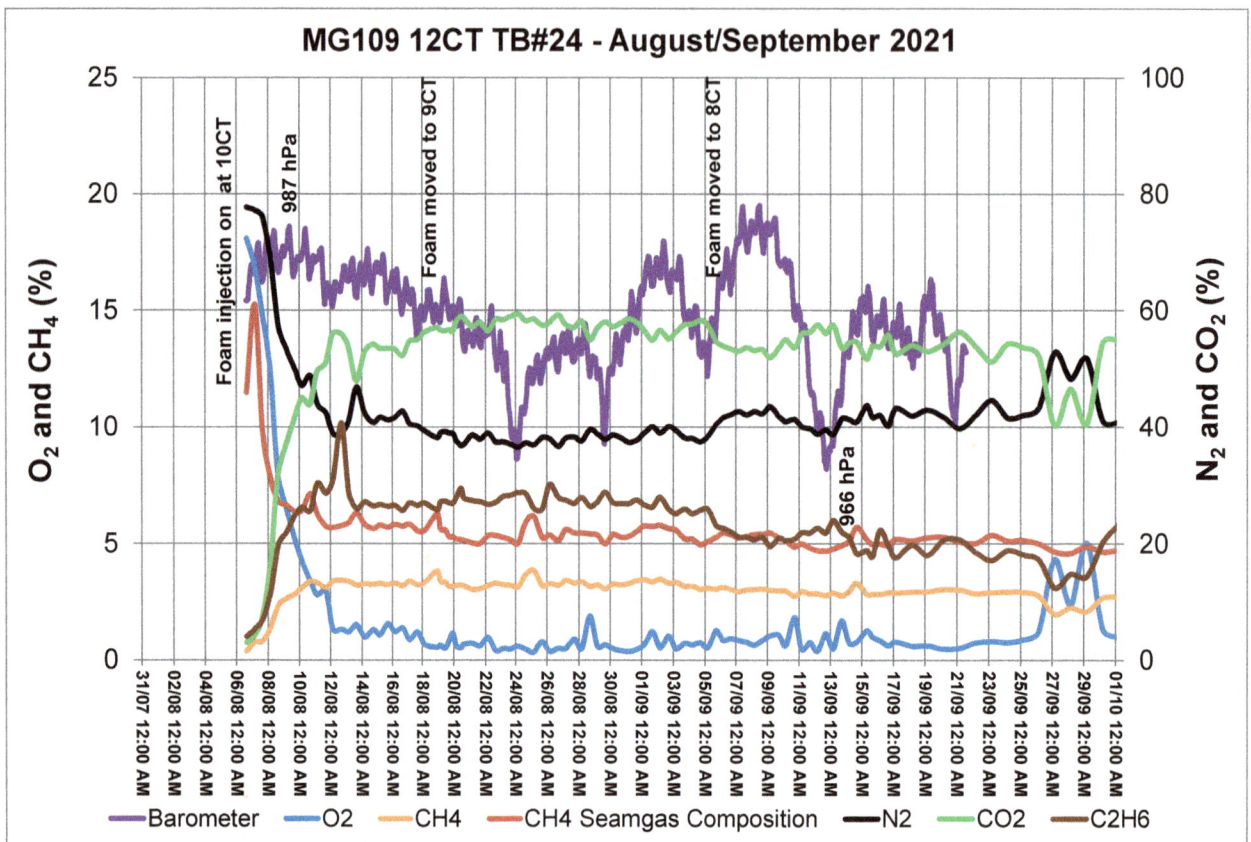

FIG 6 – Gas monitoring results for MG109 12CT showing drop in oxygen concentration.

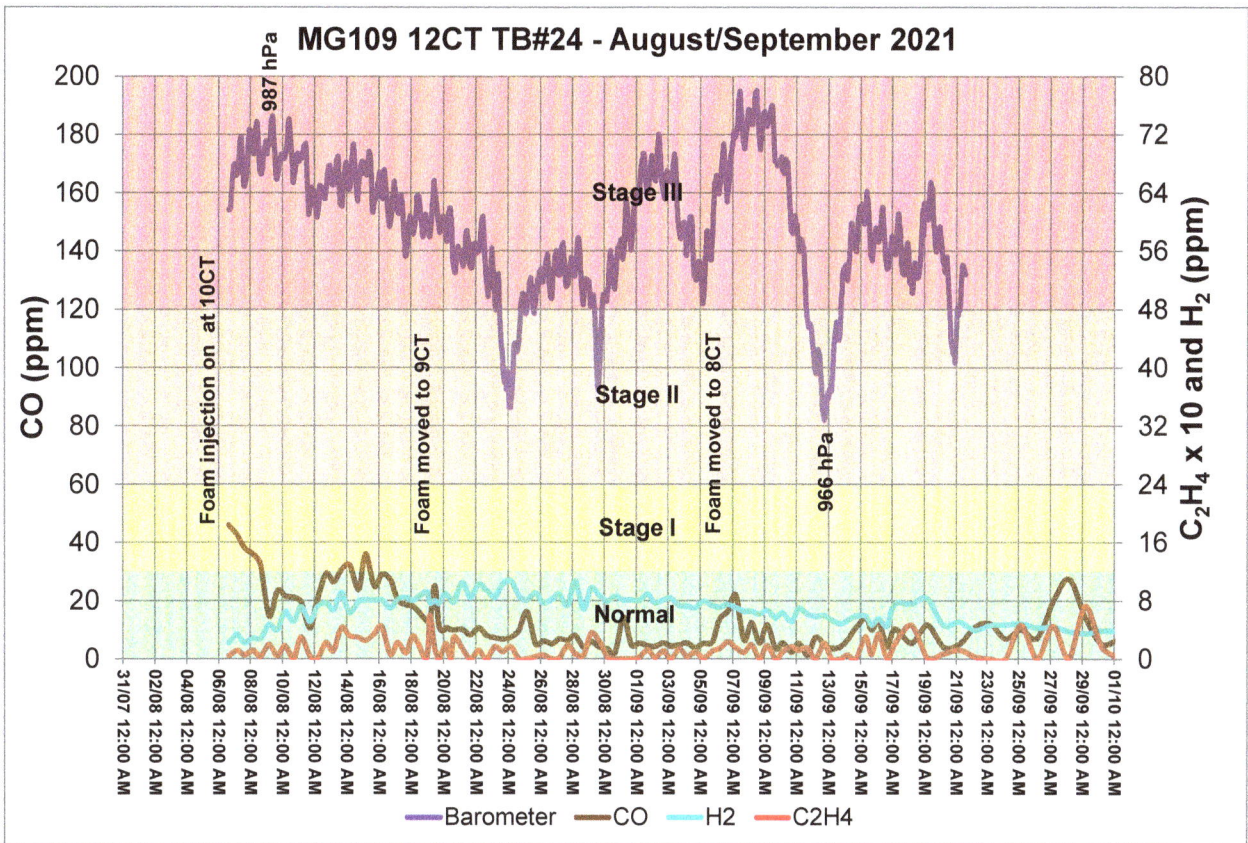

FIG 7 – Gas monitoring results for MG109 12CT showing drop in carbon monoxide concentration.

Whilst foaming operations were being undertaken tracer gas testing was carried out into LW109 goaf. At various times tracer gas was released into the goaf at the MG corner and surface boreholes. The results confirmed the effectiveness of the foam to stop air ingress along the MG goaf rib line as shown by Figure 8.

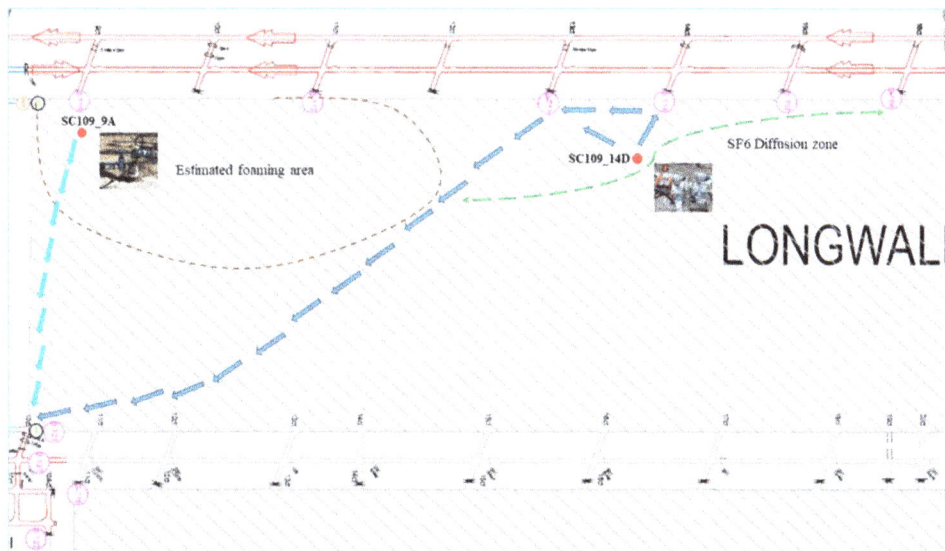

FIG 8 – Tracer gas flow paths.

CONCLUSIONS

The application of N_2 gas may slow the escalation of an event in an open goaf environment, but it will not extinguish a heating due to the opportunity to disperse throughout the goaf and flow to the tailgate return. When applied in a sealed goaf environment it is an effective control and management tool. Rigid barriers do not have enough flexibility to prevent air diverting around them into the goaf.

The use of N_2 foam, however, has a number of benefits as the foam confines the N_2 such that it can be directed to the hotspot. In addition, the foam creates an effective flexible barrier to the air pathway, cools the hot coal and releases N_2 as the foam eventually degrades.

REFERENCE

Salisbury, O, Linde, G and Beamish, B, 2022. Some operational perspectives on spontaneous combustion management of a longwall goaf, in *Proceedings of the 2022 Resource Operators Conference*, University of Wollongong, pp 145–152, https://ro.uow,edu.au/coal/830.

The experimental study on propagation of pressure wave in underground mine ventilation system

Y Shougu[1], V Tenorio[2], L Shugang[3] and Z Qinghua[4]

1. Associate Professor, Xi'an University of Science and Technology, Xi'an Shaanxi 710054, China. Email: yangsg@xust.edu.cn
2. Professor, University of Arizona, Tucson AZ 85721, USA. Email: vtenorio@email.arizona.edu
3. Professor, Xi'an University of Science and Technology, Xi'an Shaanxi 710054, China. Email: lisg@xust.edu.cn
4. Researcher, Chongqing Institute of China Coal Technology and Engineering Group, Chongqing 400037, China. Email: zqhcqmky@163.com

ABSTRACT

Many mine accidents will affect the mine ventilation system when they occur. At present, there is a lack of monitoring means to capture the occurrence of accidents in time by monitoring the ventilation system, so as to take emergency measures to reduce the loss of personnel and property. Pressure wave is an elastic wave existing in air. Most accidents in mine will cause obvious changes of pressure wave. It is an innovative idea to monitor the occurrence of accidents in mine by measuring pressure wave. The development and functions of the pressure monitoring device and related software are introduced in detail. The pressure wave monitoring test is carried out in San Xavier underground mining laboratory, and the D-wave and S-wave waveform data are collected. The waveform characteristics are analysed, and the main frequency distribution of the pressure wave is found to be within 1 Hz by using FFT. At the same time, the influence of the propagation distance of the pressure wave on the waveform is analysed. The propagation velocity of D-wave and S-wave are roughly calculated, and they are 279.6 m/s and 397.3 m/s respectively. S-wave is slightly faster than D-wave. Although there are errors and uncertainties in the test, the test verified the feasibility of the existence and collection of pressure waves in the ventilation roadway, and provides a method for the future study of the propagation of pressure waves in the mine ventilation system.

INTRODUCTION

Pressure wave is a kind of elastic wave existing in fluid. When the pressure changes in the fluid, the corresponding pressure waveform will be generated, and the pressure wave will propagate forward or backward along the fluid flow direction. Pressure wave was applied earlier to monitor the waveform of human blood in arteries, so as to diagnose the working condition of the heart (Lee, Soojeong, 2018)[1]. In the pipeline transportation industry, pressure wave is applied to the monitoring of pipe explosion accident and valve opening and closing (Cui Xiuguo *et al.*, 2010; Chen Qian *et al.*, 2017; GUO Xinlei *et al.*, 2011; Zhang Xuanyi, 2014)[2–5]. Some scholars also use pressure wave to monitor gas flow in pipelines (Huang Liang *et al.*, 2008)[6].

As a closed space, underground mine ventilation system is similar to pipeline transportation system. In the mine ventilation system, specific events will also produce obvious pressure wave phenomenon, which is generally manifested in the form of severe shock wave, such as coal and gas outburst, gas explosion, roof collapse and so on. Cao Jie *et al.*, (2018)[7] Conducted experimental research on the shock wave generated by coal and gas outburst, and concluded that the propagation speed of the shock wave was 405.36 m/s. Qu Zhiming, Hao Gangli and Wu Huige (2006)[8] Established the physical model and mathematical model for the propagation of shock wave generated during blasting operation in the driving roadway. Liu Chang, Qin Min and Peng Yun (2014)[9] pointed out that a large area of roof caving in the goaf can produce extremely destructive air shock waves, and analysed the characteristics of velocity and energy changes in the process of air shock waves from formation to extinction. The above results are aimed at the shock wave generated when the relevant mine disasters occur. Monitoring the pressure wave in the mine ventilation system can not only identify and locate the disaster events, but also monitor the operation of the mine ventilation system, which is of great significance to the mine safety production. In this paper, the pressure wave in the mine ventilation system is taken as the research object. Through the developed pressure wave monitoring device, the pressure wave generated when the damper is opened and closed is monitored in the San Xavier mine Lab at the University of Arizona, and the wave characteristics, frequency spectrum and propagation speed of the pressure wave are analysed.

THEORETICAL PRINCIPLES

Pressure Waves propagate in the air in two ways. One is compressional wave, which is caused by pressure alternative in the direction of propagation. Compressional wave also can be considered as the alternative of dynamical pressure, and follows the same direction of velocity of the air in a tunnel. Compressional wave is called D-Wave for short in this paper. D-Wave is same as alternative of velocity, and so it can be measured as differential pressure with a pitot tube. Another wave is transversal wave, which is pressure alternative at right angles to the direction of its propagation, also can be considered as the alternative of static pressure, so that it is called S-Wave for short. S-Wave can be measured with barometer.

The propagation velocity of pressure wave in pipeline medium can be calculated by the following formula (Liu Jun, 2015)[10]:

$$v = \sqrt{\frac{K}{\left(1+\frac{DK}{\delta E}\right)\rho}} \qquad (1)$$

that:

v	is the velocity of pressure wave, m/s
K	is volumetric elastic modulus of the media in the pipe, MPa
D	is the diameter of the pipe, m
δ	is the thickness of the pipe wall, m
E	is Elastic modulus of material for pipe wall, GPa

If the above formula is applied to the mine ventilation roadway, the transmission medium is air, without considering the influence of temperature change, and the volume elastic modulus of air is equal to the local atmospheric pressure. The pipe diameter is the equivalent diameter of the roadway. The thickness of the pipe wall is taken as infinity. The elastic modulus of the pipe wall material is taken as the elastic modulus of the surrounding rock of the roadway. According to the formula and the initial values, the propagation velocity of pressure wave in the tunnel is about the propagation velocity of sound with the corresponding conditions.

DEVELOPMENT OF THE TOOL

Development of pressure monitoring devices

The functions of pressure wave monitoring device include static pressure wave monitoring and dynamic pressure wave monitoring. During device development, MS5611–01 atmospheric pressure sensor and SDP810–125 differential pressure sensor are used respectively. In order to obtain the air temperature and humidity required to calculate the air speed, HTU21D sensor is also used. See Table 1 for the detail parameters of each sensor. Each sensor receives control commands and uploads data through I²C digital interface.

TABLE 1

Performances of sensors.

Sensor name	Operating range	Accuracy	Resolution	Update rate	Digital interface
MS5611–01	1–120 kPa	150 Pa	1.2–6.5 Pa	120–1800 Hz	SPI, I²C
SDP810–125	0–125 Pa	0.08 Pa	0.004 Pa	2000 Hz	I²C
HTU21D	0–100% RH (-40)-125°C	2% RH 0.3°C	0.04% RH 0.01°C	70–500 Hz 20–160 Hz	I²C

The microprocessor is MCU STC15F2S60, a type of single chip microcomputer made by SysTem Chip Ltd. In order to reduce the peripheral circuits, the development board is directly adopted here. The MCU of STC15F2S64 has 60K flash RAM, 2K SRAM, two UART interfaces, eight A/D ports, one SPI interface, and which running frequency is up to 28 MHz. The development board uses a UART port and ch340g chip to realise USB communication with the upper computer. In order to realise time synchronisation among multiple pressure wave monitoring devices, DS3231 clock module with I²C interface is also used here. The RTC module has a clock accuracy of ±5 ppm, that is, an hourly error of ±18 ms. Each electronic module uses I²C interface to communicate with MCU, and the device data circuit diagram is shown in Figure 1.

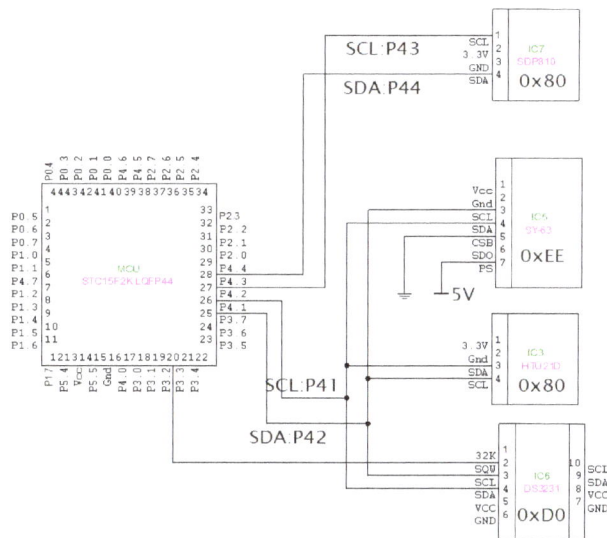

FIG 1 – Electronic diagrams of pressure monitor.

According to the circuit designed, connects each module to the MCU development board with jumpers to build a pressure wave monitoring device, as shown in Figure 2. Then the control code of the monitoring device is written, mainly including the control communication between MCU and each electronic module, data reading, data processing, data uploading and other functions. Since MCU does not provide I²C interface, code simulation is required. In the control code, the data reading and uploading cycle is set to 20 ms.

FIG 2 – Device of pressure monitor.

Development of data collecting software

The data acquisition software has been developed with c#.Net language, and the software interface is shown in Figure 3. The software mainly realises serial communication, real-time receiving data, calculating air velocity and dynamic displaying of curve. All data is stored in the Microsoft Access database in real time for upper data exporting and processing.

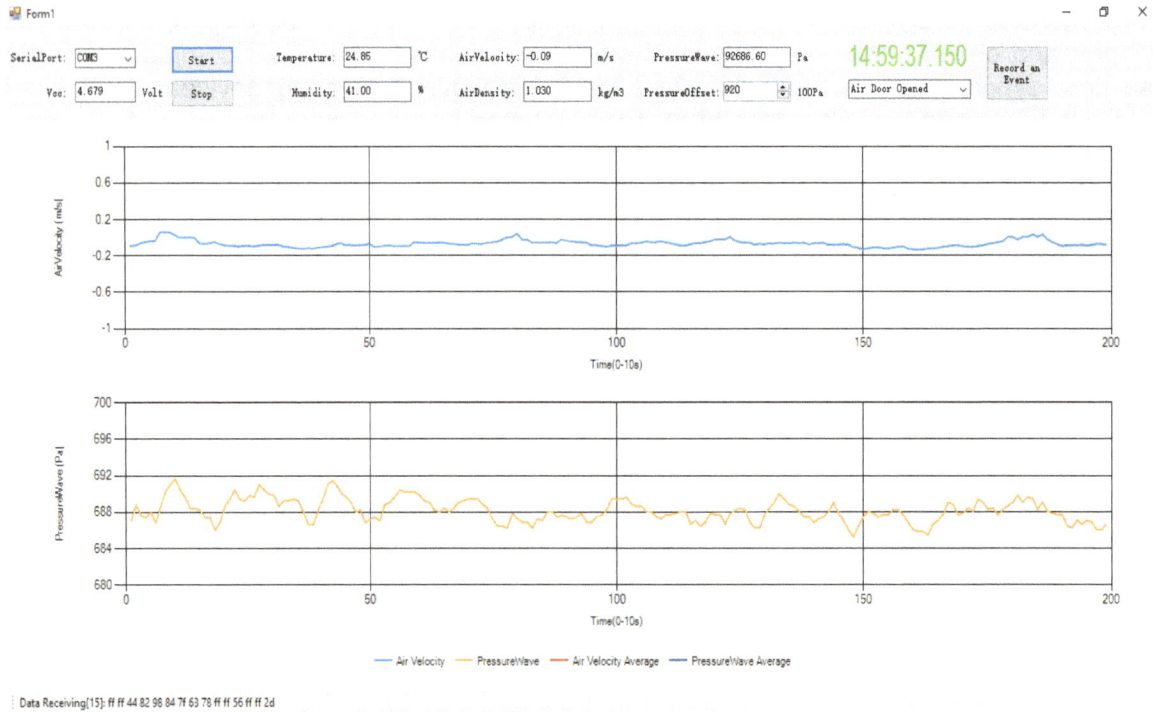

FIG 3 – User interface of data collecting software.

EXPERIMENTS OF PRESSURE WAVE MEASURING

Testing condition and points position

The field test was conducted in the San Xavier underground mining Laboratory at the University of Arizona. The Laboratory is a copper mine with a working vertical shaft and three levels which are adit level, 100 feet level and 150 level. The mine adopts positive forced ventilation. Fresh air is blown into the mine from VentRaise, enters 100 feet level through ScrewRaise, and then 150 feet level from LonyRaise. Two measuring points are arranged for this test. Measuring point 1 is arranged at the entrance of LonyRaise of 100 feet level, and measuring point 2 is arranged in the drift of 150 feet level, as shown in Figure 4.

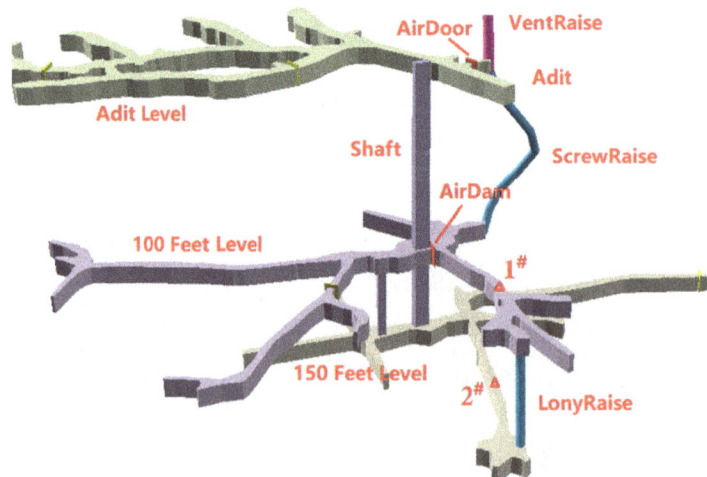

FIG 4 – Mine map and the location of measure points.

Pitot static tube is used for dynamic pressure monitoring. The pitot static tube is installed in the centre of the roadway section, and its head is facing the direction of air flow. The port of static pressure is connected to the negative input port of the differential pressure sensor, and the port of total pressure is connected to the positive input port of the sensor. The atmospheric pressure sensor is located on the circuit board and directly placed on the ground of the roadway to monitor the change of static pressure. As shown in Figure 5.

FIG 5 – Instillation of monitor device.

Testing procedures

The pressure wave is produced by the shock caused by the instantaneous change of pressure in the air flow. In order to cause such shock, this test uses the sudden closing of the air door, which located at the conjunction of the adit level and the VentRaise, to generate pressure wave. Firstly, open the air door, and part of the air flow flows through the air door to the adit level. The air quantity in the system decreases and the air flow pressure decreases. Then suddenly close the air door, the air flow to the adit level is suddenly cut-off, and the air pressure in the system rises sharply, causing a shock impact in the ventilation system. Under the action of such impact, dynamic pressure wave and static pressure wave will be generated. The pressure wave propagates forward at a speed far greater than the wind velocity and will be captured by the pressure wave monitoring devices at the measuring points, as shown in Figure 6.

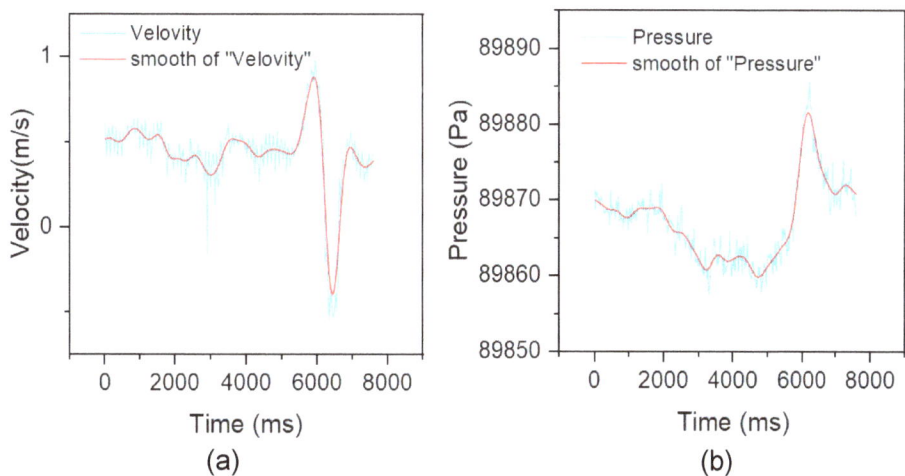

FIG 6 – Pressure waves produced by air-door shock; (a) compressional wave(D-Wave); (b) transversal wave(S-Wave).

DATA ANALYSIS

Analysis of wave characters

It can be seen from Figure 6 that when the air door is closed quickly, the wind velocity in the roadway increases sharply, reaches a peak and then drops to a low value far lower than the normal wind speed (even the wind flow is reversed), and then the wind speed returns to normal. Therefore, the D-wave caused by shock is a sine wave composed of a peak and a valley. At the same time, the static pressure in the roadway also rises sharply. After reaching the peak, it drops at the same speed, but only drops to a slightly lower normal value and then returns to normal. S-wave can be regarded as a half-sine wave with only one wave peak.

Comparing the D-waves and S-waves (shown in Figure 7) monitored at point 1 and point 2, it can also be found that the wave amplitude of D-waves at point 2 is greater than that at point 1, and the wave width at point 2 is greater than that at point 1. The amplitude of S-waves at point 2 is greater than that at point 1, and the wave width at point 2 is greater than that at point 1. It can be seen that the amplitude and width of pressure waves (D-waves and S-waves) will decrease with the increase of propagation distance.

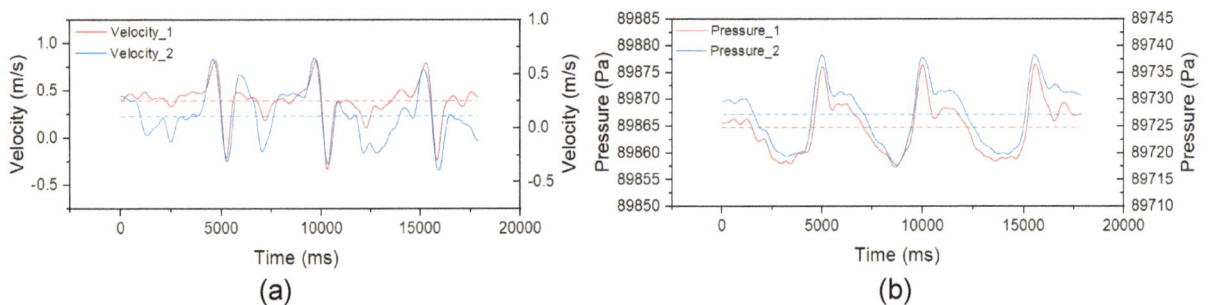

FIG 7 – Compare of Pressure Waves at Point1 and Point2: (a) D-Waves; (b) S-Waves.

Spectral analysis

According to the Fourier principle, any waveform can be combined of multiple sine waves, and also the pressure wave can be (Gao Lin *et al.*, 2017)[11]. The pressure wave signal is processed by fast Fourier transform to obtain the amplitudes of different frequencies, as shown in Figure 8. It can be seen from the figure that the main frequency of the pressure wave signal is concentrated below 1 Hz. With D-wave, the average amplitude of 0.2 Hz is 0.3753 m/s, the average amplitude of 0.4 Hz is 0.2613 m/s, the average amplitude of 0.6 Hz is 0.0903 m/s and the amplitude of other frequencies is less than 0.1 ms/s, which can be regarded as noise. With S-wave, the average amplitude of 0.2 Hz is 7.57 Pa, the average amplitude of 0.4 Hz is 2.64 Pa, the average amplitude of 0.6 Hz is 1.38 Pa and the amplitude of other frequencies is less than 1.0 Pa, which can be regarded as noise.

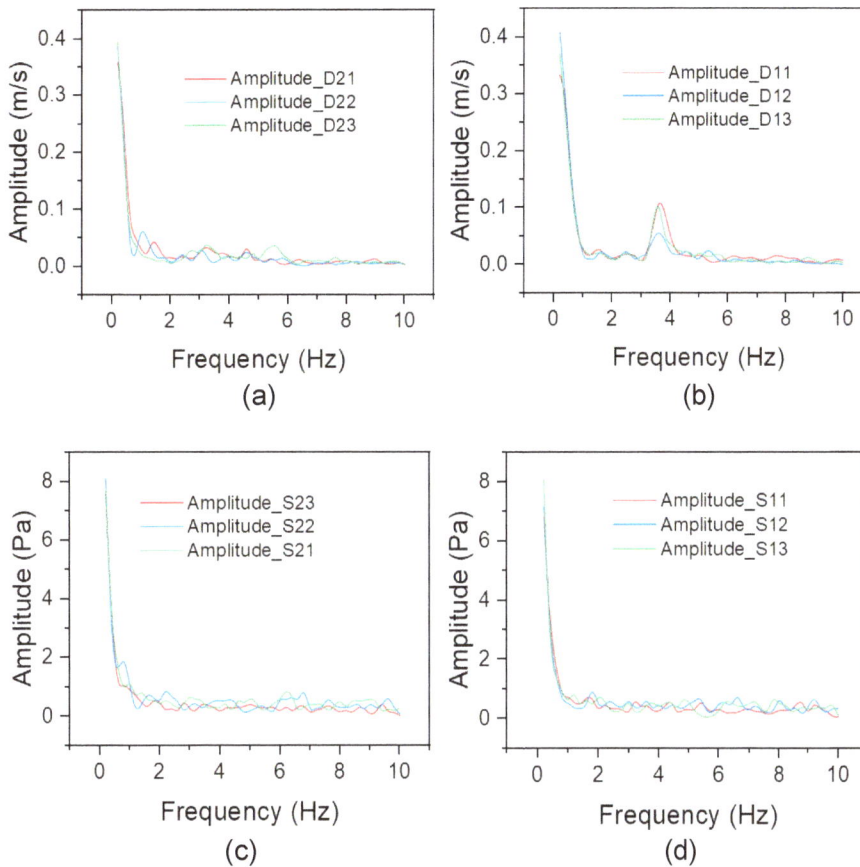

FIG 8 – FFT Result of Pressure Waves: (a) D-Wave at Point 2; (b) D-Wave at Point 2; (c) S-Wave at Point 2; (d) S-Wave at Point 2.

In addition, when the results of D-wave and the results of S-wave are drawn in the same figure respectively, it can be seen that the spectrum curve of D-wave at point 2 is basically located below the spectrum curve at point 1, while the spectrum curve of S-wave at points 1 and 2 is not largely distinguished. It shows that the waveform of D-wave is greatly affected by propagation distance, while the waveform of S-wave is less affected by propagation distance.

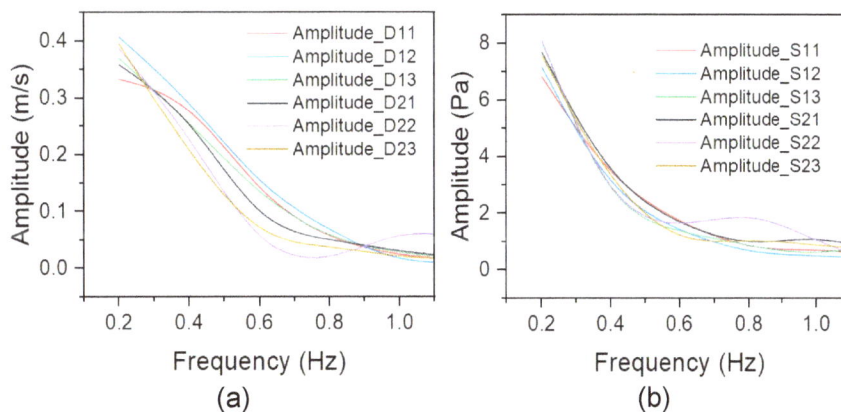

FIG 9 – FFT Result of Pressure Waves within 1 Hz: (a) D-Wave; (b) S-Wave.

Calculation of wave speed

According to Formula 1, the propagation velocity of pressure wave in the roadway is about equal to the sound velocity, 340 m/s. In this test, the measured distance from point 2 to air door generating pressure wave is 232 feet, and the distance from point 2 to point 1 is 80 feet. Using the time difference between the pressure wave from point 2 to point 1, the propagation speed of D-wave and S-wave is calculated to be 279.6 m/s and 397.3 m/s respectively. The calculated speed of S-wave is slightly faster than that of D-wave.

DISCUSS

Due to the influence of experimental conditions and the instrument accuracy, the experimental results may not be the response of real phenomena, and some phenomena may not be truly explained. In addition to the influence of atmospheric conditions, the pressure wave propagation is also affected by the roadway section roughness, cross-sectional area change, roadway elevation change, flatness and other factors. The amplitude and width of the pressure wave measured this time are compressed with the increase of propagation distance, but more tests need to be done under other conditions. At the same time, it is undeniable that during pressure wave propagation, new pressure wave will be generated by reflection, and its frequency is different from that of the original wave. As shown in Figure 8b, the frequency of the newly generated D-wave is about 3.5 Hz. In addition, when calculating the propagation velocity of pressure wave, the error caused by the acquisition frequency and the time synchronisation between the two measuring points may make the calculation result greatly different from the actual situation.

Although the accuracy of the test results is insufficient due to the above reasons. However, in this test, the complete D-wave and S-wave data were successfully collected, the characteristics of the waveform were obtained, and the main frequency distribution was obtained by FFT, which verified the existence and acquisition feasibility of the pressure wave in the ventilation tunnel, and provided ideas for the future research on the propagation of the pressure wave in the ventilation system. It also provides a new method for underground ventilation disaster monitoring.

CONCLUSIONS

This paper attempts to use pressure wave monitoring to monitor mine disasters and accidents. The development and functions of the pressure monitoring device and related software are introduced in detail. The pressure wave monitoring test is carried out in San Xavier underground mining laboratory, and the D-wave and S-wave waveform data are collected. The waveform characteristics are analysed, and the main frequency distribution of the pressure wave is found to be within 1 Hz by using FFT. At the same time, the influence of the propagation distance of the pressure wave on the waveform is analysed. The propagation velocity of D-wave and S-wave are roughly calculated, and they are 279.6 m/s and 397.3 m/s respectively. S-wave is slightly faster than D-wave. Although there are errors and uncertainties in the test, the test verified the feasibility of the existence and collection of pressure waves in the ventilation roadway, and provides a method for the future study of the propagation of pressure waves in the mine ventilation system.

ACKNOWLEDGEMENTS

The experiments had done in San Xavier Underground Mining Laboratory at University of Arizona during my visiting scholar period. I would like to thank Dr. Victor Tenorio, Dr. Moe Momayez, Steve Gravley, James Werner and Nathan Kraft for their extraordinary support in this process.

REFERENCES

[1] Lee, Soojeong, 2018. Combining Bootstrap Aggregation with Support Vector Regression for Small Blood Pressure Measurement. *Journal of Medical Systems*, 42(4):1–7.

[2] Cui Xiuguo *et al.*, 2010. Commercial Loop Test-based Transfer Characteristic of Startup Pressure Wave, *J Oil and Gas Storage and Transportation*, 29(07):491–493, p 474.

[3] Chen Qian *et al.*, 2017. Transient Simulation and Analysis of Gas Pipeline Leakage Based on Characteristics Method, *J Science Technology and Engineering*, 17(30):217–222.

[4] GUO Xinlei *et al.*, 2011. Transient Pressure Rise Method for Leak Detection in Pipeline Systems, *Journal of Basic Science and Engineering*, 19(01):20–28.

[5] Zhang Xuanyi, 2014. Testing and Locating of the Blockage in Water Supply Network Based on Pressure Wave, *Water and Wastewater Engineering*, 40(12):109–113.

[6] Huang Liang *et al.*, 2008. A New Measuring Method for Gas Flow in Pipeline Based on Pressure Wave, *J Machine Tool and Hydraulics*, 36(12):138–140.

[7] Cao Jie *et al.*, 2018. Simulation Research on Dynamic Effect of Coal and Gas Outburst, *Journal of China University of Mining and Technology*, 47(1):100–107.

[8] Qu Zhiming, Hao Gangli and Wu Huige, 2006. The Damage Effects on Ventilation Structures in Mine Excavation Roadway during Blasting Operation, *China Mine Magazine*, 15(4):83–87.

[9] Liu Chang, Qin Min and Peng Yun, 2014. Study on Time-history Characteristics of air Blast Wave due to Great Extent of Roof Instability, *Journal of Safety Science and Technology*, 10(5):49–55.

[10] Liu Jun, 2015. Continuous Pressure Fluctuation Propagation in Pipeline and Its Application in Measurement-While-Drilling and Hanger Pressure Test, *Harbin Institute of Technology*.

[11] Gao Lin *et al.*, 2017. Spectral Characteristic Simulation of Leakage Pressure Wave in Gas Pipeline, *Industrial Safety and Environmental Protection*, 43(01):65–68.

A coupled gas flow-geomechanical model based on energy analysis in coal and gas outburst simulation

F Soleimani[1], G Si[2] and H Roshan[3]

1. PhD candidate, School of Minerals and Energy Resources Engineering, University of New South Wales, Sydney NSW 2052. Email: f.soleimani@unsw.edu.au
2. Senior Lecturer, School of Minerals and Energy Resources Engineering, University of New South Wales, Sydney NSW 2052. Email: g.si@unsw.edu.au
3. Senior Lecturer, School of Minerals and Energy Resources Engineering, University of New South Wales, Sydney NSW 2052. Email: h.roshan@unsw.edu.au

ABSTRACT

Coal and gas outburst is a common hazard in coalmines and due to the complexity of its nature, efficaciously predicting the possibility of its occurrence is intricate. The mechanical and gas flow properties of coal, *in situ* and mining-induced stresses and geological structure are involved in outburst occurrence, whereas suggested thresholds for its prediction do not consider them all. Numerical modelling is known as an effective method that can not only consider couple the effect of different physics but also analyse the effect of associated parameters. In this paper, a coupled fluid flow and geomechanics model has been developed in COMSOL Multiphysics to simulate outbursts by considering the effect of gas pressure and composition, damage mechanics on related coal properties, and progressive excavation and mining-induced stress. For the proposed fully coupled outburst model, the effect of pore pressure and sorption are applied to the stress while the influence of stress changes and damage are considered in updating coal porosity and permeability in each time step. In addition, to study the effect of mining-induced stress, the excavation process is simulated by changing the material properties over time according to the schedule of heading development. Then, outburst initiation is analysed from the energy point of view by using the concept of energy conservation law. The simulation results are used in calculating energy sources of the outburst initiation, also, the effect of particle size and ejection velocity are studied, and the significance of particle size effect on the energy analysis and deformed coal on outburst initiation is discussed.

METHODOLOGY AND GOVERNING EQUATIONS

Coupling of the geomechanical and fluid flow equations

Coal and gas outburst happens in the underground coalmines and refers to the immediate emission of coal and gas during the mining activities. This is a complicated phenomenon due to the wide range of involved parameters. Numerical modelling is suggested as a method to study this phenomenon by considering all the involved physics. To that end, geomechanical and fluid flow behaviour in coal should be studied using equations in a double porosity medium, along with the sorption effect. Equation 1 is used to study the stress effect in which ε_s indicates the sorption induced stress and should be calculated using Equation 2. In Equations 1–2, σ_{ij} indicates the component of stress tensor, ε_{ij} indicates the component of strain tensor, ε_{kk} is the volumetric strain, G and λ are Lamé Parameters, δ_{ij} is Kronecker Delta, α is Biot Coefficient, and P represents pressure, K is the bulk modulus, and ε_L and P_L are the Langmuir volumetric strain and Langmuir pressure constant. Subscripts of g, m and f indicate gas, the coal matrix and fracture, respectively (Wilson and Aifantis, 1982; Wu *et al*, 2010):

$$\sigma_{ij} = 2G\varepsilon_{ij} + \lambda\varepsilon_{kk}\delta_{ij} - \alpha_m P_{g,m}\delta_{ij} - \alpha_f P_{g,f}\,\delta_{ij} - K\varepsilon_s\,\delta_{ij} \tag{1}$$

$$\varepsilon_s = \varepsilon_L \frac{P_m}{P_m + P_L} \tag{2}$$

The fluid flow in the fractures is studied through Equations 3–4, due to the higher permeability in the fractures compared to the matrix. Also, Fick's law of diffusion is used in the matrix by considering the summation of adsorbed and free gas as the total amount of gas (Equation 5) (Langmuir, 1918).

$$\frac{\partial(\rho_{g,f}\varphi_f)}{\partial t} + \nabla \cdot \left(\rho_{g,f}u\right) = (1 - \varphi_f)Q_m \tag{3}$$

$$u = -\frac{k}{\mu}\nabla P_{g,f} \tag{4}$$

$$m_{g,m} = \frac{M_W}{RT}P_{g,m}\varphi_m + \frac{V_L P_L}{P_L + P_{g,m}}\frac{M_W}{V_M}\rho_c \tag{5}$$

Q_m is the mass transfer between the matrix and fracture per unit volume expressed as the change in the gas mass in the coal matrix (Equation 6). In Equations 5–6, D_e is the effective diffusion coefficient, c is the concentration of free gas, φ is the porosity, M_W and V_M are gas molecular weight and volume, R is the gas constant, T is temperature, ρ_c is the coal density and V_L represents the Langmuir volume constant.

$$Q_m = D_e\left(c_{g,m} - c_{g,f}\right) = \frac{\partial m_{g,m}}{\partial t} \tag{6}$$

Equation 1 coupled stress with the fracture and matrix pressure. To couple the equations the other way around, the effect of stress on the porosity and permeability is considered using the equations Liu *et al* (2010) suggested. In this method, porosity and permeability are related to the cleats strain, and directional permeability is related to the directional strains using R_m (the elastic modulus reduction ratio) and considering the gas sorption ratio (Equations 7–8).

$$\frac{\varphi_f}{\varphi_{f0}} = 1 + \frac{3(1-R_m)}{\varphi_{f0}}(\Delta\varepsilon_{kk} - \Delta\varepsilon_s) \tag{7}$$

$$\frac{k_{f,i}}{k_{f,io}} = \sum_{i \neq j}\frac{1}{2}\left(1 + \frac{3(1-R_m)}{\varphi_{f0}}\left(\Delta\varepsilon_{jj} - \Delta\varepsilon_{sj}\right)\right)^3 \ where \ i,j = x,y,z \tag{8}$$

The effect of the damage variable (D) was added to Equation 8 by Zhu and Wei (2011) using a coefficient (α_k) to apply the damage degree effect on the permeability (Equation 9).

$$\frac{k_{f,i}}{k_{f,io}} = \sum_{i \neq j}\frac{1}{2}\left(1 + \frac{3(1-R_m)}{\varphi_0}\left(\Delta\varepsilon_{jj} - \Delta\varepsilon_{sj}\right)\right)^3 \exp(\alpha_k D) \tag{9}$$

To define tensile and shear damage variables, maximum tensile stress (Equation 10) and Mazars damage model (Equation 11) were used respectively (Zheng *et al*, 2018; Zhou *et al*, 2021). σ_{tr} is the residual tensile strength, ε_{tu} is the maximum tensile strain, ε_{t0} is the ultimate elastic strains of coal under tensile, A_c and B_c are damage evolution parameters, ε_f is strain threshold, ε^* is the equivalent strain under the general stress state which calculates using the principal strains in the three directions (Equation 12), κ indicates a state variable expressed as Equation 13.

$$D = \begin{cases} 0 \ (\varepsilon < \varepsilon_{t0}) \\ 1 - \frac{\sigma_{tr}}{E_0\varepsilon} \ (\varepsilon_{t0} \leq \varepsilon < \varepsilon_{tu}) \\ 1 \ (\varepsilon_{tu} \leq \varepsilon) \end{cases} \tag{10}$$

$$D = 1 - \frac{\varepsilon_f(1-A_c)}{\kappa} - A_c e^{-B_c(\kappa-\varepsilon_f)} \tag{11}$$

$$\varepsilon^* = \sqrt{\langle\varepsilon_1\rangle^2 + \langle\varepsilon_2\rangle^2 + \langle\varepsilon_3\rangle^2} \tag{12}$$

$$\kappa = \max(\varepsilon^*, \varepsilon_f) \tag{13}$$

Equations regarding outburst energy conservation

Being able to calculate the moment of the outburst initiation is of great importance in the study of the probability of the phenomenon and finding thresholds. The previous section, *Coupling of the geomechanical and fluid flow equations*, under *Methodology and governing equations*, focuses on the pre-initiation step and calculating the deformation and failure in the coal seam which can lead to the outburst occurrence. The result of the overmentioned equations is a step of the energy analysis of the event (Cai and Xiong, 2005; Gray, 1980; Hodot, 1966; Jiang and Yu, 1996; Li *et al*, 2012; Valliappan and Wohua, 1999; Wen *et al*, 2002; Zhao *et al*, 2016).

In the outburst phenomenon, there are energies that bring about its occurrence (herein supplier energies), and these energies are consumed after the outburst occurrence. The supplier energies refer to the elastic energy and gas expansion energy, and they will be consumed to crush the coal particles, transfer them, and there will be some energy loss (Lu et al, 2019a; Nie et al, 2019; Wu et al, 2020). Outburst can be expressed as Equation 14 from the energy viewpoint in which E_e is the elastic strain energy, and E_g represents the gas expansion energy, W_B, W_K and ΔE_g indicate the breakage, kinetic energy, and the residual kinetic energy of the gas.

$$E_e + E_g = W_B + W_K + \Delta E_g \qquad (14)$$

The stored energy that releases after any immediate changes in the stress state refers to the elastic strain energy and can be calculated as per unit volume of coal by Equation 15. In this equation, σ_1, σ_2, σ_3 are principal stresses, E is Young's modulus and ϑ represents Poisson's ratio (Nie et al, 2019; Wu et al, 2020).

$$E_e = \frac{1}{2E}(\sigma_1^2 + \sigma_2^2 + \sigma_3^2 - 2\vartheta(\sigma_1\sigma_2 + \sigma_2\sigma_3 + \sigma_1\sigma_3)) \qquad (15)$$

The other supplier energy is the gas expansion energy that is provided by the free and desorbed gas (Equation 16). Equation 17 calculates the free gas energy where P_{atm}, P_i and V_i are the mine atmospheric pressure, gas pressure and gas volume per unit mass, respectively. The process index (γ) is assumed 1.31 (as an adiabatic process) for the outburst due to the fast nature of the phenomenon. Equation 18 calculates the gas volume emitted per unit mass of coal during the outburst at the mine atmospheric pressure (V_i^*), in which ξ_i, T and T_0 are a pressure-dependent coefficient, the temperature before and after outburst occurrence, respectively.

$$E_g = E_{fg} + E_{dg} \qquad (16)$$

$$E_{fg} = \rho_c \sum_{i=m,f} \frac{P_{atm}V_i^*}{\gamma-1}\left[\left(\frac{P_i}{P_{atm}}\right)^{\left(\frac{\gamma-1}{\gamma}\right)} - 1\right] \qquad (17)$$

$$V_i^* = \frac{1}{\rho_c}\frac{P_i\varphi_i T}{P_{atm}T_0\xi_i} \qquad (18)$$

Equation 19 which can be applied in adiabatic processes, along with assuming that all the free gas is involved in the event, Equation 20 can be concluded from Equation 17 (An et al, 2019; Lu et al, 2019a).

$$P_{atm}(V_i^*)^\gamma = P_iV_i^\gamma \text{ where } i = m, f \qquad (19)$$

$$E_{fg} = \sum_{i=m,f}\frac{P_i\varphi_i}{\gamma-1}\left[1 - \left(\frac{P_i}{P_{atm}}\right)^{\left(\frac{1-\gamma}{\gamma}\right)}\right] \qquad (20)$$

The desorbed gas volume while the outburst occurs (V_t) determines the desorbed gas energy (Equation 21). The amount of desorbed gas is a function of coal particles' desorption rate (α_t) and the ultimate volume of the desorbed gas (V_∞). These two parameters can be calculated through Equations 22–25 (Bai and Elsworth, 2000; Busch and Gensterblum, 2011; Crosdale et al, 1998; Ruckenstein et al, 1971; Smith and Williams, 1984):

$$E_{dg} = \rho_c \frac{P_{atm}V_t}{\gamma-1}\left[\left(\frac{P_m}{P_{atm}}\right)^{\left(\frac{\gamma-1}{\gamma}\right)} - 1\right] \qquad (21)$$

$$V_t = \alpha_t V_\infty \qquad (22)$$

$$\alpha_t = 1 - \frac{6}{\pi^2}\sum_{n=1}^{\infty}\frac{1}{n^2}e^{-Dn^2\pi^2 t/r^2} \qquad (23)$$

$$\alpha_t = \frac{12}{d}\sqrt{\frac{D\,t}{\pi}} \qquad (24)$$

$$V_\infty = \left(\frac{V_L P_m}{P_L + P_m} - \frac{V_L P_{atm}}{P_L + P_{atm}}\right)(1 - M_{ad} - A_{ad}) \qquad (25)$$

If the diffusion process happens only in the micropores, Equation 23 can be derived along with assuming the diffusion happens over a short period through a homogeneous medium induces Equation 24. In these equations, r indicates the length of the diffusion path, D represents the diffusion coefficient, and d is the weighted average of coal particles after the outburst occurrence. M_{ad}, A_{ad}, V_L and P_L are moisture content, ash content, Langmuir volume constant and Langmuir pressure constant, respectively (Lu *et al*, 2015; Nandi and Walker, 1975; Ruckenstein *et al*, 1971; Zhao *et al*, 2016, 2020).

When the summation of all supplier energies reaches the least amount of energy to crush and move the coal particles, the surface theory can be used to calculate the breakage energy shown as W_B. Based on this theory, the energy to break the particles is equal to the energy required to generate new surfaces and relates the energy to the particles' size before (d_0) and after (d) the breakage (Equation 26).

$$W_B = C\left(\frac{1}{d} - \frac{1}{d_0}\right) \text{ where } C = 3A\psi \tag{26}$$

Equation 26, A represents the energy to enhance a unit of the fracture surface, and ψ is between 1.2 to 1.7, known as the uniform coefficient. In heterogeneous media, particles need to be categorised based on their size. The breakage energy would be equal to the weighted average of the required energy to break different groups.

In the outburst process, the step after the particles breakage is to supply the kinetic energy to eject them. In Equation 27, the minimum velocity to transfer the particles (V_{e_c}) is used to that end which is complicated to be found. Therefore, by assuming the transfer is only in the direction of inclination and categorising the particles into N different groups based on the ejection distance, Equation 28 can be used for the calculation. x_i and m_i represent the ejection distance and mass of each mass group, ϕ indicates the coal seam inclination angle, and f is the friction coefficient (Hodot, 1966; Nie *et al*, 2019; Wen, 2003; Wu *et al*, 2020; Zhao *et al*, 2016):

$$W_K = \frac{1}{2}\rho_c V_{e\_c}^{\ 2} \tag{27}$$

$$W_K = \sum_{i=1}^{N} x_i f(\cos\phi \pm \sin\phi)m_i \tag{28}$$

Some of the supplied energies will remain as the emitted gas energy that cannot transfer to the coal particles. Equation 29 calculates the amount of the gas flow energy loss (ΔE) using the mass of gas at the atmospheric pressure (m_g) and the critical velocity to carry the coal particles (v_c). In Equation 30 which calculates the critical velocity, $Q_{M,coal}$ indicates the mass flow rate of coal, and A_{cs} represents the roadway cross-section area (Nie *et al*, 2019; Xiong *et al*, 2009; Zhao *et al*, 2016):

$$\Delta E = \frac{1}{2}m_g v_c^2 = \frac{1}{2}V_{g,atm}\rho_{g,atm}v_c^2 \tag{29}$$

$$v_c = \frac{Q_{M,coal}}{A_{cs}\rho_c} \tag{30}$$

NUMERICAL MODELLING

In this study, COMSOL Multiphysics software is used to solve the coupled equations using the finite element method (FEM). A 3D model with two layers of rock ($100 \times 100 \times 20$ m³ each) and a layer of coal ($100 \times 100 \times 4.8$ m³) in the middle is generated. A 30 m long roadway is considered in the coal layer, and the roadway is excavated as 15×2 m length segments to accelerate the running time and increase the model convergence. Three of these segments (18–24 m) have different properties (Table 1) to be considered deformed coal with a higher chance of outburst occurrence (Figure 1).

TABLE 1

Parameters used in the model.

Parameter	Normal coal	Deformed coal	Rock	
Bulk modulus of matrix (GPa)	10	1.07		An *et al*, 2013
Young's modulus of body (GPa)	4.054	0.8	30	Lu *et al*, 2019b
Initial fracture porosity	0.01	0.005		Lu *et al*, 2019b
Initial matrix porosity	0.035	0.05		Lu *et al*, 2019b
Viscosity of the gas (Pa.s)	1.84E-5	1.84E-5		An *et al*, 2013
Poisson's ratio	0.26	0.35	0.3	Lu *et al*, 2019b
Bulk density (kg/m^3)	1500	1300	2500	Lu *et al*, 2019b
Maximum sorption-induced strain	0.011	0.0301		Lu *et al*, 2019a
Langmuir-type pressure of sorption strain (MPa)	5.14	4.125		Lu *et al*, 2019a
Maximum gas adsorption capacity (m^3/kg)	5.28E-02	4.71E-02		Lu *et al*, 2019a
Adsorption constant (1/Pa)	1.36E-06	1.32E-06		Lu *et al*, 2019a
Effective diffusion coefficient (1/s)	2.26E-06	3.79E-05		Lu *et al*, 2019b
Adsorption time (d)	11.7	5.85		Lu *et al*, 2019b
Initial permeability (m^2)	1.28E-18	6.42E-16		Lu *et al*, 2019b
Strain threshold	0.005	0.005		Zhou *et al*, 2021
Damage evolution parameter (A_c)	1.2	1.2		Zhou *et al*, 2021
Damage evolution parameter (B_c)	1000	1000		Zhou *et al*, 2021

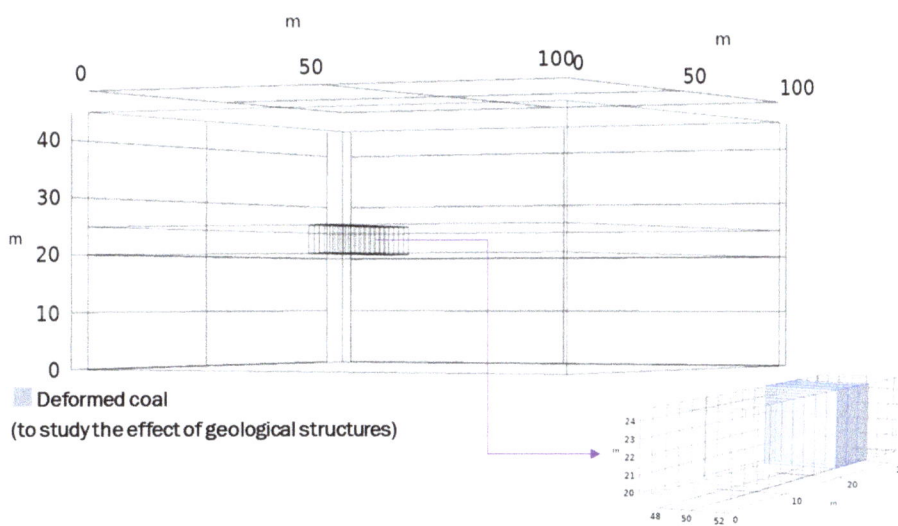

FIG 1 – Schematic of the geometric model.

For each segment, a stationary study is conducted to solve the stress-strain equation through the solid mechanic module of COMSOL, and then a time-dependent fluid flow model (based on the mining rate which is 4 m/h in this case) is done for each segment to calculate the pressure change.

Figure 2 shows the boundary conditions applied in the solid mechanic study part, in which the lower side is a fixed boundary, the back, right, and left sides are roller, the front side is symmetric, and

10 MPa overburden pressure is applied on the upper side. In the fluid flow study, all sides are at the zero-flux condition, and the roadway faces are at the atmospheric condition after the excavation.

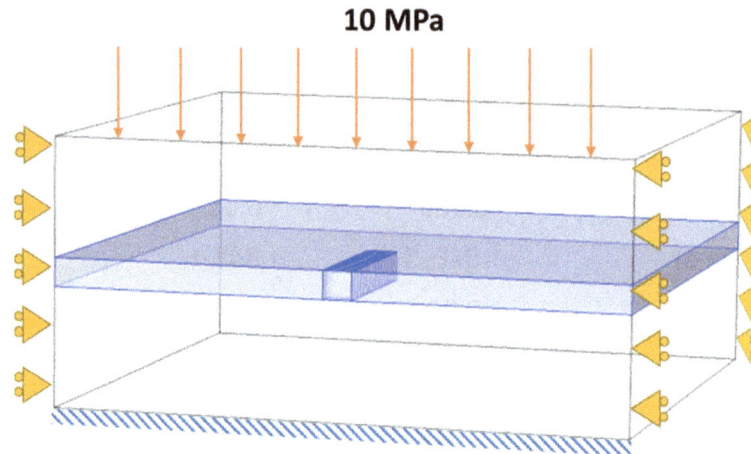

FIG 2 – solid mechanics boundary condition.

The initial pressure for fluid flow calculations is 0.5 MPa, and the stress initial condition is based on the *in situ* stress calculations of the numerical model.

RESULTS AND DISCUSSION

After simulating the discussed scenario, the results of fracture pressure, matrix pressure and horizontal stress are as Figures 3–5. The different stress distribution in the deformed coal is considerable in Figure 5. The results were used to calculate the elastic, desorbed and free gas energies to determine the possibility of the outburst occurrence. Before starting mining the coal seam, the contribution of each supplier energy was as Figure 6. As this figure shows, the role of gas expansion energy and more specifically the desorbed gas energy is more significant in the deformed coal.

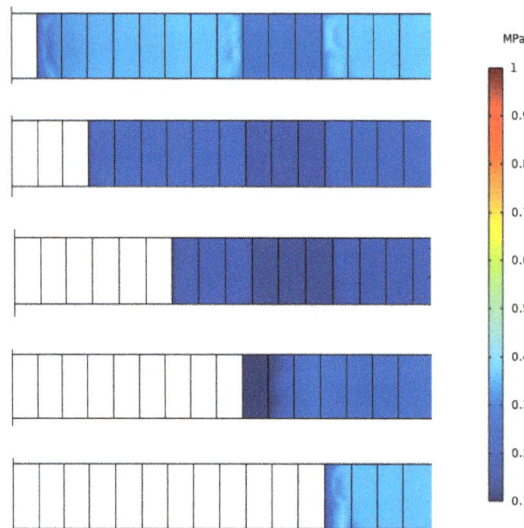

FIG 3 – The fracture pressure distribution after 2, 6, 12, 18 and 24 m of excavation.

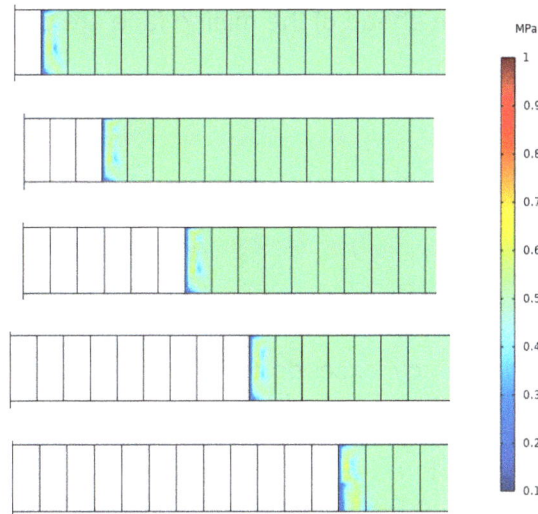

FIG 4 – The matrix pressure distribution after 2, 6, 12, 18 and 24 m of excavation.

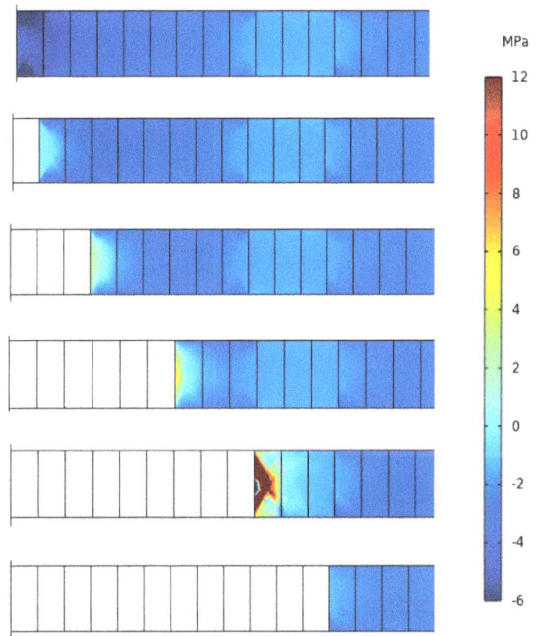

FIG 5 – The distribution of horizontal stress in front of the roadway after 0, 2, 6, 12, 18 and 24 m of excavation.

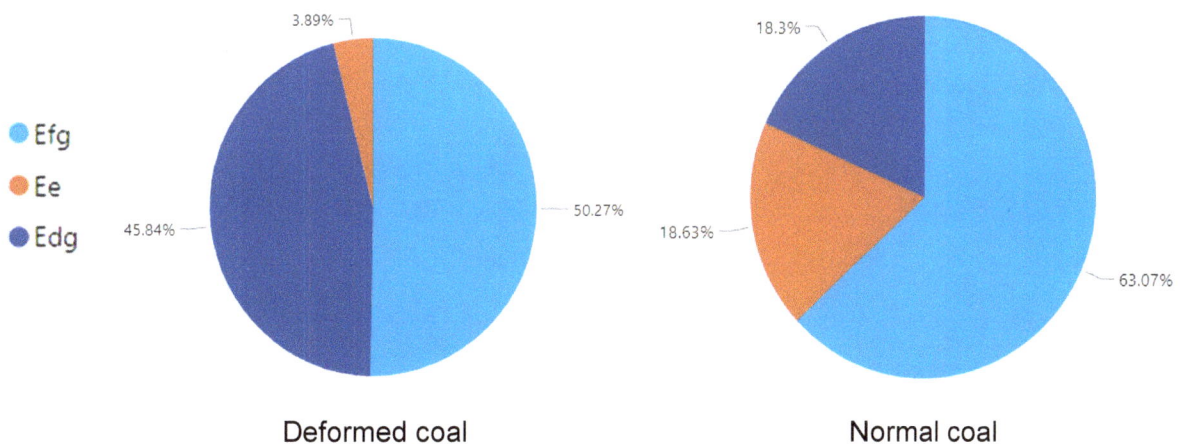

FIG 6 – Comparison of supplier energies values after 2, 12, 20 and 22 m of excavation.

After the excavation starts the new stress and pressure distribution changes the contribution of the stress values. The energy values 2 m ahead of the mining face are categorised as:

1. 0–0.1 MJ/m^3

2. 0.1–0.2 MJ/m^3

3. 0.2–0.5 MJ/m^3

4. 0.5–0.7 MJ/m^3

5. 0.7–1 MJ/m^3

6. 1–2 MJ/m^3.

As Figure 7a shows, the elastic energy exceeds the desorbed and free gas energies when the excavation gets close to deformed coal. This indicates the importance of mechanical properties in the outburst phenomenon (which is aligned with the high-stress values observed in Figure 5). The comparison of the summation of all three energies with the consumer energies indicates that the increase of the elastic energy leads to the outburst occurrence (the moment that the elastic strain and gas expansion energy exceeds the breakage and kinetic energy).

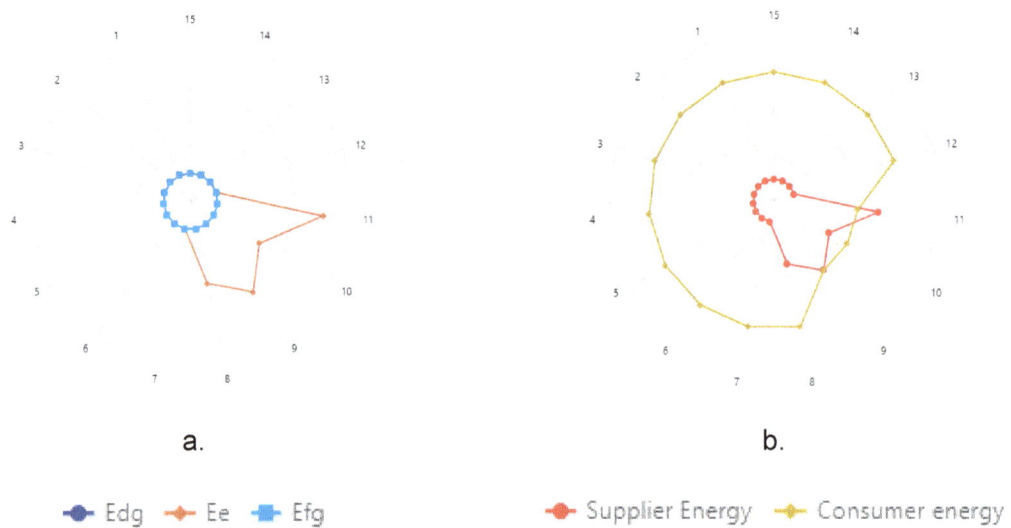

a.   b.

FIG 7 – Comparison of involved energies in the outburst based on the number of mined segments.

The comparison between outburst prone zone (red zone in Figure 8) and the damaged area indicates that the larger the damaged area, the more chance of outburst initiation (Figure 8).

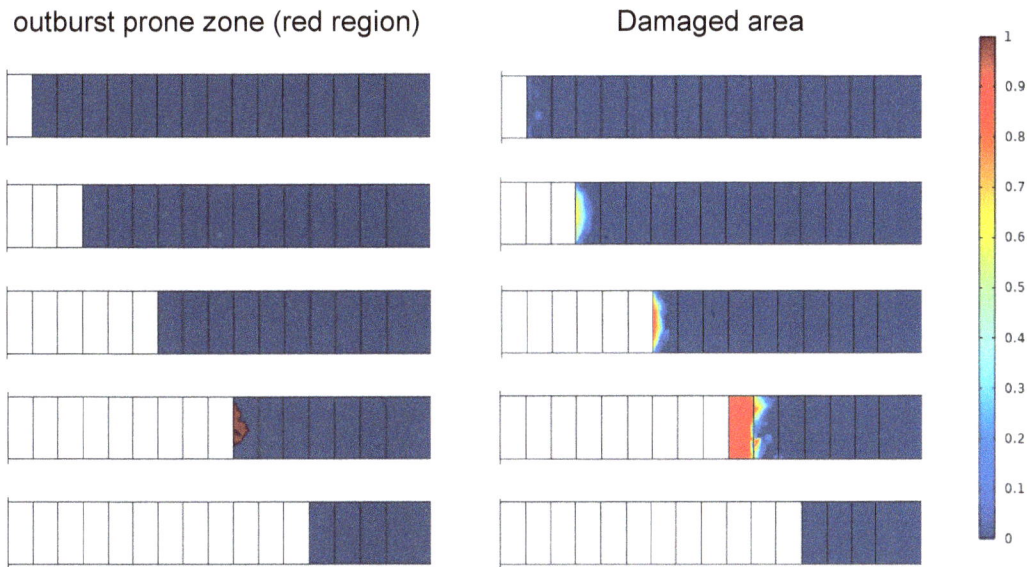

| outburst prone zone (red region) | Damaged area |

FIG 8 – Comparison between the damaged area and the outburst prone zone.

CONCLUSIONS

In this study, outburst occurrence is analysed by conducting a numerical model. The geomechanical and fluid flow equations were solved using the FEM method by COMSOL Multiphysics. The results were used to calculate the involved energies in the phenomenon. The final results showed that the magnitude of the gas expansion energy is more than the elastic energy before the excavation starts in the coal seam. The difference is more significant in the deformed coal, and compared to the normal coal, the desorbed gas contribution is more. The presence of deformed coal (geological structures) increased the possibility of outburst occurrence during mining of the coal seam, and the main energy source that leads to the outburst occurrence is the elastic strain energy in the deformed coal. Also, the damaged area can be a representative of the outburst occurrence.

REFERENCES

An, F, Yuan, Y, Chen, X, Li, Z and Li, L, 2019. Expansion energy of coal gas for the initiation of coal and gas outbursts, *Fuel*, 235:551–557.

An, F-H, Cheng, Y-P, Wang, L and Li, W, 2013. A numerical model for outburst including the effect of adsorbed gas on coal deformation and mechanical properties, *Computers and Geotechnics*, 54:222–231.

Bai, M and Elsworth, D, 2000. *Coupled Processes in Subsurface Deformation, Flow, and Transport*, American Society of Civil Engineers.

Busch, A and Gensterblum, Y, 2011. CBM and CO2-ECBM related sorption processes in coal: A review, *International Journal of Coal Geology*, 87:49–71.

Cai, C G and Xiong, Y X, 2005. Theoretical and experimental study on crushing energy of outburst-proneness coal, *Meitan Xuebao/Journal of the China Coal Society*, 30:63–66.

Crosdale, P J, Beamish, B B and Valix, M, 1998. Coalbed methane sorption related to coal composition, *International Journal of Coal Geology*, 35:147–158.

Gray, I, 1980. The Mechanism of, and Energy Release Associated with Outbursts, *Symposium on the Occurrence, Prediction and Control of Outbursts in Coal Mines* (The Australasian Institute of Mining and Metallurgy: Southern Queensland Branch).

Hodot, B B, 1966. Outburst of Coal and Coalbed Gas (Chinese Translation), *China Coal Industry Press, Beijing*, 318–320.

Jiang, C and Yu, Q, 1996. Rules of energy dissipation in coal and gas outburst, *Journal of China Coal Society*, 21:173–178.

Langmuir, I, 1918. The adsorption of gases on plane surfaces of glass, mica and platinum, *Journal of the American Chemical Society*, 40:1361–1403.

Li, C W, Xie, B J, Cao, J L, Wang, T T and Wang, X Y, 2012. The energy evaluation model of coal and gas outburst intensity, *Meitan Xuebao/Journal of the China Coal Society*, 37:1547–1552.

Liu, J, Chen, Z, Elsworth, D, Miao, X and Mao, X, 2010. Linking gas-sorption induced changes in coal permeability to directional strains through a modulus reduction ratio, *International Journal of Coal Geology*, 83:21–30.

Lu, S, Cheng, Y, Qin, L, Li, W, Zhou, H and Guo, H, 2015. Gas desorption characteristics of the high-rank intact coal and fractured coal, *International Journal of Mining Science and Technology*, 25:819–825.

Lu, S, Wang, C, Liu, Q, Zhang, Y, Liu, J, Sa, Z and Wang, L, 2019a. Numerical assessment of the energy instabilty of gas outburst of deformed and normal coal combinations during mining, *Process Safety and Environmental Protection*, 132:351–366.

Lu, S, Zhang, Y, Sa, Z, Si, S, Shu, L and Wang, L, 2019b. Damage-induced permeability model of coal and its application to gas predrainage in combination of soft coal and hard coal, *Energy Science & Engineering*, 7:1352–1367.

Nandi, S P and Walker, P L, 1975. Activated diffusion of methane from coals at elevated pressures, *Fuel*, 54:81–86.

Nie, B, Ma, Y, Hu, S and Meng, J, 2019. Laboratory Study Phenomenon of Coal and Gas Outburst Based on a Mid-scale Simulation System, *Scientific Reports*, 9:15005.

Ruckenstein, E, Vaidyanathan, A S and Youngquist, G R, 1971. Sorption by solids with bidisperse pore structures, *Chemical Engineering Science*, 26:1305–1318.

Smith, D M and Williams, F L, 1984. Diffusion models for gas production from coal, Determination of diffusion parameters, *Fuel*, 63:256–261.

Valliappan, S and Wohua, Z, 1999. Role of gas energy during coal outbursts, *International Journal for Numerical Methods in Engineering*, 44:875–895.

Wen, G, 2003. Study of coal and gas outburst energy, *Mining Safety & Environmental Protection*, 30:6–9.

Wen, G, Zhou, J and Liu, S, 2002. Study on intrinsic gas energy to do work in outburst, *Mining Safety & Environmental Protection*, 29:1–3.

Wilson, R K and Aifantis, E C, 1982. On the theory of consolidation with double porosity, *International Journal of Engineering Science*, 20:1009–1035.

Wu, X, Peng, Y, Xu, J, Yan, Q, Nie, W and Zhang, T, 2020. Experimental study on evolution law for particle breakage during coal and gas outburst, *International Journal of Coal Science & Technology*, 7:97–106.

Wu, Y, Liu, J, Elsworth, D, Chen, Z, Connell, L and Pan, Z, 2010. Dual poroelastic response of a coal seam to CO2 injection, *International Journal of Greenhouse Gas Control*, 4:668–678.

Xiong, Y, Guo, X, Gong, X, Huang, W, Zhao, J and Lu, H, 2009. Blockage critical state of pulverized coal dense-phase pneumatic conveying in horizontal pipe, *Huagong Xuebao/CIESC Journal*, 60:1421–1426.

Zhao, W, Cheng, Y, Jiang, H, Jin, K, Wang, H and Wang, L, 2016. Role of the rapid gas desorption of coal powders in the development stage of outbursts, *Journal of Natural Gas Science and Engineering*, 28:491–501.

Zhao, W, Wang, K, Cheng, Y, Liu, S and Fan, L, 2020. Evolution of gas transport pattern with the variation of coal particle size: Kinetic model and experiments, *Powder Technology*, 367:336–346.

Zheng, C, Kizil, M S, Chen, Z and Aminossadati, S M, 2018. Role of multi-seam interaction on gas drainage engineering design for mining safety and environmental benefits: Linking coal damage to permeability variation, *Process Safety and Environmental Protection*, 114:310–322.

Zhou, H W, Zhao, J W, Su, T, Zhang, L, Zhong, J C and Liu, Z L, 2021. Characterization of gas flow in backfill mining-induced coal seam using a fractional derivative-based permeability model, *International Journal of Rock Mechanics and Mining Sciences*, 138:104571.

Zhu, W C and Wei, C H, 2011. Numerical simulation on mining-induced water inrushes related to geologic structures using a damage-based hydromechanical model, *Environmental Earth Sciences*, 62:43–54.

Tube bundle analyser flow rate, why it is important

M Tsai[1] and M Watkinson[2]

1. Senior Computer Systems Engineer, Simtars, Redbank Qld 4301.
    Email: martin.tsai@simtars.com.au
2. Executive Mining Engineer, Simtars, Redbank Qld 4301.
    Email: martin.watkinson@simtars.com.au

ABSTRACT

Tube bundle gas monitoring systems have been used in the mining industry for over 50 years. The initial design was based on four individual gas analysers along with a series of solenoid valves, gas sample conditioning and flow control mechanisms to ensure the correct gas flow stream volume to the analyser.

Simtars staff have identified flow control issues when providing operational support to coalmines on numerous occasions. The accuracy of the oxygen analysis is dependent on control of the gas flow to the analyser. Oxygen is one of the parameters used in calculating Graham's ratio which is used for the detection of spontaneous combustion activity in coalmines.

This paper will provide data which identifies the flow control issues by comparing mine site records for the two surface sample lines in the analyser room and the gas cage. Mine sites will be deidentified within the paper. The paper will discuss some of the changes that have been made in modern tube bundle systems and compare the change to the initial design specification developed by the UK National Coal Board, investigate possible operational impacts and provide a technical fix to solve the current flow issues being experienced at some mines.

INTRODUCTION

Tube bundle systems were developed in the United Kingdom in the 1960s. Chamberlain *et al* (1971) states that seven installations of tube bundles had been completed following on from his presentation at Harrogate in 1970. In the paper they recommend that pumps are to be diaphragm type and both the purge pump and sample pump should be the same size to avoid issues where the airflow temporally reverses due to the different pressures. They also recommended the use of regulators on each tube to equalise the flow rates by having the longest tube unregulated. One other option discussed was to use an individual vacuum pump on each line.

Tube bundle systems are built around infra-red analysers for carbon monoxide (CO), methane (CH_4) and carbon dioxide (CO_2) and a paramagnetic analyser for oxygen (O_2). All the systems in Australia have one sample pump and several purge pumps. Purge pumps now draw on ten tubes with a manufacturers recommendation to put tubes of similar length on the same pump. One supplier controls back pressure on the sample pump by using check/non-return valves in each line to prevent the reversal of the sample pump. Another supplier uses line regulation to control the flow and stabilise the pressure in each tube. The control of the analyser flow rate is maintained by using a mass flow controller with the excess volume purged to atmosphere. The initial designs in the 1990s had matched (pressure) sample pumps and purge pumps with line regulators on each tube. (There was one purge pump to five underground tubes).

Paramagnetic analysers are flow dependant whereas infra-red analysers are not. Any variation inflow to the analyser can compromise the accuracy of the oxygen analysis. Two key uses of the oxygen analysis are in the Graham ratio calculation for the detection of spontaneous combustion activity and the monitoring of oxygen levels behind seals.

$$\text{Graham ratio} = 100 * \frac{(CO_f - CO_i)}{(O_{2i} - O_{2f})}$$

Where:

CO$_f$ is the final CO reading in %

CO$_i$ is the initial CO reading in %

O_{2i} is the initial oxygen reading in %

O_{2f} is the final oxygen reading in %

Six mines provided data for analysis, five in Queensland and one in NSW.

Flow data was reviewed for the analyser room tube and the pump room tube as well as six underground tubes. The two surface tubes are short lengths and do not have a purge pump connected to them. The mines are deidentified in this paper. Where potential issues were identified they were communicated to the mine site. Figure 1 shows the schematic for a 40-point system with four purge pumps and one sample pump.

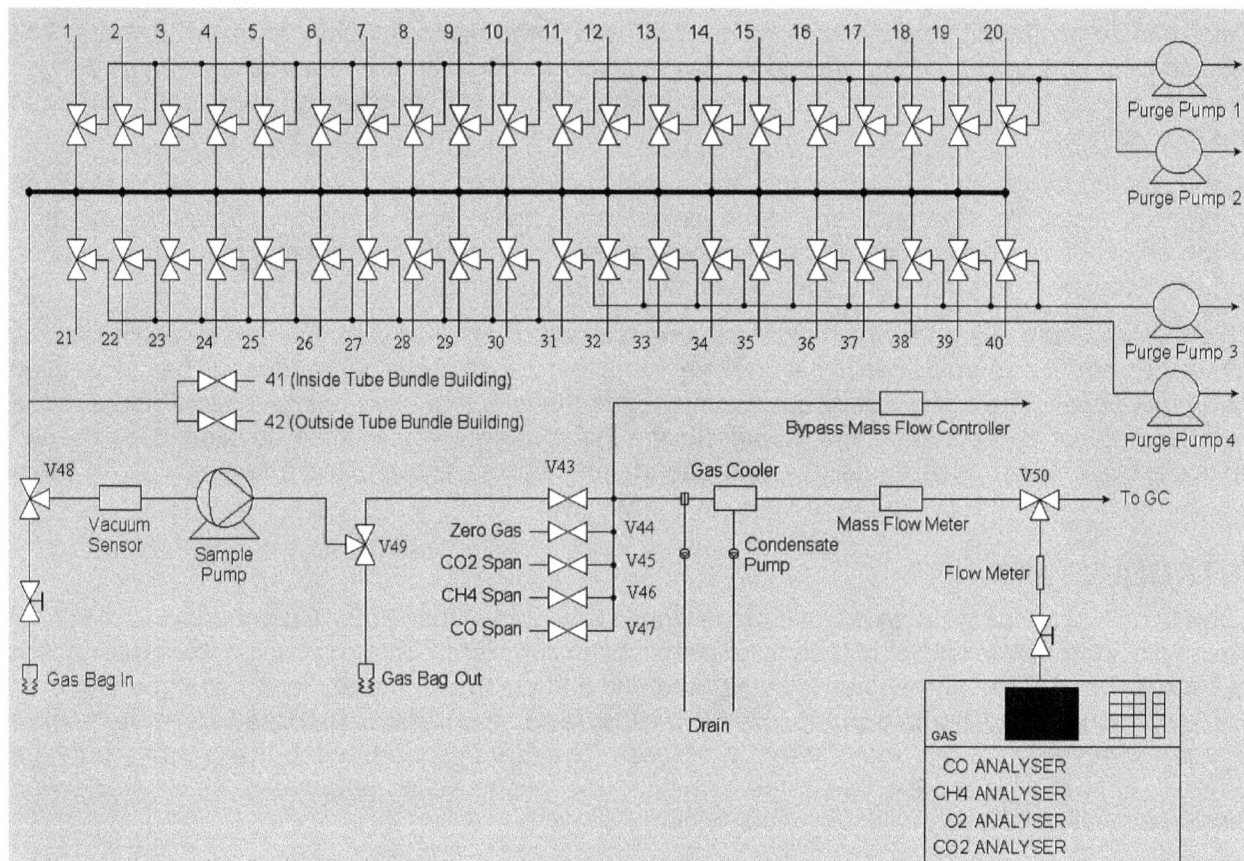

FIG 1 – Schematic of tube bundle system.

This shows the current design for a 40-point system supplied by one of the original equipment manufacturers (OEM 1). Five of the mines studied had systems supplied by OEM 1. This system has ten tubes connected to one purge pump (vane) with the advice given that all long tubes should be connected to the same purge pump. The sample pump is a diaphragm pump. The other OEM's offer a similar design.

One of the mines who supplied data has two tube bundle systems supplied by OEM 2. The National Coal Board (NCB, 1974) recommend that both vane type and diagram type pumps can be used in the system. They do however recommend that the vacuum pressure generated by the purge pump should not be greater than the pressure that the sample pump can generate. This suggestion is aimed to address vacuum imbalance as the imbalance may lead to flow reversal when system switches to next sampling point.

Installations supplied by OEM 1 encountered this problem and the issue was resolved by the insertion of check valves to prevent the sample pump from reversing. These check valves became a major issue at one mine when they were dealing with a mine fire.

The original design supplied by OEM 1 in 1997 was one purge pump per five tube lines and the sample pump and the purge pumps were the same model and were diaphragm type. This design did not have a mass flow controller and all the tubes had individual line regulation.

FLOW RATE DATA

Mine 1 has two tube bundle sheds supplied by OEM 2. The supplied design had needle valves to control and balance the flows in the tubes. Mine site personnel requested the inclusion of a mass flow controller. Only the longer tubes are regulated using the needle valves.

Figure 2 shows the flow rate for Mine 1 tube bundle system 1. This varies between 0.7 L/min and 0.69 L/min. This is an example of very good flow control on both the two surface tubes and the underground tubes.

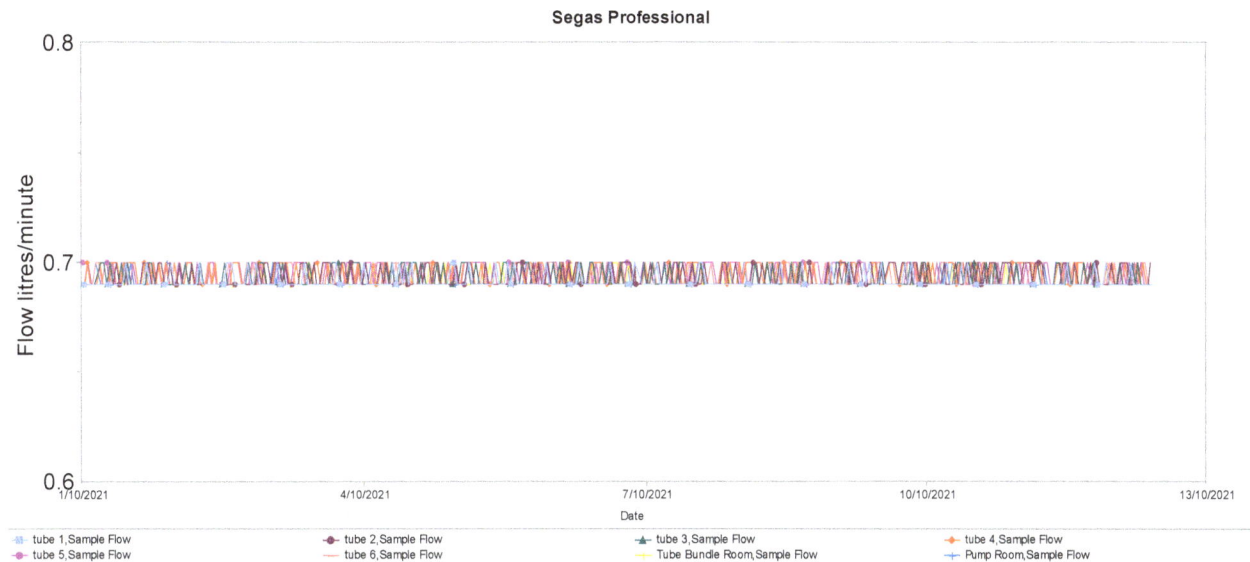

FIG 2 – Mine 1 tube 1 flow rate for eight tubes.

The vacuum pressures for Mine 1 tube bundle system 1 values can be seen in Figure 3. There are some individual variations in the vacuum pressures but overall the pressures for each tube are fairly constant. The two surface tubes are around -49 kPa.

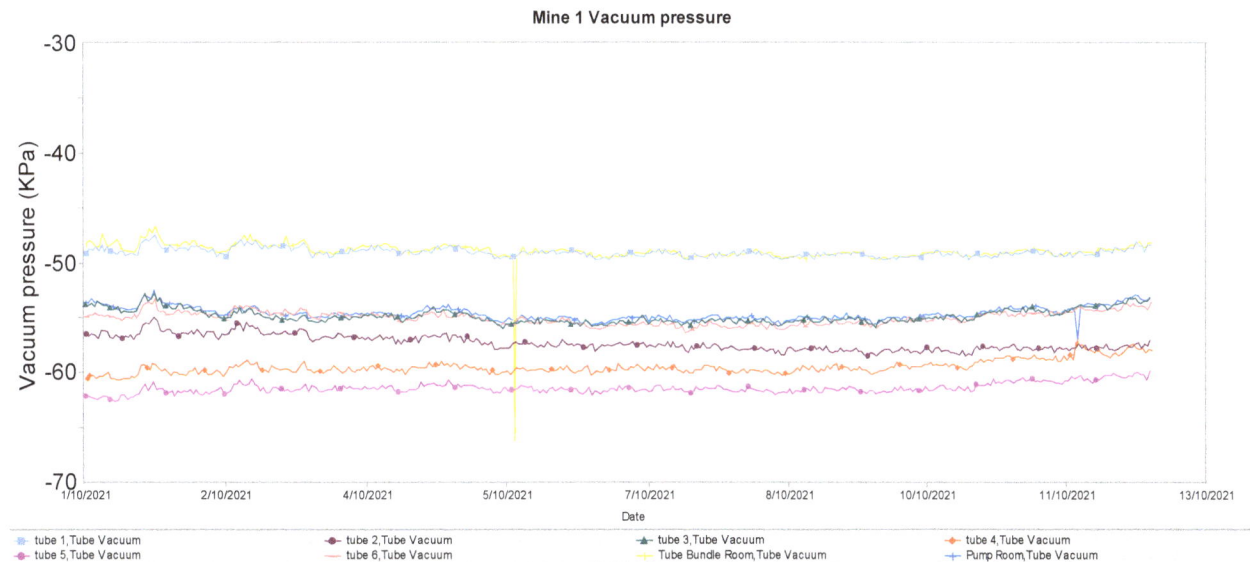

FIG 3 – Mine 1 tube 1 vacuum pressures.

Figure 4 shows the variation inflow rate being experienced by Mine 2 with variations of ±8 L/hr or 0.13 L/min. In comparison Mine 1 only varied by 0.01 L/min.

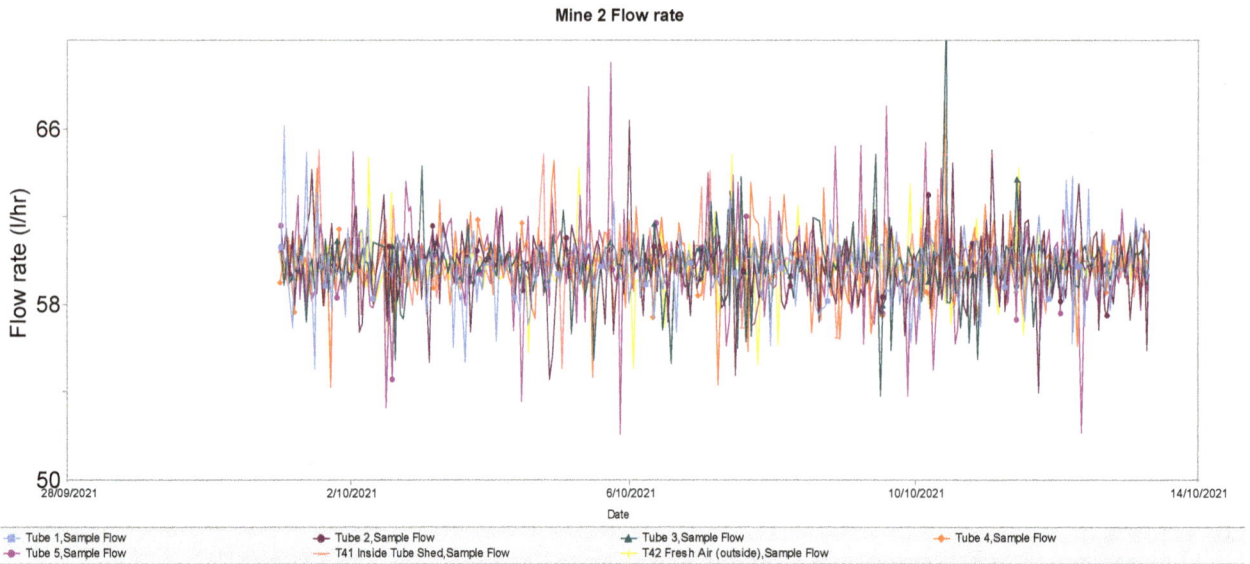

FIG 4 – Mine 2 sample flow.

Figure 5 shows the variation on the two surface tubes ±5 L/hr or 0.83 L/min.

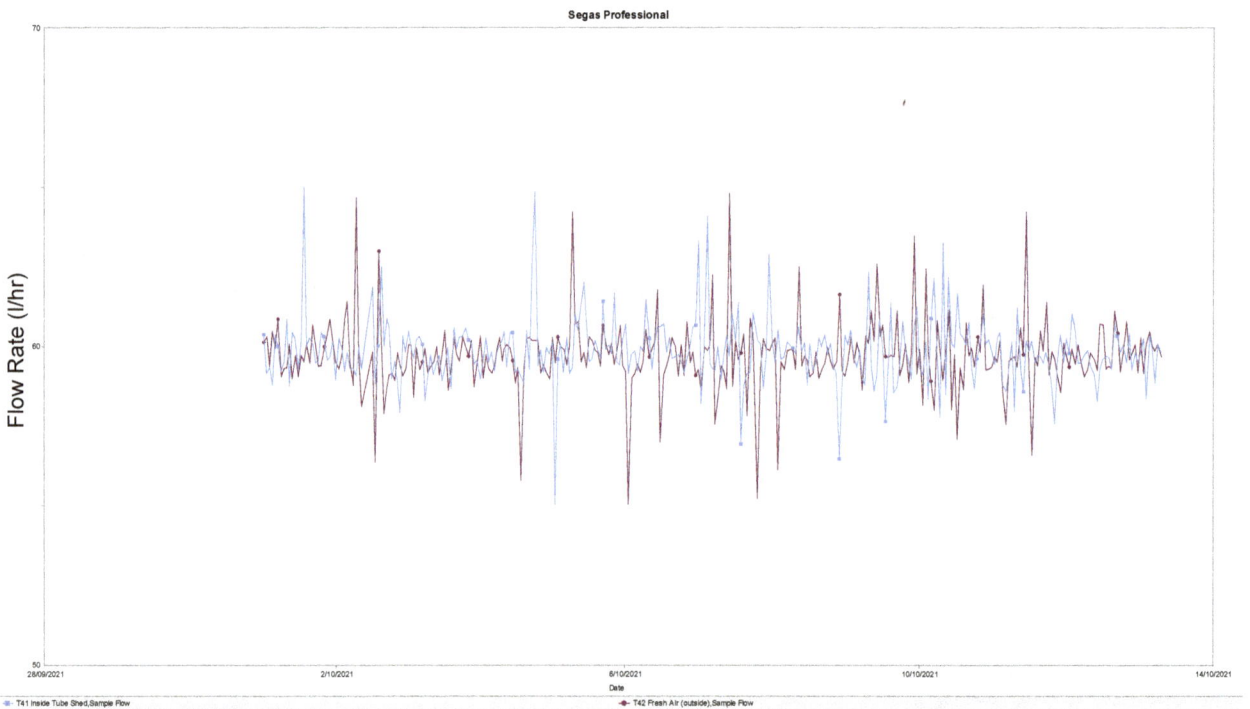

FIG 5 – Mine 2 flow rate surface tubes.

The variation in tube vacuums is like Mine 1 with the vacuum pressure on the two surface tubes being around -30 kPa (see Figure 6).

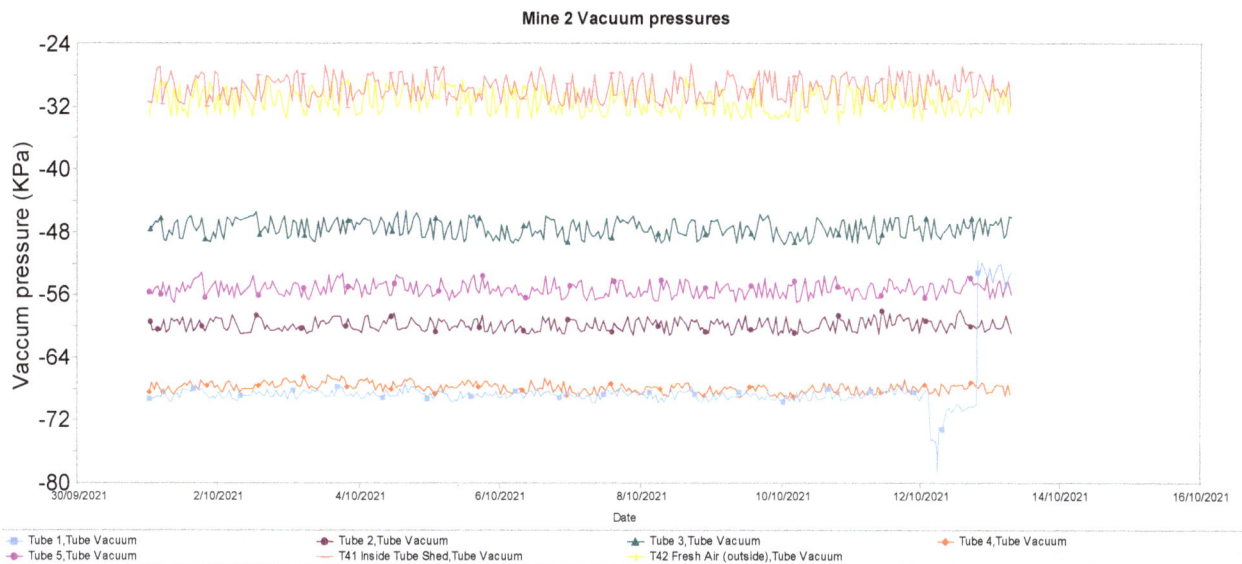

FIG 6 – Mine 2 vacuum pressures.

Figure 7 shows the analyser flow readings for Mine 5 when viewing the flows on eight tubes there is a change in response around 17 September 2021.

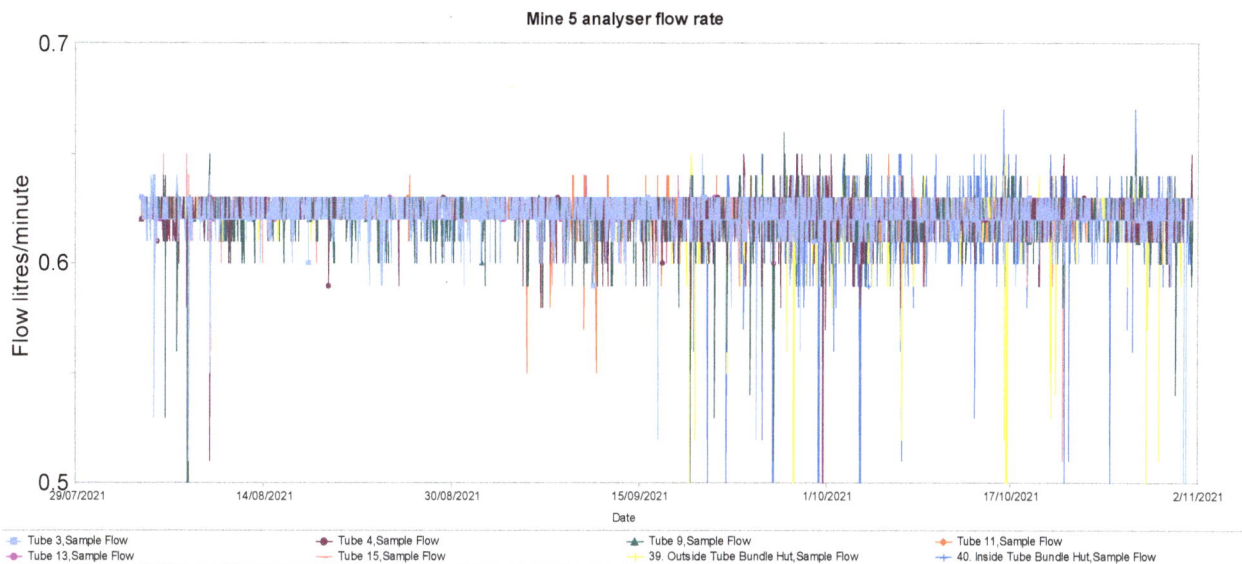

FIG 7 – Mine 5 analyser flow.

Figure 8 shows the response change on 17 September 2021 for the two surface tubes, where the change can be seen more clearly. The flow control on the surface tubes is very good prior to that date (±0.01 L/min). The question is what changed around 17 September 2021 and why couldn't the system control the analyser flows?

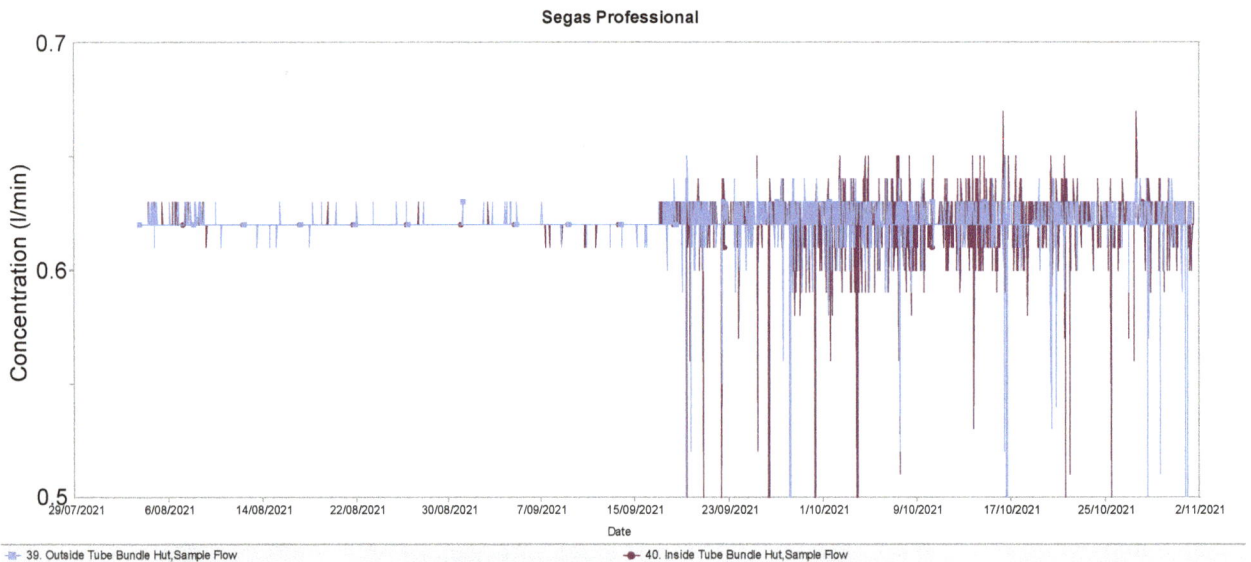

FIG 8 – Mine 5 analyser flow two surface tubes.

Figure 9 shows the vacuum pressure for eight tubes from Mine 5. This graph shows some variation over the three months that the data is available. The mine site staff reported that they have issues with material clogging the bypass valve on occasions. When this is cleared the system goes back to normal. The attribute the change on 17 September 2021 to this issue.

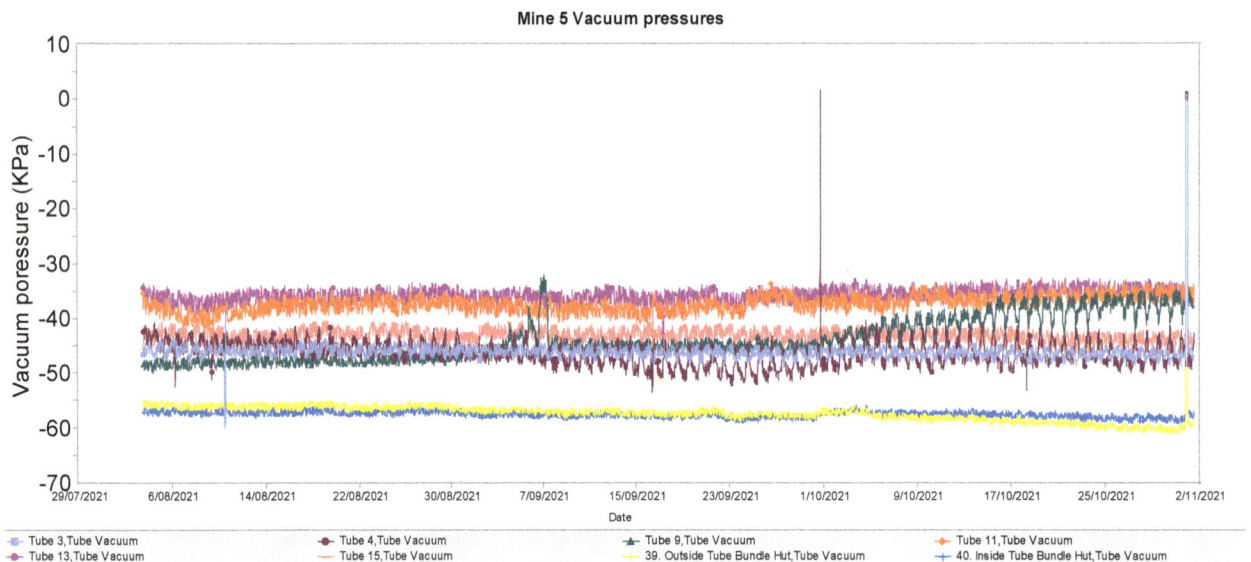

FIG 9 – Mine 5 vacuum pressures.

DISCUSSION

Most of the tube bundle systems were supplied by OEM 1 and use a mass flow meter and bypass valve to control the flow and ensure the system delivers the required flow to the analyser. This approach is known as a continuous proportional derivative (PD) control loop in automation systems. The PD control loop offers improved system stability during rapid load changes if the load change compensation error is within the acceptable range.

Factors which affect the monitoring effectiveness

Condition of the tube – the system is under a constant change in the operational environment. The overall resistance of each tube changes due to many factors including tube leakage, water inside the tube, incorrect seal in water trap and different length of the tubes. The tube bundle system needs to rebalance the entire sampling line when system selects a new sampling point. These resistance changes are part of the load changes on the system.

Control system response time

A real-time sensor used in an underground coalmine has a T_{90} requirement which is the time to reach 90 per cent of the test gas value. A standard control system should have a similar standard applied to the expected response time. A typical sampling time per location in the tube bundle system is between 90 to 120 seconds. Therefore, it is normal to expect that the system should achieve stable sample flow within 30–45 seconds. The rest of the sampling time is to enable the analyser to achieve an accurate reading under stable sample flow conditions.

One of the original tube bundle system design specifications is to supply a conditioned sample gas in a controlled environment to achieve the best accuracy result.

The original scope of this review was just to look at the sample variations on the two surface tubes and compare all the mines. When reviewing the data from Mine 1 it could be seen that very good sample flow was achieved on all tubes. Mine 1 has both needle valves and a mass flow controller.

Mine 2 has flow control issues on all tubes (Figures 4 and 5).

Mine 5 has intermittent issues as can be seen in Figures 7 and 8. Given that these two mines (2 and 5) have a mass flow bypass valve arrangement the issue appears to be with the control of the bypass valve and is independent of the needle valves.

This raises questions on how the control system performs on with standard tube condition (1~8 km tube with water, dust, various joints etc). If an operator observed an unstable system performance during a mine emergency, how much confidence would the operator have on the reported values?

System behaviour

Field observations of tube bundle systems with non-ideal control have demonstrated a sample flow oscillation see Figures 4 and 7.

Figure 10 shows three different control system behaviours. Mines 2 and 5 flow responses is behaving like an underdamped control system where it has a massive overshot (±5 L/min) then oscillates nonstop until the set time configured by the mass flow controller. If this set time is longer than the sampling frequency of the tubes there will not be stable flow to the analyser which can impact on the oxygen analysis.

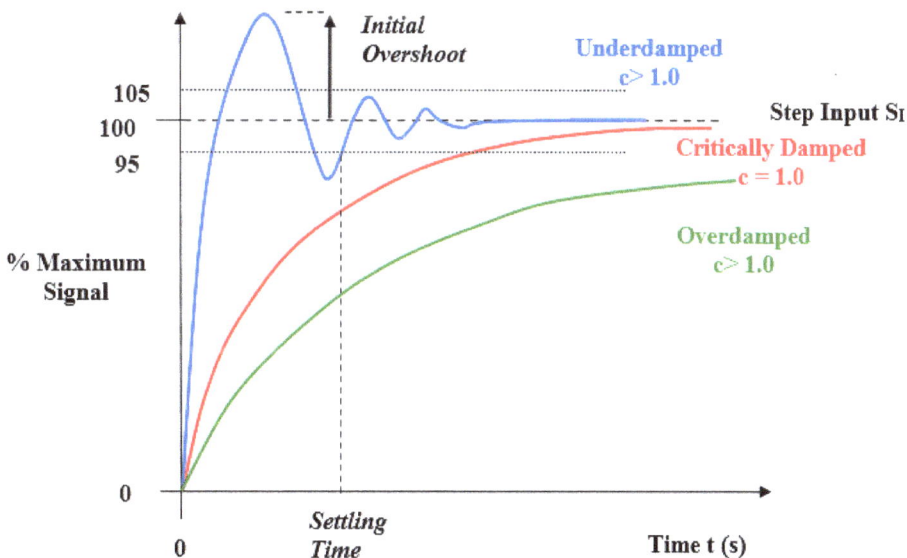

FIG 10 – Control system behaviours.

Both Mines 2 and 5 systems cannot control the flow on the shortest tubes on the surface therefore there must be a possible design flaw with the control system.

All control systems should be configured with set baseline for the shortest tubes in the factory acceptance testing (FAT) then verified on-site when longer tubes are added to the system. The analyser building contains the two short tubes for the analyser room and pump room, and these can be tested during FAT. If the delivered system fails to control the flow in the short tubes, there is a possibility that the baseline was incorrectly set.

Similar undamped behaviour was observed on the Simtars mobile laboratory (20 point tube bundle system) when the tube diameter was switched from ½ inch (12.7 mm) to 5/8 inch (15.88 mm).

The aim of smart control in a tube bundle system is managing the various factors presented in the tube and control it within an acceptable range. If tube bundle system fails to maintain control, one of the options is to reduce the variables inside the system.

If the system maintains control on the short tubes but cannot maintain control on longer tubes the load change on the system needs to be optimised to ensure that flow control is maintained. Mine 5 did achieve flow control in August and September 2021 see Figure 8.

The current configuration of tube bundle systems is to have two purge pumps, one for every ten tubes and one sample pump. The purge pumps can generate a larger vacuum pressure than the sample pump. Once a new tube has been selected to be sampled and the system energised, the sample pump must balance out the higher vacuum condition that already existed in the line. The initial switch will cause a sample flow overshot condition (underdamped Figure 10), the control system will the over correct and cause an undershoot condition; the time taken to stabilise the flow to an acceptable condition is likely to be longer that then current sample time frequency. Needle valves will help to regulate the rate of tube balance and assist tube sample condition (critical damping Figure 10). Mine 1 (Figure 2) has a needle valve installed in every tube and shows that good control of the analyser flow is being maintained.

CONCLUSIONS

- Incorrect flow to the analyser impacts the accuracy of the oxygen analysis and hence the calculation of Graham's ratio.

- Mine 1 has the best performing system and this mine uses both needle valves for line regulation as well as a mass flow controller to control the final supply to the analyser.

- The readings provided on the analyser flow are at the end of the stabilising period and additional dampening of the signal is required. The graphs do not record the dampening control of the system.

- Both mines with issues have systems supplied by OEM 1.

- Mine 5 has an issue with debris or 'gunk' as they name it in the bypass valve at the analyser.

- Both OEM's have modified the design from the initial design prepared by NCB (1974).

- Further investigation is required into PD control loop configurations used by OEM 1 and OEM 2.

REFERENCES

Chamberlain, E, Donaghue, W, Hall, D and Scott, C, 1971. The Continuous Monitoring of Mine Gases: The Development of a 'Tube Bundle' Technique, presented at General Meeting of North of England Branch of the Institution of Mining Engineers, published in *The Mining Engineer,* 1975, pp 239–251.

National Coal Board (NCB), 1974. The Tube Bundle Technique for the Continuous Monitoring of Mine Air (622.412.1).

A theoretical goaf resistance model based on goaf gas drainage data

Y Wang[1], G Si[2], B Belle[3], J Oh[4] and Z Xiang[5]

1. PhD student, School of Minerals and Energy Resources Engineering, University of New South Wales, Sydney, NSW 2052. Email: yuehan.wang@student.unsw.edu.au
2. Senior Lecturer, School of Minerals and Energy Resources Engineering, University of New South Wales, Sydney, NSW 2052. Email: g.si@unsw.edu.au
3. Adjunct Associate Professor, Anglo American Metallurgical Coal, Brisbane, Qld 4001. Email: bharath.belle@angloamerican.com
4. Associate Professor, School of Minerals and Energy Resources Engineering, University of New South Wales, Sydney, NSW 2052. Email: joung.oh@unsw.edu.au
5. PhD student, School of Minerals and Energy Resources Engineering, University of New South Wales, Sydney, NSW 2052. Email: zizhuo.xiang@student.unsw.edu.au

ABSTRACT

Inadequate goaf gas control may lead to gas exceedances in the ventilation return air, which will trip power in longwalls and cause production stoppage. To effectively control gas emissions, Australian mines heavily rely on drilling vertical goaf holes to capture coalmine methane before it migrates to mine openings. The wide application of goaf gas drainage may change the goaf pressure distribution, lead to more ventilation air leakage, and consequently increase gas explosion and spontaneous combustion risks. This paper focuses on analysing goaf gas drainage production data from a case study longwall mine in Australia. The correlation between suction pressure, total flow rate, gas flow rate and air leakage rate were analysed in detail as the face-to-hole distance changes. Suction pressure is found to be positively correlated with total flow rate, and their correlation also varies at different goaf locations due to various goaf compaction and flow resistance. This paper also proposed a theoretical method to calculate the goaf resistance of ventilation air leakage pathways. Based on several basic assumptions and different goaf drainage scenarios, the leakage airflow pathway resistance in the goaf at different face-to-hole distances can be calculated. The model results suggested that the resistance of air leakage pathways from the working face to individual goaf holes increased with the face-to-hole distance. The resistance results can also be used to calculate the corresponding permeability at the same position, which can be used as input data for CFD or Ventsim models.

INTRODUCTION

Coalmine methane is the methane released from the target coal seam or the strata near the mining coal seam during the longwall mining process (Karacan *et al*, 2011). Large amounts of methane released from the overlying strata enter the mined area (goaf), which may vary in the explosive range and pose a threat to mine safety (Coward and Jones, 1952). To control goaf gas emissions, ventilation systems together with goaf gas drainage are needed to provide fresh air to the longwall panel. However, the residual coal in the goaf would react with the leaked ventilation air and accumulate heat gradually due to low airflow rate, which may cause spontaneous combustion risks in the goaf (Wu and Liu, 2011; Yang *et al*, 2014).

Goaf gas drainage utilises a series of vertical goaf holes drilled at certain distance above the goaf to capture gas emissions from the goaf and overlying fractured strata, as shown in Figure 1. The application of goaf gas drainage may change the goaf pressure distribution and lead to more ventilation air leakage, and consequently increase the gas explosion and spontaneous combustion risks (Ren and Edwards, 2002). Thus, it is necessary to closely monitor goaf holes data and carefully manage goaf gas atmosphere to avoid excessive air leakage. Goaf gas drainage performance is related to many factors, such as gas emission sources, ventilation air leakage, suction pressure, and goaf permeability.

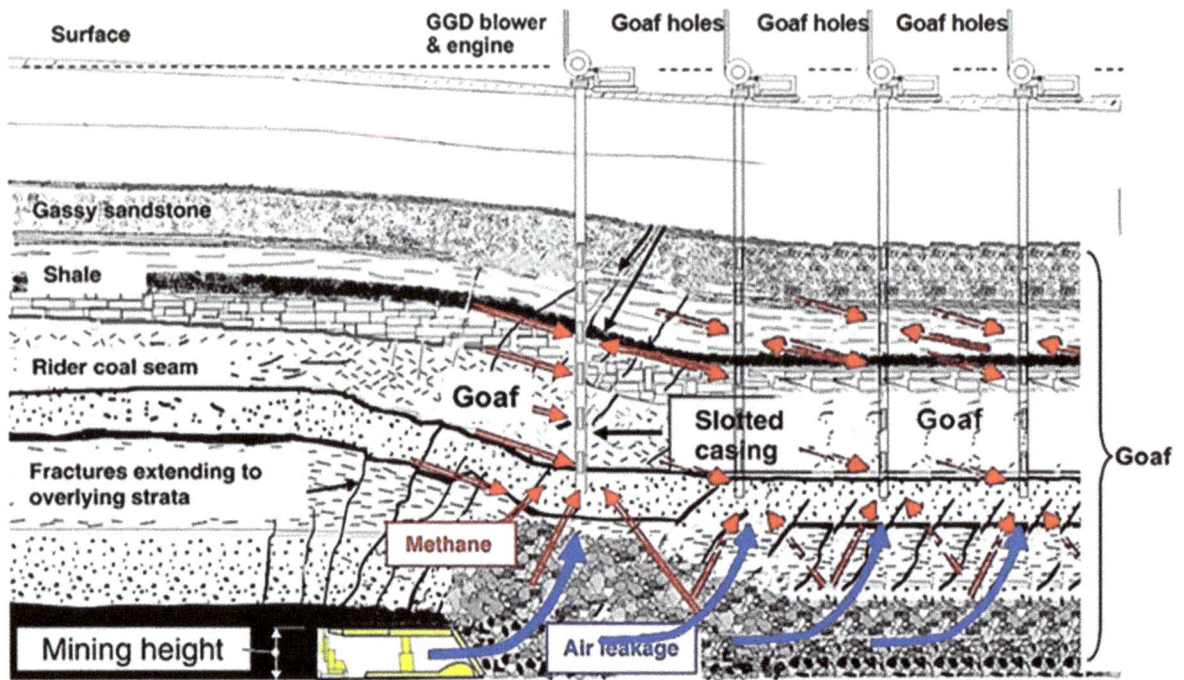

FIG 1 – The cross-section of longwall mining with goaf gas drainage (modified from Karacan, 2009a, 2009b).

Many researchers used numerical models to analyse goaf gas drainage performance and spontaneous combustion risks in the goaf. Through comparing different CFD models, goaf gas flow pathways and high-risk zones of spontaneous combustion are obtained, which can be used to improve ventilation and goaf drainage designs (Balusu *et al*, 2002; Ren and Balusu, 2005; Yuan and Smith, 2008). The CFD model developed by Ren and Edwards (2002) suggested that the increase in suction pressure would aggravate air leakage. Low suction pressure regimes can create goaf pressure sinks, which would then impact ventilation air pathways. Mining geomechanical models are also usually developed to illustrate the permeability and the progressively goaf compaction process. The modelling results are then used as the input for reservoir models to evaluate goaf gas drainage performance (Esterhuizen and Karacan, 2005, 2007; Esterhuizen *et al* 2010; Karacan *et al*, 2006, 2007; Karacan, 2009a, 2009b).

However, these models normally lacked adequate in-site monitoring data for validation, especially goaf gas drainage data as widely applied in modern high-production mines. Based on a large amount of field data of goaf hole production, this paper analysed the goaf gas emission and air leakage trends. The impact of goaf hole suction pressure on gas flow rate was also discussed, and their linear correlation suggests a laminar flow regime in goaf. This study also developed a theoretical resistance model to calculate the ventilation airflow resistance in the goaf at various goaf hole locations.

MINE SITE INFORMATION

This study used extensive goaf drainage data collected from an Australian underground coalmine, Mine A. The longwall panels in Mine A operated under the depth of cover from 250 m to 500 m with 2.8 m seam thickness. LW9 in Mine A was selected as the study longwall panel. The width of LW9 is 350 m, and its length was up to 3600 m. The planned production rate for each longwall panel in Mine A is approximately 200 000 t/week. The seam gas composition measured from the local reservoir condition is approximately 96 per cent methane with the rest is carbon dioxide (Si and Belle, 2019).

Goaf holes in the study mine are typical goaf holes with 250 mm diameter, and the bottom is around 10–20 m above the targeted coal seam. There are 98 goaf holes drilled in LW9, 74 of which were located on the tailgate side (named TG1 to TG74), and the rest were on the main gate side. The TG goaf holes were approximately spaced at 50 m intervals and 30–40 m offset away from the TG, while the MG goaf holes were drilled 50 m away from the MG side with 150 m spacing. The goaf holes started operating at around 15 m after the longwall working face passed-by. A comprehensive gas monitoring scheme was applied to individual goaf holes, which have goaf gas concentration, gas flow rate, and suction pressure measurement data. Mine site engineers collected measurement data every four hours, and the daily average values were used in this research.

FIELD DATA ANALYSIS

The relationship between suction pressure and flow rate

The differential negative pressure between the barometric and static pressure on each goaf hole, ie suction pressure, uses the absolute value in this study. Suction pressure was used as the vertical axis and flow rate was used as the horizontal axis in the following scatter plot (Figure 2). Due to a large amount of data for LW9, to examine the relationship between suction pressure and flow rate, this research focused on the first 24 boreholes (TG1 to TG24) in the tailgate side goaf. It can be found that the relationship between the suction pressure and flow rate on the tailgate side showed a roughly positive linear correlation.

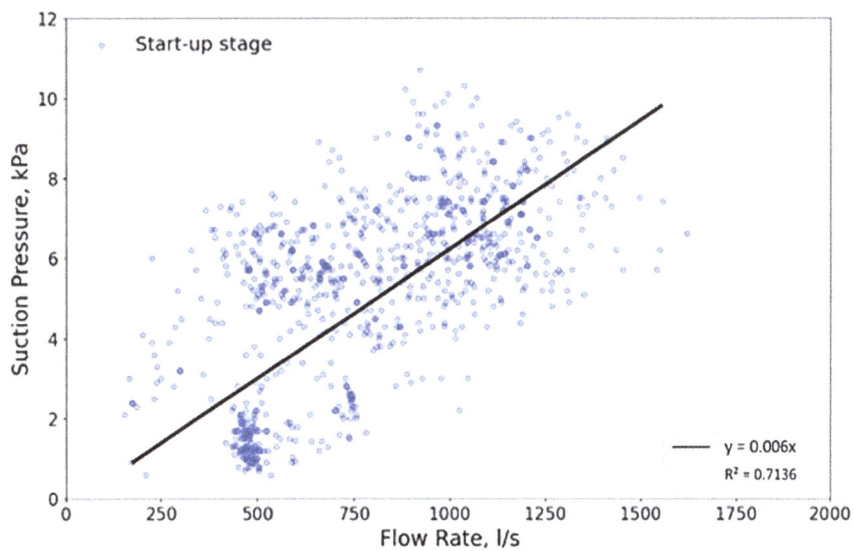

FIG 2 – The relationship between suction pressure and flow rate for start-up stage goaf holes (TG1 – TG24) in LW9.

There are a couple of goaf holes with very long production periods (>100 days) in LW9. Goaf hole TG11 was selected as an example for analysing the correlation between suction pressure and flow rate over the gas production lifespan. Figure 3 shows the positive effect of suction pressure on flow rate, and high gas purity was obtained by applying high suction pressure at the first 15 days after the initial opening. As the longwall moved away and goaf compaction developed, the elevated suction pressure incurred much air ingress, and the purity declined with more air presenting in goaf holes. Therefore, the goaf hole TG11 closed at around 30 days based on the Trigger Action Response Plans (TARPs) applied at the study site (Xiang *et al*, 2021). After 230 days, TG11 was reopened with higher purity gas being extracted from this goaf hole, and the suction pressure and flow rate are also positively correlated after that.

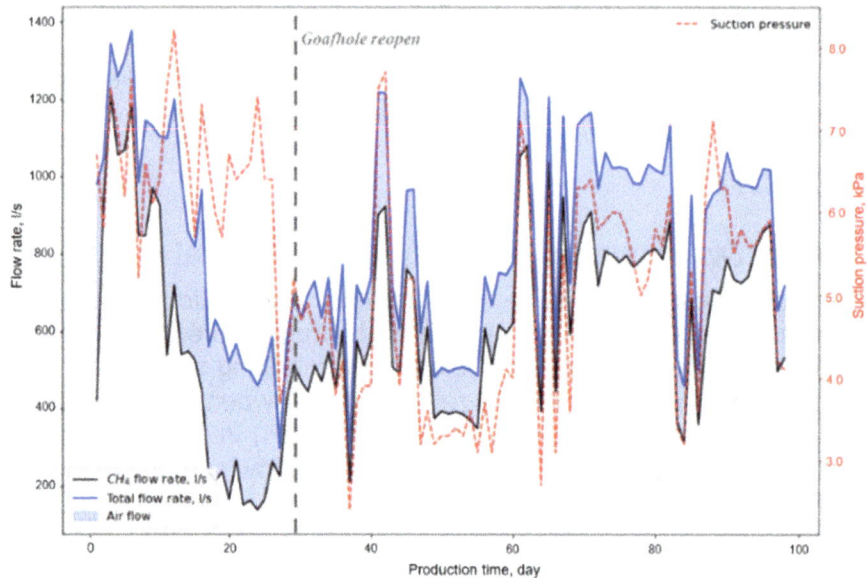

FIG 3 – Variation of flow rate against suction pressure in goaf holes TG11 at LW9.

Gas emission and air leakage rate in the goaf

The flow rates of total, seam gas, and air leakage captured by goaf holes on both LW9 TG and MG sides, with respect to different face-to-hole distances, are shown in Figure 4. As shown in Figure 4a, the total flow rate of TG goaf holes shows an upward trend and reaches a peak of 1100 L/s at about 100 m from the face, regardless of a slight increase of suction pressure by 1 kPa, after which it decreases to 600 L/s. Then the total flow rate slowly rises back to 800 L/s, driven by the continuous increase of suction pressure from 6–10 kPa, which is indicated by the red dotted line. The air leakage rate at the TG side gradually increases in the first 100 m, and then there is nearly no change until 800 m back in the goaf. Over the entire studied length of the TG side goaf, the normalised flow rate (flow rate divided by the suction pressure) trend is similar to the raw flow rate, but its variation is marginal. From Figure 4b, the normalised total flow rate varies between 100–250 L/s/kPa, and the normalised seam gas flow rate ranges between 50 to 200 L/s/kPa. Besides, even though the air leakage remains stable after removing the effect of suction pressure, the O_2/N_2 ratio declines from 0.25 to nearly 0.05 over the studied goaf length, indicating a less reactive goaf atmosphere as the consumption of O_2 component. A 'sweet-spot' distance with a high flow rate and low suction requirement can be identified within the 0–200 m behind the face, where goaf holes tend to have the best gas capture performance.

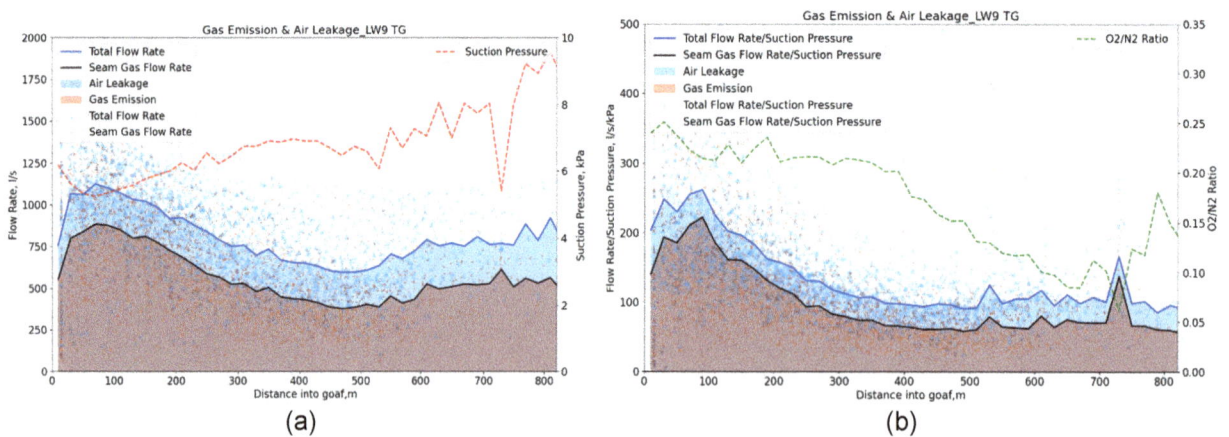

(a)

(b)

FIG 4 – (a) Total flow and seam gas flow rate profiles along the TG side goaf; (b) Total flow and seam gas flow rate profiles normalised by suction pressure along the TG side goaf in LW9.

A THEORETICAL GOAF RESISTANCE MODEL

Theoretical background and assumptions

The gas flow pathway is another important control factor for the goaf atmosphere. As the gas emission source in the goaf is rather complex, the air leakage was assumed to be only from the ventilation air at the working face. With the movement of hydraulic supports, the goaf filled with collapsed rocks and some remaining coal formed a porous media. Due to the inertia of ventilation air in the main gate side, the ventilation airflow may migrate from the working face to the gradually compacted goaf with a relatively high velocity, resulting in turbulent in this area. As the air leakage penetrates deeper into the goaf, its kinetics energy diminishes, resulting in the flow regime change from turbulent to laminar (Schmal *et al*, 1985; Akgun and Essenhigh, 2001; Wu *et al*, 2007). Based on the linear correlation between suction pressure and flow rate in the above section *The relationship between suction pressure and flow rate*, the laminar flow assumption will be used in the following theoretical resistance model. The Poiseuille equation (Equation 1) is suitable for the incompressible laminar fluid, which is used to calculate pressure drop and flow rate in a given pipe size. This equation can be applied and the flow viscosity here is not affected by the velocity (McPherson, 2009).

$$\Delta P = \frac{8\mu L}{\pi r^4} Q = R_L Q \tag{1}$$

Where, ΔP (Pa) is the pressure drop; μ ($N \cdot s/m^2$) is the fluid viscosity; L (m) is the length of the pipe; Q (m^3/s) is the volumetric flow rate; r (m) is the pipe radius; R_L ($N \cdot s/m^5$) is the laminar flow resistance.

To simplify the theoretical resistance model calculation, the following assumptions are proposed:

1. Air leakage can migrate from the working face to each goaf hole, and the working face pressure is assumed as the barometrical pressure. Then the leaked air can flow autonomously from the working face to goaf holes driven by the differential pressure.

2. The airflow into each goaf hole is categorised as two types of pathways: (1) the air directly leaked from the face ventilation, without being affected by other goaf holes, and (2) the air migrating from the face and passing through multiple goaf holes that are closer to the face. A goaf hole can only receive air passing from its nearest neighbouring goaf hole.

3. The air passing between neighbouring goaf holes can only travel in one-direction from the face to deep goaf. In addition, the differential pressure between two neighbouring goaf holes need to be positive to facilitate the migration of leaked air.

4. The air resistance in the first type of pathway is equal to the combined resistances of air flowing through the second type of pathway originated from the face.

Calculation methods

According to Kirchhoff's laws and above assumptions, a simplified model for calculating air leakage pathway resistance can be obtained. The calculation methods for different active goaf holes from simple to complex cases (two goaf holes, three goaf holes, and n goaf holes) are presented below.

The goaf air migration network with two goaf holes is shown in Figure 5, and each blue arrow represents one airflow pathway from the face to a goaf hole or the neighbouring goaf hole. The following parameters can be obtained from the field goaf gas drainage data: Q_1 (L/s) and Q_2 (L/s) are air leakage flow rate from goaf holes, BH_1 and BH_2, respectively; ΔP_{01}, ΔP_{02}, and ΔP_{12} (kPa) are the differential pressure of three airflow pathways, from the working face to BH_1, the working face to BH_2, and BH_1 to BH_2, respectively. Using the above goaf gas drainage data, R_{01}, R_{02}, R_{12}, Q_{01}, Q_{02}, and Q_{12} can be calculated simultaneously from the following equations (Equations 2–7). Under the laminar flow assumption, R_{01}, R_{02}, and R_{12} (kPa/(L/s) $= 10^6$ $N \cdot s/m^5$) represent the resistance from the working face to BH_1, the working face to BH_2, and BH_1 to BH_2, respectively. Besides, Q_{01}, Q_{02}, and Q_{12} (L/s) are the air leakage flow rate from three different pathways, the working face to BH_1, the working face to BH_2, and BH_1 to BH_2, respectively.

$$Q_{01} = Q_1 + Q_{12} \tag{2}$$

$$Q_2 = Q_{12} + Q_{02} \tag{3}$$

$$\Delta P_{01} = R_{01} Q_{01} \tag{4}$$

$$\Delta P_{02} = R_{02} Q_{02} \tag{5}$$

$$\Delta P_{12} = \Delta P_{02} - \Delta P_{01} = R_{12} Q_{12} \tag{6}$$

$$R_{02} = R_{01} + R_{12} \tag{7}$$

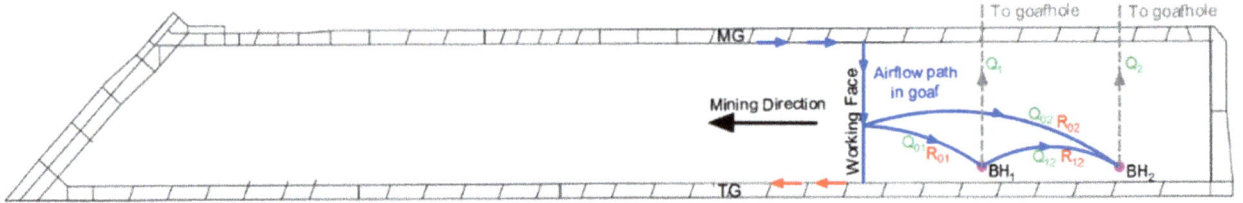

FIG 5 – A schematic of the simplified goaf ventilation network (two goaf holes case).

For the goaf air migration network with three working goaf holes (Figure 6), the goaf holes BH_1 and BH_2 are combined into an equivalent goaf hole $BH_{1,2}$ to simplify the calculation. The equivalent goaf hole $BH_{1,2}$ related parameters Q_{01-EQ}, Q_{2-EQ}, ΔP_{01-EQ}, ΔP_{02-EQ} and ΔP_{12-EQ} can be calculated from Equations 8–12. Q_1, Q_2, Q_3, ΔP_{01}, ΔP_{02} and ΔP_{03} are field monitoring data that can be easily obtained from each goaf hole. Then different airflow pathway resistances (R_{01-EQ}, R_{03}, R_{23}) and flow rate (Q_{01-EQ}, Q_{03}, Q_{23}) can be calculated following the steps in the case of two goaf holes using Equations 2–7. Once R_{01-EQ} is solved, R_{02} can be solved using Equation 13 and so as R_{01}. When ΔP_{23} is negative, there is no leakage air flowing between BH_2 and BH_3 based on the third assumption above. Thus, R_{23} is assumed to be infinity large in this case, and R_{03} is equal to $\Delta P_{03}/Q_3$. Besides, the other air pathway resistances (R_{01}, R_{02}, R_{12}) are calculated as the two goaf holes case.

$$Q_{01-EQ} = Q_{1-EQ} + Q_{23} = Q_1 + Q_2 + Q_{23} \tag{8}$$

$$Q_{2-EQ} = Q_3 \tag{9}$$

$$\Delta P_{01-EQ} = \Delta P_{02} \tag{10}$$

$$\Delta P_{02-EQ} = \Delta P_{03} \tag{11}$$

$$\Delta P_{12-EQ} = \Delta P_{23} = \Delta P_{03} - \Delta P_{02} \tag{12}$$

$$R_{01-EQ} = \cfrac{1}{\cfrac{1}{R_{01} + R_{12}} + \cfrac{1}{R_{02}}} = \cfrac{1}{\cfrac{1}{R_{02}} + \cfrac{1}{R_{02}}} = \cfrac{R_{02}}{2} \tag{13}$$

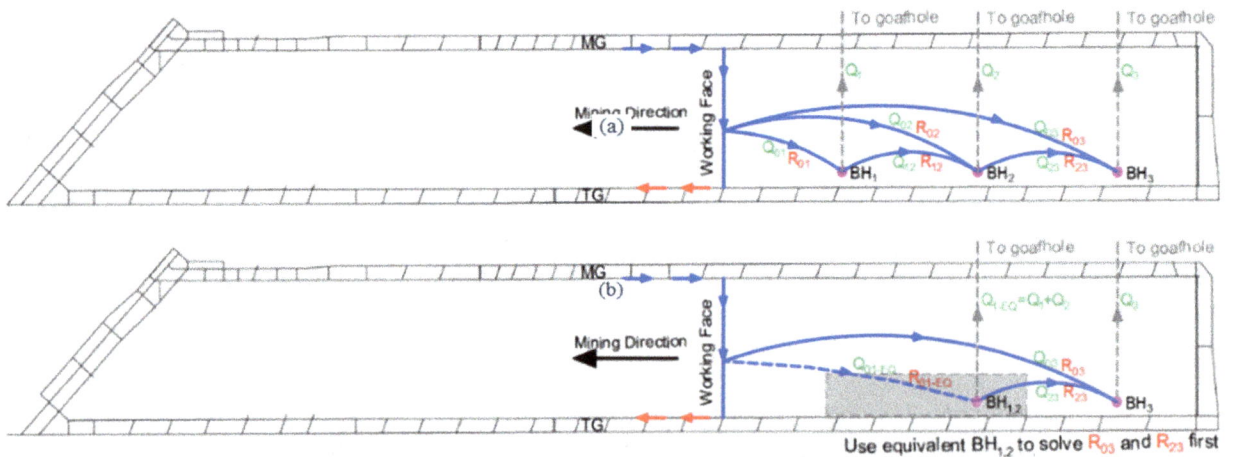

FIG 6 – A schematic of the simplified goaf ventilation network (three goaf holes case): (a) Base model; (b) Equivalent goaf hole ($BH_{1,2}$) model.

For the goaf air migration network with n (n>3) working goaf holes, the first (n-1) goaf holes are first equivalent to a new goaf hole ($BH_{1,n-1}$) as shown in Figure 7a and 7b. The relevant parameters of the new equivalent goaf hole can be solved by Equations 14–18, and then use the two goaf holes case Equations 2–7 to calculate R_{01-EQ}, $R_{0,n}$, and $R_{n-1,n}$. If $\Delta P_{n-1,n}$ is negative, there is no air leakage between BH_{n-1} and BH_n. Therefore, $R_{n-1,n}$ is assumed to be infinity large, and $R_{0,n}$ is equal to $\Delta P_{0,n}/Q_n$. Once $R_{0,n}$ and $R_{n-1,n}$ are determined, the same calculation methods can be repeated by merging the first (n-2) goaf holes into a new equivalent goaf hole $BH_{1,n-2}$ (Figure 7c), and then $R_{0,n-1}$ and $R_{n-2,n-1}$ can be calculated by repeating the previous steps. Thus, after applying the equivalent goaf hole assumption and Equations 2–7 multiple times, the resistances from working face to a number of active goaf holes can be back-propagated from BH_n to BH_2.

$$Q_{01-EQ} = Q_{1-EQ} + Q_{n-1,n} = \sum_{1}^{n-1} Q_i + Q_{n-1,n} \tag{14}$$

$$Q_{2-EQ} = Q_n \tag{15}$$

$$\Delta P_{01-EQ} = \Delta P_{0,n-1} \tag{16}$$

$$\Delta P_{02-EQ} = \Delta P_{0,n} \tag{17}$$

$$\Delta P_{12-EQ} = \Delta P_{n-1,n} = \Delta P_{0,n} - \Delta P_{0,n-1} \tag{18}$$

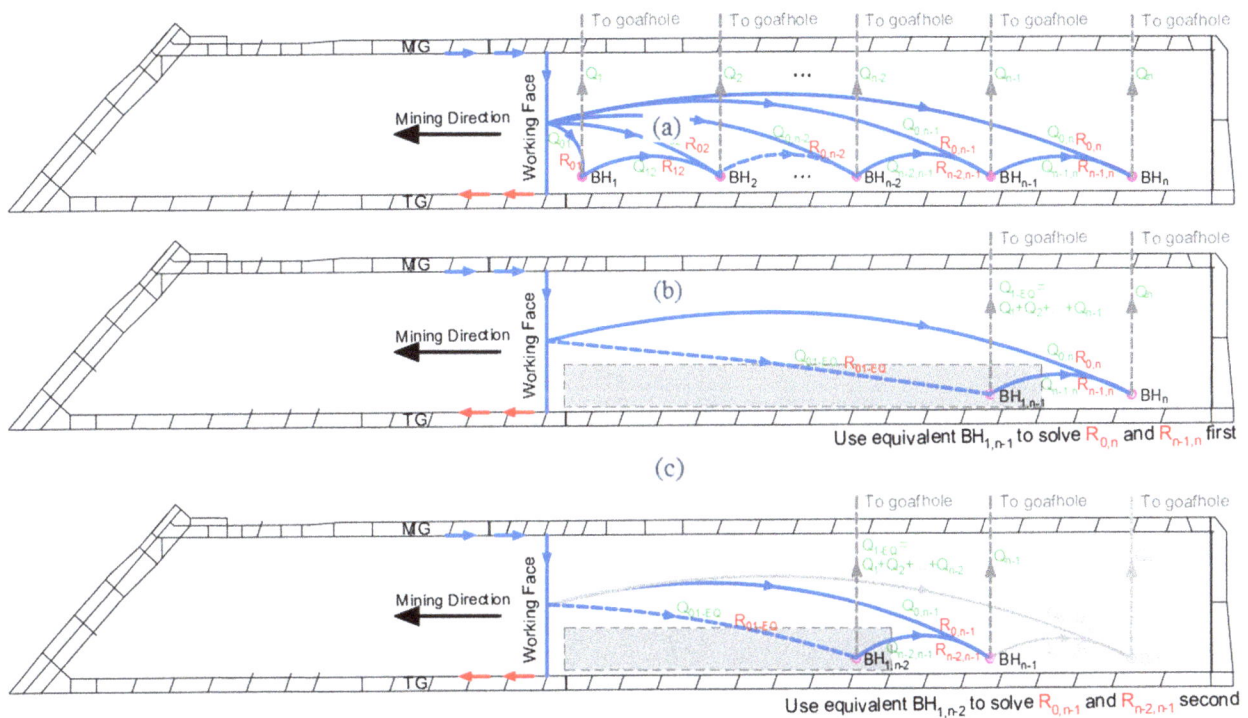

FIG 7 – A schematic of the simplified goaf ventilation network (n goaf holes case): (a) Base model; (b) Equivalent goaf hole ($BH_{1,n-1}$) model; (c) Equivalent goaf hole ($BH_{1,n-2}$) model.

Results analysis

The laminar flow assumption was proposed because the positive linear correlation between suction pressure and the flow rate was found in LW9 TG1-TG24. The resistance changes along the TG side goaf at LW9 can be obtained by applying the above theoretical resistance model. As Figure 8 shows, the resistance of several selected goaf holes is represented by dots in different colours, which refers to the resistance that the air transported from the working face to a single operating goaf hole at a specific face-to-hole position. As the airflow resistance in the goaf increases with the increase of the face-to-hole distance, it is more difficult for deep goaf holes to capture oxygen from the working face. All calculated resistances from the data set of TG1 to TG24 in LW9 are represented by blue triangles in Figure 9, and these discreated triangles were rolling-averaged per 10 m at the TG side to obtain the smooth resistance trend (the continuous line). The goaf resistance at the LW9 TG side increases rapidly to about $0.02 \times 10^6 \text{ N} \cdot \text{s/m}^5$ at the first 100 m of the goaf, and then the resistance raises slowly to $0.03 \times 10^6 \text{ N} \cdot \text{s/m}^5$ till 600 m from the working face.

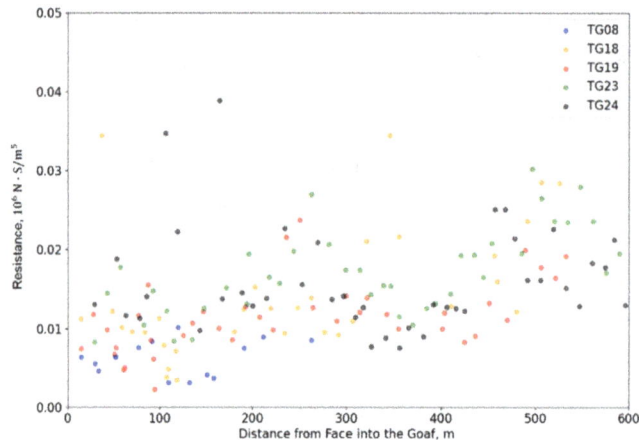

FIG 8 – Resistance profiles calculated on selected goaf holes along the TG side goaf at LW9.

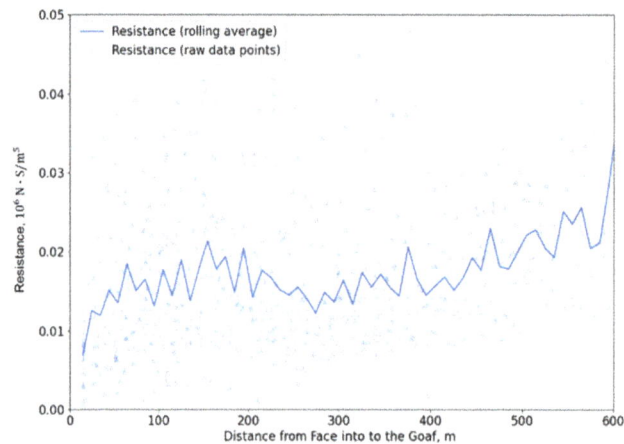

FIG 9 – Resistance profiles along the TG side goaf at LW9.

DISCUSSIONS

Applications of the theoretical resistance model

In addition to understanding the evolution of goaf resistance with respect to face-to-hole distance change, the theoretical resistance model can also predict the goaf permeability trend along the working face retreat direction. Based on the above goaf resistance calculation results, the goaf permeability can also be calculated by Equation 19 at the given goaf holes location.

$$Q = \frac{kA\Delta P}{\mu L} \tag{19}$$

Where, Q (m^3/s) is the volumetric flow rate; k (m^2) is the permeability; $A = 10 \times 2r$ (m^2) is the cross-sectional area; ΔP (Pa) is the pressure drop; μ $(\text{N} \cdot \text{s/m}^2)$ is the fluid viscosity, equals to $1.75 \times 10^{-5} \text{ N} \cdot \text{s/m}^2$ in the following calculation; L (m) is the length of the flow, equals to the face-to-hole distance.

The cross-sectional area (A in Equation 19) is calculated under the assumption that the vertical height of airflow area is 10 m, which is roughly three time of mining height. As shown in Figure 10, the horizontal length of the cross-sectional area equals to $2r_1$, $2r_2$, $2r_3$, $2r_n$ (r_1, r_2, r_3, and r_n are calculated using Equation 1) for goaf hole BH_1, BH_2, BH_3, BH_n, respectively. The volumetric flow rate (Q in Equation 19) at a goaf hole location is the cumulative value of the air leakage rate in all outby goaf holes operating simultaneously on a same working day. Besides, the pressure drop (ΔP in

Equation 19) is collected from the goaf hole drainage data, ie the differential pressure between the working face and each goaf hole. As a result, the permeability at different face-to-hole distances is displayed in Figure 11. The calculated permeability values are in a reasonable range: the maximum value is around 1×10^{-8} m^2 immediately after the longwall face, and the minimum permeability is around 0.1×10^{-8} m^2. The average permeability followed by a decreasing trend along the TG side goaf, declined from 0.6×10^{-8} m^2 to 0.2×10^{-8} m^2 at 600 m face-to-hole distance. The goaf permeability in this paper follows the inverse relationship with the related resistance. Similar observations that the goaf permeability tends to decrease first and then maintain stable in the deep goaf was also reported by Zhang *et al* (2016).

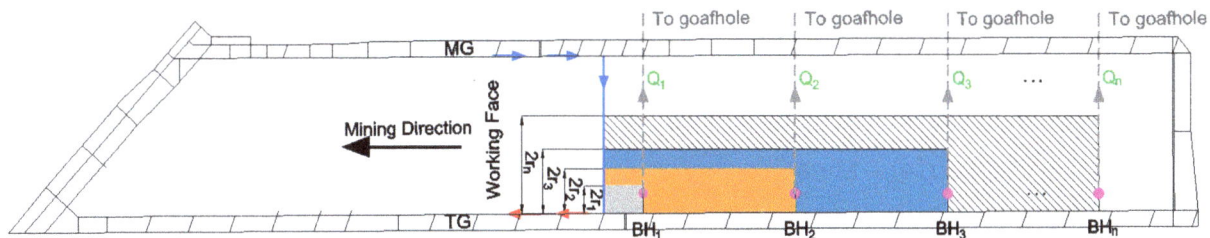

FIG 10 – A schematic of the simplified goaf air leakage area (n goaf holes case).

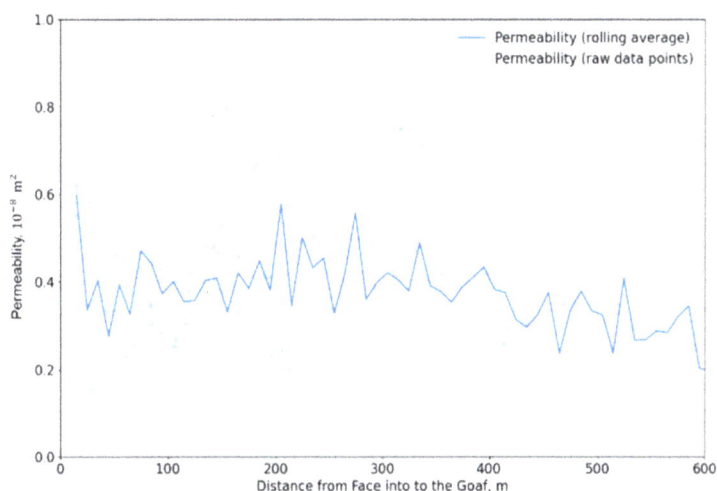

FIG 11 – Permeability profiles along the TG side goaf at LW9.

On-site measurement of goaf permeability is not widely implemented in mines due to its high implementation cost and low feasibility. Therefore, most previous studies rely on developing geomechanical models to estimate goaf permeability. Table 1 compares goaf permeability values obtained from previous investigations, which verified that the permeability values obtained using the proposed theoretical resistance method are in a reasonable range. Compared with other numerical models, results obtained from the theoretical resistance method were directly calculated from field monitoring data, which can rapidly predict goaf permeability distribution. The calculated goaf permeability distribution can be validated against other geomechanical model results or used as input for later goaf CFD models.

TABLE 1
Comparison of goaf permeability ranges with previous studies (m^2).

Previous studies	Maximum permeability ($\mathbf{m^2}$)	Minimum permeability ($\mathbf{m^2}$)
Szlązak (2001)	1.00×10^{-6}	5.10×10^{-9}
Esterhuizen and Karacan (2007)	1.00×10^{-6}	1.00×10^{-9}
Karacan (2009a) – High Bulking Factor	3.60×10^{-8}	1.50×10^{-8}
Karacan (2009a) – Low Bulking Factor	1.30×10^{-8}	5.10×10^{-9}
Ren et al (2011)	2.00×10^{-6}	2.00×10^{-9}
Marts et al (2014) – Mine C	5.10×10^{-6}	2.00×10^{-7}
Marts et al (2014) – Mine E	6.90×10^{-6}	2.00×10^{-6}
Zhang et al (2016, 2019)	9.00×10^{-5}	3.50×10^{-5}
Mine A (LW9) in this study	6.00×10^{-9}	2.00×10^{-9}

Limitations of the theoretical resistance model

Based on the linear correlation between suction pressure and flow rate (as per above section *The relationship between suction pressure and flow rate*), this theoretical model was built upon the assumption that all airflows in the goaf are laminar flows. However, after the total flow and seam gas flow rate profiles are normalised by suction pressure in Figure 4b, the goaf airflow in the first 100 m is not completely laminar.

To simplify the calculation of airflow pathways in the goaf, it assumes that the air transport can only occur from the working face to each goaf hole and between adjacent goaf holes. The ventilation branches on the TG and MG sides are only one-dimensional along the goaf, without considering the 3D air migration in the goaf or across multiple goaf holes. As per above section *Theoretical background and assumptions*, it also assumed that the pressure on the working face is roughly equal to the barometric pressure, but the actual pressure would be lower due to the general suction of mine body ventilation. Additionally, the pressure loss calculations in the theoretical model only considered head pressure losses. Based on the Darcy–Weisbach equation (Equation 20), the pressure loss due to viscous effects can be calculated. For the laminar flow in a circular pipe, the Darcy friction factor equals 64/Re, and the Darcy–Weisbach equation can be rewritten as Equation 21. As a result, the maximum friction pressure loss value is about 22 Pa (0.022 kPa) at a maximum flow rate of 1500 L/s, which is quite low for the pressure difference between two adjacent goaf holes. In this way, the friction pressure loss was ignored in this resistance model results calculation. Besides, the airflow in the goaf can only move from the higher pressure to the lower pressure regime, and there are some conditions would result some infinite large resistance values between two adjacent goaf holes.

$$\Delta P_{fric} = f_D \frac{\rho v^2 L}{2D} \tag{20}$$

$$\Delta P_{fric} = \frac{128 \mu Q L}{\pi D^4} \tag{21}$$

where, ΔP_{fric} (Pa) is friction pressure loss; f_D is the Darcy friction factor (also called flow coefficient); ρ (kg/m^3) is the fluid density; v (m/s) is the mean flow velocity, experimentally measured as the volumetric flow rate Q (m^3/s) per unit cross-sectional wetted area; L (m) is the length of the flow; D (m) is the hydraulic diameter of the pipe; μ (N · s/m^2) is the fluid viscosity.

CONCLUSIONS

This research focuses on analysing goaf gas drainage data from the study underground Mine A in Australia. To develop the correlation between suction pressure and total flow rate, the goaf seam gas flow rate and air leakage rate was analysed in detail as the face-to-hole distance increased. In general, the total flow rate along the TG side increased rapidly to the peak value until about 100 m into the goaf, and then slowly decreased and levelled off eventually. In addition, suction pressure is found to be positively correlated with total flow rate. As the suction pressure increases, gas capture and air leakage increase. Besides, this paper proposed a theoretical model to calculate the goaf resistance of ventilation air leakage pathways under the effect of intensive goaf drainage. The model results suggested that the resistance of air leakage pathways from the working face to individual goaf holes increased with the face-to-hole distance. The results from the resistance can be used to calculate goaf permeability, which is consistent with goaf permeability distributions proposed by other researchers. The goaf resistance calculation can be used as input for other numerical models, such as CFD or Ventsim, to provide field data for model calibration and validation.

ACKNOWLEDGEMENT

The authors would like to thank Australian Coal Research Association Project (ACARP) C29017 for sponsoring this research.

REFERENCES

Akgun, F and Essenhigh, R H, 2001. Self-ignition characteristics of coal stockpiles: theoretical prediction from a two-dimensional unsteady-state model, *Fuel*, 80(3):409–415.

Balusu, R, Deguchi, G, Holland, R, Moreby, R, Xue, S, Wendt, M and Mallett, C, 2002. Goaf gas flow mechanics and development of gas and Sponcom control strategies at a highly gassy mine, *Coal and Safety*, 20(3):35–45.

Coward, H F and Jones, G W, 1952. *Limits of flammability of gases and vapors*, Washington US Gov. Print. Off.

Esterhuizen, E, Mark, C and Murphy, M M, 2010. Numerical Model Calibration for Simulating Coal Pillars, Gob and Overburden Response, *Proceedings of the 29th International Conference on Ground Control in Mining*, pp 46–57.

Esterhuizen, G and Karacan, C Ö, 2007. A Methodology for Determining Gob Permeability Distributions and its Application to Reservoir Modeling of Coal Mine Longwalls, *2007 SME Annual Meeting*, Denver, CO.

Esterhuizen, G S and Karacan, C Ö, 2005. Development of numerical models to investigate permeability changes and gas emission around longwall mining panel, in *Alaska Rocks 2005, The 40th US Symposium on Rock*.

Karacan, C Ö, 2009a. Prediction of Porosity and Permeability of Caved Zone in Longwall Gobs, *Transport in Porous Media*, 82(2):413–439.

Karacan, C Ö, 2009b. Reconciling longwall gob gas reservoirs and venthole production performances using multiple rate drawdown well test analysis, *International Journal of Coal Geology*, 80(3–4):181–195.

Karacan, C Ö, Diamond, W P, Schatzel, S J and Garcia, F, 2006. *Development and Application of Reservoir Models for the Evaluation and Optimization of Longwall Methane Control Systems*, National Institute for Occupational Safety and Health, Pittsburgh Research Laboratory.

Karacan, C Ö, Esterhuizen, G S, Schatzel, S J and Diamond, W P, 2007. Reservoir simulation-based modeling for characterizing longwall methane emissions and gob gas venthole production, *International Journal of Coal Geology*, 71(2–3):225–245.

Karacan, C Ö, Ruiz, F A, Cotè, M and Phipps, S, 2011. Coal mine methane: A review of capture and utilization practices with benefits to mining safety and to greenhouse gas reduction, *International Journal of Coal Geology*, 86(2–3):121–156.

Marts, J A, Gilmore, R C, Brune, J F, Bogin Jr, G E, Grubb, J W and Saki, S, 2014. Dynamic gob response, *Mining Engineering*, 66(12):59–66.

Mcpherson, M J, 2009. *Subsurface ventilation and environmental engineering*, Chapman and Hall.

Ren, T and Balusu, R, 2005. CFD modelling of gob gas migration to improve the control of spontaneous combustion in longwalls, in *Proceedings of the 2005 Coal Operators' Conference*, University of Wollongong and The Australasian Institute of Mining and Metallurgy, pp 259–264.

Ren, T and Edwards, J S, 2002. Goaf gas modelling techniques to maximise methane capture from surface gob wells, in *Proceedings of the 9th North American/US Mine Ventilation Symposium*, pp 8–12.

Ren, T, Balusu, R and Claassen, C, 2011. Computational fluid dynamics modelling of gas flow dynamics in large longwall goaf areas, in *Proceedings of 35th Application of Computers and Operations Research in the Minerals Industry Symposium*, pp 603–613.

Schmal, D, Duyzer, Jan, H and van Heuven, J W, 1985. A model for the spontaneous heating of coal, *Fuel*, 64(7):963–972.

Si, G and Belle, B, 2019. Performance analysis of vertical goaf gas drainage holes using gas indicators in Australian coal mines, *International Journal of Coal Geology*, 216, p.103301.

Szlązak, J, 2001. The determination of a co-efficient of longwall gob permeability, *Archives of Mining Sciences*, 46(4):451–468.

Wu, J and Liu, X, 2011. Risk assessment of underground coal fire development at regional scale, *International Journal of Coal Geology*, 86(1):87–94.

Wu, Z, Jiang, S, He, X, Wang, L and Lin, B, 2007. Study of 3-D Numerical Simulation for Gas Transfer in the Goaf of the Coal Mining, *Journal of China University of Mining and Technology*, 17(2):152–157.

Xiang, Z, Si, G, Wang, Y, Belle, B and Webb, D, 2021. Goaf gas drainage and its impact on coal oxidation behaviour: A conceptual model, *International Journal of Coal Geology*, 248:103878.

Yang, Y, Li, Z, Tang, Y, Liu, Z and Ji, H, 2014. Fine coal covering for preventing spontaneous combustion of coal pile, *Natural Hazards*, 74(2):603–622.

Yuan, L and Smith, A C, 2008. Numerical study on effects of coal properties on spontaneous heating in longwall gob areas, *Fuel*, 87(15–16):3409–3419.

Zhang, C, Tu, S and Zhao, Y, 2019. Compaction characteristics of the caving zone in a longwall goaf: a review, *Environmental Earth Sciences*, 78(1).

Zhang, C, Tu, S, Zhang, L, Bai, Q, Yuan, Y and Wang, F, 2016. A methodology for determining the evolution law of gob permeability and its distributions in longwall coal mines, *Journal of Geophysics and Engineering*, 13(2):181–193.

Characterisation and calculation method of friction resistance based on fractal theory of roadway rough surface

B Wu[1], C Zhao[2] and B Lei[3]

1. Professor, School of Emergency Management and Safety Engineering, China University of Mining and Technology, Beijing 100083. Email: wbelcy@vip.sina.com
2. Doctoral candidate, School of Emergency Management and Safety Engineering, China University of Mining and Technology, Beijing 100083. Email: zcgzgkydx12@163.com
3. Associate professor, School of Emergency Management and Safety Engineering, China University of Mining and Technology, Beijing 100083. Email: leibws@163.com

ABSTRACT

At present, the friction factor is calculated by measuring pressure drop and airway configurations. However, the workload of on-site measurement is heavy, and the measurement results are easily disturbed by the external environment. In order to meet the need of intelligent ventilation to quickly and accurately obtain the friction resistance of roadway. Based on the important of roughness in Atkinson friction resistance formula, in this paper, fractal theory is used to characterise the 3D rough surface, and a method based on fractal dimension to describe the roughness of roadway surface and calculate the friction resistance was proposed. A computational model of friction resistance for fractal characterisation of roadway 3D rough surface was established. The parameters of the model were deduced and verified by using Nikuradse experimental data. The fractal dimension of roadway surface was calculated by 3D laser scanning data of Kailuan roadway, the established friction resistance calculation method was compared with the field resistance measurement results. It is considered that the calculation method of friction resistance based on fractal theory is reliable and can well represent the influence of 3D rough surface characteristics of roadway. The result show that the new established friction resistance formula compared with the traditional Atkinson formula in Nikuradse experimental, the average relative error is 6.874 per cent, and compared with the measured results of ventilation resistance in Kailuan experimental roadway, the relative error is 6.667 per cent. Comprehensive evaluation and analysis suggest that the new formula can well reflect the influence of 3D rough surface on friction resistance, combined with 3D laser scanning technology, the friction resistance of roadway can be calculated quickly and accurately. This research provided a reference for the key technology and theoretical development of intelligent ventilation parameter measurement.

INTRODUCTION

At present, Chinese mine ventilation system is facing the major demand of upgrading to information and intelligence. Rapid and accurate acquisition of resistance is the basis of real-time calculation of wind network and accurate adjustment of air volume (Zhou *et al*, 2020). Atkinson formula (McPherson, 1993) shows that the resistance is only related to the roadway characteristics. When the length, cross-sectional area, perimeter and wall relative roughness of a roadway are determined, the frictional wind resistance is also determined. However, the wall roughness is difficult to measure directly. At present, there are only two methods for coalmine technicians to obtain the resistance. One is to measure the wind volume and pressure of the roadway and calculate it by using the resistance square formula. The other is to obtain the roughness or friction resistance coefficient of similar roadway types according to the manual and calculate the resistance by using Atkinson formula. However, in the above two methods, the former has a large workload and cumbersome, and the measurement results are easy to be disturbed by the external environment, while the latter is difficult to find the exactly corresponding roughness or friction resistance coefficient value in the manual for a certain roadway, resulting in large error in the calculation results. So there is an urgent need for a method to quickly obtain roadway friction wind resistance. In recent years, 3D laser scanning technology has developed rapidly in the fields of architectural modelling and cultural relics protection because it can quickly obtain the three-dimensional shape of object surface (Yao *et al*, 2020). However, it is rarely used in mine roadway. Bråtveit, Lia and Olsen (2012) used a 3D laser scanner to scan the point cloud of water conservancy tunnel, and proposed a method to obtain the wall roughness based on the point cloud data. Watson and Marshall (2018) used point cloud data to obtain mine roadway wall roughness and calculate resistance. However, the roughness parameter in the Atkinson formula is 1D roughness, which cannot suitable for the characterisation of 3D rough surface. Therefore, it is necessary to establish a new method to calculate the resistance, which can characterise the 3D rough surface of roadway.

Fractal geometry, which was produced in the late 1970s, is a new subject with irregular geometry as the research object. Because the fractal characterisation of rough surface can well avoid the problem of too many roughness parameters in statistical model (Ge and Chen, 1999a), fractal theory has been widely popularised and applied in the characterisation of rough surface (Gagnepain and Roques-Carmes, 1986; Berry and Lewis, 1980). Fractal geometry is applied to geotechnical mechanics, and the fractal characterisation of rock roughness is discussed in detail (Xie, 1992). Sun *et al* (2019) divides the rock joint roughness into first-order large protrusion and second-order small protrusion, and uses the double-order fractal dimension to refine the characterisation of roughness. Based on the fractal idea, Ban *et al* (2019) proposed a roughness evaluation system which can reflect the anisotropy of joint surface and is not affected by the measurement scale. Ge and Chen (1999b) combines fractal dimension with scale coefficient, and proposes a characteristic roughness parameter to characterise the surface roughness.

It can be seen from the above that fractal theory has been widely used in characterising 3D rough surface, while 1D roughness is still used to calculate the roughness in the field of mine ventilation. Therefore, based on the fractal theory, this paper develop a method of resistance which can characterise the 3D roughness of roadway. Combined with the 3D laser scanning technology, the resistance of roadway can be calculated quickly and accurately, which provides a reference for the theoretical development of resistance and the development of key technologies of intelligent ventilation parameter measurement.

THEORETICAL MODEL OF RESISTANCE BASED ON FRACTAL DIMENSION

Derivation of Atkinson formula

In order to quantify the relationship between wind volume and resistance, Atkinson summarised Atkinson formula (McPherson, 1993) Based on Darcy experiment. When gas flows along the roadway, the wall shear stress is τ, The total resistance of the wall must be equal to the pressure that moves the gas, as shown in Equation 1.

$$\tau S = PA , \tag{1}$$

Where τ is the wall shear stress per unit area, pa; S is the total wall surface area, m²; P is the pressure, pa; A is the cross-sectional area of the roadway, m².

If the flow is completely turbulent, the shear stress applied to the roadway wall is also proportional to the kinetic energy of the air flow, as shown in Equation 2.

$$\tau = f\rho \frac{v^2}{2} , \tag{2}$$

Where ρ is the air density, kg/m³; v is the flow velocity of gas in the roadway, m/s; f is the friction factor. Substitute Equation 2 into Equation 1 to obtain the calculation formula of wind resistance P.

$$P = f\rho \frac{Sv^2}{2A} , \tag{3}$$

For the convenience of calculation, the roadway surface area is treated as a smooth surface. And $S = LU$ and wind volume $Q = vA$ are substituted into Equation 3.

$$P = f\rho \frac{S}{2A^3} Q^2 , \tag{4}$$

Where Q is the roadway wind volume, m³/s. L is the roadway length, m; U is the perimeter of roadway section, m.

$$R_t = f\rho \frac{S}{2A^3} = f\rho \frac{LU}{2A^3}, \tag{6}$$

Where R_t is the Atkinson resistance, kg/m^7. Predecessors (McPherson, 1993) deduced the relationship between friction factor f, Reynolds number Re and relative roughness e/d. Based on Nikuradse experiment. In the case of complete turbulence, the friction factor is only related to the roughness. As shown in Equation 7.

$$f = \left[4\log_{10}(\frac{e/d}{3.7})\right]^{-2}, \tag{7}$$

Where e is the height of the rough element, d is the equivalent diameter. For non-circular section, $d = 4A/U$.

The relative roughness in Equation 7 is the height of the rough element, which cannot reflect the change of 3D rough surface. However, both roadway surface area s and cross-sectional area a are related to roughness. Based on this, the inner surface area of roadway can be characterised by fractal theory.

Fractal theory analysis of wall roughness element

According to the calculation method of box dimension, compared with the fractal expression of pore distribution in porous media (Wang *et al*, 2019), the number distribution function expression of wall roughness element size greater than d_r is shown as Equation 8.

$$N(\geq d_r) = (d_{max}/d_r)^D, \tag{8}$$

Where d_r is size of rough element, D is the fractal dimension of rough surface, d_{max} is size of maximum roughness element when $N(d_{max})=1$. $N(d_r)$ is the number of rough elements with size greater than or equal to d_r. In Equation 8, the number of rough elements can be regarded as a continuous differentiable function, and the number of rough elements with a diameter of $d_r \sim (d_r + dd_r)$ is shown as Equation 9.

$$-dN = Dd_{max}^D d_r^{-(D+1)}dd_r. \tag{9}$$

Calculation formula of resistance based on fractal dimension

Assuming that the rough element of the inner surface of the roadway is replaced by a hemisphere (Wang *et al*, 2019), as shown in Figure 1a, the whole rough surface area is the sum of the surface areas of all hemispheres and the sum of the smooth surface.

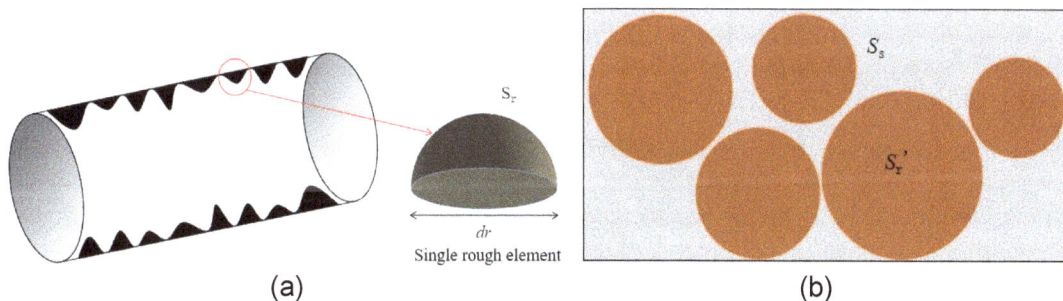

FIG 1 – Simplified structural model of wall roughness element.

The surface area of all hemispheres in a fractal set can be obtained by summing the integral from the minimum rough element to the maximum rough element. The calculation formula of the surface area of the rough element is:

$$S_r = -\int_{d\,min}^{d\,max} 2\pi(\frac{d}{2})^2 dN. \tag{10}$$

Introducing variables $\varepsilon = \dfrac{d_{min}}{d_{max}}$, Substitute Equation 9 into Equation 10 to obtain:

$$S_r = \frac{\pi D d_{max}^D}{2} \frac{1}{2-D}\left(1 - \varepsilon^{2-D}\right).$$

(11)

The smooth surface area S_s is:

$$S_s = UL - \left(-\int_{d\,min}^{d\,max} \pi \left(\frac{d}{2}\right)^2 dN\right)$$

$$= UL - \frac{\pi D d_{max}^D}{4} \frac{1}{2-D}\left(1 - \varepsilon^{2-D}\right)$$

(12)

The total rough surface area is:

$$S_t = S_r + S_s = UL + \frac{\pi D d_{max}^D}{4} \frac{1}{2-D}\left(1 - \varepsilon^{2-D}\right)$$

(13)

UL is the surface area of the smooth wall of the roadway, and Sr' is the projected area of the rough element on the plane, as shown in Figure 1b. Introduce variables φ is the percentage of the total area of the bottom surface of all hemispheres in the total area of the whole fractal set unit, then UL is:

$$UL = \frac{S_r'}{\varphi} = \frac{\pi}{4\varphi'} \frac{D d_{max}^D}{4} \frac{1}{2-D}\left(1 - \varepsilon^{2-D}\right),$$

(14)

Substitute Equation 14 into Equation 13:

$$S_t = \left(\frac{\pi}{4\varphi} + \frac{\pi}{4}\right) \frac{D d_{max}^D}{4} \frac{1}{2-D}\left(1 - \varepsilon^{2-D}\right)$$

(15)

Substituting Equation 15 into Equation 14, the friction wind resistance calculation formula Rf based on fractal dimension is:

$$R_f = f\rho \frac{\left(\dfrac{\pi}{4\varphi} + \dfrac{\pi}{4}\right) \dfrac{D d_{max}^D}{4} \dfrac{1}{2-D}\left(1 - \varepsilon^{2-D}\right)}{2A^3}.$$

(16)

For the cross-sectional area, as shown in Figure 1a, the influence of roughness change on the cross-sectional area of roadway is small and can be ignored. Therefore, the calculation formula of resistance based on fractal dimension is shown in Equation 16.

In the real mine, the roadway is usually hundreds of metres long, and the workload of calculating the fractal dimension of the whole roadway surface is huge and unnecessary. Therefore, the characteristic length l_c and the characteristic section scale u_c are introduced, make the fractal dimension of the local area of $u_c \times l_c$ close to the fractal dimension of the whole roadway surface. As shown in Figure 2, the length coefficient k and section scale coefficient m are introduced, $k = L/l_c$, $m = U/u_c$. In addition, the frictional wind resistance of roadways with different section shapes is also different. Therefore, the section shape coefficient c needs to be introduced to characterise the influence of different section shapes on resistance.

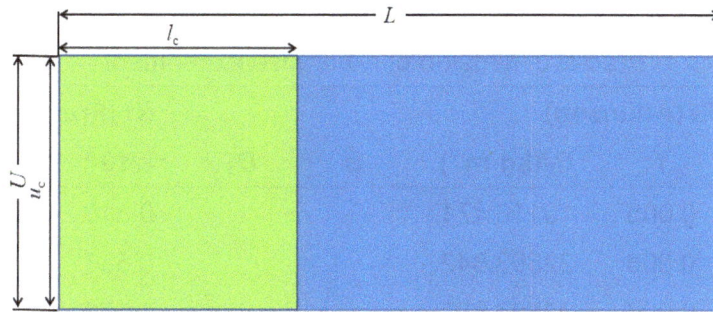

FIG 2 – Influencing factors of friction coefficient.

Therefore, f is related to the length coefficient k, the scale coefficient m, the shape coefficient c and ε.

$$f = f(k,m,c,\varepsilon) = k \times m \times c \times f'(\varepsilon) \tag{17}$$

Where f' is the roughness coefficient, only related to ε. Substitute Equation 17 into Equation 16.

$$R_f = f'\rho\frac{kmc(\frac{\pi}{4\varphi'}+\frac{\pi}{4})\frac{Dd_{\max}^D}{4}\frac{1}{2-D}\left(1-\varepsilon^{2-D}\right)}{2A^3} \tag{18}$$

F' CALCULATION AND EXPERIMENTAL VERIFICATION

It can be seen from Equation 18 that only the roughness coefficient f' in the resistance model is unknown. In order to obtain the calculation expression of f', this paper uses Nikuradse experimental data for calculation and verification. The diameter of Nikuradse experimental pipe is 0.1 m. During the experiment, sand particles were pasted in the pipeline, and the diameter of sand particles were 0.010×10^{-2}, 0.020×10^{-2}, 0.040×10^{-2}, 0.083×10^{-2}, 0.160×10^{-2}, 0.330×10^{-2} m. Nikuradse finally obtained the relationship between friction coefficient f, Reynolds number Re and relative roughness e/d through a series of experiments on pipeline velocity, as shown in Figure 3.

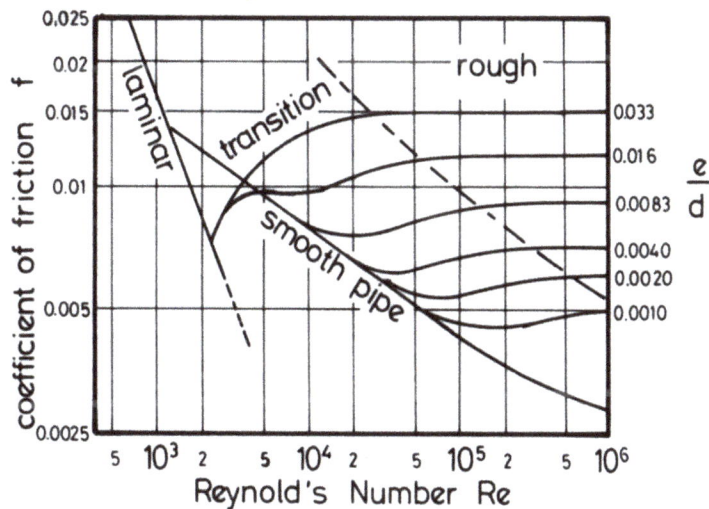

FIG 3 – The result of Nikuradse experiment.

Assuming that the length of the experimental pipeline is 10 m, according to Equation 6 and Equation 7, the resistance R_t of six roughness are calculated respectively, as shown in Table 1. In order to facilitate the calculation of the fractal dimension of the pipe surface, the arrangement of sand particles on the pipe wall is defined as the uniform arrangement as shown in Figure 4a. For the rough surface of sand particles with different diameters, the fractal dimension Ds should be the same. The relationship between 3D fractal dimension D_S and 2D fractal dimension D_L is $D_S = D_L + 1$. Therefore, this study establishes 3.14 cm × 3.14 cm rough surface, as shown in Figure 4, the coefficient φ of the bottom surface of the rough unit in the total surface is calculated by MATLAB.

TABLE 1

Fractal dimension of Nikuradse experiment.

Number	Rt (Atkinson)					Rf (Fractal)			
	e/d	f	$R_t/(kg \cdot m^{-7})$	φ	D_S	$\varepsilon/10^{-3}$	k	m	$f'/10^{-3}$
1	0.001	0.005	19118.471			0.318			0.027
2	0.002	0.006	22803.842			0.637			0.052
3	0.004	0.007	27667.387	0.785	2.667	1.270	31.85	1	0.100
4	0.008	0.009	34685.463			2.640			0.210
5	0.016	0.011	43553.892			5.090			–
6	0.033	0.015	57939.737			0.0105			–

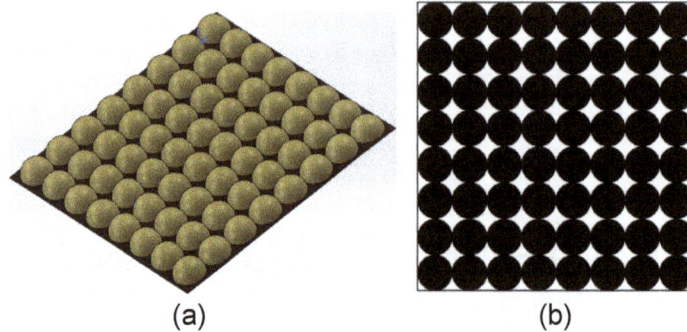

(a)   (b)

FIG 4 – Rough element diagram of Nikuradse experiment.

The 3D fractal dimension D_S of rough surface is 2.667. The value of parameters φ, ε, k, m are as shown in Table 1. The first four groups of resistance Rt were substituted into Equation 18 and calculated the roughness coefficient f'. By linear fitting ε and f', the calculation expression of roughness coefficient f' is obtained, shown as Figure 5.

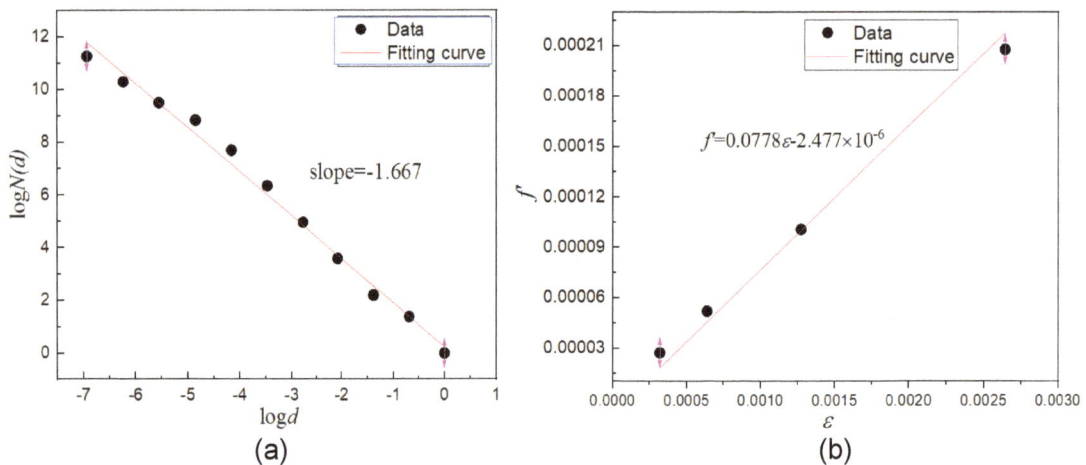

(a)   (b)

FIG 5 – Fitting of roughness coefficient f'.

Calculating the friction wind resistance R_f of groups 5–6 through Equation 19. As shown in Table 2, comparing the calculation results of R_t and R_f, when the relative roughness is 0.016 and 0.033, the relative errors of Rf and Rt are 3.963 per cent and 9.775 per cent respectively, both less than 10 per cent, and the average relative error is 6.874 per cent, which proves that the empirical formula of linear fitting is feasible. The above research is based on regular rough surface. For the irregular surface of real roadway, the feasibility of this method needs to be further verified.

TABLE 2

Comparison of fractal dimension friction resistance Rf and Atkinson friction resistance Rt.

Rt					Rf						error/%
e/d	f	R_t / (kg·m^{-7})	φ	D_S	ε / 10^{-3}	k	m	f' / 10^{-3}	R_t / (kg·m^{-7})	$|R_f - R_t|/R_t$	
0.016	0.011	43553.892	0.785	2.667	5.090	31.85	1	0.410	41827.770	3.963	
0.033	0.015	57939.737			1.050			0.900	52276.042	9.775	

FIELD VERIFICATION

In order to verify the feasibility of this calculation method in real roadway, a full-scale roadway study is carried out. The full-size roadway is located in Kailuan team of national mine emergency rescue in Tangshan City, Hebei Province. The inner surface of the roadway is supported by shotcrete, as shown in Figure 6a. By scanning the return air roadway of Kailuan roadway with a 3D laser scanner, the point cloud data is obtained, as shown in Figure 6b. Based on the point cloud data, the roadway parameters are obtained: the length L is 81.200 m, the section perimeter u is 9.370 m, and the section area a is 8.300 m^2. By using JFY-4 ventilation multi parameter detector to measure the roadway air volume Q and pressure P, the resistance R_t is calculated, as shown in Table 4.

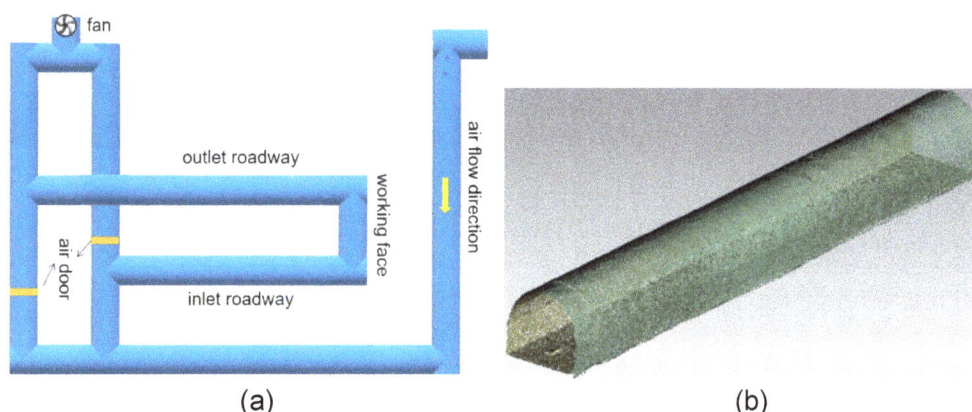

(a)   (b)

FIG 6 – 3D scanning point cloud data of Kailuan airway.

Since the roadway section shape is circular arch, the section shape coefficient C obtained according to Table 3 is 1.080.

TABLE 3

Roadway section shape coefficient (McPherson, 1993).

	Section shape	Shape coefficient
	Circular	1.000
Circular arch	Straight arm arch	1.080
	Semicircular arch	1.090
	Square	1.130
	Aspect ratio=1.5:1	1.150
Rectangle	2:1	1.200
	3:1	1.300
	4:1	1.410

As shown in Figure 7a, a rough surface of 2.7 m × 2.7 m is selected at the bottom of the roadway, and the binary Figure 7b is obtained through MATLAB. Calculating the D_S is 2.706 and the φ is 38.584 per cent.

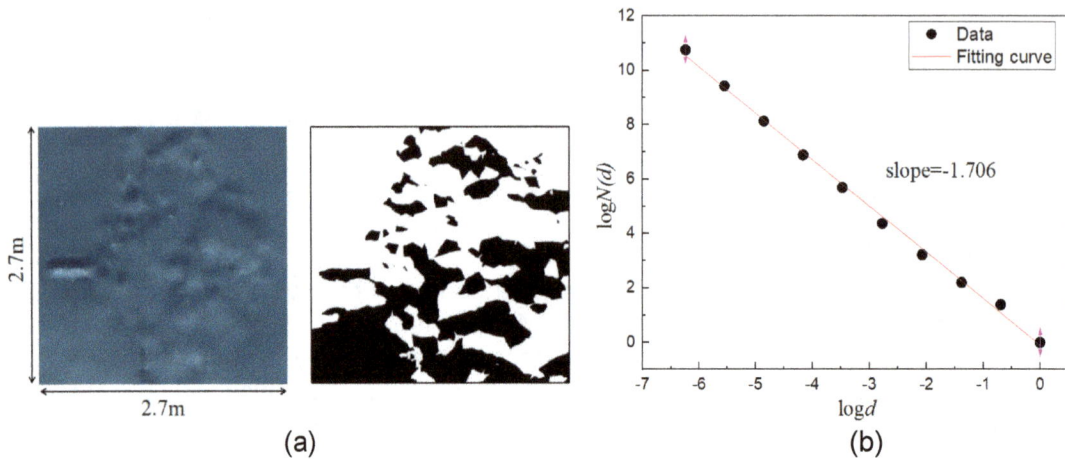

FIG 7 – Rough surface fractal dimension of return airway.

The resistance calculated by the square formula of resistance and fractal theory are shown in Table 4.

TABLE 4

Comparison of calculation and measurement results of friction resistance in Kailuan experimental roadway.

Resistance square formula			Calculation of resistance by fractal dimension									error/%
Q / $(m^3 \cdot s^{-1})$	P / Pa	R_t / $(kg \cdot m^{-7})$	k	m	c	D	d_{max} /m	d_{min}/m	f / 10^{-3}	ε / 10^{-3}	R_f / $(kg \cdot m^{-7})$	$|R_f - R_t|/R_t$ ×100
7.3	0.8	0.015	30.1	3.5	1.08	2.706	2.7	2.2×10^{-4}	0.004	0.083	0.014	6.667

It can be seen from Table 4 that the relative error between the Rf based on fractal dimension and the R_t calculated by the resistance square formula is 6.7 per cent, the R_f is smaller compared with the calculation results. The error is acceptable, which proves the reliability of the calculation method.

CONCLUSION

- The calculation formula of resistance based on Fractal representation is derived through fractal theory, as follows:

$$R_f = f' \rho \frac{kmc(\frac{\pi}{4\varphi'} + \frac{\pi}{4}) \frac{Dd_{max}^D}{4} \frac{1}{2-D} \left(1 - \varepsilon^{2-D}\right)}{2A^3}$$

The new friction formula includes length coefficient k, section scale coefficient m, shape coefficient c and roughness coefficient f'. It can better characterise the influence of the concave convex characteristics of the three-dimensional rough surface of the roadway on the resistance.

- The empirical expression of roughness coefficient f' is calculated by using 1–4 groups of experimental data of Nikuradse, and 5–6 groups of data are selected to calculate the Rf by the new method. Compared with Atkinson resistance Rt, the relative errors are 3.963 per cent and 9.775 per cent respectively, and the average relative error is 6.874 per cent.

- The point cloud data of return air roadway in Kailuan roadway are obtained by 3D laser scanner, and the fractal dimension of roadway rough surface is calculated. By comparing the Rf calculated by the new method with the Rt calculated by the resistance square formula, it is found that the relative error between them is 6.667 per cent. the result indicate that the resistance calculation method based on fractal theory is reliable, and the roadway resistance can be obtained quickly combined with 3D laser scanning technology.

ACKNOWLEDGEMENTS

The authors would like to acknowledge the financial supports by National Key R&D Program of China (Grant No. 2018YFC0808100). And thanks for the support of Kailuan rescue team.

REFERENCES

Ban, L, Qi, C, Yan, F *et al.* 2019. A new method for determining the JRC with new roughness parameters, *Journal of China Coal Society*, 44(4):1059–1065.

Berry, M V and Lewis, Z V, 1980. On the Weierstrass-Mandelbrot fractal function, *Proceedings of the Royal Society of London A Mathematical and Physical Sciences*, 370(1743):459–484.

Bråtveit, K, Lia, L and Olsen, N R B, 2012. An efficient method to describe the geometry and the roughness of an existing unlined hydro power tunnel, *Energy Procedia*, 20:200–206.

Gagnepain, J J and Roques-Carmes, C, 1986. Fractal approach to two-dimensional and three-dimensional surface roughness, *Wear*, 109(1/2/3/4):119–126.

Ge, S and Chen, G, 1999b. Characterization of surface topography changes during running-in process with characteristic roughness parameter, *Journal of China University of Mining and Technology*, 28(3):4–7.

Ge, S R and Chen, G A, 1999a. Fractal prediction models of sliding wear during the running-in process, *Wear*, 231(2):249–255.

McPherson, M J, 1993. *Subsurface Ventilation Engineering*, pp 124–160.

Sun, S, Li, Y, Tang, C *et al.* 2019. Dual fractal features of the surface roughness of natural rock joints, *Chinese Journal of Rock Mechanics and Engineering*, 38(12):2502–2511.

Watson, C and Marshall, J, 2018. Estimating underground mine ventilation friction factors from low density 3D data acquired by a moving LiDAR, *International Journal of Mining Science and Technology*, 28(4):657–662.

Xie, H, 1992. Fractal geometry and its application to rock and soil materials, *Chinese Journal of Geotechnical Engineering*, 14(1):14–24.

Yao, X, Zhou, Y, Jia, S *et al.* 2020. Realization method and application of building reverse modeling based on 3d laser scanning, *Industrial Construction*, 50(3):178–181.

Zhou, F, Wei, L, Xia, T *et al.* 2020. Principle key technology and preliminary realization of mine intelligent ventilation, *Journal of China Coal Society*, 45(06):2225–2235.

Laboratory investigation of relationships between zeta potential and coal dust suppression efficiency

Z Zhao[1], P Chang[2], G Xu[3] and A Ghosh[4]

1. PhD candidate, WA School of Mines: Minerals, Energy and Chemical Engineering, Curtin University, Kalgoorlie WA 6430. Email: zidong.zhao@postgrad.curtin.edu.au
2. Lecturer, WA School of Mines: Minerals, Energy and Chemical Engineering, Curtin University, Kalgoorlie WA 6430. Email: ping.chang@curtin.edu.au
3. Associate Professor, Department of Mining Engineering, Missouri University of Science and Technology, Rolla MO 65409, United States. Email: guang.xu@mst.edu
4. Senior Lecturer, WA School of Mines: Minerals, Energy and Chemical Engineering, Curtin University, Kalgoorlie WA 6430. Email: apurna.ghosh@curtin.edu.

ABSTRACT

Coal dust is hazardous to underground coal workers and water spray is the most common and economic method for controlling coal dust. Previous studies proved that the addition of surfactants can improve the performance of water spray on coal dust suppression. The evaluation methods for surfactants vary from different studies. The zeta potential test is a common method for evaluating surfactants with ionicity. Generally, the suspension of coal dust with pure water usually showed a negative value of zeta potential. The addition of cationic surfactants can increase the zeta potential to a positive value with a small dosage, while the addition of anionic surfactants can only lower the zeta potential. However, there is a lack of comprehensive study to investigate the relationship between the zeta potential and the suppression performance for coal dust. The aim of this study is to reveal the consistency and discrepancy between the zeta potential and the suppression efficiency of a surfactant. Firstly, the mixture of cationic surfactant solutions and coal dust would be evaluated by the zeta potential test. The wind tunnel tests would then be conducted for the same surfactant solutions to measure and calculate the suppression efficiency. It was found that the addition of surfactant can increase the zeta potential and the suppression efficiency and it reached the greatest efficiency when the zeta potential approached to zero. The equilibrium of zeta potential can benefit the suppression performance of surfactants in the wind tunnel test. The result of this study can achieve a better understanding of the mechanism for coal dust suppression. The selection of surfactants will also be suggested for researchers or industries in terms of coal dust control. It would be more economic if a similar performance could be achieved by less consumption of surfactants.

INTRODUCTION

Coal dust is a fine powdered form of coal, which is usually generated during coal production. It has been marked as hazardous in recent decades, especially to the underground safety and the health of coal workers. Long-term exposure to coal dust can cause scarring in the lungs and impair the ability to breathe and thus result in the Coal Workers' Pneumoconiosis (CWP), which is the most famous occupational disease (Ayoglu *et al*, 2014; Castranova and Vallyathan, 2000; Liu *et al*, 2009). Currently, there is no effective treatment for CWP. Although it has been considered that the CWP has been controlled significantly in the past decades, there were still some new cases diagnosed with CWP in many countries. In Australia, there were only 82 cases of mine-dust lung disease that had been reported before 2019 (Hendry, 2019). However, 20 more cases were confirmed within two weeks in 2019 and 31 cases were diagnosed in 2020 (Burt and McGhee, 2020; Hendry, 2019). In China, more than 750,300 cases were diagnosed with pneumoconiosis in 2013 and 50 per cent of total cases were considered as CWP (World Health Organization, 2013; Yan *et al*, 2018). In 2016, CWP accounted for over 60 per cent of 800,000 cases of occupational pneumoconiosis (Doney *et al*, 2020; Kurth *et al*, 2020; Zhang *et al*, 2020). In the US, 20.6 per cent of long-tenured underground workers have CWP as estimated in 2016, compared to only 10 per cent in the 1990s, which proved that the CWP shows a resurgence in recent years (Blackley *et al*, 2016; Blackley, Halldin and Laney, 2018). Due to the significant growth of CWP prevalence, improvement of coal dust control is of necessity in mine industries.

Many studies have investigated dust controlling technologies and water spray with surfactant is one of the most common methods for capturing coal dust in underground coalmines. Surfactants can significantly improve the performance of capturing coal dust by water spraying. In Zhou *et al's* (2019) study, sodium dodecyl benzene sulfonate (SDBS) based surfactant solution was tested at a coalmine in China. The surfactant solution showed 76 per cent efficiency of underground dust suppression compared to 55 per cent of untreated water. In another study by the same research team, an SDBS based compound surfactant was evaluated at Lu Wa Coalmine, China (Zhou *et al*, 2017). This surfactant showed around 85 per cent efficiency of coal dust suppression, which is 45 per cent higher maximum than untreated water. In Wang *et al's* (2014a) study, a complex wetting agent was tested in an underground roadway. It showed around 94 per cent efficiency of dust suppression at road-head and 68 per cent at 100 m away from the heading face, compared to 85 per cent and 58 per cent by spraying water only. In Chang *et al's* (2020) study, four surfactants were studied in the wind tunnel test. The highest coal dust suppression efficiency was over 60 per cent, shown by cetyltrimethylammonium bromide (CTAB), while plain water can only achieve less than 40 per cent on dust suppression. A similar study can be found in Jin *et al's* (2019) research that a compound surfactant was evaluated in the wind tunnel test. The result showed that a 5 per cent concentration of the surfactant could suppress 99 per cent of the coal dust. All the above studies illustrate that the addition of surfactants into water spraying can critically improve the performance of coal dust suppression.

Although surfactants can significantly raise the suppressing efficiency, the result could differ from evaluating methods. In previous studies, the zeta potential test is commonly used for evaluating surfactants in terms of coal dust suppression. The zeta potential is a physical property that is exhibited in particles within the suspension. It generally shows a negative value when coal dust suspends into deionised water (Yang *et al*, 2007). Adding surfactants into the suspension can change the value more negative or become positive depending on the ionicity. Usually, surfactants can be grouped into four types, including cationic surfactants, anionic surfactants, amphoteric surfactants and non-ionic surfactants. Adding anionic surfactants can lower the zeta potential of the suspension; by contrast, cationic surfactants can increase the zeta potential even from negative to positive (Crawford and Mainwaring, 2001; Guo *et al*, 2018; Maršálek, 2008; Yang *et al*, 2007). In some studies, it has been considered that the larger change of zeta potential, the better performance the surfactants can achieve. In another word, surfactants with a large difference between absence and presence usually show great absorbability on coal dust (Kost, Shirey and Ford, 1980). For example, Chen, Xu and Albijanic (2017) evaluated SDBS surfactant in the zeta potential test. The result illustrated that 0.4 per cent SDBS had the greatest performance when the greatest difference of zeta potential can be observed, because large difference of zeta potential represents greater absorbability of coal. However, some researchers illustrated that a higher value of zeta potential represents greater suppression performance of surfactants. For instance, Chang *et al* (2016) evaluated a cationic surfactant in the zeta potential of coal pitch and the result showed that cationic surfactant could charge the zeta potential of suspension from negative to positive then keep stable. Before zeta potential approached zero, the concentration of surfactant was too low to absorb coal pitch, which is obtained by the distillation of coal. This is because an adsorption film is not formed yet. When the zeta potential was increased from zero, the absorbability was improved and the surface tension was reduced. The critical micelle concentration was achieved when the zeta potential was kept stable. Finally, a bi-layer or multi-layer adsorption film is formed to control the coal pitch. In other studies, it has been considered the greater performance of surfactants can be achieved when zeta potential reaches zero. When the zeta potential of the suspension reaches zero, the charges of coal particles and surfactants are basically equal, which means the suspension can coagulate rapidly (Kumar and Dixit, 2017). For example, Guo *et al* (2018) investigated the relationship between the zeta potential and the agglomeration efficiency of coal and ash particles. The result showed that both cationic and non-ionic surfactants can increase the zeta potential of suspension, whereas anionic surfactants decrease it. Moreover, it has been considered the highest particle removal efficiency can be achieved when the pH equals 4.3 and zeta potential equals 0 mV. The author illustrated that the particle suspension with a lower absolute value of zeta potential can form aggregation with high chance, which can be effectively suppressed by the dust collector.

Although the zeta potential test can evaluate the suppression ability of surfactants, results from different studies are inconclusive. There is a lack of comprehensive studies evaluating surfactants by zeta potential tests. Moreover, the results should be compared and verified with dynamic methods such as the field test or the laboratory wind tunnel test. The objective of this study is to investigate the surfactant performance and the relationship between zeta potential and suppression efficiency. In this study, a cationic surfactant, DTAB ranging from 0 per cent w/v to 0.005 per cent w/v, is examined in both the zeta potential test and wind tunnel test. The zeta potential of DTAB is firstly evaluated from negative to a positive value, then the suppression efficiency of each concentration is measured by wind tunnel test. The result of both tests is analysed and compared by Analysis of Variance (ANOVA). The findings of this study present a clear demonstration of the relationship between zeta potential and suppression efficiency for coal dust control and it is useful to improve the evaluation methods of surfactants, thus optimising the performance.

METHOD

Materials

The coal sample in this study is made by the medium volatile sub-bituminous coal, which is provided by the Premier Coal Company in Collie, Western Australia. The sample preparation follows the standard procedures of the American Society for Testing and Materials international standard (ASTM International, 2012, 2017). The raw coal sample is firstly crushed to small by a Jaw Crusher and ground to powders by a pulveriser. Then coal powders are dehydrated until the weight loss is less than 0.1 per cent per hour around 35°C in a sample drying oven. After dehydrating, the coal powders are sieved to less than 38 µm by a power sifter. Then coal samples are stored in sealed containers and prepared for experiments.

An anionic surfactant, DTAB, is tested in this study. Six concentrations of surfactant solution are selected for analysis including 0.000 per cent w/v, 0.001 per cent w/v, 0.002 per cent w/v, 0.003 per cent w/v, 0.004 per cent w/v and 0.005 per cent w/v. All surfactants are made with deionised water. These concentrations are selected based on the zeta potential ranges from negative to positive, while the concentration of DTAB ranges from 0.000 per cent w/v to 0.005 per cent w/v.

Zeta potential test

The zeta potential is a physical character that is exhibited by particles in suspension. In this test, 1 g coal samples are placed into 1000 mL surfactant solution then stirred for 15 min until the suspension reaches an equilibrium state, as shown in Figure 1. The test samples made with the suspension is tested then by the Zetasizer, model Nano-Z. The sample of each concentration is measured six times.

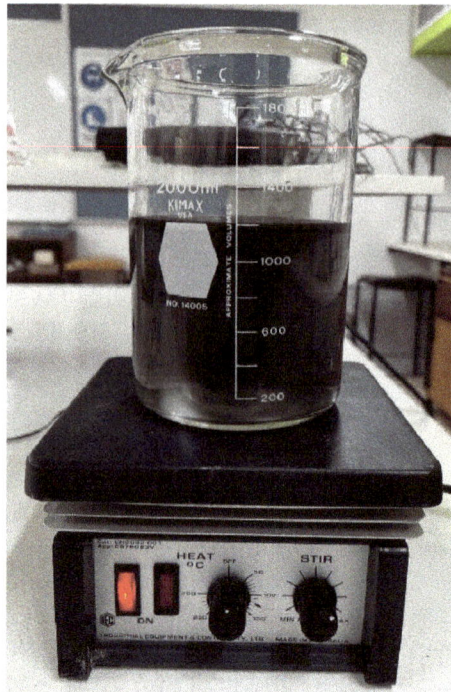

FIG 1 – Zeta potential sample preparation.

Laboratory wind tunnel test

The wind tunnel test is conducted by a laboratory wind tunnel apparatus, which is presented in our previous studies, as shown in Figure 2 (Chang *et al*, 2020; Zhao *et al*, 2021). In this study, the air velocity in the wind tunnel is 0.68 m/s. The spray rate of surfactant solutions is set to be 4.97 L/min. The main procedures of the wind tunnel test are shown as follows: (a) turn on all main power, after airflow in the tunnel keeps stable, coal particles are spread into the wind tunnel through dust generator; (b) once the airflow keeps steady, the dust concentration C_1 is measured and recorded for 120s; (c) the surfactant solution is sprayed out through the nozzle; (d) the second dust concentration C_2 is measured during spraying. The whole experiment is conducted twice. The suppression efficiency can be calculated by the following Equation (1):

$$\eta = \frac{C_1 - C_2}{C_1} \tag{1}$$

Where η represents the suppression efficiency, C_1 represents the initial coal dust concentration, C_2 represents the suppressed coal dust concentration.

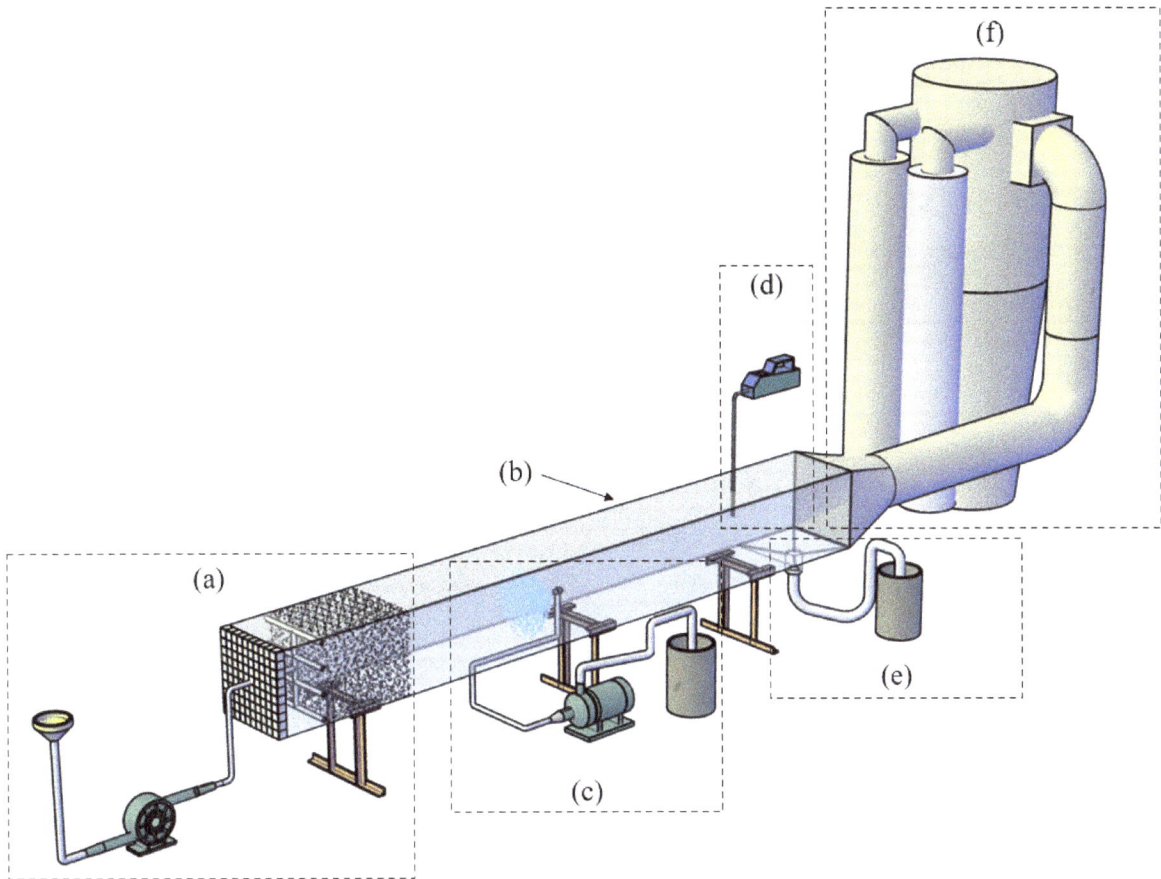

FIG 2 – Laboratory wind tunnel apparatus (Chang *et al*, 2020; Zhao *et al*, 2021): (a) dust generation section; (b) main wind tunnel; (c) surfactant solution spray section; (d) aerosol concentration measurement; (e) disposal section; (f) dust collector.

RESULTS

Table 1 shows the results of the zeta potential test and wind tunnel test. In this study, the zeta potential of each surfactant concentration was measured six times. The average value and standard deviation are also shown in Table 1. Additionally, the wind tunnel test was conducted duplicated. The average suppression efficiency is also shown in Table 1. Because there was only one dependent factor in each test, the one-factor ANOVA was applied to analyse the significance of zeta potential and suppression efficiency with the increase of surfactant concentration. The summary of ANOVA is shown in Tables 2 and 3.

TABLE 1

Results of zeta potential test and wind tunnel test.

Concentration (% w/v)	Zeta potential test		Wind tunnel test
	Zeta potential (mV)	STD	Efficiency (%)
0	-25.97	1.59	38.61
0.001	-14.18	2.74	42.10
0.002	-1.58	0.35	47.19
0.003	1.20	0.49	48.70
0.004	3.85	0.50	46.23
0.005	8.51	0.82	47.37

STD: standard deviation.

TABLE 2

ANOVA result for the zeta potential test.

	Sum of squares	df	Mean square	F	Sig.
Between groups	5005.865	5	1001.173	529.847	.000
Within groups	56.687	30	1.890		
Total	5062.551	35			

TABLE 3

ANOVA result for the wind tunnel test.

	Sum of squares	df	Mean square	F	Sig.
Between groups	149.543	5	29.909	4.570	.046
Within groups	39.265	6	6.544		
Total	188.809	11			

As can be seen from Table 2, the P-value of zeta potential is less than 0.05, which means the zeta potential can be significantly impacted by the change of surfactant concentration. Figure 3 illustrates the tendency of zeta potential with the increase of surfactant concentration. It can be clearly seen that the zeta potential of the suspension, which was made by coal dust and deionised water only, was -25.97 mV. With the increase of surfactant concentration gradually, the zeta potential raised to zero between 0.002 per cent w/v and 0.003 per cent w/v, then reached the highest value of 8.51 mV at 0.005 per cent w/v. Moreover, the zeta potential changed greater at lower concentrations less than 0.002 per cent w/v. After the suspension reached an equilibrium of zeta potential, it raised slowly with the increase of surfactant concentration.

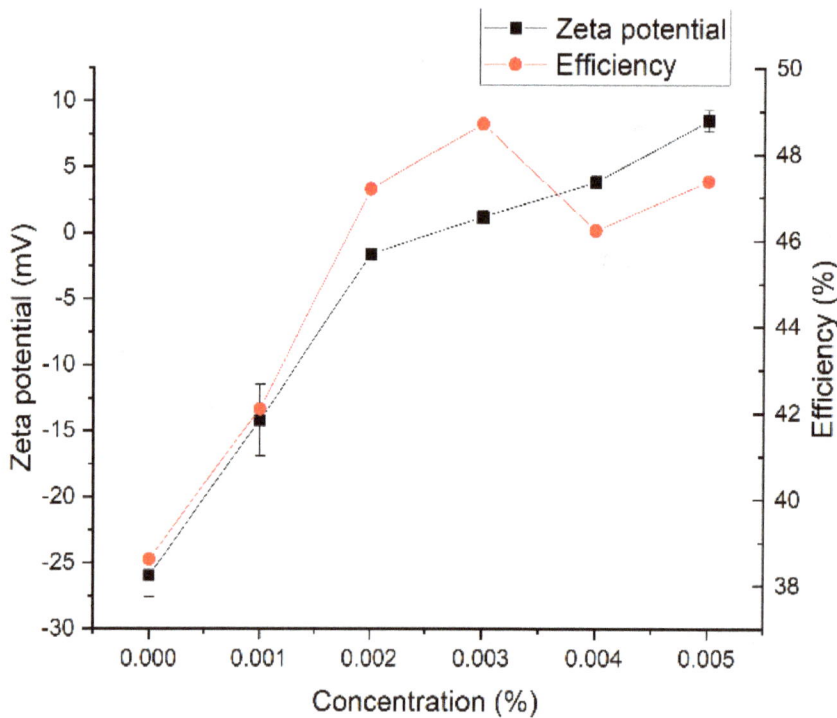

FIG 3 – The result of the zeta potential test and the wind tunnel test.

As can be seen from Table 3, the P-value of suppression efficiency is less than 0.05, which means the suppression efficiency in the wind tunnel test can be affected by the change of surfactant concentration. Usually, the surfactant performance of coal dust suppression can be improved with the increase of concentration, which is shown in Figure 3. The suppression efficiency without surfactant was 38.61 per cent and It increased greater at lower concentrations. As shown in Figure 3, the addition of surfactant into the solution can improve the efficiency to 42.10 per cent and 47.19 per cent at 0.001 per cent w/v and 0.002 per cent w/v, respectively. After 0.002 per cent w/v concentration, the suppression efficiency almost keeps stable with slight fluctuations. The efficiency peaks the highest value of 48.70 per cent when the concentration is 0.003 per cent w/v, followed by 47.37 per cent at 0.002 per cent w/v concentration.

DISCUSSIONS

The objective of this study is to investigate the relationship between zeta potential and suppression efficiency for evaluating the suppression performance of surfactants. From the ANOVA result, both the zeta potential and the suppression efficiency can be impacted significantly by the change of surfactant concentration. For better analysing the result of this study, further investigation and comparison are of importance.

In this study, both the zeta potential test and wind tunnel test shows some general results which are consistent with previous studies. In the zeta potential test, the suspension of coal and deionised water was negative, while it became positive with increasing the surfactant concentration. This behaviour also can be found in other cationic surfactants in some previous studies (Chang *et al*, 2016; Guo *et al*, 2018; Maršálek and Navrátilová, 2011; Maršálek, Pospisil and Taraba, 2011). The suspension reached an equilibrium between 0.002 per cent w/v and 0.003 per cent w/v. Although the presence of surfactants can dramatically increase the zeta potential before the concentration of 0.002 per cent w/v, the growth trended to be stable after the equilibrium point. As for the result of the wind tunnel test, although the plain water can reduce around 39 per cent of coal dust, the addition of surfactants can still improve the efficiency even with a small dosage. The greatest suppression efficiency of 48.7 per cent can be observed at 0.003 per cent w/v concentration.

From the results of this study, both the zeta potential test and the wind tunnel test have shown some consistency. Firstly, the curve tendency of zeta potential and suppression efficiency are similar. With the presence of surfactant at lower concentrations, both values of zeta potential and suppression efficiency were improved greater than higher concentrations. The zeta potential increased 24.39 mV from -25.97 mV to -1.58 mV with the increase of surfactant concentration from 0 per cent w/v to 0.002 per cent w/v, while it only increased 10.09 mV with the concentration from 0.002 per cent w/v to 0.005 per cent w/v. Moreover, in the wind tunnel test, the suppression efficiency raised 8.58 per cent and 0.18 per cent with the concentration range from 0 per cent w/v to 0.002 per cent w/v and 0.002 per cent w/v to 0.005 per cent w/v, respectively. Indeed, this similar tendency can be further proved by the Pearson Correlation analysis, as shown in Table 4. The correlation coefficient of zeta potential and efficiency was 0.936, which indicated a strong correlation between the two factors. Therefore, the difference of zeta potential is highly correlated with the performance of coal dust suppression. The larger difference of zeta potential, the better performance of the surfactant. Secondly, the suppression efficiency could take advantage when the zeta potential is around zero. The suppression efficiency had the highest growth rate from 0.001 per cent w/v to 0.002 per cent w/v concentration when the zeta potential approached equilibrium. At 0.003 per cent w/v concentration, the suppression efficiency peaked the highest value of 48.70 per cent. However, as predicted in the efficiency curve from 0.001 per cent w/v to 0.005 per cent w/v, it should be around 45 per cent at 0.003 per cent w/v concentration as shown in Figure 4, compared to 48.70 per cent as tested. After the equilibrium point, the suppression efficiency showed some fluctuations, which was dropped 2.47 per cent, followed by an increase of 1.14 per cent. Hence, the equilibrium of zeta potential can improve the performance of surfactants for controlling coal dust.

TABLE 4

Pearson Correlations of zeta potential and efficiency.

		Zeta Potential	Efficiency
Zeta Potential	Pearson Correlation	1	.936
	Sig. (2-tailed)		.006
	N	6	6
Efficiency	Pearson Correlation	.936	1
	Sig. (2-tailed)	.006	
	N	6	6

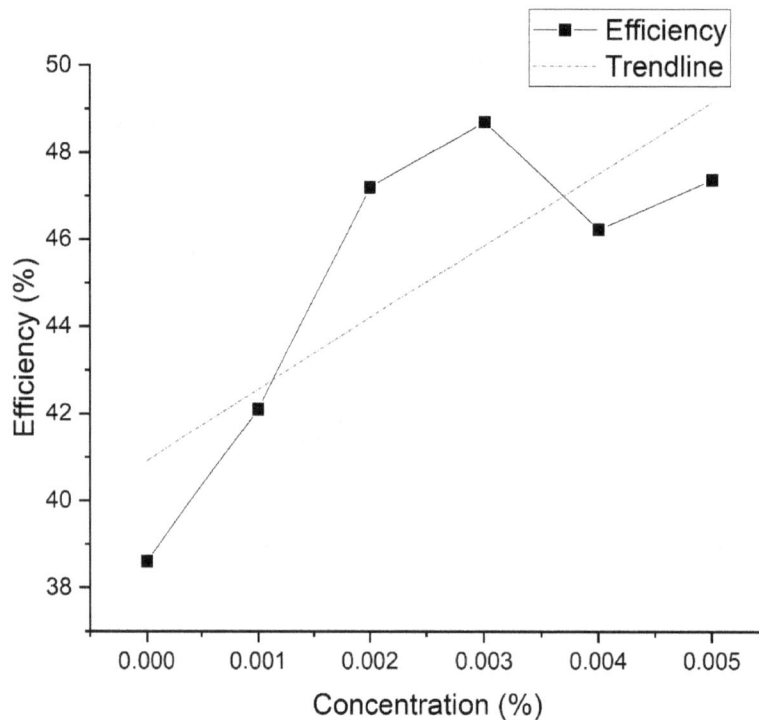

FIG 4 – Trendline of suppression efficiency.

Although the result of zeta potential test is highly correlated with that of the wind tunnel test, there are still some limitations of the zeta potential test. Firstly, the concentrations of DTAB ranging from 0.000 per cent w/v to 0.005 per cent w/v were selected based on the zeta potential test considering to the purpose of this study. The suspension of DTAB and Collie coal dust reached the equilibrium of zeta potential when the concentration was around 0.003 per cent w/v, which is relatively lower when compared with other surfactants with other evaluation methods in some previous studies. For instance, Maršálek, Pospisil and Taraba (2011) tested the performance of CTAB by zeta potential test and surface adsorption test. The result showed that the best concentration for controlling coal dust is around 0.2 per cent. Moreover, In Chen, Xu and Albijanic (2017)'s study, the largest difference of zeta potential was obtained at 0.4 per cent of SDBS, which presented the greatest performance for coal dust control. Secondly, the increase of suppression efficiency at the equilibrium point is relatively lower than the advantage of increasing the concentration. Although the equilibrium of zeta potential can benefit around 3 per cent of suppression efficiency as estimated, the total efficiency was still lower than 50 per cent. With the increase of surfactant concentration, the suppression efficiency could be further enhanced. In previous studies with a similar experimental set-up without compound surfactants, the dust control efficiency could be higher than 60 per cent or improve the performance of water spray over 40 per cent (Chang *et al*, 2020; Wang *et al*, 2014b; Zhou *et al*, 2020). Nevertheless, this finding could be beneficial to compound suppressants.

Compound suppressants usually are made with low-concentrated surfactants and electrolytes for improving the suppression performance. The equilibrium of zeta potential could further boost the performance for better controlling the underground coal dust.

CONCLUSION

In this study, the performance of DTAB ranging from 0 per cent w/v to 0.005 per cent w/v was investigated on coal dust suppression by using the zeta potential test and the wind tunnel test. Based on the experiment result, the relationship between the zeta potential and the suppression efficiency have been analysed. Firstly, it was found that both the zeta potential and suppression efficiency can be improved significantly with the presence of DTAB. With the increase of surfactant concentration, the zeta potential and suppression efficiency raised gradually. In this study, the greater difference the zeta potential can achieve, the better performance of coal dust suppression the surfactant can obtain. Moreover, the greatest efficiency can be found at the 0.003 per cent w/v concentration when the zeta potential approached zero, which means the equilibrium of zeta potential can benefit the suppression performance in the wind tunnel test. However, the advantage of equilibrium is limited compared to the improvement obtained from the increase of concentration. Although the result of this study illustrated the relationship between the zeta potential and the suppression efficiency, there was only one cationic surfactant was investigated in this study. In the future research, more surfactants should be investigated and more data should be collected to solidly prove the findings of this study. In conclusion, this study is significant because a better understanding of the relationship between the zeta potential and the suppression efficiency has been presented. The finding of this study can benefit the evaluation methods of surfactants in terms of coal dust suppression for both researchers and industries.

REFERENCES

ASTM International, 2012. *D3302/D3302M-12: Standard Test Method for Total Moisture in Coal* (inactive), West Conshohocken.

ASTM International, 2017. *D4749: Standard test method for performing the sieve analysis of coal and designating coal size,* West Conshohocken.

Ayoglu, F N, Acikgoz, B, Tutkun, E and Gebedek, S, 2014. Descriptive characteristics of coal workers' pneumoconiosis cases in Turkey. *Iranian Journal of Public Health, 43*(3):389.

Blackley, D J, Crum, J B, Halldin, C N, Storey, E and Laney, A S, 2016. Resurgence of progressive massive fibrosis in coal miners—Eastern Kentucky, *Morbidity and Mortality Weekly Report, 65*(49):1385–1389.

Blackley, D J, Halldin, C N and Laney, A S, 2018. Continued increase in prevalence of coal workers' pneumoconiosis in the United States, 1970–2017, *American Journal of Public Health, 108*(9):1220–1222.

Burt, J and McGhee, R, 2020. Cases of mine-dust lung disease and silicosis increasingly found in Queensland coal mine workers, *ABC News,* https://www.abc.net.au/news/2020–02–25/silicosis-and-black-lung-cases-rise-queensland-workers/11998404.

Castranova, V and Vallyathan, V, 2000. Silicosis and coal workers' pneumoconiosis, *Environmental Health Perspectives, 108*(suppl 4):675–684.

Chang, H, Zhang, H, Jia, Z, Li, X, Gao, W and Wei, W, 2016. Wettability of coal pitch surface by aqueous solutions of cationic Gemini surfactants, *Colloids and Surfaces A: Physicochemical and Engineering Aspects, 494*:59–64.

Chang, P, Zhao, Z, Xu, G, Ghosh, A, Huang, J and Yang, T, 2020. Evaluation of the coal dust suppression efficiency of different surfactants: A factorial experiment, *Colloids and Surfaces A: Physicochemical and Engineering Aspects,* 124686.

Chen, Y, Xu, G and Albijanic, B, 2017. Evaluation of SDBS surfactant on coal wetting performance with static methods: Preliminary laboratory tests, *Energy Sources, Part A: Recovery, Utilisation and Environmental Effects, 39*(23):2140–2150.

Crawford, R and Mainwaring, D, 2001. The influence of surfactant adsorption on the surface characterisation of Australian coals, *Fuel, 80*(3):313–320.

Doney, B C, Blackley, D, Hale, J M, Halldin, C, Kurth, L, Syamlal, G and Laney, A S, 2020. Respirable coal mine dust at surface mines, United States, 1982–2017, *American Journal of Industrial Medicine, 63*(3):232–239.

Guo, Y, Zhao, Y, Wang, S, Jiang, C and Zhang, J, 2018. Relationship between the zeta potential and the chemical agglomeration efficiency of fine particles in flue gas during coal combustion, *Fuel, 215:*756–765.

Hendry, M, 2019. Black lung advocates say 20 Queenslanders diagnosed with coal dust diseases in a fortnight, *ABC News,* https://www.abc.net.au/news/2019–02–26/dozens-of-new-black-lung-cases-qld-advocates-say/10851482.

Jin, H, Nie, W, Zhang, H, Liu, Y, Bao, Q, Wang, H and Huang, D, 2019. Preparation and characterisation of a novel environmentally friendly coal dust suppressant, *Journal of Applied Polymer Science, 136*(17):47354. doi:10.1002/app.47354

Kost, J, Shirey, G and Ford, C, 1980. *In-mine Tests for Wetting Agent Effectiveness*, Bureau of Mines, US Department of the Interior, Minerals Health and Safety.

Kumar, A and Dixit, C K, 2017. Methods for characterisation of nanoparticles, *Advances in Nanomedicine for the Delivery of Therapeutic Nucleic Acids*, pp 43–58, Elsevier.

Kurth, L, Laney, A S, Blackley, D J and Halldin, C N, 2020. Prevalence of spirometry-defined airflow obstruction in never-smoking working US coal miners by pneumoconiosis status, *Occupational and Environmental Medicine, 77*(4):265–267.

Liu, H, Tang, Z, Yang, Y, Weng, D, Sun, G, Duan, Z and Chen, J, 2009. Identification and classification of high risk groups for Coal Workers' Pneumoconiosis using an artificial neural network based on occupational histories: a retrospective cohort study, *BMC Public Health, 9*(1):366.

Maršálek, R, 2008. The influence of surfactants on the zeta potential of coals, *Energy Sources, Part A: Recovery, Utilisation and Environmental Effects, 31*(1):66–75.

Maršálek, R and Navrátilová, Z, 2011. Comparative study of CTAB adsorption on bituminous coal and clay mineral, *Chemical Papers, 65*(1):77–84.

Maršálek, R, Pospisil, J and Taraba, B, 2011. The influence of temperature on the adsorption of CTAB on coals, *Colloids and Surfaces A: Physicochemical and Engineering Aspects, 383*(1–3):80–85.

World Health Organization (WHO), 2013. National Health and Family Planning Commission of the People's Republic of China, Paper presented at the 9th Global Conference on Health Promotion, Shanghai.

Wang, K, Jiang, S, Wu, Z, Shao, H and Pei, X, 2014a. Experimental study on complex wetting agent enhancing the dusting effect and its application, *Electron. J Geotechnical Eng., 19*:9784–9798.

Wang, N, Nie, W, Cheng, W, Liu, Y, Zhu, L and Zhang, L, 2014b. Experiment and research of chemical de-dusting agent with spraying dust-settling, *Procedia Engineering, 84*:764–769.

Yan, J, Fan, J G, Jing, P, Ke, W, Zhang, P Y, Wang, H Q and Tao, L, 2018. Risk of active pulmonary tuberculosis among patients with coal workers' pneumoconiosis: a case-control study in China, *Biomedical and Environmental Sciences, 31*(6):448–453.

Yang, J, Tan, Y, Wang, Z, Shang, Y and Zhao, W, 2007. Study on the coal dust surface characteristics and wetting mechanism, *Mei T'an Hsueh Pao (Journal of China Coal Society), 32*.

Zhang, Y, Zhang, Y, Liu, B and Meng, X, 2020. Prediction of the length of service at the onset of coal workers' pneumoconiosis based on neural network, *Archives of Environmental & Occupational Health, 75*(4):242–250.

Zhao, Z, Chang, P, Xu, G, Ghosh, A, Li, D and Huang, J, 2021. Comparison of the coal dust suppression performance of surfactants using static test and dynamic test, *Journal of Cleaner Production*, 129633.

Zhou, Q, Qin, B, Ma, D and Jiang, N, 2017. Novel technology for synergetic dust suppression using surfactant-magnetised water in underground coal mines, *Process Safety and Environmental Protection, 109*:631–638.

Zhou, Q, Qin, B, Wang, F and Wang, H, 2019. Experimental investigation on the performance of a novel magnetised apparatus used to improve the dust suppression ability of surfactant-magnetised water, *Powder Technology, 354*:149–157.

Zhou, Q, Xu, G, Chen, Y, Qin, B, Zhao, Z and Guo, C, 2020. The development of an optimised evaluation system for improving coal dust suppression efficiency using aqueous solution sprays, *Colloids and Surfaces A: Physicochemical and Engineering Aspects, 602*:125104.

AUTHOR INDEX